The RF and Microwave Handbook

Second Edition

Editor-in-Chief

Mike Golio

HVVi Semiconductors, Inc.
Phoenix, Arizona, U.S.A.

Managing Editor

Janet Golio

RF and Microwave Applications and Systems

RF and Microwave Circuits, Measurements, and Modeling

RF and Microwave Passive and Active Technologies

The Electrical Engineering Handbook Series

Series Editor
Richard C. Dorf
University of California, Davis

Titles Included in the Series

RF AND MICROWAVE PASSIVE AND ACTIVE TECHNOLOGIES

Editor-in-Chief

Mike Golio

HVVi Semiconductors, Inc.
Phoenix, Arizona, U.S.A.

Managing Editor

Janet Golio

CRC Press
Taylor & Francis Group
Boca Raton   London   New York

CRC Press is an imprint of the
Taylor & Francis Group, an **informa** business

CRC Press
Taylor & Francis Group
6000 Broken Sound Parkway NW, Suite 300
Boca Raton, FL 33487-2742

© 2008 by Taylor & Francis Group, LLC
CRC Press is an imprint of Taylor & Francis Group, an Informa business

Library of Congress Cataloging-in-Publication Data

RF and microwave passive and active technologies / [Ed.] Mike Golio.
    p. cm.
  Includes bibliographical references and index.
  ISBN 978-0-8493-7220-9 (alk. paper)
  1. Microwave circuits. 2. Radio circuits. 3. Wireless communication systems. I. Golio, John Michael, 1954- II. Title.

TK7876.R488 2008
621.381'32--dc22
                                       2007016318

Visit the Taylor & Francis Web site at
http://www.taylorandfrancis.com

and the CRC Press Web site at
http://www.crcpress.com

Contents

SECTION I Passive Technologies

SECTION II   Active Device Technologies

SECTION III   Materials Properties

Acknowledgments

Although the topics and authors for this book were identified by the editor-in-chief and the advisory board, they do not represent the bulk of the work for a project like this. A great deal of the work involves tracking down those hundreds of technical experts, gaining their commitment, keeping track of their progress, collecting their manuscripts, getting appropriate reviews/revisions, and finally transferring the documents to be published. While juggling this massive job, author inquiries ranging from, "What is the required page length?", to, "What are the acceptable formats for text and figures?", have to be answered and re-answered. Schedules are very fluid. This is the work of the Managing Editor, Janet Golio. Without her efforts there would be no second edition of this handbook.

The advisory board has facilitated the book's completion in many ways. Board members contributed to the outline of topics, identified expert authors, reviewed manuscripts, and authored several of the chapters for the book.

Hundreds of RF and microwave technology experts have produced the chapters that comprise this second edition. They have all devoted many hours of their time sharing their expertise on a wide range of topics.

I would like to sincerely thank all of those listed above. Also, it has been a great pleasure to work with Jessica Vakili, Helena Redshaw, Nora Konopka, and the publishing professionals at CRC Press.

Editors

Editor-in-Chief

Mike Golio, since receiving his PhD from North Carolina State University in 1983, has held a variety of positions in both the microwave and semiconductor industries, and within academia. As Corporate Director of Engineering at Rockwell Collins, Dr. Golio managed and directed a large research and development organization, coordinated corporate IP policy, and led committees to achieve successful corporate spin-offs.

As an individual contributor at Motorola, he was responsible for pioneering work in the area of large signal microwave device characterization and modeling. This work resulted in over 50 publications including one book and a commercially available software package. The IEEE recognized this work by making Dr. Golio a Fellow of the Institute in 1996.

He is currently RF System Technologist with HVVi Semiconductor, a start-up semiconductor company. He has contributed to all aspects of the company's funding, strategies, and technical execution.

Dr. Golio has served in a variety of professional volunteer roles for the IEEE MTT Society including: Chair of Membership Services Committee, founding Co-editor of *IEEE Microwave Magazine*, and MTT-Society distinguished lecturer. He currently serves as Editor-in-chief of *IEEE Microwave Magazine*. In 2002 he was awarded the N. Walter Cox Award for exemplary service in a spirit of selfless dedication and cooperation.

He is author of hundreds of papers, book chapters, presentations and editor of six technical books. He is inventor or co-inventor on 15 patents. In addition to his technical contributions, Dr. Golio recently published a book on retirement planning for engineers and technology professionals.

Managing Editor

Janet R. Golio is Administrative Editor of *IEEE Microwave Magazine* and webmaster of www.golio.net. Prior to that she did government, GPS, and aviation engineering at Motorola in Arizona, Rockwell Collins in Iowa, and General Dynamics in Arizona. She also helped with the early development of the personal computer at IBM in North Carolina. Golio holds one patent and has written six technical papers. She received a BS in Electrical Engineering Summa Cum Laude from North Carolina State University in 1984.

When not working, Golio actively pursues her interests in archaeology, trick roping, and country western dancing. She is the author of young adult books, *A Present from the Past* and *A Puzzle from the Past*.

Advisory Board

Peter A. Blakey

Peter A. Blakey obtained a BA in applied physics from the University of Oxford in 1972, a PhD in electronic engineering from the University of London in 1976, and an MBA from the University of Michigan in 1989. He has held several different positions in industry and academia and has worked on a wide range of RF, microwave, and Si VLSI applications. Between 1991 and 1995 he was the director of TCAD Engineering at Silvaco International. He joined the Department of Electrical Engineering at Northern Arizona University in 2002 and is presently an emeritus professor at that institution.

Nick Buris

Nick Buris received his Diploma in Electrical Engineering in 1982 from the National Technical University of Athens, Greece, and a PhD in electrical engineering from the North Carolina State University, Raleigh, North Carolina, in 1986. In 1986 he was a visiting professor at NCSU working on space reflector antennas for NASA. In 1987 he joined the faculty of the Department of Electrical and Computer Engineering at the University of Massachusetts at Amherst. His research work there spanned the areas of microwave magnetics, phased arrays printed on ferrite substrates, and broadband antennas. In the summer of 1990 he was a faculty fellow at the NASA Langley Research Center working on calibration techniques for dielectric measurements (space shuttle tiles at very high temperatures) and an ionization (plasma) sensor for an experimental reentry spacecraft.

In 1992 he joined the Applied Technology organization of Motorola's Paging Product Group and in 1995 he moved to Corporate Research to start an advanced modeling effort. While at Motorola he has worked on several projects from product design to measurement systems and the development of proprietary software tools for electromagnetic design. He currently manages the Microwave Technologies Research Lab within Motorola Labs in Schaumburg, Illinois. Recent and current activities of the group include V-band communications systems design, modeling and measurements of complex electromagnetic problems, RF propagation, Smart Antennas/MIMO, RFID systems, communications systems simulation and modeling, spectrum engineering, as well as TIA standards work on RF propagation and RF exposure.

Nick is a senior member of the IEEE, and serves on an MTT Technical Program Committee. He recently served as chair of a TIA committee on RF exposure and is currently a member of its Research Division Committee.

Lawrence P. Dunleavy

Dr. Larry Dunleavy, along with four faculty colleagues established University of South Florida's innovative Center for Wireless and Microwave Information Systems (WAMI Center—http://ee.eng.usf.edu/WAMI).

In 2001, Dr. Dunleavy co-founded Modelithics, Inc., a USF spin-off company, to provide a practical commercial outlet for developed modeling solutions and microwave measurement services (www.modelithics.com), where he is currently serving as its president.

Dr. Dunleavy received his BSEE degree from Michigan Technological University in 1982 and his MSEE and PhD in 1984 and 1988, respectively, from the University of Michigan. He has worked in industry for E-Systems (1982–1983) and Hughes Aircraft Company (1984–1990) and was a Howard Hughes Doctoral Fellow (1984–1988). In 1990 he joined the Electrical Engineering Department at the University of South Florida. He maintains a position as professor in the Department of Electrical Engineering. His research interests are related to microwave and millimeter-wave device, circuit, and system design, characterization, and modeling. In 1997–1998, Dr. Dunleavy spent a sabbatical year in the noise metrology laboratory at the National Institute of Standards and Technology (NIST) in Boulder, Colorado. Dr. Dunleavy is a senior member of IEEE and is very active in the IEEE MTT Society and the Automatic RF Techniques Group (ARFTG). He has authored or co-authored over 80 technical articles.

Jack East

Jack East received his BSE, MS, and PhD from the University of Michigan. He is presently with the Solid State Electronics Laboratory at the University of Michigan conducting research in the areas of high-speed microwave device design, fabrication, and experimental characterization of solid-state microwave devices, nonlinear and circuit modeling for communications circuits and low-energy electronics, and THz technology.

Patrick Fay

Patrick Fay is an associate professor in the Department of Electrical Engineering at the University of Notre Dame, Notre Dame, Indiana. He received his PhD in Electrical Engineering from the University of Illinois at Urbana-Champaign in 1996 after receiving a BS in Electrical Engineering from Notre Dame in 1991. Dr. Fay served as a visiting assistant professor in the Department of Electrical and Computer Engineering at the University of Illinois at Urbana-Champaign in 1996 and 1997, and joined the faculty at the University of Notre Dame in 1997.

His educational initiatives include the development of an advanced undergraduate laboratory course in microwave circuit design and characterization, and graduate courses in optoelectronic devices and electronic device characterization. He was awarded the Department of Electrical Engineering's IEEE Outstanding Teacher Award in 1998–1999. His research interests include the design, fabrication, and characterization of microwave and millimeter-wave electronic devices and circuits, as well as high-speed optoelectronic devices and optoelectronic integrated circuits for fiber optic telecommunications. His research also includes the development and use of micromachining techniques for the fabrication of microwave components and packaging. Professor Fay is a senior member of the IEEE, and has published 7 book chapters and more than 60 articles in refereed scientific journals.

David Halchin

David Halchin has worked in RF/microwaves and GaAs for over 20 years. During this period, he has worn many hats including engineering and engineering management, and he has worked in both academia and private industry. Along the way, he received his PhD in Electrical Engineering from North Carolina State University. Dave currently works for RFMD, as he has done since 1998. After a stint as a PA designer, he was moved into his current position managing a modeling organization within RFMD's Corporate Research and Development organization. His group's responsibilities include providing compact models for circuit simulation for both GaAs active devices and passives on GaAs. The group also provides compact models for a handful of Si devices, behavioral models for power amplifier assemblies, and physics-based simulation for GaAs device development. Before joining RFMD, Dave spent time at Motorola and Rockwell working in the GaAs device development and modeling areas. He is a member of the IEEE-MTT and EDS. Dave is currently a member of the executive committee for the Compound IC Symposium.

Alfy Riddle

Alfy Riddle is vice president of Engineering at Finesse. Before Finesse, Dr. Riddle was the principal at Macallan Consulting, a company he founded in 1989. He has contributed to the design and development of a wide range of products using high-speed, low-noise, and RF techniques. Dr. Riddle developed and marketed the Nodal circuit design software package that featured symbolic analysis and object-oriented techniques. He has co-authored two books and contributed chapters to several more. He is a member of the IEEE MTT Society, the Audio Engineering Society, and the Acoustical Society of America. Dr. Riddle received his PhD in Electrical Engineering in 1986 from North Carolina State University. When he is not working, he can be found on the tennis courts, hiking in the Sierras, taking pictures with an old Leica M3, or making and playing Irish flutes.

Robert J. Trew

Robert J. Trew received his PhD from the University of Michigan in 1975. He is currently the Alton and Mildred Lancaster Distinguished Professor of Electrical and Computer Engineering and Head of the ECE Department at North Carolina State University, Raleigh. He previously served as the Worcester Professor of Electrical and Computer Engineering and Head of the ECE Department of Virginia Tech, Blacksburg, Virginia, and the Dively Distinguished Professor of Engineering and Chair of the Department of Electrical Engineering and Applied Physics at Case Western Reserve University, Cleveland, Ohio. From 1997 to 2001 Dr. Trew was director of research for the U.S. Department of Defense with management responsibility for the $1.3 billion annual basic research program of the DOD. Dr. Trew was vice-chair of the U.S. government interagency group that planned and implemented the U.S. National Nanotechnology Initiative. Dr. Trew is a fellow of the IEEE, and was the 2004 president of the Microwave Theory and Techniques Society. He was editor-in-chief of the *IEEE Transactions on Microwave Theory and Techniques* from 1995 to 1997, and from 1999 to 2002 was founding co-editor-in-chief of the *IEEE Microwave Magazine*. He is currently the editor-in-chief of the *IEEE Proceedings*. Dr. Trew has twice been named an IEEE MTT Society Distinguished Microwave Lecturer. He has earned numerous honors, including a 2003 Engineering Alumni Society Merit Award in Electrical Engineering from the University of Michigan, the 2001 IEEE-USA Harry Diamond Memorial Award, the 1998 IEEE MTT Society Distinguished Educator Award, a 1992 Distinguished Scholarly Achievement Award from NCSU, and an IEEE Third Millennium Medal. Dr. Trew has authored or co-authored over 160 publications, 19 book chapters, 9 patents, and has given over 360 presentations

Contributors

David Anderson
Freescale Semiconductor
Tempe, Arizona

Mike Carroll
RF Micro Devices
Greensboro, North Carolina

Prashant Chavarkar
Infinera
Sunnyvale, California

Julio Costa
RF Micro Devices
Greensboro, North Carolina

John C. Cowles
Analog Devices-Northwest Labs
Beaverton, Oregon

Lawrence P. Dunleavy
Modelithics, Inc.
Tampa, Florida

Jack R. East
The University of Michigan
Ann Arbor, Michigan

K. F. Etzold
IBM T. J. Watson Research Center
Yorktown Heights, New York

Patrick Fay
University of Notre Dame
Notre Dame, Indiana

S. Jerry Fiedziuszko
Lockheed Martin CSS
Sunnyvale, California

Mike Golio
HVVi Semiconductor
Phoenix, Arizona

H. Mike Harris
Georgia Tech Research Institute
Atlanta, Georgia

Thomas K. Ishii
Marquette University
Milwaukee, Wisconsin

William Liu
Maxim Integrated Products
Beaverton, Oregon

Donald C. Malocha
University of Central Florida
Orlando, Florida

Michael E. Majerus
Freescale Semiconductor
Tempe, Arizona

Tom McKay
RF Micro Devices
Greensboro, North Carolina

Imran Mehdi
Jet Propulsion Laboratory
California Institute of
    Technology
Pasadena, California

Umesh K. Mishra
University of California
Santa Barbara, California

Jeanne S. Pavio
Rockwell Collins, Inc.
Scottsdale, Arizona

G. Ali Rezvani
RF Micro Devices
Greensboro, North Carolina

Alfy Riddle
Finesse, LLC
Santa Clara, California

Michael S. Shur
Rensselaer Polytechnic Institute
Troy, New York

Jan Stake
Chalmers University of
    Technology
Göteborg, Sweden

Michael B. Steer
North Carolina State University
Raleigh, North Carolina

Robert J. Trew
North Carolina State University
Raleigh, North Carolina

Karl R. Varian
Raytheon Company
Dallas, Texas

Robert R. Weirather
Harris Corporation
Quincy, Illinois

John P. Wendler
Tyco Electronics Wireless Network
    Solutions
Lowell, Massachusetts

James B. West
Rockwell Collins—Advanced
    Technology Center
Cedar Rapids, Iowa

Jerry C. Whitaker
Advance Television Systems
    Committee
Washington, DC

James C. Wiltse
Georgia Tech Research Institute
Georgia Institute of Technology
Atlanta, Georgia

Introduction to Microwaves and RF

Patrick Fay
University of Notre Dame

I.1   Introduction to Microwave and RF Engineering

Modern microwave and radio frequency (RF) engineering is an exciting and dynamic field, due in large part to the symbiosis between recent advances in modern electronic device technology and the explosion in demand for voice, data, and video communication capacity that started in the 1990s and continues through the present. Prior to this revolution in communications, microwave technology was the nearly exclusive domain of the defense industry; the recent and dramatic increase in demand for communication systems for such applications as wireless paging, mobile telephony, broadcast video, and tethered as well as untethered computer networks has revolutionized the industry. These communication systems are employed across a broad range of environments, including corporate offices, industrial and manufacturing facilities, infrastructure for municipalities, as well as private homes. The diversity of applications and operational environments has led, through the accompanying high production volumes, to tremendous advances in cost-efficient manufacturing capabilities of microwave and RF products. This in turn has lowered the implementation cost of a host of new and cost-effective wireless as well as wired RF and microwave services. Inexpensive handheld GPS navigational aids, automotive collision-avoidance radar, and widely available broadband digital service access are among these. Microwave technology is naturally suited for these emerging applications in communications and sensing, since the high operational frequencies permit both large numbers of independent channels for the wide variety of uses envisioned as well as significant available bandwidth per channel for high-speed communication.

Loosely speaking, the fields of microwave and RF engineering together encompass the design and implementation of electronic systems utilizing frequencies in the electromagnetic spectrum from approximately 300 kHz to over 100 GHz. The term "RF" engineering is typically used to refer to circuits and systems having frequencies in the range from approximately 300 kHz at the low end to between 300 MHz and 1 GHz at the upper end. The term "microwave engineering," meanwhile, is used rather loosely to refer to design and implementation of electronic systems with operating frequencies in the range from 300 MHz to 1 GHz on the low end to upwards of 100 GHz. Figure I.1 illustrates schematically the electromagnetic spectrum

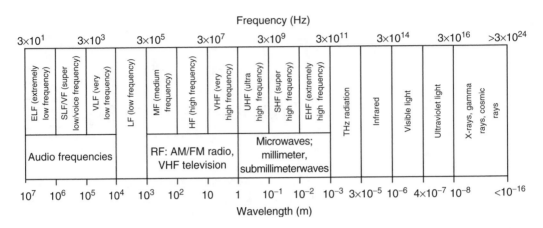

FIGURE I.1 Electromagnetic frequency spectrum and associated wavelengths.

from audio frequencies through cosmic rays. The RF frequency spectrum covers the medium frequency (MF), high frequency (HF), and very high frequency (VHF) bands, while the microwave portion of the electromagnetic spectrum extends from the upper edge of the VHF frequency range to just below the THz radiation and far-infrared optical frequencies (approximately 0.3 THz and above). The wavelength of free-space radiation for frequencies in the RF frequency range is from approximately 1 m (at 300 MHz) to 1 km (at 300 kHz), while those of the microwave range extend from 1 m to the vicinity of 1 mm (corresponding to 300 GHz) and below.

The boundary between "RF" and "microwave" design is both somewhat indistinct as well as one that is continually shifting as device technologies and design methodologies advance. This is due to implicit connotations that have come to be associated with the terms "RF" and "microwave" as the field has developed. In addition to the distinction based on the frequency ranges discussed previously, the fields of RF and microwave engineering are also often distinguished by other system features as well. For example, the particular active and passive devices used, the system applications pursued, and the design techniques and overall mindset employed all play a role in defining the fields of microwave and RF engineering. These connotations within the popular meaning of microwave and RF engineering arise fundamentally from the frequencies employed, but often not in a direct or absolute sense. For example, because advances in technology often considerably improve the high frequency performance of electronic devices, the correlation between particular types of electronic devices and particular frequency ranges is a fluid one. Similarly, new system concepts and designs are reshaping the applications landscape, with mass market designs utilizing ever higher frequencies rapidly breaking down conventional notions of microwave-frequency systems as serving "niche" markets.

The most fundamental characteristic that distinguishes RF engineering from microwave engineering is directly related to the frequency (and thus the wavelength, λ) of the electronic signals being processed. This distinction arises fundamentally from the finite speed of propagation of electromagnetic waves (and thus, by extension, currents and voltages). In free space, $\lambda = c/f$, where f is the frequency of the signal and c is the speed of light. For low-frequency and RF circuits (with a few special exceptions such as antennae), the signal wavelength is much larger than the size of the electronic system and circuit components. In contrast, for a microwave system the sizes of typical electronic components are often comparable to (i.e., within approximately 1 order of magnitude of) the signal wavelength. A schematic diagram illustrating this concept is shown in Figure I.2. As illustrated in Figure I.2, for components much smaller than the wavelength (i.e., $\ell < \lambda/10$), the finite velocity of the electromagnetic signal as it propagates through the component leads to a modest difference in phase at opposite ends of the component. For components comparable to or larger than the wavelength, however, this end-to-end phase difference becomes increasingly significant. This gives rise to a reasonable working definition of the two

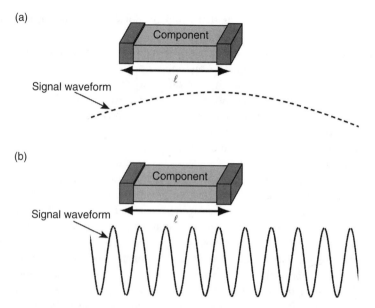

FIGURE I.2 Schematic representation of component dimensions relative to signal wavelengths. Conventional lumped-element analysis techniques are typically applicable for components for which $\ell < \lambda/10$ (a) since the phase change due to electromagnetic propagation across the component is small, while for components with $\ell > \lambda/10$ (b) the phase change is significant and a distributed circuit description is more appropriate.

design areas based on the underlying approximations used in design. Since in conventional RF design, the circuit components and interconnections are generally small compared to a wavelength, they can be modeled as lumped elements for which Kirchoff's voltage and current laws apply at every instant in time. Parasitic inductances and capacitances are incorporated to accurately model the frequency dependencies and the phase shifts, but these quantities can, to good approximation, be treated with an appropriate lumped-element equivalent circuit. In practice, a rule of thumb for the applicability of a lumped-element equivalent circuit is that the component size should be less than $\lambda/10$ at the frequency of operation. For microwave frequencies for which component size exceeds approximately $\lambda/10$, the finite propagation velocity of electromagnetic waves can no longer be as easily absorbed into simple lumped-element equivalent circuits. For these frequencies, the time delay associated with signal propagation from one end of a component to the other is an appreciable fraction of the signal period, and thus lumped-element descriptions are no longer adequate to describe the electrical behavior. A distributed-element model is required to accurately capture the electrical behavior. The time delay associated with finite wave propagation velocity that gives rise to the distributed circuit effects is a distinguishing feature of the mindset of microwave engineering.

An alternative viewpoint is based on the observation that microwave engineering lies in a "middle ground" between traditional low-frequency electronics and optics, as shown in Figure I.1. As a consequence of RF, microwaves, and optics simply being different regimes of the same electromagnetic phenomena, there is a gradual transition between these regimes. The continuity of these regimes results in constant re-evaluation of the appropriate design strategies and trade-offs as device and circuit technology advances. For example, miniaturization of active and passive components often increases the frequencies at which lumped-element circuit models are sufficiently accurate, since by reducing component dimensions the time delay for propagation through a component is proportionally reduced. As a consequence, lumped-element components at "microwave" frequencies are becoming increasingly common in systems previously based on distributed elements due to significant advances in miniaturization, even though the operational frequencies remain unchanged. Component and circuit miniaturization also leads to tighter packing of interconnects and components, potentially introducing new parasitic coupling and distributed-element effects into circuits that could previously be treated using lumped-element RF models.

The comparable scales of components and signal wavelengths has other implications for the designer as well, since neither the ray-tracing approach from optics nor the lumped-element approach from RF circuit design are valid in this middle ground. In this regard, microwave engineering can also be considered to be "applied electromagnetic engineering," as the design of guided-wave structures such as waveguides and transmission lines, transitions between different types of transmission lines, and antennae all require analysis and control of the underlying electromagnetic fields.

Guided wave structures are particularly important in microwave circuits and systems. There are many different approaches to the implementation of guided-wave structures; a sampling of the more common options is shown in Figure I.3. Figure I.3a shows a section of coaxial cable. In this common cable type, the grounded outer conductor shields the dielectric and inner conductor from external signals and also prevents the signals within the cable from radiating. The propagation in this structure is controlled by the dielectric properties, the cross-sectional geometry, and the metal conductivity. Figure I.3b shows a rectangular waveguide. In this structure, the signal propagates in the free space within the structure, while the rectangular metal structure is grounded. Despite the lack of an analog to the center conductor in the coaxial line, the structure supports traveling-wave solutions to Maxwell's equations, and thus can be used to transmit power along its length. The lack of a center conductor does prevent the structure from providing any path for dc along its length. The solution to Maxwell's equations in the rectangular waveguide also leads

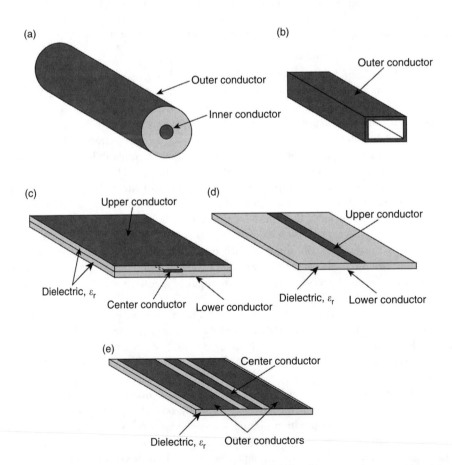

FIGURE I.3 Several common guided-wave structures. (a) coaxial cable, (b) rectangular waveguide, (c) stripline, (d) microstrip, and (e) coplanar waveguide.

to multiple eigenmodes, each with its own propagation characteristics (e.g., characteristic impedance and propagation constant), and corresponding cutoff frequency. For frequencies above the cutoff frequency, the mode propagates down the waveguide with little loss, but below the cutoff frequency the mode is evanescent and the amplitude falls off exponentially with distance. Since the characteristic impedance and propagation characteristics of each mode are quite different, in many systems the waveguides are sized to support only one propagating mode at the frequency of operation. While metallic waveguides of this type are mechanically inflexible and can be costly to manufacture, they offer extremely low loss and have excellent high-power performance. At W-band and above in particular, these structures currently offer much lower loss than coaxial cable alternatives. Figures I.3c through I.3e show several planar structures that support guided waves. Figure I.3c illustrates the stripline configuration. This structure is in some ways similar to the coaxial cable, with the center conductor of the coaxial line corresponding to the center conductor in the stripline, and the outer shield on the coaxial line corresponding to the upper and lower ground planes in the stripline. Figures I.3d and I.3e show two planar guided-wave structures often encountered in circuit-board and integrated circuit designs. Figure I.3d shows a microstrip configuration, while Figure I.3e shows a coplanar waveguide. Both of these configurations are easily realizable using conventional semiconductor and printed-circuit fabrication techniques. In the case of microstrip lines, the key design variables are the dielectric properties of the substrate, the dielectric thickness, and the width of the top conductor. For the coplanar waveguide case, the dielectric properties of the substrate, the width of the center conductor, the gap between the center and outer ground conductors, and whether or not the bottom surface of the substrate is grounded control the propagation characteristics of the lines.

For all of these guided-wave structures, an equivalent circuit consisting of the series concatenation of many stages of the form shown in Figure I.4 can be used to model the transmission line. In this equivalent circuit, the key parameters are the resistance per unit length of the line (R), the inductance per unit length (L), the parallel conductance per unit length of the dielectric (G), and the capacitance per unit length (C). Each of these parameters can be derived from the geometry and material properties of the line. Circuits of this form give rise to traveling-wave solutions of the form

$$V(z) = V_0^+ e^{-\gamma z} + V_0^- e^{\gamma z}$$

$$I(z) = \frac{V_0^+}{Z_0} e^{-\gamma z} - \frac{V_0^-}{Z_0} e^{\gamma z}$$

In these equations, the characteristic impedance of the line, which is the constant of proportionality between the current and voltage associated with a particular traveling-wave mode on the line, is given by $Z_0 = \sqrt{(R + j\omega L)/(G + j\omega C)}$. For lossless lines, $R = 0$ and $G = 0$, so that Z_0 is real; even in many practical cases the loss of the lines is small enough that the characteristic impedance can be treated as real. Similarly, the propagation constant of the line can be expressed as $\gamma = \alpha + j\beta = \sqrt{(R + j\omega L)(G + j\omega C)}$. In this expression, α characterizes the loss of the line, and β captures the wave propagation. For lossless lines, γ is pure imaginary, and thus α is zero. The design and analysis of these guided-wave structures is treated in more detail in Chapter 30 of the companion volume *RF and Microwave Applications and Systems* in this handbook series.

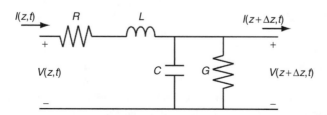

FIGURE I.4 Equivalent circuit for an incremental length of transmission line. A finite length of transmission line can be modeled as a series concatenation of sections of this form.

The distinction between RF and microwave engineering is further blurred by the trend of increasing commercialization and consumerization of systems using what have been traditionally considered to be microwave frequencies. Traditional microwave engineering, with its historically military applications, has long been focused on delivering performance at any cost. As a consequence, special-purpose devices intended solely for use in high performance microwave systems and often with somewhat narrow ranges of applicability were developed to achieve the required performance. With continuing advances in silicon microelectronics, including Si bipolar junction transistors, SiGe heterojunction bipolar transistors (HBTs) and conventional scaled CMOS, microwave-frequency systems can now be reasonably implemented using the same devices as conventional low-frequency baseband electronics. These advanced silicon-based active devices are discussed in more detail in Chapters 16–19. In addition, the commercialization of low-cost III–V compound semiconductor electronics, including ion-implanted metal semiconductor field-effect transistors (MESFETs), pseudomorphic and lattice-matched high electron mobility transistors (HEMTs), and III–V HBTs, has dramatically decreased the cost of including these elements in high-volume consumer systems. These compound-semiconductor devices are described in Chapters 17 and 20–22. This convergence, with silicon microelectronics moving ever higher in frequency into the microwave spectrum from the low-frequency side and compound semiconductors declining in price for the middle of the frequency range, blurs the distinction between "microwave" and "RF" engineering, since "microwave" functions can now be realized with "mainstream" low-cost electronics. This is accompanied by a shift from physically large, low-integration-level hybrid implementations to highly-integrated solutions based on monolithic microwave integrated circuits (MMICs) (see Chapters 24–25 and the companion volume *RF and Microwave Circuits, Measurements, and Modeling*, Chapters 25–26). This shift has a dramatic effect not only on the design of systems and components, but also on the manufacturing technology and economics of production and implementation as well. A more complete discussion of the active device and integration technologies that make this progression possible is included in Section II, while modeling of these devices is described in Section III of the *RF and Microwave Circuits, Measurements, and Modeling* companion volume in this handbook series.

Aside from these defining characteristics of RF and microwave systems, the behavior of materials is also often different at microwave frequencies than at low frequencies. In metals, the effective resistance at microwave frequencies can differ significantly from that at dc. This frequency-dependent resistance is a consequence of the skin effect, which is caused by the finite penetration depth of an electromagnetic field into conducting material. This effect is a function of frequency; the depth of penetration is given by $\delta_s = (1/\sqrt{\pi f \mu \sigma})$, where μ is the permeability, f is the frequency, and σ is the conductivity of the material. As the expression indicates, δ_s decreases with increasing frequency, and so the electromagnetic fields are confined to regions increasingly near the surface as the frequency increases. This results in the microwave currents flowing exclusively along the surface of the conductor, significantly increasing the effective resistance (and thus the loss) of metallic interconnects. Further discussion of this topic can be found in Chapter 26 of this volume and Chapter 28 of the *RF and Microwave Applications and Systems* volume in this handbook series. Dielectric materials also exhibit frequency-dependent characteristics that can be important. The permeability and loss of dielectrics arises from the internal polarization and dissipation of the material. Since the polarization within a dielectric is governed by the response of the material's internal charge distribution, the frequency dependence is governed by the speed at which these charges can redistribute in response to the applied fields. For ideal materials, this dielectric relaxation leads to a frequency-dependent permittivity of the form $\varepsilon(\omega) = \varepsilon_\infty + (\varepsilon_{dc} - \varepsilon_\infty)/(1 + j\omega\tau)$, where ε_{dc} is the low-frequency permittivity, ε_∞ is the high-frequency (optical) permittivity, and τ is the dielectric relaxation time. Loss in the dielectric is incorporated in this expression through the imaginary part of ε. For many materials the dielectric relaxation time is sufficiently small that the performance of the dielectric at microwave frequencies is very similar to that at low frequencies. However, this is not universal and some care is required since some materials and devices exhibit dispersive behavior at quite low frequencies. Furthermore, this description of dielectrics is highly idealized; the frequency response of many real-world materials is much more complex than this idealized model would suggest. High-value capacitors and semiconductor devices are among the classes of devices that are particularly likely to exhibit complex dielectric responses.

In addition to material properties, some physical effects are significant at microwave frequencies that are typically negligible at lower frequencies. For example, radiation losses become increasingly important as the signal wavelengths approach the component and interconnect dimensions. For conductors and other components of comparable size to the signal wavelengths, standing waves caused by reflection of the electromagnetic waves from the boundaries of the component can greatly enhance the radiation of electromagnetic energy. These standing waves can be easily established either intentionally (in the case of antennae and resonant structures) or unintentionally (in the case of abrupt transitions, poor circuit layout, or other imperfections). Careful attention to transmission line geometry, placement relative to other components, transmission lines, and ground planes, as well as circuit packaging is essential for avoiding excessive signal attenuation and unintended coupling due to radiative effects.

A further distinction in the practice of RF and microwave engineering from conventional electronics is the methodology of testing and characterization. Due to the high frequencies involved, the capacitance and standing-wave effects associated with test cables and the parasitic capacitance of conventional test probes make the use of conventional low-frequency circuit characterization techniques impractical. Although advanced measurement techniques such as electro-optic sampling can sometimes be employed to circumvent these difficulties, in general the loading effect of measurement equipment poses significant measurement challenges for debugging and analyzing circuit performance, especially for nodes at the interior of the circuit under test. In addition, for circuits employing dielectric or hollow guided-wave structures, voltage and current often cannot be uniquely defined. Even for structures in which voltage and current are well-defined, practical difficulties associated with accurately measuring such high-frequency signals make this difficult. Furthermore, since a dc-coupled time-domain measurement of a microwave signal would have an extremely wide noise bandwidth, the sensitivity of the measurement would be inadequate. For these reasons, components and low-level subsystems are characterized using specialized techniques.

One of the most common techniques for characterizing the linear behavior of microwave components is the use of s-parameters. While z-, y-, and h-parameter representations are commonly used at lower frequencies, these approaches can be problematic to implement at microwave frequencies. The use of s-parameters essentially captures the same information as these other parameter sets, but instead of directly measuring terminal voltages and currents, the forward and reverse traveling waves at the input and output ports are measured instead. While perhaps not intuitive at first, this approach enables accurate characterization of components at very high frequencies to be performed with comparative ease. For a two-port network, the s-parameters are defined by:

$$\left[\begin{array}{c} V_1^- \\ V_2^- \end{array} \right] = \left[\begin{array}{cc} s_{11} & s_{12} \\ s_{21} & s_{22} \end{array} \right] \left[\begin{array}{c} V_1^+ \\ V_2^+ \end{array} \right]$$

where the V^- terms are the wave components traveling away from the two-port, and the V^+ terms are the incident terms. These traveling waves can be thought of as existing on "virtual" transmission lines attached to the device ports. From this definition,

$$s_{11} = \left. \frac{V_1^-}{V_1^+} \right|_{V_2^+ = 0}$$

$$s_{12} = \left. \frac{V_1^-}{V_2^+} \right|_{V_1^+ = 0}$$

$$s_{21} = \left. \frac{V_2^-}{V_1^+} \right|_{V_2^+ = 0}$$

$$s_{22} = \left. \frac{V_2^-}{V_2^+} \right|_{V_1^+ = 0}$$

To measure the s-parameters, the ratio of the forward and reverse traveling waves on the virtual input and output transmission lines is measured. To achieve the $V_1^+ = 0$ and $V_2^+ = 0$ conditions in these expressions, the ports are terminated in the characteristic impedance, Z_0, of the virtual transmission lines. Although in principle these measurements can be made using directional couplers to separate the forward and reverse traveling waves and phase-sensitive detectors, in practice modern network analyzers augment the measurement hardware with sophisticated calibration routines to remove the effects of hardware imperfections to achieve accurate s-parameter measurements. A more detailed discussion of s-parameters, as well as other approaches to device and circuit characterization, is provided in Section I in the companion volume *RF and Microwave Circuits, Measurements, and Modeling* in this handbook series.

I.2 General Applications

The field of microwave engineering is currently experiencing a radical transformation. Historically, the field has been driven by applications requiring the utmost in performance with little concern for cost or manufacturability. These systems have been primarily for military applications, where performance at nearly any cost could be justified. The current transformation of the field involves a dramatic shift from defense applications to those driven by the commercial and consumer sector, with an attendant shift in focus from design for performance to design for manufacturability. This transformation also entails a shift from small production volumes to mass production for the commercial market, and from a focus on performance without regard to cost to a focus on minimum cost while maintaining acceptable performance. For wireless applications, an additional shift from broadband systems to systems having very tightly-regulated spectral characteristics also accompanies this transformation.

For many years the driving application of microwave technology was military radar. The small wavelength of microwaves permits the realization of narrowly-focused beams to be achieved with antennae small enough to be practically steered, resulting in adequate resolution of target location. Long-distance terrestrial communications for telephony as well as satellite uplink and downlink for voice and video were among the first commercially viable applications of microwave technology. These commercial communications applications were successful because microwave-frequency carriers (f_c) offer the possibility of very wide absolute signal bandwidths (Δf) while still maintaining relatively narrow fractional bandwidths (i.e., $\Delta f / f_c$). This allows many more voice and data channels to be accommodated than would be possible with lower-frequency carriers or baseband transmission.

Among the current host of emerging applications, many are based largely on this same principle, namely, the need to transmit more and more data at high speed, and thus the need for many communication channels with wide bandwidths. Wireless communication of voice and data, both to and from individual users as well as from users and central offices in aggregate, wired communication including coaxial cable systems for video distribution and broadband digital access, fiber-optic communication systems for long- and short-haul telecommunication, and hybrid systems such as hybrid fiber-coax systems are all designed to take advantage of the wide bandwidths and consequently high data carrying capacity of microwave-frequency electronic systems. The widespread proliferation of wireless Bluetooth personal-area networks and WiFi local-area networks for transmission of voice, data, messaging and online services operating in the unlicensed ISM bands is an example of the commoditization of microwave technology for cost-sensitive consumer applications. In addition to the explosion in both diversity and capability of microwave-frequency communication systems, radar systems continue to be of importance with the emergence of nonmilitary and nonnavigational applications such as radar systems for automotive collision avoidance and weather and atmospheric sensing. Radar based noncontact fluid-level sensors are also increasingly being used in industrial process control applications. Traditional applications of microwaves in industrial material processing (primarily via nonradiative heating effects) and cooking have recently been augmented with medical uses for microwave-induced localized hyperthermia for oncological and other medical treatments.

In addition to these extensions of "traditional" microwave applications, other fields of electronics are increasingly encroaching into the microwave-frequency range. Examples include wired data networks based on coaxial cable or twisted-pair transmission lines with bit rates of over 1 Gb/s, fiber-optic communication systems with data rates well in excess of 10 Gb/s, and inexpensive personal computers and other digital systems with clock rates of well over 1 GHz. The continuing advances in the speed and capability of conventional microelectronics are pushing traditional circuit design ever further into the microwave-frequency regime. These advances have continued to push digital circuits into regimes where distributed circuit effects must be considered. While system- and board-level digital designers transitioned to the use of high-speed serial links requiring the use of distributed transmission lines in their designs some time ago, on-chip transmission lines for distribution of clock signals and the serialization of data signals for transmission over extremely high-speed serial buses are now an established feature of high-end designs within a single integrated circuit. These trends promise to both invigorate and reshape the field of microwave engineering in new and exciting ways.

I.3   Frequency Band Definitions

The field of microwave and RF engineering is driven by applications, originally for military purposes such as radar and more recently increasingly for commercial, scientific, and consumer applications. As a consequence of this increasingly diverse applications base, microwave terminology and frequency band designations are not entirely standardized, with various standards bodies, corporations, and other interested parties all contributing to the collective terminology of microwave engineering. Figure I.5 shows graphically the frequency ranges of some of the most common band designations. As can be seen from the complexity of Figure I.5, some care must be exercised in the use of the "standard" letter designations; substantial differences in the definitions of these bands exist in the literature and in practice. While the IEEE standard for radar bands [8] expressly deprecates the use of radar band designations for nonradar applications, the convenience of the band designations as technical shorthand has led to the use of these band designations in practice for a wide range of systems and technologies. This appropriation of radar band designations for other applications, as well as the definition of other letter-designated bands for other applications (e.g., electronic countermeasures) that have different frequency ranges is in part responsible for the complexity of Figure I.5. Furthermore, as progress in device and system performance opens up new system possibilities and makes ever-higher frequencies useful for new systems, the terminology of microwave engineering is continually evolving.

Figure I.5 illustrates in approximate order of increasing frequency the range of frequencies encompassed by commonly-used letter-designated bands. In Figure I.5, the dark shaded regions within the bars indicate the IEEE radar band designations, and the light cross-hatching indicates variations in the definitions by different groups and authors. The double-ended arrows appearing above some of the bands indicate other non-IEEE definitions for these letter designations that appear in the literature. For example, multiple distinct definitions of L, S, C, X, and K band are in use. The IEEE defines K band as the range from 18 to 27 GHz, while some authors define K band to span the range from 10.9 to 36 GHz, encompassing most of the IEEE's K_u, K, and K_a bands within a single band. Both of these definitions are illustrated in Figure I.5. Similarly, L band has two substantially different, overlapping definitions, with the IEEE definition of L band including frequencies from 1 to 2 GHz, with an older alternative definition of 390 MHz–1.55 GHz being found occasionally in the literature. Many other bands exhibit similar, though perhaps less extreme, variations in their definitions by various authors and standards committees. A further caution must also

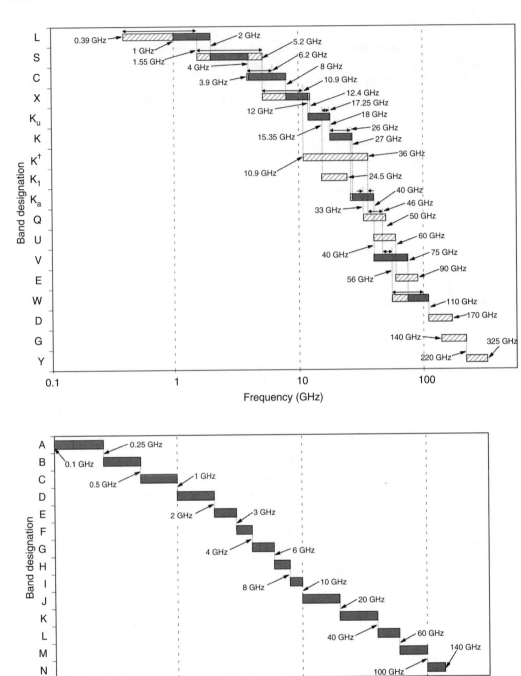

FIGURE I.5   Microwave and RF frequency band designations [1–7]. (a) Industrial and IEEE designations. Diagonal hashing indicates variation in the definitions found in literature; dark regions in the bars indicate the IEEE radar band definitions [8]. Double-ended arrows appearing above bands indicate alternative band definitions appearing in the literature, and K† denotes an alternative definition for K band found in Reference [7]. (b) U.S. military frequency band designations [2–5].

be taken with these letter designations, as different standards bodies and agencies do not always ensure that their letter designations are not used by others. As an example, the IEEE and U.S. military both define C, L, and K bands, but with very different frequencies; the IEEE L band resides at the low end of the microwave spectrum, while the military definition of L band is from 40 to 60 GHz. The designations (L–Y) in Figure I.5a are presently used widely in practice and the technical literature, with the newer U.S. military designations (A–N) shown in Figure I.5b having not gained widespread popularity outside of the military community.

I.4   Overview of The RF and Microwave Handbook

The field of microwave and RF engineering is inherently interdisciplinary, spanning the fields of system architecture, design, modeling, and validation; circuit design, characterization, and verification; active and passive device design, modeling, and fabrication, including technologies as varied as semiconductor devices, solid-state passives, and vacuum electronics; electromagnetic field theory, atmospheric wave propagation, electromagnetic compatibility and interference; and manufacturing, reliability and system integration. Additional factors, including biological effects of high-frequency radiation, system cost, and market factors also play key roles in the practice of microwave and RF engineering. This extremely broad scope is further amplified by the large number of technological and market-driven design choices faced by the practitioner on a regular basis.

The full sweep of microwave and RF engineering is addressed in this three-volume handbook series. Active and passive device technologies, as well as the behavior of materials at high frequencies, are discussed in this volume. Section I includes coverage of key passive device technologies, including radiating elements, cables and connectors, and packaging technology, as well as in-circuit passive elements including resonators, filters, and other components. The fundamentals of active device technologies, including semiconductor diodes, transistors and integrated circuits as well as vacuum electron devices, are discussed in Section II. Important device technologies including varactor and Schottky diodes, as well as bipolar junction transistors and heterojunction bipolar transistors in both the SiGe and III-V material systems are described, as are Si MOSFETs and III-V MESFETs and HEMTs. A discussion of the fundamental physical properties at high frequencies of common materials, including metals, dielectrics, ferroelectric and piezoelectric materials, and semiconductors, is provided in Section III.

A companion volume in this handbook series, *RF and Microwave Circuits, Measurements, and Modeling*, includes coverage of the unique difficulties and challenges encountered in accurately measuring microwave and RF devices and components. Section I of this companion volume discusses linear and non-linear characterization approaches, load-pull and large-signal network analysis techniques, noise measurements, fixturing and high-volume testing issues, and testing of digital systems. Consideration of key circuits for functional blocks in a wide array of system applications is addressed in Section II, including low-level circuits such as low-noise amplifiers, mixers, oscillators, power amplifiers, switches, and filters, as well as higher-level functionalities such as receivers, transmitters, and phase-locked loops. Section III of this companion volume discusses technology computer-aided design (TCAD) and nonlinear modeling of devices and circuits, along with analysis tools for systems, electromagnetics, and circuits.

Discussion of system-level considerations for high-frequency systems is featured in a third companion volume in this handbook series, *RF and Microwave Applications and Systems*. Section I of this companion volume focuses on system-level considerations with an application-specific focus. Typical applications, ranging from nomadic communications and cellular systems, wireless local-area networks, analog fiber-optic links, satellite communication networks, navigational aids and avionics, to radar, medical therapies, and electronic warfare applications are examined in detail. System-level considerations from the viewpoint of system integration and with focus on issues such as thermal management, cost modeling, manufacturing, and reliability are addressed in Section II of this volume in the handbook series, while the fundamental physical principles that govern the operation of devices and microwave and RF systems generally are discussed in Section III. Particular emphasis is placed on electromagnetic field theory through Maxwell's equations, free-space and guided-wave propagation, fading and multipath effects in wireless channels, and electromagnetic interference effects.

References

1. Chang, K., Bahl, I., and Nair, V., *RF and Microwave Circuit and Component Design for Wireless Systems*, John Wiley & Sons, New York, 2002.
2. Collin, R. E., *Foundations for Microwave Engineering*, McGraw-Hill, New York, 1992, 2.
3. Harsany, S. C., *Principles of Microwave Technology*, Prentice Hall, Upper Saddle River, 1997, 5.
4. Laverghetta, T. S., *Modern Microwave Measurements and Techniques*, Artech House, Norwood, 1988, 479.
5. Misra, D. K., *Radio-Frequency and Microwave Communication Circuits: Analysis and Design*, John Wiley & Sons, New York, 2001.
6. Rizzi, P. A., *Microwave Engineering*, Prentice-Hall, Englewood Cliffs, 1988, 1.
7. *Reference Data for Radio Engineers*, ITT Corp., New York, 1975.
8. IEEE Std. 521-2002.

1

Overview of Microwave Engineering

Mike Golio
HVVi Semiconductor

1.1 Semiconductor Materials for RF and Microwave Applications

In addition to consideration of unique properties of metal and dielectric materials, the radio frequency (RF) and microwave engineer must also make semiconductor choices based on how existing semiconductor properties address the unique requirements of RF and microwave systems. Although semiconductor materials are exploited in virtually all electronics applications today, the unique characteristics of RF and microwave signals requires that special attention be paid to specific properties of semiconductors which are often neglected or of second-order importance for other applications. Two critical issues to RF applications are (a) the speed of electrons in the semiconductor material and (b) the breakdown field of the semiconductor material.

The first issue, speed of electrons, is clearly important because the semiconductor device must respond to high frequency changes in polarity of the signal. Improvements in efficiency and reductions in parasitic losses are realized when semiconductor materials are used which exhibit high electron mobility and velocity. Figure 1.1 presents the electron velocity of several important semiconductor materials as a

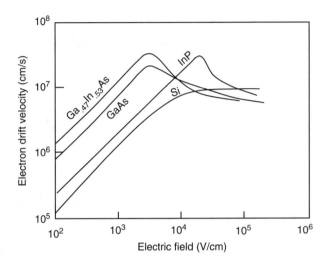

FIGURE 1.1   The electron velocity as a function of applied electric field for several semiconductor materials which are important for RF and microwave applications.

TABLE 1.1   Mobility and Breakdown Electric Field Values for Several Semiconductors Important for RF and Microwave Transmitter Applications

Property	Si	SiC	InP	GaAs	GaN
Electron mobility (cm^2/Vs)	1900	40–1000	4600	8800	1000
Breakdown field (V/cm)	3×10^5	20×10^4 to 30×10^5	5×10^5	6×10^5	$>10 \times 10^5$

function of applied electric field. The carrier mobility is given by

$$\mu_c = \frac{v}{e} \quad \text{for small values of } E \tag{1.1}$$

where v is the carrier velocity in the material and E is the electric field.

Although Silicon is the dominant semiconductor material for electronics applications today, Figure 1.1 illustrates that III–V semiconductor materials such as GaAs, GaInAs, and InP exhibit superior electron velocity and mobility characteristics relative to Silicon. Bulk mobility values for several important semiconductors are also listed in Table 1.1. As a result of the superior transport properties, transistors fabricated using III–V semiconductor materials such as GaAs, InP, and GaInAs exhibit higher efficiency and lower parasitic resistance at microwave frequencies.

From a purely technical performance perspective, the above discussion argues primarily for the use of III–V semiconductor devices in RF and microwave applications. These arguments are not complete, however. Most commercial wireless products also have requirements for high yield, high volume, low cost, and rapid product development cycles. These requirements can overwhelm the material selection process and favor mature processes and high volume experience. The silicon high volume manufacturing experience base is far greater than that of any III–V semiconductor facility.

The frequency of the application becomes a critical performance characteristic in the selection of device technology. Because of the fundamental material characteristics illustrated in Figure 1.1, Silicon device structures will always have lower theoretical maximum operation frequencies than identical III–V device structures. The higher the frequency of the application, the more likely the optimum device choice will be a III–V transistor over a Silicon transistor. Above some frequency, $f_{\text{III–V}}$, compound semiconductor devices dominate the application space, with Silicon playing no significant role in the microwave portion of the product. In contrast, below some frequency, f_{Si}, the cost and maturity advantage of Silicon provide little opportunity for III–V devices to compete. In the transition spectrum between these two frequencies Silicon and III–V devices coexist. Although Silicon devices are capable of operating above frequency f_{Si}, this operation is often gained at the expense of DC current drain. As frequency is increased above f_{Si} in the transition spectrum, efficiency advantages of GaAs and other III–V devices provide competitive opportunities for these parts. The critical frequencies, f_{Si} and f_{III-V} are not static frequency values. Rather, they are continually being moved upward by the advances of Silicon technologies—primarily by decreasing critical device dimensions.

The speed of carriers in a semiconductor transistor can also be affected by deep levels (traps) located physically either at the surface or in the bulk material. Deep levels can trap charge for times that are long compared to the signal period and thereby reduce the total RF power carrying capability of the transistor. Trapping effects result in frequency dispersion of important transistor characteristics such as transconductance and output resistance. Pulsed measurements as described in Section 1.4.4 (especially when taken over temperature extremes) can be a valuable tool to characterize deep level effects in semiconductor devices. Trapping effects are more important in compound semiconductor devices than in silicon technologies.

The second critical semiconductor issue listed in Table 1.1 is breakdown voltage. The constraints placed on the RF portion of radio electronics are fundamentally different from the constraints placed on digital circuits in the same radio. For digital applications, the presence or absence of a single electron can theoretically define a bit. Although noise floor and leakage issues make the practical limit for bit signals larger than this, the minimum amount of charge required to define a bit is very small. The bit charge minimum is also independent of the radio system architecture, the radio transmission path or the external environment. If the amount of charge utilized to define a bit within the digital chip can be reduced, then operating voltage, operating current, or both can also be reduced with no adverse consequences for the radio.

In contrast, the required propagation distance and signal environment are the primary determinants for RF signal strength. If 1 W of transmission power is required for the remote receiver to receive the signal, then reductions in RF transmitter power below this level will cause the radio to fail. Modern radio requirements often require tens, hundreds, or even thousands of Watts of transmitted power in order for the radio system to function properly. Unlike the digital situation where any discernable bit is as good as any other bit, the minimum RF transmission power must be maintained. A Watt of RF power is the product of signal current, signal voltage and efficiency, so requirements for high power result in requirements for high voltage, high current and high efficiency.

The maximum electric field before the onset of avalanche breakdown, *breakdown field*, is the fundamental semiconductor property that often limits power operation in a transistor. Table 1.1 presents breakdown voltages for several semiconductors that are commonly used in transmitter applications. In addition to Silicon, GaAs and InP, two emerging widebandgap semiconductors, SiC and GaN are included in the table. Interest from microwave engineers in these less mature semiconductors is driven almost exclusively by their attractive breakdown field capabilities. Figure 1.2 summarizes the semiconductor material application situation in terms of the power–frequency space for RF and microwave systems.

1.2 Propagation and Attenuation in the Atmosphere

Many modern RF and microwave systems are wireless. Their operation depends on transmission of signals through the atmosphere. Electromagnetic signals are attenuated by the atmosphere as they propagate from source to target. Consideration of the attenuation characteristics of the atmosphere can be critical in the design of these systems. In general, atmospheric attenuation increases with increasing frequency. As shown in Figure 1.3, however, there is significant structure in the atmospheric attenuation versus frequency plot. If only attenuation is considered, it is clear that low frequencies would be preferred for long range communications, sensor, or navigation systems in order to take advantage of the low attenuation of the atmosphere. If high data rates or large information content is required, however, higher frequencies

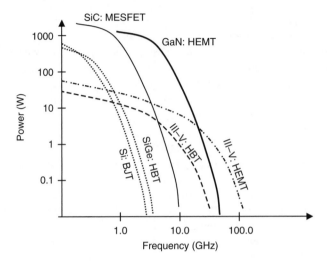

FIGURE 1.2 Semiconductor choices for RF applications are a strong function of the power and frequency required for the wireless application.

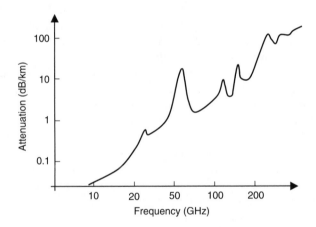

FIGURE 1.3 Attenuation of electromagnetic signals in the atmosphere as a function of frequency.

are needed. In addition to the atmospheric attenuation, the wavelengths of microwave systems are small enough to become effected by water vapor and rain. Above 10 GHz these effects become important. Above 25 GHz, the effect of individual gas molecules becomes important. Water and oxygen are the most important gases. These have resonant absorption lines at ~23, ~69, and ~120 GHz. In addition to absorption lines, the atmosphere also exhibits "windows" that may be used for communication, notably at ~38 and ~98 GHz.

RF and microwave signal propagation is also affected by objects such as trees, buildings, towers, and vehicles in the path of the wave. Indoor systems are affected by walls, doors, furniture, and people. As a result of the interaction of electromagnetic signals with objects, the propagation channel for wireless communication systems consists of multiple paths between the transmitter and receiver. Each path will experience different attenuation and delay. Some transmitted signals may experience a deep fade (large attenuation) due to destructive multipath cancellation. Similarly, constructive multipath addition can produce signals of large amplitude. Shadowing can occur when buildings or other objects obstruct the line-of-site path between transmitter and receiver.

The design of wireless systems must consider the interaction of specific frequencies of RF and microwave signals with the atmosphere and with objects in the signal channel that can cause multipath effects.

1.3 Systems Applications

There are four important classes of applications for microwave and RF systems: communications, navigation, sensors, and heating. Each of these classes of applications benefits from some of the unique properties of high-frequency electromagnetic fields.

1.3.1 Communications

Wireless communications applications have exploded in popularity over the past decade. Pagers, cellular phones, radio navigation, and wireless data networks are among the RF products that consumers are likely to be familiar with. Prior to the growth of commercial wireless communications, RF and microwave radios were in common usage for communications satellites, commercial avionics communications, and many government and military radios. All of these systems benefit from the high frequencies that offer greater bandwidth than low frequency systems, while still propagating with relatively low atmospheric losses compared to higher frequency systems.

Cellular phones are among the most common consumer radios in use today. Analog cellular (first generation or 1G cellular) operates at 900 MHz bands and was first introduced in 1983. Second generation (2G) cellular using TDMA, GSM TDMA, and CDMA digital modulation schemes came into use more than 10 years later. The 2G systems were designed to get greater use of the 1.9 GHz frequency bands than their analog predecessors. Emergence of 2.5G and 3G systems operating in broader bands as high as 2.1 GHz is occurring today. These systems make use of digital modulation schemes adapted from 2G GSM and CDMA systems. With each advance in cellular phones, requirements on the microwave circuitry have increased. Requirements for broader bandwidths, higher efficiency and greater linearity have been coupled with demands for lower cost, lighter, smaller products, and increasing functionality. The microwave receivers and transmitters designed for portable cellular phones represent one of the highest volume manufacturing requirements of any microwave radio. Fabrication of popular cell phones has placed an emphasis on manufacturability and yield for microwave radios that was unheard of prior to the growth in popularity of these products.

Other microwave-based consumer products that are growing dramatically in popularity are the wireless local area network (WLAN) or Wi-Fi and the longer range WiMAX systems. These systems offer data rates more than five times higher than cellular-based products using bandwidth at 2.4, 3.5, and 5 GHz. Although the volume demands for Wi-Fi and WiMAX components are not as high as for cellular phones, the emphasis on cost and manufacturability is still critical to these products.

Commercial communications satellite systems represent a microwave communications product that is less conspicuous to the consumer, but continues to experience increasing demand. Although the percentage of voice traffic carried via satellite systems is rapidly declining with the advent of undersea fiber-optic cables, new video and data services are being added over existing voice services. Today satellites provide worldwide TV channels, global messaging services, positioning information, communications from ships and aircraft, communications to remote areas, and high-speed data services including internet access. Allocated satellite communication frequency bands include spectrum from as low as 2.5 GHz to almost 50 GHz. These allocations cover extremely broad bandwidths compared to many other communications systems. Future allocation will include even higher frequency bands. In addition to the bandwidth and frequency challenges, microwave components for satellite communications are faced with reliability requirements that are far more severe than any earth-based systems.

Avionics applications include subsystems that perform communications, navigation, and sensor applications. Avionics products typically require functional integrity and reliability that are orders of magnitude more stringent than most commercial wireless applications. The rigor of these requirements is matched or exceeded only by the requirements for space and/or certain military applications. Avionics must function in environments that are more severe than most other wireless applications as well. Quantities of products required for this market are typically very low when compared to commercial wireless applications, for example, the number of cell phones manufactured every single working day far exceeds the number of aircraft that are manufactured in the world in a year. Wireless systems for avionics applications cover an extremely wide range of frequencies, function, modulation type, bandwidth, and power. Due to the number of systems aboard a typical aircraft, Electromagnetic Interference (EMI) and Electromagnetic Compatibility (EMC) between systems is a major concern, and EMI/EMC design and testing is a major factor in the flight certification testing of these systems. RF and microwave communications systems for avionics applications include several distinct bands between 2 and 400 MHz and output power requirements as high as 100 Watts.

In addition to commercial communications systems, military communication is an extremely important application of microwave technology. Technical specifications for military radios are often extremely demanding. Much of the technology developed and exploited by existing commercial communications systems today was first demonstrated for military applications. The requirements for military radio applications are varied but will cover broader bandwidths, higher power, more linearity, and greater levels of integration than most of their commercial counterparts. In addition, reliability requirements for these systems are stringent. Volume manufacturing levels, of course, tend to be much lower than commercial systems.

1.3.2 Navigation

Electronic navigation systems represent a unique application of microwave systems. In this application, data transfer takes place between a satellite (or fixed basestation) and a portable radio on earth. The consumer portable product consists of only a receiver portion of a radio. No data or voice signal is transmitted by the portable navigation unit. In this respect, electronic navigation systems resemble a portable paging system more closely than they resemble a cellular phone system. The most widespread electronic navigation system is GPS. The nominal GPS constellation is composed of 24 satellites in six orbital planes, (four satellites in each plane). The satellites operate in circular 20,200 km altitude (26,570 km radius) orbits at an inclination angle of 55°. Each satellite transmits a navigation message containing its orbital elements, clock behavior, system time, and status messages. The data transmitted by the satellite are sent in two frequency bands at 1.2 and 1.6 GHz. The portable terrestrial units receive these messages from multiple satellites and calculate the location of the unit on the earth. In addition to GPS, other navigation systems in common usage include NAVSTAR, GLONASS, and LORAN.

1.3.3 Sensors (Radar)

Microwave sensor applications are addressed primarily with various forms of radar. Radar is used by police forces to establish the speed of passing automobiles, by automobiles to establish vehicle speed and danger of collision, by air traffic control systems to establish the locations of approaching aircraft, by aircraft to establish ground speed, altitude, other aircraft and turbulent weather, and by the military to establish a multitude of different types of targets.

The receiving portion of a radar unit is similar to other radios. It is designed to receive a specific signal and analyze it to obtain desired information. The radar unit differs from other radios, however, in that the signal that is received is typically transmitted by the same unit. By understanding the form of the transmitted signal, the propagation characteristics of the propagation medium, and the form of the received (reflected) signal, various characteristics of the radar target can be determined including size, speed, and distance from the radar unit. As in the case of communications systems, radar applications benefit from the propagation characteristics of RF and microwave frequencies in the atmosphere. The best frequency to use for a radar unit depends upon its application. Like most other radio design decisions, the choice of frequency usually involves trade-offs among several factors including physical size, transmitted power, and atmospheric attenuation.

The dimensions of radio components used to generate RF power and the size of the antenna required to direct the transmitted signal are, in general, proportional to wavelength. At lower frequencies where wavelengths are longer, the antennae and radio components tend to be large and heavy. At the higher frequencies where the wavelengths are shorter, radar units can be smaller and lighter.

Frequency selection can indirectly influence the radar power level because of its impact on radio size. Design of high power transmitters requires that significant attention be paid to the management of electric field levels and thermal dissipation. Such management tasks are made more complex when space is limited. Since radio component size tends to be inversely proportional to frequency, manageable power levels are reduced as frequency is increased.

As in the case of all wireless systems, atmospheric attenuation can reduce the total range of the system. Radar systems designed to work above about 10 GHz must consider the atmospheric loss at the specific frequency being used in the design.

Automotive radar represents a large class of radars that are used within an automobile. Applications include speed measurement, adaptive cruise control, obstacle detection, and collision avoidance. Various radar systems have been developed for forward-, rear-, and side-looking applications.

V-band frequencies are exploited for forward looking radars. Within V-band, different frequencies have been used in the past decade, including 77 GHz for U.S. and European systems, and 60 GHz in some Japanese systems. The choice of V-band for this application is dictated by the resolution requirement, antenna size requirement and the desire for atmospheric attenuation to insure the radar is short range. The frequency requirement of this application has contributed to a slow emergence of this product into mainstream use, but the potential of this product to have a significant impact on highway safety continues to keep automotive radar efforts active.

As in the case of communications systems, avionics and military users also have significant radar applications. Radar is used to detect aircraft both from the earth and from other aircraft. It is also used to determine ground speed, establish altitude, and detect weather turbulence.

1.3.4 Heating

The most common heating application for microwave signals is the microwave oven. These consumer products operate at a frequency that corresponds to a resonant frequency of water. When exposed to electromagnetic energy at this frequency, all water molecules begin to spin or oscillate at that frequency. Since all foods contain high percentages of water, electromagnetic energy at this resonant frequency interacts with all foods. The energy absorbed by these rotating molecules is transferred to the food in the form of heat.

RF heating can also be important for medical applications. Certain kinds of tumors can be detected by the lack of electromagnetic activity associated with them and some kinds of tumors can be treated by heating them using electromagnetic stimulation.

The use of RF/microwaves in medicine has increased dramatically in recent years. RF and microwave therapies for cancer in humans are presently used in many cancer centers. RF treatments for heartbeat irregularities are currently employed by major hospitals. RF/microwaves are also used in human subjects for the treatment of certain types of benign prostrate conditions. Several centers in the United States have been utilizing RF to treat upper airway obstruction and alleviate sleep apnea. New treatments such as microwave aided liposuction, tissue joining in conjunction with microwave irradiation in future endoscopic surgery, enhancement of drug absorption, and microwave septic wound treatment are continually being researched.

1.4 Measurements

The RF/microwave engineer faces unique measurement challenges. At high frequencies, voltages and currents vary too rapidly for conventional electronic measurement equipment to gauge. Conventional curve tracers and oscilloscopes are of limited value when microwave component measurements are needed. In addition, calibration of conventional characterization equipment typically requires the use of open and short circuit standards that are not useful to the microwave engineer. For these reasons, most commonly exploited microwave measurements focus on the measurement of power and phase in the frequency domain as opposed to voltages and currents in the time domain.

1.4.1 Small Signal

Characterization of the linear performance of microwave devices, components and boards is critical to the development of models used in the design of the next higher level of microwave subsystem. At lower frequencies, direct measurement of y-, z-, or h-parameters is useful to accomplish linear characterization. As discussed in Chapter 1, however, RF and microwave design utilizes s-parameters for this application. Other small signal characteristics of interest in microwave design include impedance, VSWR, gain, and attenuation. Each of these quantities can be computed from two-port s-parameter data.

The s-parameters defined in Chapter 1 are complex quantities normally expressed as magnitude and phase. Notice that S_{11} and S_{22} can be thought of as complex reflection ratios since they represent the magnitude and phase of waves reflected from port 1 (input) and 2 (output), respectively. It is common to measure the quality of the match between components using the *reflection coefficient* defined as

$$\Gamma = |S_{11}| \tag{1.2}$$

for the input reflection coefficient of a two-port network, or

$$\Gamma = |S_{22}| \tag{1.3}$$

for the output reflection coefficient.

Reflection coefficient measurements are often expressed in dB and referred to as *return loss* evaluated as

$$L_{\text{return}} = -20\log(\Gamma). \tag{1.4}$$

Analogous to the reflection coefficient, both a forward and reverse *transmission coefficient* can be measured. The forward transmission coefficient is given as

$$T = |S_{21}| \tag{1.5}$$

while the reverse transmission coefficient is expressed

$$T = |S_{12}|. \tag{1.6}$$

As in the case of reflection coefficient, transmission coefficients are often expressed in dB and referred to as *gain* given by

$$G = 20\log(T). \tag{1.7}$$

Another commonly measured and calculated parameter is the *standing wave ratio* or the *voltage standing wave ratio* (VSWR). This quantity is the ratio of maximum to minimum voltage at a given port. It is commonly expressed in terms of reflection coefficient as

$$\text{VSWR} = \frac{1 + \Gamma}{1 - \Gamma}. \tag{1.8}$$

The vector network analyzer (VNA) is the instrument of choice for small signal characterization of high-frequency components. Figure 1.4 illustrates a one-port VNA measurement. These measurements

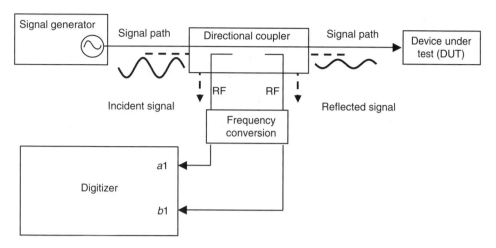

FIGURE 1.4 Vector network analyzer measurement configuration to determine *s*-parameters of a high-frequency device, component, or subsystem.

use a source with well-defined impedance equal to the system impedance and all ports of the device under test (DUT) are terminated with the same impedance. This termination eliminates unwanted signal reflections during the measurement. The port being measured is terminated in the test channel of the network analyzer that has input impedance equal to the system characteristic impedance. Measurement of system parameters with all ports terminated minimizes the problems caused by short-, open-, and test-circuit parasitics that cause considerable difficulty in the measurement of *y*- and *h*-parameters at very high frequencies. If desired, *s*-parameters can be converted to *y*- and *h*-parameters using analytical mathematical expressions.

The directional coupler shown in Figure 1.4 is a device for measuring the forward and reflected waves on a transmission line. During the network analyzer measurement, a signal is driven through the directional coupler to one port of the DUT. Part of the incident signal is sampled by the directional coupler. On arrival at the DUT port being measured, some of the incident signal will be reflected. This reflection is again sampled by the directional coupler. The sampled incident and reflected signals are then downconverted in frequency and digitized. The measurement configuration of Figure 1.4 shows only one-half of the equipment required to make full two-port *s*-parameter measurements. The *s*-parameters as defined in Chapter 1 are determined by analyzing the ratios of the digitized signal data.

For many applications, knowledge of the magnitude of the incident and reflected signals is sufficient (i.e., Γ is all that is needed). In these cases, the scalar network analyzer can be utilized in place of the VNA. The cost of the scalar network analyzer equipment is much less than VNA equipment and the calibration required for making accurate measurements is easier when phase information is not required. The scalar network analyzer measures reflection coefficient as defined in Equations 2.1 and 2.2.

1.4.2 Large Signal

Virtually all physical systems exhibit some form of nonlinear behavior and microwave systems are no exception. Although powerful techniques and elaborate tools have been developed to characterize and analyze linear RF and microwave circuits, it is often the nonlinear characteristics that dominate microwave engineering efforts. Nonlinear effects are not all undesirable. Frequency conversion circuitry, for example, exploits nonlinearities in order to translate signals from one frequency to another. Nonlinear performance characteristics of interest in microwave design include harmonic distortion, gain compression, intermodulation distortion (IMD), phase distortion, and adjacent channel power. Numerous other nonlinear phenomena and nonlinear figures-of-merit are less commonly addressed, but can be important for some microwave systems.

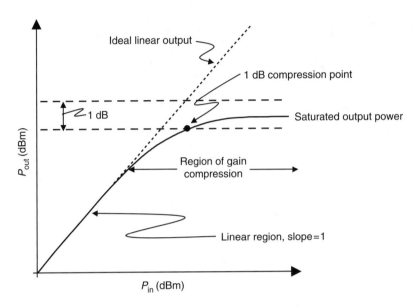

FIGURE 1.5   Output power versus input power at the fundamental frequency for a nonlinear circuit.

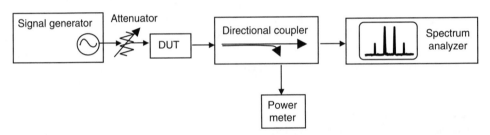

FIGURE 1.6   Measurement configuration to characterize gain compression and harmonic distortion. By replacing the signal generator with two combined signals at slightly offset frequencies, the configuration can also be used to measure intermodulation distortion.

1.4.2.1 Gain Compression

Figure 1.5 illustrates gain compression characteristics of a typical microwave amplifier with a plot of output power as a function of input power. At low power levels, a single frequency signal is increased in power level by the small signal gain of the amplifier ($P_{out} = G * P_{in}$). At lower power levels, this produces a linear P_{out} versus P_{in} plot with slope = 1 when the powers are plotted in dB units as shown in Figure 1.5. At higher power levels, nonlinearities in the amplifier begin to generate some power in the harmonics of the single frequency input signal and to compress the output signal. The result is decreased gain at higher power levels. This reduction in gain is referred to as *gain compression*. Gain compression is often characterized in terms of the power level when the large signal gain is 1 dB less than the small signal gain. The power level when this occurs is termed the *1dB compression point* and is also illustrated in Figure 1.5.

The microwave spectrum analyzer is the workhorse instrument of nonlinear microwave measurements. The instrument measures and displays power as a function of swept frequency. Combined with a variable power level signal source (or multiple combined or modulated sources), many nonlinear characteristics can be measured using the spectrum analyzer in the configuration illustrated in Figure 1.6.

1.4.2.2 Harmonic Distortion

A fundamental result of nonlinear distortion in microwave devices is that power levels are produced at frequencies which are integral multiples of the applied signal frequency. These other frequency components are termed *harmonics* of the fundamental signal. Harmonic signal levels are usually specified and measured relative to the fundamental signal level. The harmonic level is expressed in dBc, which designates dB relative to the fundamental power level. Microwave system requirements often place a maximum acceptable level for individual harmonics. Typically third and second harmonic levels are critical, but higher-order harmonics can also be important for many applications. The measurement configuration illustrated in Figure 1.6 can be used to directly measure harmonic distortion of a microwave device.

1.4.2.3 Intermodulation Distortion

When a microwave signal is composed of power at multiple frequencies, a nonlinear circuit will produce IMD. The IMD characteristics of a microwave device are important because they can create unwanted interference in adjacent channels of a radio or radar system. The intermodulation products of two signals produce distortion signals not only at the harmonic frequencies of the two signals, but also at the sum and difference frequencies of all of the signal's harmonics. If the two signal frequencies are closely spaced at frequencies f_c and f_m, then the IMD products located at frequencies $2f_c - f_m$ and $2f_m - f_c$ will be located very close to the desired signals. This situation is illustrated in the signal spectrum of Figure 1.7. The IMD products at $2f_c - f_m$ and $2f_m - f_c$ are third-order products of the desired signals, but are located so closely to f_c and f_m that filtering them out of the overall signal is difficult.

The spectrum of Figure 1.7 represents the nonlinear characteristics at a single power level. As power is increased and the device enters gain compression, however, harmonic power levels will grow more quickly than fundamental power levels. In general, the nth-order harmonic power level will increase at n times the fundamental. This is illustrated in the P_{out} versus P_{in} plot of Figure 1.8 where both the fundamental and the third-order products are plotted. As in the case of the fundamental power, third-order IMD levels will compress at higher power levels. IMD is often characterized and specified in terms of the *third-order intercept point*, IP3. This point is the power level where the slope of the small signal gain and the slope of the low power level third-order product characteristics cross as shown in Figure 1.8.

1.4.2.4 Phase Distortion

Reactive elements in a microwave system give rise to time delays that are nonlinear. Such delays are referred to as *memory effects* and result in *AM–PM distortion* in a modulated signal. AM–PM distortion creates

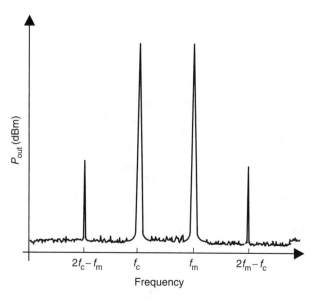

FIGURE 1.7 An illustration of signal spectrum due to intermodulation distortion from two signals at frequencies f_c and f_m.

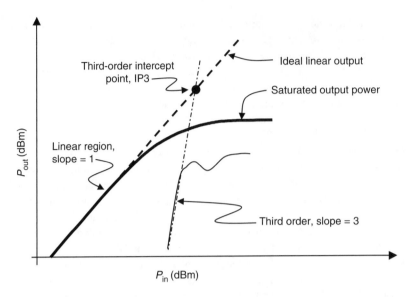

FIGURE 1.8   Relationship between signal output power and intermodulation distortion product levels.

sidebands at harmonics of a modulating signal. These sidebands are similar to the IMD sidebands, but are repeated for multiple harmonics. AM–PM distortion can dominate the out-of-band interference in a radio. At lower power levels, the phase deviation of the signal is approximately linear and the slope of the deviation, referred to as the modulation index, is often used as a figure-of-merit for the characterization of this nonlinearity. The *modulation index* is measured in degrees per volt using a VNA. The phase deviation is typically measured at the 1 dB compression point in order to determine modulation index. Because the VNA measures power, the computation of modulation index, k_ϕ, uses the formula

$$k_\phi = \frac{\Delta\Phi(P_{1\text{dB}})}{2Z_0\sqrt{P_{1\text{dB}}}} \tag{1.9}$$

where $\Delta\Phi(P_{1\text{dB}})$ is the phase deviation from small signal at the 1 dB compression point, Z_0 is the characteristic impedance of the system and $P_{1\text{dB}}$ is the 1 dB output compression point.

1.4.2.5   Adjacent Channel Power Ratio

Amplitude and phase distortion affect digitally modulated signals resulting in gain compression and phase deviation. The resulting signal, however, is far more complex than the simple one or two carrier results presented in Sections 1.4.2.2 through 1.4.2.4. Instead of IMD, *adjacent channel power ratio* (ACPR) is often specified for digitally modulated signals. ACPR is a measure of how much power leaks into adjacent channels of a radio due to the nonlinearities of the digitally modulated signal in a central channel. Measurement of ACPR is similar to measurement of IMD, but utilizes an appropriately modulated digital test signal in place of a single tone signal generator. Test signals for digitally modulated signals are synthesized using an *arbitrary waveform generator*. The output spectrum of the DUT in the channels adjacent to the tested channel are then monitored and power levels are measured.

1.4.2.6   Error Vector Magnitude

Adjacent channel power specifications are not adequate for certain types of modern digitally modulated systems. *Error vector magnitude* (EVM) is used in addition to, or instead of adjacent channel power for these systems. EVM specifications have already been written into system standards for GSM, NADC, and PHS, and they are poised to appear in many important emerging standards.

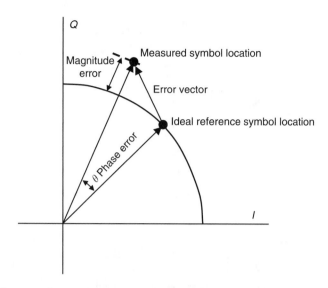

FIGURE 1.9 *I–Q* diagram indicating the error vector for EVM measurements.

The EVM measurement quantifies the performance of a radio transmitter against an ideal reference. A signal sent by an ideal transmitter would have all points in the *I–Q* constellation fall precisely at the ideal locations (i.e., magnitude and phase would be exact). Nonideal behavior of the transmitter, however, causes the actual constellation points to fall in a slightly scattered pattern that only approximates the ideal *I–Q* location. EVM is a way to quantify how far the actual points are from the ideal locations. This is indicated in Figure 1.9.

Measurement of EVM is accomplished using a vector signal analyzer (VSA). The equipment demodulates the received signal in a similar way to the actual radio demodulator. The actual *I–Q* constellation can then be measured and compared to the ideal constellation. EVM is calculated as the ratio of the root mean square power of the error vector to the RMS power of the reference.

1.4.3 Noise

Noise is a random process that can have many different sources such as thermally generated resistive noise, charge crossing a potential barrier, and generation–recombination (G–R) noise. Understanding noise is important in microwave systems because background noise levels limit the sensitivity, dynamic range and accuracy of a radio or radar receiver.

1.4.3.1 Noise Figure

At microwave frequencies noise characterization involves the measurement of noise power. The noise power of a linear device can be considered as concentrated at its input as shown in Figure 1.10. The figure considers an amplifier, but the analysis is easily generalized to other linear devices.

All of the amplifier noise generators can be lumped into an equivalent noise temperature with an equivalent input noise power per Hertz of $N_e = kT_e$, where k is Boltzmann's constant and T_e is the equivalent noise temperature. The noise power per Hertz available from the noise source is $N_S = kT_S$ as shown in Figure 1.10. Since noise limits the system sensitivity and dynamic range, it is useful to examine noise as it is related to signal strength using a signal-to-noise ratio (SNR). A figure-of-merit for an amplifier, *noise factor* (F), describes the reduction in SNR of a signal as it passes through the linear device illustrated in Figure 1.10. The noise factor for an amplifier is derived from the figure to be

$$F = \frac{\text{SNR}_{\text{IN}}}{\text{SNR}_{\text{OUT}}} = 1 + \frac{T_e}{T_S} \tag{1.10}$$

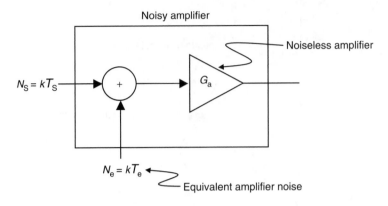

FIGURE 1.10   System view of amplifier noise.

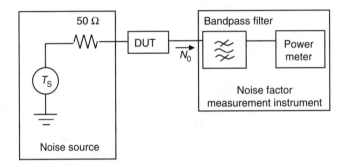

FIGURE 1.11   Measurement configuration for noise factor measurement.

Device noise factor can be measured as shown in Figure 1.11. To make the measurement, the source temperature is varied resulting in variation in the device noise output, N_0. The device noise contribution, however, remains constant. As T_S changes the noise power measured at the power meter changes providing a method to compute noise output.

In practice, the noise factor is usually given in decibels and called the *noise figure*,

$$NF = 10 \log F \tag{1.11}$$

1.4.3.2   Phase Noise

When noise is referenced to a carrier frequency it modulates that carrier and causes amplitude and phase variations known as phase noise. Oscillator phase modulation (PM) noise is much larger than amplitude modulation (AM) noise. The phase variations caused by this noise result in *jitter* which is critical in the design and analysis of digital communication systems.

Phase noise is most easily measured using a spectrum analyzer. Figure 1.12 shows a typical oscillator source spectrum as measured directly on a spectrum analyzer. Characterization and analysis of phase noise is often described in terms of the power ratio of the noise at specific distances from the carrier frequency. This is illustrated in Figure 1.12.

1.4.4   Pulsed I–V

Although most of the measurements commonly utilized in RF and microwave engineering are fre-quency domain measurements, pulsed measurements are an important exception used to characterize

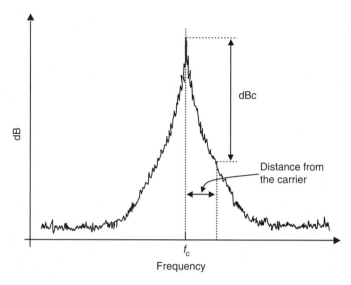

FIGURE 1.12 Typical phase noise spectrum observed on a spectrum analyzer.

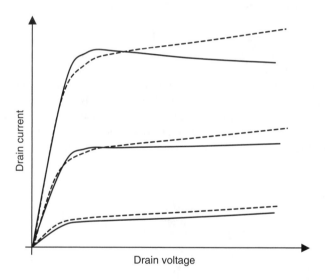

FIGURE 1.13 Pulsed *I–V* characteristics of a microwave FET. Solid lines are DC characteristics while dashed lines are pulsed.

high-frequency transistors. At RF and microwave frequencies, mechanisms known as *dispersion effects* become important to transistor operation. These effects reveal themselves as a difference in *I–V* characteristics obtained using a slow sweep as opposed to *I–V* characteristics obtained using a rapid pulse. The primary physical causes of *I–V* dispersion are thermal effects and carrier traps in the semiconductor. Figure 1.13 illustrates the characteristics of a microwave transistor under DC (solid lines) and pulsed (dashed lines) stimulation. In order to characterize dispersion effects, pulse rates must be shorter than the thermal and trapping time constants that are being monitored. Typically, for microwave transistors, that requires a pulse on the order of 100 ns or less. Similarly, the quiescent period between pulses should be long compared to the measured effects. Typical quiescent periods are on the order of 100 ms or more. The discrepancy between DC and pulsed characteristics is an indication of how severely the semiconductor traps and thermal effects will impact device performance.

Another use for pulsed *I–V* measurement is the characterization of high power transistors. Many high power transistors (greater than a few dozen Watts) are only operated in a pulsed mode or at a bias level far below their maximum currents. If these devices are biased at higher current levels for a few milliseconds, the thermal dissipation through the transistor will cause catastrophic failure. This is a problem for transistor model development, since a large range of *I–V* curves—including high current settings—is needed to extract an accurate model. Pulsed *I–V* data can provide input for model development while avoiding unnecessary stress on the part being characterized.

1.5 Circuits and Circuit Technologies

Figure 1.14 illustrates a generalized radio architecture that is typical of the systems used in many wireless applications today. The generalized diagram can apply to either communications or radar applications. In a wired application, the antenna of Figure 1.14 can be replaced with a transmission line. The duplexer of Figure 1.14 will route signals at the transmission frequency from the PA to the antenna while isolating that signal from the low noise amplifier (LNA). It will also route signals at the receive frequency from the antenna to the LNA. For some systems, input and output signals are separated in time instead of frequency. In these systems, an RF switch is used instead of a duplexer. Matching elements and other passive frequency selective circuit elements are used internally to all of the components shown in the figure. In addition, radio specifications typically require the use of filters at the ports of some of the components illustrated in Figure 1.14.

A signal received by the antenna is routed via the duplexer to the receive path of the radio. An LNA amplifies the signal before a mixer downconverts it to a lower frequency. The downconversion is accomplished by mixing the received signal with an internally generated local oscillator (LO) signal. The ideal receiver rejects all unwanted noise and signals. It adds no noise or interference and converts the signal to a lower frequency that can be efficiently processed without adding distortion.

On the transmitter side, a modulated signal is first upconverted and then amplified by the PA before being routed to the antenna. The ideal transmitter boosts the power and frequency of a modulated signal to that required for the radio to achieve communication with the desired receiver. Ideally, this

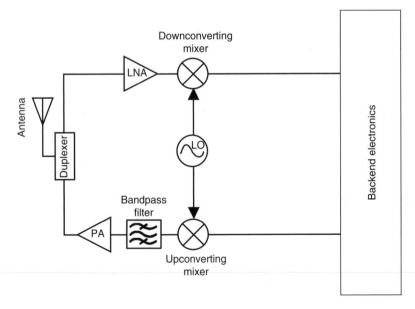

FIGURE 1.14 Generalized microwave radio architecture illustrating the microwave components in both the receiver and transmitter path.

process is accomplished efficiently (minimum DC power requirements) and without distortion. It is especially important that the signal broadcast from the antenna include no undesirable frequency components.

To accomplish the required transmitter and receiver functions, RF and microwave components must be developed either individually or as part of an integrated circuit. The remainder of this section will examine issues related to individual components that comprise the radio.

1.5.1 Low Noise Amplifier

The LNA is often most critical in determining the overall performance of the receiver chain of a wireless radio. The noise figure of the LNA has the greatest impact of any component on the overall receiver noise figure and receiver sensitivity. The LNA should minimize the system noise figure, provide sufficient gain, minimize nonlinearities, and assure stable 50 Ω impedance with low power consumption. The two performance specifications of primary importance to determine LNA quality are gain and noise figure.

In many radios, the LNA is part of a single chip design that includes a mixer and other receiver functions as well as the LNA. In these applications, the LNA may be realized using Silicon, SiGe, GaAs or another semiconductor technology. Si BJTs and SiGe HBTs dominate the LNA business at frequencies below a couple of GHz because of their tremendous cost and integration advantages over compound semiconductor devices. Compound semiconductors are favored as frequency increases and noise figure requirements decrease. For applications that require extremely low noise figures, cooled compound semiconductor HEMTs are the favored device.

1.5.2 Power Amplifier

A PA is required at the output of a transmitter to boost the signal to the power levels necessary for the radio to achieve a successful link with the desired receiver. PA components are almost always the most difficult and expensive part of microwave radio design. At high power levels, semiconductor nonlinearities such as breakdown voltage become critical design concerns. Thermal management issues related to dissipating heat from the RF transistor can dominate the design effort. Efficiency of the amplifier is critical, especially in the case of portable radio products. PA efficiency is essential to obtain long battery lifetime in a portable product. Critical primary design specifications for PAs include output power, gain, linearity, and efficiency.

For many applications, PA components tend to be discrete devices with minimal levels of on-chip semiconductor integration. The unique semiconductor and thermal requirements of PAs dictate the use of unique fabrication and manufacturing techniques in order to obtain required performance. The power and frequency requirements of the application typically dictate what device technology is required for the PA. At frequencies as low as 800 MHz and power levels of 1 Watt, compound semiconductor devices often compete with Silicon and SiGe for PA devices. As power and frequency increase from these levels, compound semiconductor HBTs and HEMTs dominate in this application. Vacuum tube technology is still required to achieve performance for some extremely high-power or high-frequency applications.

1.5.3 Mixer

A mixer is essentially a multiplier and can be realized with any nonlinear device. If at least two signals are present in a nonlinear device, their products will be produced at the device output. The mixer is a frequency translating device. Its purpose is to translate the incoming signal at frequency, f_{RF}, to a different outgoing frequency, f_{IF}. The LO port of the mixer is an input port and is used to *pump* the RF signal and create the IF signal.

Mixer characterization normally includes the following parameters:

- Input match (at the RF port)
- Output match (at the IF port)

- LO to RF leakage (from the LO to RF port)
- LO to IF leakage (from LO to IF port)
- Conversion Loss (from the RF port to the IF port)

The first four parameters are single frequency measurements similar to s-parameters S_{11}, S_{22}, S_{13}, and S_{23}. Conversion loss is similar to s-parameter S_{21}, but is made at the RF frequency at the input port and at the IF frequency at the output port.

Although a mixer can be made from any nonlinear device, many RF/microwave mixers utilize one or more diodes as the nonlinear element. FET mixers are also used for some applications. As in the case of amplifiers, the frequency of the application has a strong influence on whether Silicon or compound semiconductor technologies are used.

1.5.4 RF Switch

RF switches are control elements required in many wireless applications. They are used to control and direct signals under stimulus from externally applied voltages or currents. Phones and other wireless communication devices utilize switches for duplexing and switching between frequency bands and modes.

Switches are ideally a linear device so they can be characterized with standard s-parameters. Since they are typically bi-directional, $S_{21} = S_{12}$. Insertion loss (S_{21}) and reflection coefficients (S_{11} and S_{22}) are the primary characteristics of concern in an RF switch. Switches can be reflective (high impedance in the off state) or absorptive (matched in both on-and off-state).

The two major classes of technologies used to implement switches are PIN diodes and FETs. PIN diode switches are often capable of providing superior RF performance to FET switches but the performance can come at a cost of power efficiency. PIN diodes require a constant DC bias current in the on state while FET switches draw current only during the switching operation. Another important emerging technology for microwave switching is the micro-electro-mechanical systems (MEMS) switch. These integrated circuit devices use mechanical movement of integrated features to open and close signal paths.

1.5.5 Filter

Filters are frequency selective components that are central to the operation of a radio. The airwaves include signals from virtually every part of the electromagnetic spectrum. These signals are broadcast using various modulation strategies from TV and radio stations, cell phones, base stations, wireless LANs, police radar, and so on. An antenna at the front end of a radio receives all these signals. In addition, many of the RF components in the radio are nonlinear, creating additional unwanted signals within the radio. In order to function properly, the radio hardware must be capable of selecting the specific signal of interest while suppressing all other unwanted signals. Filters are a critical part of this selectivity. An ideal filter would pass desired signals without attenuation while suppressing signals at all other frequencies to elimination.

Although Figure 1.14 shows only one filter in the microwave portion of the radio, filters are typically required at multiple points along both the transmit and receive signal paths. Further selectivity is often accomplished by the input or output matching circuitry of amplifier or mixer components.

Filter characteristics of interest include the bandwidth or passband frequencies, the insertion loss within the passband of the filter, as well as the signal suppression outside of the desired band. The quality factor, Q, of a filter is a measure of how sharply the performance characteristics transition between passband and out-of-band behavior.

At lower frequencies, filters are realized using lumped inductors and capacitors. Typical lumped components perform poorly at higher frequencies due to parasitic losses and stray capacitances. Special manufacturing techniques must be used to fabricate lumped inductors and capacitors for microwave applications. At frequencies above about 5 or 6 GHz, even specially manufactured lumped element components are often incapable of producing adequate performance. Instead, a variety of technologies are exploited to accomplish frequency selectivity. Open- and short-circuited transmission line segments are often realized in stripline, microstrip, or coplanar waveguide forms to achieve filtering. Dielectric resonators, small pucks of high dielectric material, can be placed in proximity to transmission lines to achieve frequency selectivity. Surface acoustic wave (SAW) filters are realized by coupling the electromagnetic signal into piezoelectric materials and tapping the resulting surface waves with appropriately spaced contacts. Bulk acoustic wave (BAW) filters make use of acoustic waves flowing vertically through bulk piezoelectric material. MEMS are integrated circuit devices that combine both electrical and mechanical components to achieve both frequency selectivity and switching.

1.5.6 Oscillator

Oscillators deliver power either within a narrow bandwidth, or over a frequency range (i.e., they are tunable). Fixed oscillators are used for everything from narrowband power sources to precision clocks. Tunable oscillators are used as swept sources for testing, FM sources in communication systems, and the controlled oscillator in a phase-locked loop (PLL). Fixed tuned oscillators will have a power supply input and the oscillator output, while tunable sources will have one or more additional inputs to change the oscillator frequency. The output power level, frequency of output signal and power consumption are primary characteristics that define oscillator performance. The quality factor, Q, is an extremely important figure-of-merit for oscillator resonators. Frequency stability (jitter) and tunability can also be critical for many applications.

The performance characteristics of an oscillator depend on the active device and resonator technologies used to fabricate and manufacture the component. Resonator technology primarily affects the oscillator's cost, phase noise (jitter), vibration sensitivity, temperature sensitivity, and tuning speed. Device technology mainly affects the oscillator maximum operating frequency, output power, and phase noise (jitter).

Resonator choice is a compromise of stability, cost, and size. Generally the quality factor, Q, is proportional to volume, so cost and size tend to increase with Q. Technologies such as quartz, SAW, yttrium-iron-garnet (YIG) and dielectric resonators allow great reductions in size while achieving high Q by using acoustic, magnetic, and dielectric materials, respectively. Most materials change size with temperature, so temperature stable cavities have to be made of special materials. Quartz resonators are an extremely mature technology with excellent Q, temperature stability, and low cost. Most precision microwave sources use a quartz crystal to control a high-frequency tunable oscillator via a PLL. Oscillator noise power and jitter are inversely proportional to Q^2, making high resonator Q the most direct way to achieve a low noise oscillator.

Silicon bipolar transistors are used in most low noise oscillators below about 5 GHz. Heterojunction bipolar transistors (HBTs) are common today and extend the bipolar range to as high as 100 GHz. These devices exhibit high gain and superior phase noise characteristics over most other semiconductor devices. For oscillator applications, CMOS transistors are poor performers relative to bipolar transistors, but offer levels of integration that are superior to any other device technology. Above several GHz, compound semiconductor MESFETs and HEMTs become attractive for integrated circuit applications. Unfortunately, these devices tend to exhibit high phase noise characteristics when used to fabricate oscillators. Transit time diodes are used at the highest frequencies where a solid-state device can be used. IMPATT and Gunn diodes are the most common types of transit-time diodes available. The IMPATT diode produces power at frequencies approaching 400 GHz, but the avalanche breakdown mechanism inherent to its operation causes the device to be very noisy. In contrast, Gunn diodes tend to exhibit very clean signals at frequencies as high as 100 GHz.

1.6 CAD, Simulation, and Modeling

The unique requirements of RF and microwave engineering establish a need for design and analysis tools that are distinct from conventional electrical engineering tools. Simulation tools that work well for a digital circuit or computational system designer fail to describe RF and microwave behavior adequately. Component and device models must include detailed descriptions of subtle parastic effects not required for digital and low-frequency design. Circuit CAD tools must include a much wider range of components than for traditional electrical circuit design. Transmission line segments, wires, wire bonds, connector transitions, specialized ferrite, and acoustic wave components are all unique to microwave circuit design. In addition, the impact of the particular package technology utilized as well as layout effects must be considered by the microwave engineer. Electromagnetic simulators are also often required to develop models for component transitions, package parasitics, and complex board layouts.

In the microwave design environment, a passive component model for a single chip resistor requires a model that uses several ideal circuit elements. Shunt capacitances are required to model parasitic displacement currents at the input and output of the component. Ideal capacitors, inductors, or transmission line segments must be used internal to the desired component to model phase-shift effects. Nonideal loss mechanisms require the inclusion of additional resistor elements. Similar complexities are required for chip capacitors and inductors. The complexity is compounded by the fact that the ideal element values required to model a single component will change depending on how that component is connected to the circuit board and the kind of circuit board used.

Device models required for microwave and RF design are significantly more detailed than those required for digital or low-frequency applications. In digital design, designers are concerned primarily with two voltage–current states and the overall transition time between those states. A device model that approximately predicts the final states and timing is adequate for many of these applications. For analog applications, however, a device model must describe not only the precise $I–V$ behavior of the device over all possible transitions, but also accurately describe second and third derivatives of those transitions for the model to be able to predict second-and third-order harmonics. Similar accuracy requirements also apply to the capacitance-voltage characteristics of the device. In the case of PA design, the device model must also accurately describe gate leakage and breakdown voltage—effects that are considered second order for many other types of circuit design. Development of a device model for a single microwave transistor can require weeks or months of detailed measurement and analysis.

Another characteristic that distinguishes RF microwave design is the significant use of electromagnetic simulation. A microwave circuit simulation that is completely accurate for a set of chips or components in one package may fail completely if the same circuitry is placed in another package. The transmission line-to-package transitions and proximity effects of package walls and lid make these effects difficult to determine and model. Often package effects can only be modeled adequately through the use of multi-dimensional electromagnetic simulations. Multilevel circuit board design also requires use of such simulations. Radiation effects can be important contributors of observed circuit behavior but cannot be captured without the use of electromagnetic simulation.

Because of the detail and complexity required to perform adequate RF and microwave design, the procedure used to develop such circuits and systems is usually iterative. Simple ideal-element models and crude models are first used to determine preferred topologies and approximate the final design. Nonideal parasitics are then included and the design is reoptimized. Electromagnetic simulators are exercised to determine characteristics of important transitions and package effects. These effects are then modeled and included in the simulation. For some circuits, thermal management can become a dominant concern. Modeling this behavior requires more characterization and simulation complexity. Even after all of this characterization and modeling has taken place, most RF and microwave circuit design efforts require multiple passes to achieve success.

References

1. Collin, R. E., *Foundations for Microwave Engineering*, McGraw-Hill, New York, 1992, p. 2.
2. Adam, S. F., *Microwave Theory and Applications*, Prentice Hall, Englewood Cliffs, NJ, 1969.
3. Halliday, D., Resnick, R. and Walker, J., *Fundamentals of Physics*, 7th ed., John Wiley & Sons, Hoboken, NJ, 2005.

4. Schroder, D. K., *Semiconductor Material and Device Characterization*, John Wiley & Sons, Hoboken, NJ, 1990.

5. Plonus, M. A., *Applied Electromagnetics*, McGraw-Hill, New York 1978.

6. Jonscher, A. K., *Dielectric Relaxation in Solids*, Chelsea Dielectric Press, London, 1983.

7. Kittel, C., *Introduction to Solid State Physics*, 3rd ed. John Wiley & Sons, Hoboken, NJ, 1967.

8. Tsui, J. B., *Microwave Receivers and Related Components*, Peninsula Publishing, Los Altos Hills, CA, 1985.

9. Pearce, C. W. and Sze, S. M., *Physics of Semiconductors*, John-Wiley & Sons, New York, 1981.

10. Bahl, I. and Bhartia P., *Microwave Solid State Circuit Design*, John Wiley & Sons, New York, 1988.

11. Lee, W. C. Y., *Mobile Communications Engineering*, McGraw-Hill Book Company, New York, 1982.

12. Jakes, W. C., Jr. (ed.), *Microwave Mobile Communications*, John Wiley & Sons, Inc., New York, 1974.

13. Evans, J., *Network Interoperability Meets Multimedia*, Satellite Communications, pp. 30–36, February 2000.

14. Laverghetta, T. S., *Modern Microwave Measurements and Techniques*, Artech House, Dedham, MA, 1989.

15. Ramo, S., Winnery, J. R., and Van Duzer, T., *Fields and Waves in Communications Electronics*, 2nd ed., John Wiley & Sons, New York, 1988.

16. Matthaei, G. L., Young, L., and Jones, E. M. T., *Microwave Filters, Impedance-Matching Networks, and Coupling Structures*, Artech House, Dedham, MA, 1980.

17. Ambrozy, A., *Electronic Noise*, McGraw-Hill, New York, 1982.

18. Watson, H. A., *Microwave Semiconductor Devices and their Circuit Applications*, McGraw-Hill, New York, 1969.

19. Vendelin, G., Pavio, A. M., and Rohde, U. L., *Microwave Circuit Design*, John Wiley & Sons, Hoboken, NJ, 1990.

I

Passive Technologies

2

Passive Lumped Components

Alfy Riddle

Finesse, LLC

Lumped components such as resistors, capacitors, and inductors make up most of the glue that allows microwave discrete transistors and integrated circuits to work. Lumped components provide impedance matching, attenuation, filtering, DC bypassing, and DC blocking. More advanced lumped components such as chokes, baluns, directional couplers, resonators, and EMI filters are a regular part of RF and microwave circuitry. Figure 2.1 shows examples of lumped resistors, capacitors, inductors, baluns, and directional couplers. Surface mount techniques and ever-shrinking package sizes now allow solderable lumped components useful to 10 GHz.[1]

FIGURE 2.1   Assorted surface mount lumped components.

As lumped components are used at higher and higher frequencies, the intrinsic internal parasitics, as well as the package and mounting parasitics play a key role in determining overall circuit behavior both in and out of the desired frequency range of use. Mounting parasitics come from excess inductance caused by traces between the component soldering pad and a transmission line, and excess capacitance from relatively large mounting pads and the component body.

2.1 Resistors

Figure 2.2 shows a typical surface mount resistor. These resistors use a film of resistive material deposited on a ceramic substrate with solderable terminations on the ends of the component. Individual surface mount resistors are available in industry standard sizes from over 2512 to as small as 0201. The package size states the length and width of the package in hundredths of an inch. For example an 0805 package is 8 hundredths long by 5 hundredths wide or 80 mils by 50 mils, where a mil is 1/1000th of an inch. Some manufacturers use metric designations rather than English units, so an 0805 package would be 2 mm by 1.2 mm or a 2012 package. The package size determines the intrinsic component parasitics to the first order, and in the case of resistors determines the allowable dissipation. For resistors the most important specifications are power dissipation and tolerance of value.

2.1.1 Resistance

Chip resistors have plated, wrap-around end contacts that overlap the resistive material deposited on the top surface of the ceramic carrier. Circuits built from film on alumina can have multiple thin or thick film resistors on one substrate. The same principles for determining resistance, heat dissipation limits, parasitic inductance, and parasitic capacitance apply to both chips and thin film circuits.[2] Standard chip resistors come in 10% and 1% tolerances, with tighter tolerances available.

The resistive material is deposited in a uniform thickness, t_R, and has a finite conductivity, σ_R. The material is almost always in a rectangle that allows the resistance to be calculated by Equation 2.1.

$$R = l_R / \left(\sigma_R \, w_R \, t_R \right) = R_p \, l_R / w_R \tag{2.1}$$

Increasing the conductivity, material thickness, or material width lowers the resistance, and the resistance is increased by increasing the resistor length, l_R. Often it is simplest to combine the conductivity and thickness into $R_p = 1/(\sigma_R \, t_R)$, where any square resistor will have resistance R_p. Resistor physical size determines the heat dissipation, parasitic inductance and capacitance, cost, packing density, and mounting difficulty.

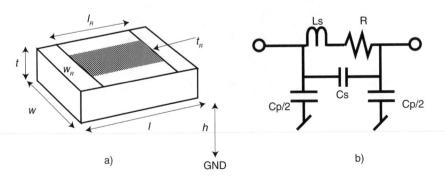

FIGURE 2.2 Surface mount resistor: a) component, b) schematic.

2.1.1.1 Heat Dissipation

Heat dissipation is determined mostly by resistor area, although a large amount of heat can be conducted through the resistor terminations.[3] Generally, the goal is to keep the resistor film below 150 degrees C and the PCB temperature below 100 degrees C. Derating curves are available from individual manufacturers with a typical curve shown in Figure 2.3. Table 2.1 shows maximum power dissipation versus size for common surface mount resistors.

2.1.1.2 Intrinsic Inductive Parasitics

The resistor length and width determine its effective series inductance. This inductance can be computed from a transmission line model or as a ribbon inductor if the resistor is far enough above the ground plane. The ribbon inductor equation is given in Equation 2.2, with all units MKS.[20]

$$L_S = 0.125 \, l_R \left(\ln\left[2 \, l_R / w_R \right] + 0.5 + 2 \, w_R / 9 \, l_R \right) \mu H \tag{2.2}$$

For example, an 0805 package has an inductance of about 0.7 nH. If the resistor width equals the transmission line width, then no parasitic inductance will be seen because the resistor appears to be part of the transmission line.

2.1.1.3 Intrinsic Capacitive Parasitics

There are two types of capacitive parasitics. First, there is the shunt distributed capacitance to the ground plane, C_P. The film resistor essentially forms an RC transmission line and is often modeled as such. In this case, a first order approximation can be based on the parallel plate capacitance plus fringing of the resistor area above the ground plane, as shown in Equation 2.3. The total permittivity, ε, is a product of the permittivity of free space, ε_O, and the relative permittivity, ε_R. An estimate for effective width, w_E, is

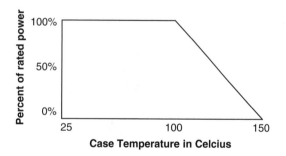

FIGURE 2.3   Surface mount resistor derating curve.

TABLE 2.1   Maximum Resistor Dissipation vs. Size

Resistor Size	P_{Max}(Watts)
2512	1
2510	0.5
1210	0.25
1206	0.125
805	0.1
603	0.0625
402	0.0625

$1.4^* w_R$, but this depends on the ground plane distance, h, and the line width, w. A more accurate method would be to just use the equivalent microstrip line capacitance.

$$C_p = \varepsilon \, w_E \, l_R / h \qquad (2.3)$$

For an 0805 resistor on 0.032" FR4 the capacitance will be on the order of 0.2 pF. As with the inductance, when the resistor has the same width as the transmission line, the line C per unit length absorbs the parasitic capacitance. When the resistor is in shunt with the line, the parasitic capacitance will be seen, often limiting the return loss of a discrete pad.

An additional capacitance is the contact-to-contact capacitance, C_S. This capacitance will typically only be noticed with very high resistances. Also, this capacitance can be dominated by mounting parasitics such as microstrip gap capacitances. C_S can be approximated by the contact area, the body length, and the body dielectric constant as in a parallel plate capacitor.

2.2 Capacitors

Multilayer chip capacitors are available in the same package styles as chip resistors. Standard values range from 0.5 pF up to microfarads, with some special products available in the tenths of picofarads. Parallel plate capacitors are available with typically lower maximum capacitance for a given size. Many different types of dielectric materials are available, such as NPO, X7R, and Z5U. Low dielectric constant materials, such as NPO, usually have low loss and either very small temperature sensitivity, or well-defined temperature variation for compensation, as shown in Table 2.2. Higher dielectric constant materials, such as X7R and Z5U, vary more with temperature than NPO. Z5U will lose almost half its capacitance at very low and very high temperatures. Higher dielectric constant materials, such as X7R and Z5U, also have a reduction in capacitance as voltage is applied.[5,23] The critical specification for these capacitors is the voltage rating. Secondary specifications include temperature stability, Q, tolerance, and equivalent series resistance (ESR).

2.2.1.1 Parallel Plate

Parallel plate capacitors, as shown in Figure 2.4a, can use a thin dielectric layer mounted on a low resistance substrate such as silicon, or they can be a thick ceramic with plated terminations on top and bottom. These capacitors can be attached by soldering or bonding with gold wire. Some capacitors come with several pads, each pad typically twice the area of the next smaller, which allows tuning. These capacitors obey the parallel plate capacitance equation below.

$$C = \varepsilon w l / t_d \qquad (2.4)$$

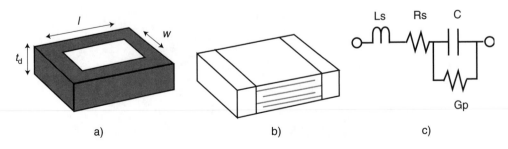

FIGURE 2.4   Capacitor styles and schematic: (a) parallel plate, (b) multilayer, and (c) schematic.

Parasitic resistances, R_S, for these capacitors are typically small and well controlled by the contact resistance and the substrate resistance. Parasitic conductances, G_P, are due to dielectric loss. These capacitors have limited maximum values because of using a single plate pair. The voltage ratings are determined by the dielectric thickness, t_d, and the material type. Once the voltage rating and material are chosen, the capacitor area determines the maximum capacitance. The parasitic inductance of these capacitors, which determines their self-resonance frequency, is dominated by the wire connection to the top plate. In some cases these capacitors are mounted with tabs from the top and bottom plate. When this occurs, the parasitic inductance will be the length of the tab from the top plate to the transmission line, as well as the length of the capacitor acting as a coupled transmission line due to the end launch from the tab.

2.2.1.2 Multilayer Capacitors

Multilayer chip capacitors are a sandwich of many thin electrodes between dielectric layers. The end terminations connect to alternating electrodes, as shown in Figure 2.4b. A wide variety of dielectric materials are available for these capacitors and a few typical dielectric characteristics are given in Table 2.2.[23]

Table 2.2 shows that critical capacitors will have to use NPO dielectric (or alumina) because of tolerance and stability. DC blocking and bypass capacitors will often use X7R or Z5U dielectrics, but these dielectrics must be used with the knowledge that their capacitance varies significantly with temperature and has a voltage dependent nonlinearity.

$$C_P = (n-1)\varepsilon w_P l/t_L \tag{2.5}$$

Multilayer capacitors have a more complicated structure than parallel plate capacitors. Their capacitance is given by Equation 2.5, where t_L is the dielectric layer thickness and w_P is the plate width. The series resistance of the capacitor, Rs, is determined by the parallel combination of all the plate resistance and the conductive loss, Gp, is due to the dielectric loss. Often the series resistance of these capacitors dominates the loss due to the very thin plate electrodes.

By using the package length inductance, as was given in Equation 2.2, the first series resonance of a capacitor can be estimated. The schematic of Figure 2.4c only describes the first resonance of a multilayer capacitor. In reality many resonances will be observed as the multilayer transmission line cycles through quarter- and half-wavelength resonances due to the parallel coupled line structure, as shown in Figure 2.5.[4] The 1000 pF capacitor in Figure 2.5 has a series inductance of 0.8 nH and a series resistance of 0.8 ohms at the series resonance at 177 MHz. The first parallel resonance for the measured capacitor is at 3 GHz. The equivalent circuit in Figure 2.4c does not show the shunt capacitance to ground, C_P, caused by mounting the chip capacitor on a PCB. C_P may be calculated just as it was for a chip resistor in the section on Intrinsic Capacitive Parasitics. The Qs for multilayer capacitors can reach 1000 at 1 GHz.

2.2.1.3 Printed Capacitors

Printed capacitors form very convenient and inexpensive small capacitance values because they are printed directly on the printed circuit board (PCB). Figure 2.6 shows the layout for gap capacitors and interdigital

TABLE 2.2 Chip Capacitor Dielectric Material Comparison

Type	ε_R	Temp. Co. (ppm/degC)	Tol (%)	Range (pF in 805)	Voltage Coeff. (%)
NPO	37	0+/−30	1–20	0.5 p–2200 p	0
4	205	−1500+/−250	1–20	1 p–2200 p	0
7	370	−3300+/−1000	1–20	1 p–2200 p	0
Y	650	−4700+/−1000	1–20	1 p–2200 p	0
X7R	2200	+/−15%	5–20	100 p–1 μ	+0/−25
Z5U	9000	+22/−56%	+80/−20	0.01 μ–0.12 μ	+0/−80

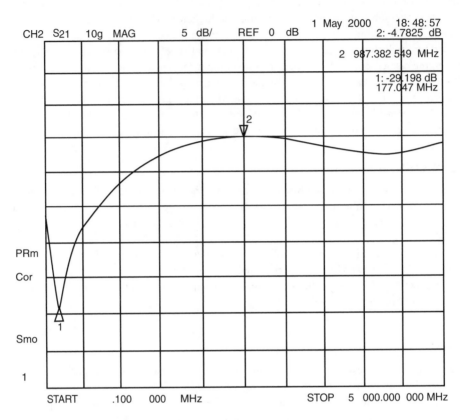

FIGURE 2.5   S21 of a shunt 0805 1000 pF multilayer capacitor, as if used for bypassing.

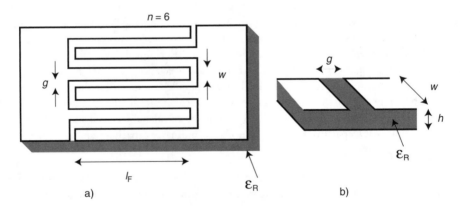

FIGURE 2.6   Printed capacitors: (a) Gap capacitor; (b) interdigital capacitor.

capacitors. The capacitance values for a gap capacitor are very low, typically much less than 1 pF, and estimated by Equation 2.6.[6] Gap capacitors are best used for very weak coupling and signal sampling because they are not particularly high Q. Equation 2.6 can also be used to estimate coupling between two circuit points to make sure a minimum of coupling is obtained.

$$C_G = \varepsilon_O \, \varepsilon_R \, w \left(\ln\left[0.25 + \left(h/g \right)^2 \right] + g/h \, \mathrm{Tan}^{-1}\left[2h/g \right] \right) \bigg/ \left(2\pi \right) \qquad (2.6)$$

Interdigital capacitors, shown in Figure 2.6a are a planar version of the multilayer capacitor. These capacitors have medium Q, are accurate, and are typically less than 1 pF. These capacitors can also be tuned by cutting off fingers. Equation 2.7 gives an estimate of the interdigital capacitance using elliptic functions, K.[7] Because interdigital capacitors have a distributed transmission line structure, they will show multiple resonances as frequency increases. The first resonance occurs when the structure is a quarter wavelength. The Q of this structure is limited by the current crowding at the thin edges of the fingers.

$$C_I = \varepsilon_O \left(1 + \varepsilon_R\right)\left(n - 1\right) l_F \, K\!\left[\mathrm{Tan}\!\left(w\,\pi\big/\!\left(4\!\left(w+g\right)\right)\right)\right]^4 \Big/ K\!\left[1 - \mathrm{Tan}\!\left(w\,\pi\big/\!\left(4\!\left(w+g\right)\right)\right)\right]^4 \qquad (2.7)$$

2.3 Inductors

Inductors are typically printed on the PCB or surface mount chips. Important specifications for these inductors are their Q, their self-resonance frequency, and their maximum current. While wire inductors have their maximum current determined by the ampacity of the wire or trace, inductors made on ferrite or iron cores will saturate the core if too much current is applied. Just as with capacitors, using the largest inductance in a small area means dealing with the parasitics of a nonlinear core material.[8,22]

2.3.1.1 Chip Inductors

Surface mount inductors come in the same sizes as chip resistors and capacitors, as well as in air-core "springs." "Spring" inductors have the highest Q because they are wound from relatively heavy gauge wire, and they have the highest self-resonance because of their air core. Wound chip inductors, shown in Figure 2.7a, use a fine gauge wire wrapped on a ceramic or ferrite core. These inductors have a mediocre Q of 10 to 100 and a lowered self-resonance frequency because of the dielectric loading of the ceramic or ferrite core. However, these inductors are available from 1 nH to 1 mH in packages from 402 to 1812. The chip inductors shown in Figure 2.7b use a multilayer ceramic technology, although some planar spiral inductors are found in chip packages. Either way, these inductors typically have lower Q than wound inductors, but Qs can still reach 100. Multilayer chip inductors use 805 and smaller packages with a maximum inductance of 470 nH. Although the self-resonance frequency of these inductors is high because of the few turns involved, the dielectric loading of the sandwich makes the resonance lower than that of an equivalent "spring" or even a wound inductor, as shown in Table 2.3 and Figure 2.8. The solenoid formula given in Equation 2.8 uses MKS units and describes both wound and "spring" inductors.[10] In Equation 2.8, n is the number of turns, d is the coil diameter, and l is the coil length.

$$L = 9.825 \, n^2 \, d^2 \big/ \left(4.5 \, d + 10 \, l\right) \mu H \qquad (2.8)$$

Figure 2.8 shows an S21 measurement of several series inductors. The inductors are mounted in series to emphasize the various resonances. The 470 nH wound inductor used a 1008 package and had a first

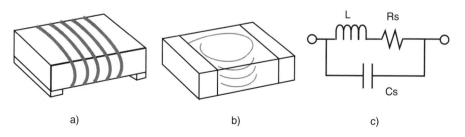

a) b) c)

FIGURE 2.7 Chip inductors: (a) wound; (b) multilayer; and (c) schematic.

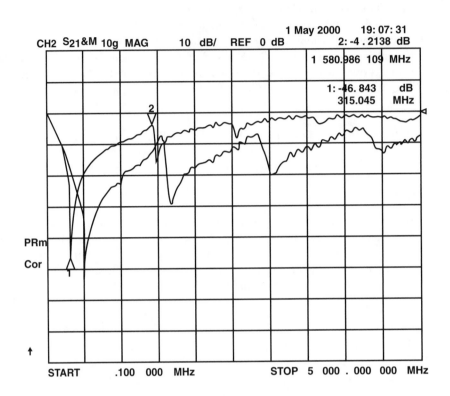

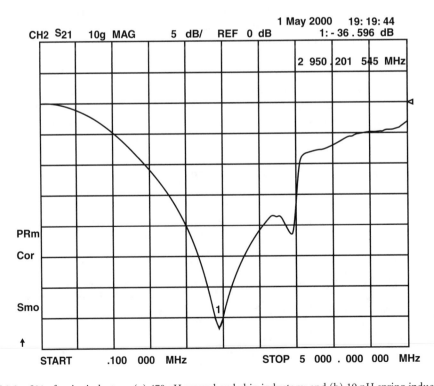

FIGURE 2.8 S21 of series inductors: (a) 470 nH wound and chip inductors; and (b) 19 nH spring inductor.

parallel resonance at 600 MHz. The 470 nH chip inductor used an 805 package and had a first parallel resonance of 315 MHz. The chip inductor shows a slightly shallower first null, indicating a higher series resistance and lower Q. Both inductors had multiple higher order resonances. The chip inductor also showed less rejection at higher frequencies. Figure 2.8b shows the S21 of a series "spring" inductor with a first parallel resonance of 3 GHz. No higher modes of the "spring" inductor can be seen because of the limited bandwidth of the network analyzer, only a fixture resonance is shown around 4 GHz. Table 2.3 shows a matrix of typical "spring," wound, and chip inductor capability.

2.3.1.2 Current Capability

Chip inductors have limited current-carrying capability. At the very least, the internal conductor size limits the allowable current. When the inductor core contains ferrite or iron, magnetic core saturation will limit the useful current capability of the inductor, causing the inductance to decrease well before conductor fusing takes place.

$$Q = \omega \, L/Rs \tag{2.9}$$

2.3.1.3 Parasitic Resistance

The inductor Q is determined by the frequency, inductance, and effective series resistance as shown in Equation 2.9. The effective series resistance, Rs, comes from the conductor resistance and the core loss when a magnetic core is used. The conductor resistance is due to both DC and skin effect resistance as given by Equations 2.1 and 2.10.[9]

$$R_{SKIN} = l_R/\left(\sigma_R \rho_R \delta\right), \tag{2.10}$$

where ρ_R is the perimeter of the wire and δ is the skin depth,

$$\delta = \mathrm{Sqrt}\left[\left(2.5 \; E \; 6\right)/\left(\sigma_R f \pi^2\right)\right]. \tag{2.11}$$

When the skin depth is less than half the wire thickness, Equation 2.10 should be used. For very thin wires the DC resistance is valid up to very high frequencies. For example, 1.4-mil thick copper PCB traces will show skin effect at 14 MHz, while 5 microns of gold on a thin film circuit will not show skin effect until almost 1 GHz.

TABLE 2.3   Inductor Matrix

Type	Size	Inductance Range	Q Range	Current Range	Resonance Range
Spring					
	Micro	1.6–13 nH	50–225	NA	4.5–6 GHz
	Mini	2.5–43 nH	90–210	NA	1.5–3 GHz
Wound					
	603	1.8–120 nH	10–90	0.3–0.7 A	1.3–6 GHz
	805	3.3–220 nH	10–90	0.3–0.8 A	0.8–6 GHz
	1008	3.3–10,000 nH	10–100	0.24–1 A	0.06–6 GHz
	1812	1.2–1000 µH	10–60	0.05–0.48 A	1.5–230 MHz
Multilayer					
	402	1–27 nH	10–200	0.3 A	1.6–6 GHz
	603	1.2–100 nH	5–45	0.3 A	0.83–6 GHz
	805	1.5–470 nH	10–70	0.1–0.3 A	0.3–6 GHz

2.3.1.4 Parasitic Capacitance

The broadband inductor sweep of Figure 2.8 shows resonances in wound, chip, and spring inductors. Manufacturers, and the schematic of Figure 2.7, only show the first parallel resonance. Wound inductors can be modeled as a helical transmission line.[10] The inductance per unit length is the total solenoid inductance divided by the length of the inductor. The capacitance to ground can be modeled as a cylinder of diameter equal to the coil diameter. The first quarter wave resonance of this helical transmission line is the parallel resonance of the inductor, while the higher resonances follow from transmission line theory. When the inductor is tightly wound, or put on a high dielectric core, the interwinding capacitance increases and lowers the fundamental resonance frequency.

$$L_{WIRE} = 200 \, l_w \left(\ln\left[4 \, l_w / d \right] + 0.5 \, d/l_w - 0.75 \right) nH \qquad (2.12)$$

2.3.1.5 Bond Wires and Vias

The inductance of a length of wire is given by Equation 2.12.[1] This equation is a useful first order approximation, but rarely accurate because wires usually have other conductors nearby. Parallel conductors such as ground planes or other wires reduce the net inductance because of mutual inductance canceling some of the flux. Perpendicular conductors, such as a ground plane terminating a wire varies the inductance up to a factor of 2 as shown by the Biot-Savart law.[11]

2.3.1.6 Spiral Inductors

Planar spiral inductors are convenient realizations for PCBs and ICs. They are even implemented in some low-value chip inductors. These inductors tend to have low Q because the spiral blocks the magnetic flux; ground planes, which reduce the inductance tend to be close by; and the current crowds to the wire edges, which increases the resistance. With considerable effort Qs of 20 can be approached in planar spiral inductors.[12] Excellent papers have been written on computing spiral inductance values.[13,14] In some cases, simple formulas can produce reasonable approximations; however, the real challenge is computing the parallel resonance for the spiral. As with the solenoid, the best formulas use the distributed nature of the coil.[14] A first order approximation is to treat the spiral as a microstrip line, compute the total capacitance, and compute the resonance with the coupled wire model of the spiral.

2.4   Chokes

Chokes are an important element in almost all broadband RF and microwave circuitry. A choke is essentially a resonance-free high impedance. Generally, chokes are used below 2 GHz, although bias tees are a basic choke application that extends to tens of GHz. Schematically, a choke is an inductor. In the circuit a choke provides a high RF impedance with very little loss to direct current so that supply voltages and resistive losses can be minimized. Most often the choke is an inductor with a ferrite core used in the frequency range where the ferrite is lossy. A ferrite bead on a wire is the most basic form of choke. It is important to remember that direct current will saturate ferrite chokes just as it does ferrite inductors. At higher frequencies, clever winding tricks can yield resonance-free inductance over decades of bandwidth.[15] Ferrite permeability can be approximated by Equation 2.13.[16]

$$\mu\left(f_{MHz} \right) = 1 / \left(j \, f_{MHz} / 2250 + 1 / \left(\mu_i \left(1 + j \, 0.8 \, \mu_i \, f_{MHz} / 2250 \right) \right) \right) \qquad (2.13)$$

For frequencies below 2250 MHz/μ_i, the ferrite makes a low loss inductor, but above this frequency the ferrite becomes lossy and appears resistive.[17] Figure 2.9 shows a comparison of a 4.7 µH choke and a 4.7 µH wound chip inductor. The high frequency resonances present in the wound inductor are absorbed by the ferrite loss. The best model for a ferrite choke is an inductor in parallel with a resistor. The resistance comes from the high frequency core loss, and the inductor comes from the low frequency

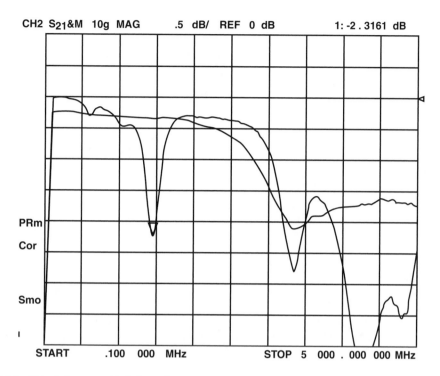

FIGURE 2.9 S21 of shunt 4.7 μH choke and 4.7 μH wound chip inductor, as if used in a DC bias application.

ferrite permeability, core shape, and wire turns. As can be seen in Equation 2.13, the real part of the permeability decreases as frequency increases, causing the loss term to dominate.

2.5 Ferrite Baluns

Ferrite baluns are a cross between transmission lines and chokes. While transmission line baluns can be used at any frequency, ferrite baluns are most useful between 10 kHz and 2 GHz. Basically, the balun is made from twisted or bifilar wire with the desired differential mode impedance, such as 50 or 75 ohms. The wire is wound on a ferrite core so any common mode currents see the full choke impedance, as shown in Figure 2.10a. Two-hole balun cores are used because they provide the most efficient inductance per turn in a small size. The ferrite core allows the balun to work over decades of frequency, for example 10 MHz to 1000 MHz is a typical balun bandwidth. Also, the ferrite allows the baluns to be small enough to fit into T0-8 cans on a 0.25-in. square SMT substrate as shown in Figure 2.1. Extremely high performance baluns are made from miniature coax surrounded by ferrite beads.[18]

Manufacturers design baluns for specific impedance levels and bandwidths. In order to understand the performance trade-offs and limitations, some discussion will be given. Figure 2.10a shows a typical ferrite and wire balun. Figure 2.10b shows a schematic representation of the balun in Figure 2.10a. The balun is easiest to analyze using the equivalent circuit of Figure 2.10c. The equivalent circuit shows that the inverting output is loaded with the even-mode choke impedance, Zoe. For low frequency symmetry in output impedance and gain, a choke is often added to the non-inverting output as shown in Figure 2.10d. The configuration of Figure 2.10d is especially useful when coaxial cable is used instead of twisted pair wire. This is because the coax shields the center conductor from the ferrite so the center conductor has infinite even-mode impedance to ground. For twisted pair baluns, the even-mode impedance to ground is finite and creates a lossy transmission line on the ferrite that attenuates the high frequencies. For broadest band operation, ferrite and wire baluns use the configuration of Figure 2.10b

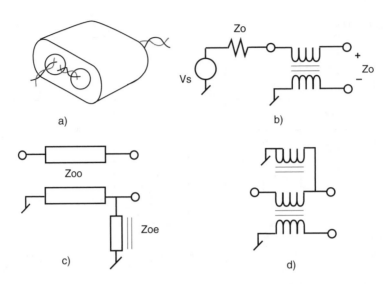

FIGURE 2.10   Ferrite balun (a), with schematic (b), equivalent circuit (c), and symmetric version *(d).

because Figure 2.10d causes excessive high frequency attenuation. Finally, transitions into and out of the balun, as well as imperfect wire impedance contribute to excess series inductance for the differential-mode impedance. The inductive S11 increases with frequency and is often tuned out with small capacitors across the twisted pair.[19]

References

1. Bahl, I, and Bhartia, P., *Microwave Solid State Circuit Design*, John Wiley & Sons, New York, 1988.
2. Chin, S., RLC Components, *Electronic Products*, 31–33, July 1989.
3. Florida RF Labs, *Component Reference Catalog*, Florida RF Labs, Stuart, FL, 2000.
4. ATC, *The RF Capacitor Handbook*, ATC, Huntington Station, NY, 1979.
5. Ingalls, M.W., Perspectives on Ceramic Chip Capacitors, *RF Design*, 45–53, Nov. 1989.
6. Pavlov, V.I., Gap Capacity in Microstrip, *Radioelectronica and Communication Systems*, 85–87, 1981.
7. Wolff, I., and Kibuuka, G., Computer Models for MMIC Capacitors and Inductors, *14th Euorpean Microwave Conference*, 853–859, 1984.
8. Grover, F.W., *Inductance Calculations*, Dover, New York, 1962.
9. Ramo, S., Whinnery, J.R., and van Duzer, T., *Fields and Waves in Communication Electronics*, John Wiley & Sons, New York, 1984.
10. Rhea, R.W., A Multimode High-Frequency Inductor Model, *Applied Microwaves and Wireless*, 70–80, Nov/Dec 1997.
11. Goldfard, M.E., and Pucel, R.A., Modeling Via Hole Grounds in Microstrip, *IEEE Microwave Guided Letters*, 135–137, June 1991.
12. Gecorgyan, S., Aval, O., Hansson, B., Jaconsson, H., and Lewin, T., Loss Considerations for Lumped Inductors in Silicon MMICs, *IEEE MTT-S Digest*, 859–862, 1999.
13. Remke, R.L., and Burdick, G.A., Spiral Inductors for Hybrid and Microwave Applications, *24th Electronic Components Conference*, 152–161, May 1974.
14. Lang, D., Broadband Model Predicts S-Parameters of Spiral Inductors, *Microwaves & RF*, 107–110, Jan. 1988.
15. Piconics, *Inductive Components for Microelectronic Circuits*, Piconics, Tyngsboro, 1998.
16. Riddle, A., Ferrite and Wire Baluns with under 1 dB Loss to 2.5 GHz, *IEEE MTT-S Digest*, 617–620, 1998.

17. Trans-Tech, *Microwave Magnetic and Dielectric Materials*, Trans-Tech, Adamstown, MD, 1998.
18. Barabas, U., On an Ultrabroadband Hybrid Tee, *IEEE Trans MTT*, 58-64, Jan. 1979.
19. Hilbers, A.H., High-Frequency Wideband Power Transformers, *Electronic Applications*, Philips, 30, 2, 65–73, 1970.
20. Boser, O., and Newsome, V., High Frequency Behavior of Ceramic Multilayer Capacitors, *IEEE Trans CHMT*, 437–439, Sept. 1987.
21. Ingalls, M., and Kent, G., Monolithic Capacitors as Transmission Lines, *IEEE Trans MTT*, 964–970, Nov. 1989.
22. Grossner, N.R., *Transformers for Electronic Circuits*, McGraw Hill, New York, 1983.
23. AVX, RF Microwave/Thin-Film Products, AVX, Kyocera, 2000.

3

Passive Microwave Devices

Michael B. Steer
North Carolina State University

Wavelengths in air at microwave and millimeter-wave frequencies range from 1 m at 300 MHz to 1 mm at 300 GHz. These dimensions are comparable to those of fabricated electrical components. For this reason circuit components commonly used at lower frequencies, such as resistors, capacitors, and inductors, are not readily available. The relationship between the wavelength and physical dimensions enables new classes of distributed components to be constructed that have no analogy at lower frequencies. Components are realized by disturbing the field structure on a transmission line, which results in energy storage and thus reactive effects. Electric (E) field disturbances have a capacitive effect and the magnetic (H) field disturbances appear inductive. Microwave components are fabricated in waveguide, coaxial lines, and strip lines. The majority of circuits are constructed using strip lines as the cost is relatively low since they are produced using photolithography techniques. Fabrication of waveguide components requires precision machining, but they can tolerate higher power levels and are more easily realized at millimeter-wave frequencies (30 to 300 GHz) than can either coaxial or microstrip components.

3.1 Characterization of Passive Elements

Passive microwave elements are defined in terms of their reflection and transmission properties for an incident wave of electric field or voltage. In Figure 3.1(a) a traveling voltage wave with phasor V_1^+ is incident at port 1 of a two-port passive element. A voltage V_1^- is reflected and V_2^- is transmitted. In the

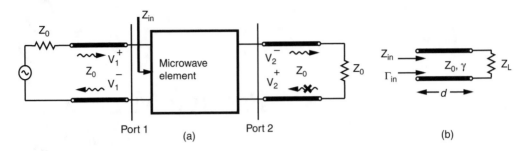

FIGURE 3.1   Incident, reflected and transmitted traveling voltage waves at (a) a passive microwave element, and (b) a transmission line.

absence of an incident voltage wave at port 2 (the voltage wave V_2^- is totally absorbed by Z_0), at port 1 the element has a voltage reflection coefficient

$$\Gamma_1 = \mathbf{V}_1^- / \mathbf{V}_1^+$$

and transmission coefficient

$$T = \mathbf{V}_2^- / \mathbf{V}_1^+ .$$

More convenient measures of reflection and transmission performance are the return loss and insertion loss as they are relative measures of power in transmitted and reflected signals. In decibels

$$\text{RETURN LOSS} = -20 \log \Gamma \ (\text{dB}) \text{ and RETURN LOSS} = -20 \log T \ (\text{dB})$$

The input impedance at port 1, Z_{in}, is related to Γ by

$$Z_{\text{in}} = Z_0 \frac{1 + \Gamma_1}{1 - \Gamma_1} \text{ or by } \ldots \ \Gamma = \frac{Z_L - Z_0}{Z_L + Z_0} \tag{3.1}$$

The reflection characteristics are also described by the voltage standing wave ratio (VSWR), a quantity that is more easily measured. The VSWR is the ratio of the maximum voltage amplitude $|V_1^+| + |V_1^-|$ on the input transmission line to the minimum voltage amplitude $|V_1^+| - |V_1^-|$. Thus

$$\text{VSWR} = \frac{1 + |\Gamma|}{1 - |\Gamma|}$$

These quantities will change if the loading conditions are changed. For this reason scattering (S) parameters are used that are defined as the reflection and transmission coefficients with a specific load referred to as the reference impedance. Thus

$$S_{11} = \Gamma_1 \text{ and } S_{21} = T$$

S_{22} and S_{12} are similarly defined when a voltage wave is incident at port 2. For a multiport $S_{pq} = \mathbf{V}_q^- / \mathbf{V}_p^+$ with all of the ports terminated in the reference impedance. Simple formulas relate the S parameters to other network parameters [1], [2]. S parameters are the most convenient network parameters to use with distributed circuits as a change in line length results in a phase change. As well they are the only network parameters that can be measured directly at microwave and millimeter-wave frequencies. Most passive devices, with the notable exception of ferrite devices, are reciprocal and so $S_{pq} = S_{qp}$. A lossless passive device also satisfies the unitary condition: $\sum_p |S_{pq}|^2$ which is a statement of power conservation indicating that all power is either reflected or transmitted. A passive element is fully defined by its S parameters

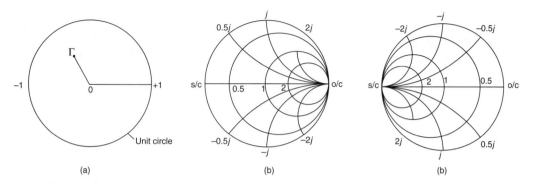

FIGURE 3.2 Smith charts: (a) polar plot of reflection coefficient; $\Gamma = -0.3 + j05$; (b) normalized impedance Smith chart; and (c) normalized admittance Smith chart.

together with its reference impedance, here Z_0. In general the reference impedance at each port can be different.

Circuits are designed to minimize the reflected energy and maximize transmission at least over the frequency range of operation. Thus the return loss is high and the VSWR ≈ 1 for well-designed circuits. However, individual elements may have high reflections.

A terminated transmission line such as that in Figure 3.1b has an input impedance

$$Z_{in} = Z_0 \frac{Z_L + jZ_0 \tanh \gamma d}{Z_0 + jZ_L \tanh \gamma d}$$

Thus a short section ($\gamma d \ll 1$) of short circuited ($Z_L = 0$) transmission line looks like an inductor, and looks like a capacitor if it is open circuited ($Z_L = \times$). When the line is a quarter wavelength long, an open circuit is presented at the input to the line if the other end is short circuited.

3.2 The Smith Chart

Scattering parameters can be conveniently displayed on a polar plot as shown in Figure 3.2a. For passive devices $|S_{pq}| \le 1$ and only for active devices can a scattering parameter be greater than 1. Thus the location of a scattering parameter, plotted as a complex number on the polar plot with respect to the unit circle indicates a basic physical property. As shown earlier in Equation 3.4, there is a simple relationship between load impedance and reflection coefficient. With the reactance of a load held constant and the load resistance varied, the locus of the reflection coefficent is a circle as shown. Similarly, arcs of circles result when the resistance is held constant and the reactance varied. These lines result in the normalized impedance Smith chart of Figure 3.2b.

3.3 Transmission Line Sections

The simplest microwave circuit element is a uniform section of transmission line that can be used to introduce a time delay or a frequency-dependent phase shift. More commonly it is used to interconnect other components. Other line segments used for interconnections include bends, corners, twists, and transitions between lines of different dimensions (see Figure 3.3).

The dimensions and shapes are designed to minimize reflections and so maximize return loss and minimize insertion loss.

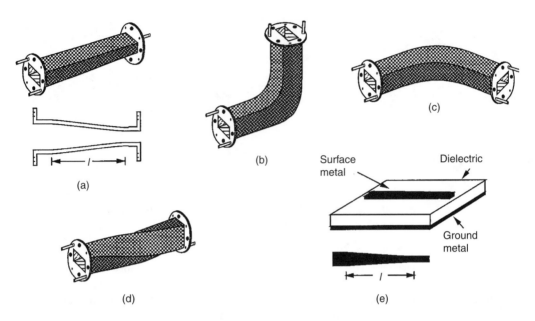

FIGURE 3.3   Sections of transmission lines used for interconnecting components: (a) waveguide tapered section, (b) waveguide E-plane bend, (c) waveguide H-plane bend, (d) waveguide twist, and (e) microstrip tapered line.

3.4   Discontinuities

The rectangular waveguide discontinuities shown in Figure 3.4a–f illustrate most clearly the use of E and H field disturbances to realize capacitive and inductive components. An E plane discontinuity, Figure 3.4a, is modeled approximately by a frequency-dependent capacitor. H plane discontinuities (Figures 3.4b and 3.4c) resemble inductors as does the circular iris of Figure 3.4d. The resonant waveguide iris of Figure 3.4e disturbs both the E and H fields and can be modeled by a parallel LC resonant circuit near the frequency of resonance. Posts in waveguide are used both as reactive elements (Figure 3.4f), and to mount active devices (Figure 3.4g). The equivalent circuits of microstrip discontinuities (Figures 3.4j–o), are again modeled by capacitive elements if the E-field is interrupted and by inductive elements if the H field (or current) is disturbed. The stub shown in Figure 3.4j presents a short circuit to the through transmission line when the length of the stubs is $\lambda_g/4$. When the stub is electrically short $\ll\lambda_g/4$ it introduces a shunt capacitance in the through transmission line.

3.5   Impedance Transformers

Impedance transformers interface two sections of line of different characteristic impedance. The smoothest transistion and the one with the broadest bandwidth is a tapered line as shown in Figures 3.3a and 3.3e. This element tends to be very long as $l > \lambda_g$ and so step terminations called quarter-wave impedance transformers (see Figure 3.4h) are sometimes used, although their bandwidth is relatively small centered on the frequency at which $l = \lambda_g$. Ideally $Z_{0,2} = \sqrt{Z_{0,1}Z_{0,3}}$.

3.6   Terminations

In a termination, power is absorbed by a length of lossy material at the end of a shorted piece of transmission line (Figures 3.5a and 3.5c). This type of termination is called a matched load as power is absorbed and reflections are small irrespective of the characteristic impedance of the transmission line.

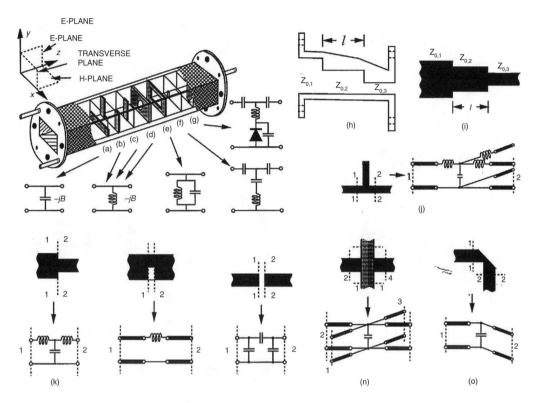

FIGURE 3.4 Discontinuities. Waveguide discontinuities: (a) capacitive E-plane discontinuity, (b) inductive H-plane discontinuity, (c) symmetrical inductive H-plane discontinuity, (d) inductive post discontinuity, (e) resonant window discontinuity, (f) capacitive post discontinuity, (g) diode post mount, and (h) quarter-wave impedance transformer; microstrip discontinuities: (i) quarter-wave impedance transformer, (j) open microstrip stub, (k) step, (l) notch, (m) gap, (n) crossover, and (o) bend.

This is generally preferred as the characteristic impedance of transmission lines varies with frequency — particularly so for waveguides. When the characteristic impedance of a line does not vary much with frequency, as is the case with a coaxial line or microstrip, a simpler and smaller termination can be realized by placing a resistor to ground (Figure 3.5b).

3.7 Attenuators

Attenuators reduce the signal level traveling along a transmission line. The basic design is to make the line lossy but with characteristic impedance approximating that of the connecting lines so as to reduce reflections. The line is made lossy by introducing a resistive vane in the case of a waveguide, Figure 3.5d, replacing part of the outer conductor of a coaxial line by resistive material, Figure 6.5e, or covering the line by resistive material in the case of a microstrip line Figure 3.5f. If the amount of lossy material introduced into the transmission line is controlled, a variable attenuator is obtained, e.g., Figure 3.5g.

3.8 Microwave Resonators

In a lumped element resonant circuit, stored energy is transfered between an inductor, which stores magnetic energy, and a capacitor, which stores electric energy, and back again every period. Microwave

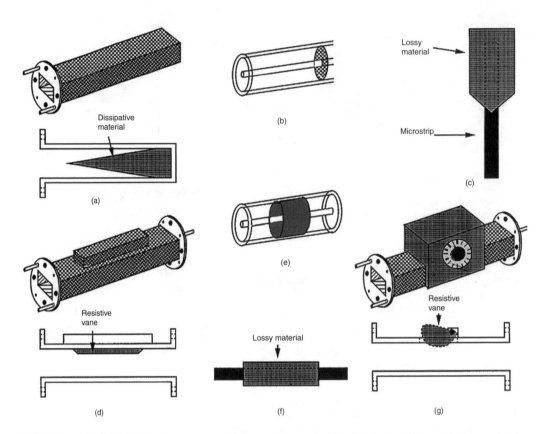

FIGURE 3.5   Terminations and attenuators: (a) waveguide matched load, (b) coaxial line resistive termination, (c) microstrip matched load, (d) waveguide fixed attenuator, (e) coaxial fixed attenuator, (f) microstrip attenuator, and (g) waveguide variable attenuator.

resonators function the same way, exchanging energy stored in electric and magnetic forms but with the energy stored spatially. Resonators are described in terms of their quality factor

$$Q = 2\pi f_0 \left| \frac{\text{Maximum energy stored in the cavity at } f_0}{\text{Power lost in the cavity}} \right|$$

where f_0 is the resonant frequency. The Q is reduced and thus the resonator bandwidth is increased by the power lost to the external circuit so that the loaded Q

$$Q_L = 2\pi f_0 \left| \frac{\text{Maximum energy stored in the resonator at } f_0}{\text{Power lost in the cavity and to the external circuit}} \right| = \frac{1}{1/Q + 1/Q_{ext}}$$

where Q_{ext} is called the external Q, Q_L accounts for the power extracted from the resonant circuit and is typically large. For the simple response shown in Figure 3.6a the half power (3 dB) bandwidth is f_0/Q_L.

Near resonance, the response of a microwave resonator is very similar to the resonance response of a parallel or series *RLC* resonant circuit, Figures 3.6f and 3.6g. These equivalent circuits can be used over a narrow frequency range.

Several types of resonators are shown in Figure 3.6. Figure 3.6b is a rectangular cavity resonator coupled to an external coaxial line by a small coupling loop. Figure 3.6c is a microstrip patch reflection resonator.

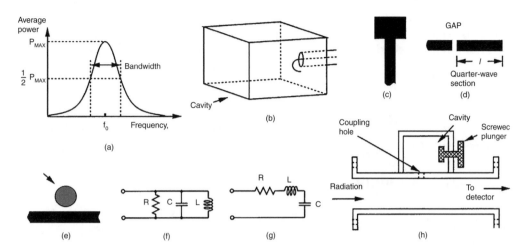

FIGURE 3.6   Microwave resonators: (a) resonator response, (b) rectangular cavity resonator, (c) microstrip patch resonator (d) microstrip gap-coupled reflection resonator, (e) transmission dielectric transmission resonator in microstrip, (f) parallel equivalent circuits, (g) series equivalent circuits, and (h) waveguide wavemeter.

This resonator has large coupling to the external circuit. The coupling can be reduced and photolithographically controlled by introducing a gap as shown in Figure 3.6d for a microstrip gap-coupled transmission line reflection resonator. The Q of a resonator can be dramatically increased by using a high dielectric constant material as shown in Figure 3.6e for a dielectric transmission resonator in microstrip.

One simple application of cavity resonance is the waveguide wavemeter, Figure 3.6h. Here the resonant frequency of a rectangular cavity is varied by changing the the physical dimensions of the cavity with a null of the detector indicating that the frequency corresponds to the cavity resonant frequency.

3.9   Tuning Elements

In rectangular waveguide the basic adjustable tuning element is the sliding short shown in Figure 3.7a. Varying the position of the short will change resonance frequencies of cavities. It can be combined with hybrid tees to achieve a variety of tuning functions. The post in Figure 3.9f can be replaced by a screw to obtain a screw tuner, which is commonly used in waveguide filters.

Sliding short circuits can be used in coaxial lines and in conjunction with branching elements to obtain stub tuners. Coaxial slug tuners are also used to provide matching at the input and output of active circuits. The slug is movable and changes the characteristic impedance of the transmission line. It is more difficult to achieve variable tuning in passive microstrip circuits. One solution is to provide a number of pads as shown in Figure 3.5c which, in this case, can be bonded to the stub to obtain an adjustable stub length.

Variable amounts of phase shift can be inserted by using a variable length of line called a line stretcher, or by a line with a variable propagation constant. One type of waveguide variable phase shifter is similar to the variable attenuator of Figure 3.5g with the resistive material replaced by a low-loss dielectric.

3.10   Hybrid Circuits and Directional Couplers

Hybrid circuits are multiport components that preferentially route a signal incident at one port to the other ports. This property is called directivity. One type of hybrid is called a directional coupler the schematic of which is shown in Figure 3.8a. Here the signal incident at port 1 is coupled to ports 2 and 4 while very little is coupled to port 3. Similarly a signal incident at port 2 is coupled to ports 1 and 3, but very little power appears at port 4. The feature that distinguishes a directional coupler from other types of hybrids is that the power at the output ports are different. The performance of a directional coupler is specified by two parameters:

Coupling Factor $= P_1/P_4$
Directivity $\quad = P_4/P_3$

Microstrip and waveguide realizations of directional couplers are shown in Figures 3.8b and 3.8c.

The power at the output ports of the hybrids shown in Figures 3.9 and 3.10 are equal and so the hybrids serve to split a signal in half (in terms of power) as well having directional sensitivity.

3.11   Ferrite Components

Ferrite components are nonreciprocal in that the insertion loss for a wave traveling from port A to port B is not the same as that from port B to port A.

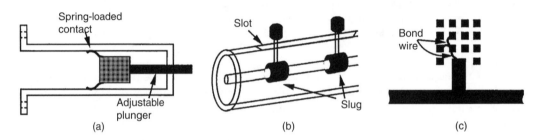

FIGURE 3.7   Tuning elements: (a) waveguide sliding short circuit, (b) coaxial line slug tuner, (c) microstrip stub with tuning pads.

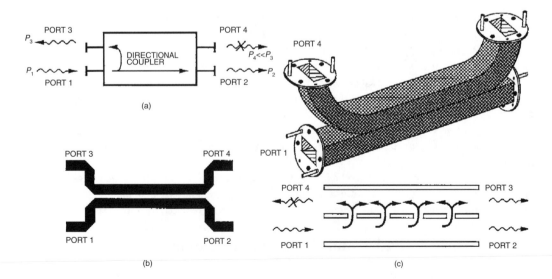

FIGURE 3.8   Directional couplers: (a) schematic, (b) microstrip directional coupler, (c) waveguide directional coupler.

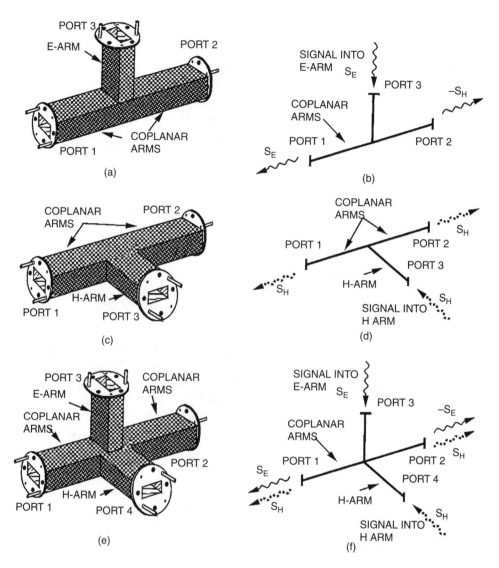

FIGURE 3.9 Waveguide hybrids:(a) E-plane tee and (b) its signal flow; (c) H-plane tee and (d) its signal flow; and (e) magic tee and (f) its signal flow. The negative sign indicates 180° phase reversal.

3.11.1   Circulators and Isolators

The most important ferrite component is a circulator, Figures 3.11a and 3.11b. The essential element of a circulator is a piece of ferrite which, when magnetized, becomes nonreciprocal preferring progression of electromagnetic fields in one circular direction. An ideal circulator has the scattering matrix

$$S = \begin{vmatrix} 0 & 0 & S_{13} \\ S_{21} & 0 & 0 \\ 0 & S_{32} & 0 \end{vmatrix}$$

In addition to the insertion and return losses, the performance of a circulator is described by its isolation, which is its insertion loss in the undesired direction. An isolator is just a three-port circulator with one

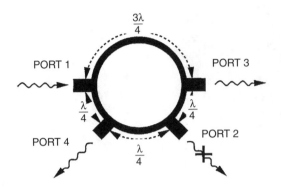

FIGURE 3.10 Microstrip rat race hybrid.

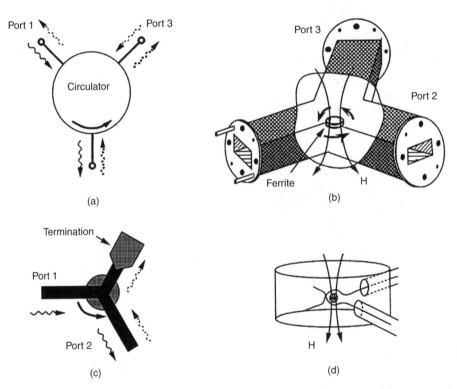

FIGURE 3.11 Ferrite Components: (a) Schematic of a circulator, (b) a waveguide circulator, (c) a microstrip isolator, and (d) a YIG tuned band-pass filter.

of the ports terminated in a matched load as shown in the microstip realization of Figure 3.11c. It is used in a transmission line to pass power in one direction but not in the reverse direction. It is commonly used to protect the output of equipment from high reflected signals. A four-port version is called a duplexer and is used in radar systems and to separate the received and transmitted signals in a transciever.

3.11.2   YIG Tuned Resonator

A magnetized YIG (Yttrium Iron Garnet) sphere shown in Figure 3.11d, provides coupling between two lines over a very narrow band. The center frequency of this bandpass filter can be adjusted by varying the magnetizing field.

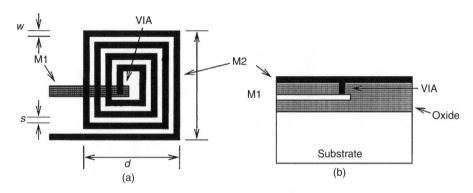

FIGURE 3.12 A spiral inductor fabricated ona monolithic integrated circuit: (a) top view; and (b) cross-section.

3.12   Filters and Matching Networks

Filters are combinations of microwave passive elements designed to have a specified frequency response. Typically a topology of a filter is chosen based on established lumped element filter design theory. Then computer-aided design techniques are used to optimize the response of the circuit to the desired response. Matching networks contain reactive elements and have the essential purpose of realizing maximum power transfer. They often interface a complex impedance termination to a resistance source. On a chip, and at RF frequencies, the matching network consists of parallel plate capacitors and spiral inductors such as that shown in Figure 3.12.

3.13   Passive Semiconductor Devices

A semiconductor diode is modeled by a voltage dependent resistor and capacitor in shunt. Thus an applied DC voltage can be used to change the value of a passive circuit element. Diodes optimized to produce a voltage variable capacitor are called varactors. In detector circuits, a diode's voltage variable resistance is used to achieve rectification and, through design, produce a DC voltage proportional to the power of an incident microwave signal. The controllable variable resistance is used in a PIN diode to realize an electronically controllable switch.

Defining Terms

Characteristic impedance: Ratio of the voltage and current on a transmission line when there are no reflections.
Insertion loss: Power lost when a signal passes through a device.
Reference impedance: Impedance to which scattering parameters are referenced.
Return loss: Power lost upon reflection from a device.

References

1. G.D. Vendelin, A.M. Pavio, and U.L. Rohde, *Microwave Circuit Design Using Linear and Nonlinear Techniques*, Wiley: New York, 1990.
2. T.C. Edwards and M.B. Steer, *Foundations of Interconnect and Microstrip Design*, 3rd edition, Wiley: Chichester, 2000.

Further Information

The following books provide good overviews of passive microwave components: *Foundations of Interconnect and Microstrip Design* by T.C. Edwards and M.B. Steer, 3rd edition, Wiley, Chichester, 2000, *Microwave*

Engineering Passive Circuits by P.A. Rizzi, Prentice Hall, Englewood Cliffs, N.J., 1988; *Microwave Devices and Circuits* by S.Y. Liao, 3rd edition, Prentice Hall, Englewood Cliffs, N.J., 1990; *Microwave Theory, Components and Devices* by J.A. Seeger, Prentice Hall, Englewood Cliffs, N.J., 1986; *Microwave Technology* by E. Pehl, Artech House, Dedham, MA, 1985; *Microwave Engineering and Systems Applications* by E.A. Wolff and R. Kaul, Wiley, New York, 1988; and Microwave Engineering by T.K. Ishii, 2 nd edition, Harcourt Brace Jovanovich, Orlando, Florida, 1989. *Microwave Circuit Design Using Linear and Nonlinear Techniques* by G.D. Vendelin, A.M. Pavio and U.L. Rohde, Wiley, New York, 1990, also provides a comphrensive treatment of computer-aided design techniques for both passive and active microwave circuits.

The monthly journals, *IEEE Transactions on Microwave Theory and Techniques*, *IEEE Microwave and Guided Wave Letters*, and *IEEE Transactions on Antennas and Propagation* publish articles on modeling and design of microwave passive circuit components. Articles in the first two journals are more circuit and component oriented while the third focuses on field theoretic analysis. These are published by The Institute of Electrical and Electronics Engineers, Inc. For subscription or ordering contact: IEEE Service Center, 445 Hoes Lane, PO Box 1331, Piscataway, New Jersey 08855-1331, U.S.A.

Articles can also be found in the biweekly magazine, *Electronics Letters*, and the bimonthly magazine, *IEE Proceedings Part H — Microwave, Optics and Antennas*. Both are published by the Institute of Electrical Engineers and subscription enquiries should be sent to IEE Publication Sales, PO Box 96, Stenage, Herts. SG1 2SD, United Kingdom. Telephone number (0438) 313311.

The *International Journal of Microwave and Millimeter-Wave Computer-Aided Engineering* is a quarterly journal devoted to the computer-aided design aspects of microwave circuits and has articles on component modeling and computer-aided design techniques. It has a large number of review-type articles. For subscription information contact John Wiley & Sons, Inc., Periodicals Division, PO Box 7247-8491, Philadelphia, Pennsylvania 19170-8491, U.S.A.

4

Dielectric Resonators

S. Jerry Fiedziuszko
Lockheed Martin CSS

Resonating elements are key to the function of most microwave circuits and systems. They are fundamental to the operation of filters and oscillators, and the quality of these circuits is basically limited by the resonator quality factor. Traditionally, microwave circuits have been encumbered by large, heavy, and mechanically complex waveguide structures that are expensive and difficult to adjust and maintain. Dielectric resonators, which can be made to perform the same functions as waveguide filters and resonant cavities, are, in contrast very small, stable, and lightweight. The popularization of advanced dielectric resonators roughly coincides with the miniaturization of many of the other associated elements of most microwave circuits. When taken together, these technologies permit the realization of small, reliable, lightweight, and stable microwave circuits.

Historically, guided electromagnetic wave propagation in dielectric media received widespread attention in the early days of microwaves. Surprisingly, substantial effort in this area predates 1920 and includes such famous scientists as Rayleigh, Sommerfeld, J.C. Bose, and Debye.[1] The term "dielectric resonator" first appeared in 1939 when R.D. Richtmyer of Stanford University showed that unmetalized dielectric objects (sphere and toroid) can function as microwave resonators.[2] However, his theoretical work failed to generate significant interest, and practically nothing happened in this area for more than 25 years. In 1953, a paper by Schlicke[3] reported on super high dielectric constant materials (~1,000 or more) and their applications at relatively low RF frequencies. In the early 1960s, researchers from Columbia University, Okaya and Barash, rediscovered dielectric resonators during their work on high dielectric materials (rutile), paramagnetic resonance and masers. Their papers[4,5] provided the first analysis of modes and resonator design. Nevertheless, the dielectric resonator was still far from practical applications. High dielectric constant materials such as rutile exhibited poor temperature stability causing correspondingly large resonant frequency changes. For this reason, in spite of high Q factor and small size, dielectric resonators were not considered for use in microwave devices.

In the mid-1960s, S. Cohn and his co-workers at Rantec Corporation performed the first extensive theoretical and experimental evaluation of the dielectric resonator.[6] Rutile ceramics were used for experiments that had an isotropic dielectric constant in the order of 100. Again, poor temperature stability prevented development of practical components.

A real breakthrough in ceramic technology occurred in the early 1970s when the first temperature stable, low-loss, Barium Tetratitanate ceramics were developed by Raytheon.[7] Later, a modified Barium Tetratitanate with improved performance was reported by Bell Labs.[8] These positive results led to the actual implementations of dielectric resonators as microwave components. The materials, however, were in scarce supply and not commercially available.

The next major breakthrough came from Japan when Murata Mfg. Co. produced $(Zr-Sn)TiO_4$ ceramics.[9] They offered adjustable compositions so that the temperature coefficient could be varied between +10 and −12 ppm/degree C. These devices became commercially available at reasonable prices. Afterward, the theoretical work and use of dielectric resonators expanded rapidly.

4.1 Microwave Dielectric Resonator

What is it? A dielectric resonator is a piece of high dielectric constant material usually in the shape of a disc that functions as a miniature microwave resonator (Figure 4.1).

How does it work? The dielectric element functions as a resonator because of the internal reflections of electromagnetic waves at the high dielectric constant material/air boundary. This results in confinement of energy within, and in the vicinity of, the dielectric material which, therefore forms a resonant structure.

Why use it? Dielectric resonators can replace traditional waveguide cavity resonators in most applications, especially in MIC structures. The resonator is small, lightweight, high Q, temperature stable, low cost, and easy to use. A typical Q exceeds 10,000 at 4 GHz.

4.2 Theory of Operation

A conventional metal wall microwave cavity resonates at certain frequencies due to the internal reflections of electromagnetic waves at the air (vacuum)/metal boundary. These multiple reflections from this highly conductive boundary (electrical short) form a standing wave in a cavity with a specific electromagnetic field distribution at a unique frequency. This is called a "mode." A standard nomenclature for cavity

FIGURE 4.1   Dielectric resonators.

modes is based on this specific electromagnetic field distribution of each mode. Since a metal wall cavity has a very well-defined boundary (short) and there is no field leaking through the wall, the associated electromagnetic field problem can be easily solved through exact mathematical analysis and modes for various cavity shapes (e.g., rectangular cavity or circular cavity) are precisely defined.

The TE (transverse electric) and TM (transverse magnetic) mode definitions are widely used. Mode indices e.g., TE_{113} (rectangular cavity analyzed in Cartesian coordinates) indicate how many of the electromagnetic field variations we have along each coordinate (in this case 1 along x and y, and 3 along z). The case of a dielectric resonator situation is more complicated. An electromagnetic wave propagating in a high dielectric medium and impinging on a high dielectric constant medium/air boundary will be reflected. However, contrary to a perfectly conducting boundary (e.g., highly conductive metal) this is a partial reflection and some of the wave will leak through the boundary to the other, low dielectric constant medium (e.g., air or vacuum). The higher the dielectric constant is of the dielectric medium, more of the electromagnetic wave is reflected and this boundary can be modeled not as a short (metal) but as an "open." As in a metal wall cavity, these internal reflections form a resonant structure called a dielectric resonator.

As in a conventional metal wall cavity, an infinite number of modes can exist in a dielectric resonator. To a first approximation, a dielectric resonator can be explained as a hypothetical magnetic wall cavity, which is the dual case of a metal (electric) wall cavity. The magnetic wall concept (on which the normal component of the electric field and tangential component of a magnetic field vanish at the boundary) is well known and widely used as a theoretical tool in electromagnetic field theory. In a very crude approximation, the air/high dielectric constant material interface can be modeled as such a magnetic wall (open circuit). Hence, the field distribution and resonant frequencies for such a resonator can be calculated analytically.

To modify this model we have to take into consideration that in actuality some of the electromagnetic field leaks out of the resonator and eventually decays exponentially in its vicinity. This leaking field portion is described by a mode subscript δ. Mode subscript δ is always smaller than one and varies with the field confinement in a resonator. If the dielectric constant of the resonator increases, more of the electromagnetic field is confined in the resonator and the mode subscript δ starts approaching one. The first modification of the magnetic wall model to improve accuracy was to remove two xy plane magnetic walls (Figure 4.2), and to create a magnetic wall waveguide below cut off filled with the dielectric.[5,6,10]

This gave a calculated frequency accuracy for the $TE_{01\delta}$ mode of about 6%. Figure 4.2 shows the magnetic wall waveguide below cutoff with a dielectric resonator inside. Later, the circular wall was also removed (dielectric waveguide model) and the accuracy of calculations of resonant frequency was

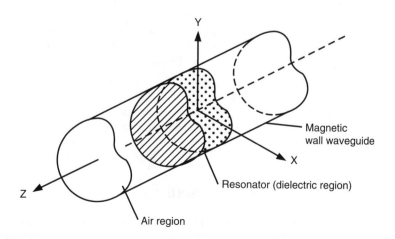

FIGURE 4.2 Magnetic wall waveguide below cut-off model of a dielectric resonator.

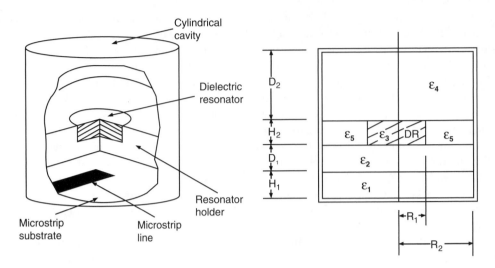

FIGURE 4.3   Practical configurations of a dielectric resonator and corresponding theoretical model.

improved to 1 to 2%.[11,12] In an actual resonator configuration, usually some sort of metal wall cavity or housing is necessary to prevent radiation of the electromagnetic field and resulting degradation of resonator Q. This is illustrated in Figure 4.3. Taking this into consideration, the model of the dielectric resonator assembly was modified, and accurate formulas for resonant frequency and electromagnetic field distribution in the structure were obtained through the mode matching method.[13]

In advanced models, additional factors such as dielectric supports, tuning plate, and microstrip substrate, can also be taken into account. The resonant frequency of the dielectric resonator in these configurations can be calculated using mode matching methods with accuracy much better than 1%.

The most commonly used mode in a dielectric resonator is the $TE_{01\delta}$ (in cylindrical resonator) or the $TE_{11\delta}$ (in rectangular resonator). The $TE_{01\delta}$ mode for certain Diameter/Length (D/L) ratios has the lowest resonant frequency, and therefore is classified as the fundamental mode. In general, mode nomenclature in a dielectric resonator is not as well defined as for a metal cavity (TE and TM modes). Many mode designations exist[14-16] and this matter is quite confusing as is true for the dielectric waveguide. In the authors opinion, the mode designation proposed by Y. Kobayashi[14] is the most promising and should be adopted as a standard (this was addressed by MTT Standards Committee and presently two designations including Kobayashi's are recommended). Some of the modes and their field distributions are presented in Figure 4.4.

The $TE_{01\delta}$ mode is the most popular and is used in single mode filters and oscillators. $HE_{11\delta}$ (HE indicates a hybrid mode) is used in high performance, dual mode filters,[17] directional filters and oscillators. TM mode is being used in cavity combiners and filters. Hybrid modes have all six components of the electromagnetic field. Also, low frequency filters (~800 MHz) for cellular telephone base stations were designed using the triple mode TM_{010}.[18]

Very high order modes called "whispering gallery modes" are finding applications at millimeter wave frequencies. These modes were first observed by Rayleigh in a study of acoustic waves.

Analogous propagation is possible in dielectric cylinders. Since these modes are well confined near the surface of the dielectric resonator, very high Q values at millimeter frequencies are posssible.

4.3   Coupling to Microwave Structures

An advantage of dielectric resonators is the ease with which these devices couple to common transmission lines such as waveguides and microstrip. A typical dielectric resonator in the $TE_{01\delta}$ mode can be transversely inserted into a rectangular waveguide. It couples strongly to the magnetic field, and acts as a

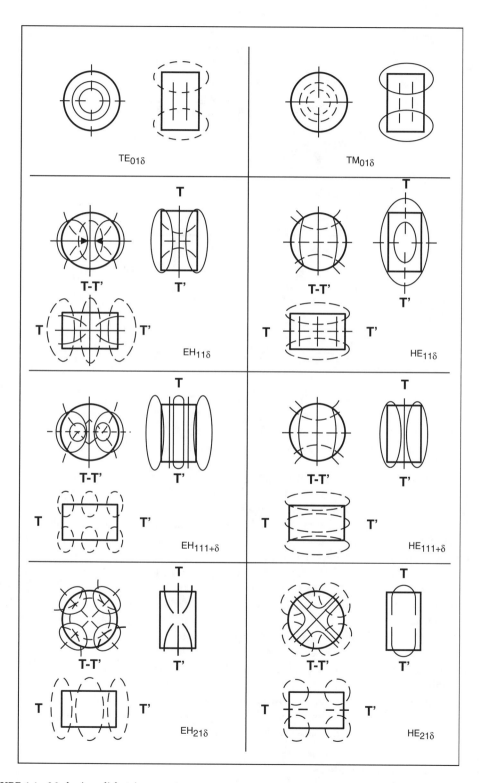

FIGURE 4.4 Modes in a dielectric resonator.

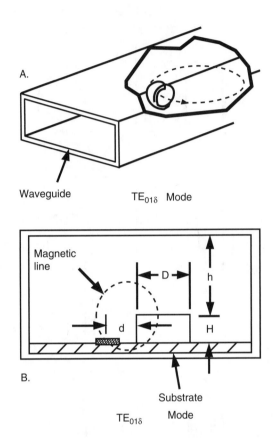

A.

Waveguide $TE_{01\delta}$ Mode

B.

Substrate

$TE_{01\delta}$ Mode

FIGURE 4.5   Magnetic field coupling of a dielectric to rectangular waveguide (top) and Microstrip line (bottom).

simple bandstop filter. Figure 4.5 illustrates the magnetic field coupling into the microstrip. Coupling to a magnetic field in the waveguide can be adjusted by either rotating/tilting the resonator or moving a resonator toward the side of the waveguide. In microstrip line applications, a dielectric $TE_{11\delta}$ resonator couples magnetically and forms a bandstop filter. This is shown in Figure 4.5. The coupling can be easily adjusted by either moving the resonator away (or toward center) from the microstrip or by lifting the resonator on a special support above the microstrip.

The resonant frequency of a dielectric resonator in this very practical case can be calculated using an equation derived in Reference 13. The resonant frequency in this topology can be adjusted to higher frequency with a metal screw or plate located above the resonator and perturbing the magnetic field, or down in frequency by lifting the resonator (moving it away from the ground plane). A typical range is in the order of 10%. Extra care must be taken, however, not to degrade the Q factor or temperature performance of the resonator by the closely positioned metal plate.

An interesting modification of the dielectric resonator is the so-called double resonator. This configuration is shown in Figure 4.6. In this configuration, two halves of the ceramic disc or plate act as one resonator. Adjustment of the separation between the two halves of the resonator results in changes of the resonant frequency of the structure (Figure 4.6). A much wider linear tuning range can be obtained in this configuration without degradation of the Q factor.[19-21]

4.4   Ceramic Materials

The major problem with previously available high Q materials, such as rutile or rutile ceramics, was the poor temperature stability of the dielectric constant and the resulting instability of the resonant frequency

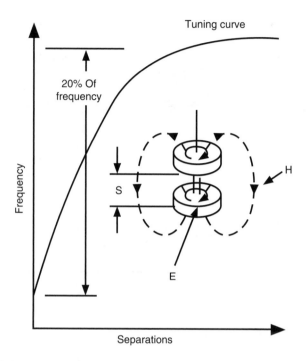

FIGURE 4.6   Double dielectric resonator configuration and tuning curve showing 20% tuning range.

TABLE 4.1   Basic Properties of High Quality Ceramics

Material Composition	Manufacturer	ε	Loss Tangent @ 4 GHz	Temp. Coeff. ppm/degree C
$Ba\,Ti_4O_9$	Raytheon, Transtech	38	0.0001	+4
$Ba_2Ti_9O_{20}$	Bell Labs	40	0.0001	+2
$(Zr\text{-}Sn)\,TiO_4$	Murata Tekelec Siemens Transtech	38	0.0001	−4 to +10 adj.
$Ba(Zn_{1/3}\,Nb_{2/3})O_2-$ $Ba(Zn_{1/3}\,Ta_{2/3})O_2$	Murata	30	0.00004	0 to +10 adj.

of the dielectric resonators. Newly developed high Q Ceramics, however, have excellent temperature stability and an almost zero temperature coefficient is possible. The most popular materials are composed of a few basic, high Q compounds capable of providing negative and positive temperature coefficients. By adjusting proportions of the compounds and allowing for the linear expansion of the ceramic, perfect temperature compensation is possible. Basic properties of high quality ceramics developed for dielectric resonator applications are presented below in Table 4.1.[22–23]

4.5   Applications

The miniaturization of microwave components began with the introduction of microwave integrated circuits (MIC) and general advances in semiconductor technology, especially gallium arsenide. MIC components have become very common and presently more advanced, monolithic circuits (MMICs) are being used in many applications. MIC/MMIC structures have suffered, however, from a lack of high Q miniature elements that are required to construct high performance, narrowband filters, and highly stable, fundamental frequency oscillators.Expensive and bulky coaxial and waveguide resonators made

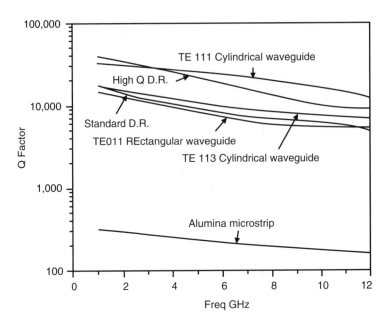

FIGURE 4.7   Achievable quality factor (Q) of various microwave resonator types (copper waveguide — 60 to 70% of theoretical Q is assumed).

out of temperature-stable materials such as INVAR or graphite composites were the only solution in the past (Figure 4.7). With the dielectric resonator described above, a very economical alternative, which also satisfies very stringent performance requirements, was introduced. Dielectric resonators find use as probing devices to measure dielectric properties of materials as well as the surface resistance of metals, and more recently high temperature superconductors (HTS). Additional applications include miniature antennas, where strongly radiating lower order resonator modes are successfully used.[24]

4.6   Filters

Simultaneously with advances in dielectric resonator technology, significant advances were made in microwave filter technology. More sophisticated, high performance designs (such as elliptic function filters) are now fairly common. Application of dielectric resonators in high quality filters is most evident in bandpass and bandstop filters. There are some applications in directional filters and group delay equalizers, but bandpass and bandstop applications using dielectric resonators dominate the filter field.

Bandpass and bandstop filter fields can be subdivided according to the dielectric resonator mode being used. The most commonly used mode is the $TE_{01\delta}$. The $HE_{11\delta}$ (degenerate) hybrid mode finds applications in sophisticated elliptic function filters and high frequency oscillators. This particular mode offers the advantage of smaller volume and weight (approximately $\frac{1}{2}$) when compared to a single mode device. This is possible since each resonator resonates in two orthogonal, independent modes.

4.7   Single Mode Bandpass Filters

The basic bandpass filter topology element generally can be described as a section of an evanescent-mode waveguide (waveguide below cutoff) in which the dielectric resonators are housed. This particular configuration was originally proposed by Okaya and Barash[5] and later expanded by Cohn[6] and Harrison.[25] The orientation of dielectric resonators can be either transverse or coaxial.

The transverse configuration yields a larger filter, but is presently preferred in practical designs because these can be more conveniently tuned with screws concentric with the resonators. Typical configurations

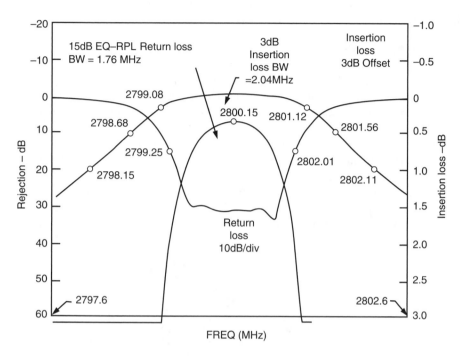

FIGURE 4.8 Typical performance of a four-pole, single made dielectric resonator filter showing insertion loss corresponding to Q factor of 9000.

FIGURE 4.9 Single mode dielectric resonator filters (2.8 GHz and 5.6 GHz).

of such filters are presented in Figure 4.9. Actual performance of one of the filters is shown in Figure 4.8. This particular design suffers from spurious responses on the high frequency side of the filter and suppressing techniques for these frequencies are necessary. The situation is worse when a microstrip transmission line is used to couple between the resonators.

The equivalent Q-factor of the filter is degraded and mounting of the dielectric resonator on special supports is usually necessary. Also, extra care must be taken to select proper dimensions of the dielectric resonator (e.g., D/L ratio) to place spurious modes as far as possible from the operating frequency.

Sufficient spacing from metal walls of the housing is also important since close proximity of conductive walls degrades the high intrinsic Q of the dielectric resonator.

4.8 Dual Mode Filters

After reviewing literature in the dielectric resonator area, it is obvious that most attention has been directed toward analysis and applications of the fundamental $TE_{01\delta}$ mode. Higher order modes and the $HE_{11\delta}$ mode, which for certain ratios of diameter/length has a lower resonant frequency than that of the $TE_{01\delta}$ mode, are considered spurious and hard to eliminate. Even for a radially symmetrical mode like $TE_{01\delta}$, which has only 3 components of the electromagnetic field, rigorous analysis is still a problem, and various simplifying assumptions are required. The situation is much more complex for higher modes that are generally hybrid, are usually degenerate, and have all six components of the electromagnetic field.

A typical filter configuration using the $HE_{11\delta}$ mode (in line) is presented in Figure 4.10.[26] Coupling between modes within a single cavity is achieved via a mode-coupling screw with an angular location of 45 degrees with respect to orthogonal tuning screws.

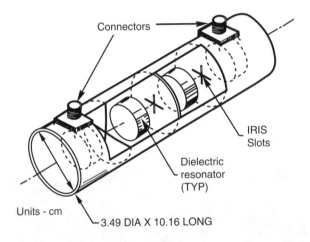

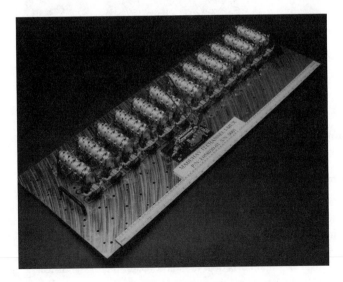

FIGURE 4.10 Dual mode bandpass filter configuration and multiplexer assembly using dual mode dielectric resonator filters.

Intercavity coupling is provided by polarization-sensitive coupling slots. This arrangement is similar to that presently used in metal-cavity filters. The design is identical and the standard filter synthesis method can be used. Dielectric resonators are mounted axially in the center of each evanescent, circular cavity. A low-loss stable mounting is required to ensure good electrical and temperature performance.

Size comparison between traditional cavity filters and a dielectric resonator filter is shown in Figure 4.11. Weight reduction by a factor of five and volume reduction by a factor of > 20 can be achieved. Spurious response performance of the 8-pole filters is similar to the TE_{111} mode cavity filter.

It was found that selection of diameter/length ratios greater than two yields optimum spacing of spurious responses. The $TE_{01\delta}$ mode is not excited because of the axial orientation of the resonator in the center of a circular waveguide.

One of the factors in evaluating a filter design, that is equal in importance to its bandpass characteristics, is its temperature stability. Since most of the electromagnetic field of a dielectric resonator is contained in the material forming the resonator, temperature properties of the filter are basically determined by

FIGURE 4.11 Size comparisons between traditional cavity filters and dielectric resonator filter (single mode rectangular waveguide filter, dual mode circular cavity filter, and dual mode dielectric resonator filter are shown).

properties of the ceramics. Typical temperature performance of the filters is in the order of ±1ppm/°C with almost perfect temperature compensation possible.[28]

4.9 Dielectric Resonator Probe

The dielectric resonator probe configuration illustrated in Figure 4.12 employs a dielectric resonator sandwiched between two conductive metal plates in a "post resonator" configuration. The "post resonator" is a special configuration of the dielectric resonator. In a "post resonator," the xy surfaces of the resonator are conductive (e.g., metalized).

The measured Q factor from this configuration is dominated by the losses from the conductive plates directly above and below the dielectric resonator and the dielectric loss (loss tangent) of the dielectric

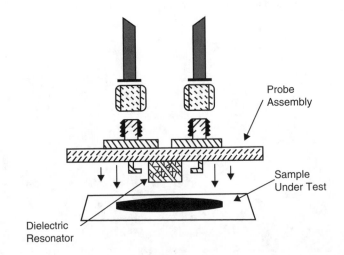

FIGURE 4.12   Illustration and photograph of the dielectric resonator probe.

resonator materials. Either one of these two loss contributors can be calibrated out and the other one can be determined with a great accuracy. If we calibrate out the conductive loss (in metal plates) the dielectric loss can be determined.[27,28] In the other case, the loss tangent of the dielectric resonator is known and the conductivity of the nearby metal or superconductor is unknown.[29] As in the case of the other measurement, surface resistance is calculated from a measured Q value.

The TE_{011} mode (post resonator) is used for these measurements since it is easily identified, relatively insensitive to small gaps between the dielectric and the test sample, and has no axial currents across any possible discontinuities in the probe fixture.

4.10 Diode Oscillators

Dielectric resonators have been used with Gunn diode oscillators to reduce frequency variations resulting from temperature or bias changes and to lower the oscillator phase noise. A typical configuration of such an oscillator is shown in Figure 4.13. Stability in order of 0.5 ppm/degree C is achievable at 10 GHz.[30]

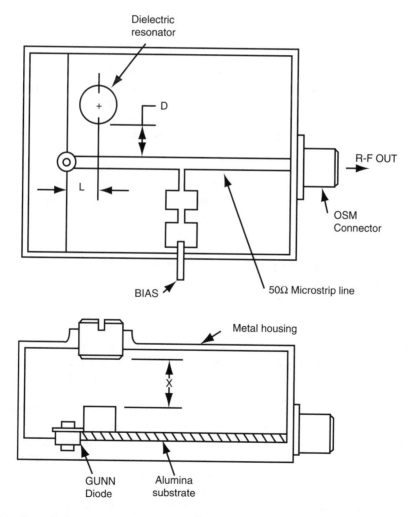

FIGURE 4.13 Gunn diode oscillator stabilized by a dielectric resonator.

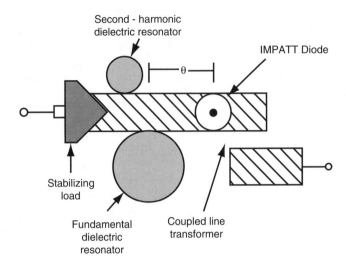

FIGURE 4.14   IMPATT diode oscillator stabilized by dielectric resonators.

Gunn diode oscillators exhibit lower phase noise at higher frequencies when compared to FET oscillators. Their performance is inferior, however, when factors such as efficiency and required bias levels are taken into consideration. An interesting application of dielectric resonators, which were used for stabilization of a 4 GHz high power IMPATT oscillator, is shown in Figure 4.14.[36] The novel oscillator configuration uses two dielectric resonators. This technique allows independent control of fundamental and harmonic frequencies of the oscillator.

At very high millimeter wave frequencies, dielectric resonators are too small to be effectively controlled. Therefore, much larger resonators utilizing whispering gallery dielectric resonator modes are preferred.[32-38] An additional advantage of these resonators (higher order modes) is better confinement of the electromagnetic field inside the dielectric resonator and consequently, the higher Q factor.

4.11   Field Effect Transistor and Bipolar Transistor Oscillators

FET (or bipolar) oscillators using dielectric resonators are classified as reflection or feedback oscillators (Figure 4.15).

For a reflection oscillator, initial design starts with either an unstable device or external feedback (low Q) to obtain negative resistance and reflection gain at the desired frequency. Next, a properly designed dielectric resonator is placed approximately one-half wavelength away from the FET or bipolar device. In this configuration, the dielectric resonator acts as a weakly coupled bandstop filter with a high external Q. Part of the output energy is reflected toward the device and such a self-injected oscillator will generate a signal at the resonant frequency of the dielectric resonator.

Typical reflection oscillators exhibit very good phase noise characteristics and frequency stability of approximately 1.5 ppm/degree C. Because of the reflective mode of operation, however, these designs are sensitive to load changes and require an output isolator or buffer amplifier.

Feedback oscillators can be divided into two classes: shunt feedback and series feedback (see Figure 4.15). In these examples, a dielectric resonator actually forms the feedback circuit of the amplifying element, usually a FET or bipolar transistor. In the shunt feedback arrangement, the resonator is placed between the output and input of the device (e.g., between gate and source, or gate and drain circuits).

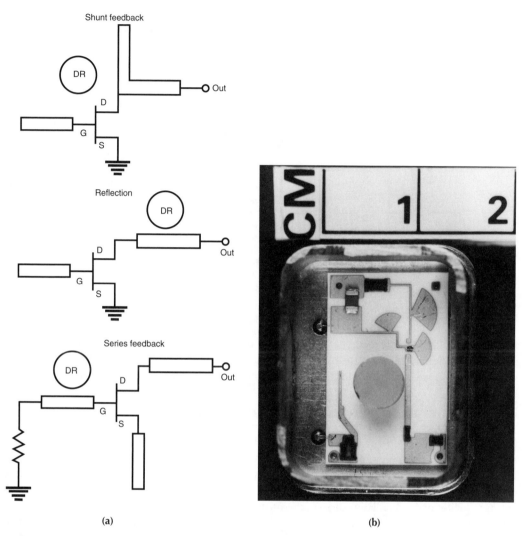

FIGURE 4.15 Basic configurations and photograph of dielectric resonator oscillator.

The conditions for oscillations are met at the resonant frequency of the dielectric resonator. In the shunt feedback scheme, however, the resonator is strongly coupled to the drain and gate transmission lines. Therefore, the loaded Q of the circuit is quite low and phase noise performance is degraded.

Another circuit that yields high stability and low phase noise is the series feedback oscillator.[39] This circuit consists of a high-gain, low-noise FET or bipolar transistor, a 50 ohm transmission line connected to the FET gate (bipolar-base) terminated with a 50 ohm resistor for out-of-band stability, a dielectric resonator coupled to the line and located at the specific distance from the gate (base), a shunt reactance connected to the FET source or drain (collector), and matching output impedance. Critical to the performance of this circuit is the placement of the dielectric resonator on the gate port, where it is isolated from the output circuits by the very low drain to gate capacitance inherent in the device. This isolation minimizes interaction between the output and input ports, which allows the resonator to be lightly coupled to the gate, resulting in a very high loaded Q, and therefore, minimum phase noise. A photograph of the typical dielectric resonator oscillator is shown in Figure 4.15. The phase noise performance, which demonstrates suitability of such oscillators for stringent communication systems is presented in Figure 4.16.

To further increase stability of the oscillator and improve close-in phase noise, phase-locked systems are getting increasingly popular. In such systems, a tunable (e.g., varactor) dielectric resonator oscillator

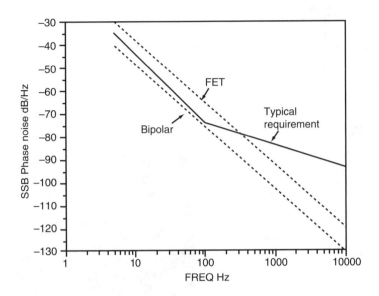

FIGURE 4.16   Phase noise characteristics of a dielectric resonator oscillator.

is phase locked to a low frequency crystal source or digital synthesizer (typically in ASIC form). This approach significantly enhances the overall performance of DROs allowing their use in all (even most stringent and demanding) communication systems (including satellite receivers).

4.11   Conclusions and Future Outlook

Applications of dielectric resonators in various microwave components are very cost effective and lead to significant miniaturization, particularly when MIC or MMIC structures are used. Excellent performance in filters and oscillators is currently being achieved. Dielectric resonators are widely used in wireless communication systems. Additional applications include dielectric or superconductor testing and antenna applications and radiating dielectric resonators. Miniature dielectric filled coaxial resonators are commonly used in wireless headsets (cellular and PCS phones). Recently available very high Q materials will extend commercial applications of dielectric resonators to much higher frequencies. Applications as high as 100 GHz are being reported.

Development of higher dielectric materials (80 to 100) has had a significant impact on lower frequency microwave devices (1 GHz region). Such dielectric resonators are used in practically all cellular and PCS base stations.

However, further material development is needed, mostly in dielectric materials with lower dielectric constants, which are used to mount dielectric resonators. New, low-loss plastics and adhesives should be developed to ensure that the excellent properties of dielectric resonator ceramics are not degraded.

Dielectric resonators are here to stay, and a wide variety of commercial wireless components using these elements is readily available. With the advent of new materials and improved circuit techniques, the field of dielectric resonators will continue to develop and will certainly be exciting in the future.

References

1. K.S. Packard, The Origin of Waveguides: A Case of Multiple Rediscovery, *IEEE Trans. Microwave Theory & Tech.*, MTT-32, 961–969, September 1984.
2. R.D. Richtmyer, Dielectric Resonator, *J. Appl. Phys.*, 10, 391–398, June 1939.

3. H.M. Schlicke, Quasi-degenerate Modes in High –ε Dielectric Cavities, *J. Appl. Phys.*, 24, 187–191, February 1953.

4. A. Okaya, The Rutile Microwave Resonator, *Proc. IRE*, 48, 1921, November 1960.

5. A. Okaya, L.F. Barash, The Dielectric Microwave Resonator, *Proc. IRE*, 50, 2081–2092, October 1962.

6. S.B.Cohn, Microwave Bandpass Filters Containing High Q Dielectric Resonators, *IEEE Trans. Microwave Theory & Tech.*, MTT-16, 218–227, April 1968.

7. D.J. Masse et al., A New Low Loss High-k Temperature Compensated Dielectric for Microwave Applications, *Proc. IEEE*, 59, 1628–1629, November 1971.

8. J. K. Plourde, D.F. Linn, H.M. O'Bryan Jr., and J. Thompson Jr, $Ba_2Ti_9O_{20}$ as a Microwave Dielectric Resonator, *J. Amer. Ceram. Soc.*, 58, 418–420, October-November 1975.

9. T. Nishikawa, Y. Ishikawa, and H. Tamura, Ceramic Materials for Microwave Applications, *Electronic Ceramics*, Spring Issue, Special Issue on Ceramic Materials for Microwave Applications, Japan, 1979.

10. S. J. Fiedziuszko and A. Jelenski, The Influence of Conducting Walls on Resonant Frequencies of the Dielectric Resonator, *IEEE Trans. Microwave Theory & Tech.*, MTT-19, 778, September 1971.

11. T. Itoh and R. Rudokas, New Method for Computing the Resonant Frequencies for Dielectric Resonators, *IEEE Trans. Microwave Theory & Tech.*, MTT-25, 52–54, January 1977.

12. M. W. Pospieszalski Cylindrical Dielectric Resonators and Their Applications in TEM Line Microwave Circuits, *IEEE Trans. Microwave Theory & Tech.*, MTT-27, 233–238, March 1979.

13. R.Bonetti and A. Atia, Resonant Frequency of Dielectric Resonators in Inhomogeneous Media, *1980 IEEE MTT-S Int. Symposium Digest*, 376–378, May 1980.

14. Y. Kobayashi, N. Fukuoka, and S. Yoshida, Resonant Modes in a Shielded Dielectric Rod Resonator, *Electronics & Communications in Japan*, 64-B, 11, 44–51, 1981.

15. K.A. Zaki, C.Chen, Field Distribution of Hybrid Modes in Dielectric Loaded Waveguides, *1985 IEEE MTT-S, Int. Symposium Digest*, 461–464, June 1985.

16. P. Wheless Jr. and D. Kajfez, The Use of Higher Resonant Modes in Measuring the Dielectric Constant of Dielectric Resonators, *IEEE MTT-S Int. Microwave Symposium Digest*, 473–476, June 1985.

17. S. J. Fiedziuszko, Dual Mode Dielectric Resonator Loaded Cavity Filters, *IEEE Trans. Microwave Theory & Tech.*, MTT-30, 1311–1316, September 1982.

18. T. Nishikawa, K. Wakino, H. Wada, and Y. Ishikawa, 800 MHz Band Dielectric Channel Dropping Filter Using TM_{010} Triplet Mode Resonance, *IEEE MTT-S Int. Microwave Symposium Digest*, 289–292, June 1985.

19. M. Stiglitz, Frequency Tuning of Rutile Resonators, *Proc. IEEE*, 54, 413–414, March 1966.

20. S. J. Fiedziuszko and A. Jelenski, Double Dielectric Resonator, *IEEE Trans. Microwave Theory and Tech.*, MTT-19, 79–780, September 1971.

21. S. Maj and M. Pospieszalski, A Composite Multilayered Cylindrical Dielectric Resonator, *IEEE MTT-S Int. Microwave Symposium Digest*, 190–192, June 1984.

22. Murata Mfg.Co — catalog

23. Transtech- catalog

24. D. Kajfez and P. Guillon, Editors, Dielectric Resonators, Artech House, 1986, available from Vector Fields, PO Box 757, University, MS 38677.

25. W.H. Harrison, A Miniature High Q Bandpass Filter Employing Dielectric Resonators, *IEEE Trans. Microwave Theory & Tech.*, MTT-16, 210–218, April 1968.

26. S. J. Fiedziuszko, Dielectric Resonator Design Shrinks Satellite Filters and Resonators, *Microwave Systems News*, August 1985.

27. B.W. Hakki, P.D. Coleman, Dielectric Resonator Method of Measuring Inductive Capacities in the Millimeter Range, *IRE Trans. Microwave Theory and Tech.*, MTT-8, 402–410, July 1960.

28. Y. Kobayashi and M. Katoh, Microwave Measurement of Dielectric Properties of Low-Loss Materials by the Dielectric Rod Resonator Method, *IEEE Trans. Microwave Theory and Tech.*, MTT-33, 586–592, July 1987.

29. S.J. Fiedziuszko and P.D. Heidemann, Dielectric Resonator Used as a Probe for High Tc Superconductor Measurements, *IEEE MTT-S International Microwave Symposium Digest,* Long Beach, CA, 1989.

30. T. Makino, Temperature Dependence and Stabilization Conditions of an MIC Gunn Oscillator Using Dielectric Resonator, *Trans. IECE Japan,* E62, 262–263, April 1979.

31. M. Dydyk, H. Iwer, Planar IMPATT Diode Oscillator Using Dielectric Resonator, *Microwaves & RF,* October 1984.

32. C. Chen and K.A. Zaki, Resonant Frequencies of Dielectric Resonators Containing Guided Complex Modes, *IEEE Trans. on Microwave Theory and Tech.,* MTT-36, 1455–1457, October 1988.

33. D. Cros and P. Guillon, Whispering Gallery Dielectric Resonator Modes for W-Band Devices, *IEEE Trans. on Microwave Theory and Tech.,* MTT-38, 1657–1674, November 1990.

34. D.G. Santiago, G. J. Dick, and A. Prata, Jr., Mode Control of Cryogenic Whispering Gallery Mode Sapphire Dielectric Ring Resonators, *IEEE Trans. on Microwave Theory and Tech.,* MTT-42, 52–55, January 1994.

35. E.N. Ivanov, D.G. Blair, and V.I. Kalinichev, Approximate Approach to the Design of Shielded Dielectric Disk Resonators with Whispering-Gallery Modes, *IEEE Trans. on Microwave Theory and Tech.,* MTT-41, 632–638, April 1993.

36. J. Krupka, D. Cros, M. Aubourg, and P. Guillon, Study of Whispering Gallery Modes in Anisotropic Single-Crystal Dielectric Resonators, *IEEE Trans. on Microwave Theory and Tech.,* MTT-42, 56–61, January 1994.

37. D. Cros, C. Tronche, P. Guillon, and B. Theron, W Band Whispering Gallery Dielectric Resonator Mode Oscillator, *1991 MTT-S Int. Microwave Symposium Digest,* 929–932, June 1991.

38. J. Krupka, K. Derzakowski, A. Abramowicz, M. Tobar, and R.G. Geyer, Complex Permittivity Measurements of Extremely Low Loss Dielectric Materials using Whispering Gallery Modes, *1997 MTT-S Int. Microwave Symposium Digest,* 1347–1350, June 1997.

39. K.J. Anderson, A.M. Pavio, FET Oscillators Still Require Modeling But Computer Techniques Simplify the Task, *Microwave Systems News,* September 1983.

5

RF MEMS

Karl Varian
Raytheon Company

5.1 Introduction

Microelectromechanical Systems (MEMS) are integrated circuit (IC) devices or systems that combine both electrical and mechanical components. MEMS are fabricated using typical IC batch-processing techniques with characteristic sizes ranging from nanometers to millimeters. RF MEMS are microelectromechanical systems that interact with a radio frequency (RF) signal. The integration/implementation of RF MEMS provides engineers with an additional integration option for better performance, smaller size, and lower cost in their designs.

Figure 5.1 includes a selection of RF MEMS devices and application of those devices in circuits that will be discussed later in this chapter. Figure 5.1 includes static devices, such as transmission lines and resonators; active devices, such as switches and variable capacitors; and circuits, such as oscillators (fixed frequency and voltage controlled) and tunable filters. The range of frequencies covered by these circuits is from a few MHz to the millimeter wave region. Both static and active RF MEMS devices will be discussed, but the chapter emphasis will be on the active devices.

RF MEMS provide microwave and RF engineers with low-insertion loss, high-Q, small size, very low-current consumption, and potentially low-cost options to solving their design problems. The low-insertion loss is obtained by replacing the moderate losses associated with semiconductors with lower metallic losses. Cost and size reduction is the result of utilizing semiconductor batch-processing techniques in RF MEMS manufacturing. Like existing semiconductor devices, the RF MEMS circuitry must be protected from

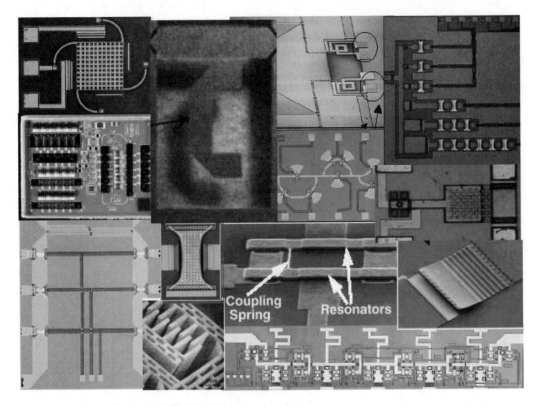

FIGURE 5.1   Examples of typical RF MEMS devices and applications.

the environment. But unlike semiconductors, the environmental protection is required due to either the mechanical movement and/or the mechanical fragility of the parts.

This chapter is an introduction to the terminology and technologies involved in the batch-processing, operation, and implementation of RF MEMS. The first section provides an RF MEMS technical overview, based on a switch. The operational theory of a switch is conceptually similar to most RF MEMS devices. This section also includes a switch equivalent circuit model, actuation methods, and various performance issues (e.g., power handling, switching time, reliability, and packaging).

After the RF MEMS overview, the following sections will cover fabrication, circuit elements, and typical circuit implementations. The fabrication section introduces the IC batch-processing technology, micromachining, and wafer bonding, utilized in the fabrication of RF MEMS. This section provides the background required to understand some of the manufacturing constraints and how RF MEMS devices are constructed. The next section covers RF MEMS circuit elements, describing various RF MEMS switches, variable capacitors, resonators, and transmission lines. The final section provides several typical RF MEMS implementation examples, such as phase shifters, tunable filters, oscillators, and reconfigurable elements.

5.2   RF MEMS Technical Overview

The RF MEMS switch is a conceptually simple device, with actuation mechanisms common to RF MEMS technology. The most common actuation mechanism is electrostatic. A simple electromechanical model will be used to introduce basic switch concepts of "on" and "off" states. An electrostatic model for the switch will then be described introducing the concept of pull-down voltage. An alternative switch actuation mechanism, electrothermal, will be discussed. The section will conclude with a discussion of several RF MEMS performance issues, such as power handling, switching time, reliability, and packaging.

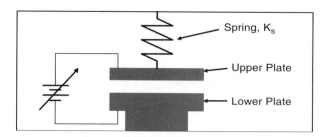

FIGURE 5.2 One-dimensional electromechanical model of an RF MEMS switch.

5.2.1 One-Dimensional Electromechanical Model

Switch actuation can be conceptualized with the following one-dimensional electromechanical example in Figure 5.2: two parallel plates (with zero mass); one supported above the other with a gap between them. The upper plate is held above the lower plate by some force, such as an ideal linear spring. This initial condition can be thought of as the switch "off" condition. When an electrostatic potential is applied between the two parallel plates, there is a resulting attractive electrostatic force. The spring force will balance out this attractive electrostatic force as the gap between the plates is reduced. The upper plate movement continues until the attractive electrostatic force overcomes the spring force, bringing the plates together. When the plates are together, the switch is in the "on" condition or activated. When the electrostatic force is removed, the spring restores the upper plate to the original noncontacting "off" position.

5.2.2 Electromechanical Switch Activation Characteristics [1]

The one-dimensional electromechanical model just described approximates the switch electromechanical motion. The model approximates the switch as a single rigid, but moveable, parallel plate (switch body) suspended above a fixed ground plate by an ideal linear spring (*note*: a classical representation of a capacitor is as two parallel plates separated by a dielectric gap). This model has a single degree of freedom, which is the gap between the movable top plate and the fixed bottom plate. An important feature of this model is its ability to correctly predict the pull-in of the membrane as a function of applied voltage. This switch motion can be described by the pressure balance equation:

$$P(g) = K_s(g_0 - g) - \left(\frac{\varepsilon V^2}{2g^2}\right), \tag{5.1}$$

where P is the total pressure on the mechanical body of the switch, g is the height of the switch body above the bottom plate, g_0 is the initial height of g with no applied field, and V is the applied electrostatic potential. The spring constant of the switch body, K_s, is determined by the Young's modulus and Poisson ratio of the membrane metal and the residual stress within the switch body. The gap permittivity, ε, is the permittivity of the dielectric material located between the upper and lower plates.

As the electrostatic field is applied to the switch, the switch membrane (switch body) starts to deflect downward, decreasing the gap g and increasing the electrostatic pressure on the membrane. At a critical height of 2/3 g_0, this mechanical system becomes unstable, causing the membrane to suddenly snap down onto the bottom plate. A graph (Figure 5.3) of the gap height as a function of applied voltage is obtained by solving the above equation.

A key parameter associated with RF MEMS switches is pull-down voltage. The pull-down voltage is the voltage at which the membrane suddenly snaps down onto the bottom plate. The pull-down voltage for the simple one-dimensional system can be solved as

$$V_p = \sqrt{\left(\frac{8K_s g_0^3}{27\varepsilon}\right)}. \tag{5.2}$$

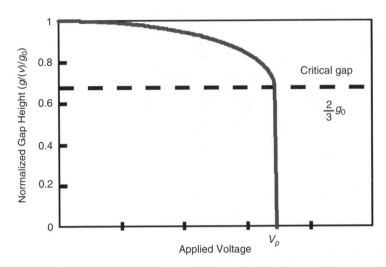

FIGURE 5.3   Gap height as a function of applied voltage. (From Goldsmith, C. et al. In *1996 IEEE MTT-S International Microwave Digest*, pp. 1141–1144, 1996.)

FIGURE 5.4   Electrothermal variable capacitor. (From Feng, Z. et al. In *1999 IEEE MTT-S International Microwave Digest*, pp. 1507–1510, 1999.)

From this equation, one can see that pull-down voltage for the simple one-dimensional system is related to the switch body spring constant, permittivity of the material in the gap, and to the initial membrane gap.

When the electrostatic force is removed from the switch, the tension in the metal membrane, K_s, pulls it back into the unactuated state.

5.2.3   Electrothermal Switch Actuation Characteristics

Besides the electrostatic mechanism just described there also exists electrothermal actuators [2–4] (Figure 5.4). Electrothermal actuators use the thermal expansion mismatch of different beam materials to produce a force that moves the plate up and down. The advantages of the electrothermal actuator are avoidance of static charges from collecting on the plates, approximately linear capacitance tuning, and activation with lower driving voltage (below 5 V). Disadvantages of electrothermal actuator are a slower tuning speed, requirement for current draw, small displacements, and additional space requirements. This actuation method can be used when precise physical displacement is required. Variable capacitors utilizing the electrothermal actuation have demonstrated capacitive tuning ratios of around 5:1 with a 1.8 μm gap change.

5.2.4 Switching Time

RF MEMS (mechanical) switches are inherently slower than electronic switches, with switching speeds in the microsecond to millisecond range, depending on the material and switch construction. Membrane switches, with relatively smaller mass and spring constant, have the fastest demonstrated switching times, with rise times of 6 μs and fall times of 8 μs [1]. The best switching times reported for cantilever beams switches on the order of 20 μs [5].

5.2.5 Power Handling

The RF power handling capability of RF MEMS switches is constrained by two major factors. One constraint is due to the classical switch current and/or voltage handling capability of the contact and contact area. The classical solution to this constraint has been to increase the contact area, use of special contact metals, sequential DC/RF switching, and so on.

The other major constraining factor is inadvertent switch actuation due to the RF power at the switch. The inadvertent switching occurs since RF MEMS switches are activated by an electrostatic potential. The average RF power carried on the line can generate enough electrostatic potential to cause the switch to be activated. On the basis of Equation 15.2, the pull-down voltage (and hence the switch power handling capability) can be increased by increasing the spring constant associated with either the beam or membrane, increasing the gap between the contact and RF line, or changing the permittivity in the gap. Since RF MEMS switches are mechanical switches, increasing either the spring constant or the gap would probably decrease the switching speed, while the impact on size would be nominal.

Potentially, the cantilever beam switch has a power handling advantage over the capacitive switch, since the bias electrode is separate from the contact region. This allows the cantilever beam switch designer to design the switch contact area separate from the bias electrode area. Typical power handling capability of cantilever beam switches have not been reported, while the power handling capability of the membrane switch is in the 2–9 W range.

5.2.6 Reliability

Outside of physical damage to the mechanical structure, the major reliability concern of RF MEMS devices occurs when the desired device movement is restricted. This lack of movement is usually referred to as stiction. Stiction occurs when the restoring force associated with an activated microstructure is insufficient to overcome the force that is holding the microstructure in the activated position. Possible causes of stiction are as follows:

1. *Metal-to-metal*: caused by van der Waal's forces between the two clean, smooth metal surfaces
2. *Microwelding*: caused by high-current density within the device during hot switching of signals
3. *Dielectric charging*: created by tunneling of electric charges into the dielectric that subsequently become trapped in the dielectric and screen the applied electric field
4. *Humidity*: surface tension of water can exert enough force between films to cause unwanted sticking
5. *ESD*: static electricity can cause unwanted MEMS actuation
6. *Inadequate micromachining*: contaminants and residues left from an incomplete undercut can cause sticking of the MEMS

Some, but not all, of these stiction sources may be catastrophic and irreversible. Sometimes, "stuck" RF MEMS devices can be freed by temperature cycling the device.

5.2.7 Packaging

Owing to the sensitivity of RF MEMS devices to environmental damage, handling, and humidity, packaging is a key technology. The MEMS packaging technology of choice is referred to as "Wafer Level" packaging. Wafer level packaging is the permanent bonding of two wafers (one wafer contains the MEMS devices while the other wafer is micromachined with pockets) before the wafer dicing operation. This packaging approach has many packaging advantages, but the main advantage as far as MEMS devices are concerned is that the MEMS devices are never exposed to environmental influences outside of the MEMS manufacturing process.

5.2.8 Actuation Voltage

The actuation voltage for MEMS devices (in particular MEMS switches) is in the range of 5 V to over 100 V with nominal voltages in the 20–40 V region. The selected actuation voltage is part of a design trade between spring stiffness, life, restoring force, temperature effects, and power handling. The use of laminate structures may expand this trade space.

5.2.9 Effects of Temperature

Electrical characteristics of MEMS devices tend to be sensitive to temperature variations. The temperature sensitivity is due to the dissimilar material properties of the substrate and the MEMS device materials and the small mechanical dimensions required for these devices. The differences in thermal expansion coefficient can change the actuation voltage from 0 V at elevated temperature or deform the MEMS device at lower temperatures. These issues are presently solved by either design or material selection.

5.2.10 Fabrication Technology [6–8]

The major fabrication technologies used in RF MEMS batch processing are micromachining and wafer bonding. Micromachining is the process of building three-dimensional structures either on or into a supporting wafer material (substrate). Wafer bonding is the building of three-dimensional structures by the stacking and attachment of wafers. Both micromachining and wafer bonding utilize typical batch-processing techniques used in traditional semiconductor wafer processing. Lesser-used technologies in RF MEMS fabrication such as laser micromachining, three-dimensional lithography, and so forth will not be discussed. RF MEMS devices have been fabricated on a number of different substrate materials: silicon, gallium arsenide, quartz, and so on with a majority of the work utilizing silicon. Several different metallizations (gold, aluminum, copper, etc.) have been used. By leveraging the silicon batch-process technology, rapid progress has been made in RF MEMS advancement. The term silicon and substrate will be used interchangeably in the rest of this chapter, but not restricted to silicon exclusively.

5.2.11 Micromachining Processes

The micromachining processes used in the fabrication of RF MEMS structures are bulk micromachining, surface micromachining, and LIGA (a refinement of surface micromachining). To build an RF MEMS structure with micromachining, the wafer could be processed using conventional processes to create transmission lines, capacitors, resistors, inductors, transistors, and so on that are required before the RF MEMS processing starts. A resist layer is then deposited, patterned, and cured in the areas not requiring RF MEMS process (to protect this part of the circuit from the RF MEMS processing steps). This layer can be removed at the completion of RF MEMS fabrication.

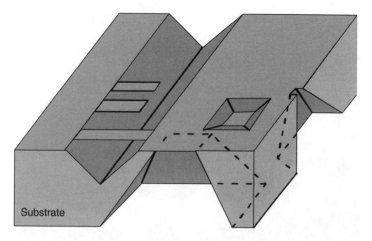

FIGURE 5.5 Example of bulk micromachining; note removal of substrate material and different etch depths.

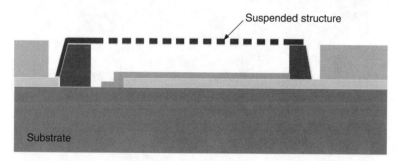

FIGURE 5.6 Example of surface micromachining. The structure is built up on the substrate surface. (Courtesy of Raytheon.)

Bulk micromachining involves the selective removal of the bulk silicon substrate material with either an anisotropic wet etch or by deep reactive ion etching (DRIE) as shown in Figure 5.5. Wet etching normally produces features with sloped sides at an angle of 54.7° while DRIE produces nearly vertical walls. For bulk micromachining the wafer is coated with a resist layer, patterned and cured. The areas that are not protected by a resist layer are then etched, either wet or DRIE, to the required depth. If the circuit requires features of different depths then the process of applying a resist, patterning, and etching are repeated until all features are defined. The wafer surface is metallized, as required, after wafer etching using a resist, patterning, and depositioning process. Upon RF MEMS process completion on this surface, the remaining resist is removed. If "back-side" processing is required, the wafer can be turned over and the processing steps repeated. This technique has been used to generate low-loss transmission lines and resonant cavities (with wafer-bonding techniques to be described later).

Surface micromachining involves the selective adding and removing of metal, dielectric, and sacrificial layers on the substrate surface. A resist layer is deposited, patterned, and cured on the wafer. Depending on the step, a metal, dielectric, or sacrificial layer is then deposited, patterned, and etched. This sequence of steps is repeated until the required RF MEMS three-dimensional structure is completed. The structures that are to be suspended are then "released" by the removal of the sacrificial material under the "to be suspended structure". Surface micromachining has been used to generate three-dimensional suspended structures such as cantilever beams and membranes (see Figure 5.6).

A refined form of surface micromachining is the process referred to as LIGA [9,10]. LIGA is an acronym that comes from the German name for the process, Lithographie Galvanoformung Abformung (lithography, electroplating, and molding). One of the first steps in the LIGA process is locating sacrificial

FIGURE 5.7 These three gears are examples of structures that were built with the LIGA process. See http://mems.engr.wisc.edu/images/gears/.

metallic pads where the LIGA microstructures will occur. A resist layer is then deposited to the required depth, patterned with x-rays, and cured. Metal is then electroplated into the patterned areas of the resist layer making contact with the sacrificial metal layer. The microstructure metal is released by removing the resist layer and original metallic pad. This process has demonstrated finely defined microstructures of up to 1000 μm high and the major structures formed with the LIGA process are gears (see Figure 5.7 [11]).

Comparisons of the three RF MEMS micromachining processes are shown in Table 5.1. The majority of the RF MEMS devices are fabricated with surface and bulk micromachining processes. The LIGA process is used at present primarily for micromachine structures.

5.2.12 Wafer Bonding

Wafer bonding is used to generate three-dimensional RF MEMS structures that are buried in a substrate. Wafer bonding is the permanent bonding of two or more wafers. Wafer bonding techniques that will be discussed are fusion, anodic, and eutectic. Typical applications are resonant cavities and packaging.

Silicon fusion bonding [12] occurs when pressure is applied to smooth, flat, clean, and hydrated surfaces. The resulting bond is a silicon-to-silicon bond with a water by-product (Figure 5.8). The completed assemblies can then be annealed at temperatures in the range of 800–1200°C to increase the bond strength by an order of magnitude. The resulting bond strength reaches the bond strength of crystalline silicon: 10–20 MPa.

TABLE 5.1 RF MEMS Process Technology Comparison

Capability	Bulk (Wet and DRIE)	Surface	LIGA
Maximum structural thickness	Wafer(s) thickness	$<50\,\mu m$	$1000\,\mu m$
Planar geometry	Wet—rectangular	Unrestricted	Unrestricted
	DRIE—unrestricted		
Minimum planar feature size	$\sqrt{2}\times$ depth	$1\,\mu m$	$3\,\mu m$
Side wall features	Wet—54.74° slope	Limited by dry etch	$0.2\,\mu m$ runout over $400\,\mu m$
	DRIE—Limited by the dry etch		
Surface and edge definitions	Excellent	Mostly adequate	Very good
Material properties	Very well controlled	Mostly adequate	Very good
Integration with electronics	Demonstrated	Demonstrated	Difficult
Capital investment and costs	Low	Moderate	High
Published knowledge	Very high	High	Moderate

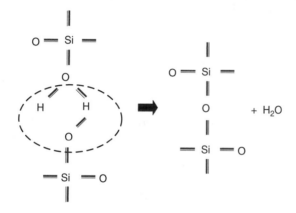

FIGURE 5.8 Chemical reaction during silicon-to-silicon fusion bonding.

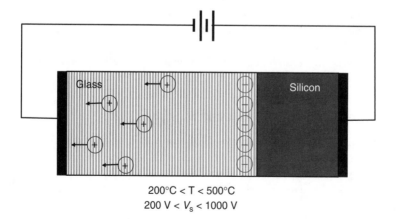

$200°C < T < 500°C$
$200\,V < V_s < 1000\,V$

FIGURE 5.9 Schematic representation of anodic bonding.

Anodic bonding occurs when an electrostatic potential of between 200 and 1000 V is applied across a smooth glass to silicon interface at elevated temperatures of 200–500°C (Figure 5.9). The resulting chemical reaction results in a trapped electrical charge in the glass at the glass–silicon interface. This trapped charge electrically pulls the silicon wafer into intimate contact with the glass substrate. The resulting bond strength is 2–3 MPa.

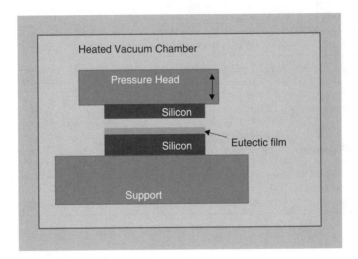

FIGURE 5.10   Schematic representation of eutectic bonding.

In eutectic bonding, a eutectic film is deposited on either of the wafers that are to be bonded together. The assembly is then placed in a vacuum chamber and heated to the eutectic temperature (Au–Si is 363°C) with pressure applied (Figure 5.10). Under the appropriate conditions, a eutectic bond is formed between the two substrates. The resulting bond strength (with a Au–Si bond) has measured 148 MPa.

5.3 Devices, Components, and Circuits

This section will cover various RF MEMS devices. Emphasis will be on switches, followed by examples of various types of variable capacitors, transmission lines, and resonators. The section begins with a description of the two different types of switches, cantilever beam and membrane, followed by a discussion of both ohmic and capacitive switch contacts. Equivalent circuits and a comparison between the different contact approaches will then be presented. On the basis of the equivalent circuits, several figures-of-merit will be introduced, which are used to compare various switches and technologies. The section concludes with a discussion of some typical physical switch dimensions.

5.3.1 Cantilever Beam and Membrane Switches

A cantilever beam switch [4,13–15] has one end of a "beam" anchored to a fixed point(s) while the other end of the beam is free to move, Figures 5.11 through 5.13. With no electrostatic force applied, the cantilever beam is up, the beam is in a minimum force state. The beam is actuated when a sufficient electrostatic potential is applied to a bias electrode (bias pad or control contact) located under the beam. With a sufficient electrostatic potential, the cantilever beam is pulled into contact with the bias pad, and consequently the end of the cantilever beam is pulled into contact in the contact region. For a solid metal cantilever beam, the beam will be electrically connected to the contact region. A variation on this approach is for the end of the cantilever beam to be electrically isolated from the rest of the cantilever beam and the contact region to actually be two contact regions (Figure 5.13) similar to a relay [16]. Therefore, when the switch is actuated, the two contact regions are connected together. The cantilever beam is restored to its original position when the electrostatic potential is removed. In most cases, the bias electrode and contact region are physically different as shown in Figures 5.11 through 5.13.

A membrane switch consists of a thin membrane anchored on more than one side and allowed to flex in the middle as shown in Figure 5.14. In the unactuated state, the membrane switch exhibits a high

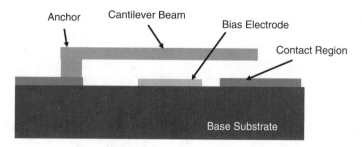

FIGURE 5.11 Cross-sectional view, cantilever beam switch.

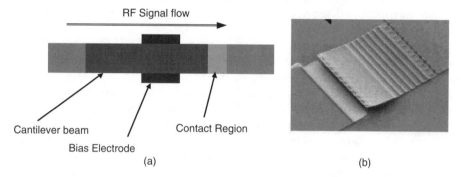

FIGURE 5.12 Cantilever beam switch, (a) overhead illustration and (b) perspective picture. (From Bozler, C. et al. In *2000 IEEE MTT-S International* Microwave Symposium Digest, pp. 153–156, 2000.)

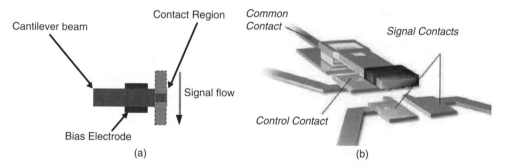

FIGURE 5.13 Cantilever beam relay, (a) overhead and (b) perspective view. (From K.R. Varian and Micromachined-relay illustration, Analog Devices, http://www.analog.com/industry/umic/relay.html.) *Note:* Control Contact and Bias electrode perform the same function.

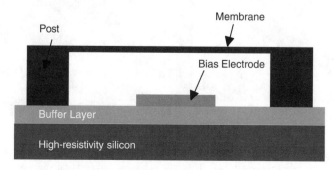

FIGURE 5.14 Cross-sectional view, membrane switch.

impedance due to the air gap between the membrane and bias electrode. Application of an electrostatic potential between the membrane and bias electrode causes the thin upper membrane to deflect downwards due to the electrostatic attraction. When the electrostatic force is greater than the restoring tensile force, the top membrane deflects into the actuated position. In this state, the membrane is in contact with the bias electrode. The membrane is restored to its original position when the electrostatic potential is removed. In most cases, the bias electrode and contact area are the same.

Although not shown in Figures 5.11 through 5.14, a feature that is usually added to the switches is a structure preventing the cantilever beam or membrane from shorting out to the bias pad. This structure, such as "bumps", can occur on either the cantilever beam, membrane, or substrate.

Besides the two different types of switches, there are two different contact mechanisms: ohmic and capacitive contacts. An ohmic contact occurs when there is metal-to-metal contact, whereas a capacitive contact occurs when there is a dielectric layer located between the potentially contacting areas, preventing metal-to-metal contact.

When in the off condition, both contact types are characterized by an "off" capacitance, C_{off}. The "off" capacitance is proportional to the distance and dielectric properties of the material between the contact pad and the contact, the contact shape, and the contact area.

The contacts illustrated in Figures 5.11 through 5.14 were ohmic contacts. Examples of an ohmic cantilever beam switch and an ohmic cantilever beam relay are shown in Figure 5.15 [5,17]. When actuated, an ohmic contact switch makes direct metal-to-metal contact. The "on" characteristic of an ohmic contact switch is characterized by the metal-to-metal contact resistance and is referred to as an "on" resistance, R_{on}. The combined simplified equivalent circuit for an ohmic contact switch is shown in Figure 5.16. For metal-to-metal contacts, switching with RF present at the contacts (hot switching) may lead to degradation of the metal in the contact region. This issue is addressed either in the design of the switch requiring the RF signal to remain below a critical level or by removing the RF signal before switching (cold switching).

The frequency range of ohmic switch is from dc to the upper frequency limit. The upper frequency limit is set by the reactance of C_{off}.

To explain a capacitive switch, a capacitive membrane switch, Figure 5.17, is used [18–21]. When a capacitive switch is actuated, the upper contact surface (membrane) is pulled into intimate contact with a dielectric layer that is located on the lower contact bias electrode, as shown in Figure 5.17. The large area formed by the metallic membrane–dielectric interface acts as a capacitor, preventing the metal-to-metal contact. Examples of capacitive membrane switches are shown in Figure 5.18 [18,22].

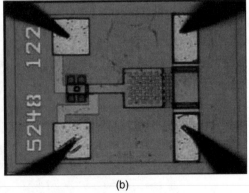

(a)   (b)

FIGURE 5.15   Photograph of ohmic contact (a) cantilever beam switch. (From Bozler, C. et al. In *2000 IEEE MTT-S International Microwave Symposium Digest*, pp. 153–156, 2000), and (b) cantilever beam relay switch. (From Hyman, D. et al., *Electronics Letters*, Vol. 35, No. 3, pp. 224–226, February 4, 1999).

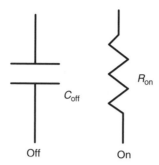

FIGURE 5.16 Simplified equivalent circuit of an ohmic contact switch.

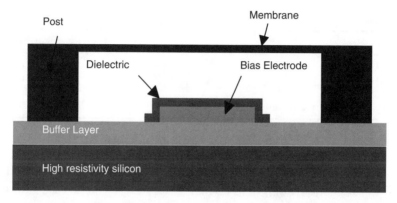

FIGURE 5.17 Cross-sectional view, membrane capacitive switch. (Courtesy of Raytheon.)

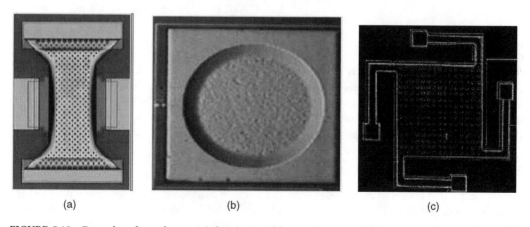

(a)   (b)   (c)

FIGURE 5.18 Examples of membrane switches (or variable capacitors), to different scales. Figures (a) and (b) courtesy of Raytheon. (From Young, D.J. and Boser, B.E. In *Proceedings of 1996 Solid-State Sensor and Actuator Workshop*, Hilton Head Island, pp. 86–89, 1996.)

In the design of a capacitive switch, the electrostatic field strength required to hold the membrane in "on" condition must not exceed the dielectric breakdown voltage.

An "on" capacitance, C_{on}, and an effective series "on" resistance, R_{on}, characterize the actuated capacitive switch. C_{on} is related to the membrane-dielectric-electrode area while R_{on} is related to the electrode and membrane ohmic losses. A simplified equivalent circuit of a capacitive switch is shown in Figure 5.19. Capacitive switches are inherently band-limited switches. The reactance of C_{on} sets the lower frequency

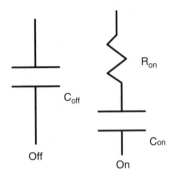

FIGURE 5.19 Simplified equivalent circuit of a capacitive switch.

TABLE 5.2 Comparison of Switch Technologies

Device Type	$R_{ON}(\Omega)$	C_{OFF} (fF)	f_c (GHz)
GaAs MESFET	2.3	249	280
GaAs pHEMT	4.7	80	420
GaAs P-I-N Diode	5.6	39	730
Capacitive Membrane MEMS	0.25	35	18,000

range of the capacitive switch while the upper frequency range is set by the reactance of C_{off}. A capacitive switch figure-of-merit is the ratio of C_{on}/C_{off} with acceptable ratios in the range of 40–100 presently being reported for RF MEMS capacitive switches. The larger the C_{on}/C_{off} ratio the broader the frequency range over which the switch can work.

A general switch figure-of-merit is cut-off frequency, f_c, and is defined as

$$f_c = \frac{1}{2\pi R_{on} C_{off}}. \tag{5.3}$$

This figure-of-merit is a measure of two critical switch characteristics, R_{on} and C_{off}, both of which should be as small as possible, consequently, the larger the cut-off frequency the better the switch. As shown in Table 5.2, a table comparing the f_c of various switch technologies, the RF MEMS switch is more than a factor of 20 better than the next best switch technology, P-I-N diodes. An additional benefit over the P-I-N diode is negligible power consumption.

5.3.2 Typical Switch Construction Details

The ohmic cantilever beam switch has four basic components: a cantilever beam, an anchor point, a bias electrode region, and the RF contact (Figures 5.11 through 5.13). The cantilever beam is anchored at one point, the anchor point, while the contact point is normally at the opposite end with the pull-down electrode located in between. The distance between the pull-down electrode and the anchor point and the cantilever beam spring constant determines the restoring force. Located between the pull-down electrode and the bias electrode, there is normally either a dielectric or a mechanical stop. This prevents the bias electrode (the cantilever beam) from shorting out to the pull-down electrode, maintaining an electrical separation between the RF and control circuitry. A dc ground return is required.

The capacitive membrane switch consists of an anchored, RF-grounded membrane suspended above a dielectrically protected pull-down electrode (Figure 5.17). The membrane thickness is in the range of 0.2–2.0 μm. In the actuated state, the membrane is pulled down onto the dielectrically protected pull-down electrode. As in the cantilever beam case, the restoring force is related to the distance between the anchor points and the pull-down electrode. In RF capacitive switch design, the pull-down electrode is usually an RF line and a dc ground return is required.

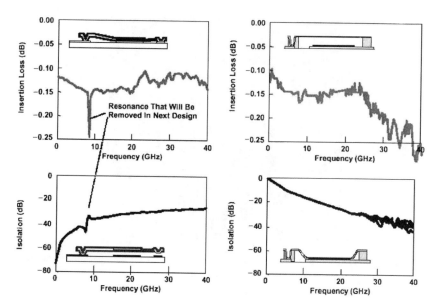

FIGURE 5.20 Insertion loss and isolation comparison between ohmic and capacitive switches versus frequency. (From B. Pierce, personal communication.)

The membrane height for capacitive switches and the contact height for cantilever beam switches are in the range of 1.5–5 μm. Owing to the normally larger contact area associated with capacitive switches, the membrane height is generally greater than cantilever beam contact height in order to minimize the off capacitance.

Figure 5.20 has typical response curves of insertion loss and isolation illustrating the RF electrical performance differences between an ohmic switch and a capacitive switch [23]. Recent measurements of the capacitive membrane switch indicate that its loss is less than the ohmic contact switch up to about 30 GHz. The loss of the ohmic contact switch is due to contact and "beam" losses. The major loss mechanism for the capacitive membrane switch is related to the electrode ohmic losses. Above 25 GHz, the loss of the capacitive switch increases due to modeling of the CPW structure, mismatch losses, or dielectric losses. Recent measurements of a capacitive membrane switch have indicated combined ohmic and dielectric losses of less than 0.1 dB up to 30 GHz [24].

The fundamental difference between these two switch losses is shown in the isolation measurements. For the ohmic switch, isolation is infinite at dc increasing to an isolation corresponding to the off capacitance at high frequencies. Meanwhile, the capacitive switch has a near-zero isolation at low frequencies, limited by electrode ohmic losses, and the isolation increases, corresponding to the voltage ratio between the C_{off} reactance and the load.

5.3.3 Variable Capacitors

Many different types of RF MEMS variable capacitors have been developed. They cover both the digital and analog spectrum. A few conceptually different ones will be described, one digital and three analog variable capacitors.

The digital approach utilizes switches to select a discrete capacitor that is in series with the switch. If ideal ohmic switches were used, then activating a switch(s) would add the discrete capacitor(s) to the circuit. The capacitive tuning range over which this type of variable capacitor can work is limited by the series R_{on} and C_{off}.

If an ideal capacitive switch is used in place of the ohmic switch, then activating the switch(es) would add the series combination of the discrete capacitor(es) and C_{on} of the switch(es), as shown in Figure 5.21 [25]. The capacitive tuning range over which this type of variable capacitor can work is limited by the

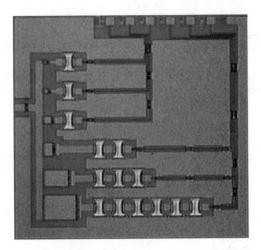

FIGURE 5.21   Six-bit variable capacitor, MEMS switches are in series with a fixed capacitor. (From Goldsmith, C.L. et al. *International Journal of RF & Microwave Computer-Aided Engineering*, Vol. 9, No. 4, pp. 362–374, 1999.)

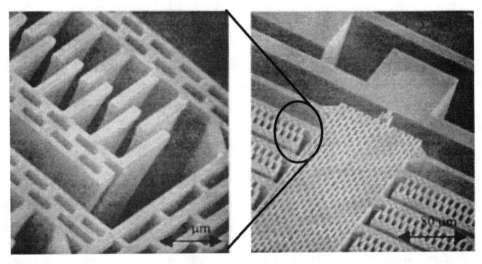

FIGURE 5.22   Variable capacitor using interdigitated metallic fingers. (From Yao, J.J. et al. In *Proceedings of 1998 Solid-State Sensor and Actuator Workshop*, Hilton Head Island, pp. 124–127, 1998.)

series combination of $C_{on} - R_{on}$ and C_{off}. This latter approach to building a digital variable capacitor with capacitive switches was demonstrated with a tunable filter (see Section 5.4.2), where the tunable elements of the filter are the digital variable capacitor.

Remembering the one-dimensional description of how a switch works, one recalls that as a potential is applied to the plates, the plates move closer together. This moving of the top plate continues until the height is reduced to two-thirds of the static state. When two-thirds of the static state height is reached, the switch becomes unstable and is pulled (collapses) into the fully activated state. In the region before the collapse, the movable plate (membrane) is a variable capacitor [22]. This type of variable capacitor has a limited tuning range and has been demonstrated with both electrostatic potential as described above and with electrothermal switches. The theoretical maximum capacitive tuning ratio is 1.5:1.

Another analog variable capacitor approach is to use interdigitated metallic fingers (see Figure 5.22) to provide the capacitance [26,27]. One set of fingers, similar to a comb, is fixed while the other set is

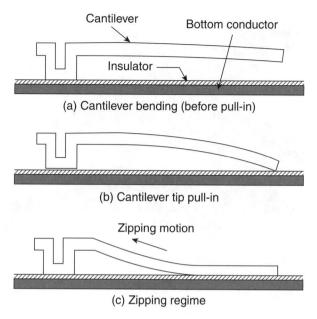

FIGURE 5.23 Capacitive cantilever beam "zipper" variable capacitor. (From Hung, E. and Senturia, S. In *Proceedings of 1998 Solid-State Sensor and Actuator Workshop*, Hilton Head Island, pp. 292–295, 1998.)

constrained to movement in only one direction. The movable comb structure is moved by the electrostatic force generated by fringing fields at the ends (both fixed and stationary) of the comb structure, and thus is independent of and not limited by the gap spacing between the comb structures. This relatively uniform force as the overlap of the variable capacitor interdigital fingers increases results in a relatively linear voltage–capacitance curve. This mechanical configuration for an interdigitated variable capacitor allows a motion on the order of tens of microns, thus a large tuning range for the variable capacitor, with relative small electrostatic fields. The interdigital variable capacitor has demonstrated a maximum capacitance of approximately 6 pF, a capacitance tuning ratio of greater than 4.55:1 over a voltage tuning range of 0–5.2 V, and a series resonance of over 5 GHz.

A capacitive cantilever beam electrostatic "zipper" switch (Figure 5.23) can also be used as a variable capacitor [28]. A cantilever beam zipper switch is constructed similar to a normal cantilever beam switch except that the bias electrode (bottom electrode) is shaped to enhance a zipper action on the beam. When an increasing voltage is applied between the beam and the bias electrode, the beam first bends downward, then collapses toward the substrate. At first, only the beam tip contacts the substrate, but as additional voltage is applied, the tip flattens and the beam zips along the substrate toward the cantilever beam anchor point. A specific capacitance–voltage characteristic can be tailored by lithographically "programming" the bias electrode geometry. This particular variable capacitor has demonstrated a capacitive tuning ratio of 1.7:1. A ratio of available gap sizes and large capacitances per unit area can be used to increase the capacitive tuning range.

5.3.4 Transmission Lines [29–31]

Obtaining transmission lines with low loss has always been a very challenging goal. Micromachining has enabled the fabrication of transmission lines on less than 2-μm-thick dielectric layers. Examples of some transmission lines that have been fabricated are shown and described in Figure 5.24. These transmission lines and resulting circuits have demonstrated zero dispersion, very low loss, and very small parasitics.

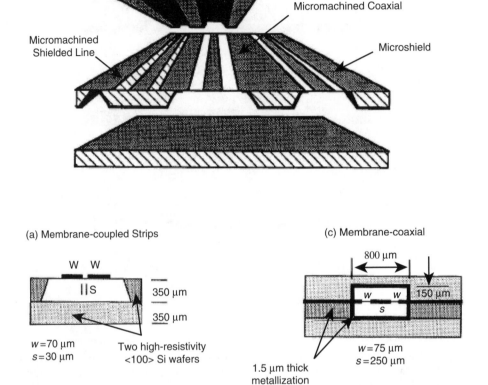

FIGURE 5.24 Different styles of demonstrated low loss transmission lines. (From Katehi, L.P.B. and Rebeiz, G.M. In *1996 IEEE MTT-S International Microwave Symposium Digest*, pp. 1145–1148, 1996.)

The circuitry is patterned on a gold film located on a stress-compensated 1.4-m membrane layer consisting of SiO_2–Si_3N_4–SiO_2 layer on a high-resistivity silicon substrate using thermal oxidation and low-pressure chemical vapor deposition. The silicon is completely etched away from under the circuitry until the circuitry is left on a thin dielectric membrane. When applicable, the mating wafers are patterned, etched, plated, and mounted to the main wafer. Besides low-loss transmission lines, inductors [32–34], resonators, and filters have been demonstrated based on this technology.

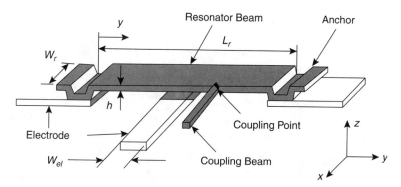

FIGURE 5.25 Mechanical MEMS resonator, a single pole. (From Nguyen, C.T.-C. In *1998 IEEE Aerospace Conference*, 1, pp. 445–460, 1998.)

5.3.5 Resonators (Filters)

Resonators are a basic building block in frequency selective systems. Owing to the diverse technologies involved and the low insertion loss associated with MEMS technology, several different resonator types exist. Three types of resonant structures, demonstrated over widely different frequency ranges, will be discussed: mechanical (300 KHz to 100 MHz), cavity (greater than 20 GHz), and piezoelectric film (1.5–7.5 GHz) resonators.

Mechanical resonators utilize the small feature sizes of the microstructures to create a mechanically resonant structure in the 300 KHz to 100 MHz range [35–42]. The mechanically resonant structure is a micromechanical "clamped–clamped" (clamped at both ends) beam resonator (Figure 5.25). A bias electrode is located centrally under the beam. These structures are then electrostatically coupled with RF lines to create electromechanical resonances. Bandpass filters can then be created by coupling two or more of these structures together.

A perspective view and a schematic representation of a mechanically resonant structure is shown in Figure 5.26. In this figure, key mechanical structures are identified, along with the required bias and excitation scheme for proper operation.

A dc bias voltage, V_P, is applied to the resonator. The ac input signal, v_i, is applied to the bias electrode. When the applied ac signal frequency enters the passband of the beam resonator, the microstructure will vibrate in a direction perpendicular to the electrode, creating a dc-biased time varying capacitor. An ac current, i_{ac}, will thus flow

$$i_{ac} = V_P \left(\frac{\partial C}{\partial t} \right), \tag{5.4}$$

where C is the bias electrode to beam capacitance. The time varying current is then ac coupled-off of the beam resonator (when more than one resonator is in the circuit, the resonators are coupled together and the signal is coupled-off of the second resonator with a sense electrode) and converted to an output voltage signal.

An example of a typical transmission spectrum for a second order (two resonator) (Figure 5.27) maximally flat bandpass filter centered at 14.54 MHz is shown in Figure 5.28. The resulting filter Q is 1000 with a 13 dB insertion loss and 24 dB of stop band rejection.

Cavity resonators are created by using bulk micromachining, plating, and epoxy bonding the wafers together to generate buried cavities. The cavity can be either a typical waveguide structure [43–45] (with and without obstructions) or a cavity containing transmission line resonating structures [46–49]. The

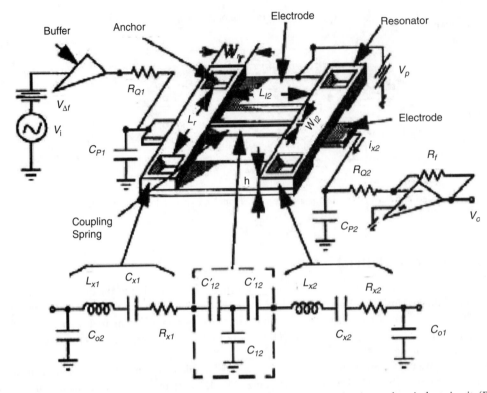

FIGURE 5.26 Mechanical MEMS resonator, 2-pole filter illustrating support circuitry and equivalent circuit. (From Bannon, F.D. III, et al. In *1996 International Electron Devices Meeting,* pp. 773–776, 1996.)

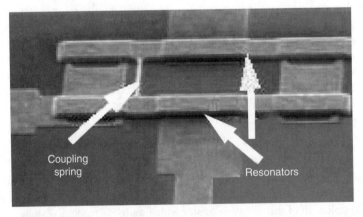

FIGURE 5.27 Two resonator, two-pole, MEMS bandpass filter. (From Nguyen, C.T.-C. *IEEE Transactions on Microwave Theory and Techniques,* Vol. 47, No. 8, pp. 1486–1503, August 1999.)

structure shown in Figure 5.29 is made by bonding three wafers together with epoxy. The central wafer has the desired circuitry while the other two wafers form a shield cavity around the circuitry. Micromachined low-loss transmission lines with gold metallization are used to define the circuitry in the cavity. Silicon is completely etched away from under the circuitry, until the circuitry is left on a thin dielectric membrane. At the same time, grooves are opened all around the circuitry to ensure complete shielding. This completes

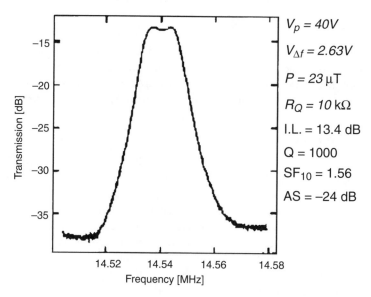

FIGURE 5.28   Transmission response of a two-resonator, two-pole, MEMS bandpass filter. (From Clark, J.R. et al. In *Proceedings of 1997 International Conference on Solid-State Sensor and Actuators*, Chicago, pp. 1161–1164, 1997.)

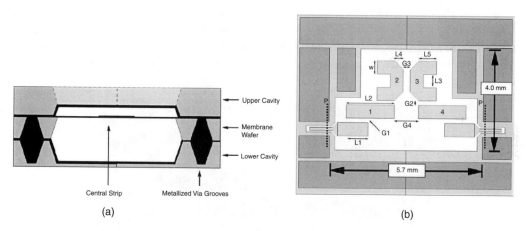

FIGURE 5.29   MEMS cavity resonator, (a) cross-sectional view, (b) a typical circuitry that would be placed on the central strip, a 4-pole elliptic function filter. (From Blondy, P. et al. *IEEE Transactions on Microwave Theory and Techniques*, Vol. 46, No. 12, pp. 2283–2288, 1998.)

the processing of the central wafer. The upper and lower wafers are patterned, etched, and plated to form the required cavities. The resulting wafers are then joined together with silver epoxy as shown in Figure 5.29. The four-pole, 60-GHz elliptic function filter demonstrated an insertion loss of 1.5 dB and rejection of better than −40 dB.

The final resonator structure to be considered is deposited piezoelectric films sandwiched between conductors in a similar fashion to quartz crystals [50–52]. Such resonators are constructed using aluminum nitride piezoelectric films with demonstrated Qs of over 1000 and resonance frequencies of 1.5–7.5 GHz. Although very promising, improvements in process and frequency trimming (tuning) are required.

5.4   Typical RF MEMS Applications

The utilization of RF MEMS devices and some implementation techniques will be discussed. Examples were chosen to illustrate the variety of applications in which RF MEMS can be used. Circuits included are phase shifters, tunable filters, oscillators (both fixed and voltage controlled), and reconfigurable matching elements and antennas. Some of the examples include how the RF MEMS devices were packaged.

5.4.1   Phase Shifters (Time Delay)

Phase shifters and time delay networks are similar. A phase shifter adds a required phase shift to a signal relative to a reference path, while a time delay network shifts (delays) the signal in time relative to a reference path. At a fixed frequency, a time delay corresponds to a phase shift. As the frequency changes the time delay remains approximately the same whereas the phase shift changes linearly with frequency. An example of a time delay network is a transmission line of fixed length (distributed). An example of a phase shift circuit is a series capacitor or inductor (lumped elements). The following RF MEMS phase shifter circuits are a hybrid combination of lumped (RF MEMS switches and other discontinuities) and distributed (transmission lines) elements. Consequently, they are thought of more as phase shifters, although in a limited range they are also time delay networks.

Three circuits will be described: a switched path transmission line phase shifter, a reflection transmission line phase shifter, and a capacitively loaded transmission line phase shifter. Of these, the capacitively loaded transmission line phase shifter may be a true time delay phase shifter over the widest frequency range.

The phase shift in a switched path transmission line phase shifter occurs when an additional length of transmission line, corresponding to the required phase shift, is added to the signal path (see Figure 5.30) [53,54]. The switching is accomplished with RF MEMS capacitive membrane switches that are shunted to ground using shunt RF MEMS capacitively coupled membrane switches and quarter-wave length ($\lambda/4$)

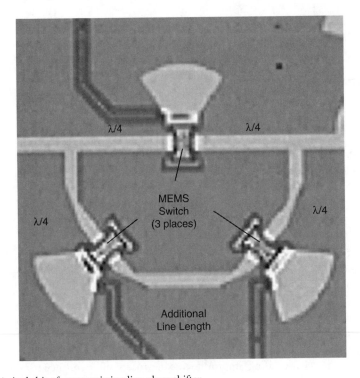

FIGURE 5.30   A single bit of a transmission line phase shifter.

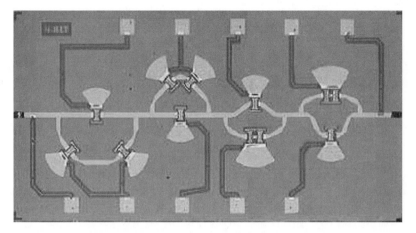

FIGURE 5.31 Ka-band 4-bit RF MEMS phase shifter. (From Pillans, B. et al. *IEEE Microwave and Guided Wave Letters*, Vol. 9, No. 12, pp. 520–522, December 1999.)

FIGURE 5.32 Ka-bond 3-bit phase shifter with an epoxy bonded lid. (From Pillans, B. et al. In *2000 IEEE Radio Frequency Integrated Circuits Symposium Digest*, pp. 195–198, 2000.)

transmission lines. To turn off a section of line, the switch in that line is activated. Consequently, an open will appear at the tee junction for the activated line due to the two quarter-wave transformations, occurring from the tip of the resonant stub ($\lambda/4$ long) to the tee junction. In the other path (the desired path), the switches are not activated (off) and the RF signal follows the selected path. An RF MEMS Ka-band, four-bit switched path transmission line phase shifter is shown in Figure 5.31. The circuit has an average insertion loss of 2.25 dB with better than a 15 dB return loss.

A three-bit version of this same phase shifter with two phase shifters on a single die was fabricated. The die size was increased to permit a transparent glass lid (with a cavity over the circuitry) to be epoxy bonded onto the die (Figure 5.32). The epoxy glass lid permitted the phase shifter to function in a noncontrolled environment for a limited length of time.

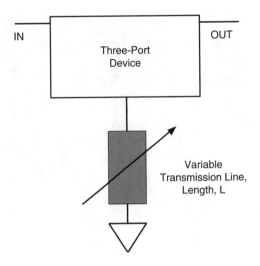

FIGURE 5.33   Basic reflection phase shifter.

A reflection transmission line phase shifter consists of a nonreciprocal device, a variable length transmission line, and a short as shown in Figure 5.33 [55]. The signal enters through one port of the nonreciprocal device (such as a Lange coupler or circulator). A variable length transmission line connects the nonreciprocal device to a short, which reflects the signal back to the nonreciprocal device and out the exit port. The phase shift occurs when the length of the transmission line changes and the phase shift corresponds to twice the change in electrical length of the transmission line. A 10 GHz, 4-bit RF MEMS reflection phase shifter is shown in Figure 5.34. The 4-bit phase shifter consists of two different cascaded 2-bit phase shifters. One phase shifter steps the phase in 90° steps from 0E to 270E, while the other phase shifter varies the phase between the 0E, 22.5E, 45E, and 67.5E states. A Lange coupler is used as the nonreciprocal device and RF MEMS capacitive membrane switches are used to change the line length by changing the location of the short. After the Lange coupler, one of four different phase states can be selected from either phase shifter. The combined 4-bit phase shifter has an average insertion loss of 1.4 dB with less than 11 dB of return loss at 10 GHz.

A distributed phase shifter can be realized by capacitively loading a transmission line (see Figure 5.35) [56,57]. The capacitors can be either an analog or digital variable capacitor, as described earlier. Changing the shunt capacitance changes the line impedance and hence the phase shift of the line. This type of phase shifter has demonstrated, in a single bit configuration, insertion loss of 0.6–0.7 dB for roughly a 180E of phase shift.

5.4.2 Tunable Filters [18]

The RF MEMS variable capacitors (both analog and digital versions) have been used to fabricate RF tunable bandpass filters utilizing both discrete and distributed elements. The frequency range covered by the filters ranged from 100 MHz to 30 GHz, with tunability bandwidths of 2.5% to greater than 35%. An example of a five-pole 1-GHz discrete element bandpass filter tunable over a 20% band is shown in Figure 5.36. The filter utilizes the digital variable capacitors and contains 44 capacitive switches.

This filter is part of a filter bank of seven filters covering the frequency range of 100–2800 MHz. The filters were packaged into a 1.25 in. × 1.75 in. ball grid array hermetic package (Figure 5.37). The package contained around 450 capacitive membrane switches, both discrete and integral (part of the RF MEMS-integrated circuit) inductors, and a controller to convert an incoming data stream into the control signals

FIGURE 5.34   10 GHz, 4-bit reflection phase shifter using Lange couplers as the nonreciprocal device. (From Malczewski, A. et al. *IEEE Microwave and Guided Wave Letters*, Vol. 9, No. 12, pp. 517–519, 1999.)

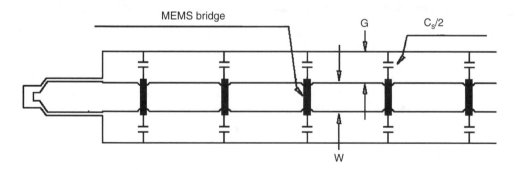

FIGURE 5.35   Capacitively loaded transmission line phase shifter. MEMS bridge refers to a MEMS switch, G the gap between the transmission line and ground, W the width of the transmission line, and C2 the shunt capacitor. (From Hayden, J.S. and Rebeiz, G.M. In *2000 IEEE MTT-S International Microwave Symposium Digest*, pp. 161–164, 2000.)

required for band selection and filter tuning. Hermetic packaging permits assembly testing for an extended time in an uncontrolled environment.

In Figures 5.38 and 5.39, a photograph and the corresponding responses of K- and Ka-band, bandpass filters are shown [58]. These filters used MEMS variable capacitors to tune either a lumped element filter

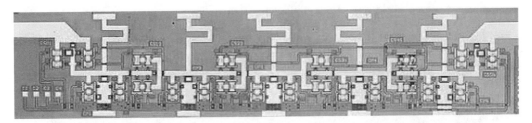

FIGURE 5.36 Five-pole, 1 GHz tunable bandpass filter. (From Goldsmith, C. *2000 IEEE MTT-S International Microwave Symposium*, 2000.)

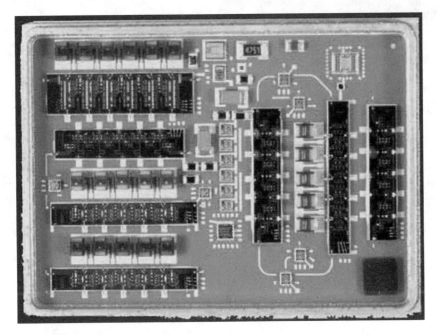

FIGURE 5.37 Seven-filter, filter bank in a hermetic package, plus control circuitry. (From Goldsmith, C. *2000 IEEE MTT-S International Microwave Symposium*, 2000.)

centered at 26 GHz or a distributed, half-wave resonator, filter at 30 GHz. Respective minimum insertion losses, 4.9 dB and 3.8 dB, were measured with corresponding tuning ranges of 4.2% and 2.5%.

5.4.3 Oscillators

A key oscillator parameter is phase noise. Phase noise is directly related to the circuit losses (inversely related to the square of the unloaded Q). RF MEMS provides a means of reducing circuit losses. The RF MEMS structures that have been used in oscillator design are low-loss transmission lines, resonators, and variable capacitors.

Low-loss transmission lines have been used to provide stronger coupling into a dielectric resonator [59,60]. The stronger coupling permits the dielectric resonator to be decoupled more from the circuit, which in turn increases the unloaded Q of the resonator and lowers the phase noise of the oscillator. An unloaded Q of 1600 was obtained for a whispering gallery mode dielectric resonator oscillator at 35 GHz with micromachined coplanar transmission lines.

An additional use for a low-loss transmission line has been to form a low-loss resonating structure (this scheme has also been used for fixed tuned filters) (see Figure 5.40) [61,62]. This resonating structure is

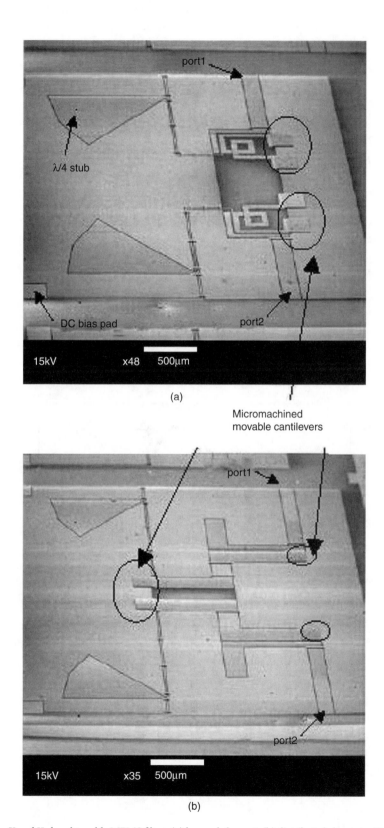

FIGURE 5.38   K and Ka band tunable MEMS filters (a) lumped element, (b) distributed elements. (From Kim, H.-T. et al. In *1999 IEEE MTT-S International Microwave Symposium Digest*, Vol. 3, pp. 1235–1238, 1999.)

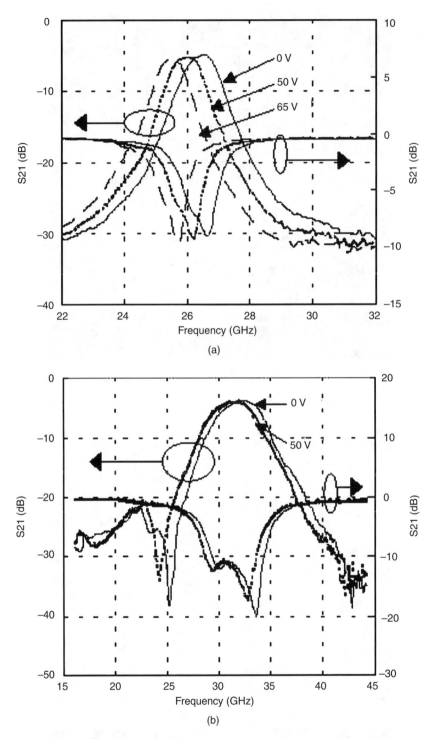

FIGURE 5.39 Response of the K and Ka band tunable MEMS filters in Figure 5.38, (a) lumped element, (b) distributed elements. (From Kim, H.-T. et al. In *1999 IEEE MTT-S International Microwave Symposium Digest*, Vol. 3, pp. 1235–1238, 1999.)

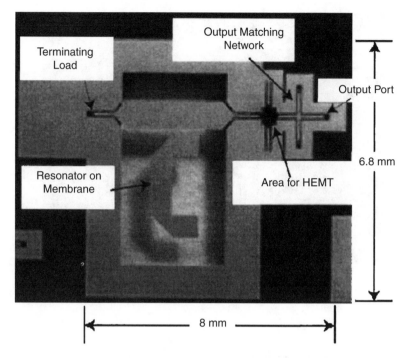

FIGURE 5.40 Photograph of a 28.65 GHz oscillator utilizing a low-loss transmission line resonator. (From Brown, A.R. and Rebeiz, G.M. *IEEE Transactions on Microwave Theory and Techniques*, Vol. 47, No. 8, pp. 1504–1508, August 1999.)

then coupled with an active device in such a way that the circuit will resonate. An unloaded Q of 460 have been measured with oscillators at 28.65 GHz, with corresponding phase noise of -92 dBc/Hz at a 100-kHz offset frequency and -122 dBc/Hz at a 1-MHz offset frequency. The RF MEMS resonating structure resulted in a 10 dB improvement in phase noise.

RF MEMS variable capacitors have been used as the tunable element in a voltage-controlled oscillator (VCO) [22,63–66]. The frequency range of VCOs report to date has been less than 2.4 GHz and tuning ranges of 3.4% or less. A photograph of the RF MEMS variable capacitor used is shown in Figure 5.41 and a schematic utilizing the RF MEMS variable capacitor is shown in Figure 5.42. This oscillator had an output power of -14 dBm at 2.4 GHz and a measured phase noise of -93 dBc/Hz at a 100-kHz offset and -122 dBc/Hz at a 1-MHz offset.

5.4.4 Reconfigurable Elements

RF MEMS devices have also been used to actively reconfigure matching elements and antennas. Three different examples will be discussed. Two are using RF MEMS switches to activate different portions of an antenna/array. The other example mechanically changes the shape of a radiating element.

The approach consisting of an array of MEMS switches is illustrated in Figure 5.43 [17]. The array of RF MEMS cantilever beam ohmic switches activates metal patches. By appropriately selecting the desired row and column actuator control lines, a specific metal patch will be selected. When this array concept was applied to a patch antenna (see Figure 5.44), the patch demonstrated the capability of changing its radiating frequency from roughly 10 to 20 GHz. This approach was also used to vary the input and output matching network of an amplifier from an operation center frequency of 10–20 GHz.

Another conceptually straightforward concept is shown in Figure 5.45 [67,68]. By using a MEMS cantilever beam ohmic switches, the length of a half-wave dipole antenna is changed. This reconfigurable

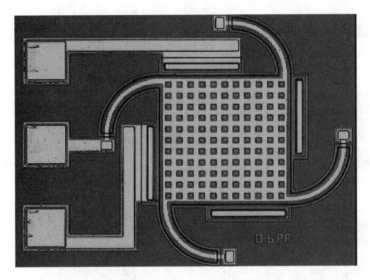

FIGURE 5.41 Photograph of MEMS variable capacitor used in VCO. (From Dec, A. and Suyama, K. In *1999 IEEE MTT-S International Microwave Symposium Digest*, Vol. 1, pp. 78–82, 1999.)

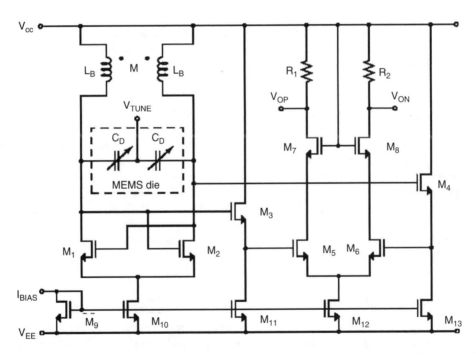

FIGURE 5.42 Schematic of a 2.4 GHz MEMS VCO. (From Dec, A. and Suyama, K. In *1999 IEEE MTT-S International Microwave Symposium Digest*, Vol. 1, pp. 79–82, 1999.)

antenna element was demonstrated with the antenna response changing from being centered at 12 GHz to above 18 GHz.

A MEMS reconfigurable Vee antenna (Figure 5.46) has demonstrated the capability of beam steering and beam shaping [69]. The mechanical arrangement utilizes metal-coated polysilicon and

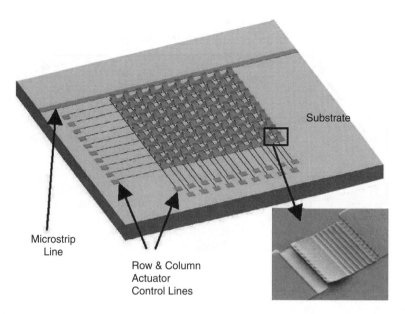

FIGURE 5.43 RF MEMS switches used in a reconfigurable array. (From Bozler, C. et al. In *2000 IEEE MTT-S International Microwave Symposium Digest*, pp. 153–156, 2000.)

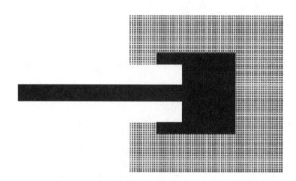

FIGURE 5.44 RF MEMS switch array layout for a patch antenna. (From Bozler, C. et al. In *2000 IEEE MTT-S International Microwave Symposium Digest*, pp. 153–156, 2000.)

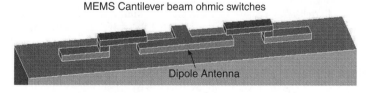

FIGURE 5.45 Reconfigurable dipole antenna using MEMS cantilever switches.

metal-to-metal contacts to construct three-dimensional reconfigurable radiating and wave-guiding structures. The illustrated example contains fixed and moveable rotating hinges activated by scratch drive actuators [70]. The resulting structure demonstrated beam steering up to 48E off bore site along with the ability to reshape the beam.

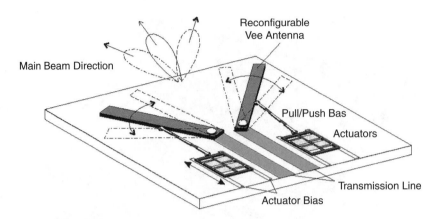

FIGURE 5.46 A MEMS reconfigurable vee antenna. (From Chiao, J.-C. et al. In *1999 IEEE MTT-S International Microwave Symposium Digest*, Vol. 4, pp. 1515–1518, 1999.)

References

1. Goldsmith, C., Randall, J., Eshelman, S., Lin, T. H., Denniston, D., Chen, S., Norvell, B. Characteristics of micromachined switches at microwave frequencies. In *1996 IEEE MTT-S International Microwave Digest*, pp. 1141–1144, 1996.

2. Feng, Z., Zhang, H., Zhang, W., Su, B., Gupta, K. C., Bright, V. M., Lee, Y. C. MEMSbased variable capacitor for millimeter-wave applications. In *2000 Solid-State Sensor and Actuator Workshop*, Hilton Head Island, pp. 255–258, 2000.

3. Feng, Z., Zhang, W., Su, B., Harsh, K. F., Gupta, K. C., Bright, V., Lee, Y. C. Design and modeling of RF MEMS tunable capacitors using electro-thermal actuators. In *1999 IEEE MTT-S International Microwave Digest*, pp. 1507–1510, 1999.

4. Wu, H. D., Harsh, K. F., Irwin, R. S., Zhang, W., Mickelson, A. R., Lee, Y. C. MEMS designed for tunable capacitors. In *1998 IEEE MTT-S International Microwave Digest*, pp. 127–129, 1999.

5. Hyman, D., Lam, J., Warneke, B., Schmitz, A., Hsu, T. Y., Brown, J., Schaffner, J., Walston, A., Loo, R. Y., Mehregany, M., Lee, J. Surface-micromachined RF MEMS switches on GaAs Substrates. *International Journal of RF & Microwave Computer-Aided Engineering*, Vol. 9, No. 4, pp. 348–361, 1999.

6. Madou, M. *Fundamentals of Microfabrication*. CRC Press, 1997.

7. Wolf, S., Tauber, R. N. *Silicon Processing for the VLSI Era*. Lattice Press, 1986.

8. Tang, W. C. Overview of MEMS fabrication technology in microwave and photonic applications of MEMS. Presented at *Proceedings of the 2000 IEEE MTT-S International Microwave Symposium*, Boston, MA, 2000.

9. Willke, T. L., Gearhart, S. S. LIGA micromachined planar transmission lines and filters. *IEEE Transactions on Microwave Theory and Techniques*, Vol. 45, No. 10, pp. 1681–1688, October 1997.

10. Becker, E. W., Ehrfeld, W., Hagmann, P., Maner, A., Munchmeyer, D. Fabrication of microstructures with high-aspect ratios and great structural heights by synchrotron radiation lithography, galvanoforming, and plastic moulding (LIGA process). *Microelectronic Engineering*, Vol. 4, pp. 35–36, 1986.

11. Gear picture from University of Wisconson–Madison Web site, http://mems.engr.wisc.edu/MEMS.html

12. Harendt, C., Appel, W., Graf, H.-G., Hfflinger, B., Penteker, E. Wafer fusion bonding and its application to silicon-on-insulator fabrication. *Journal of Micromechanical Microengineering*, Vol. 1, pp. 145–151, 1991.

13. Hyman, D., Schmitz, A., Warneke, B., Hsu, T. Y., Lam, J., Brown, J., Schaffner, J., Walston, A., Loo, R. Y., Tangonan, G. L., Mehregany, M., Lee, J. GaAs-compatible surface-micromachined RF MEMS switches. *Electronics Letters*, Vol. 35, No. 3, pp. 224–226, February 1999.

14. Zavracky, P. M., McGruer, N. E., Morrison, R. H., Potter, D. Microswitches and microrelays with a view toward microwave applications. In *International Journal of RF & Microwave Computer-aided Engineering*, pp. 338–347, Vol. 9, No. 4, 1999.

15. McGruer, N. E., Zavracky, P. M., Majumder, S., Morrison, R., Adams, G. G. Electrostatically actuated microswitches; scaling properties in Proceedings of 1998 Solid-State Sensor and Actuator Workshop, Hilton Head Island, pp. 132–135, 1998.

16. Micromachined-relay illustration courtesy of Analog Devices, http://www.analog.com/industry/umic/relay.html

17. Bozler, C., Drangmeister, R., Duffy, S., Gouker M., Knecht, J., Kushner, L., Parr, R., Rabe, S., Travis, L. MEMS Microswitch arrays for reconfigurable distributed microwave components. In *2000 IEEE MTT-S International Microwave Symposium Digest*, pp. 153–156, 2000.

18. Goldsmith, C. RF MEMS devices and circuits for radar and receiver applications at the Microwave and photonic applications of MEMs Workshop, at the *2000 IEEE MTT-S International Microwave Symposium*, 2000.

19. Goldsmith, C. L., Yao, Z., Eshelman, S., Denniston, D. Performance of low-loss RF MEMS capacitive switches. In *IEEE Microwave and Guided Wave Letters*, Vol. 8, No. 8, pp. 517–519, August 1998.

20. Yao, Z. J., Chen, S., Eshelman, S., Denniston, D., Goldsmith, C. Micromachined low-loss microwave switches. In *IEEE Journal of Microelectromechanical Systems*, Vol. 8, No. 2, pp. 129–134, June 1999.

21. Peroulis, D., Pacheco, S., Sarabandi, K., Katehi, L. P. B. MEMS devices for high isolation switching and tunable filtering. In *2000 IEEE MTT-S International Microwave Symposium Digest*, pp. 1217–1220, 2000.

22. Young, D. J. Boser, B. E. A micromachine-based RF low-noise voltage-controlled oscillator. In *Proceedings of IEEE 1997 Custom Integrated Circuits Conference*, pp. 431–434, 1997.

23. Pierce, B. Private Communication

24. Malczewski, A. Private Communication

25. Goldsmith, C. L., Malczewski, A., Yao, Z. J., Chen, S., Ehmke, J., Hinzel, D. H. RF MEMS variable capacitors for tunable filters. In *International Journal of RF & Microwave Computer-aided Engineering*, pp. 362–374, Vol. 9, No. 4, 1999.

26. Yao, J. J., Park, S., Anderson, R., DeNatale, J. A low power / low voltage electrostatic actuator for RF MEMS applications. In *Proceedings of 2000 Solid-State Sensor and Actuator Workshop*, Hilton Head Island, pp. 246–249, 2000.

27. Yao, J. J., Park, S., DeNatale, J. High tuning-ratio MEMS-based tunable capacitors for RF communications applications. In *Proceedings of 1998 Solid-State Sensor and Actuator Workshop*, Hilton Head Island, pp. 124–127, 1998.

28. Hung, E, Senturia, S. Tunable capacitors with programmable capacitance-voltage characteristics. In *Proceedings of 1998 Solid-State Sensor and Actuator Workshop*, Hilton Head Island, pp. 292–295, 1998.

29. Katehi, L. P. B., Rebeiz, G. M. Novel micromachined approaches to MMICs using low-parasitic, high-performance transmission media and environments. In *1996 IEEE MTT-S International Microwave Symposium Digest*, pp. 1145–1148, 1996.

30. Ayon, A. A., Kolias, N. J., MacDonald, N. C. Tunable, micromachined parrallel-plate transmission lines. In *IEEE/Cornell Conference Proceedings on Advanced Concepts in High Speed Semiconductor Devices and Circuits*, pp. 201–208, 1995.

31. Kudrie, T. D., Neves, H. P., MacDonald, N. C. Microfabricated single crystal silicon transmission lines. In *RAWCON'98 Proceedings*, pp. 269–272, 1998.

32. Burghartz, J. N., Edelstein, D. C., Ainspan, H. A., Jenkins, K. A. RF Circuit design aspects of spiral inductors on silicon. In *IEEE J. Solid-State Circuits*, Vol. 33, No. 12, Dec. 1998.

33. Chang, J. Y. C., Abidi, A. A., Gaitan M. Large suspended inductors on silicon and their use in a 2-mm CMOS RF amplifier. In *IEEE Electron Device Letters*, Vol. 14, No. 5, pp. 246–248, 1993.

34. Chi, C. Y., Rebeiz, G. M. Planar microwave and milimeter-wave lumped elements and coupled-line filters using micro-machining techniques. In *IEEE Transactions on Microwave Theory and Techniques*, Vol. 43, No. 4, pp. 730–738, 1995.

35. Clark, J. R., Bannon, F. D. III, Wong, A.-C., Nguyen, C. T.-C. Parrallel-resonator HF micromechanical bandpass filters. In *Proceedings of 1997 International Conference on Solid-State Sensor and Actuators*, Chicago, pp. 1161–1164, 1997.

36. Bannon, F. D. III, Clark, J. R., Nguyen, C. T.-C. High frequency microelectromenchanical IF filters. In *1996 International Electron Devices Meeting*, pp. 773–776, 1996.

37. Bannon, F. D. III, Clark, J. R., Nguyen, C. T.-C. High-Q HF microelectromechanical filters. In *IEEE Journal of Solid-State Circuits*, Vol. 35, No. 4, pp. 512–526, April 2000.

38. Nguyen, C. T.-C. Frequency-selective MEMS for miniaturized communication devices. In *1998 IEEE Aerospace Conference*, Vol. 1, pp. 445–460, 1998.
39. Nguyen, C. T.-C. Frequency-selective MEMS for miniaturized low-power communication devices. In *IEEE Transactions on Microwave Theory and Techniques*, Vol. 47, No. 8, pp. 1486–1503, August 1999.
40. Nguyen, C. T.-C., Wong, A.-C., Ding, H. Tunable, switchable, high-Q VHF microelectromechanical bandpass filters. In *1999 IEEE International Solid-State Circuits Conference Digest*, pp. 78–79, 448, 1999.
41. Nguyen, C. T.-C., Howe, R. T. An integrated CMOS micromechanical resonator high-Q oscillator. In *IEEE Journal of Solid-State Circuits*, Vol. 34, No. 4, pp. 440–454, April 1999.
42. Wang, K., Bannon, F. D. III, Clark, J. R., Nguyen, C. T.-C. Q-enhancement of microelectromechanical filters via low-velocity spring coupling. In *1997 IEEE Ultrasonics Symposium Digest*, pp. 323, 1997.
43. Becker, J. P., Katehi, L. P. B. Toward a novel planar circuit compatible silicon micromachined waveguide. In *1999 Electrical Performance and Electronic Packaging Digest*, pp. 221–224, 1999.
44. Katehi, L. P. B., Rebeiz, G. M., Nguyen, C. T.-C. MEMS and Si-micromachined components for low-power, high-frequency communications systems. In *1998 IEEE MTT-S International Microwave Symposium Digest*, Vol. 1, pp. 331–333, 1998.
45. Papapolymerou, J., Jui-Ching Cheng, East, J., Katehi, L. P. B. A micromachined high-Q X-band resonator. In *IEEE Microwave and Guided Wave Letters*, Vol. 7, No. 6, pp. 168–170, June 1997.
46. Blondy, P., Brown, A. R., Cros, D., Rebeiz, G. M. Low-loss micromachined filters for millimeter-wave communication systems. In *IEEE Transactions on Microwave Theory and Techniques*, Vol. 46, No. 12, Dec., pp. 2283–2288, 1998.
47. Brown, A. R., Rebeiz, G. M. Micromachined micropackaged filter banks and tunable bandpass filters. In *1997 Wireless Communications Conference Digest*, pp. 193–197, 1997.
48. Robertson, S. V., Katehi, L. P. B., Rebeiz, G. M. Micromachined self-packaged W-band bandpass filters. In *1995 IEEE MTT-S International Microwave Symposium Digest*, pp. 1543–1546, 1995.
49. Robertson, S. V., Katehi, L. P. B., Rebeiz, G. M. Micromachined W-band filters, *IEEE Transactions on Microwave Theory and Techniques*, Vol. 44, No. 4, pp. 598–606, April 1996.
50. Nguyen, C. T.-C., Katehi, L. P. B., Rebeiz, G. M. Micromachined devices for wireless communications. In *Proceedings of the IEEE*, Vol. 86, No. 8, August, pp. 1756–1768, 1998
51. Krishnaswamy, S. V., Rosenbaum, J, Horwitz, S., Yale, C., Moore, R. A. Compact FBAR filters offer low-loss performance in Microwave & RF, pp. 127–136, Sept., 1991.
52. Rudy, R., Merchant, P. Micromachined thin film bulk acoustic resonators. In Proceedings of the *1994 IEEE International Frequency Control Symposium*, Boston, MA, June 1–3, 1994, pp. 135–138, 1994.
53. Pillans, B., Eshelman, S., Malczewski, A., Ehmke, J., Goldsmith, C. Ka-band RF MEMS phase shifters. In *IEEE Microwave and Guided Wave Letters*, Vol. 9, No. 12, pp. 520–522, December 1999.
54. Pillans, B., Eshelman, S., Malczewski, A., Ehmke, J., Goldsmith, C. Ka-band RF MEMS phase shifters for phased array applications. In *2000 IEEE Radio Frequency Integrated Circuits Symposium Digest*, pp. 195–198, 2000.
55. Malczewski, A., Eshelman, S., Pillans, B., Ehmke, J., Goldsmith, C.L. X-band RF MEMS phase shifters for phased array applications. In *IEEE Microwave and Guided Wave Letters*, Vol. 9, No. 12, pp. 517–519, Dec. 1999.
56. Hayden, J. S., Rebeiz, G. M. One and two-bit low-loss cascadable MEMS distributed X-band phase shifters. In *2000 IEEE MTT-S International Microwave Symposium Digest*, pp. 161–164, 2000.
57. Barker, N. S., Rebeiz, G. M. Optimization of distributed MEMS phase shifters. In 1999 *IEEE MTT-S International Microwave Symposium Digest*, pp. 299–302, Anaheim, CA, 1999.
58. Kim, H.-T., Park, J.-H., Kim, Y.-K., Kwon, Y. Millimeter-wave micromachined tunable filters. In *1999 IEEE MTT-S International Microwave Symposium Digest*, Vol. 3, pp. 1235–1238, 1999.
59. Guillon, B., Cros, D., Pons, P., Gazaux, J.L., Lalaurie, J.C., Plana, R., Graffeuil, J. Ka band micromachined dielectric resonator oscillator. In *Electronics Letters*, Vol. 35, No. 11, pp. 909–910, May 27, 1999.
60. Guillon, B., Cros, D., Pons, P., Grenier, K., Parra, T., Cazaux, J. L., Lalaurie, J. C., Graffeuil, J., Plana, R. Design and realization of high Q millimeter-wave structures through micromachining techniques. In *1999 IEEE MTT-S International Microwave Symposium Digest*, Vol. 4 , pp. 1519 –1522, 1999.
61. Brown, A. R., Rebeiz, G. M. Micromachined high-Q resonators, low-loss diplexers, and low phase-noise oscillators for a 28 GHz front-end. In *1999 IEEE Radio and Wireless Conference*, pp. 247–253, 1999.

62. Brown, A. R., Rebeiz, G. M. A Ka-band micromachined low-phase-noise oscillator. In *IEEE Transactions on Microwave Theory and Techniques*, Vol. 47, No. 8, pp. 1504–1508, August 1999.

63. Young, D. J., Malba, V., Ou, J.-J., Bernhardt, A. E., Boser, B. E. A low-noise RF voltage-controlled oscillator using on-chip high-Q three-dimensional coil inductor and micromachined variable capacitor. In *Proceedings of 1998 Solid-State Sensor and Actuator Workshop*, Hilton Head Island, pp. 128–131, 1998.

64. Young, D. J., Boser, B. E. A micromachined variable capacitor for monolithic low-noise VCOs. In *Proceedings of 1996 Solid-State Sensor and Actuator Workshop*, Hilton Head Island, pp. 86–89, 1996.

65. Dec, A., Suyama, K. A 2.4 GHz CMOS LC VCO using micromachined variable capacitors for frequency tuning. In *1999 IEEE MTT-S International Microwave Symposium Digest*, Vol. 1, pp. 79–82, 1999.

66. Dec, A., Suyama, K. A 1.9 GHz micromachined-based low-phase-noise CMOS VCO. In *1999 IEEE International Solid-State Circuits Conference Digest*, pp. 80–81, 449, 1999.

67. Tangonan, G., Loo, R., Schaffner, J., Lee, J. J. Microwave photonic applications of MEMS technology. In *1999 International Topical Meeting on Microwave Photonics*, Vol. 1, pp. 109–112, 1999.

68. Izadpanah, H., Warneke, B., Loo, R., Tangonan, G. Reconfigurable low power, light weight wireless system based on the RF MEM switches. In *1999 IEEE MTT-S Symposium on Technologies for Wireless Applications*, pp. 175–180, 1999.

69. Chiao, J.-C., Fu, Y., Chio, I. M., DeLisio, M., Lin, L.-Y. MEMS reconfigurable Vee antenna. In *1999 IEEE MTT-S International Microwave Symposium Digest*, Vol. 4, pp. 1515–1518, 1999.

70. Akiyama, T., Shono, K. Controlled step-wise motion in polysilicon microstructures. In *Journal of MEMS*, Vol. 2, No. 3, pp. 106, Sept. 1993.

6

Surface Acoustic Wave (SAW) Filters

Donald C. Malocha
University of Central Florida

A **surface acoustic wave (SAW)**, also called a Rayleigh wave, is composed of a coupled compressional and shear wave in which the SAW energy is confined near the surface. There is also an associated electrostatic wave for a SAW on a piezoelectric substrate which allows electroacoustic coupling via a transducer. SAW technology's two key advantages are its ability to electroacoustically access and tap the wave at the crystal surface and that the wave velocity is approximately 100,000 times slower than an electromagnetic wave, assuming an electromagnetic wave velocity of 3×10^8 m/s and an acoustic wave velocity of 3×10^8 m/s, Table 6.1 compares relative dimensions versus frequency and delay. The SAW wavelength is on the same order of magnitude as line dimensions that can be photolithographically produced and the lengths for both small and long delays are achievable on reasonable size substrates. The corresponding E&M transmission lines or waveguides would be impractical at these frequencies.

TABLE 6.1   Comparison of SAW and E&M Dimensions versus Frequency and Delay, Where Assumed Velocities are $v_{SAW} = 3000$ m/s and $v_{EM} = 3 \times 10^8$ m/s

Parameter	SAW	E&M
$F_0 = 10$ MHz	$\lambda_{SAW} = 300$ μm	$\lambda_{EM} = 30$ m
$F_0 = 2$ GHz	$\lambda_{SAW} = 1.5$ μm	$\lambda_{EM} = 0.15$ m
Delay = 1 ns	$L_{SAW} = 3$ μm	$L_{EM} = 0.3$ m
Delay = 10 μs	$L_{SAW} = 30$ mm	$L_{EM} = 3000$ m

Because of SAWs' relatively high operating frequency, linear delay, and tap weight (or sampling) control, they are able to provide a broad range of signal processing capabilities. Some of these include linear and dispersive filtering, coding, frequency selection, convolution, delay line, time impulse response shaping, and others. There are very broad ranges of commercial and military system applications that include components for radars, front-end and IF filters, CATV and VCR components, cellular radio and pagers, synthesizers and analyzers, navigation, computer clocks, tags, and many, many others [Campbell, 1989; Matthews, 1977].

There are four principal SAW properties: transduction, reflection, regeneration, and nonlinearities. Nonlinear elastic properties are principally used for convolvers and will not be discussed. The other three properties are present, to some degree, in all SAW devices, and these properties must be understood and controlled to meet device specifications.

A finite-impulse response (FIR) or transversal filter is composed of a series of cascaded time delay elements that are sampled or "tapped" along the delay line path. The sampled and delayed signal is summed at a junction which yields the output signal. The output time signal is finite in length and has no feedback. A schematic of an FIR filter is shown in Figure 6.1.

A SAW transducer is able to implement an FIR filter. The electrodes or fingers provide the ability to sample or "tap" the SAW and the distance between electrodes provides the relative delay. For a uniformly sampled SAW transducer, the delay between samples, Δt, is given by $\Delta t = \Delta L/v_a$, where ΔL is the electrode period and v_a is the acoustic velocity. The typical means for providing attenuation or weighting is to vary the overlap between adjacent electrodes which provides a spatially weighted sampling of a uniform wave. Figure 6.1 shows a typical FIR time response and its equivalent SAW transducer implementation. A SAW filter is composed of a minimum of two transducers and possibly other SAW components. A schematic of a simple SAW bidirectional filter is shown in Figure 6.2. A **bidirectional transducer** radiates energy equally from each side of the transducer (or port). Energy not being received is absorbed to eliminate spurious reflections.

6.1 SAW Material Properties

There are a large number of materials currently being used for SAW devices. The most popular single-crystal piezoelectric materials are quartz, lithium niobate ($LiNbO_3$), and lithium tantalate ($LiTa_2O_5$). The

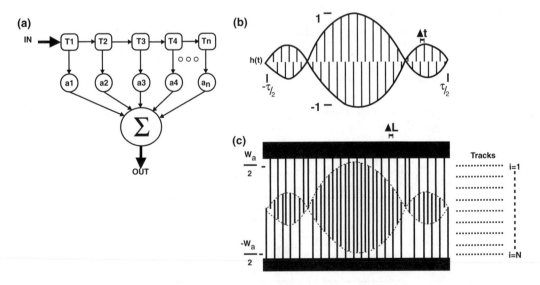

FIGURE 6.1 (a) Schematic of a finite-impulse response (FIR) filter. (b) An example of a sampled time function; the envelope is shown in the dotted lines. (c) A SAW transducer implementation of the time function $h(t)$.

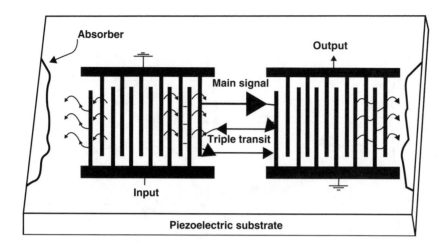

FIGURE 6.2 Schematic diagram of a typical SAW bidirectional filter consisting of two interdigital transducers. The transducers need not be identical. The input transducer launches waves in either direction and the output transducer converts the acoustic energy back to an electrical signal. The device exhibits a minimum 6-dB insertion loss. Acoustic absorber damps unwanted SAW energy to eliminate spurious reflections that could cause distortions.

TABLE 6.2   Common SAW Material Properties

Parameter/Material	ST-Quartz	YZ LiNbO$_3$	128° YX LiNbO$_3$	YZ LiTa$_2$O$_3$
k^2 (%)	0.16	4.8	5.6	0.72
C_s (pf/cm-pair)	0.05	4.6	5.4	4.5
v_0 (m/s)	3,159	3,488	3,992	3,230
Temp. coeff. of delay (ppm/°C)	0	94	76	35

materials are anisotropic, which will yield different material properties versus the cut of the material and the direction of propagation. There are many parameters that must be considered when choosing a given material for a given device application. Table 6.2 shows some important material parameters for consideration for four of the most popular SAW materials [Datta, 1986; Morgan, 1985].

The coupling coefficient, k^2, determines the electroacoustic coupling efficiency. This determines the fractional bandwidth versus minimum insertion loss for a given material and filter. The static capacitance is a function of the transducer electrode structure and the dielectric properties of the substrate. The values given in the table correspond to the capacitance per pair of electrodes having quarter wavelength width and one-half wavelength period. The free surface velocity, v_0, is a function of the material, cut angle, and propagation direction. The temperature coefficient of delay (TCD) is an indication of the frequency shift expected for a transducer due to a change of temperature and is also a function of cut angle and propagation direction.

The substrate is chosen based on the device design specifications and includes consideration of operating temperature, fractional bandwidth, and insertion loss. Second-order effects such as diffraction and beam steering are considered important on high-performance devices [Morgan, 1985]. Cost and manufacturing tolerances may also influence the choice of the substrate material.

6.2   Basic Filter Specifications

Figure 6.3 shows a typical time domain and frequency domain device performance specification. The basic frequency domain specification describes frequency bands and their desired level with respect to a given reference. Time domain specifications normally define the desired impulse response shape and any spurious time responses. The overall desired specification may be defined by combinations of both time

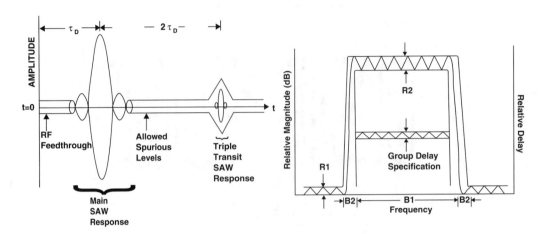

FIGURE 6.3 Typical time and frequency domain specification for a SAW filter. The filter bandwidth is B_1, the transition bandwidth is B_2, the inband ripple is R_2 and the out-of-band sidelobe level is R_1.

and frequency domain specifications. Since time, $h(t)$, and frequency, $H(\omega)$, domain responses form unique Fourier transform pairs, given by

$$h(t) = 1/2\pi \int_{-\infty}^{\infty} H(\omega) e^{j\omega t} d\omega \qquad (6.1)$$

$$H(\omega) = \int_{-\infty}^{\infty} h(t) e^{-j\omega t} dt \qquad (6.2)$$

it is important that combinations of time and frequency domain specifications be self-consistent.

The electrodes of a SAW transducer act as sampling points for both transduction and reception. Given the desired modulated time response, it is necessary to sample the time waveform. For symmetrical frequency responses, sampling at twice the center frequency, $f_s = 2f_0$, is sufficient, while nonsymmetric frequency responses require sampling at twice the highest frequency of interest. A very popular approach is to sample at $f_s = 4f_0$. The SAW frequency response obtained is the convolution of the desired frequency response with a series of impulses, separated by f_s, in the frequency domain. The net effect of sampling is to produce a continuous set of harmonics in the frequency domain in addition to the desired response at f_0. This periodic, time-sampled function can be written as

$$g(t_n) = \sum_{-N/2}^{N/2} a_n \cdot \delta(t - t_n) \qquad (6.3)$$

where a_n represents the sample values, $t_n = n\Delta t$, $n = n$th sample, and $\Delta t =$ time sample separation. The corresponding frequency response is given by

$$G(f) = \sum_{-N/2}^{N/2} g(t_n) e^{-j2\pi f t_n} = \sum_{-N/2}^{N/2} g(t_n) e^{-j2\pi f/f_s} \qquad (6.4)$$

where $f_s = 1/\Delta t$. The effect of sampling in the time domain can be seen by letting $f = f + mf_s$, where m is an integer, which yields $G(f + mf_s) = G(f)$, which verifies the periodic harmonic frequency response.

Before leaving filter design, it is worth noting that a SAW filter is composed of two transducers that may have different center frequencies, bandwidth, and other filter specifications. This provides a great deal of flexibility in designing a filter by allowing the product of two frequency responses to achieve the total filter specification.

6.3 SAW Transducer Modeling

The four most popular and widely used models include the transmission line model, the coupling of modes model, the impulse response model, and the superposition model. The superposition model is an extension of the impulse response model and is the principal model used for the majority of SAW bidirectional and multiphase filter synthesis which do not have inband, interelectrode reflections. As is the case for most technologies, many models may be used in conjunction with each other for predicting device performance based on ease of synthesis, confidence in predicted parameters, and correlation with experimental device data.

6.3.1 The SAW Superposition Impulse Response Transducer Model

The impulse response model was first presented by Hartmann et al. [1973] to describe SAW filter design and synthesis. For a linear causal system, the Fourier transform of the device's frequency response is the device impulse time response. Hartmann showed that the time response of a SAW transducer is given by

$$h\left(t\right)=4k\sqrt{C_s}f_i^{3/2}\left(t\right)\sin\left[\theta\left(t\right)\right] \text{ where } \theta\left(t\right)=2\pi\int_0^t f_i\left(\tau\right)d\tau \tag{6.5}$$

where the following definitions are k^2 = SAW coupling coefficient, C_s = electrode pair capacitance per unit length (pf/cm-pair), and $f_i(t)$ = instantaneous frequency at a time, t. This is the general form for a uniform beam transducer with arbitrary electrode spacing. For a uniform beam transducer with periodic electrode spacing, $f_i(t) = f_0$ and $\sin \theta(t) = \sin \omega t$. This expression relates a time response to the physical device parameters of the material coupling coefficient and the electrode capacitance.

Given the form of the time response, energy arguments are used to determine the device equivalent circuit parameters. Assume a delta function voltage input, $v_{in}(t) = \delta(t)$, then $V_{in}(\omega) = 1$. Given $h(t)$, $H(\omega)$ is known and the energy launched as a function of frequency is given by $E(\omega) = 2 \cdot |H(\omega)|^2$. Then

$$E\left(\omega\right)=V_{in}^2\left(\omega\right)\cdot G_a\left(\omega\right)=1\cdot G_a\left(\omega\right) \tag{6.6}$$

or

$$G_a\left(\omega\right)=2\cdot\left|H\left(\omega\right)\right|^2 \tag{6.7}$$

There is a direct relationship between the transducer frequency transfer function and the transducer conductance. Consider an **interdigital transducer** (IDT) with uniform overlap electrodes having N_p interaction pairs. Each gap between alternating polarity electrodes is considered a localized SAW source. The SAW impulse response at the fundamental frequency will be continuous and of duration τ, where $\tau = N \cdot \Delta t$, and $h(t)$ is given by

$$h\left(t\right)=\kappa\cdot\cos\left(\omega_0 t\right)\cdot rect\left(t/\tau\right) \tag{6.8}$$

where $\kappa = 4k\sqrt{C_s}\,f_0^{3/2}$ and f_0 is the carrier frequency. The corresponding frequency response is given by

$$H\left(\omega\right)=\frac{\kappa\tau}{2}\left\{\frac{\sin\left(x_1\right)}{x_1}+\frac{\sin\left(x_2\right)}{x_2}\right\} \tag{6.9}$$

where $x_1 = (\omega - \omega_0) \cdot \tau/2$ and $x_2 = (\omega + \omega_0) \cdot \tau/2$.

This represents the ideal SAW continuous response in both time and frequency. This can be related to the sampled response by a few substitutions of variables. Let

$$\Delta t = \frac{1}{2 \cdot f_0}, \quad t_n = n \cdot \Delta t, \quad N \cdot \Delta t = \tau, \quad N_p \cdot \Delta t = \tau/2 \tag{6.10}$$

Assuming a frequency bandlimited response, the negative frequency component centered around $-f_0$ can be ignored. Then the frequency response, using Equation 6.9, is given by

$$H(\omega) = \kappa \left\{ \frac{\pi N_p}{\omega_0} \right\} \cdot \frac{\sin(x_n)}{x_n} \tag{6.11}$$

where

$$x_n = \frac{(\omega - \omega_0)}{\omega_0} \pi N_p = \frac{(f - f_0)}{f_0} \pi N_p$$

The conductance, given using Equations 6.6 and 6.10, is

$$G_a(f) = 2\kappa^2 \left\{ \frac{\pi N_p}{2\pi f_0} \right\}^2 \frac{\sin^2(x_n)}{x_n^2} = 8k^2 f_0 C_s N_p^2 \cdot \frac{\sin^2(x_n)}{x_n^2} \tag{6.12}$$

This yields the frequency-dependent conductance per unit width of the transducer. Given a uniform transducer of width, W_a, the total transducer conductance is obtained by multiplying Equation 6.12 by W_a. Defining the center frequency conductance as

$$G_a(f_0) = G_0 = 8k^2 f_0 C_s W_a N_p^2 \tag{6.13}$$

the transducer conductance is

$$G_a(f_0) = G_0 \cdot \frac{\sin^2(x_n)}{x_n^2} \tag{6.14}$$

The transducer electrode capacitance is given as

$$C_e = C_s W_a N_p \tag{6.15}$$

Finally, the last term of the SAW transducer's equivalent circuit is the frequency-dependent susceptance. Given any system where the frequency-dependent real part is known, there is an associated imaginary part that must exist for the system to be real and causal. This is given by the Hilbert transform susceptance, defined as B_a, where [Datta, 1986]

$$B_a(\omega) = \frac{1}{\pi} \int_{-\infty}^{\infty} \frac{G_a(u)}{(u - \omega)} du = G_a(\omega) * 1/\omega \tag{6.16}$$

where "*" indicates convolution.

These three elements compose a SAW transducer equivalent circuit. The equivalent circuit, shown in Figure 6.4, is composed of one lumped element and two frequency-dependent terms that are related to the substrate material parameters, transducer electrode number, and the transducer configuration. Figure 6.5 shows the time and frequency response for a uniform transducer and the associated frequency-dependent conductance and Hilbert transform susceptance. The simple impulse model treats each electrode as an ideal

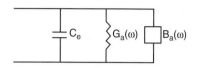

FIGURE 6.4 Electrical equivalent circuit model.

impulse; however, the electrodes have a finite width that distorts the ideal impulse response. The actual SAW potential has been shown to be closely related to the electrostatic charge induced on the transducer by the input voltage. The problem is solved assuming a quasi-static and electrostatic charge distribution, assuming a semi-infinite array of electrodes, solving for a single element, and then using superposition and convolution. The charge distribution solution for a single electrode with all others grounded is defined as the basic charge distribution function (BCDF). The result of a series of arbitrary voltages placed on a series of electrodes is the summation of scaled, time-shifted BCDFs. The identical result is obtained if an array factor, $a(x)$, defined as the ideal impulses localized at the center of the electrode or gap, is convolved with the BCDF, often called the element factor. This is very similar to the analysis of antenna arrays. Therefore, the ideal frequency transfer function and conductance given by the impulse response model need only be modified by multiplying the frequency-dependent element factor. The

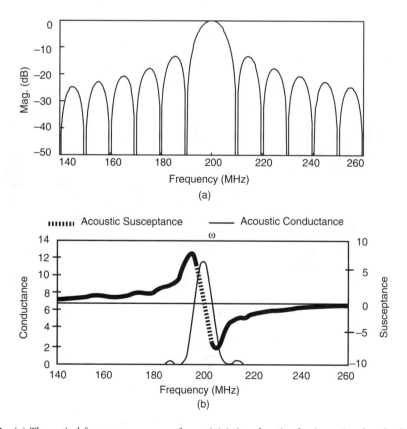

FIGURE 6.5 (a) Theoretical frequency response of a *rect*(t/τ) time function having a time length of 0.1 μs and a 200-MHz carrier frequency. (b) Theoretical conductance and susceptance for a SAW transducer implementing the frequency response. The conductance and susceptance are relative and are given in millisiemens.

analytic solution to the BCDF is given in Datta [1986] and Morgan [1985], and is shown to place a small perturbation in the form of a slope or dip over the normal bandwidths of interest. The BCDF also predicts the expected harmonic frequency responses.

6.3.2 Apodized SAW Transducers

Apodization is the most widely used method for weighting a SAW transducer. The desired time-sampled impulse response is implemented by assigning the overlap of opposite polarity electrodes at a given position to a normalized sample weight at a given time. A tap having a weight of unity has an overlap across the entire beamwidth while a small tap will have a small overlap of adjacent electrodes. The time impulse response can be broken into tracks which have uniform height but whose time length and impulse response may vary. Each of these time tracks is implemented spatially across the transducer's beamwidth by overlapped electrode sections at the proper positions. This is shown in Figure 6.1. The smaller the width of the tracks, the more exact the approximation of uniform time samples. There are many different ways to implement the time-to-spatial transformation; Figure 6.1 shows just one such implementation.

The impulse response can be represented, to any required accuracy, as the summation of uniform samples located at the proper positions in time in a given track. Mathematically this is given by

$$h(t) = \sum_{i=1}^{I} h_i(t) \tag{6.17}$$

and

$$H(\omega) = \sum_{i=1}^{I} H_i(\omega) = \sum_{i=1}^{I} \left\{ \int_{-\tau/2}^{\tau/2} h_i(t) e^{-j\omega t} dt \right\} \tag{6.18}$$

The frequency response is the summation of the individual frequency responses in each track, which may be widely varying depending on the required impulse response. This spatial weighting complicates the calculations of the equivalent circuit for the transducer. Each track must be evaluated separately for its acoustic conductance, acoustic capacitance, and acoustic susceptance. The transducer elements are then obtained by summing the individual track values yielding the final transducer equivalent circuit parameters. These parameters can be solved analytically for simple impulse response shapes (such as the rect, triangle, cosine, etc.) but are usually solved numerically on a computer [Richie et al., 1988].

There is also a secondary effect of apodization when attempting to extract energy. Not all of the power of a nonuniform SAW beam can be extracted by an a uniform transducer, and reciprocally, not all of the energy of a uniform SAW beam can be extracted by an apodized transducer. The transducer efficiency is calculated at center frequency as

$$E = \frac{\left| \sum_{i=1}^{I} H(\omega_0) \right|^2}{I \cdot \sum_{i=1}^{I} H^2(\omega_0)} \tag{6.19}$$

The apodization loss is defined as

$$\text{apodization loss} = 10 \times \log(E) \tag{6.20}$$

Typical apodization loss for common SAW transducers is 1 dB or less.

Finally, because an apodized transducer radiates a nonuniform beam profile, the response of two cascaded apodized transducers is not the product of each transducer's individual frequency responses, but rather is given by

$$H_{12}(\omega) = \sum_{i=1}^{I} H_{1i}(\omega) \cdot H_{2i}(\omega) \neq \sum_{i=1}^{I} H_{1i}(\omega) \cdot \sum_{i=1}^{I} H_{2i}(\omega) \tag{6.21}$$

In general, filters are normally designed with one apodized and one uniform transducer or with two apodized transducers coupled with a spatial-to-amplitude acoustic conversion component, such as a multistrip coupler [Datta, 1986].

6.4 Distortion and Second-Order Effects

In SAW devices there are a number of effects that can distort the desired response from the ideal response. The most significant distortion in SAW transducers is called the **triple transit echo** (**TTE**) which causes a delayed signal in time and an inband ripple in the amplitude and delay of the filter. The TTE is primarily due to an electrically regenerated SAW at the output transducer, which travels back to the input transducer, where it induces a voltage across the electrodes, which in turn regenerates another SAW which arrives back at the output transducer. This is illustratedtically in Figure 6.2. Properly designed and matched **unidirectional transducers** have acceptably low levels of TTE due to their design. Bidirectional transducers, however, must be mismatched in order to achieve acceptable TTE levels. To first order, the TTE for a bidirectional two-transducer filter is given as

$$\text{TTE} \approx 2 \cdot IL + 6 \text{ dB} \tag{6.22}$$

where IL = filter insertion loss, in dB [Matthews, 1977]. As examples, the result of TTE is to cause a ghost in a video response and intersymbol interference in data transmission.

Another distortion effect is electromagnetic feedthrough which is due to direct coupling between the input and output ports of the device, bypassing any acoustic response. This effect is minimized by proper device design, mounting, bonding, and packaging.

In addition to generating a SAW, other spurious acoustic modes may be generated. Bulk acoustic waves (BAW) may be both generated and received, which causes passband distortion and loss of out-of-band rejection. BAW generation is minimized by proper choice of material, roughening of the crystal backside to scatter BAWs, and use of a SAW track changer, such as a multistrip coupler.

Any plane wave that is generated from a finite aperture will begin to diffract. This is exactly analogous to light diffracting through a slit. Diffraction's principal effect is to cause effective shifts in the filter's tap weights and phase which results in increased sidelobe levels in the measured frequency response. Diffraction is minimized by proper choice of substrate and filter design.

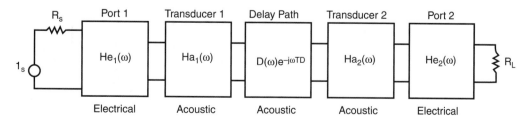

FIGURE 6.6 Complete transfer function of a SAW filter including the acoustic, electrical, and delay line transfer functions. The current generator is I_s, and R_s and R_L are the source and generator resistances, respectively.

Transducer electrodes are fabricated from thin film metal, usually aluminum, and are finite in width. This metal can cause discontinuities to the surface wave which cause velocity shifts and frequency-dependent reflections. In addition, the films have a given sheet resistance which gives rise to a parasitic electrode resistance loss. The electrodes are designed to minimize these distortions in the device.

6.5 Bidirectional Filter Response

A SAW filter is composed of two cascaded transducers. In addition, the overall filter function is the product of two acoustic transfer functions, two electrical transfer functions, and a delay line function, as illustrated in Figure 6.6. The acoustic filter functions are as designed by each SAW transducer. The delay line function is dependent on several parameters, the most important being frequency and transducer separation. The propagation path transfer function, $D(\omega)$, is normally assumed unity, although this may not be true for high frequencies ($f > 500$ MHz) or if there are films in the propagation path. The electrical networks may cause distortion of the acoustic response and are typically compensated in the initial SAW transducer's design.

The SAW electrical network is analyzed using the SAW equivalent circuit model plus the addition of packaging parasitics and any tuning or matching networks. Figure 6.7 shows a typical electrical network that is computer analyzed to yield the overall transfer function for one port of the two-port SAW filter [Morgan, 1985]. The second port is analyzed in a similar manner and the overall transfer function is obtained as the product of the electrical, acoustic, and propagation delay line effects.

6.6 Multiphase Unidirectional Transducers

The simplest SAW transducers are single-phase bidirectional transducers. Because of their symmetrical nature, SAW energy is launched equally in both directions from the transducer. In a two-transducer configuration, half the energy (3 dB) is lost at the transmitter, and reciprocally, only half the energy can be received at the receiver. This yields a net 6-dB loss in a filter. However, by adding nonsymmetry into the transducer, either by electrical multiphases or nonsymmetry in reflection and regeneration, energy can be unidirectionally directed yielding a theoretical minimum 0-dB loss.

The most common SAW UDTs are called the three-phase UDT (3PUDT) and the group type UDT (GUDT). The 3PUDT has the broadest bandwidth and requires multilevel metal structures with cross-overs. The GUDT uses a single-level metal but has a narrower unidirectional bandwidth due to its structure. In addition, there are other UDT or equivalent embodiments that can be implemented but will not be discussed [Morgan, 1985]. The basic structure of a 3PUDT is shown in Figure 6.8. A unit cell

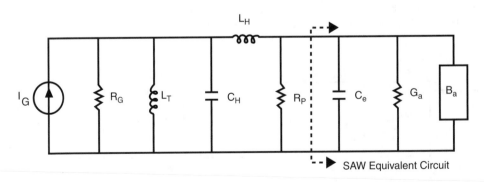

FIGURE 6.7 Electrical network analysis for a SAW transducer. I_G and R_G represent the generator source and impedance, L_T is a tuning inductor, C_H and L_H are due to the package capacitance and bond wire, respectively, and R_P represents a parasitic resistance due to the electrode transducer resistance. The entire network, including the frequency-dependent SAW network, is solved to yield the single-port transfer function.

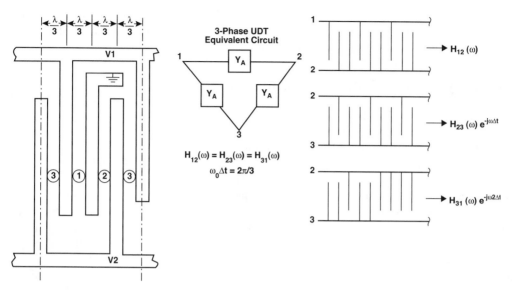

FIGURE 6.8 Schematic of a unit cell of a 3PUDT and the basic equivalent circuit. The 3PUDT can be analyzed as three collinear transducers with a spatial offset.

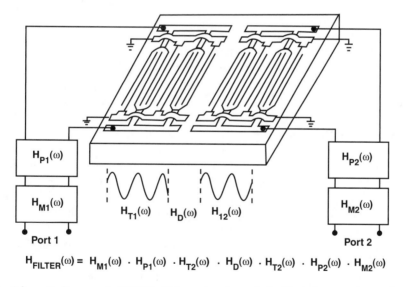

$$H_{FILTER}(\omega) = H_{M1}(\omega) \cdot H_{P1}(\omega) \cdot H_{T2}(\omega) \cdot H_D(\omega) \cdot H_{T2}(\omega) \cdot H_{P2}(\omega) \cdot H_{M2}(\omega)$$

FIGURE 6.9 Schematic diagram of a 3PUDT which requires the analysis of both the acoustic transducer responses as well as electrical phasing and matching networks.

consists of three electrodes, each connected to a separate bus bar, where the electrode period is $\lambda_0/3$. One bus bar is grounded and the other two bus bars will be driven by an electrical network where $V_1 = V_2 \angle 60°$. The transducer analysis can be accomplished similar to a simple IDT by considering the 3PUDT as three collinear IDTs with a spatial phase shift, as shown in Figure 6.8. The electrical phasing network, typically consisting of one or two reactive elements, in conjunction with the spatial offset results in energy being launched in only one direction from the SAW transducer. The transducer can then be matched to the required load impedance with one or two additional reactive elements. The effective unidirectional bandwidth of the 3PUDT is typically 20% or less, beyond which the transducer behaves as a normal bidirectional transducer. Figure 6.9 shows a 3PUDT filter schematic consisting of two transducers and their associated matching and phasing networks. The overall filter must be analyzed with all external electrical components in place for accurate prediction of performance. The external components can be miniaturized and may be fabricated using only printed circuit board material and area. This type of device has demonstrated as low as 2 dB insertion loss.

6.7 Single-Phase Unidirectional Transducers

Single-phase unidirectional transducers (SPUDT) use spatial offsets between mechanical electrode reflections and electrical regeneration to launch a SAW in one direction. A reflecting structure may be made of metal electrodes, dielectric strips, or grooved reflectors that are properly placed within a transduction structure. Under proper design and electrical matching conditions, the mechanical reflections can exactly cancel the electrical regeneration in one direction of the wave over a moderate band of frequencies. This is schematically illustrated in Figure 6.10 which shows a reflector structure and a transduction structure merged to form a SPUDT. The transducer needs to be properly matched to the load for optimum operation. The mechanical reflections can be controlled by modifying the width, position, or height of the individual reflector. The regenerated SAW is primarily controlled by the electrical matching to the load of the transduction structure. SPUDT filters have exhibited as low as 3 dB loss over fractional bandwidths of 5% or less and have the advantage of not needing phasing networks when compared to the multiphase UDTs.

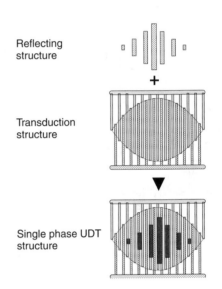

FIGURE 6.10 Schematic representation of a SPUDT which is a combination of transduction and reflecting structures to launch a SAW in one direction over moderate bandwidths.

6.8 Dispersive Filters

SAW filters can also be designed and fabricated using nonuniformly spaced electrodes in the transducer. The distance between adjacent electrodes determines the "local" generated frequency. As the spacing between the electrodes changes, the frequency is slowly changed either up (decreasing electrode spacing) or down (increasing electrode spacing) as the position progresses along the transducer. This slow frequency change with time is often called a "chirp." Figure 6.11 shows a typical dispersive filter consisting of a chirped transducer in cascade with a uniform transducer. Filters can be designed with either one or two chirped transducers and the rate of the chirp is variable within the design. These devices have found wide application in radar systems due to their small size, reproducibility, and large time bandwidth product.

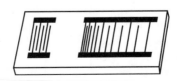

FIGURE 6.11 A SAW dispersive filter consisting of a uniform transducer and a "down chirp" dispersive transducer. The high frequencies have a shorter delay than the low frequencies in this example.

6.9   Coded SAW Filters

Because of the ability to control the amplitude and phase of the individual electrodes or taps, it is easy to implement coding in a SAW filter. Figure 6.12 shows an example of a coded SAW filter implementation. By changing the phase of the taps, it is possible to generate an arbitrary code sequence. These types of filters are used in secure communication systems, spread spectrum communications, and tagging, to name a few [Matthews, 1977].

SAW devices can also be used to produce time impulse response shapes for use in modulators, equalizers, and other applications. An example of a SAW modulator used for generating a cosine envelope for a minimum shift keyed (MSK) modulator is shown in Figure 6.13 [Morgan, 1985].

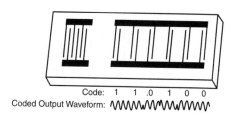

Code: 1   1 .0   1   0   0
Coded Output Waveform: ᳶᳶᳶᳶᳶᳶᳶᳶ

FIGURE 6.12   Example of a coded SAW tapped delay line.

6.10   Resonators

Another very important class of devices is SAW resonators. Resonators can be used as frequency control elements in oscillators, as notch filters, and as narrowband filters, to name a few. Resonators are typically fabricated on piezoelectric quartz substrates due to its low TCD which yields temperature-stable devices. A resonator uses one or two transducers for coupling energy in/out of the device and one or more distributed reflector arrays to store energy in the device. This is analogous to an optical cavity with the distributed reflector arrays acting as the mirrors. A localized acoustic mirror, such as a cleaved edge, is not practical for SAW because of spurious mode coupling at edge discontinuities which causes significant losses.

A distributive reflective array is typically composed of a series of shorted metal electrodes, etched grooves in the substrate, or dielectric strips. There is a physical discontinuity on the substrate surface due to the individual reflectors. Each reflector is one-quarter wavelength wide and the periodicity of the array is one-half wavelength. This is shown schematically in Figure 6.14. The net reflections from all the individual array elements add synchronously at center

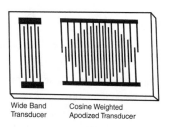

Wide Band   Cosine Weighted
Transducer   Apodized Transducer

FIGURE 6.13   A SAW filter for implementing an MSK waveform using a wideband input transducer and a cosine envelope apodized transducer.

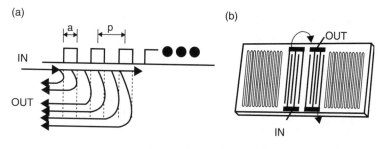

FIGURE 6.14   (a) SAW reflector array illustrating synchronous distributed reflections at center frequency. Individual electrode width (a) is 1/4 wavelength and the array period is 1/2 wavelength at center frequency. (b) A schematic of a simple single-pole, single-cavity two-port SAW resonator.

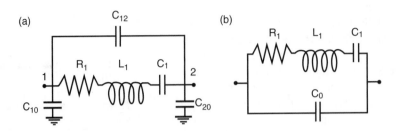

FIGURE 6.15 (a) Two-port resonator equivalent circuit and (b) one-port resonator equivalent circuit.

frequency, resulting in a very efficient reflector. The reflection from each array element is small and very little spurious mode coupling results.

Figure 6.14 shows a typical single-pole, single-cavity, two-port SAW resonator. Resonators can be made multipole by addition of multiple cavities, which can be accomplished by inline acoustic coupling, transverse acoustic coupling, and by electrical coupling. The equivalent circuit for SAW two-port and one-port resonators is shown in Figure 6.15. SAW resonators have low insertion loss and high electrical Qs of several thousand [Campbell, 1989; Datta, 1986; Morgan, 1985].

Defining Terms

Bidirectional transducer: A SAW transducer which launches energy from both acoustic ports which are located at either end of the transducer structure.

Interdigital transducer: A series of collinear electrodes placed on a piezoelectric substrate for the purpose of launching a surface acoustic wave.

Surface acoustic wave (SAW):   A surface acoustic wave (also known as a Rayleigh wave) is composed of a coupled compressional and shear wave. On a piezoelectric substrate there is also an electrostatic wave which allows electroacoustic coupling. The wave is confined at or near the surface and decays away rapidly from the surface.

Triple transit echo (TTE):   A multiple transit echo received at three times the main SAW signal delay time. This echo is caused due to the bidirectional nature of SAW transducers and the electrical and/or acoustic mismatch at the respective ports. This is a primary delayed signal distortion which can cause filter distortion, especially in bidirectional transducers and filters.

Unidirectional transducer (UDT):   A transducer that is capable of launching energy from primarily one acoustic port over a desired bandwidth of interest.

References

D.S. Ballintine, *Acoustic Wave Sensors*, San Diego, Calif.: Academic Press, 1995.

D.S. Ballantine, R.M. White, S.J. Martin, A.J. Ricco, E.T. Zllers, G.C. Frye, and H. Wohltjen, *Acoustic Wave Sensors, Theory, Design, Physico-Chemical Applications*, San Diego, CA: Academic Press, 1997.

C. Campbell, *Surface Acoustic Wave Devices and their Signal Processing Applications*, San Diego, Calif.: Academic Press, 1989.

S. Datta, *Surface Acoustic Wave Devices*, Englewood Cliffs, N.J.: Prentice-Hall, 1986.

A.J. DeVries and R. Adler, Case history of a surface wave TV IF filter for color television receivers, *IEEE Proceedings*, 64, 5, 671–676, 1976.

C.S. Hartmann, D.T. Bell, and R.C. Rosenfeld, Impulse model design of acoustic surface wave filters, *IEEE Transactions on Microwave Theory and Techniques*, 21, 162–175, 1973.

C.S. Hartmann, J.C. Andle, and M.B. King, SAW notch filters, *1987 IEEE Ultrasonics Symposium*, 131–138.

J. Machui, J. Baureger, G. Riha, and I. Schropp, SAW devices in cellular and cordless phones, *1995 Ultrasonics Symposium Proceedings*, 121–130.

D.C. Malocha, Surface Acoustic Wave Applications, *Wiley Encyclopedia of Electrcial and Electronics Engineering,* 21, 117–127, 1999.

H. Matthews, *Surface Wave Filters,* New York: Wiley Interscience, 1977.

D.P. Morgan, *Surface Wave Devices for Signal Processing,* New York: Elsevier, 1985.

D.P. Morgan, Surface Acoustic Wave, *Wiley Encyclopedia of Electrical and Electronics Engineering,* 21, 127–139, 1999.

S.M. Richie, B.P. Abbott, and D.C. Malocha, Description and development of a SAW filter CAD system, *IEEE Transactions on Microwave Theory and Techniques,* 36, 2, 1988.

C.C.W. Ruppel, R. Dill, A Fischerauer, W. Gawlik, J. Machui, F. Mueller, L. Reindl, G. Scholl, I. Schropp, and K. Ch. Wagner, SAW devices for consumer applications, *IEEE Trans. on Ultrasonics, Ferroelectrics, and Frequency Control,* 40, 5, 438–452, 1993.

Special Issue on Surface Acoustic Wave Devices and Applications, *IEEE Proceedings,* 64, 5, May 1976.

Special Issue on Applications, *IEEE Trans. on Ultrasonics, Ferroelectrics, and Frequency Control,* 40, 5, Sept. 1993.

K. Tsubouchi, H. Nakase, A. Namba, and K. Masu, Full duplex transmission operation of a 2.45 GHz asynchronous spread spectrum using a SAW convolver, *IEEE Trans. on Ultrasonics, Ferroelectrics, and Frequency Control,* 40, 5, 478–482, 1993.

Further Information

The *IEEE Transactions on Ultrasonics, Ferroelectrics, and Frequency Control* provides excellent information and detailed articles on SAW technology.

The *IEEE Ultrasonics Symposium Proceedings* provides information on ultrasonic devices, systems, and applications for that year. Articles present the latest research and developments and include invited articles from eminent engineers and scientists.

The *IEEE Frequency Control Symposium Proceedings* provides information on frequency control devices, systems, and applications (including SAW) for that year. Articles present the latest research and developments and include invited articles from eminent engineers and scientists.

For additional information, see the following references:

IEEE Transaction on Microwave Theory and Techniques, vol. 21, no. 4, 1973, special issue on SAW technology.

IEEE Proceedings, vol. 64, no. 5, special issue on SAW devices and applications.

Joint Special Issue of *IEEE Transaction on Microwave Theory and Techniques* and *IEEE Transactions on Sonics and Ultrasonics,* MTT-vol. 29, no. 5, 1981, on SAW device systems.

M. Feldmann and J. Henaff, *Surface Acoustic Waves for Signal Processing,* Norwood, Mass.: Artech House, 1989.

B.A. Auld, *Acoustic Fields and Waves in Solids,* New York: Wiley, 1973.

V.M. Ristic, *Principles of Acoustic Devices,* New York: Wiley, 1983.

A. Oliner, *Surface Acoustic Waves,* New York: Springer-Verlag, 1978.

7
RF Coaxial Cables

Michael E. Majerus
Freescale Semiconductor

RF/microwave cables come in many different forms and types. The one thing they all have in common is that they are used to convey RF energy between locations. For RF/microwave frequencies (500 MHz to 50 GHz) the coaxial cable is the most widely used medium for information transfer. The following sections present a short abbreviated history of the coaxial cable. This is followed by sections on the basic characteristics of a coaxial cable and some of the materials used to make a modern coaxial cable.

7.1 History of the Coaxial Cable

Heinrich Hertz used the coaxial cable in experiments to prove Faraday's and Maxwell's theories about the wave nature of RF energy. His use of the coaxial structure helped him demonstrate standing waves by producing minimum and maximum wave voltages at alternating quarter wavelength intervals along a coaxial line. Hertz published his investigations from 1887 to 1891.[1] Through out the 1920s many improvements in the design of coaxial lines were realized. The state-of-the-art in the mid 1920s was a coaxial cable that used a ridged outer conductor and an air dielectric with bead supports for the center conductor as illustrated in Figure 7.1.[2-4] In the 1930s, coaxial cable was used almost exclusively in low frequency VHF and UHF radio applications. Primarily, it was used because its excellent shielding properties reduced static interference, which is a common problem at low VHF and UHF frequencies. For most high frequency applications, 1 GHz and up, a waveguide structure was used. The waveguide was the preferred option because it had lower losses, and reducing the required power directly reduces cost.

With the start of World War II, the use and development of higher frequency coaxial components and the production of coaxial cable increased at a rate that had never been seen before. This was due to the size reduction achievable with coaxial cable compared to waveguides. There was no real plan for miniaturization at the time; in the field, coaxial cable was easier to work with than waveguide. The development of radar

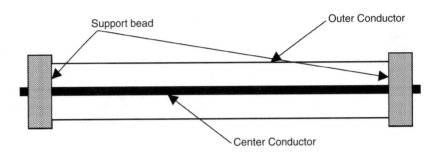

FIGURE 7.1   Air line with dielectric bead supports.

during World War II was the driving force behind the need for higher frequency components and cable. To manage this development and procurement, the joint Army-Navy Cable Coordinating Committee was established. This committee published the specifications for cables and connectors for the Army and Navy.

7.1.1   Why 50 Ohms?

Why was 50 ohms selected as the reference impedance of radar and radio equipment? There are many misconceptions surrounding this bit of history. "The first coaxial dimensions just happened to define the 50 ohm reference impedance." "Fifty ohms matched up well with the antennas in use 60 years ago." In actuality, the 50 ohm reference impedance was selected from a trade-off between the lowest loss and maximum power-handling dimension for an air line coaxial cable. The optimum ratio of the outer conductor to inner conductor, for minimum attenuation in a coaxial structure with air as the dielectric, is 3.6. This corresponds to an impedance, Zo, of 77 ohms.[5] Although this yields the best performance from a loss standpoint, it does not provide the maximum peak power handling before dielectric break-down occurs. The best power performance is achieved when the ratio of the outer conductor to inner conductor is 1.65. This corresponds to a Zo of 30 ohms.[5] The geometric mean of 77 ohms and 30 ohms is approximately 50 ohms [Equation 7.1]; thus, the 50 ohm standard is a compromise between best attenuation performance and maximum peak power handling in the coaxial cable.

$$50 \approx \sqrt{30 * 77}$$ (7.1)

7.2   Characteristics of Coaxial Cable

7.2.1   TEM Mode of Propagation

A key feature of coaxial cable is that its characteristic impedance is very broadband. The fundamental mode setup in a coaxial cable is the TEM mode. This means that the electric and magnetic fields are transverse to the direction of propagation. Figure 7.2 shows a snapshot in time for a propagating wave.

7.2.2   Characteristic Impedance and Cutoff Frequency

In the TEM mode of propagation, below the cutoff frequency, the characteristic impedance of the coaxial cable is independent of frequency. The Zo of the cable is defined by the ratio of outer conductor "D" to the inner conductor "d" (see Figure 7.3), and the relative dielectric constant (ε_r) of the dielectric material. The relationship is shown in Equation 7.2:

$$Z_o = \frac{60}{\sqrt{\varepsilon_r}} * \ln\frac{D}{d}$$ (7.2)

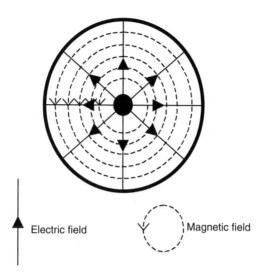

FIGURE 7.2 TEM Mode in coax.

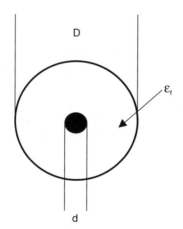

FIGURE 7.3 Cross-section of a coaxial cable showing outer diameter and inner diameter dimensions.

The approximate cutoff frequency for a cable, the frequency at which the first non-TEM mode of propagation begins, can be calculated using Equation 7.3. At frequencies above F_{cutoff}, other propagation modes dominate and the characteristic impedance becomes frequency dependent.

$$F_{cutoff} \, GHz \approx \frac{7.51}{\sqrt{\varepsilon_r}} * \frac{1}{D+d} \tag{7.3}$$

7.2.3 Electrical Length

The electrical length of a cable can be expressed in degrees of phase shift. These units are useful when using a Smith chart to synthesize a matching network. To find the electrical length, the dielectric material properties need to be known. The most common dielectric is Polytetrafluoroethylene (PTFE) with a ε_r

of 2.10. Using Equation 7.4 the velocity factor (υ_f) may be calculated: The value of υ_f represents the velocity of propagation of the

$$\upsilon_f = \frac{1}{\sqrt{\varepsilon_r}} = \frac{1}{\sqrt{2.1}} = 0.69 \tag{7.4}$$

RF signal down the cable relative to the velocity in free space. The velocity of propagation may be calculated by multiplying the free space velocity "c" (300 million meters/sec) by the υ_f. This is shown in Equation 7.5. Then Equation 7.6 is used to calculate the electrical length. To get the desired cable length, take the ratio of the desired

$$\upsilon_f = \frac{1}{\sqrt{\varepsilon_r}} * c \tag{7.5}$$

$$\frac{1}{\text{Frequency}} * \left(\upsilon_f * 3.0 \times 10^8\right) = \text{length in meters} \tag{7.6}$$

angle or phase shift to 360° and multiply by the cable length found in Equation 7.6. This will yield the length for the desired phase shift [(Equation 7.7]. As an example, a phase shift of 45° at 500 MHz, using a PTFE dielectric $\upsilon_f = 0.69$, is desired.

$$\frac{\text{Phase shift needed in deg.}}{360} * \text{wavelength} = \text{cable length} \tag{7.7}$$

Using Equation 7.6 a full wavelength or 360° of phase shift at 500 MHz is determined to be ~41.1 cm. Since we only need 45° of electrical length, we need to take the ratio of 45° to 360° and multiply that by 41.1 cm as shown in Equation 7.8. This will give us the cable length of 5.137 cm, which is needed for an electrical length of phase shift of 45°.

$$\frac{45}{360} * 41.1 \text{ cm} = 5.137 \text{ cm} \tag{7.8}$$

7.2.4 Cable Attenuation

The losses in a coaxial cable arise from two sources: one is from the resistance of the conductors and the currents flowing in the center and outer conductor; the second is from the conduction current in the dielectric. The conductor losses are ohmic and due to the skin effect in the conductor's increase with square root of the frequency. The skin effect loss increases proportional to the square root of the frequency.[6] Equations 7.9 and 7.10 can be used to calculate this loss. This represents the key trade-off in coaxial cable selection. The trade-off of cable cutoff frequency and the skin effect loss implies that the larger the diameter of the cable used for the frequency of operation, the better. The dielectric loss is the resistance of the dielectric material to the conduction currents, is linear with frequency, and is calculated with Equation 7.11. The losses in the center conductor, outer conductor, and dielectric added together to give the total loss of the cable, as shown in Equation 7.12. A third type of loss in a coaxial cable is

$$L_O\left(dB/100_{ft}\right) = \frac{.435 * \sqrt{F_{MHz}}}{Z_O * d} \qquad L_O = \text{Loss in outer conductor} \tag{7.9}$$

$$L_O\left(dB/100_{ft}\right) = \frac{.435 * \sqrt{F_{MHz}}}{Z_O * D} \qquad L_O = \text{Loss in outer conductor} \tag{7.10}$$

$$L_D\left(dB/100_{ft}\right) = 2.78\rho * \sqrt{\varepsilon_r} * F_{MHZ} \quad L_D = \text{Dielectric loss} \qquad (7.11)$$

$\varepsilon_r = 2.1$ for solid PTFE and $\qquad\qquad F_{MHz} = $ Frequency in MHz
1.6 for expanded PTFE

$$(7.12)$$

$$\text{Loss}_{Total} = L_C + L_O + L_D$$

due to radiation. This loss is usually minimal in the coaxial cable because the outer conductor confines.[1] The losses from the major contributors add to give the approximate total losses in a coaxial cable.[7]

7.2.5 Maximum Peak Power

The maximum peak voltage that a cable can sustain is that voltage at which dielectric breakdown occurs. In an air filled coaxial cable, this happens when the maximum electric field, E_m, reaches about 2.9×10^4 volts per cm.[5] The maximum power (Pmax) under this condition is given by Equation 7.13. The maximum power in an air dielectric coaxial cable is realized with a characteristic impedance of 30 ohms.

$$P_{max} = \frac{E_m^2}{480} * D^2 * \frac{\ln D/d}{\left(D/d\right)^2} \qquad (7.13)$$

7.3 Cable Types

7.3.1 Semirigid

The outer conductor as well as the center conductor can be made of different materials. The outer conductor will be discussed first. A type of cable called a semirigid cable has an outer conductor that is a solid sheath of extruded metal such as copper (see Figure 7.4). This cable is the hardest to form into

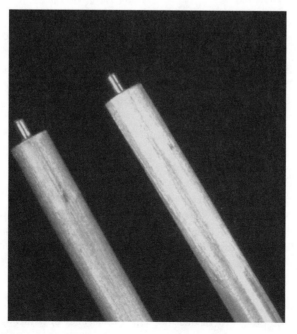

FIGURE 7.4 Photograph of copper- and tin-plated copper semi-rigid coaxial cable.

complex shapes and care needs to be taken when bending. The cable must be cut to size and then bent into the shape required. After shaping, heat is applied to the cable to expand the dielectric and relax the stress in the dielectric. After this step, the desired connector may be attached.

7.3.2 Pliable Semirigid

A variation to the semirigid cable is a type manufactured with a soft outer conductor, such as dead soft aluminum or an un-annealed copper. These types of cables are easier to form, and can often be formed without special tooling.

7.3.3 Flexible Cable

Another type of coaxial cable is the flexible cable, which uses a braided outer conductor, as shown in Figure 7.5 (upper right). This cable has less phase stability with flexure than semirigid because of the limited dimensional rigidity around the dielectric material, but can be much easier to use. This type of cable typically uses a mechanical means of attaching the connector, such as a crimp-on fitting or a screw-on fitting.

A variation on the braided outer conductor cable has the exterior braid coated with solder as shown in the lower half of Figure 7.5. This makes this cable similar to a semirigid cable. The use of the soft solder sheath makes the cable very easy to shape. The drawback is that only a limited number of bends are permitted before the cable fails.

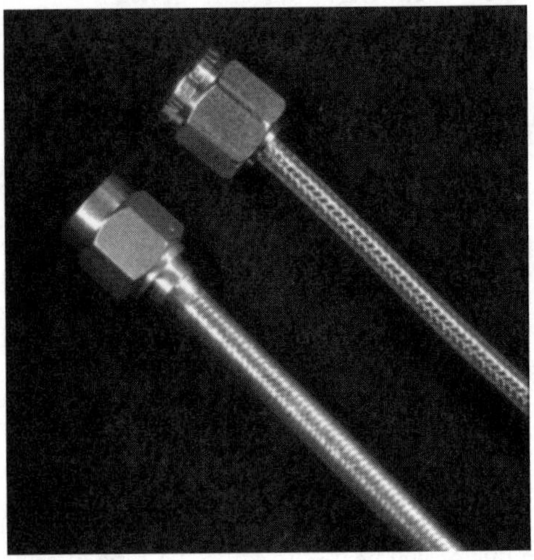

FIGURE 7.5 Photograph of a solder-coated and braided coaxial cable.

7.3.4 Center Conductors

The center conductor of a coaxial cable can take different forms. The most common are solid and stranded. The solid center conductor is the most prevalent. It commonly is made of copper, beryllium copper, and aluminum. Very often, the center conductor is plated with silver or tin. The stranded center conductor is not as prevalent because its performance advantage, reduced attenuation, is at the lower frequencies, 1 GHz and below. Above 1 GHz, the performance is the same as the solid center conductor.

7.4 Dielectric Materials

7.4.1 Solid PTFE

The dielectric used in most modern coaxial cable is polytetrafluoroethylene (PTFE), also known as TeflonS (a Dupont registered trademark). Solid PTFE is an extruded form, which is relatively sensitive to temperature changes. The solid PTFE has a negative phase shift with temperature,[6] i.e., as the temperature increases the electrical length decreases. Both the phase and characteristic impedance solid PTFE coaxial will change with changes in temperature.

7.4.1.1 Expanded PTFE

To improve the temperature performance of the coaxial cable, a more stable dielectric is PTFE that has been expanded with air. Another benefit due to a smaller dialectic constant of expanded PTFE is the reduction of the dielectric losses (ε_r is approximately 1.60). See Table 7.1 for examples.

TABLE 7.1   Common Cable Types

Cable RG #	Dielectric Type	Outer Conductor Dia.	Dielectric Dia.	Center Conductor Dia.	Attenuation per 100 ft	Average Power Max
.405/U	Solid PTFE	.0865	.0658	.0201	19 dB	.1 Kwatts
.402/U	Solid PTFE	.141	.1175	.036	11 dB	.3 Kwatts
.401/U	Solid PTFE	.250	.208	.0641	6.5 dB	.7 Kwatts
N/A	Air expanded PTFE	.141	.116	.043	9.5 dB	.55 Kwatts
N/A	Air expanded PTFE	.250	.210	.074	.55 dB	12 Kwatts

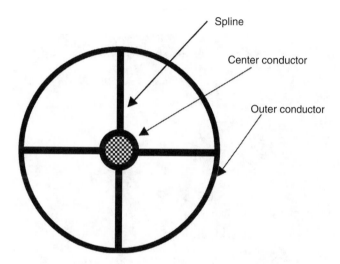

FIGURE 7.6   Cross-section of a spline coaxial cable.

7.4.2   Splined PTFE

To further reduce the dielectric losses, a unique cable configuration was developed: the spline cable. This dielectric sheath is made such that ridges or splines run the length of the cable, supporting the center conductor within the outer conductor, as shown in Figure 7.6. The spline structure reduces the dielectric loss to nearly that of air. In addition, this configuration has a positive phase shift with temperature.[6]

7.5   Standard Available Coaxial Cable Types

Coaxial cable can be made in many sizes and impedances. Table 7.1 is a lists some of the standard commercially available cables.[8,9] The power maximums listed are average power ratings. This is the maximum average power that can be maintained without damaging the dielectric.

7.6   Connectors

The coaxial cable is of little use if the RF energy is not efficiently coupled into it, thus, high quality connectors are necessary. The connector usually defines the usable frequency range of a coaxial cable. In most, but not all cases, the connector will have a cutoff frequency lower than the cable itself. Table 7.2 is a list of common RF connectors and their cutoff frequencies.

Each type of connector can be attached to a cable in several ways: direct solder, crimp-on, and screw-on. Figure 7.7 indicates a direct solder connection where the connector body is soldered onto the outer

TABLE 7.2   Common Coaxial Connectors

Connector Type	Cutoff Frequency	Mating Torque
BNC	4.0 GHz	N/A
SMB	4 GHz	N/A
SMC	10 GHz	30–50 in-oz
TNC	15 GHz	12–15 in-lbs
Type-N	18 GHz	12–15 in-lbs
7 mm	18 GHz	12–15 in-lbs
SMA	18 GHz	7–10 in-lbs
3.5 mm	26.5 GHz	7–12 in-lbs
2.9 mm	46 GHz	8–10 in-lbs
2.4 mm	50 GHz	8–10 in-lbs

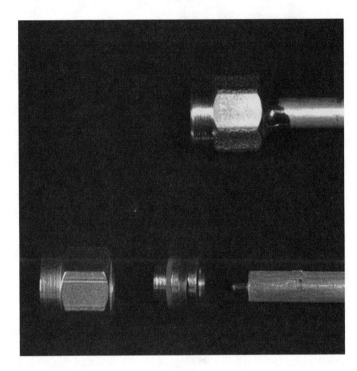

FIGURE 7.7   Photograph of a solder-on, connector-to-cable assembly.

conductor.[9] This makes the most reliable connection. The drawback to this method is that it is the most demanding on assembly techniques to prevent damage to the dielectric. A second method is "crimp-on" as shown in Figure 7.8.[8,9] This type of connection is easy to make, but is not durable under hard usage. The last method, "screw-on" is used for lower frequency cables (under 1 GHz). This method uses a swage that tightens onto the outer conductor (typically a braided outer conductor), as shown in Figure 7.9.[10] It is important to understand the proper methods of connector attachment to realize optimum performance. Proper assembly tools and techniques must be used.

7.7   Summary

The coaxial cable remains the backbone for signal interconnects between equipment and systems since the early 1920s due to its desirable features. The ease of use, relatively low cost, and broad operating bandwidth ensure its continued popularity. The wireless world is made possible through the lowly coaxial cable. This basic component brings the wireless technology to us.

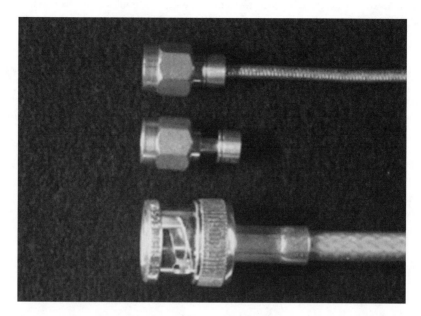

FIGURE 7.8   Photograph of a crimp-on, connector-to-cable assembly.

FIGURE 7.9   Photograph of a swage-on, connector-to-cable assembly.

References

1. J.H. Bryant, Coaxial Transmission Lines, Related Two-conductor Transmission lines, Connectors, and Components: A U.S. Historical Perspective, *IEEE Trans. Microwave Theory and Techniques,* MTT-32, 970–983, Sept. 1984.

2. R.S. Ohl, Means for transferring high frequency energy, U.S. Patent 1 619 882, Mar. 8 1927 (filed 1924).
3. H.A. Affel, E.I. Green, Concentric conducting system, U.S. Patent 1 781 092, Nov 11 1930 (filed 1929).
4. L. Espenscheid, H.A. Affel, Concentric conducting system,, U.S. Patent 1 835 031, Dec. 8 1931 (filed 1929).
5. G. Matthaei, L. Young, E.M.T. Jones, Properties of some common microwave filter elements, in *Microwave Filters, Impedance-Matching Networks, and Coupling Structures,* Artech House, Norwood, MA, 1985, 165–168.
6. P.H. Smith, *Electronic Applications of the Smith Chart,* 2nd edition, Noble, Atlanta, 1995, 38.
7. Astrolab Catalog.
8. Precision Tube Co. Coaxitube Division Catalog, 1993.
9. M/A-COM Coaxial Connectors, Adapters, Tools and Accessories. Catalog, 1995.
10. Pasternack Catalog.

8

Coaxial Connectors

David Anderson
Freescale Semiconductor

Coaxial connectors are one of the fundamental tools of microwave technology and yet they are taken for granted in many instances. A good understanding of coaxial connectors, both electrically and mechanically, is required to utilize them fully and derive their full benefits.

8.1 History*

The only coaxial connector in general use in the early 1940s was the UHF connector, which is still manufactured and used today. E. C. Quackenbush of the American Phenolic Company (later Amphenol) in Chicago developed it. The UHF connector was not deemed suitable for higher frequency use, and the

---

*Bryant, John H., Coaxial Transmission Lines, Related Two-Conductor Transmission Lines, Connectors, and Components: A U.S. Historical Perspective, *IEEE MTT Transactions*, 9, 970, 1990.

connector committee undertook to develop one in 1942. The result was the original type N (Navy) connector. This connector does not have a constant impedance from interface to interface. The three main coaxial parts, center conductor, dielectric, and outer conductor or shell were held together by a system of steps and shoulders. These mechanical discontinuities represent electrical discontinuities and limit performance. However, modified designs were made in an effort to improve microwave performance and still maintain interface-mating capability.

Several others, none of which appears to have received microwave design attention, followed the type N connector design. The type BNC, a "baby type N," is somewhat scaled down in diameter and provides a twist-lock coupling mechanism. The TNC followed the BNC in 1956.

Development of the SMA connector was driven by the need to provide compact, low-reflection connections for microwave components. By September 1962, the market for connectors as well as other microwave/RF components was emerging. Omni Spectra was manufacturing connectors for use in its components. Based on demand, the decision was made to offer connectors to help promote the use of their components. Thus, a microwave company was in the business of selling connectors. OSM, for Omni Spectra Miniature, was adopted as the trademark for this first microwave connector. The name SMA was adopted in 1968 as the military designation for this connector under specification MIL-C-39012.

8.2 Definitions*

There are two distinct categories of connectors, sexless and sexed. A sexless connector has two equal mating halves. A sexed connector has either a female or male configuration to form a mated pair. All sexed connectors are of the pin and socket type, with a male contact or pin and a female contact or socket. Two basic types of female sockets are used for precision connectors, either slotted or slotless.

To insure mechanical mating compatibility and electrical repeatability, the mechanical configuration, dimensions, and tolerances of a connector and its mating properties must be clearly defined. This set of data defines the connector's interface.

A connector's reference plane is defined as the outer conductor-mating surface of a coaxial connector. It is desirable to have both the outer and center conductors mating surface coplanar. When this occurs, the connectors are referred to as coplanar. Typical coplanar connectors are 14 mm, 7 mm, 3.5 mm, and SMA connectors. Typical noncoplanar connectors are Type-N, TNC, and BNC.

Connector coupling describes how the outer conductors are connected when the connector is mated. Coupling is either threaded, bayonet, or snap on. Precision connectors generally use threaded coupling. The best example of the bayonet or twist coupling is the BNC.

Dielectric styles in connectors are of two types, air or solid dielectric. Air dielectric simplifies connector construction and generally is used on precision connectors so accurate standards can be created. Solid dielectrics such as Teflon are used two configurations, flush and overlapping. Overlapping dielectrics are generally used for higher power applications and to prevent voltage breakdown.

8.3 Design†

Two equations define the general design parameters for coaxial connectors. One defines the coaxial characteristic impedance and the other defines the TE_{11} mode cutoff frequency.

Characteristic Impedance:

$$Z_o = 60 * \left(\sqrt{\mu_r} \big/ \sqrt{\varepsilon_r} \right) * \ln b/a \, \Omega \tag{8.1}$$

*Maury Jr., Mario A., Microwave Coaxial Connector Technology: A Continuing Evolution, *Microwave Journal*, State of the Art Reference, 39, 1990.

†Rizzi, Peter A., *Microwave Engineering Passive Circuits*, Prentice-Hall, Englewood Cliffs, 1988, chap. 5.

Cutoff Frequency TE$_{11}$ mode:

$$f_c = \frac{c}{\pi(a+b)\sqrt{\mu_r \varepsilon_r}} Hz$$

(8.2)

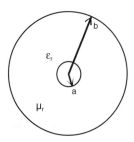

where: b is the inner radius of the outer conductor.
$\quad\quad a$ is the radius of the center conductor.
$\quad\quad \varepsilon_r$ is the relative permittivity of the transmission medium.
$\quad\quad \mu_r$ is the relative permeability of the transmission medium.
$\quad\quad c$ is the velocity of light.

The frequency range of any connector is limited by the occurrence of the first circular waveguide mode in the coaxial structure. Decreasing the diameter of the outer conductor increases the highest usable frequency. Filling the air space with dielectric lowers the highest usable frequency and increases the system loss.*

8.4   Connector Care†

Dimensions of microwave connectors are small and some of the mechanical tolerances are very precise. Seemingly minor defects, damage and dirt, can significantly degrade repeatability and accuracy. In addition, the mating surfaces of most precision connectors are gold plated over a beryllium-copper alloy. This makes them very susceptible to mechanical damage due to the comparative softness of the metals.

Among the most important general care and usage recommendations for all types of microwave connectors are:

1. Connectors must be kept clean and the mating plane surfaces protected from harm during storage.
2. Connectors should be inspected visually before every connection and damaged connectors discarded immediately.
3. Connectors should be cleaned first with compressed air. Solvent should never be sprayed into a connector; use a cotton swab or lint-free cloth and the least amount of solvent possible. The plastic support beads should not come in contact with solvent.
4. Connectors should be inspected mechanically with a connector gage before being used for the first time and periodically thereafter.
5. Connectors should be aligned carefully when mated and the preliminary connection should only be made by lightly turning the connector nut to pull the connectors together. The final connection should be made by using a torque wrench.
6. Turning one connector body relative the other should never be done to make connections and disconnections. This is extremely harmful and can occur whenever the connector body rather than the connector nut alone are turned.

8.4.1   Handling

Microwave connectors must be handled carefully, inspected before each use, and stored for maximum protection.

The connector mating surface should never be touched and should never be placed in contact with a foreign surface, such as a table. Natural skin oils and microscopic particles of dirt are easily transferred to the connector interface and are very difficult to remove. In addition, damage to the plating and to the

---

*Microwave Test Accessories Catalog, Hewlett Packard, 1991, 80.
†Microwave Connector Care, Hewlett Packard, April 1986, Part One.

mating plane surfaces occurs readily when the interface comes into contact with any surface. Connectors should not be stored with the contact end exposed. Plastic end caps are provided with precision connectors and these should be retained after unpacking and placed over the connector ends before storage. Connectors should never be stored loose in a box or in a desk or bench drawer. Careless handling of this kind is the most common cause of connector damage during storage.

8.4.2 Visual Inspection

Visual inspection and, if necessary, cleaning should be done every time a connection is made. Metal and metal by-product particles from the connector threads often find their way onto the mating plane surfaces when a connection is disconnected, and even one connection made with a dirty or damaged connector can damage both connectors beyond repair.

The connectors should be examined prior to use for obvious defects or damage, e.g., badly worn plating, deformed threads, or bent, broken, or misaligned center conductors. Connector nuts should turn smoothly and be free of burrs or loose metal particles. Connectors with these types of problems should be discarded or repaired.

The mating plane surfaces of connectors should also be examined. Flat contact between the connectors at all points on the mating plane surfaces is essential for a good connection. Therefore, particular attention should be paid to deep scratches or dents and to dirt and metal or metal by-product particles on the connector mating plane surfaces. The mating plane surfaces of the center and outer conductors should be examined for bent or rounded edges and also for any signs of damage due to excessive or uneven wear or misalignment.

If a connector displays deep scratches or dents, or particles on the mating plane surfaces, clean it and inspect it again. Damage or defects of these kinds — dents or scratches deep enough to displace metal on the mating plane surface of the connector — may indicate the connector is damaged and should not be used.

8.4.3 Cleaning

Careful cleaning of all connectors is essential to assure long, reliable connector life, to prevent accidental damage to connectors, and to obtain maximum measurement accuracy and repeatability. However, this one step is most often neglected.

Loose particles on the connector mating plane surfaces can usually be removed with a quick blast of compressed, dry, air. Clean dry air cannot damage the connectors or leave particles or residues behind and should therefore be employed as the first attempt at connector cleaning. Dirt and other contaminants that cannot be removed with compressed air can often be removed with either a cotton swab or lint-free cleaning cloth and a solvent. The least possible amount of solvent should be used and neither the bead nor dielectric supports in the connectors should come in contact with the solvent. A very small amount of solvent should be applied to a cotton swab or a lint-free cleaning cloth. The connector should then be wiped as gently as possible. The precaution should be taken to always use solvents in a well-ventilated area, avoid prolonged breathing of solvent vapors, and avoid contact of solvents with the skin.

The threads of the connectors should be cleaned first since a small amount of metal wears off the threads every time a connection or a disconnection is made, and this metal often finds its way onto the mating plane surfaces of the connectors. Then, wet a clean swab, and clean the mating plane surfaces of the connector.

8.4.4 Mechanical Inspection

Even a perfectly clean unused connector can cause trouble if it is mechanically out of specification. Since the critical tolerances in precision microwave connectors are very small, using a connector gage is essential.

Before using any connector for the first time, inspect it for mechanical tolerance by using a connector gage. Connectors should be gaged after that depending upon their usage. In general, connectors should be gaged whenever visual inspection or electrical performance suggests that the connector interface may be out of specification, for example due to wear or damage. Connectors on calibration and verification devices should also be gaged whenever they have been used on another system or piece of equipment.

A different gage is required for each type of connector. See Figure 8.1 for a typical connector gage and Figure 8.2 for a master gage. Sexed connectors require two gages for female and male. Every connector gage requires a master gage for zeroing the gage. Care is necessary in selecting a connector gage to measure precision microwave connectors. Some have a very strong plunger springs, strong enough, in some cases, to push the center conductor back through the connector, damaging the connector itself.

The critical dimension to be measured is the position, recession or setback, of the center conductor relative to the outer conductor mating plane. Mechanical specifications for connectors specify a maximum and minimum distance the center conductor can be positioned with respect to the outer conductor-mating plane. Before gauging any connector, consult the mechanical specifications provided with the connector or the device itself.

Before using any connector gage it must be inspected, cleaned and zeroed. Inspect and clean the gage and master gage for dirt and particles exactly as the connector was cleaned and inspected. Dirt on the gage or master gage will make the gage measurements of the connectors incorrect and can transfer dirt to the connectors themselves, damaging the connectors during gauging or connection. Zero the gage using the correct connector gage master gage. Hold the gage by the plunger barrel only. Slip the gage master gage into the circular housing of the connector gage. Carefully bring

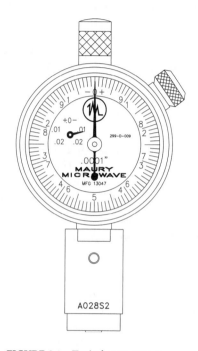

FIGURE 8.1   Typical connector gauge.

the gage and the master gage together applying only enough pressure to the gage and the master gage to result in the dial indicator settling at a reading. Gently rock the two surfaces together to verify they have come together flatly. The gage indicator should now line up exactly with the zero mark of the gage. If it does not, inspect and clean both the gage and master gage again.

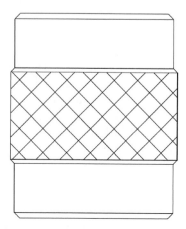

FIGURE 8.2   Typical connector master gauge.

Measuring the recession of the center conductor behind the outer conductor mating plane in a connector is performed in exactly the same manner as zeroing the gage. If the connector has a retractable sleeve or sliding connector nut, extend the sleeve or nut fully. This makes it easier to keep the gage centered in the connector. Hold the gage by the plunger barrel and slip the gage into the connector so that the gage plunger rests against the center conductor. Carefully bring the gage into firm contact with the outer conductor mating plane. Apply only enough pressure to the gage as results in the gage indicator settling at a reading. Gently rock the connector gage within the connector, to insure the gage and the outer conductor have come together flatly. Read the recession or protrusion from the gage dial. See Figure 8.3 for a typical 7 mm example. For maximum accuracy, measure the connector several times and average the readings. Rotate the gage relative to the connector between each measurement.

8.4.5 Connections

Skill is essential in making good connections. The sensitivity of modern test instruments and the mechanical tolerances of precision microwave connectors are such that slight errors that once went unnoticed now have a significant effect on measurements.

Before making any connections, inspect all connectors visually, clean them if necessary, and use a connector gage to verify that all the center conductors are within specifications. If connections are made to static-sensitive devices, avoid electrostatic discharge by wearing a grounded wrist strap and grounding yourself and all devices before making any connections.

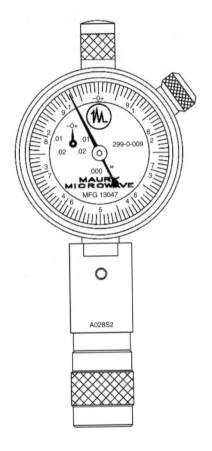

FIGURE 8.3 Typical 7 mm connector gauge inspection.

Careful alignment of the connectors is critical in making a good connection, both to avoid damaging connectors and devices and to assure accurate measurements. As one connector is connected to the other, care should be taken to notice any signs that the two connectors are not aligned perfectly. If misalignment has occurred, stop and begin again.

Alignment is especially important for sexed connectors to avoid bending or breaking the contact pins. The center pin on the male connector must slip concentrically into the contact fingers of the female connector and requires great care aligning the two connectors as they are mated.

When they have been aligned, the center conductors must be pushed straight together, not twisted or screwed together and only the connector nut, not the connector body, should be rotated to make the connection. Slight resistance is generally felt as the center conductors mate; very light finger pressure, 2 inch-ounces of torque, is enough.

When the preliminary connection has been made, use a torque wrench to make the final connection. The connection should be tightened only until the break point of the wrench is reached, when the handle gives way at its internal pivot point.

8.4.6 Disconnections

Disconnect connectors by grasping the body firmly to prevent it from rotating. Then loosen the connector nut that was tightened to make the connection. If necessary, use the torque wrench or an open-end wrench to start the process, but leave the connection finger tight. At all times support the connectors and the connection to avoid putting lateral or bending forces on the connector mating planes.

Complete the process by disconnecting the connector nut completely. As in making the connections, turn only the connector nut. Never allow the connectors to rotate relative to each other. Twisting the connection can damage the connector by damaging the center conductors or the interior components. It can also scrape the plating off the male contact pin or even unscrew the male or female contact pin slightly from its interior mounting.

8.5   Type N

Precision Type-N connectors are similar in size to 7 mm connectors but relatively inexpensive. They are more rugged, and were developed for severe operating environments or applications in which many connections and disconnections must be made. They are among the most popular general-purpose connectors used in the DC to 18 GHz frequency range.

Unlike precision 7 mm connectors, Type-N connectors are sexed connectors. The male contact pin slides into the female contact fingers and electrical contact is made by the inside surfaces of the tip of the female contact on the sides of the male contact pin. The position of the center conductor in the male connector is defined as the position of the shoulder of the male contact pin, not the position of the tip.

Type-N connectors differ from other connectors in that the outer conductor mating plane is offset from the mating plane of the center conductors. The outer conductor sleeve in the male connector extends in front of the shoulder of the male contact pin. When the connection is made, this outer conductor sleeve fits into a recess in the female outer conductor behind the tip of the female contact fingers.

Type-N connectors should not be used if there is a possibility of interference between the shoulder of the male contact pin and the tip of the female contact fingers when the connectors are mated. In practice, this means that no Type-N connector pair should be mated when the separation between the tip of the female contact fingers and the shoulder of the male contact pin could be less than zero when the connectors are mated. Care should be taken when gauging Type-N connectors to avoid damage.

As Type-N connectors wear, the protrusion of the female contact fingers generally increases, due to wear of the outer conductor mating plane inside the female connector. This decreases the total center conductor contact separation and must be monitored carefully. At lower frequencies the effects of wide contact separation are small; only at higher frequencies does the contact separation become important.

Type-N connectors are available in both 50 Ω and 75 Ω impedance. However 75 Ω Type-N connectors differ from 50 Ω Type-N connectors most significantly in that the center conductor, male contact pin, and female contact hole are smaller. Therefore, mating a male 50 Ω Type-N connector to a female 75 Ω Type-N connector will destroy the female 75 Ω connector by spreading the female contact fingers apart permanently or even breaking them. If both 50 Ω and 75 Ω Type-N connectors are among those on the devices you are using, mark the 75 Ω Type-N connectors to insure they are never mated with any 50 Ω Type-N connectors.

8.5.1   Type-N Specifications

Frequency Range:	DC to 18 GHz
Impedance:	50 Ω and 75 Ω
Mating Torque:	12 in-lbs.
Female Socket:	0.207 +0.000 −0.010 inches
Male Pin:	0.207 +0.010 −0.000 inches

8.5.2   Type-N Dimensions

See Figures 8.4 through 8.7 for details.

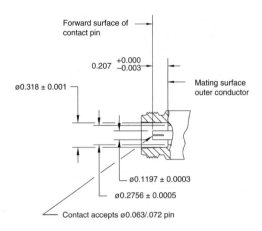

FIGURE 8.4   Type-N female 50 Ω interface.

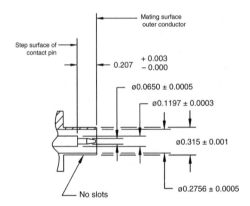

FIGURE 8.5   Type-N male 50 Ω interface.

8.6   BNC

The BNC is a general-purpose connector for low frequency uses. It is a dielectrically loaded, sexed connector. The male contact pin slides into the female contact fingers and electrical contact is made by the inside surfaces of the tip of the female contact fingers on the sides of the male contact pin. BNC connectors are available in both 50 Ω and 75 Ω versions, and the two versions will mate successfully with each other.

8.6.1   BNC Specifications

Frequency Range:	DC to 4 GHz		
Impedance:	50 Ω and 75 Ω		
Mating Torque:	None		
Female Socket:	0.206	+0.000	−0.003 inches
Female Dielectric Top:	0.208	+0.000	−0.008 inches
Female Dielectric Bottom:	0.000	+0.008	−0.000 inches
Male Pin:	0.209	+0.003	−0.000 inches
Male Dielectric Top:	0.008	+0.004	−0.000 inches
Male Dielectric Bottom:	0.212	+0.006	−0.000 inches

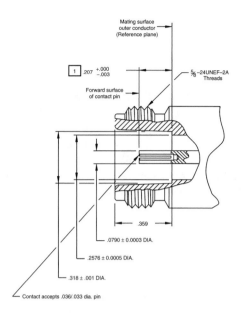

FIGURE 8.6   Type-N female 75 Ω interface.

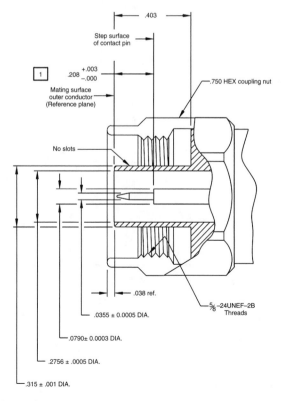

FIGURE 8.7   Type-N male 75 Ω interface.

8.6.2 BNC Dimensions

See Figures 8.8 through 8.11 for details.

8.7 TNC

The TNC is a dielectric loaded sexed connector with many different configurations. The male contact pin slides into the female contact fingers and electrical contact is made by the inside surfaces of the tip of the female contact fingers on the sides of the male contact pin.

Female and male TNC connectors of the same specification are designed to provide the best-matched condition when mated together. When female and male TNC connectors of different specifications are mated, less than optimum electrical performance can be experienced even though they are mechanically compatible. Table 8.1 displays TNC connector compatibility.*

8.7.1 TNC Specifications

Frequency Range:	DC to 18 GHz		
Impedance:	50 Ω		
Mating Torque:	12 in-lbs.		
Female Socket:	0.208	+0.000	–0.005 inches
Female Dielectric Top:	0.208	+0.000	–0.008 inches
Female Dielectric Bottom:	0.000	+0.000	–0.004 inches
Male Pin:	0.209	+0.005	–0.000 inches
Male Dielectric Top:	0.000	+0.004	–0.000 inches
Male Dielectric Bottom:	0.209	+0.005	–0.000 inches

8.7.2 TNC Dimensions

See Figures 8.12 and 8.13 for details.

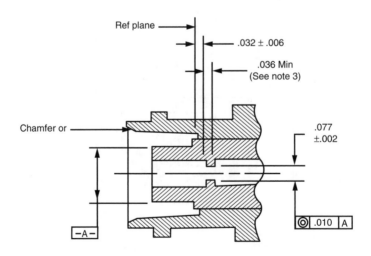

FIGURE 8.8   BNC female 50 Ω outer conductor.

---

* TNC Compatibility Chart, 5E-057, Maury Microwave, Dec. 1998.

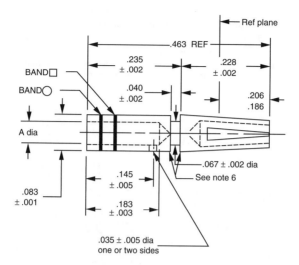

FIGURE 8.9   BNC female 50 Ω center conductor.

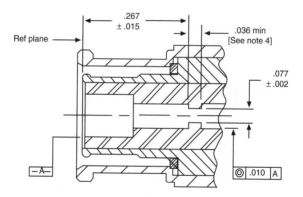

FIGURE 8.10   BNC male 50 Ω outer conductor.

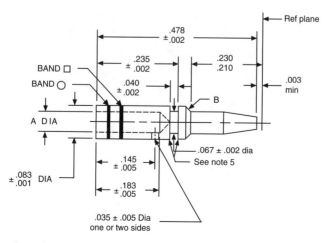

FIGURE 8.11   BNC male 50 Ω center conductor.

Table 8.1   TNC Connector Compatibility

A.   Mating these TNC connectors together will result in non-contacting outer conductors.
B.   These TNC connectors should not be mixed except in cases where one connector has been chosen as a test connector
      and it is characterized on a network analyzer for error corrected measurements.
C.   The male contact pin interface of this TNC connector specification has not been fully defined.

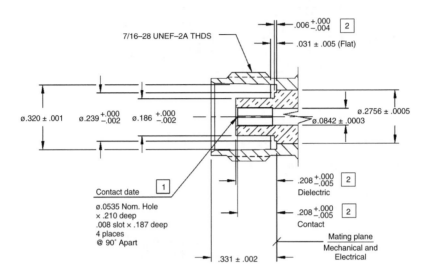

FIGURE 8.12   Typical TNC female interface.

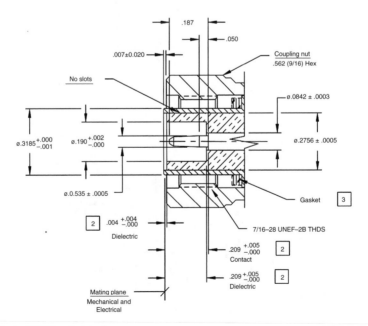

FIGURE 8.13   Typical TNC male interface.

8.8 SMA

SubMiniature Type-A or SMA is one of the most popular microwave connectors. Because of their smaller size, SMA connectors can be used at higher frequencies than Type-N connectors. The most common application is for semirigid cable and components that are connected only a few times because of the fragility of the outer connector wall.

SMA connectors are not precision devices. They are not designed for repeated connections and disconnections. SMA connectors wear out quickly and are often found to be out of specification even before they have been used. They are used most often as "one-time-only" connectors in internal component assemblies and in similar applications in which few connections or disconnections will be made.

The SMA is a dielectrically loaded, sexed connector. The male contact pin slides into the female contact fingers and electrical contact is made by the inside surfaces of the tip of the female contact fingers on the sides of the male contact pin. The mechanical specifications for the SMA connector gives a maximum and a minimum recession of the shoulder of the male contact pin and a maximum and minimum recession of the tip of the female contact fingers behind the outer conductor mating plane. An SMA connector will mate with 3.5 mm and 2.92 mm connectors without damage.

8.8.1 SMA Specifications

Frequency Range:	DC to 18 GHz
Impedance:	50 Ω
Mating Torque:	5 in-lbs.
Female Socket:	0.000 +0.005 −0.000 inches
Female Dielectric:	0.000 ±0.002 inches
Male Pin:	0.000 +0.005 −0.000 inches
Male Dielectric:	0.000 ±0.002 inches

8.8.2 SMA Dimensions

See Figures 8.14 through 8.17 for details.

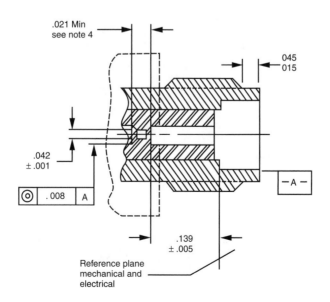

FIGURE 8.14 SMA female outer conductor.

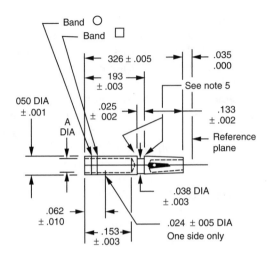

FIGURE 8.15   SMA female center conductor.

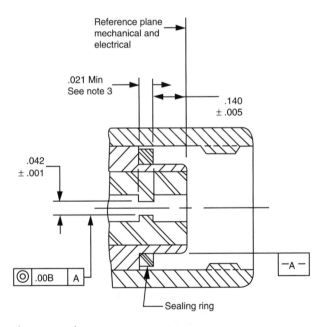

FIGURE 8.16   SMA male outer conductor.

8.9 7-16

The 7-16 connector is a air dielectric loaded, sexed connector. The male contact pin slides into the female contact fingers and electrical contact is made by the inside surfaces of the tip of the female contact fingers on the sides of the male contact pin. The 7-16 is intended as a replacement connector for Type-N connectors in high power applications.

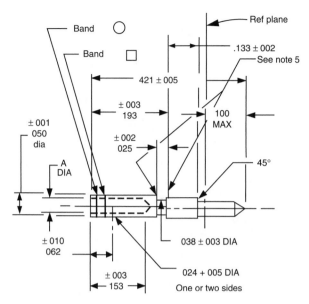

FIGURE 8.17 SMA male center conductor.

8.9.1 7-16 Specifications

Frequency Range: DC to 7 GHz
Impedance: 50 Ω
Mating Torque: 20 in-lbs.
Female Socket: 0.0697 +0.012 −0.000 inches
Male Pin: 0.0697 +0.000 −0.010 inches

8.9.2 7-16 Dimensions

See Figures 8.18 and 8.19 for details.

8.10 7 mm

Precision 7 mm connectors, among them APC-7® (Amphenol Precision Connector-7 mm) connectors, are used in the DC to 18 GHz frequency band and offer the lowest SWR and the most repeatable connections of any 7 mm connector type. Development of these connectors was begun by Hewlett Packard (Agilent Technologies) in the mid-1960s and improved upon Amphenol Corporation.

Precision 7 mm connectors are air dielectric devices. Only a plastic support bead inside the connector body supports the center conductor. The conductors are generally made of beryllium copper alloy plated with gold.

Precision 7 mm connectors are durable and are suitable for many connections and disconnections. Therefore they are widely used in test and measurement applications requiring a high degree of accuracy and repeatability.

Precision 7 mm connectors are generally designed for use as sexless connectors, able to mate with all other precision 7 mm connectors. There is no male or female and contact between the center is made by replaceable inserts called collets designed to make spring-loaded butt contact when the connector is torqued. Small mechanical differences do sometimes exist between precision 7 mm connectors made by different manufacturers and occasionally these differences can cause difficulty in making connections. Always inspect all connectors mechanically, using a precision connector gage, to insure the connectors meet their critical interface specifications.

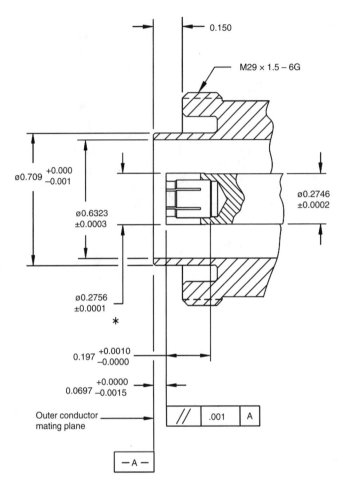

FIGURE 8.18   7-16 female interface.

In precision 7 mm connectors, contact between the center conductors is made by spring-loaded contacts called collets. These protrude slightly in front of the outer conductor mating plane when the connectors are disconnected. When the connection is tightened, the collets are compressed into the same plane as the outer conductors. For this reason two mechanical specifications are generally given for precision 7 mm connectors:

1. The maximum and minimum allowable recession of the center conductor behind the outer conductor mating plane with the center conductor collet removed. The critical mechanical specification is the recession or setback of the center conductor relative to the outer conductor mating plane with the center conductor collet removed.
2. The maximum and minimum allowable protrusion of the center conductor collet in front of the outer conductor mating plane with collet in place. No protrusion of the center conductor in front of the outer conductor mating plane is ever allowable and sometimes a minimum recession is required.

The center conductor collet should also spring back immediately when pressed with a blunt plastic rod or with the rounded plastic handle of the collet removing tool.

Nominal specifications for precision 7 mm connectors exist, but the allowable tolerances differ from manufacturer to manufacturer and from connector to connector. Before gaging any precision 7 mm connector, consult the manufacturer's mechanical specifications provided with the connector.

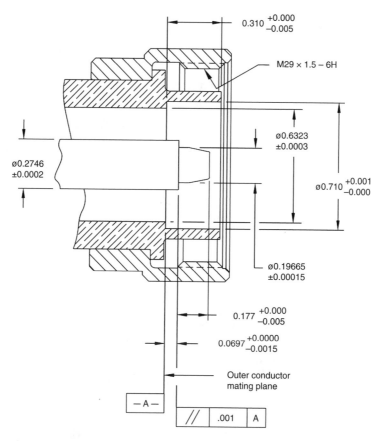

FIGURE 8.19   7-16 male interface.

8.10.1   7 mm Specifications

Frequency Range: DC to 18 GHz
Impedance: 50 Ω
Mating Torque: 12 in-lbs.
Contact: 0.0000   +0.0000   −0.0015 inches

8.10.2   7 mm Dimensions

See Figures 8.20 and 8.21 for details.

8.11   3.5 mm

Precision 3.5 mm connectors, also known as APC-3.5® (Amphenol Precision Connector-3.5 mm) connectors, were developed during the early 1970s jointly by Hewlett Packard (Agilent Technologies) and Amphenol Corporation. The goal was to produce a durable high-frequency connector that could mate with SMA connectors, exhibit low SWR and insertion loss, and be mode free up to 34 GHz.

Unlike SMA connectors, precision 3.5 mm connectors are air dielectric devices. That is, air is the insulating dielectric between the center and outer conductors. A plastic support bead inside the connector body supports the center conductor. APC- 3.5 mm connectors are precision devices. Therefore, they are more expensive than SMA connectors, but are durable enough to permit repeated connections and disconnections. A 3.5 mm connector will mate with either a SMA or 2.92 mm connectors.

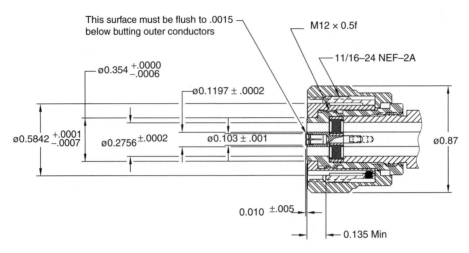

FIGURE 8.20   7 mm interface (sleeve retracted).

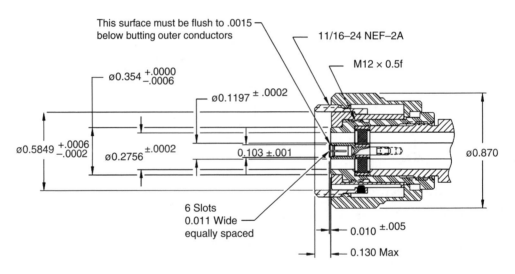

FIGURE 8.21   7 mm interface (sleeve extended).

8.11.1   3.5 mm Specifications

Frequency Range: DC to 34 GHz
Impedance: 50 Ω
Mating Torque: 8 in.-lbs.
Female Socket: 0.000   +0.003   −0.000 inches
Male Pin: 0.000   +0.003   −0.000 inches

8.11.2   3.5 mm Dimensions

See Figures 8.22 and 8.23 for details.

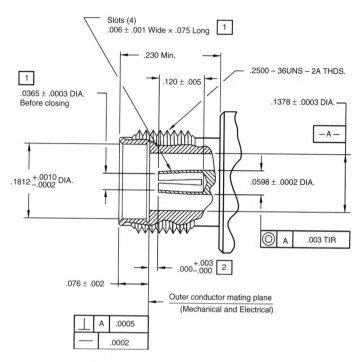

FIGURE 8.22   3.5 mm female interface.

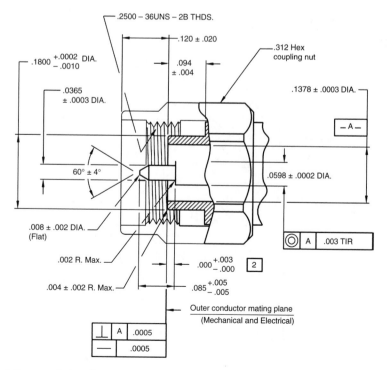

FIGURE 8.23   3.5 mm male interface.

8.12   2.92 mm

Precision 2.92 mm connectors are air dielectric loaded, sexed connectors. The male contact pin slides into the female contact fingers and electrical contact is made by the inside surfaces of the tip of the female contact fingers on the sides of the male contact pin. Air is the insulating dielectric between the center and outer conductors. A plastic support bead inside the connector body supports the center conductor. A 2.92 mm connector will mate with SMA, 3.5 mm, or 2.4 mm connectors.

8.12.1   2.92 mm Specifications

Frequency Range:	DC to 40 GHz
Impedance:	50 Ω
Mating Torque:	8 in-lbs.
Female Socket:	0.000 +0.003 −0.000 inches
Male Pin:	0.000 +0.003 −0.000 inches

8.12.2   2.92 mm Dimensions

See Figures 8.24 and 8.25 for details.

8.13   2.4 mm

Precision 2.4 mm connectors are air dielectric loaded, sexed connectors. The male contact pin slides into the female contact fingers and electrical contact is made by the inside surfaces of the tip of the female contact fingers on the sides of the male contact pin. Air is the insulating dielectric between the center and outer conductors. A plastic support bead inside the connector body supports the center conductor. A 2.4 mm connector will mate with SMA, 3.5 mm, or 2.92 mm connectors.

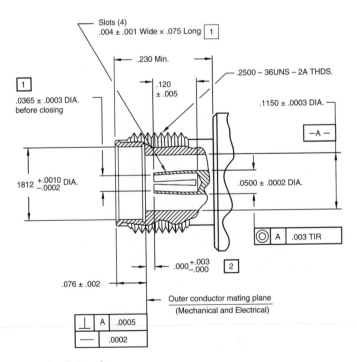

FIGURE 8.24   2.92 mm female interface.

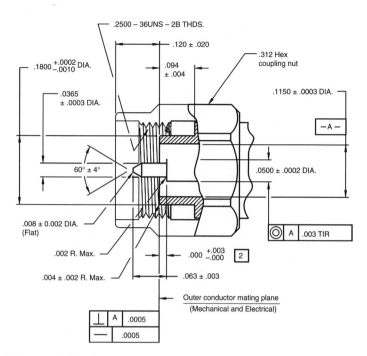

FIGURE 8.25 2.92 mm male interface.

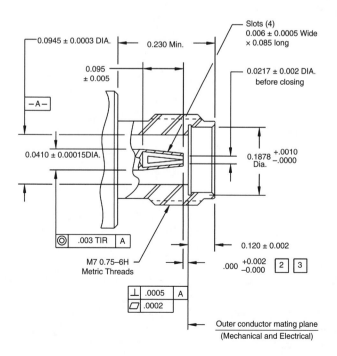

FIGURE 8.26 2.4 mm female interface.

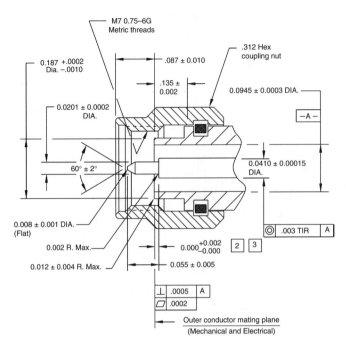

FIGURE 8.27   2.4 mm male interface.

8.13.1   2.4 mm Specifications

Frequency Range:	DC to 50 GHz
Impedance:	50 Ω
Mating Torque:	8 in-lbs.
Female Socket:	0.000 +0.002 –0.000 inches
Male Pin:	0.000 +0.002 –0.000 inches

8.13.2   2.4 mm Dimensions

See Figures 8.26 and 8.27 for details.

References

1. Bryant, John H., Coaxial Transmission Lines, Related Two-Conductor Transmission Lines, Connectors, and Components: A U.S. Historical Perspective, *IEEE MTT Transactions*, 9, 970, 1990.
2. Maury Jr., Mario A., Microwave Coaxial Connector Technology: A Continuing Evolution, *Microwave Journal*, State of the Art Reference, 39, 1990.
3. Rizzi, Peter A., *Microwave Engineering Passive Circuits*, Prentice-Hall, Englewood Cliffs, NJ, 1988, chap. 5.
4. Microwave Test Accessories Catalog, Hewlett Packard, 1991, 80.
5. Microwave Connector Care, Hewlett Packard, April 1986, Part One.
6. TNC Compatibility Chart, 5E-057, Maury Microwave, Dec. 1998.

9

Antenna Technology

James B. West
Rockwell Collins—Advanced
Technology Center

9.1 Introduction

This chapter is a brief overview of contemporary antenna types used in cellular, communication links, satellite communication, radar, and other microwave and millimeter wave systems. In this presentation, microwave is presumed to cover the frequency spectrum from 800 MHz to 94 GHz. A discussion of Maxwell's equations of electromagnetic theory and the wave equation, which together mathematically describe nonionizing electrodynamic wave propagation and radiation, are described in detail in Chapters 28 and 29 of the companion volume, *RF and Microwave Applications and Systems*, in this handbook series. The reader is referred to the literature for detailed electromagnetic analysis of the multitude of antennas used in contemporary microwave and millimeter wave systems.

9.1.1 Fundamental Antenna Parameter Definitions [1–5]

An antenna is fundamentally a device that translates guided wave energy into radiating energy. Antenna physics, unlike RF/microwave transmission lines, simultaneously exhibits spatial and frequency (time) dependencies. Schantz provides an excellent phenomenological description of an antenna device, where he characterizes antenna functionality from multiple perspectives [1].

Electromagnetic Radiation: The emission of energy from a device in the form of electromagnetic waves.

Radiation Pattern: A graphical or mathematical description of the radiation properties of an antenna as a function of space coordinates. The standard (r, θ, ϕ) spherical coordinate system is typically used, as shown Figure 9.1. Radiation properties include radiation intensity (Watt/solid angle), radiation density (Watt/m^2), gain, directivity, radiated phase, and polarization parameters, all of which are discussed

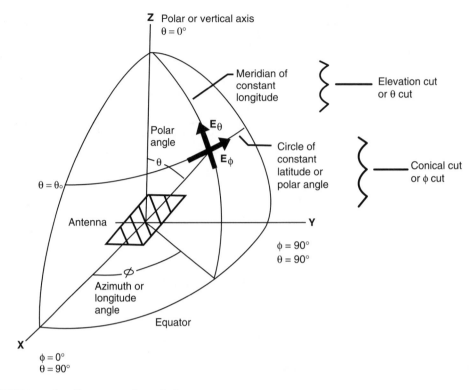

FIGURE 9.1   Coordinate system for radiation-pattern measurement.

in subsequent paragraphs. Radiated power, rather than electric field, is commonly used at microwave frequencies since radiated power is readily measured with contemporary antenna metrology equipment. As an example, a *power radiation pattern* is an expression of the variation of received or transmitted power density as quantified at a constant radius about the antenna. Power patterns are often graphically depicted as normalized to their main beam peak.

Radiation Pattern cuts are two-dimensional cross sections of the three-dimensional power radiation pattern. A cut in the plane of the theta unit vector is the *theta cut*, while a pattern cut in the phi unit vector direction is the *phi cut*, as shown in Figure 9.2. The *principal plane cuts* are two dimensional, orthogonal antenna cuts taken in the E field and H field planes of the power radiation patterns, as illustrated in Figure 9.3.

Isotropic Radiator: A hypothetical lossless antenna having equal radiation in all directions. The three-dimensional radiation pattern of an isotropic radiator is a sphere.

Directional Antenna: An antenna that has a peak sensitivity that is a function of direction, as illustrated in Figure 9.2 and 9.3.

Omnidirectional Antenna: An antenna that has essentially a nondirectional radiation pattern in one plane and a directional pattern in any orthogonal plane.

Radiation Pattern Lobes: A radiation lobe is a portion of the radiation pattern that is bound by regions of lesser radiation intensity. Typical directional antennas have one major *main lobe* that contains the antenna's radiation peak, and several *minor lobes*, or *side lobes*, which are any lobes other than the major lobe. A *back lobe* is a radiation lobe that is located approximately 180° from the main lobe. The *side lobe level* is a measure of the power intensity of a minor lobe, usually referenced in dB to the main lobe peak of the power radiation pattern. The *front-to-back ratio* is measure of the power of the main beam to that of the back lobe. A *pattern null* is an angular position in the radiation pattern where the power radiation pattern is at a minimum. The parameters are depicted in Figure 9.2.

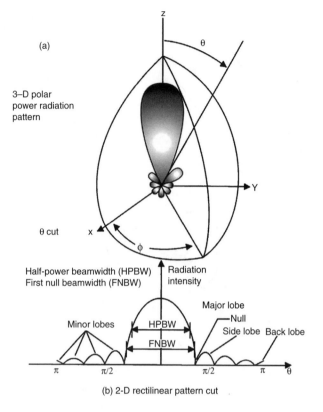

FIGURE 9.2 Antenna parameters.

Field Regions: The three-dimensional space around an antenna is divided into three regions. The region closest to the antenna is called the *reactive near field*, the intermediate region is the *radiating near field (Fresnel)* region, and the farthest region is the *far field (Fraunhofer)* region. The reactive near field region is where reactive fields dominate. The Fresnel region is where radiation fields dominate, but the angular radiation pattern variation about the antenna is a function of distance away from the antenna. The far field is the region where the angular variation of the radiation pattern is essentially independent of distance. Various criteria have been used to quantify the boundaries of these radiation regions. The commonly accepted boundaries between the field regions are

$$\text{Reactive near field/Fresnel boundary: } R_{\text{RNF}} = 0.62 \cdot \sqrt{\frac{D^3}{\lambda}} \tag{9.1}$$

$$\text{Fresnel/far field boundary: } R = \frac{2D^2}{\lambda}, \tag{9.2}$$

where

D = the largest dimension of the antenna
λ = the wavelength in free space

The far field formulation assumes a 22.5° phase error across a circular aperture designed for a −25 dB side lobe level, which creates a ± 1.0 dB error at the 25 dB side lobe level [6]. More stringent far field criterion is applicable to lower side lobe levels [7]. Antennas operate in the far field for typical communication systems and radar applications.

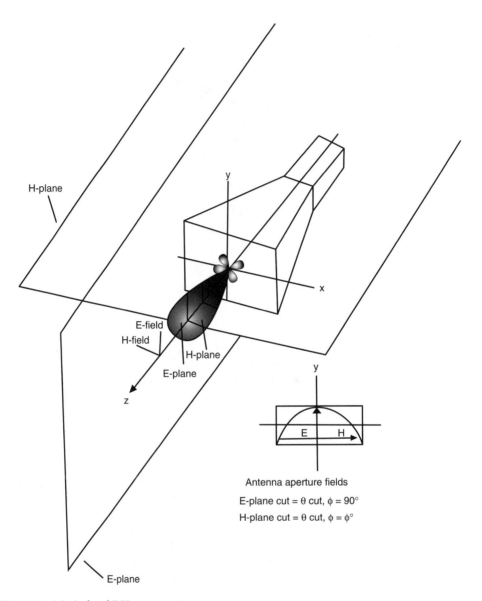

FIGURE 9.3 Principal and E/H pattern cuts.

Directivity: The directivity of an antenna is the ratio of the radiation intensity in a given direction to that of the radiation intensity of the antenna averaged over all directions. This ratio is usually expressed in decibels. The term directivity most often implicitly refers to the *maximum directivity*.

Directivity is a measure of an antenna's ability to focus power density (transmitting) or to preferentially receive an incoming wave's power density as a function of spatial coordinates. A passive, lossless antenna does not amplify its input signal due to conservation of energy considerations, but rather, it redistributes input energy as a function of spatial coordinates. Consider an isotropic radiator, and a highly directive antenna such as a parabolic reflector. Assume that each antenna is lossless, impedance and polarization matched, and has the same power at their input terminal. The matched isotropic radiator couples the same total input power into free space as that of the parabolic reflector, but the power density as a function of spatial coordinates is drastically different. The parabola focuses the majority of the total input power into a very narrow spatial sector, and is commensurately more "sensitive" in this region. In contrast,

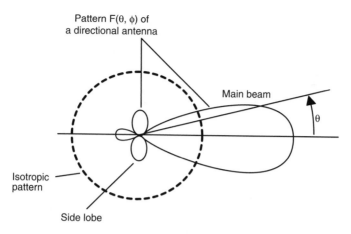

FIGURE 9.4 Comparison of 2-D isotropic and directional pattern cuts.

the isotropic radiator uniformly distributes the input power over all directions and lacks the directional sensitivity. These concepts are illustrated in Figure 9.4.

Directivity does not account for resistive loss mechanisms, polarization mismatch, or impedance mismatch factors in an antenna. As previously noted, a high value of directivity means that a high percentage of antenna input power is focused in a small angular region. By way of example, parabolic reflector antennas used on radio astronomy can have directivities exceeding 80 dB. An isotropic radiator, in contrast, has a directivity of 0 dB (1.0 numeric) since it radiates equally in all directions. A half-wave resonant dipole radiating in free space has a theoretical directivity of 2.14 dB.

Half-Power Beam Width is the angle between two directions in the maximum lobe of the power radiation pattern where the directivity is one half the peak directivity, or 3 dB down, as illustrated in Figure 9.2. The half-power beam width is a measure of the spatial selectivity of an antenna. Low beam width antennas are very sensitive over a small angular region. Half-power beam width and directivity are inversely proportional for antennas in which the major portion of the radiation resides within the main beam.

Power Gain: The gain of an antenna is the ratio of the radiation intensity in a given direction to the total input power accepted by the antenna input port. This ratio is usually expressed in decibels above an isotropic radiator (**dBi** for linear polarization and **dBic** for circular polarization).

Gain is related to directivity through the *radiation efficiency* parameter. Radiation efficiency is a measure of the insertion dissipative power losses internal to the antenna:

$$G = \eta_{cd} \times D, \tag{9.3}$$

where

G = power gain
η_{cd} = antenna efficiency due to conductor and dielectric dissipative losses
D = the antenna's directivity

Power gain usually implicitly refers to the peak power gain. The gain parameter includes the effects of dissipative losses, but does not include the effects of polarization and impedance mismatches.

Antenna Polarization: The polarization of an antenna refers to the electric field polarization properties of the propagating wave received by the antenna. Wave polarization is a description of the contour that the radiating electric field vector traces as the wave propagates through space. The most general wave polarization is *elliptical. Circular and linear polarizations* are special cases of elliptical polarization. Examples of elliptical, circular, and linear polarized propagating electric field vectors are illustrated in Figure 9.5. *Vertical and horizontal polarizations* are sometimes used as well, and loosely refer to the

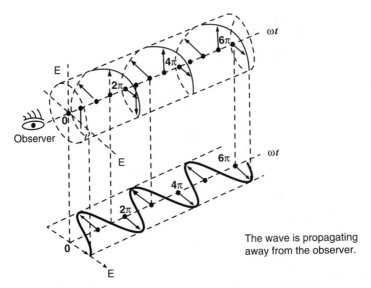

The wave is propagating away from the observer.

(a) Elliptically polarized radiating wave

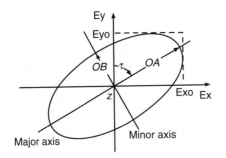

Rotation of a plane electromagnetic wave and its polarization ellipse at $z = 0$ as a function of time.

(b) Polarization ellipse

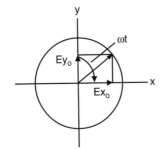

Exo = Eyo, 90° phase shift between Exo and Eyo

(c) Circular polarization

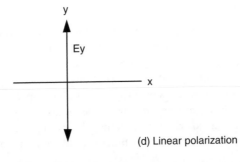

(d) Linear polarization

FIGURE 9.5   Antenna polarization.

orientation of the linear polarized electric field vector at the plane of the antenna. The elliptically polarized wave is described by its *polarization ellipse*, which is the planar projection of the contour that its electric field vector sweeps out as the wave propagates through space. The orientation of the polarization ellipse is the *tilt angle*, and the ratio of the *major diameter* to the *minor diameter* of the ellipse is the *axial ratio*. The direction of rotation of the radiated field is expressed either as left handed or right handed as the wave propagates from a reference point. Left-hand circular polarization (LHCP) and right-hand circular polarization (RHCP) are commonly used in contemporary microwave systems.

In most applications, it is desirable to have the receive antenna's polarization properties match those of the transmitting antennas, and vice versa. If an antenna is polarized differently than the wave it is attempting to receive, then power received by the antenna will be less than the maximum, and the effect is quantified by a *polarization mismatch loss*, which is defined by

$$\text{PML} = |\boldsymbol{\rho_r} \cdot \boldsymbol{\rho_t}|^2 = |\cos \Psi_p|^2, \tag{9.4}$$

where

$\boldsymbol{\rho_r} = $ the receive polarization unit vector
$\boldsymbol{\rho_t} = $ the transmit polarization unit vector
$\Psi_p = $ the angle between the two unit vectors.

This formulation is applicable to elliptical, circular, and linear polarizations.

Some commonly encountered polarization mismatch losses are depicted in Figure 9.6. It is apparent from the figure that a pure RHCP wave is completely isolated from a LHCP wave. This property is exploited in polarization diversity systems, for example, satellite communications systems, and cellular radio systems.

The *axial ratio*, usually expressed in dB, is a measure of the ellipticity of a polarized wave. Zero decibel axial ratio describes pure circular polarization. Six decibel axial ratio is typically used as a rule of thumb

Wave polarization						
	Vertical	Horizontal	Right hand circular	Left hand circular	45° Right linear	45° Left linear
Vertical	0 dB	∝	3 dB	3 dB	3 dB	3 dB
Horizontal	∝	0 dB	3 dB	3 dB	3 dB	3 dB
Right hand circular	3 dB	3 dB	0 dB	∝	3 dB	3 dB
Left hand circular	3 dB	3 dB	∝	0 dB	3 dB	3 dB
45° Right linear	3 dB	3 dB	3 dB	3 dB	0 dB	∝
45° Left linear	3 dB	3 dB	3 dB	3 dB	∝	0 dB

(Left axis label: Wave polarization. Right axis label: Polarization loss.)

NOTE: Direction of propagation is into the page.

FIGURE 9.6 Polarization loss between receive and transmit antennas.

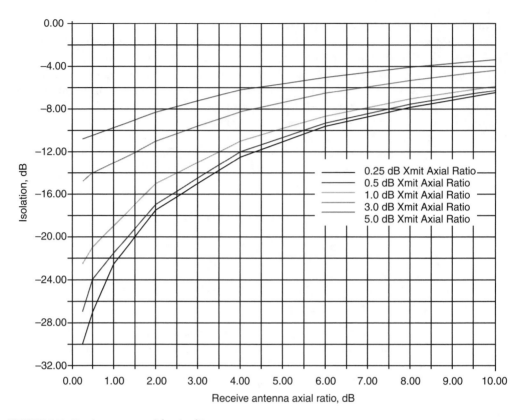

FIGURE 9.7   Receive antenna axial ratio, dB.

for the maximum axial ratio in which an elliptical wave is considered circular. Axial ratio does not describe the sense of the propagating wave. The axial ratio of an antenna is a three-dimensional function of spatial coordinates.

An arbitrary elliptically polarized wave can be mathematically decomposed in to a LHCP component and a RHCP component. Consider two circular antennas of opposite polarization sense, one in the transmit mode and the other in the receiving mode. As each antenna deviates from pure circular polarization toward elliptical polarization, the orthogonal (undesired) polarization components of each antenna becomes increasingly influential, which in turn reduces the isolation between the antennas. Similarly, a deviation from pure circular polarizations between receiving and transmitting antennas of the same polarization sense manifests itself as an increased loss. Figure 9.7 quantifies the minimum polarization isolation given the axial ratio and polarization sense of each antenna [8].

Radiation Efficiency accounts for losses at the input terminal of the antenna and the losses within the structure. Efficiency due to conductor and dielectric losses is defined as

$$\eta_{cd} = \eta_c \times \eta_d, \tag{9.5}$$

where

$\eta_c =$ the efficiency due to conduction loss
$\eta_d =$ the efficiency due to dielectric loss

Conduction loss is a function of the conductivity of the metal used. The dielectric loss is typically specified in terms of the loss tangent parameter, and can be anisotropic. Tabulations of both parameters are available in the literature [9]. Conduction and dielectric loss efficiencies can also be determined experimentally.

Input Impedance: The input impedance of an antenna is the impedance at the input terminal reference plane. *Impedance* is the ratio of the voltage to the current at the terminal plane, which is related to ratio of the electric and magnetic fields at the terminal plane. An antenna can be thought of as a *mode translator* of electromagnetic waves. Guided waves propagating through the input transmission line from the input generator to the antenna input terminal are transformed into unbounded (radiation) electromagnetic waves. Thus, the input port of an antenna radiating into free space can be thought of as the impedance with real and imaginary components that are a function of frequency.

The real or resistive component of the input impedance is required for the transfer of real power from the input generator into the radiated electrical and magnetic fields to initiate an average power flow out of the antenna. This resistive component is called the *radiation resistance*. The reactive near fields in the immediate vicinity of the antenna structure, that is, the reactive near field region, are manifested as a reactance at the antenna input terminal. A *matched impedance condition* over the desired frequency range is required to optimally transfer input generator power into radiated waves. Both the real and imaginary components of the input impedance can vary with frequency, further complicating the impedance matching problem. Standard transmission line and broad band circuit matching techniques can be applied to ensure the optimal transfer of energy between the input transmission line and the antenna's input terminal [10,11].

Impedance Mismatch Loss is a measure of the amount of power from the input generator that passes through the antenna's input terminals. The input reflection coefficient is a measure of the amount of power that is reflected from the antenna input back into the generator. Mismatch loss is not a dissipative, or resistive loss, but rather a loss due to reflections of guided electromagnetic waves back to the input generator.

The impedance mismatch loss is defined as

$$\text{MML} = (1 - |\Gamma|^2), \tag{9.6}$$

where

Γ = the voltage reflection coefficient at the antenna input terminal.

The voltage reflection coefficient is usually specified and measured as either a return loss or a voltage standing wave ratio (VSWR):

$$\text{Return loss} = 10\log(|\Gamma|^2) \tag{9.7}$$

and

$$\text{VSWR} = \frac{1 + |\Gamma|}{1 - |\Gamma|}. \tag{9.8}$$

Overall Antenna Efficiency: The overall efficiency of an antenna is minimally related to the previously mentioned parameters of conductor/dielectric losses, polarization loss, and impedance mismatch factors. In addition, there are other efficiency parameters that are specific to certain classes of antennas. These additional terms include

- Illumination efficiencies (reflector and lens antennas)
- Spill over efficiencies (reflector and lens antennas)
- Blockage efficiencies (reflector antennas)
- Random phase and amplitude errors (arrays, lens, apertures, and reflector antennas)
- Aperture taper efficiencies (arrays, lens, apertures, and reflector antennas)
- Others

Effective Area is the ratio of the available power at the antenna input terminals due to a polarization matched plane wave incident on the antenna from a given direction to the power density of the same plane wave incident on the antenna. Effective area is a measure of an antenna's power capturing properties under plane wave illumination. The effective area of an aperture antenna is related to its physical area by its *aperture efficiency* as shown below:

$$\eta_{ap} = \frac{A_{em}}{A_p} = \frac{\text{Maximum effective area}}{\text{Physical area}}. \tag{9.9}$$

The effective area is related to its directivity by

$$A_{em} = \frac{\lambda^2}{4\pi} \times D_0 \eta_{cd} \left(1 - |\Gamma^2|\right) \times |\rho_r \cdot \rho_t|^2. \tag{9.10}$$

The effective area parameter is useful in first order array and aperture antenna directivity calculations.

Bandwidth is a frequency range in which a particular electrical parameter of an antenna conforms to a specified performance. Examples of parameters include gain, beam width, side lobe level, efficiency, axial ratio, input impedance, and beam direction or squint. Since all of the above mentioned parameters have different frequency dependencies, it is impossible to uniquely specify a single bandwidth parameter.

Electrically Small Antennas: An electrically small antenna features dimensions that are small relative to their operating wavelengths. Conventional antenna structures typically operate most efficiency when their largest dimension is an appreciable fraction of a wavelength (i.e., $\lambda/2$ or $\lambda/4$), which is unacceptably large for many applications, particularly those in the HF through lower UHF regions. It is therefore highly desirable to determine that maximum theoretical bandwidth of an antenna that encloses a given volume in space. The fundamental limits of electrically small antennas have been examined by several researchers [12–14]. Their findings can be verbally summarized as follows:

An antenna that can be enclosed within a sphere of radius r has maximum bandwidth when its geometric structure optimally fills the volume of the sphere.

This implies that three-dimensional (3D) radiating elements will exhibit larger bandwidths than thin wire radiators, which had been verified within the research community through several examples [1].

The bandwidth of the antenna can be expressed by its quality factor, or "Q," which is the ratio of reactive power to real power. Chu has shown that the Q of an electrically small antenna can be expressed as [13]:

$$Q = \frac{1 + 2(kr)^2}{(kr)^3[1 + (kr)^2]} \cong \frac{1}{(kr)^3} \text{ for } kr \ll 1 \tag{9.11}$$

where

$k = 2\pi/\lambda$, the wave propagation constant
$\lambda =$ the free space wavelength
$r =$ the radius of the enclosing sphere

For antenna with $Q > 2$, the fractional bandwidth is related to the quality factor by the following expression:

$$\text{Fractional band width} = \frac{\Delta f}{fo} = \frac{1}{Q} \tag{9.12}$$

Equation 9.12 shows that lower values of Q translate to larger bandwidths.

Equation 9.13 is known as the fundamental limit on the electrical size of the antenna. Figure 9.8 illustrates this limit for 100% radiation efficiency (no dissipative loss) as a function of kr. Note that Q proportionately lowers with decreasing radiation efficiency. The classic resonant $\lambda/2$ resonant dipole and

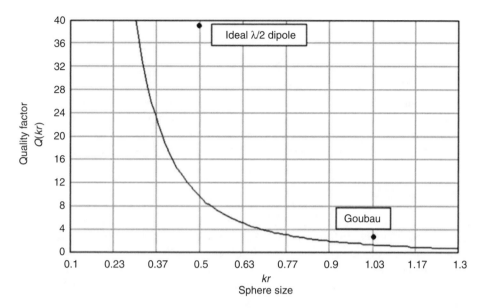

FIGURE 9.8 Fundamental Q limit for 100% radiation efficiency.

the Goubau electrically small multielement monopole are compared to the theoretical limit [15]. The Goubau antenna has traditionally been the standard for broad band electrically small antennas. Excellent discussions of bandwidth properties for several new antenna elements are available in the literature [16,17].

Reciprocity is a theorem of electromagnetics that requires receive and transmit properties to be identical for linear and reciprocal antenna structures. This concept is very useful in antenna measurements, since it is normally more convenient to measure the properties of an antenna under test (AUT) in the receive mode. Also, it can be easier to understand some antenna concepts through a receive mode formulation, while others are easier through a transmit mode interpretation.

Friis Transmission Formula describes the coupling of electromagnetic energy between two antennas under far field radiation conditions. The power received by the receiving antenna is related to the power transmitted through the transmitting antenna by the following expression:

$$\frac{P_r}{P_t} \eta_{cdr} \cdot \eta_{cdt} \cdot \left(1 - |\Gamma_r|^2\right) \cdot \left(1 - |\Gamma_r|^2\right) \cdot \left(\frac{\lambda}{4\pi}\right)^2 D_{or} \cdot D_{ot} \cdot |\bar{\rho}_r \cdot \bar{\rho}_t|^2, \tag{9.13}$$

where θ and ϕ are the usual spherical coordinate angles, and the r and t subscripts refer to the receive and transmit modes, respectively.

Antenna Noise Temperature: All objects with a physical temperature greater than 0 K radiate energy. Antenna noise temperature is an important parameter in radio astronomy and satellite communications systems since the antenna, ground, and sky background noise contributes to the total system noise. This system noise ultimately sets a limit to the system signal to noise ratio. Satellite system link performance is typically specified in terms of the ratio of system gain to the system noise temperature, which is called the **G/T** figure of merit. The noise temperature of an antenna is defined as follows:

$$T_A = \frac{\int_0^{2\pi} \int_0^{\pi} T_B(\theta, \phi) G(\theta, \phi) \sin\theta \, d\theta \, d\phi}{\int_0^{2\pi} \int_0^{\pi} G(\theta, \phi) \sin\theta \, d\theta \, d\phi} \tag{9.14}$$

where

$$T_B = \varepsilon(\theta, \phi) T_m = \left(1 - |\Gamma|^2\right) T_m = \text{the equivalent brightness temperature,} \tag{9.15}$$

where

$\varepsilon(\theta, \phi) =$ emissivity
$T_m =$ molecular (physical) temperature (K)
$G(\theta, \phi) =$ power gain of the antenna

The brightness temperature is a function of frequency, polarization of the emitted radiation, and the atomic structure of the object. In the microwave frequency range, the ground has an equivalent temperature of about 300 K, and the sky temperature is about 5 K when observing toward the zenith (straight up, perpendicular to the ground), and between (100–150 K) when looking toward the horizon.

9.2 Radiating Element Types

Several of the most prominent radiating elements used in contemporary microwave and millimeter wave systems are shown in Table 9.1, which is an extension of the work of Salati [1]. The intent is to catalog the salient features of these radiating elements in terms of broad classifications and to summarize the most important electrical parameters useful for system design. Detailed analysis and design methodologies are beyond the scope of this presentation and the reader is referred to the extensive literature for further information.

9.2.1 Wire Antennas [2–4]

Wire antennas are arguably the most common antenna type and they are used extensively in contemporary microwave systems. The classic *dipole, monopole,* and *whip* antennas are used extensively throughout the microwave frequency bands. Broadband variations of the classic wire antenna include the *electrically thick monopole and dipole, biconical dipole, bow tie, coaxial dipole and monopole, the folded dipole, discone, and conical skirt monopole.* Various techniques to reduce the size of this type of antenna for a given resonant frequency include foreshortening the antenna's electrical length and compensating with lumped-impedance loading, top hat capacitive loading, dielectric material loading, helical winding of monopole element to retain electrical length, but with foreshortened effective height. An excellent discussion on electrically small antennas is found in Fujimoto et al. [5].

9.2.2 Loop Antennas [2,6]

Loop antennas are used extensively in the HF through UHF band, and have application at L band and above as field probes. The circular loop is the most commonly used, but other geometric contours, such as rectangular, are used as well. Loops are typically classified as *electrically small*, where the overall wire length (circumference multiplied by the number of turns) is less than 0.1 wavelength and electrically large where the loop circumference is approximately 1 wavelength. Electrically small antennas suffer from low radiation efficiency, but are used in portable radio receivers and pagers and as field probes for electromagnetic field strength measurements. The ubiquitous AM portable radio antenna is a ferrite material-loaded multiturn loop antenna. *Electrically large* loops are commonly used as array antenna radiating elements. The radiation pattern in this case is in end fire toward the axis of the loop.

9.2.3 Slot Antennas [7]

Slot antennas are used extensively in aircraft, missile, and other applications where physical low profile and ruggedness are required. Slot antennas are usually half-wave resonant and are fed by introducing an excitation electric field across their gap. *Cavity-backed slots* have application in the UHF bands. The annular slot antenna is a very low profile structure that has a monopole wire-like radiation pattern. Circular polarization is possible with crossed slots fed in phase quadrature. Stripline-fed slots are frequently used in phased array applications for the microwave bands [8].

TABLE 9.1 Common Antenna Radiating Elements

Antenna Class	Antenna Type	Configuration	Parameters	Typical Frequency Range (MHz)	Gain dB above isotropic	Beam Width at Center Frequency (Degrees)	3 dB Gain Reduction Bandwidth	Input Resistance at Center Frequency	2.0:1 VSWR Bandwidth %	Polarization	Side Lobe Levels
Ideal Isotropic	Isotropic			N/A	0	N/A	N/A	N/A	N/A		N/A
Wire Antennas	Small Dipole		$L < \lambda/2$		1.74	90			N/A	LP,CP	N/A
	Thin $\lambda/2$ Dipole		$L = \lambda/2$, $L/D = 276$	to X Band	2.14	78	34%	60 Ω		LP	N/A
	Thick $\lambda/2$ Dipole		$L = \lambda/2$, $L/D = 51$	to X Band	2.14	78	55%	49 Ω		LP	N/A

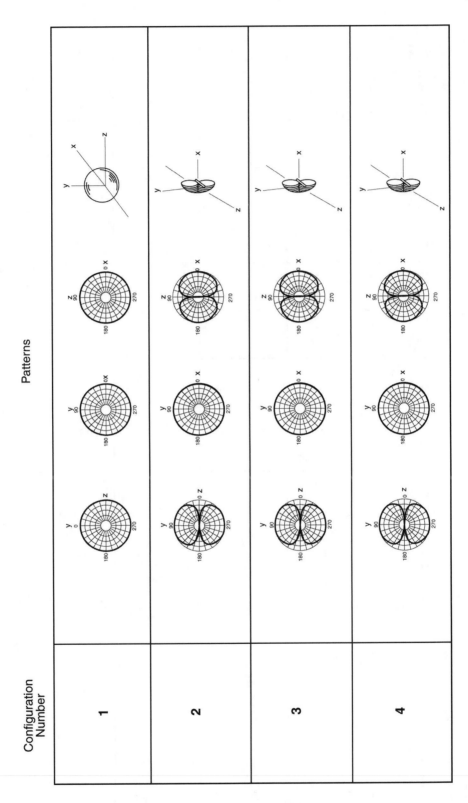

Antenna Class	Antenna Type	Configuration	Parameters	Typical Frequency Range (MHz)	Gain dB above Isotropic	Beam Width at Center Frequency (Degrees)	3 dB Gain Reduction Bandwidth	Input Resistance at Center Frequency	2.0:1 VSWR Bandwidth %	Polarization	Side Lobe Levels
Wire Antennas (cont.)	Cylindrical Dipole	5	$L = \lambda/2$, L/D = 10	to X Band	2.14	78	100%	37 Ω		LP	N/A
	Biconical Dipole	6	$L = \lambda/2$ phi = 40 deg.	to X Band	2.14	78	100%	72 Ω		LP	N/A
	Folded Dipole	7	$L = \lambda/2$, L/D = 25.5	to S Band	2.14	78	45%	300 Ω		LP	N/A
	Folded Dipole Above Perfect Ground	8	$L = \lambda/2$, L/D = 25.5	to S Band	7.14		20%	150 Ω		LP	N/A

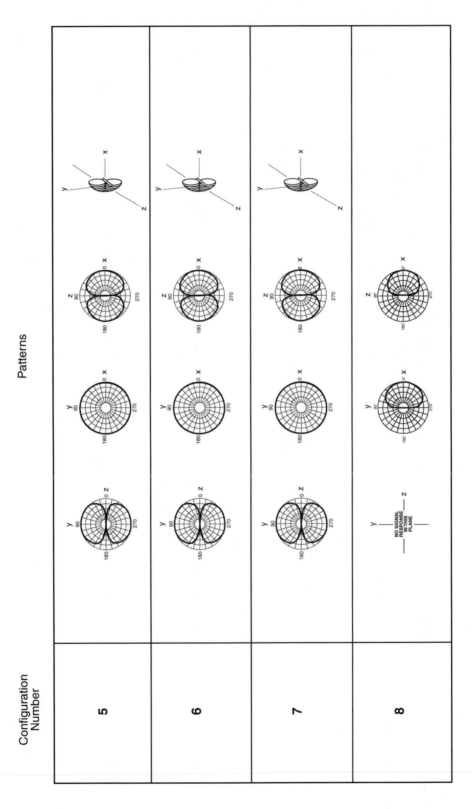

Antenna Class	Antenna Type	Configuration	Parameters	Typical Frequency Range (MHz)	Gain dB above isotropic	Beam Width at Center Frequency (Degrees)	3 dB Gain Reduction Bandwidth	Input Resistance at Center Frequency	2.0:1 VSWR Bandwidth %	Polarization	Side Lobe Levels
Wire Antennas (cont.)	Coaxial Dipole	9	$L = \lambda/4$, $L/D = 4$	to X Band	2.14	78	16%	50 Ω		LP	N/A
	Turnstile Dipole	10	$L/d = 25.5$	to L Band	-0.86		50%	150 Ω		LP	N/A
	Collinear Dipole Array	11	n, a, s	to L Band	2 to 4.5	elevation A_3 = omni				LP	N/A
	Yagi–Uda Array	12	λn, sn, n, a	to X Band	2 to 16	variable		50 Ω with match	10–60%	LP	N/A

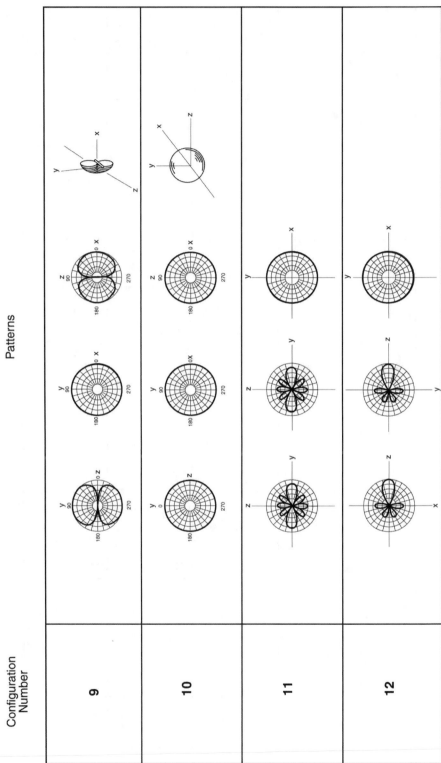

Antenna Class	Antenna Type	Configuration	Parameters	Typical Frequency Range (MHz)	Gain dB above isotropic	Beam Width at Center Frequency (Degrees)	3 dB Gain Reduction Bandwidth	Input Resistance at Center Frequency	2.0:1 VSWR Bandwidth %	Polarization	Side Lobe Levels
Wire Antennas (cont.)	$\lambda a/4$ Monopole over Infinite Ground	13	$L = \lambda/4$, thin	to X Band	5.14			36.5 Ω		LP	N/A
	$\lambda/4$ Monopole over Finite Perfect Ground	14	$L = \lambda/4$, $L/a = 53$, $D = \lambda$	to X Band	2.14	78		28 Ω		LP	N/A
	Helical Monopole over Infinite Ground	15	L total $= \lambda/2$, $h > 0.05\lambda$	to L Band	5.14	78		35 Ω	<4%	LP	N/A
	Center loaded $\lambda/4$ by $\lambda/4$ Whip over Infinite Ground	16	$L/d = 0.05$, $d =$ dia.	to X Band	8.3			50 Ω	10–60%	LP	−10

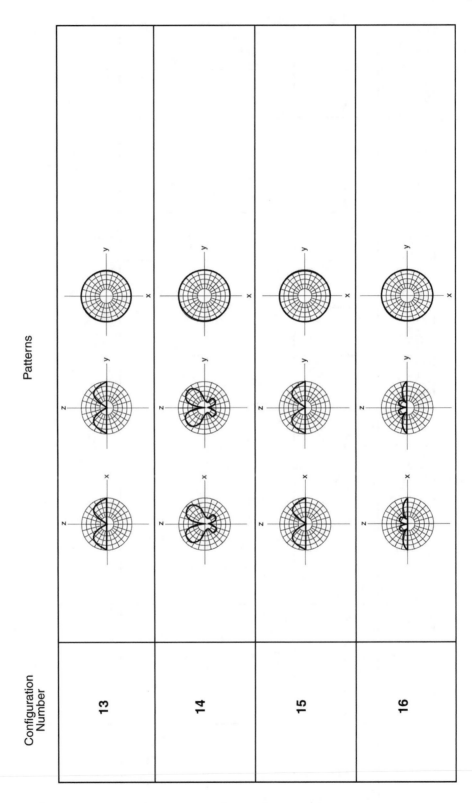

Antenna Class	Antenna Type	Configuration	Parameters	Typical Frequency Range (MHz)	Gain dB above isotropic	Beam Width at Center Frequency (Degrees)	3 dB Gain Reduction Bandwidth	Input Resistance at Center Frequency	2.0:1 VSWR Bandwidth %	Polarization	Side Lobe Levels
Wire Antennas (cont.)	Inverted F Monopole Over Infinite Ground	17	S, B, Px, Py, h	to X Band	5.14	78		50 Ω	2.50%	Linear:VP	N/A
	Discone Monopole	18	C, A, B	to S Band	2.14	78	4:01	50 Ω	Octaves	LP	N/A
	Helical Antenna	19	D, C, S, d, N, L, α, l	to S Band	6.0 to 18	Variable	<200%	approx. 140 Ω	<40%	CP	-13
Microstrip	Rectangular, Single Feed, Edge, or Probe Fed	20	εr, L, w, h	to mmWave	6	90		50 Ω	a few %	LP	N/A

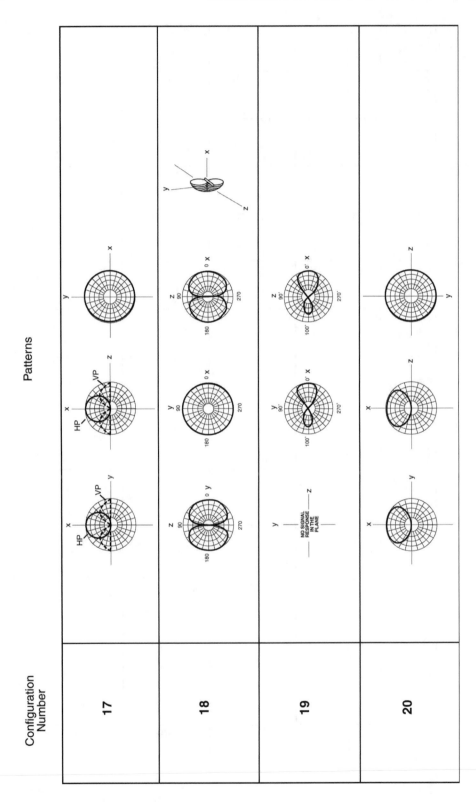

Antenna Class	Antenna Type	Configuration	Parameters	Typical Frequency Range (MHz)	Gain dB above isotropic	Beam Width at Center Frequency (Degrees)	3 dB Gain Reduction Bandwidth	Input Resistance at Center Frequency	2.0:1 VSWR Bandwidth %	Polarization	Side Lobe Levels
Microstrip (cont.)	Circular, single feed, edge or probe fed	**21**	ε_r, h, a, r	to mmWave	5.0 – 12	90		50 Ω	a few %	LP	N/A
	Circular Dual Fed	**22**	ε_r, h, r, a, 90 deg. feed	to mmWave	5 – 6	90		50 Ω	<5%	CP	N/A
	Circular, Higher Ordered Mode TM_{210}	**23**	ε_r, h, r, a, feed phase as shown		~ 5.14	78		50 Ω	a few %	LP, CP	N/A
Slot	$\lambda/2$ planer slot in infinite ground plane	**24**	$L = \lambda/2$, a	to mmWave	2.14	78		350 Ω		LP	N/A

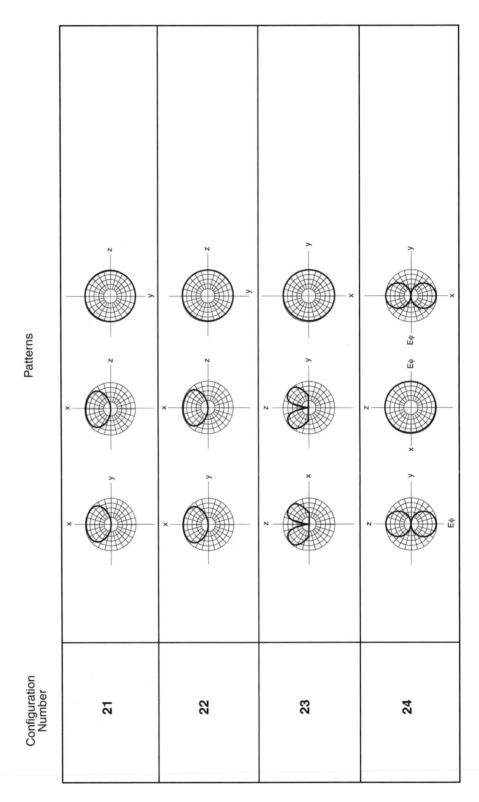

Antenna Class	Antenna Type	Configuration	Parameters	Typical Frequency Range (MHz)	Gain dB above isotropic	Beam Width at Center Frequency (Degrees)	3 dB Gain Reduction Bandwidth	Input Resistance at Center Frequency	2.0:1 VSWR Bandwidth %	Polarization	Side Lobe Levels
Slot (cont.)	Annular Slot	**25**	a & b < λa	to mmWave	1.5	70				LP	N/A
	Longitudinal Broadwall Waveguide Slot	**26**	ℓ, a, b, d, λg	to mmWave	approx. 5.0		feed dependent	high	N/A	LP	N/A
Aperture	Rectangular Aperture, Uniform Excitation	**27**	a by b	to mmWave	10 log(g), $g = 4\pi ab/(\lambda)^2$	xz plane: 50.6(λ)/b yz plane: 50.6(λ)/a		N/A	N/A	LP	−13.2 −13.2
	Rectangular Aperture, TE11 Excitation	**28**	a,b	to mmWave	10 log(g), $g = 3.24\pi ab/(\lambda)^2$	yz plane: 50.6(λ)/b xz plane: 68.8(λ)/a		N/A	N/A	LP	−13.2 −23

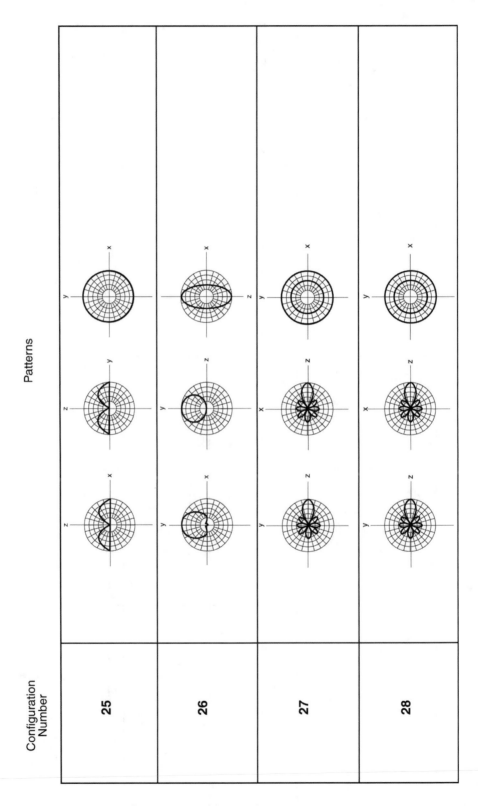

Antenna Class	Antenna Type	Configuration	Parameters	Typical Frequency Range (MHz)	Gain dB above Isotropic	Beam Width at Center Frequency (Degrees)	3 dB Gain Reduction Bandwidth	Input Resistance at Center Frequency	2.0:1 VSWR Bandwidth %	Polarization	Side Lobe Levels
Aperture	Circular Aperture, Uniform Excitation	**29**	a	to mmWave	$G = 10 \log(g)$, $g = (2\pi a/\lambda)^2$	xz plane: $29.2(\lambda)/b$ yz plane: $29.2(\lambda)/a$		N/A	N/A	LP	–17.6 –17.6
	Circular Aperture, TM11 Excitation	**30**	a	to mmWave	$G = 10 \log(g)$, $g = 10.5\pi(a/\lambda)^2$	yz plane: $29.2(\lambda)/b$ xz plane: $37.0(\lambda)/a$		N/A	N/A	LP	–17.6 –26
	Circular Aperture, Tapered Distribution	**31**	a, $1-(r/a)^2$ distribution	to mmWave	$G = 10 \log(g)$, $g = 0.75^*(2\pi^*a/\lambda)^2$	$36.4\lambda/a$		N/A	N/A	LP	–24.6
	Prime Focus Reflector (dipole feed)	**32**	$D = \dfrac{5\lambda}{2}$	to mmWave	14.75			300 Ω	30		

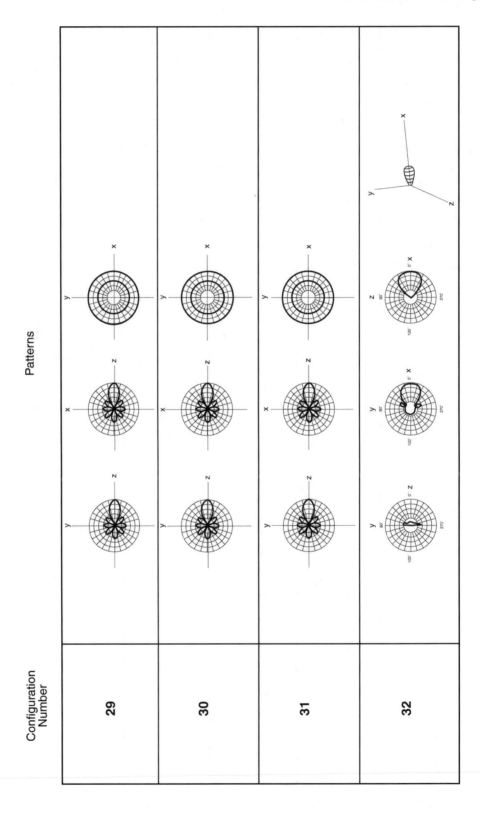

Antenna Class	Antenna Type	Configuration	Parameters	Typical Frequency Range (MHz)	Gain dB above isotropic	Beam Width at Center Frequency (Degrees)	3 dB Gain Reduction Bandwidth	Input Resistance at Center Frequency	2.0:1 VSWR Bandwidth %	Polarization	Side Lobe Levels
Aperture	Pyramidal Horn	33	L, l, b	to mmWave	< 35			50 Ω	35	LP	
Loop	Full wave loop	34	d,D	to X Band	3.14			45 Ω	13	LP	
Frequency Independent	Logperiodic Diapole Array	35	a, an, sn, ln, Rn	to X Band	7–12	> 50 deg.	N/A	50 – 100 Ω	~ 200	LP & CP	< –13 dB
	Cavity Backed Planer Sprial	36	ρ, φ mode 2 has null	to mmWave	< +4.0	~ 60 deg. Model 1	N/A	Balun required	> Decade	LP & CP	N/A

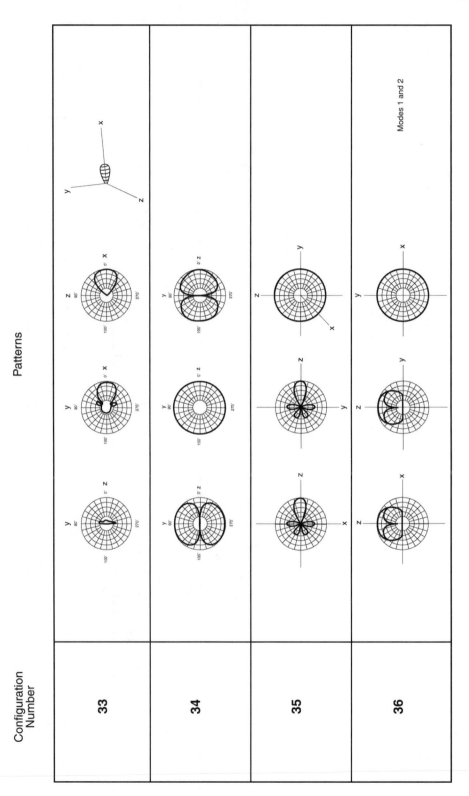

Antenna Class	Antenna Type	Configuration	Parameters	Typical Frequency Range (MHz)	Gain dB above isotropic	Beam Width at Center Frequency (Degrees)	3 dB Gain Reduction Bandwidth	Input Resistance at Center Frequency	2.0:1 VSWR Bandwidth %	Polarization	Side Lobe Levels
Frequency Independent	Conical Spiral	37	α, ρ, φ mode 2 has null	to mmWave	1 to 15	140 to 40	N/A	50 to 300 Ω Balun req.		LP & CP	N/A
	Tapered Dielectric Rod	38	d1, d2, d3, d4, P, L l1, l2, l3, l4	to mmWave	18–20	55 $(\lambda_o/L)^{0.5}$	±15%	Zo matched feed	40	LP	<–20
Leaky Wave	Long Shot in Waveguide	39	h, w, d, L, θ L = slot length	mmWave	10 log (2L/λ)	$\Delta\theta \approx \frac{L}{(\frac{L}{\lambda_o})} \cos\theta m$ θm = Angle of max. radiation	Freq, beam scan	Zo matched to feed	Broad	LP & CP	–30
	Leaky Waveguide Trough	40	w, d1, d2,	mmWave	10 log (2L/λ)	$\Delta\theta \approx \frac{L}{(\frac{L}{\lambda_o})} \cos\theta m$ θm = Angle of max. radiation	Freq, beam scan	Zo matched to feed	Broad	LP & CP	–30

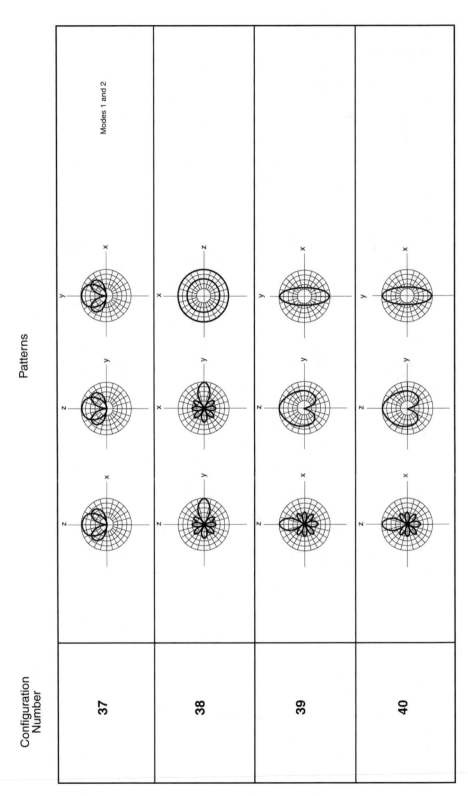

Antenna Class	Antenna Type	Configuration	Parameters	Typical Frequency Range (MHz)	Gain dB above isotropic	Beam Width at Center Frequency (Degrees)	Gain Reduction Bandwidth	Input Resistance at Center Frequency	2.0:1 VSWR Bandwidth %	Polarization	Side Lobe Levels
Lens	Dielectric	41	N, f, c, c1, o, C0, εr n = refractive index $n = \sqrt{\varepsilon_r}$ $k = \frac{2\pi}{\lambda}$ N = number of zones	x Band to mmWave	≤ 50	prop. to gain	25/(N-1) percent	N/A	N/A	LP & CP	< -20
	Waveguide	42	a, c0, c1, s, D, dn, ℓ nₙ refractive index	x Band to mmWave	< = 50	prop. to gain	25no/(1+Nno) percent	N/A	N/A	LP	< -20
	Fresnel Zone Plate Antennas	43	f, D, Z Z = number of transport zones fo = antenna freq.	mmWave	< 40	prop. to gain	fo/Z	N/A	N/A	LP	< -10
Notch	Linear Tapered Slot	44	t, W, L, teff, εr $t_{eff} = \frac{(\sqrt{\varepsilon_r}-1)t}{\lambda_o}\ \frac{\lambda_o}{} \leq 0.01$ L = 4λ to 10λ ∝ = 11.2°	X Band to mmWave	$10\log\left(\frac{4L}{\lambda_o}\right) - 2$	$77/\sqrt{\frac{L}{\lambda_o}}$ degrees	5:1 (3 dB)	50–200 Ω	N/A	LP	< -10

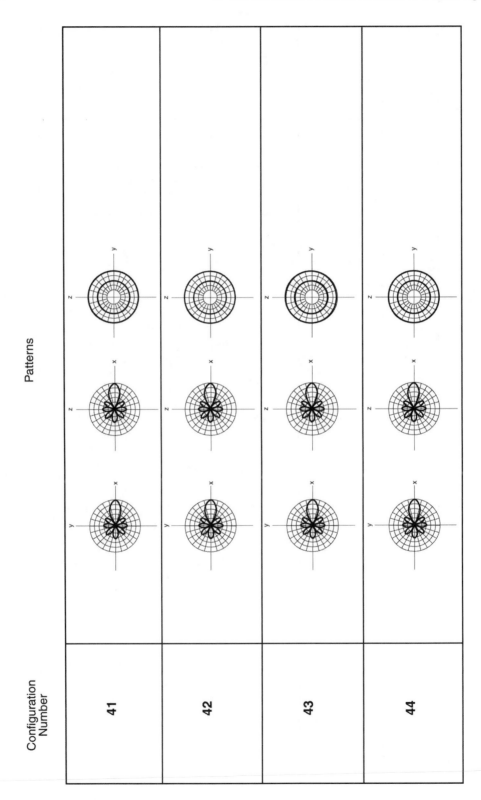

Waveguide-fed slotted array antennas find extensive use in the microwave and millimeter wave bands [9]. Linear arrays of resonant slots are formed by machining slots along the length of standard waveguide transmission lines, and a collection of these waveguide linear array "sticks" are combined together to form two-dimensional planar phased arrays. The most common waveguide slots currently in use are the *edge slot*, and the *longitudinal broad wall slot*. This array type is a cost-effective way to build high efficiency, controlled side lobe level arrays. One-dimensional electrically scanned phased arrays can be realized by introducing phase shifters in the feed manifold that excited each linear waveguide array stick within the two-dimensional aperture.

9.2.4 Helical Antennas [3]

The helical antenna has seen extensive use in the UHF through microwave frequencies, both for single radiating elements, and as phased array antenna elements. An excellent helical antenna discussion can be found in Kraus [3]. The helical antenna can operate either in the *axial (end fire) mode*, or in the *normal mode*. The axial mode results in a directional, circularly polarized pattern that can operate over a 2:1 frequency bandwidth.

The normal mode helix has found wide application for broad-beam, cardioid-shaped radiation patterns with very good axial ratio performance. Kilgus developed bifilar and quadrafilar helical antennas for satellite signal reception applications [10,11]. Circular polarization is generated by exciting each quadrafilar element of the helix in phase quadrature.

Helical antennas are also used as radiating elements for phased arrays, and as feeds for parabolic reflector antennas [12].

9.2.5 Yagi–Uda Array [2]

The Yagi–Uda array is a very common directional antenna for the VHF and UHF bands. End fire radiation is realized by a complex mutual coupling (surface wave) mechanism between the driven radiation element, a reflector element and one, or more, director elements. Dipole elements are most commonly used, but loops and printed dipoles elements are used as well. Balanis describes a detailed step-by-step design procedure for this type of antenna [2]. Method-of-Moment computer simulations are often used to verify and optimize the design.

9.2.6 Frequency-Independent Antennas [13]

Classic antenna radiator types, such as the wire dipole antenna, radiate efficiency when their physical dimensions are a certain fraction of the operating wavelength. As an example, the half-wave dipole is strictly resonant only at one frequency, and acceptable performance can be realized for finite bandwidths. The frequency-independent antenna is based on designs that are specified in terms of geometric angles. Rumsey has shown that if an antenna shape can be completely specified by angles, then it could theoretically operate over an infinite bandwidth [14]. In practice, the upper operating frequency limit is dictated by the antenna feed structure, and the lower operating frequency is limited by the physical truncation of the antenna structure. The *planar spiral, planar slot spiral, conical spiral,* and the *cavity-back planar spiral* antennas all exploit this concept.

A closely related and very useful antenna is the *log-periodic antenna*. This type of antenna has an electrical periodicity that is a logarithm of frequency. Because the antenna shape cannot be completely specified by angles, it is not truly frequency independent, but, nevertheless, is extremely broadband. The most common architecture is the log-periodic dipole array. Balanis has documented a detailed design procedure for this structure [2].

A key of frequency antennas is that the electrically active region of the antenna is a subset of the entire antenna structure, and the active region migrates to different regions of the antenna as a function of frequency [15].

9.2.7   Aperture Antennas [2,16,17]

Horn antennas are the most commonly used aperture antennas in the microwave and millimeter wave frequency bands. There is an abundance of horn antenna design information available in the literature. *E plane and H plane sectoral horns* generate fan beam radiation, that is, a narrow beamwidth in one principal plane, and a broad beamwidth in the orthogonal principal plane. The *pyramidal horn* provides a narrow beam in both principal planes. Sector and pyramidal horns are typically fed with rectangular waveguide and are linearly polarized. The *ridged horn* is a broadband variation of the pyramidal horn [18]. A *dual ridge square aperture horn* in conjunction with an orthomode transducer (OMT) can be used to generate broadband, circularly polarized radiation patterns.

Conical horns are typically fed with circular waveguide. An important variation of the basic conical horn includes the *corrugated horn*, which is typically used as a feed for high efficiency, electrically large reflector antennas systems [19]. The corrugations are used to extinguish field diffractions of the edge of the aperture plane that can lead to spurious radiation in the back lobe and side lobe regions. The *aperture-matched horn* also minimized aperture edge diffraction by blending the aperture edge to make a gradual transition into free space. *Dual mode conical horns* use a superposition of waveguide modes in the throat region to suppress beamwidth, control orthogonal beamwidths, and minimize cross-polarization. *Multimode horns* can be used in monopulse radar applications where sum and difference beam can be generated by means of mode-generating/-combining feed structure.

9.2.8   Reflector Antennas [2,20]

Reflector antennas are used extensively where very high gain, high radiation efficiency, and narrow beamwidths are required. Applications include satellite communication systems, navigation systems, terrestrial communications systems, deep space communication systems, and radio astronomy. Gain in excess of 80 dBi has been reported in radio astronomy applications [20]. *Prime focus reflector* systems have the feed antenna at the focal point of the reflector, typically a paraboloid. A disadvantage of this approach is the blockage created by the feed and its mounting struts. *Multiple reflector* systems are used in some applications for improved performance over prime focus designs. A multiple reflector system has several advantages, including (1) the feed can be mounted on the back of the main reflector, which removes the interconnecting transmission line between the feed and the receiver front end, (2) spill over radiation and side lobe radiation can be reduced, (3) a large focal length can be realized with a shorter physical length, and (4) beam shaping can be realized by shaping the main reflector and subreflector surfaces. *Cassegrain* and *offset subreflector* systems are in common use. The offset reflector has the advantage of significantly reducing feed blockage and minimizing cross-polarization.

Various reflector concepts include the planar reflector [21], the microstrip reflectarray [22], and the use of clustered feeds [23] and phased arrays as reflector feeds to generate limited scan arrays [24].

9.2.9   Microstrip Antennas [2,25,26]

Microstrip antennas offer the advantages of being low profile, light weight, and easily producible using contemporary printed circuit board materials and fabrication techniques. This technology has found extensive use in the Global Positioning System, wireless and satellite communication systems, and the cellular phone industry. In general, they have the disadvantages of narrow bandwidth and high loss at upper microwave and millimeter wave frequencies.

Linear polarized designs are realized with single feeds, either by a *microstrip transmission line feed* on the same layer as the radiating element, or by *coaxial probe feeding*. The microstrip transmission line feed has the advantage of ease of fabrication due to single layer printed circuit board construction, but the disadvantage of spurious radiation caused by the feed lines, particularly in array applications. The coaxial probe feeding method works well, but feed probe inductance must be taken into account. *Aperture coupling* is a broader band excitation method, but it increases the complexity of fabrication since multilayer printed circuit boards are required.

TABLE 9.2   Microstrip Patch Antenna Bandwidths

Microstrip Antenna Configuration	Polarization	Maximum Impedance Bandwidth (2:1 VSWR)	Axial Ratio Bandwidth
Simple patch with lower ε_r, thick substrate	LP	10%	N/A
Planar coupled	LP	20%	N/A
EM/aperture coupled	LP	70%	N/A
Stacked, multiresonator	LP	30%	N/A
Slot loaded	LP	45%	N/A
Log periodic array	LP	Multioctave	N/A
Varactor diode tuning	LP	20% around f_0. Dual band embodiments possible	N/A
Orthogonal feed patch	CP	30%	30%

Circular Polarization can be generated with a single feed using rectangular patch geometry. The resultant circularly polarized radiation is very narrow band. Square and circular patches can create circular polarization with dual feeds fed in phase quadrature. The resultant circular polarization signal is much broader band, with 6.0 dB axial ratio bandwidth of greater than 5%.

First order simple patch design can be accomplished with the *transmission line model*, and with the more accurate *cavity model* [2]. Method of Moment, Finite Element, and Finite Difference Time Domain (FDTD) techniques are used for detailed design.

Significant effort continues in the area of bandwidth optimization. Kumar and Ray review the current state-of-the-art microstrip-radiating element broad banding techniques [27,28]. *Broad banding techniques* include patch shape optimization, substrate parameter optimization (lower dielectric constant and increasing thickness), multiresonator planar configurations, stacked patch multiple resonator configurations, parasitically coupled top-hat configurations, aperture and electromagnetically coupled feed configurations, slot-loading techniques, traveling wave circularly polarized patches, quadrature-fed patch for circular polarization, broadband impedance-matching techniques, tunable ferrite material substrates, multimode excitation, and linear and planar log periodic array embodiments [29,30]. Table 9.2 summarizes the bandwidth properties of various microstrip patch architectures.

Microstrip-radiating elements are commonly used in contemporary phased array systems. Scan blindness (as discussed in the phased array antenna section) is problematic since a large dielectric slab can sustain a surface wave mode of guided wave propagation, which deteriorates the array radiation. Mutual coupling further complicates the phased array design problem. The reader is referred to the literature for detailed treatises.

9.2.10   Millimeter Wave Antennas [31]

The most widely used millimeter wave antennas in use today are the reflector antenna, the lens antenna, and the horn antenna. In addition to these structures, several lower frequency antennas can be adapted to millimeter wave frequencies, including slot, dipole, biconical dipole, and monopoles [32].

Slotted waveguide arrays have been demonstrated up to 94 GHz. The *Purcell slot array* has renewed interest due to increased slot dimensions and relaxed mechanical tolerance making it attractive in the millimeter wave regime [33].

Printed Microstrip Patch Arrays on low permitivity substrates have been demonstrated to 100 GHz. Feed loss of conventional microstrip lines is especially troublesome in the millimeter frequency bands. Several methods have been studied to reduce feed loss, including feeding microstrip-radiating element with the fringing fields associated with dielectric image guide, and E field probe coupling of the microstrip patches to a feed waveguide. Special attention must be given to mechanical tolerances and assembly techniques in the millimeter wave implementation of these structures.

Open dielectric structure antennas are attractive since most millimeter wave transmission lines are open structures. Strategically placed discontinuities in these structures initiate a *surface wave* phenomenon, which initiates radiation. Examples include *the tapered dielectric rod* and *the periodic dielectric antenna* [34,35].

Leaky Wave structures include the *long slot in waveguide, periodically loaded waveguide,* and *trough waveguide* structures [36]. Leaky waves are initiated when an open or closed waveguide is perturbed at periodic intervals, or by a continuous perturbation. Usually, the amount of leakage per waveguide length is kept low to minimize impedance mismatch with the antenna feed. The antennas are designed to be electrically long so that most of the energy radiates away before the guided wave reaches the end of the antenna. Beam scanning with input frequency is possible with this type of antenna.

The *periodically loaded waveguide antenna* utilizes a periodic surface perturbation, usually either a corrugation or a metal grating to initiate the surface wave mode. Other leaky wave antenna structures are discussed in the literature.

9.2.11   Dielectric Resonator Antennas [37]

Dielectric Resonator Antennas (DRAs) have gained recent popularity due to their compact size, additional degrees of design freedom, and low mutual coupling within an array environment. Luk and Leung provide a systematic overview of this technology [37]. A DRA is essentially a resonant structure designed to radiate electromagnetic energy, similar to a microstrip patch, and its resonance and bandwidth are strongly dependent on the size (dimensions) of the resonator and material properties. Common shapes include rectangular and body of revolution (BOR) configurations. Hemispherical DRA antennas can also be thought of as material-loaded monopoles. DRAs exhibit radiation characteristics that are mode dependent; different modes can be excited through multiport feeds. Both monopole and cardioid radiation patterns are realizable through various mode excitations. Both linear and circular polarizations are possible and instantaneous bandwidths up to 25% have been demonstrated, but the usual trade offs between dielectric loading, operating frequency, size, and bandwidth still apply. Typical feed configurations include coaxial probe and aperture-coupled microstrip.

Lens Antennas are very similar to reflector antennas since they are used to collimate spherical waves into plane waves. Reflector antennas are more versatile than lens antennas in general. Lens technology suffers from the following phenomena: surface loss from internal and external reflections, spill over loss, lens zone blockage, dielectric material loss, and tangent losses; (2) lens technology tends to be bulky and heavy; (3) and lenses must be edge supported. However, lenses are superior in some applications, such as wide angle scanning [38].

Dielectric Lenses have very strong optical analogies. If the subtended feed angle between the feed and the extremities of the lens are large, the lenses can become very bulky and heavy. Placing steps in the lens to create separate zones can reduce bulk. The reader is referred to Elliot for a detailed exposition [39]. Lenses may also be constructed of artificial dielectrics, parallel metal plates, and waveguide technology [40].

Fresnel zone plate antennas are a class of lenses that collimates spherical electromagnetic wave impinging on it by means of diffraction, rather than refraction [41–43]. The zone plate architecture offers several advantages over conventional lens technology, including a simpler, lighter, and planar construction with reduced material-induced insertion loss. Wiltse reviews the history of Fresnel zone plate antenna development in [44]. Hristov and Herben describe a Fresnel Zone plate lens with enhanced focusing capability [45]. Dual-band Fresnel zone plate antennas are discussed in [46]. Gouker and Smith described an integrated circuit version of the Fresnel zone plate antenna operating at 230 GHz [47].

9.2.12   The Fractal Antenna

Fractal antenna radiating elements have physical shapes based on fractal geometric curves popularized by Benoit Mandelbrot, and others [48–50]. An overview of fractal electromagnetic research is given by Werner [51]. There are two basic types of fractal curves, random and deterministic. Deterministic fractals utilize a repetitive motif, or pattern, that is applied to successive size scales. Fractals are mathematically described in terms of "n" iterations of the motif, where n goes to infinity. Practically, fractal structures are based on a finite number of iterations.

The general concept for a fractal-based radiating element is to introduce fractal-based impedance loading into the radiating element structure to shrink the physical size of the element for a given frequency band. Standard fractal antennas tend to exhibit nonharmonically related multiple sub-band resonances (segregated sub-band regions of lower VSWR), as opposed to low VSWR over a contiguous bandwidth. Fractal antennas typically require auxiliary impedance matching networks for true wideband operation. Cohen has demonstrated fractal loop arrays and monofilar helical antennas [52,53]. Fractal techniques have been successfully used to raise the real part of the impedance of an electrically small loop antenna over a narrow bandwidth [2]. Breden and Langley discus printed fractal technology as applied to multiband radio and television service antennas [54]. Multiband and wide-band printed fractal branched antennas are discussed by Sindduo and Sourdois [55].

Fractal antennas are one scheme to maximally fill the available area (volume) for an electrically small antenna to maximize bandwidth, as supported by the Chu-Wheeler electrically small antenna concepts [2]. Best has shown that Fractal antenna geometries offer little advantage over other space filling antenna geometries, for example, helical monopoles, spherical monopoles, and genetic algorithm derived structures and arbitrary three-dimensional wire radiators [56]. The relationship of fractal antennas to log periodic antennas, frequency-independent antennas, and genetic algorithm-based wire antenna design is described in [57] and [58].

9.2.13 Frequency Selective Surfaces [59]

Frequency selective surfaces (FSS) structures are essentially plane wave "filters" that exhibit angular and frequency dependence with either band pass or band stop characteristics. They are used in multiband, wide band phased array applications, and in applications where it is desirable to block signal transmission for one frequency band while passing another frequency band [60]. Dichroic reflector technology is another example of a FSS application [61].

9.2.14 The Genetic Algorithm

Genetic Algorithm-Based Antenna Design: Although the genetic algorithm-based optimization techniques strictly fall under the category of computational electromagnetic techniques, they are briefly mentioned here because of the nontraditional class of antennas that result from these design techniques. An introduction to genetic algorithm optimization techniques is given in Haupt and Haupt [62]. An excellent overview of genetic algorithms for electromagnetic applications, and comparisons to traditional local optimization techniques, is given by Haupt [63] and [64], and Johnson and Rahmat-Samii [65]. Traditional local-minimum-based gradient optimization techniques are detailed by Cuthbert [66]. A sampling of genetic-based, algorithm-based wire antenna designs is discussed in [67], [68], and [69].

9.2.15 Micromachined Electro Mechanical Switches [70]

Micromachined Electro Mechanical System (MEMS) Switches are miniature semiconductor switches used in a variety of antenna applications. Brown provides an excellent overview of MEMS technology for RF (radio frequency) applications [71]. The MEMS switch is physically small; has very low on-state loss; high off-state isolation; and low average dc power consumption, as compared to ferrite waveguide and semiconductor switching elements such as PIN diode and FET transistor switches.

The two most common switch topologies are the cantilever, or type 1 switch and the membrane, or type 2 switch. Phase shifters and true time-delay devices using MEMS type 2 switches have been reported at C Band, through W Band. The type 1 switch topology is superior for lower (below X Band) applications due to its high off state isolation.

An intriguing application of MEMS device technology is for reconfigurable antenna technology [72]. The desire to use multiband, multifunction, shared asset apertures for advanced radar and communications systems has led researchers to consider radiating elements whose physical properties can be dynamically adjusted. This is already demonstrated, to a simpler degree, by the use of reactive loads distributed along the length of wire antennas, such as dipoles, to give multiband performance [73]. The MEMS switch offers the potential to generalize this concept to two dimensions using a switching matrix.

MEMS technology has been successfully demonstrated within the research community and efforts are underway to commercialize the technology, including lifetime reliability studies, RF compatibility, and environmental packaging techniques [70]. Device packaging issues are particularly challenging since the MEMS switch electromechanical device that requires unrestricted mechanical motion within the package.

Micromachined antenna technology can be used to synthesize a localized artificial dielectric material with low dielectric constant for a microstrip antenna substrate from a high dielectric substrate. The lower effective dielectric constant improves the radiation efficiency and reduces surface waves for a given substrate thickness. Gauthier et al., describe a method of drilling a densely spaced two-dimensional grid of holes into the substrate to reduce the local dielectric constant of the substrate [74]. A reduction of dielectric constant to 2.3 from the intrinsic substrate dielectric constant of 10.8 was reported. Nguyen et al., describe micromachining techniques that are used to suspend dipole or slot radiators on a very thin membrane to realize a local substrate dielectric constant of approximately 1.0, which is quite useful for millimeter wave applications [75]. Veidt et al., describe a micromachined diagonal horn at 110 GHz [76].

9.3   Antenna Metrology [1–3]

Several excellent books are available on the subject of antenna metrology, so only a few of the basic concepts are briefly described herein. The reader is referred to the literature for a more detailed exposition.

Impedance, VSWR, return loss, radiation patterns, cross-polarization, axial ratio, and the polarization ellipse tilt angle are the radiation parameters most often measured. *Impedance* and *VSWR* are measured using normal network analyzer techniques with the antenna radiating into an anechoic environment [4,5]. Typically swept frequency measurements are made in small anechoic chambers designed for only impedance and VSWR testing, in a full-scale anechoic chamber where radiation patterns are measured, or in an open, reflectionless environment.

Radiation pattern cuts are measured by precisely rotating the AUT and recording the resultant phase and amplitude response as a function of position, under the condition of source antenna illumination. The amplitude and phase response are recorded as a function of two orthogonal spherical coordinates. Figure 9.9 illustrates a conceptual block diagram of a typical far field data acquisition set up appropriate for both outdoor and anechoic chamber use.

Gain is most often measured by the substitution method. A signal level is recorded from a calibrated gain standard and the source antenna. The two antennas are typically axially aligned. Then the AUT is measured using the same source antenna and power levels. The difference in the receive level of both measurements is calculated and this difference is added to the gain standard's calibrated gain number to determine the AUT's gain above isotropic at the measurement frequency. Dipoles and dipole arrays are used as gain standards in the VHF and UHF frequency bands. Pyramidal and conical horns are used above 800 MHz and on into the millimeter wave region. Axial mode helical antennas, dual mode ridged square aperture/OMT-fed waveguide horns, and OMT-fed circular horns of various types can be used directly as circularly polarized gain standards.

Cross-Polarization of a linearly polarized antenna is easily measured by rolling the source antenna so that its polarization is oriented at 90° from the AUT. Cross-polarization is typically referenced to the copolarized peak of beam signal. The cross-polarization of a circularly polarized antenna can be directly measured by comparing the received signal from the AUT under copolarized excitation with cross-polarized signal using an identical source antenna of opposite polarization. In addition, cross-polarization rejection can be calculated knowing the source antenna and AUT's axial ratio and polarization tilt angle angles.

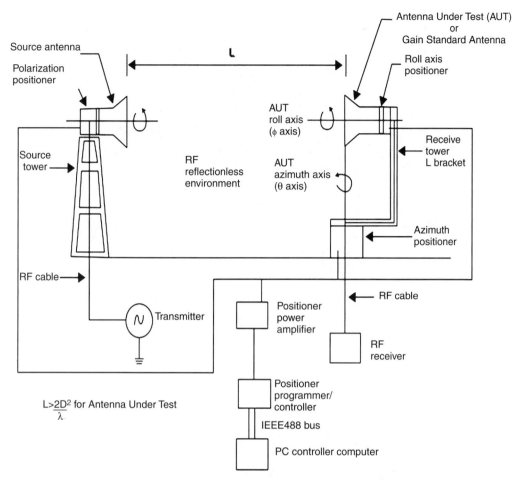

FIGURE 9.9 Basic roll/azimuth for field test setup.

Axial Ratio is usually measured with the spinning linear source antenna method. A linear source antenna with low cross-polarization is rapidly spun about its roll axis while the AUT's positioner is recording a pattern at a much slower rotation rate. The resulting pattern has a high-frequency sinusoidal-like pattern superimposed on the basic antenna pattern. If the AUT is highly linear, its response will vary drastically as the source antenna is rolled. On the other hand, if the AUT is purely circular, there will be no undulation of the signal since the circular antenna is equally sensitive to all orientations of linear polarization. It can be shown that the difference in amplitude of adjacent maxima and minima in the spinning linear plot is equivalent to the axial ratio of the AUT for that particular direction. The spinning linear measurement, therefore, provides an easy way to determine the axial ratio of the AUT across a pattern cut. An example of axial ratio pattern cut is shown in Figure 9.10.

The *axial ratio* and *polarization tilt angle* for a single point on the antenna radiation pattern can be determined by a *polarization ellipse* measurement. In this test, the AUT is parked in the position of interest and the source antenna is rolled one complete revolution. The recorded contour is the polarization ellipse and the ratio of the major to minor diameters is the axial ratio. The angle of the major diameter relative to a specified reference is related to the polarization tilt angle. A typical polarization ellipse of an axial mode helix is illustrated in Figure 9.11.

Near Field Measurements are becoming increasingly popular for phased array antenna design and diagnostics [6–8]. The basic concept of near field metrology is related to the electromagnetic theorem called

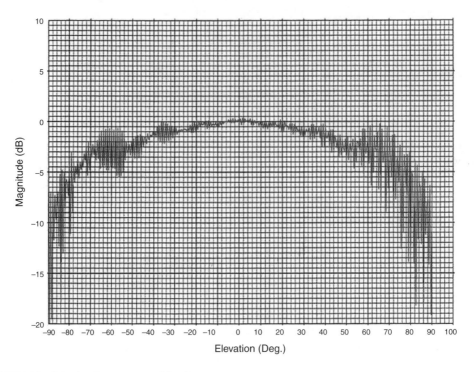

FIGURE 9.10 Spinning linear source axial ratio measurement.

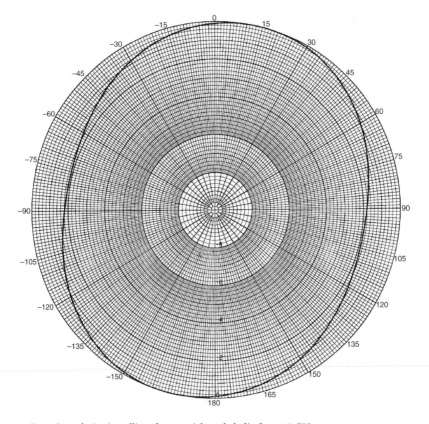

FIGURE 9.11 Bore site polarization ellipse for an axial mode helix fo = 1.5 GHz.

the *equivalence theorem* [9]. The theorem can be summarized as follows: If two orthogonal orientations of electric field are known everywhere on a closed surface, the electromagnetic field can be determined at any other observation point can be determined. In the near field measurement scenario, a field probe is used to sample two orthogonal polarizations of electrical field along the closed surface, a correction algorithm is applied to negate the perturbation of the measured field by the probe, and finally Fourier transform techniques are used to calculate the far field radiation pattern.

The three fundamental types of near field ranges are the *planar, cylindrical,* and *spherical near field,* named after the closed surfaces in which the data is sampled. In the planar near field set up, the enclosed measurement surface is approximated as a plane. This is appropriate for directional antennas such as planar arrays and reflector antennas. The cylindrical near-field range is appropriate for fan beam antennas such as linear arrays. The spherical near field is appropriate for all antenna types, including broad beam antennas. The planar near-field algorithm is the most computationally efficient, while the spherical near field is the least computationally and data acquisition inefficient.

The near-field measurement techniques offer several advantages over the traditional far-field pattern cut procedures, even though the required data acquisition and data processing time is much longer. The advantages are listed below:

1. The complete data set is available to the analysis for future use.
2. In phased array measurements, the measurement frequency and the beam position can be time multiplexed such that multiple data sets from numerous beam positions and operating frequencies can be extracted from one data acquisition session.
3. Inverse Fourier transform techniques can be used to holographically image the aperture excitation.

The ability to directly produce an *image of the array aperture excitation* is a very powerful tool in the array design process. In addition, it can be used to diagnose element or phase shifter failures, and so forth, along the array face.

9.4 Emerging Antenna Technologies

This section briefly describes recent significant advances in antenna technologies within the past 5 years. A similar discussion for phased arrays is contained in the next chapter of this handbook.

Metamaterials are one of the most significant recent advances in antenna technology that promises profound impact on the microwave/millimeter wave antenna and circuit design discipline [1–6]. The interested reader is referred to the large body of literature covering this topic. Excellent overviews are provided by Eleftheriades et al., Yeo and Mittra, and Kildal et al [1–3]. Metamaterials are engineered composites that exhibit properties not observed in nature as constituent parameters. This offers designers additional degrees of freedom over the traditional metallic, dielectric, and magnetic materials. Metamaterials can be broadly classified as *double negative materials* (DNM), or *electromagnetic band gap* (EBP) materials [2]. DNMs have feature sizes of much less than a wavelength, and their apparent (macroscopic) dielectric and magnetic permeability can both be made negative, which allows backward wave propagation. EBGs have feature sizes on the order of one half wavelength or more. EBG variations include *Tunable impedance surfaces* (TIS) and *Artificial Magnetic Conductors* (AMC), which are realizable for narrow-to-moderate bandwidths.

Metamaterials can be applied to the antenna art in the following ways: ground plane impedance tailoring (AMC), *directivity enhancement* of microstrip patch radiators, *multi-band/wide band enhancement,* radar cross-section reduction (FSS), applications of electromagnetic "hard" and "soft" surfaces to waveguides, horns and reflector antennas, series feed antenna array *beam squint mitigation, double negative index lens* antennas, *EGB analog waveguide and phase shifters* and *reflect-arrays,* and FSS-based leaky wave antennas.

AMC materials synthesize perfect "H" or Neumann boundary condition, over a finite bandwidth, where the reflection at the surface interfaces is unity with zero degree phase shift. An artificial "magnetic" material can be used at frequencies where natural ferrite materials perform poorly. This property can be exploited to realize horizontal dipole and other *low-profile antennas* very close to the artificial AMC surfaces for conformal applications. AMC properties can be used to improve the input impedance of low-profile structures. In addition, AMC can mitigate the *parasitic low elevation angle scattering*, normally associated with metallic ground planes for vertically polarized monopoles [4].

Electromagnetic "hard" and "soft" surfaces, which draw analogies from the acoustics field, have been used extensively [3]. Physically compact hard waveguides—with uniform TEM excitation—are realizable without the constraints associated with metallic waveguides. *Hard horn antennas* can be tailored for pattern symmetry, low cross-polarization and minimum aperture blockage in reflector antenna applications. The soft horn is also known as the traditional corrugated horn, a common staple in reflector antenna art. *Hard/soft reflector feed strut assemblies* are used to mitigate aperture shadowing and parasitic diffraction.

DNI lens technology offers another degree of freedom in quasioptic lens antenna design [5,6]. The optical index of refraction is directly related to its permittivity and permeability. DNI materials can therefore realize *negative index of refraction* over a finite bandwidth. The designer is therefore no longer restricted to positive index of refraction materials. DNI spherical lens can be made physically compact, can be *impedance matched to free space*, and they can have *superior focusing properties*.

References

Section 1: Antenna Parameters

1. Schantz, H., *The Art and Science of Ultrawideband Antennas*, Artech House, Inc, 685 Canton Street, Norwood, MA, 02062, 2005.
2. *IEEE Standard Definitions of Terms for Antennas*, IEEE Standard 145-1983, Institute of Electrical and Electronics Engineers, 1983.
3. Balanas, C. A., *Antenna Theory, Analysis, and Design*, 2nd Ed., John Wiley and Sons, New York, 1997.
4. Macnamara, T., *Handbook of Antennas for EMC*, Artech House, Inc., Norwood, MA, 1995.
5. *Antennas and Antenna Systems*, Watkins-Johnson Company, Catalog Number 200, September, 1990, Chapter 12, pp. 101–127.
6. Hollis, J. S., et al, *Microwave Antenna Measurements*, Scientific-Atlanta, Inc., Atlanta Georgia, November, 1985.
7. Hacker, P. S., and Schrank, H. E., "Range Distance Requirements for Measuring Low and Ultralow Side-lobe Antenna Patterns," *IEEE Transactions on Antennas and Propagation*, Vol. AP-30, No. 5, September, 1982, pp. 956–966.
8. Offutt, W. B., and DeSize, L. K., *Antenna Engineering Handbook*, 2nd Ed., Johnson, R. C., and Jasik, H., Editors, McGraw-Hill, New York, 1984, Methods of Polarization Synthesis, pp. 23-8 through 23-9.
9. Reference Data for Radio Engineers, 6th Ed., 3rd Printing, Howard K. Sams/ITT, New York, 1979, pp. 4-21 through 4-23.
10. Bowman, D. F, *Antenna Engineering Handbook*, 2nd Ed., Johnson, R. C., and Jasik, H, Editors, McGraw-Hill, New York, 1984, Chapter 23,Impedance Matching and Broadbanding, Chapter 43, pp. 43-1 through 43-32.
11. Cuthbert, T. R., *Broad Band Direct-Coupled and Matching RF Networks*, TRCPEP Publication, Greenwood Arkansas, 1999.
12. Wheeler, H. A., "Fundamental Limitations of Small Antennas," *Proceedings of the IEEE*, Institute of Radio Engineers, New York, NY, December, 1947, pp. 1479–1483.
13. Chu, L. J., "Physical Limitations of Omnidirectional Antennas," *Journal of Applied Physics*, Vol. 19, December, 1948, pp. 1163–1175.
14. Hansen, R. C., "Fundamental Limitations in Antennas," *Proceedings of the IEEE*, New York, NJ, Vol. 69, No. 2, February 1981, pp. 170–182.
15. Goubau, G., "Multi-Element Monopole Antennas," *Proceedings of the Workshop on Electrically Small Antennas*, ECOM, Ft. Monmouth, NJ, May, 1976, pp. 63–67.
16. Best, S. R., "The Performance of an electrically Small Folded Spherical Helix Antenna," IEEE International Symposium and Antennas and Propagation, 2002.
17. Best, S. R., "On the Performance of the Koch Fractal and Other Bent Wire Monopoles," IEEE Transactions on Antennas and Propagation, Vol. 51, No. 6, June, 2003, pp. 1293–1300.

Section 2: Radiating Elements

1. Salati, O. M., "Antenna Chart for Systems Designers," *Electronic Engineer*, January, 1968.
2. Balanis, C. A., *Antenna Theory, Analysis, and Design*, 2nd Ed., John Wiley and Sons, New York, 1997.
3. Kraus, J. D., *Antennas*, 2nd Ed., McGraw-Hill, New York, 1988.
4. Tai, C. T, *Antenna Engineering Handbook*, 2nd Ed., Johnson, R. C., and Jasik, H., Editors, McGraw-Hill, New York, 1984, Dipoles and Monopoles, Chapter 4, pp. 4-1 through 4-34.
5. Fujimoto, K., Henderson, A., Hirasaw, K., and James, J. R., *Small Antennas*, John Wiley and Sons, Inc., New York, 1987.
6. Macnamara, T., *Handbook of Antennas for EMC*, Artech House, Inc., Norwood, MA, 1995.
7. Blass, J., *Antenna Engineering Handbook*, 2nd Ed., Johnson, R. C., and Jasik, H., Editors, McGraw-Hill, New York, 1984, Slot Antennas, Chapter 8, pp. 8-1 through 8-26.
8. Mailloux, R. J., "On the Use of Metallized Cavities in Printed Slot Arrays with Dielectric Substrates," *IEEE Transactions on Antennas and Propagation*, Vol. Ap-35, No. 55, May, 1987, pp. 477–487.
9. Elliot, R. S., *Antenna Theory and Design*, Prentice-Hall, Inc., Englewood, New Jersey, 1981.
10. Kilgus, C. C., "Resonant Quadrafilar Helix," *IEEE Transactions on Antennas and Propagation*, Vol. Ap-23, May, 1975, pp. 392–397.
11. Donn, C., Imbraie, W. A. and Wong, G. G., "An S Band Phased Array Design for Satellite Application," *IEEE International Symposium on Antennas and Propagation*, 1977, pp. 60–63.
12. Holland, J. "Multiple Feed Antenna Covers L, S, and C Bands Segments," *Microwave Journal*, October, 1981, pp. 82–85.
13. Mayes. P. E., *Antenna Handbook, Theory, Applications, and Design*, Lo, Y. T., and Lee, S. W., Editors, Van Nostrand Reinhold Co., New York, 1988, Frequency Independent Antennas, Chapter 9, pp. 9-1 through 9-121.
14. Rumsey, V. H., *Frequency Independent Antennas*, Academic Press, New York, 1966.
15. DuHamel, R. H., A*ntenna Engineering Handbook*, 2nd Ed., Johnson, R. C., and Jasik, H., Editors, McGraw-Hill, New York, 1984, Frequency Independent Antennas, Chapter 14, pp. 14-1 through 14-44.
16. Love, A. W., *Electromagnetic Horn Antennas*, IEEE Press, The Institute of Electrical and Electronics Engineers, Inc., New York, 1976.
17. Olver, A. D., Clarricoats, P. J. B., Kishk, A. A., and Shafai, L., *Microwave Horns and Feeds*, IEEE Press, The Institute of Electrical and Electronics Engineers, Inc. New York, 1994.
18. Walton, K. L., and Sunberg, V. C., "Broadband Ridge Horn Design," The *Microwave Journal*, March, 1964, pp. 96–101.
19. Clarricoats, P. J. B., and Olver, A. D., *Corrugated Horns for Microwave Antennas*, Peter Peregrinus, London, UK, 1984.
20. Love, A. W., *Reflector Antennas*, IEEE Press, The Institute of Electrical and Electronics Engineers, Inc., New York, 1978.
21. Encinar, J. A., "Design of Two-Layer printed Reflectarrays for Bandwidth Enhancement," *Proceedings of the of the IEEE Antennas and Propagation Society 1999 International Symposium*, 1999, pp. 1164–1167.
22. Haung, J. A., "Capabilities of Printed Reflectarray Antenna," *Proceedings from the 1996 IEE International Symposium on Phased Array Systems and Technology*, Institute of Electrical and Electronics Engineers, Inc., Piscataway, NJ, 1996, pp. 131–134.
23. Ford Aerospace and Communications Corporation, "Design for Arabsat C Band Communication antenna System," Palo Alta, CL.
24. Mailloux, R. J., *Phased Array Antenna Handbook*, Artech House, Inc., Norwood, MA, 1994, pp 480–504.
25. James, J. R., and Hall, P. S., *Handbook of Microstrip Antennas*, Vols 1 and 2, Peter Peregrinus, London, UK, 1989.
26. Pozar, D. M., and Schaubert, D. H., *Microstrip Antennas: the Analysis and Design of Microstrip Antennas and Arrays*, IEEE Press, The Institute of Electrical and Electronics Engineers, Inc., New York, 1995.
27. Kumer, G., and Ray, K. P., *Broadband Microstrip Antennas*, Artech House, Inc., Norwood, MA, 02062, 2003.
28. Zurcher, J.-F., and Gardiol, F. E., *Broadband Patch Antennas*, Artech House, Boston, 1995.
29. Hall, P. S., "Multi-octave Bandwidth Log-Periodic Microstrip Antenna Array," *IEE Proceedings*, Vol. 133, Pt. H, No. 2, 1986, pp.127–136.
30. Tripp, V. K., and Papanicolopoulos, C. D., "Frequency-Independent Geometry for a Two Dimensional Array," *IEEE AP-S Symposium*, July 18, 2000, paper 33.2.

31. Schwering, F., and Oliner, A., *Antenna Handbook, Theory, Applications, and Design*, Lo, Y. T., and Lee, S. W., Editors, Van Nostrand Reinhold Co., New York, 1988, Millimeter Wave Antennas, Chapter 17, pp. 17-3 through 17-150.

32. Kay, A. F., "Millimeter-Wave Antennas," *Proceeding of the IEEE*, Piscataway, NJ, Vol. 54, pp. 641–647.

33. Silver, S. Editor, *Microwave Antenna Theory and Design*, New York: McGraw-Hill Book Co., 1949.

34. Zucker, F. J., *Antenna Engineering Handbook*, 2nd Ed., Johnson, R. C., and Jasik, H., Editors, McGraw-Hill, New York, 1984, Surface Wave Antennas and Surface-Wave-Excited Arrays, Chapter 12, pp. 12-1 through 12-36.

35. Mittra, R. *Antenna Engineering Handbook*, 2nd Ed., Johnson, R. C., and Jasik, H., Editors, McGraw-Hill, New York, 1984, Leaky Wave Antennas, Chapter 10, pp.

36. Schwering, F., and Peng, S. T., "Design of Dielectric Grating Antennas for Millimeter Wave Applications," *IEEE Transactions of Microwave Theory and Techniques*, Vol. MTT-31, February, 1983, pp. 199–209.

37. Luk, K. M., and Leung, K. W., *Dielectric Resonator Antennas*, Research Studies Press, LTD, Baldock, Hertfordshire, SG7 6AE, England.

38. Peeler, G. D. M., *Antenna Engineering Handbook*, 2nd Ed., Johnson, R. C., and Jasik, H., Editors, McGraw-Hill, New York, 1984, Lens Antennas, Chapter 16, pp. 10-1 through 10-21.

39. Elliot, R. S., *Antenna Theory and Design*, Prentice-Hall, Inc., Englewood, NJ, 1981, pp. 529–532.

40. Elliot, R. S., *Antenna Theory and Design*, Prentice-Hall, Inc., Englewood, New Jersey, 1981, pp. 538–545.

41. Wiltse, J. C., and Garrett, J. E., "The Fresnel Zone Plate Antenna," *The Microwave Journal*, January, 1991, pp. 101–114.

42. Minin & Minin, *Diffractional Optics of Millimetre Waves*, Institute of Physics, 2004.

43. Hristov, H. D., *Fresnel Zones in Wireless Links, Zone Plate Lenses and Antennas*, Artech House, Inc, 685 Canton Street, Norwood, MA, 02062, 2000.

44. Wiltse, J., "History and Evolution of Fresnel Zone Plate Antennas for Microwaves and Millimeter Waves," *IEEE Antennas and Propagation Society*, 1999 IEEE International Symposium, 1999, pp. 722–725.

45. Hristov, H. D., and Herben, M. H. A. J., "Millimeter-Wave Fresnel-Zone Plate and Lens Antenna," *IEEE Transactions on Microwave Theory and Techniques*, Vol. 43, Number 12, December, 1995, pp. 2779–2785.

46. Wiltse, J. C., "Dual Band Fresnel Zone Plate Antennas," *Proceedings SPIE Aerospace Conference*, Orlando, FL, Vol. 3062, April, 1997, pp. 181–185.

47. Gouker, M. A., and Smith, G. S., "A Millimeter-Wave Integrated-Circuit Antenna Based in the Fresnel Zone Plate," *IEEE Microwave Theory and Techniques Symposium Digest*, 1991, pp. 157–160.

48. Fractal Antenna Systems, Inc., "Fractal Antenna White Paper," July 10, 1999. This document is available on the Fractal Antenna Systems, Inc. web site.

49. Cohen, N., "Fractal Antennas, Part 1," *Communications Quarterly*, Summer, 7, 1995.

50. Mendelbrot, B. B., *The Fractal Geometry of Nature*, W.H. Freeman, New York, 1983.

51. Werner, D, H., "An Overview of Fractal Electrodynamics Research," *Proceedings of the 11th Annual Review of Progress in Applied Computational Electromagnetics*, Monterey, CL, March 20–24, 1995, pp. 964–969.

52. Cohen, N., "Simple CP Fractal Loop Array With Parasitic," 14th Annual Review of Progress on Applied Computational Electromagnetics, the Applied Computational Electromagnetics Society, March, 1998, pp. 1047–1050.

53. Cohen, N., "NEC4 Analysis of a Fractalized Monofilar Helix in an Axial Mode," 14th Annual Review of Progress on the Applied Computational Electromagnetics, Applied Computational Electromagnetics Society, March, 1998, pp. 1051–1057.

54. Breden, R., and Langley, R. J., "Printed Fractal Antennas", *IEEE National Conference on Antennas and Propagation*, 1999.

55. Sindou, M. A., and Sourdois, C., "Multiband and Wideband Properties of Printed Fractal Branched Antennas," *Electronics Letters*, Vol. 35, No. 3, February 4, 1999. pp. 181–182.

56. Best, S. R., "On the Performance of the Koch Fractal and Other Bent Wire Monopoles", *IEEE Transactions on Antennas and Propagation*, Vol. 51, No. 6, June, 2003.

57. Werner, D. H., and Werner, P. L., "Frequency Independent Features of Self-Similar Fractal Antennas," *Radio Science*, Vol. 31, No. 6, November–December 1996, pp. 1331–13343.

58. Werner, D. H., and Werner, P. L., "Fractal Radiation Pattern Synthesis," National Radio Science Meeting, Boulder Colorado, January 6–11, 1992, p. 66.

59. Munk, B. E., *Frequency Selective Surfaces, Theory and Design*, John Wiley and Sons, Inc., New York, 2000.

60. Mittra, R., Chan, C. H., and Cwik, T., "Techniques for Analyzing Frequency Selective Surfaces-A Review," *Proceeding of the IEEE*, Vol. 76, No. 12, December, 1988, pp. 1593–1615.

61. Schennum, G. H., "Frequency Selective Surfaces for Multiple-Frequency Antennas," *The Microwave Journal*, May, 1973, pp. 55–76.

62. Haupt, R. L., and Haupt, S. E., *Practical Genetic Algorithms*, John Wiley and Sons. Inc. New York, 1998.

63. Haupt, R. L., "Introduction to Genetic Algorithms for Electromagnetics," *IEEE Antennas and Propagation Magazine*, Vol. 37, No. 2, April, 1995, pp. 7–15.

64. Haupt, R. L., "Comparison Between Genetic and Gradient-Based Optimization Algorithms or Solving Electromagnetics Problems," *IEEE Transactions on Magnetics*, Vol. MAG-31, No. 3, May, 1995, pp. 1932–1935.

65. Johnson, J. M., and Rahmat-Samii, Y., "Genetic Algorithms in Engineering Electromagnetics," *IEEE Antennas and Propagation Magazine*, Vol. 39, No. 4, August, 1997, pp. 7–25.

66. Cuthbert, T. R., *Optimization Using Personal Computers with Applications to Electrical Networks*, John Wiley and Sons, Inc., New York, 1987.

67. Bahr, M., Boag, A., Michielson, E., and Mittra, R., "Design of Ultra-Broadband Loaded Monopoles," *IEEE Antenna and Propagation Society*, International Symposium Digest, Seattle, WA, USA, June, 1994, Vol. 2, pp. 1290–1293.

68. Altshuler, E. E., and Linden, D. S., "Wire-Antenna Design Using Genetic Algorithms," *IEEE Antennas and Propagation Magazine*, Vol. 39, No. 2, April, 1997, pp. 33–43.

69. Werner, P. L., Altman, Z., Mittra, R., Werner, D. H., and Ferraro, A. J., "Genetic Algorithm Optimization of Stacked Vertical Dipoles Above a Ground Plane," *IEEE Antenna and Propagation Society*, International Symposium Digest, Vol. 3, Montreal, Canada, 1997, pp. 1976–1979.

70. Rebeiz, G. M., *RF MEMS Theory, Design and Technology*, John Wiley and Sons, NJ, 2003.

71. Brown, E. R., "RF-MEMS Switches for Reconfigurable Integrated Circuits," *IEEE Transactions on Microwave Theory and Techniques*, Vol. 11, November, 1998, pp. 1868–1880.

72. Lee, J. J., Atkinson, D., Lam, J. J., Hackett, L., Lohr, R., Larson, L., Loo, R. et al., "MEMS in Antenna Systems: Concepts, Design, and Systems Implications," National Radio Science Meeting, Boulder, Colorado, 1998.

73. *The ARRL Antenna Handbook*, The American Radio Relay League, Inc.

74. Gauthier, G. P., Courtay, A., and Rebeiz, G. M., "Microstrip Antennas on Synthesized Low Dielectric-Constant Substrates," *IEEE Transactions on Antennas and Propagation*, Vol. 45, No. 8, August, 1997, pp. 1310–1314.

75. Nguyen, C T. C., Katehi, L. P. B., and Rabeiz, G. M., "Micromachined Devices for Wireless Communications," *Proceedings of the IEEE*, Institute of Electrical and Electronics Engineers, Vol. 86, No. 8, August, 1998, pp. 1756–1768.

76. Veidt, B., Kornelsen, K., Vaneldik, J. F., Rutledge, D., and Brett, M. J., "Diagonal Horn Integrated with Micromachined Waveguide for Sub-Millimetre Applications," *Electronics Letters*, Vol. 31, No. 16, August 3, 1995, pp. 1307–1309.

Section 3: Antenna Metrology

1. *IEEE Standard Test Procedures for Antennas*, IEEE Std. 149-1979, The Institute of Electrical and Electronics Engineers, 1979.

2. Hollis, J. S., et al., *Microwave Antenna Measurements*, Scientific-Atlanta, Inc., Atlanta Georgia, November, 1985.

3. Evans, G. *Antenna Metrology*, Artech House, Norwood, MA, 1990.

4. Hewlett Packard, H. P. "Understanding the Fundamental Principles of Vector Network Analysis," Application Note 1278-1, Hewlett Packard Company, No. 5965 –770E, May, 1997.

5. Hewlett Packard, H. P. "Antenna Measurements Using the HP 8753C Network Analyzer," Product Note 8753-4, Hewlett Packard Company, 5952 –2776, October, 1990.

6. Rudge, A. W., Milne, K., Olver, A. D., and Knight, P., *The Handbook of Antenna Design*, Vol. 1, Peter Peregrinus, Ltd, London, UK, 11982, pp. 594–628.

7. Slater, D., *Near Field Antenna Measurements*, Artech House, Inc., Norwood, MA, 1991.

8. Kerns, D. M., *Plane-Wave Scattering-Matrix Theory of Antenna and Antenna-Antenna Interactions*, National Bureau of Standards Monograph 162, Washington, DC, 1981.

9. Balanis, C. A., *Advanced Engineering Electromagnetics*, John Wiley and Sons, New York, 1989, pp. 329–334.

Section 4: Emerging Technologies

1. Eleftheriades, G. V., and Balmain, K. G., Editors, *Negative-Refraction Metamaterials*, IEEE Press, Wiley Interscience, John Wiley and Sons, Hoboken, NJ, 2005.

2. Yeo, J., and Mittra, R., *Performance Enhancement of Communications Antennas using Metamaterials*, short course notes, IEEE APS International Symposium and USNC/URSI Meeting, Washington, DC, 2005.

3. Kidal, P. S., Maci, S., and Sievenpiper, D. F., *Theory and Applications of PBG Structure used as Artificial Magnetic conductors and Soft and Hard Surfaces*, short course notes, IEEE APS International Symposium, Columbus, OH, 2003.

4. Balanis, C. A., *Antenna Theory, Analysis and Design*, 3rd Ed., John Wiley and Sons, Hoboken, NJ, 2005, Chapter 16, pp. 945–1000.

5. Veselago, V. G., "The Electrodynamics of Substances with Simultaneously Negative Values of ε and μ," *Soviet Physics Usp*, Vol. 10, pp. 509–514, 1968.

6. Pendry, J. B., "Negative Refraction Makes a Perfect Lens," *Physics Review Letters*, Vol. 85, pp. 3966–3969, 2000.

10

Phased Array Antenna Technology

James B. West
*Rockwell Collins—Advanced
Technology Center*

10.1 Introduction

An *array antenna* is a structure composed of a collection of individual radiation elements, each excited with a specific amplitude and phase, whose electromagnetic fields are spatially combined to realize a specific radiation pattern, usually in the far field. An array antenna can also be thought of as a continuous aperture distribution that is sampled at discrete locations within the aperture. A *phased array* antenna [1–4] is an array whose main beam can be electrically steered in one or more dimensions by changing the excitation phases on the individual radiation elements of the array. Beam scanning rates on the order of 1 µs are possible for certain array architectures, which is orders of magnitude more rapid than state-of-the-art mechanical scanning with motor technology. Rapid beam scanning is crucial in fire control and other radar systems where wide scan volumes must be rapidly interrogated and where multiple target tracking is a requirement.

Array antennas can take on one of four forms: (1) a *linear array* where the array radiating elements are in a single line, (2) a *circular array* where the radiating elements are laid out in a circular arrangement relative to a center reference location, (3) a *planar array*, where the radiating elements are laid out on in a two-dimensional grid, and (4) a *conformal array*, where the radiating elements locations are described by a three-dimensional surface. Figure 10.1 illustrates the various array architectures.

The discussion herein will focus on linear phased array technology for brevity. The fundamental phased array concepts developed for linear arrays can be extended to two-dimensional and conformal arrays. The reader is referred to the literature for more detailed expositions.

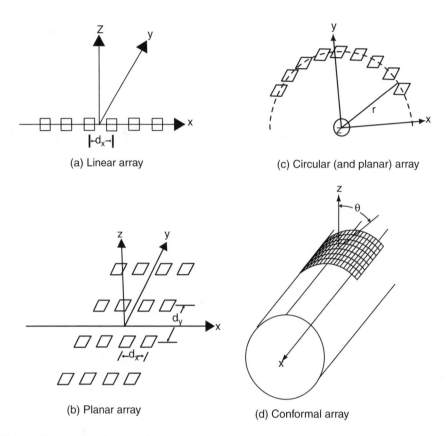

FIGURE 10.1   Various phased array architectures.

10.2   Linear Array Technology

A linear array is conceptually illustrated in Figure 10.2. This structure consists of a collection of *radiating elements* laid on in a straight line that form the transition from free space to the array feed system. Several of the antennas summarized in the antenna catalog are appropriate choices for phased array design. Dipoles; monopoles; crossed dipoles; bow tie monopoles; open sleeve dipoles; edge and broad wall waveguide-fed slots; printed Vivaldi notches; stripline- or microstrip-fed linear and crossed printed slots; circular, rectangular, and ridged open end waveguide; TEM horns; microstrip patches; and printed dipoles are commonly used. Each radiating element has a radiation pattern, which along with its element amplitude and phase excitation; contribute to the aggregate radiation pattern. The *feed system* is an apparatus that distributes the microwave energy between the radiating elements and the receiver/transmitter. The feed systems can either be constructed out of guided wave transmission lines, or be a radiation-based space feed. *The phase shifter* is a microwave/millimeter wave circuit device that outputs variable insertion phase, or time delay for a broadband array system, relative to its input. By adjusting the insertion phase of each phase shifter, the proper relative phase shift across the radiating elements is realized for beam scanning. *Amplitude weighting* is the nonuniform distribution of power across the array and is used to generate low side lobe far-field radiation patterns. The amplitude weighting is usually implemented in the feed system, but is shown conceptually as blocks A_{-N} through A_N in the figure. The beam steering computer (BSC) computes the proper insertion phase value for a given beam pointing direction and sends the appropriate commands to each phase shifter.

    The first order array analysis assumes isotropic radiating elements that are isolated from one another, that is, there is no *mutual coupling*. Mutual coupling causes a given radiating element's impedance

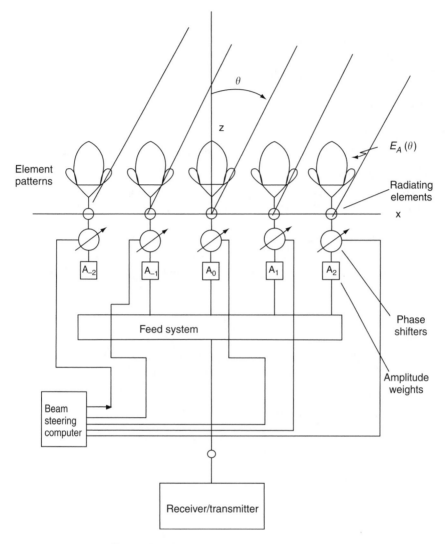

FIGURE 10.2   Center section of linear phased array.

properties, and its radiation pattern, to be effected by its proximity to neighboring radiating elements, and their individual amplitude and phase excitations. This effect will be discussed in further detail in subsequent paragraphs.

Consider a linear array in the transmit mode. In the far field, the observation point is at infinity, which causes the line-of-site rays from each radiating element and the observation point to be parallel. This approximation leads to the following expression for the far field electric field patterns as a function of θ:

$$E_A\left(\theta\right) \equiv \underbrace{\left[\frac{\exp\left(-j\frac{2\pi r_0}{\lambda}\right)}{r_0}\right]}_{\substack{\text{Isotropic} \\ \text{element} \\ \text{pattern}}} \times \underbrace{\sum_N A_n \exp\left[-j\frac{2\pi}{\lambda}n\Delta x(\sin\theta - \sin\theta_o)\right]}_{\text{Array factor}} \tag{10.1}$$

where

N = number of radiating elements
r_0 = distance from the center of the array to the far field observation point
A_n = array element amplitude weighting coefficients
θ = beam pointing, referenced to broadside ($\theta = 0$)
Δx = the array element spacing
λ = wavelength
θ_0 = the desired beam scan angle
$\Phi_n \equiv \frac{2\pi}{\lambda} n \Delta x \sin \theta_0$ = The insertion phase for the nth phase shifter required to scan the beam to θ_0.

The isotropic pattern does not influence the far field amplitude pattern of the array and is typically dropped from the expression. One can see that the number of radiating elements, their relative spacing, the operating frequency (wavelength), and phase shifter settings of the array all influence the radiation pattern.

Only one main beam is desired in most system applications. Secondary or false main beams are called *grating lobes* and they are a function of: (1) the array element spacing, and (2) the scan volume of the phased array. Consider Equation 10.1. All radiation elements add constructively in phase when the term

$$\frac{1}{\lambda} n \Delta x (\sin \theta - \sin \theta_0) \equiv m, \quad \text{where } m = 0 \pm 1, 2, 3, \ldots \tag{10.2}$$

$m = 0$ is the principal beam and the others represent grating lobes. Grating lobes, as a function of element spacing and beam scanning can be predicted by the following expression:

$$\sin \theta_{gl1} \equiv \sin \theta_0 + \frac{\lambda}{\Delta x}, \tag{10.3}$$

where $\sin \theta_{gl1}$ is the first grating lobe in visible space.

If the array is required to scan over the $\pm \theta_m$ scan volume, then the required spacing is

$$\Delta x \leq \frac{\lambda}{\sin \theta_m}. \tag{10.4}$$

If the array interelement spacing is one half free space wavelengths, then the array can scan $\pm 90°$ without grating lobes. Figure 10.3 illustrates the relationship between the main beam, grating lobes, and the region of visible space for a given array element spacing.

Amplitude weighting is used to create low side lobe patterns. The theoretical side lobe level for a uniformly excited linear array is -13.5 dB relative to the main beam. The uniformly illuminated array also has maximum gain and minimum beam width for a given number of array elements and interelement spacing. Lower side lobe level designs require a nonuniform amplitude illumination across the array. There is a trade off between gain and beam width, and side lobe level. Low side lobe level arrays exhibit lower gain, and increased beam width for a given array relative to uniform illumination.

The *directivity* of a linear array of ideal isotropic architecture elements can be expressed as

$$D_0 \equiv \frac{\left| \sum_N A_n \right|^2}{\sum_N |A_n|^2}, \tag{10.5}$$

where A_n = the element excitation coefficients.

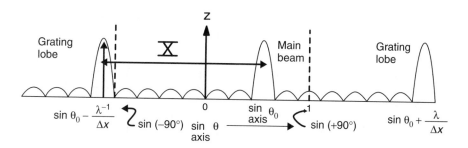

- $\mathbf{X}$ is a function of element spacing
- Grating lobes and the main Beam shift in concert along the sin θ axis as the array is scanned
- The left grating lobe is on the edge of visible space

FIGURE 10.3 Relationship of grating lobe to the main beam.

The reduction in gain due to amplitude weighting can be expressed as aperture efficiency by the following formula:

$$\eta_{ap} \equiv \frac{\left|\sum_N A_n\right|^2}{N \sum_N |A_n|^2},$$ (10.6)

where A_n is the element excitation coefficients.

Several pattern synthesis methodologies have been created for low side lobe level designs for continuous line sources and two-dimensional apertures. These procedures have been extended to array theory by continuous aperture sampling at the array element locations, or null matching techniques [5]. The most commonly used procedure is the *Taylor weighting*. The Taylor design yields a far field radiation pattern that is an optimal compromise between beam width and side lobe level. The details of the generation of the Taylor coefficients are beyond the scope of this article and the reader is referred to the literature [6]. Figure 10.4 compares uniform illumination, a -20, -25, and -30 dB Taylor side lobe level designs.

Realistic radiation elements are not isotropic, but have an element pattern, and element pattern weights the array factor, as shown in Equation 10.7:

$$E_A(\theta) \equiv \underbrace{[E_e(\theta)]}_{\text{Element pattern}} \times \underbrace{\sum_N A_n \exp\left[-j\frac{2\pi}{\lambda} n\Delta x(\sin\theta - \sin\theta_0)\right]}_{\text{Array factor}}$$ (10.7)

The isotropic element pattern is suppressed in this expression for clarity. A useful approximation for an ideal element is

$$E_e(\theta) = \cos(\theta),$$ (10.8)

where θ is referenced off the Z axis.

The ideal element pattern takes into account the aperture projection loss as the array scans off bore sight, as illustrated in Figure 10.5, but does not account for the effects of mutual coupling between the radiating elements. All linear- and planar-phased arrays suffer at least a $\cos(\theta)$ scan loss. This loss can actually be more severe, that is, $\cos^x(\theta)$, where x is a real number greater than 1.0, when mutual coupling

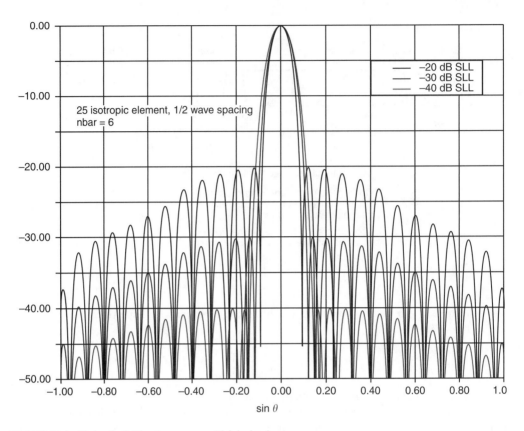

FIGURE 10.4   Taylor far field patterns versus sidelobe level.

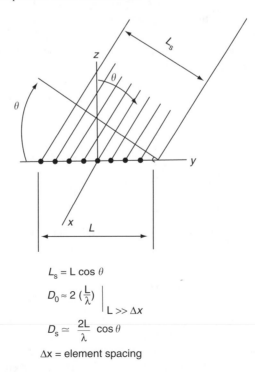

FIGURE 10.5   Aperture projection loss.

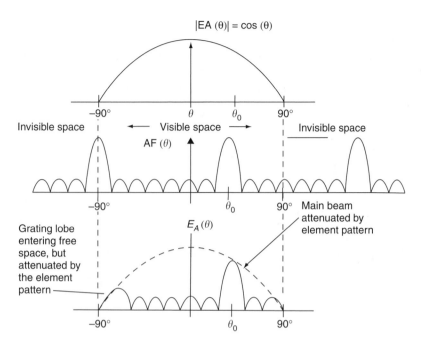

FIGURE 10.6 Phased array element pattern weighting.

is present. This effect will be described in subsequent paragraphs. Figure 10.6 illustrates the effect of the element pattern weighting on the main beam and side lobes of the array space factor. Several points are evident from this figure:

1. The main beam gain falls off at least by $\cos(\theta)$ with scan angle.
2. Grating lobes just coming into visible scan manifest themselves as high sidelobes.
3. A broad side array cannot scan to end fire since $\cos(90°) = 0$. Arrays can be designed to operate at under this "end fire" condition. The reader is referred to the literature [7].

Array bandwidth is affected by several parameters, including impedance, polarization, and radiation pattern bandwidth of the radiating elements, and mutual coupling between the radiation elements. The array factor experiences a beam squint as a function of frequency due to the frequency response of the phase shifters. For a linear array, the frequency change that causes the beam position to shift $\pm 1/2$ of its nominal 3 dB beam width can be expressed as a percent bandwidth by the following expression:

$$\text{Fractional band width} = \frac{\Delta f}{f} = \frac{BW_0}{\sin \theta_s}, \tag{10.9}$$

where

θ_s = the scan angle off bore site, in radians
BW_0 = the beam width of the array at bore site.

Broad bandwidth can be realized when the array phase shifters are replaced with true time delay (TTD) devices. Beam scanning is achieved by programming time delay into each radiating element to compensate for the difference in propagation time from each radiating element to an arbitrary observation point in the far field. In this case, the bandwidth is not limited by the array factor because there is no beam squint with frequency. The array bandwidth is ultimately determined by the radiating element bandwidth within the array environment along with feed network bandwidth.

10.3   Array Error Sources

Random amplitude and phase errors cause increased side lobe levels, pointing errors and reduced directivity [8]. Random phase and amplitude errors can be treated statistically. The phase error ϕ_n is described by a Gaussian probability distribution with zero mean and variance $\bar{\phi}_n^2$. The amplitude error δ_n has variance $\bar{\delta}_n^2$ and zero mean. The *average side lobe level due to random phase and amplitude errors* can be expressed as

$$\bar{\sigma}_I^2 = \left[\overline{\bar{\phi}^2 + \bar{\sigma}^2} \right] \times g_e, \tag{10.10}$$

where

$\overline{\sigma_I^2}$ = the average side lobe level relative to the isotropic side lobe level in the plane of the linear array axis

g_e = the element directivity.

The reduction of directivity for a linear array due to random phase and amplitude errors is given by

$$\frac{D}{D_0} = \frac{1}{1 + \bar{\phi}^2 + \overline{\sigma^2}}, \tag{10.11}$$

where

D = the directivity of the array with errors

D_0 = the directivity of the error free array.

The beam pointing error due to random phase errors is given by

$$\overline{\Delta^2} = \overline{\phi^2} \times \frac{\sum I_i^2 \frac{x_i}{\Delta x}}{\left(\sum I_i \left(\frac{x_i}{\Delta x} \right) \right)^2}, \tag{10.12}$$

where

$\frac{x_i}{\Delta x}$ = the array element position from the origin relative to the element spacing

I_i = the element amplitude weighting.

Most contemporary phased arrays incorporate digital phase shifters. The development discussed thus far has considered only a perfect linear phase front across the elements to initiate beam scanning. This linear insertion phase required for beam scanning is typically implemented with finite insertion phase values, for example, a 4-bit phase shifter has 180°, 90°, 45°, and 22.5° phase settings. In this case, the 180° bit is referred to as the most significant bit (MSB), while the 22.5° bit is referred to as the least significant bit (LSB). Any phase shifter required by the radiating element for a given beam position is approximated by linear combinations of these phase bits. The higher the bit count, the better the approximation is for a given analog phase slope across the array. Three to eight bits of insertion phase are typically used for practical phased array systems. These discrete phase approximations lead to errors in the beam pointing angle and can also deteriorate the array radiation pattern side lobe levels as a function of beam scanning.

In addition to random error effects, *quantization errors due to digital phase shifters* add additional errors. The number of bits in the phase shifters determines *the minimum beam steering increment* of the array by the following expression:

$$\theta_{\min} \cong \frac{1}{2^P} \times BW_0, \tag{10.13}$$

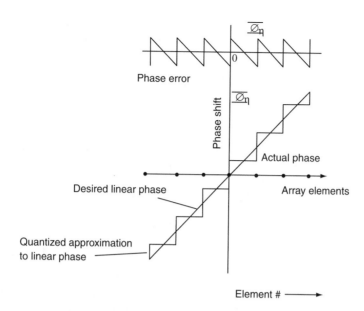

FIGURE 10.7 Quantization-induced phase error across a linear array.

where

P = the number of bits in the phase shifter,
$\theta_{\min}$ = the minimum steering increment of the array
BW_0 = the bore site 3 dB beam width of the array.

It can readily be seen that a 4-bit phase shifter allows a minimum beam steering increment of 1/16 of the 3 dB beam width.

The following development is after Miller [9]. The digital approximation to a linear phase front required to scan the main beam creates a sawtooth phase error across the array as shown in Figure 10.7. This is because a P-bit phase shifter has states separated by the least significant bit:

$$\phi_0 = \frac{2\pi}{2^P},\tag{10.14}$$

where

P = the number of phase shifter bits.

The peak side lobe level due to phase quantization can be approximated as

$$SLL_{pk\_QL} = \frac{1}{2^{2P}},\quad \text{or}\quad SLL_{pk\_DL}(dB) = -6P.\tag{10.15}$$

This result is shown in tabular form below:

TABLE 10.1 Peak Side Lobe Level vs. Phase Shifter Bit Number

Number of Bits	Peak SLL (dB), Relative to Main Beam
3	−18
4	−24
5	−30
6	−36
7	−42

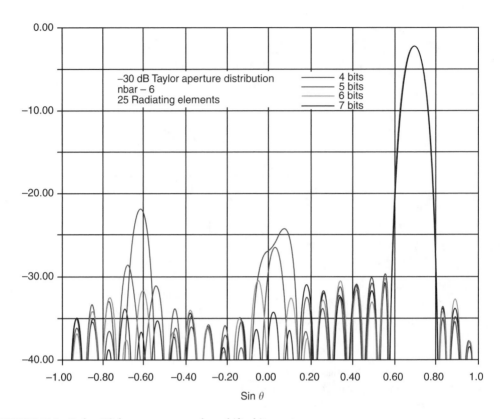

FIGURE 10.8   Taylor 45°, beam scan versus phase shifter bit count.

Figure 10.8 shows how the peak side lobe level is affected by the number of bits of the phase shifting device.

The average side lobe level due to the triangular phase error has also been analyzed. The mean squared error is given as

$$\overline{\phi^2} = \frac{1}{3} \times \frac{\pi^2}{2^{2P}}. \tag{10.16}$$

The average side lobe level due to quantization errors alone is

$$\overline{\sigma_{QL}^2} = \frac{1}{3g_A} \times \frac{\pi^2}{2^{2P}}, \tag{10.17}$$

where

  g_A = the directivity of the linear array.

The effects of phase quantization on pointing errors can be described by

$$\Delta_{QL} = \frac{\pi}{4} \times \frac{1}{2^P} \text{ beamwidths.} \tag{10.18}$$

The periodic nature of the sawtooth phase error function can be broken up by adding a random phase offset at each element [10]. The random, but known, offsets are included in the calculation of the phase shifter settings for a given beam position. The procedure tends to "whiten" the peak quantization side lobes level without changing the average phase error level [11].

10.4 True Time Delay Steering and Subarrays

Time delayed phased arrays replace the phase shifter devices in traditional devices with programmable time delay units at each radiating element. Scanning is realized by using time delay to compensate for the differences in propagation time from each radiating element to the target. The propagation delay across the phased array aperture becomes significant as the array is scanned off bore site. The instantaneous bandwidth is thus not limited by the antenna array factor, that is, there is no beam squint with frequency. TTD devices have historically been lossy, bulky, heavy, and expensive, but recent advances in Micro-machined ElectroMechanical Switches (MEMS) hold promise for high performance at moderate input power levels in compact monolithic semiconductor architectures. A time delay-steered phased array is illustrated in Figure 10.9. After a series of mathematical simplifications, the array factor for a time delay-steered phased array can be expressed as

$$E(\theta) = \sum_{n=1}^{M} A_n \exp\left[\left(\left(j \times \frac{2\pi}{\lambda}\right) \times n \times \Delta x\right) \cdot (\sin\theta - \sin\theta_0)\right]. \qquad (10.19)$$

The maximum of the electric field occurs at θ_0 independent of the wavelength (operating frequency), so no beam squint occurs. Two points should be noted: (1) the array beam width and gain are still a function

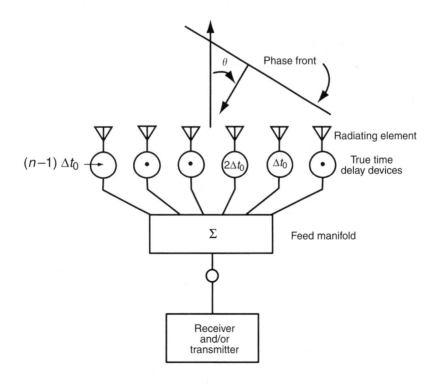

True time delay phased array architecture

FIGURE 10.9 True time delay phased array architecture.

of frequency and (2) the ultimate bandwidth of an actual array is dependent on the actual bandwidth of the radiating elements and the TTD devices used in the array.

A large phased array can be partitioned as a group of smaller array called *subarrays* [12]. Subarrays are used for the following: (1) improve the broadband performance of an array by incorporating a fewer number of TTD devices, (2) simplify the feed manifold structure, and (3) provide convenient amplifier/feed network integration.

Grouping an array into subarrays will generally increase the side lobe level of the array, as illustrated in Figure 10.10. Random errors have a larger effect at the subarray level, as opposed to the individual element level, since there are typically much fewer subarrays than individual radiation elements, and these errors tend to be statistically correlated. Peak side lobe errors are also larger because they are manifested as grating lobes since the subarray's physical implementation requires a large electrical spacing between adjacent subarrays.

The subarrayed phased array antenna, as shown in Figure 10.11, can be treated as the product of the radiation element pattern, the subarray pattern, and the array factor for the array of subarrays. This

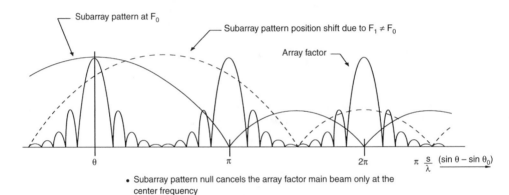

FIGURE 10.10 Subarray sidelobes as a function of frequency.

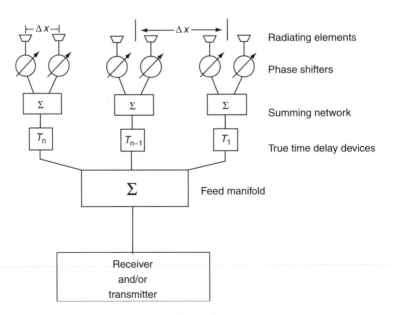

FIGURE 10.11 True time delay subarrayed architecture.

analysis neglects the effects of element pattern edge effects and mutual coupling between the radiation elements. Under these conditions, the electric field expression can be written as

$$E(\theta) = \sum_{m=1}^{M} W_m \times \exp\left[j\left(\frac{2\pi}{\lambda}\right) \times m \times \Delta s \times (\sin\theta - \sin\theta_0)\right]$$

$$\times \sum_{n=1}^{N} A_n \times \exp\left[j\left(\frac{2\pi}{\lambda}\right) \times n \times \Delta x \times \left(\frac{\sin\theta}{\lambda} - \frac{\sin\theta}{\lambda}\right)\right] \qquad (10.20)$$

where

Δx = the element spacing within the subarray
M = the number of radiating elements within each subarray
A_n = the amplitude weighting within each subarray
Δs = the subarray spacing within the array
N = the number of subarrays in the array
W_m = the amplitude weighting of the subarrays across the array

The bandwidth of this type of array is essentially the bandwidth of an individual subarray. Grating lobes generated off frequency due to the array factor ultimately sets the array bandwidth. Unequal array spacing techniques can be used to reduce these grating lobes at the expense of higher overall side lobe level for the top-level array [13].

Subarray level time delay steering without the generation of grating lobes can be accomplished by means of overlapping subarrays. The technique involves pattern synthesis of an aperture size larger than the intersubarray spacing. Flat topped, narrow, subarray patterns are possible that can suppress the array factor grating lobes, but the required feed networks become quite complex [8].

10.5 Feed Manifold Technology

The *feed manifold* in a phased array is a guided wave electromagnetic wave system that distributes the RF energy from an input and out port to each of the radiating elements in some prescribed manor. Transmission line technology is well documented in the literature and is therefore not described herein [18]. The introductory discussion herein will assume the antenna is in a transmission mode, but the principals are directly applicable to the receive case due to reciprocity. *Constrained feeds* utilize guided wave transmission line to distribute the microwave energy to the radiating elements. A *space feed* utilizes a smaller antenna system that excites the main phased array through radiation between the feed antenna and the receive elements of the phased array. The signals are then phase shifted for beam steering, perhaps amplified, and perhaps amplitude weighted for pattern control, and finally transmitted. Space feeds are much simpler to implement than constrained feeds for very large phased arrays, albeit with commensurate performance trades.

Constrained phased array feeds are generally classified as either *series*, where the radiating elements are in series with the feeding transmission line, or *shunt*, where the radiating elements are in parallel with the feeding transmission line. Series array feeds can be further subdivided into either *resonant* or *traveling wave* feeds. Parallel feeds can be further subdivided into *corporate* and *distributed* feeds. Figure 10.12 illustrates the various feed architectures.

Series feeds are typically used for waveguide linear arrays. Microstrip patches can also be series fed via interconnecting series microstrip transmission lines, but potential problems with this approach include spurious radiation off the series transmission lines, particularly for low side lobe level designs, and limitations of element amplitude tapering.

In resonant series feed design, each radiating element should be resonant (real impedance) at the operating frequency. The sum of the radiating element's resistance (or conductance) must equal the

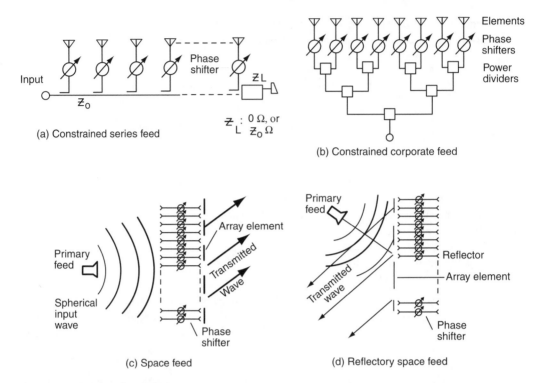

FIGURE 10.12 Phased array feed.

characteristic impedance (admittance) of the feed transmission line. Short-circuited quarter wave or half wave transmission line sections are used as the feed line loads to excite the radiating elements at voltage or current standing wave peaks. The resonant array exhibits high efficiency since all of the input power to the feed is radiated (ignoring dissipative losses in the system). The feed is inherently narrow band since the standing wave peaks must be precisely located at the radiating element, which strictly happens only at one frequency. It can be shown [11] that for a series resonant array, the relationship between the number of array elements and voltage standing wave ratio (VSWR) is

$NB = 66$ for a uniformly excited array
$NB = 50$, for a 25 dB SLL Taylor aperture distribution,

where

N = the number of array elements in the series-fed array
B = the % Bandwidth of the array.

A trade between the number of radiating element in a series feed and the percentage bandwidth is immediately apparent. Parallel feeding of smaller series-fed subarrays is sometimes used to increase bandwidth.

A *traveling wave feed* also has radiating elements distributed along a feed transmission line, but the input is at one end of the transmission line and a load is at the opposite end of the feed, as shown in Figure 10.12a. As the input wave propagates from the input toward the load, a portion of the input power is coupled to free space by each radiation slot, and if the feed is properly designed, only a small fraction of the input power is absorbed by the load. Since this structure is nonresonant, wider bandwidths can be realized, but there is beam squint as a function of frequency. Also, this type of array is less efficient than the resonant array since approximately 15% of the power is typically delivered to the load [16]. The reader is referred to the literature for series feed design details.

Corporate feeds are typically used in dipole, waveguide aperture, microstrip patch, and printed slot radiating elements. In this architecture, there are successive power subdivisions, until all of the radiating elements of the array are excited. Figure 10.12b illustrates a binary corporate feed exciting an eight-element linear array. The division ratio at each junction is typically binary, but it can be from two to five, depending on the number of radiating elements and the type of power splitters used. The performance of a corporate feed is critically dependent on the power splitter architecture used. Hybrid couplers such as the Wilkenson, Lange, or branch line hybrids are typically used since they reduce the detrimental effects of radiating element impedance mismatch on the feed, which causes amplitude and phase imbalance. This imbalance can deteriorate the array's radiation pattern, particularly in low side lobe level designs.

Distributed feeds have each radiating element, or a group of radiating elements, connected to their own receiver/transmitter module. Contemporary examples of this approach are based on MMIC Transmit/Receive (T/R) modules, where each T/R module can contain a transmit, low-noise amplifier (LNA), phase shift, attenuation, low and higher power T/R switch, and bias- and beam-steering control logic functions. Collections of distributed feeds for a large array are typically fed with a passive corporate feed. Depending on the complexity of the T/R module, the passive corporate feed can operate at an intermediate frequency (IF) with the frequency translation to the actual radiated carrier frequencies occurring within the T/R module. The reader is referred to Chapter 25 of the companion volume *RF and Microwave Circuits, Measurements, and Modeling* for a detailed description of MMIC-based T/R modules.

The actual hardware implementation of the feed manifold, phase shifter, and time delay device technology are covered in Chapter 9 of this handbook.

Space-fed arrays typically take one of two basic forms, the in-line space-fed array, and the reflect array, as shown in Figure 10.12c. The in-line space-fed phased array is analogous to exciting a lens structure. The array collects the energy radiated from the primary feed, compensates for its spherical nature, and adds the desired phase shifter for beam scanning, while the output set of radiating elements reradiate the beam into free space. This type of array has a natural amplitude pattern due to the spherical wave spreading of the primary feed's radiated energy. Low side lobe level antennas can be designed, but careful attention must be applied to the feed antenna design to minimize spill over loss. Space-fed arrays are currently the only practical alternative for extremely large arrays, for example, space-based systems.

A variation of the space-fed array is the *reflect array*, as shown in Figure 10.12d. Here the array receives the energy from the primary feed, collimates and phase shifts it to realize the desired beam position, and reradiates it from the same set of radiating elements used to receive the signal due to the reflection off the back plane of the array. This arrangement is convenient for the following reasons: (1) the phase shifters only have to realize one half of the required phase shift, since the signal passes through the phase shifter twice and (2) the control and bias circuitry of for the phase shifters can be conveniently be mounted on the back of the reflector plane. One disadvantage of this approach is the aperture blockage of the primary feed of the antenna.

10.6 Element Mutual Coupling Effects and the Small Array

The mathematical array theory development to this point has assumed ideal radiating elements. The impedance properties and radiation pattern of each element is assumed to be independent of its environment, unaffected by its neighboring elements, and identical. Array pattern multiplication is valid under these conditions. However, this is rarely the case in actual arrays. *Mutual coupling* describes the interaction between one element and all others. Mutual coupling has been analyzed as a simple current sheet model by Wheeler, and also with infinite array techniques, finite array techniques, and with measurement-based techniques [17,18]. The infinite array exploits periodicity of the array structure to reduce the problem to a "unit cell," which allows significant mathematical simplification. This technique is commonly used.

Several finite array techniques, such as direct impedance formulations (Method of Moment) and other semiinfinite array techniques can be applied to the finite array problem. Measurement-based techniques using measured scan impedance and scan element gain pattern data are commonly used. A variation of the measurement-based approach is to simulate measured data with an electromagnetic simulator data on small test arrays as though the resulting analysis were measured data.

Wheeler has derived fundamental impedance and element radiation pattern properties of planar phased arrays from an ideal phased current sheet formulation. The scan resistance of a phased array increases in one plane, and decreases in the perpendicular plane, and the reflection coefficient is given as

$$\Gamma = \tan^2(\theta), \tag{10.21}$$

where θ is the scan angle off boresite.

This analysis does not account for radiating element reactance. Another outcome of the Wheeler analysis is to derive the "ideal" element pattern:

$$E(\theta) = \sqrt{\cos(\theta)}, \tag{10.22}$$

where θ is the scan angle off boresite.

Equations 10.21 and 10.22 illustrate that the gain of on ideal phased array falls off rapidly beyond 45° scan off boresite. This is manifested both as a roll off of element pattern gain, and resistive impedance mismatch. These equations suggest that you cannot scan a broad side array to end fire. Special conditions for end fire performance exist and are described in the literature [18].

The above analysis is idealized in the sense that it does not account for the impedance and mutual coupling properties of practical radiating elements.

Array element mutual coupling can be modeled in terms of multielement S-parameter network theory. Consider Figure 10.13, which illustrates linear array in the transmit mode. The development equally applies to the receive mode through reciprocity. Each element radiates a signal that is received by neighboring signals, and is subsequently reradiated. The impedance properties of a given element are due to its self, or *driving point impedance*, which is the impedance of an isolated element, and the mutual impedance due to signals arriving from neighboring elements. Array impedance can be formulated as a multiport S-parameter network. The analysis, termed *free excitation*, assumes that the feed network is replaced by constant available power sources connected to each radiating element. Under the conditions on an electrically large array, the scan reflection coefficient for the (m,n)th array element is given by:

$$\Gamma_{mn} = \sum_{p} \sum_{q} S_{mn,pq} \cdot \frac{A_{pq}}{A_{mn}}, \tag{10.23}$$

where $S_{mn,pq}$ is the coupling coefficient between the m,nth, and p,qth radiating elements, and, A_{pq}, A_{mn} are amplitude coefficients.

It is possible for the scan reflection coefficient to have a magnitude of 1, in which case total reflection exists at the element port. This condition is termed *scan blindness* and one electromagnetic interpretation is that it is due to the generation of surface waves, traveling on the array surface, that do not radiate into free space.

The *scan impedance*, which is closely related to the scan reflection coefficient, is defined as the impedance of a radiating element as a function of scan angle, when all of the radiating elements of the array are excited with the proper magnitude and phase. The scan impedance of a nonideal radiating element is a function of the phase front across the array aperture.

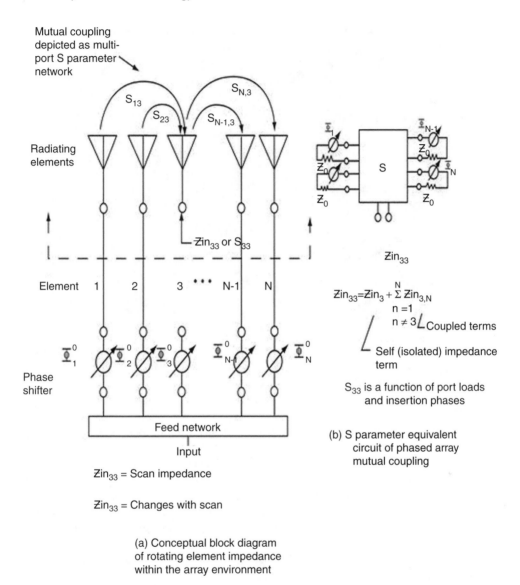

FIGURE 10.13 Mutual coupling for an N-element linear phased array.

Mutual impedance also affects the element gain pattern. The scanned element pattern can be approximated as

$$g_s(\theta) \cong \frac{4\pi A_{elem}}{\lambda^2} \times \cos(\theta) \times (1 - |\Gamma(\theta)|^2), \tag{10.24}$$

where

$g_s(\theta)$ = scanned element pattern
A_{elem} = a unit cell radiating element area
λ = the operating wavelength
θ = the scan angle off bore site.

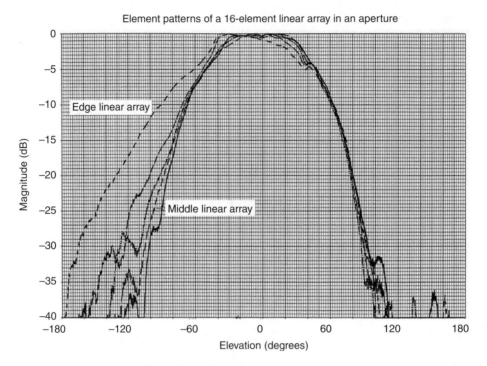

FIGURE 10.14   Element patterns of a 16-element linear array in an aperture.

This expression is applicable to electrically large arrays where no grating lobes exist, and where higher modes can be neglected.

Edge effects are present due to asymmetries at the edge of the array face. These effects are neglected in infinite array analysis, and often can be neglected in the design of electrically large arrays. There effects, however, can be significant in small array design. The radiation pattern of an edge element can be significantly different than that of an element in the center of the array. Figure 10.14 illustrates the asymmetries present in the H plane across an edge-slotted waveguide aperture consisting of 16 edge slot linear arrays spaced at 0.764 λ.

Measurement-based array design uses experimental measurements of small test arrays, or electromagnetic simulations that predict the same test array parameters. The scan reflection coefficient formulation of Equation 10.23 can predict reflection coefficient, which in turn is used to calculate an effect of mismatch loss, and worst case, scan blindness as a function of scan. Pattern measurements of individual radiation elements, with all other elements properly terminated in a matched load, will give the scanned element pattern. Nulls in the scanned element patterns are related to scan blindness at specific scan angles, assuming that these nulls are not present in an isolated element.

The scan performance of a linear array with mutual coupling can be predicted from the following expression:

$$E_A(\theta) \equiv \cdot \sum_{N} \underbrace{[E_{en}(\theta)]}_{\text{Element pattern}} \ A_n \exp\left[-j\frac{2\pi}{\lambda}n\Delta x(\sin\theta - \sin\theta_0)\right] \qquad (10.25)$$

$$\underbrace{\phantom{E_A(\theta) \equiv \cdot \sum_{N} [E_{en}(\theta)] \ A_n \exp\left[-j\frac{2\pi}{\lambda}n\Delta x(\sin\theta - \sin\theta_0)\right]}}_{\text{Array factor}}$$

In this expression each element pattern can be different to account for the effects of mutual coupling, edge array effects, and so forth.

10.7   Thinned Arrays

Large arrays of regularly spaced radiation elements are often prohibitively large, expensive, and can exhibit excessive mutual coupling. Beam width specifications often dictate the length of the array. It is possible to retain the same array length, but thin the number of elements to retain an equivalent beamwidth with a commensurate loss in gain. Periodic thinning of the array element in a uniformly spaced array will create grating lobes, but randomizing the element positions (spacing) can offset this effect. A useful probabilistic approach to thinned array design was formulated by Aggarwal and Lo [19], where the array sidelobe level threshold is described by a probability density function. The probability of power side lobe level being below a threshold is given b:

$$P = [1 - \exp(-\alpha)] \exp\left(-\frac{2}{\lambda}\sqrt{\frac{\pi\alpha}{3}}\exp(-\alpha)\right), \qquad (10.26)$$

where

α = N/SLR
L = the linear array level
N = the number of elements
SLR = the power side lobe level.

The N parameter sets an average element spacing for a given array length L. The use of directive radiation elements can reduce far out side lobe levels. An analysis by Steinberg [20] reveals that the use of directive element can reduce the overall length of the linear array described in Equation 10.26, by the following:

$$\frac{L_{\text{eff}}}{\lambda} = \frac{L/\lambda}{\sqrt{\alpha}(W/\lambda)}, \qquad (10.27)$$

where

L = the linear array length
W/λ = the electrical size of the radiating element.

10.8   Phased Array Component Technologies

The previous section addressed basic linear array phased array theoretical concepts that are readily extendable to planar and generally conformal array apertures. The following discussion emphasizes the current state of the industry in phased array hardware.

Monolithic microstrip antenna-phased arrays are becoming increasingly important due to more stringent systems needs and recent advances in integrated circuit technology, cost effective semiconductor manufacturing and test methodologies. The required dimensions of the microstrip radiating elements at millimeter wave frequencies are on the same order as the active circuit devices, thus making monolithic integration possible. The radiating elements are integrated with the T/R module subcircuits, such as power amplifiers, LNAs, phase shifters, attenuators, and T/R switches. The details of T/R module technology are described in Chapter 25 of the companion volume, *RF and Microwave Circuits, Measurements, and Modeling*. Packaging to protect the semiconductor device from mechanical damage and environmental harshness is a current research topic within the industry [21]. Additional challenges to the antenna designer include bandwidth, radiation efficiency, and surface wave/scan blindness concerns due to the electrical parameter constraints dictated by Silicon (Si) and Gallium Arsenide (GaAs) semiconductor substrate technology.

Micromachining technology is used to mitigate mutual coupling due to grounded dielectric slab substrate surface wave modes in wide beam scanned microstrip-phased array applications. This problem is particularly troublesome for substrates with a high relative dielectric constant, such as GaAs and Si, the typical semiconductor substrates used for integrated radiating element T/R module configurations. Micromachining techniques can be used to synthesize a localized artificial dielectric material with low dielectric constant for microstrip antenna substrate from a high dielectric substrate. The lower effective dielectric constant improves the radiation efficiency and reduces surface waves for a given substrate thickness. Gauthier et al. describe a method of drilling a densely spaced two-dimensional grid of holes into the substrate to reduce the local dielectric constant of the substrate [22]. A reduction of dielectric constant to 2.3 from the intrinsic substrate dielectric constant of 10.8 was reported. Nguyen et al. [23] describe micromachining techniques that are used to suspend dipole or slot radiators on a very thin membrane to realize a local substrate dielectric constant of approximately 1.0, which is quite useful for millimeter wave applications. Papapolymerou, et al. [24] have demonstrated a 64% bandwidth improvement and 28% radiation efficiency improvement with the membrane technique for substrate materials with dielectric constants similar to Si.

MEMS are miniature semiconductor switches used in a variety of antenna applications. Rebiez provides an excellent overview of MEMS technology for antenna RF (radio frequency) applications, including a state-of-the-industry assessment of switch lifetime and packaging technology [25]. The MEMS switch is physically small; has very low on state loss; high off state isolation; and low average dc power consumption, as compared to ferrite waveguide and semiconductor switching elements such as PIN diode and FET transistor switches. These features have led some researchers to propose MEMS phase shifter-based, passive electronically scanned array (PESA) architectures as a viable alternative to the monolithic T/R module-based, active electronically scanned array (AESA) architectures for very large space-based military radar and communications systems [26]. The MEMS architectures compare favorably in terms of weight, dc power consumption, and required system effective isotropic radiated power (EIRP).

An intriguing application of MEMS device technology is for reconfigurable antenna technology [27,28]. The desire to use multiband, multifunction, shared asset apertures for advanced radar and communications systems has led researchers to consider radiating elements whose physical properties can be dynamically adjusted. The MEMS switch offers the potential to generalize this concept to two dimensions using a switching matrix.

Another potential application for MEMS switches is to dynamically impedance match phased array radiating elements as a function of beam scan. Mutual coupling between radiating elements within the array environment causes the scan impedance of the radiating element to change with beam scan position. A dynamic impedance-matching network consisting of switchable banks of reactive elements and planar transmission lines can be used to dynamically tune the radiating element.

Significant progress had been made to commercialize the technology, including life time reliability studies, and RF compatible, environmental packaging techniques. The reader is again referred to Rebiez [25].

Fractal Antenna Array Technology: Fractal radiating elements were previously discussed in the antenna Chapter 9 of this handbook. Fractal array antennas can also have fractal-based radiating element shape and array element spacing. Fractal-based arrays have been studied by many researchers [29–34]. A comprehensive overview of fractal antenna array theory is given by Werner, Haupt, and Werner [33]. Fractal arrays hold potential to combine the robustness of a random array and the efficiency of a periodically spaced array, but with a reduced element count. Fractal designs have also been applied to low side lobe level microstrip arrays and multiband array designs [34].

Genetic Algorithm-Based Phased Array Design [35–37]: Genetic algorithm (GA)-based radiating elements are addressed in the antenna Chapter 9 of this handbook and have utility for phased array application. GA-based optimization techniques can be utilized at the subarray and overall array factor level. GA-based optimal array thinning (reduced element count) is discussed in References 35 and 36 and optimal subarray amplitude tapering for low sidelobe level design is detailed by Haupt [37].

Frequency Selective Surfaces (FSS): FSS devices are used in multiband and wide band phased array applications and in applications where it is desirable to block signal transmission for one frequency band while passing another frequency band. A FSS can be roughly thought of as a spatial band pass filter to radiating RF signals. FSS use has extensively proliferated for general antenna problems, since its design theory was recently declassified and Munk's seminal text books were published [38,39]. FSS techniques are used in multiband, multilayer arrays and low radar cross section (RCS) array and radome design [40]. **Multiband arrays [41–45]:** Multiband arrays feature more than one frequency band of operation. This is important for systems such as X/Ku/Ka/Q Band DOD SATCOM, current and future commercial SATCOM and reconfigurable radar/communications for robotic armored vehicles, such as the U.S. Army's Future Combat Systems concept. Multiband operation is typically realized in one of the following ways: (1) multiband radiating elements located in a conventional rectangular or hexagonal array lattice [41], (2) nested array elements [42], and (3) multilayer FSS arrays [43].

Several examples of multiband radiating elements, such as microstrip patch variations, exist in the literature [44]. In addition, recent advances in hybrid microstrip/waveguide elements have been demonstrated by Zaghloul et al. [45]. The challenge with this approach is that the overall array lattice spacing is constrained to be one-half wavelength ($\lambda o_{high}/2$) at the highest operating frequency for grating-lobe free operation, as previously discussed. This results in an overly dense array at the lower operating band and puts severe physical constraints on the radiating element, since it cannot be larger than the $\lambda o/2$ unit cell at the higher frequency band. This often results in poor low-band radiating element performance. A compromise must often be struck between radiating element performance and scan performance.

The nested array approach attempts to interpose or superimpose one array lattice suitable for first operating frequency bandwidth and second array lattice appropriate for the second operating band. Again, the performance of both bands is often compromised, depending on the band separation, and so forth. The nested approach can generate unique cross band mutual coupling issues that can affect scan performance.

The multilayer dichroic/FSS-based array structures have been successfully demonstrated in the literature. In this scenario, the "top" array is electrically transparent to the lower level array's operating frequency while being opaque to the second band of operation.

Wide band Arrays [46–50]: Wide band phased arrays are becoming increasingly important to support the industry trends of broadband, reconfigurable radio systems and the move away from federated communications/navigation systems in DOD avionic, maritime, and terrestrial systems. Broadband ESA design is very challenging due to the need for broadband apertures with wide scan performance and the realization of practical low loss TTD circuits to prevent beam squinting and low loss feed networks.

Significant research efforts have been directed towards broadband ESA aperture design. Classic log-periodic concepts have been extended to linear printed and printed planar (2-D aperture) array architectures. Tripp's log-periodic planar array (LPPA) features a constant beam width and gain over very large, multioctave bandwidths that can be a key advantage for certain system applications [46]. It does not exhibit optimal aperture illumination efficiency, since, as with any log period structure, only a percentage of the aperture is illuminated for a given subband.

There has been extensive research on variations of the Vivaldi for broadband ESA applications. Researchers at Raytheon have demonstrated a dual polarized subarray bandwidth with scanning to 60° off bore site over a decade of bandwidth [47,48]. Vivaldi arrays have the disadvantage of requiring contiguous grounds across the adjacent radiating elements in dual polarized apertures. Solder metallization, for example, is very difficult to accomplish and also makes practical field repair extremely difficult. Researchers have addressed these issues in several ways.

ElSallal and Schaubert have demonstrated a linearly polarized balanced antipodal Vivaldi antenna (BAVA) element with a 3:1 bandwidth with excellent scan impedance performance to 45° off bore site [49]. This element does not require contiguous mechanical and electrical element-to-element contacts. Therefore, this element can be the basis of an integrated "plug and play" radiating element/TTD or T/R module that is readily field repairable.

Holter has demonstrated a dielectric-free, body of revolution "ogive shaped" radiating element that has excellent scan impedance performance over a decade of bandwidth in an infinite array environment [50].

Broadband arrays are typically required to have maximum aperture efficiency across a broadband. This results in an element "compaction" problem, similar to multiband array architectures, where the low-frequency band is oversampled (too many elements) for grating lobe-free operation, because the overall array lattice spacing is constrained to be one-half wavelength at the highest operating frequency for grating lobe-free operation ($\lambda o_{high}/2$). Researchers at the Naval Research Lab have developed a nonuniform array lattice concept that is grating lobe free across the band of interest and has superior performance to a random array lattice [51]. The highest frequency portion of the array is energized at the center elements and increasingly lower frequencies are excited at elements away from the array center. The active portion of the array effectively grows and shrinks with frequency through judicious radiating element design and the array lattice effectively performs as a $\lambda o/2$ lattice spacing at any band of operation. This architecture therefore strikes an optimal compromise between lattice density and element count.

Analog Beam Formers: Simultaneous multiple beam operation is of current interest in state-of-the-art radar and communications systems. Simultaneous, multibeam operation can be realized with both analog and digital beam former topologies. Digital Beam forming (DBF) is briefing discussed in the subsequent emerging technologies section. Classic simultaneous beam feed networks include the Butler and Blass matrices, variations of Rotman and Luneburg lens feeds, passive, cross bar and active vector modulation schemes. The Butler and Blass matrices create a suite of equiangular, orthogonal staring beams. Element level phase shift or time delay is required to independently electronically steer these beams. The reader is referred to Fourikis' text for a detailed exposition [52].

Smart Antennas: Smart systems are a fusion of phased array antenna, DBF and Digital Signal Processing (DSP) technologies [53]. Smart antennas are used in communications systems that sense, adapt, and optimize performance on the basis of spatial signal processing techniques. Smart systems *exploit adaptive, or agile, directional radiation patterns. Switched beam antennas* exploit a collection of common beam width antennas that are commutated to realize narrow beam width pattern coverage over a wide field of view. Cellular telephone systems and modern wireless communications systems are exploiting these technologies at frequencies below S Band. Smart antenna systems are used to perform the following tasks: *increase system gain*, null or *reduce parasitic interfering signals, compensate for signal fading due to multipath* in electromagnetically complex environments and provide *simultaneous multiple beam operation*.

The heart of a smart antenna system is the direction of arrival (DOA) and *adaptive beam forming* (ABF) algorithms, as implemented in DSP technologies. Adaptive nulling is very powerful due to the ability to simultaneously maximize the pattern gain for a signal of interest (SOI) while minimizing signals of noninterest (SONI). Balanis provides an excellent top level exposition of smart antenna systems [54].

10.9 Emerging Phased Array Antenna Technologies

Phased array technology is a vibrant research area, particularly within the DOD, with a multitude of new technologies currently under investigation. Brookner provides an excellent and succinct overview of the current phased array state of the art [1]. This section will briefly touch on some of the important ongoing research. The reader is referred to the literature for a more detailed exposition.

Photonics systems for the phased array application offer several advantages, including electromagnetic interference (EMI) immunity, greater than one octave signal bandwidths, low-radar cross-section, and extremely low loss for digital control and RF signal interconnections [2]. The photonic system can perform one, or all of the following system functions within the phased array architectures: (1) the distribution of digital control signals, (2) the distribution of RF energy to and from the array radiating elements, and (3) optical signal processing of the array's RF signals.

The transmission of digital control signals is required to adjust the amplitude and phase weighting for each radiating element within the phased array antenna architecture. Traditional interconnect schemes such as cable bundles, printed circuit boards, and flexible circuit board technology are very cumbersome for phased arrays with hundreds or thousands of radiating elements. Kunath et al. describe an optically controlled Ka Band phased array antenna that incorporates an OptoElectronic Interface Circuit (OEIC) that allows the distribution of control signals to individual antenna elements through a single fiber optic cable [3].

Microwave/millimeter wave feed waveguide manifold technology can be replaced by modulating the RF signal to optical frequencies for the distribution of RF energy to and from the phased array's radiating elements. Traditional microwave/millimeter wave-guided wave transmission line technology is very bulky, heavy, and inherently narrow band, as compared to future optical cable technology.

Optical based space-time signal processing is feasible once the RF signal is translated to the optical domain. The wide bandwidth of the optical channel would allow the high processing burden that is required for large DBF phased arrays. One possible T/R module optical control scheme is discussed in [4]. Phased shifting and TTD required for array beam scanning can be accomplished in the optical domain. Parent [5] and Newberg [6] discuss TTD schemes using optical technology.

Optical technology holds much potential for phased arrays, but significant technology advances are still required for improved optical-to-microwave frequency conversion efficiency (dc power consumption), physically compact and further maturity of electrical-optical integrated circuits (EPIC) device packaging, and embedded fiber interconnect technologies.

10.9.1 Material Advances

There has been interest in *ferroelectric materials* for several years within the phased array community. The ferro-material's dielectric constant can be modulated by adjusting a static electric bias field. It is therefore possible to build antennas that are monolithically loaded with ferro-material and realize electronic scanning by modulating the biasing electric field [7]. This is similar in concept to the classic waveguide frequency scanned array. Both phased array and steerable, space-fed lens prototypes have been demonstrated [7].

Progress in ferroelectric arrays has been slow to date due to the inability to fabricate ferroelectric materials with the appropriate tuning and dielectric constant parameters while simultaneously exhibiting low-insertion loss at microwave frequencies [8]. Ferroelectric materials have been demonstrated in three embodiments: bulk crystal, ceramic, and thin film. Theoretically, the bulk crystals should exhibit lower loss tangent properties but required thousands of volts of dc drive voltage to adequately modulate the static electric field for material samples of practical thicknesses. Thin film is attractive to reduce the drive voltage requirements, but surface defects of the thin film process deteriorate performance.

Thin film millimeter wave coplanar waveguide ferroelectric phase shifters are commercially available, but insertion loss remains an issue. Several design issues will need to be solved to further exploit ferroelectric material advantages. There is a trade space between tunability, temperature stability, bias field, loss tangent, dielectric constant, and so forth. In addition, since the dielectric material constant is so large, impedance matching and higher order mode suppression are problematic.

Metamaterials are one of the most significant recent advances in antenna technology that promises profound impact on phased array design. The interested reader is referred to the large body of literature covering this topic. Excellent overviews are provided by Eleftheriades et al., Yeo and Mittra, and Kildal et al. [9–11]. Metamaterials are engineered composites that exhibit properties not observed in nature as constituent parameters. This offers designers additional degrees of freedom over the traditional metallic, dielectric, and magnetic materials. Metamaterials can be broadly classified as double-negative materials (DNM) or electromagnetic band gap (EBG) materials. DNMs have feature sizes of much less than a wavelength, and their apparent (macroscopic) dielectric permittivity and magnetic permeability can both be made negative, which allows backward wave propagation [9]. EBGs have feature sizes on the order of one-half wavelengths or more. EBG variations include tunable impedance surfaces (TIS) and artificial magnetic conductors (AMC), which are realizable for narrow-to-moderate bandwidths. EBG technologies are already used extensively within the phased array community.

EBG materials can be applied to the phased array antenna art in the following ways: (1) elimination of dielectric surface wave mode induced scan blindness in printed phased arrays [12]; (2) ground plane impedance tailoring (AMC) [13]; (3) *directivity enhancement* of microstrip patch radiators [14]; and (4) *EGB analog waveguide radiating element* and *phase shifters* [15,16].

AMC materials synthesize perfect "H" or Neumann boundary condition, over a finite bandwidth, where the reflection at the surface interfaces is unity with zero degrees phase shift. This property can be exploited to realize horizontal dipole and other *low-profile radiating elements* very close to the artificial AMC surfaces for conformal array applications. AMC properties can be used to improve the input impedance of low-profile structures [10].

Multifunction Reconfigurable, Shared Aperture Systems: Future DOD systems require broadband and reconfigurable phased array antenna systems to support the reconfigurable and broadband radio suite vision, reduce platform antenna count, maximally utilize limited mounting space on vehicular platforms, and minimize radar cross section. An Aegis class Navy destroyer, for example, currently has over one hundred antennas [17]. Considerable analysis has been performed within the DOD community to optimally contrast federated single radio/single antenna systems with shared aperture integrated systems [18]. The Advanced Multifunction Radio Frequency Component Concept (AMRFC), a U.S. Navy-sponsored initiative, is to advance the state-of-the-art in multifunction, shared aperture technologies [19,20].

A technology closely related to the AMRFC initiative is direct digital synthesizer (DDS)-based phase shifting [21]. The traditional phase shifter and analog or digital phase shifters of conventional array feeds are replaced with DDS-integrated circuit (IC) chips. In the transmit mode, the DDS directly generates the transmit signal, with up conversion if the operating frequency is above the operational limits of the DDS chip. In the receive mode, a DDS chip provides a carefully controlled local oscillator (LO) signal required for down conversion to I/Q base band signals where DSP techniques are used to reconstruct the receive beam. The limitations of this approach closely parallel the current state of the industry of DDS technology.

10.9.2 Digital Beam Forming Systems

Practical and inexpensive DBF systems are the ultimate phased array solution in terms of flexibility, low probability of detection (LPD), low probability of intercept (LPI), interference suppression, direction finding, real-time pattern synthesis, radiating element parasitic mutual coupling mitigation, and multi-beam generation. Space time adaptive processing (STAP) algorithms are currently a vibrant research topic within the radar community [24].

A digital beam former can perform all the functions to the classic analog beam former. DBF exploits the mathematical power of DSP to generate and steer multiple beams. The ultimate DBF systems would incorporate LNAs at the individual element level immediately proceeded with analog-to-digital and digital-to-analog (ADC/DAC) converters to translate analog RF array signals to digital data streams. DBF technology is particularly suited to conformal array applications since in this case there is a need for amplitude and phase control, array sector switching, polarization control, element pattern compensation, beam steering algorithms, and built-in-test (BIT) and *in situ* calibration.

Unlike with analog beam forming, the receive system signal-to-noise ratio is set by the LNA and is not affected by further down stream processing, which effectively creates "lossless" beamforming. Hybrid systems, where there is a first up and down conversions in analog RF, are used at carrier frequencies, where ADC/DAC technology cannot directly sample and reconstruct the analog waveforms. This digital solution places extreme requirements on the ADC/DAC technologies and on the transmission, storage, and processing of large amounts of digital data, particularly for large arrays. The hope is that the rapid advances in computer hardware and software continue at the current intense pace out into the future.

Current DBF systems consist of combinations of several subsystems and components. Imperfections and nonuniformity within these constitutive subsystems can be problematic. Receive channel imbalance, amplitude and phase errors, A/D offset errors, frequency errors, and so forth, present significant challenges. Array calibration, including mutual coupling mitigation, is an important aspect of DBF system design [25,26]. In addition, the ability of an array to self-heal after experiencing channel failures is crucial to practical systems operation. These concerns have been addressed by the U.S. Air force Rome Laboratory [27].

Several DBF systems have been demonstrated in the literature. An experimental X Band system with 570 active elements conforming to the leading edge of an aircraft wing is described in [28]. Synthesis of sectored beams with low side lobe levels (-40 dB) was demonstrated.

DBF systems are currently hindered by the state of the industry in microelectronics wafer-scale and chip-scale integration and systems-on-a-package technologies, including embedded fiber optical interconnect, 3-D packaging, embedded 3-D thermal management, and so forth. Since the unit cell for any scanning array is typically chosen as $\lambda o/2 \times \lambda o/2$, it is difficult to place the DBF circuitry and DSP computer interconnect directly behind the array aperture. This is particularly problematic for electrically large arrays of hundreds or thousands of elements, and especially for higher microwave and millimeter wave frequencies, where the $\lambda o/2 \times \lambda o/2$ unit cell is very physically small. DBF technologies must ultimately ride the wave of miniaturization as driven by the commercial wireless industry.

Recent research in DBF circuit layout and electronics packaging have been reported in the literature. Hybrid wafer-scale integrated circuits technologies were used to realize an 8-channel, 4-beam 10 MHz bandwidth beamformer in a package only 50 mm on a side [29]. Curns et al. discuss a scalable, modular architecture with a goal of high volume production economies of scale [30]. They developed a 512-element X band receive-only system that performs digital beamforming in azimuth by means of 32 columns of 16 channel beamformers. Each of the columnar digital beam formers connect to a 4-channel real-time parallel processor board. The DBF receivers are hybrid in nature; each has an RF front-end and a digital back-end that interconnect at the RF intermediate frequency. The use of a standardized IF allows the RF modules to be swapped out to realize different RF operating bands. A scalable multibeam DSP processor architecture is featured. The researchers exploited commercial off-the-shelf (COTS) components such as field-programmable gate array (FPGA) technologies to minimize cost. Continued advancement in these areas is critical to make DBF practical solutions for electrically large systems.

References

Section 1: Phased Arrays

1. Hansen, R. C., *Phased Array Antennas*, John Wiley & Sons, Inc., New York, 1997.
2. Mailloux, R. J., *Phased Array Antenna Handbook*, 2nd Ed., Artech House, Inc., Norwood, MA, 2005.
3. Fourikis, N., *Advanced Array Systems, Applications and RF Technologies*, Academic Press, London, UK, 2000.
4. Josefsson, L., and Persson, P., *Conformal Array Antenna Theory and Design*, Wiley, Interscience, Hoboken, NJ, 2006.
5. Elliot, R. S., *Antenna Theory and Design*, Prentice-Hall, Inc., Englewood Cliffs, NJ, 1981, pp. 172–180.
6. Elliot, R. S., *Antenna Theory and Design*, Prentice-Hall, Inc., Englewood Cliffs, NJ, 1981, pp. 157–165.
7. Balanis, C. A, *Antenna Theory, Analysis, and Design*, 2nd Ed., John Wiley & Sons, New York, 1997, pp. 249–338.
8. Mailloux, R. J., *Phased Array Antenna Handbook*, Artech House, Inc., Norwood, MA, 1994, pp. 394–403.
9. Miller, C. J., "Minimizing the Effects of Phase Quantization Errors in an Electronically Scanned Array," *Proc. 1964 Symposium Electronically Scanned Phase Arrays and Applications*, RADCTDR-64-225, RADC Griffins AFB, Vol. 1, pp. 17–38.
10. Smith, M. S., and Guo, Y. C., "A Comparison of Methods for Randomizing Phase Quantization Errors in Phased Arrays' *IEEE Trans, Antennas and Propagation*, Vol. AP-31, No. 6, November 1983, pp. 821–827.
11. Mailloux, R. J., "Periodic Arrays," *Antenna Handbook Theory, Applications, and Design*, Lo, Y. T., and Lee, S. W., Editors, Van Nostrand Reinhold Co., New York, 1988., Chapter 13, pp. 13.30–13.32.
12. Tang, R., "Survey of Time Delay Beam Steering Techniques," Phased Array Antennas: *Proceedings of the 1970 Phased Array Antenna Symposium*, Dedham, MA, Artech House, 1972, pp. 254–260.
13. Pozar, D. M., *Microwave Engineering*, John Wiley & Sons, Inc., New York, 1998.
14. Hansen, R. C., *Phased Array Antennas*, John Wiley & Sons, Inc., New York, 1997. p. 166.
15. Rudge, A. W., Milne, K., Olver, A. D., and Knight, P., *The Handbook of Antenna Design*, Vol. 1, Peter Peregrinus, Ltd, London, UK, 11982, pp. 87–98.
16. Hansen, R. C., *Phased Array Antennas*, John Wiley & Sons, Inc., New York, 1997. pp. 215–301.

17. Wheeler, H. A., "Simple Relations for a Phased Array Antenna Made of an Infinite Current Sheet," *IEEE Transactions on Antennas and Propagation,* Vol. AP-13, No. 4, July 1965, pp. 506–514.

18. Balanis, C. A., *Antenna Theory, Analysis, and Design,* 2nd Ed., John Wiley & Sons, New York, 1997, pp. 264–276.

19. Aggarwal, V. D., and Lo, Y. T., "Mutual Coupling in phased Arrays of Randomly Spaced Antennas," *IEEE Transactions on Antennas and Propagation,* Vol. AP-20, May, 1972, pp. 288–295.

20. Steinberg, B. D., "Comparison Between the Peak Sidelobe Level of the Random Arrays and Algorithmically Designed Aperiodic Arrays," *IEEE Transactions on Antennas and Propagation,* Vol. AAP-21, May, 1973, pp, 366–369.

21. Cohen, E. D., "Trends in the Development of MMICS and Packages for Electronically Scanned Arrays (AESAs)," *1996 Proceedings of the IEEE International Symposium on Phased Array Systems and Technology,* IEEE Press, Institute of Electrical and Electronic Engineers, Piscatway, NJ, 1996, pp. 1–4.

22. Gauthier, G. P., Courtay, A., and Rebeiz, G. M., "Microstrip Antennas on Synthesized Low Dielectric-Constant Substrates," *IEEE Transactions on Antennas and Propagation,* Vol. 45, No. 8, August, 1997, pp. 1310–1314.

23. Nguyen, C. T. C., Katehi, L. P. B., and Rebeiz, G. M., "Micromachined Devices for Wireless Communications," *Proceedings of the IEEE, Institute of Electrical and Electronics Engineers,* Vol. 86, No. 8, August, 1998, pp. 1756–1768.

24. Papapolymerou, I., Drayton, R. F., and Katehi, L. P. B., "Micromachined Patch Antennas," *IEEE Transactions on Antennas and Propagation,* Vol. 46, No. 2, February, 1998, pp. 275–283.

25. Rebeiz, G. M., *RF MEMS Theory, Design and Technology,* John Wiley & Sons, New Jersey, 2003.

26. Weedon, W. H., and Payne, W. J., "MES-Switched Reconfigurable Multi-Band Antenna: Design and Modeling," *Proceedings of the 1999 Antenna Applications Symposium,* September 15–17, 1999, Robert Allerton Park, The University of Illinois, 1999.

27. Norvell, B. R., Hancock, R. J., Smith, J. K., Pugh, M. L, Theis, S. W., and Kviatkofsky, J., "Micro Electromechanical Switch (MEMS) Technology Applied to Electronically Scanned Arrays for Space Based Radar," *Proceeding of the IEEE Aerospace Conference,* 1999, pp. 239–247.

28. Lee, J. J., Atkinson, D., Lam, J. J., Hackett, L., Lohr, R., Larson, L., Loo, R et al., "MEMS in Antenna Systems: Concepts, Design, and Systems Implications," *National Radio Science Meeting,* Boulder, Colorado, 1998.

29. Werner, D. H., and Werner, P. L., "Fractal Radiation Pattern Synthesis," *National Radio Science Meeting,* Boulder, Colorado, January 6–11, 1992, p. 66.

30. Werner, P. L., and Werner, D. H., "Fractal Arrays and Fractal Radiation Patterns," *Proceedings of the 11th Annual Review of Progress in Applied Computational Electromagnetics,* Monterey, California, March 20–24, 1995, pp. 970–978.

31. Werner, P. L., and Werner, D. H., "On the Synthesis of Fractal Radiation Patterns," *Radio Science,* Vol. 30, No. 1, 1995, pp. 29–45.

32. Werner, P. L., and Werner, D. H., "Correction to: On the Synthesis of Fractal Radiation Patterns," *Radio Science,* Vol. 30, No. 3, 1995. p. 603.

33. Werner, D. H., Haupt, R. L., and Werner, P. L., "Fractal Antenna Engineering: The Theory and Design of Fractal Antenna Arrays," *IEEE Antennas and Propagation Magazine,* Vol. 41, No. 5, October 1999, pp. 37–59.

34. Pueente-Baliarda, C., and Pous, R., "Fractal Design of Multi-band and Low Side-Lobe Arrays," *IEEE Transactions on Antennas and Propagation,* Vol. 44, No. 5, May 1996, p. 730.

35. Haupt, R. L., Menozzi, J. J., and McCormack, C. J., "Thinned Arrays Using Genetic Algorithms," *1993 International Symposium Digest,* Vol. 3, Ann Arbor, MI, 19933, pp. 712–715.

36. O'Neill, D. J., "Element Placement in Thinned Array Using Genetic Algorithms," *Proceedings OCEANS '94,* Brest, France, September 1994, II/301-3-6. pp. 712–715.

37. Haupt, R. L., "Optimization of Sub-Array Amplitude Tapers," *1995 International Symposium Digest,* Vol. 3, Newport Beach, California, June, 1995, pp. 1830–1833.

38. Munk, B. A., *Finite Antenna Arrays and FSS,* Wiley Interscience, Hoboken, NJ, 2003.

39. Munk, B. A, *Frequency Selective Surfaces Theory and Design,* Wiley Interscience, Hoboken, NJ, 2000.

40. Schennum, G. H., "Frequency Selective Surfaces for Multiple-Frequency Antennas," *The Microwave Journal,* May, 1973, pp. 55–76.

41. Bancroft, R. "Accurate Design of Dual-Band Patch Antennas," *Microwaves & RF*, September, 1988, p. 113.
42. Shively, D. G., and Stutzman, W., "Wideband Arrays with Variable Element Sizes", *IEE Proceedings*, Vol. 137, Pt. H, No. 4, August, 1990, p. 238.
43. Yunus, E. E., Kubilay, S., Roland A., Wright, D. E., and Volakas, J. L., "Frequency-selective Surfaces to Enhance Performance of Broad-Band Reconfigurable Arrays", *IEEE Transactions on Antennas and Propagation*, Vol. 50, No. 12, December, 2002, pp. 1716–1724.
44. Pozar, D. M., "A Shared Aperture Dual-Polarized Microstrip Array," *IEEE Transactions on Antennas and Propagation*, Vol. 49, No. 2, February, 2001, pp. 150–157.
45. Ravpati, C. B., Zaghloul, A. I., and Kawser, M. T., "Investigations on Microstrip/Waveguide Hybrid Antenna Element for Multiple Frequencies," *AP-S International Symposium and USNC/URSI National Science Radio Meeting*, Washington, DC, July 6, 2005.
46. Tripp, V. K., and Papanicolopoulos, C. D., "Frequency-Independent Geometry for a Two Dimensional Array," *IEEE AP-S Symposium*, July 18, 2000, paper pp. 628–631.
47. Corey, L., Jaska, E., and Guerci, J., "Phased Array Developments at DARPA," *IEEE International Symposium on Phased Array Systems and Technology*, 2003, Boston, MA, October 14–17, 2003, pp. 9–16.
48. Dover, R. T., Irion, J., and McGraph, D. T., "Ultra-Wideband Arrays," *IEEE International Symposium on Phased Array Systems and Technology*, 2003, Boston, MA, October 14–17, 2003, pp. 387–392,
49. Elsallal, W., and Schaubert, D., "Reduced-Height Array of Balanced Antipodal Vivaldi Antennas (BAVA) with Greater than Octave Bandwidth," *Allerton Antenna Applications Symposium*, 2005, pp. 226–242.
50. Holter, H., "A New Type of Antenna Element of Wide-Band Wide-Angle dual Polarized Phased Array Antennas," *IEEE International Symposium on Phased Array Systems and Technology*, 2003, Boston, MA, October 14–17, 2003, pp. 393–398,
51. Cantreel, B., Rao, J, Tavik, G., Dorsey, M., and Krichevsky, V., "Wideband Array Antenna Concept," *IEEE Aerospace and Electronics Systems Magazine*, Vol. 21, Issue 1, January, 2006, pp. 9–12.
52. Fourikis, N., *Advanced Array Systems, Applications and RF Technologies*, Academic Press, London, UK, 20000, pp. 323–335.
53. Litva, J., and Kwok-Yueng Lo, T., *Digital Beamforming in Wireless Communications*, Artech House, Norwood, MA, 1996,
54. Balanis, C.A., *Antenna Theory, Analysis and Design*, 3rd Ed., John Wiley & Sons, Hoboken, NJ, 2005, Chapter 16, pp. 945–1000.

Section 2: Emerging Phased Array Antenna Technologies

1. Brookner, E., "Phased Arrays Around the World – Progress and Future Trends," *2003 Proceedings of the IEEE International Symposium on Phased Array Systems and Technology*, IEEE Press, Institute of Electrical and Electronic Engineers, Boston, MA, 2003, pp. 1–8.
2. Van Blarium, M. L., "Photonic Systems for Antenna Applications," *IEEE Antennas and Propagation Magazine*, Vol. 36, No. 5, October 1994, pp 30–38.
3. Kunath, R. R., Lee, R. Q., Martzaklis, K. S., Shalkhauser. K. A., Downey, A. N., and Simmons, R., "An Optically controlled Ka-Band Phased Array Antenna," *Proceedings of the 1992 Antenna Applications Symposium*, Robert Allerton Park, The University of Illinois, Monticello, IL, September 23–25, 1992.
4. Herzfeld, P., "Monolithic Microwave-Photonic Integrated Circuits: A Possible Follow-up to MIMIC," *The Microwave Journal*, April 1985, pp. 121–131.
5. Parent, M. G., "A Survey of Optical Beamforming Techniques," *Proceedings of the 1995 Antenna Applications Symposium*, Robert Allerton Park, The University of Illinois, Monticello, IL, September 20–22, 1995.
6. Newberg. I. L., "Antenna True-Time-Delay Beamsteering Utilizing Fiber Optics," *Proceedings of the 1992 Antenna Applications Symposium*, September 23–25, 1992, Robert Allerton Park, The University of Illinois, 1995.
7. Rao, J. B. L., Trunk, G. V., and Patel, D. P., "Two Low-cost Arrays," *1996 Proceedings of the IEEE International Symposium on Phased Array Systems and Technology*, IEEE Press, Institute of Electrical and Electronic Engineers, Piscatway, NJ, 1996, pp. 119–124.

8. Sengupta, L. C., Stowell, S., Ngo, E., O'Day, M. E., and Lancto, R., "Investigation of the Electronic Properties of Ceramic Phase Shifting Material," *Proceedings of the Advanced Ceramic Technology for Electronic Applications Conference*, U.S. Army, CECOM, Fort Monmouth, September 28, 1993.

9. Eleftheriades, G. V., and Balmain, K. G., Editors, *Negative-Refraction Metamaterials*, IEEE Press, Wiley Interscience, John Wiley & Sons, Hobokem, NJ, 2005.

10. Yeo, J., and Mittra, R., *Performance Enhancement of Communications Antennas using Metamaterials*, short course notes, IEEE AP-S International Symposium and USNC/URSI Meeting, Washington, DC, 2005.

11. Kidal, P. S., Maci, S., and Sievenpiper, D. F., *Theory and Applications of PBG Structure used as Artificial Magnetic Conductors and Soft and Hard Surfaces*, short course notes, IEEE AP-S International Symposium, Columbus, OH, 2003.

12. Sievenpiper, D., and Yablonovitch, E., "Eliminating Surface Currents with Metallodieletric Photonic Crystals," *IEEE Antennas and Propagation International Symposium Proceedings*, 1998, pp. 663–666.

13. Kidal, P. S., Maci, S., and Sievenpiper, D. F., *Theory and Applications of PBG Structure used as Artificial Magnetic Conductors and Soft and Hard Surfaces*, short course notes, IEEE AP-S International Symposium, Columbus, OH, 2003

14. Yeo, J., and Mittra, R., *Performance Enhancement of Communications Antennas using Metamaterials*, short course notes, IEEE AP-S International Symposium and USNC/URSI Meeting, Washington, DC, 2005.

15. Xin, H., West, J. B., Mather, J. C., Doane, J. P., Higgins, J. A., Kazemi, H., and Rosker, M. J., "A Two-dimensional Millimeter Wave Phase Scanned Lens Utilizing Analog Electromagnetic Crystal (EMXT) Waveguide Phase Shifters," *IEEE Transactions on Antennas and Propagation*, Vol. 53, Issues, Part 1, January, 2005, pp. 151–159.

16. West, J. B., Herting, B. J., and Mather, J. C., "Measured Results of a Dual Band Dual Mode Millimeter Wave Analog EMXT Phase Shifter Electronically Scanned Antenna", *IEEE Antennas and Wireless Propagation Letters*, Vol. 5, Issue 1, December, 2005, pp. 7–10.

17. Fourikis, N., *Advanced Array Systems, Applications and RF Technologies*, Academic Press, London, UK, 2000, pp. 198–208.

18. Axford, R., Major, R., and Rockway, J., "An Assessment of Multi-function Phased Array Antennas for Modern Military Platforms," *2003 Proceedings of the IEEE International Symposium on Phased Array Systems and Technology*, IEEE Press, Institute of Electrical and Electronic Engineers, Boston, MA, 2003, pp. 365–370.

19. Hughes, P., and Choe, J., "Overview of Advanced Multi-function RF Systems (AMRFS)," *2000 Proceedings of the IEEE International Symposium on Phased Array Systems and Technology*, IEEE Press, Institute of Electrical and Electronic Engineers, Dana Point, CA, 2000, pp. 21–24.

20. de Graaf, J., Tavik, G., and Bottoms, M., "Calibration Overview of the AMRFC Test Bed," *2003 Proceedings of the IEEE International Symposium on Phased Array Systems and Technology*, IEEE Press, Institute of Electrical and Electronic Engineers, Boston, MA, 2003, pp. 535–540.

21. Jiaguo, L., Manqin, W., Xueming, J., and Fang, Z., "Active Phased Array Antenna Based on DDS," *2003 Proceedings of the IEEE International Symposium on Phased Array Systems and Technology*, IEEE Press, Institute of Electrical and Electronic Engineers, Boston, MA, 2003, pp. 511–517.

22. Josefsson, L., and Persson, P., *Conformal Array Antenna Theory and Design*, Wiley, Interscience, Hoboken, NJ, 2006, pp. 380–383.

23. Fourikis, N., *Advanced Array Systems, Applications and RF Technologies*, Academic Press, London, UK, 2000, pp. 335–339.

24. Wirth, W.-D., *Radar Techniques using Array Antennas*, The Institute of Electrical Engineers (IEE), London, UK, 2001.

25. Steyskal, H., "Array Error Effects in Adaptive Beam Forming," *Microwave Journal*, September, 1991, pp. 101–112.

26. Mailoux, R. J., "Phased Array Error Correction Scheme," *Electronics Letters*, Vol. 29, No 7, April, 1994, p. 577.

27. Steyskal, H., and Herd, J., "Mutual Coupling Compensation in Small Array Antennas," *IEEE Transactions on Antennas and Propagation*, Vol. 39, No. 12, December, 1990, p. 1971.

28. Kanno, M. T., Hashimure, T., Katada, T., Ato, M., Fukatini, K., and Suzuki, A., "Digital Beam Forming for Conformal Active Array Antenna," *1996 Proceedings of the IEEE International Symposium on Phased Array Systems and Technology*, IEEE Press, Institute of Electrical and Electronic Engineers, Boston, MA. 1996, pp. 21–24.

29. Lanston, J. L., and Hinman, K. A., "A Digital Beamforming Processor for Multiple Beam Antennas," *IEEE Transaction on Antennas and Propagation*, Vol. 39, No. 12, 1990, pp. 383–391.

30. Curns, D. D., Thomas, R.W., and Payne, W.J., "32-Channel Digital Beamforming Plug-and-Play Receive Array," *2003 Proceedings of the IEEE International Symposium on Phased Array Systems and Technology*, IEEE Press, Institute of Electrical and Electronic Engineers, Boston, MA 2003, pp. 205–210.

11

The Fresnel-Zone Plate Antenna

James C. Wiltse
Georgia Tech Research Institute
Georgia Institute of Technology

11.1 Introduction

The Fresnel-zone plate (FZP) is a quasi-optical component that provides the functions of a lens (or a reflective antenna when the zone plate is backed by a mirror). The zone plate has the advantages of simplicity of design and construction, planar configuration, low loss, low weight, and low cost, while giving performance comparable to a lens (sometimes better). The zone plate produces lens-like focusing and imaging of electromagnetic waves by means of diffraction, rather than refraction. Thus the field is often referred to as diffraction optics. The zone plate transforms a normally incident plane wave into a converging wave, concentrating the radiation field to a small focal region. Much of the material that follows has been adapted from the author's papers in the list of references.

There are two categories of zone plates in practice. One has been used in optical systems for decades; this is the configuration employing focal lengths (F) much larger than the diameter (D) of the plate (the small angle or high F-number case). The other is the case employing focal lengths that are comparable to the diameter (F/D near unity, often in the range between 0.3 and 2.5), referred to as the large-angle or "fast" configuration. The latter case has seen extensive investigations at microwave and millimeter wavelengths [1–8] and a few studies at terahertz frequencies [9] between 210 GHz and 1 THz. The phase-correcting zone plate is most often a stepped planar lens (Figure 11.1) made from low-loss materials such as polystyrene, Teflon, polyethylene, TPX, polycarbonate, foamed polystyrene, low-density polytetrafluroethylene (PTFE), foamed polyethylene, quartz, sapphire, ceramics, and semiconductor materials such as high-resistivity silicon. These materials range in dielectric constant from approximately 1.1 to 12. It has been shown that materials having dielectric constants below 5 are preferable because they have lower

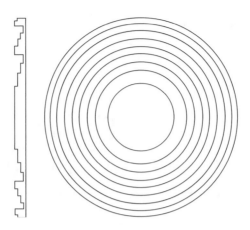

FIGURE 11.1   Quarter-wave corrected Fresnel-zone plate antenna. (From Wiltse, J.C., *Proceedings SPIE*, Vol. 5411, Orlando, April 12, 2004.)

surface and internal reflections [6]. There are nonplanar zone plate versions, such as spherical or paraboloidal types, but they are relatively rare and will not be discussed here. Figure 11.1 illustrates a stepped, quarter-wave corrected zone plate. An alternative approach would be to use rings of different dielectric constants (including air) to produce the stepped phase corrections [2–3].

At microwave or millimeter wavelengths, the steps, which are fractions of the operating wavelength, are of convenient dimensions for machining or molding of dielectrics. At terahertz frequencies, where wavelengths are on the order of a millimeter or less, the steps are inconveniently small. To cope with this, very low dielectric constant materials are recommended [9]. The other major difficulty in the terahertz region is keeping losses low.

11.2   Design Considerations

The typical geometry for a zone plate antenna is shown in Figures 11.2 and 11.3. From Figure 11.3, the equation for the radii of the Fresnel zones is given by

$$r_n = [2nF\lambda/P + (n\lambda/P)^2]^{1/2} \tag{11.1}$$

where n is the zone number, F is the focal length, λ is the free-space wavelength, and P is the number of steps to implement the phase correction (e.g., $P = 2$ is half-wave, $P = 4$ is quarter-wave, $P = 8$ is eighth-wave). If we set $r_n = D/2$ we can find the value of n at the outer edge of the zone plate, thus giving us the number of zones. Equation 11.1 is said to define interferometric zone plates which are free of spherical aberration. Note that the radii of the zones are independent of the dielectric constant, ε. Typical millimeter-wave zone plates have an F/D near unity, as illustrated in Figure 11.3. Defining θ as the angle between the optical axis and the ray from the focal point to the outer edge of the zone plate, we find that θ ranges from 11.3° for $F/D = 2.5$ to 45° for $F/D = 0.5$.

At millimeter-wavelengths the number of zones is often in the range from 5 to 80 for diameters between 10 and 50 cm, depending on the degree of phase correction. The number of zones as a function of diameter is shown in Figure 11.4 for frequencies of 10, 35, and 94 GHz. The effect of the ratio of focal length to the diameter is shown in Figure 11.5 for three diameters at 94 GHz with quarter-wave correction. Equation 11.1 can be plotted as a function of n; Figure 11.6 shows the result for half-wave correction at 30 GHz for apertures of 20 or 40 cm diameter. The curves have a similar shape for higher-order phase correction. The curvature is a reflection of the fact that the width of the zones is decreasing as n increases. However, the area of each zone increases with n. The contribution from each zone at the focal point is a

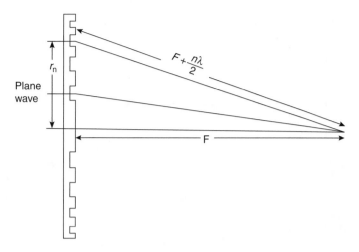

FIGURE 11.2 Ray geometry for half-wave correction. (From Wiltse, J.C., *Proceedings SPIE*, Vol. 5411, Orlando, April 12, 2004.)

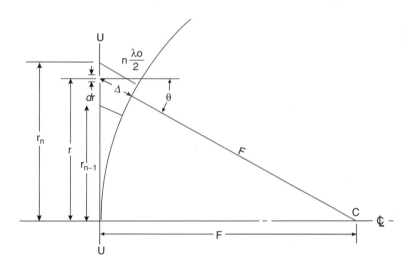

FIGURE 11.3 Geometry of the *n*th zone. (From Wiltse, J.C., *Proceedings SPIE*, Vol. 5411, Orlando, April 12, 2004.)

function of the product of the area of the zone times the cosine of the angle between the normal to the zone and the ray to the focal point. This product is nearly a constant if the aperture is uniformly illuminated.

For the grooved zone plate, the expression for the depth of the grooves is

$$d = \frac{\lambda}{P\left(\sqrt{\varepsilon} - 1\right)} \tag{11.2}$$

where the parameters are as defined above. Notice that d is independent of the focal length and the diameter of the zone plate. Figure 11.7 shows the depth of cut required as a function of frequency for dielectric constants of 1.6, 2.1, and 2.54, assuming a half-wave phase correction is used. It is obvious that the depth of the grooves is tending toward very small physical dimensions as frequency increases toward the terahertz region.

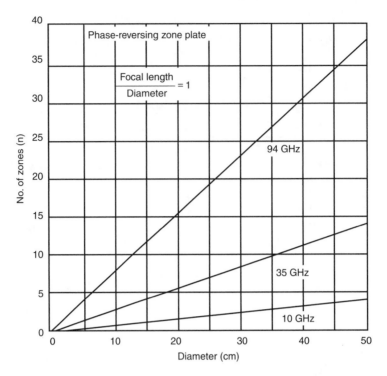

FIGURE 11.4 Number of zones as a function of diameter. (From Wiltse, J.C., *Proceedings of SPIE*, Vol. 3464, San Diego, CA, pp. 146–154, July 22–23, 1998.)

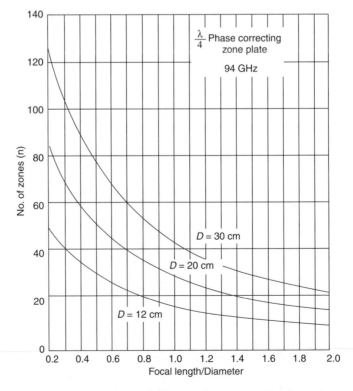

FIGURE 11.5 Number of zones as a function of *F/D*. (From Wiltse, J.C., *Proceedings of SPIE*, Vol. 3464, San Diego, CA, pp. 146–154, July 22–23, 1998.)

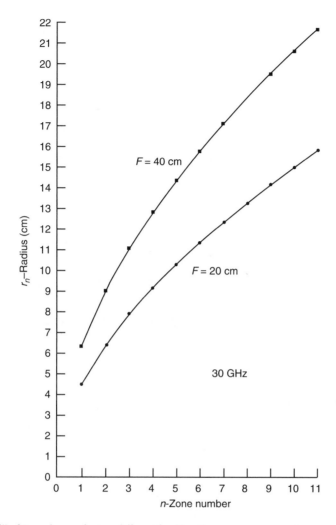

FIGURE 11.6 Radii of Fresnel zones for two different focal lengths.

11.3   Feed Considerations

The typical geometry for a FZP and feed is shown in Figures 11.2 and 11.3. Usually the feed device is a horn antenna, although dipole feeds have been used. The corrugated horn gives equal beamwidths in both principal planes, which is ideal for the circular symmetry case. If the amplitude distribution produced across the zone plate were uniform, the resulting gain is maximum, but the sidelobe level is not minimized. The feed horn should have a beam pattern wide enough to cover the zone plate with an amplitude taper that is several dB down at the edge of the plate. The amplitude taper reduces sidelobes, but also decreases the gain and broadens the beamwidth. If we assume a transmitting configuration, with a feed at the focal point, there are two factors that produce a taper across the zone plate aperture. One is the radiation pattern from the feed horn, and the other is the fact that the radiation from the focus travels a greater distance (R_{max}) to the edge than to the center of the zone plate (F). The zone plate is normally in the far field of the feed horn, and the path length dependence is inverse-squared. This type of dependence $[F/R_{max}]^2$ and the calculated angles are given in Table 11.1 for various values of F/D. It can be seen that this path length dependence does not have a significant effect on the edge illumination of the zone plate until F/D becomes very small.

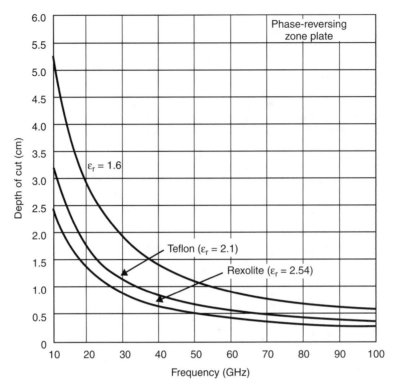

FIGURE 11.7   Depth of cut at different frequencies. (From Wiltse, J.C., *Proceedings of SPIE*, Vol. 3464, San Diego, CA, pp. 146–154, July 22–23, 1998.)

TABLE 11.1   Calculated Angles as a Function of F/D

F/D	θ	2θ	$[F/R_{max}]^2$ dB
2.5	11.31°	22.62°	
2.0	14.036	28.072	−0.263
1.5	18.435	36.87	−0.46
1.0	26.565	53.130	−0.97
0.5	45	90	−3.0
1/3	56.31	112.62	−5.13
0.2	68.199	136.397	−8.6

In addition to this path length dependence, we would like some amplitude taper due to the feed-horn pattern. We could design a feed pattern for which the first nulls would fall at the edge of the zone plate. This would improve efficiency because all the main-beam energy from the feed pattern would be included, and there would be low spill-over loss. If the feed horn (with circular aperture) were uniformly illuminated, the horn main beam would contain 83.3% of the total radiated power. Since the horn aperture undoubtedly has some amplitude taper, it will have more energy in the horn main beam, and less in the horn pattern sidelobes. Thus the horn beam efficiency may be over 90%. By contrast, for a uniformly illuminated feed-horn, only 47.6% of the total radiated horn power is included in the 3 dB beamwidth. If the horn pattern illuminated the zone plate with a 3 dB edge taper, approximately half of the radiated power would be lost, although this has been recommended by other authors. A preferred horn beam pattern would provide a truncated Gaussian distribution across the zone plate aperture [10]. High-quality scalar feed horns (corrugated horns) often produce a pattern that is nearly Gaussian. Some recently developed multimode horns produce Gaussian beams. While a nontruncated Gaussian aperture distribution has the desirable property that its Fourier transform is also Gaussian (far-field pattern with no sidelobes), real antenna patterns have sidelobes resulting from the truncation of the aperture distribution. In addition, a zone plate pattern is affected by nonconstant phase across the aperture; nonetheless, the far-field pattern is primarily determined by the amplitude distribution in the aperture. Goldsmith [10] has carried out an analysis to show that the aperture efficiency for a circular aperture has a maximum value of 81.5% for a Gaussian distribution with an edge taper of 10.9 dB. Aperture efficiency is defined as the product of spillover efficiency and the taper efficiency (resulting from the effect of the taper compared to uniform illumination). Obviously this means that both spillover and taper efficiencies are high (~90%) [10]. The first sidelobe level peak is at approximately −24 dB below the main beam compared to the value of −17.6 dB for uniform illumination. In an actual application, there are still the questions of how much gain is needed or how narrow the main beam should be, and given the conditions above, the answer is determined by how large the aperture must be. The gain is related to the efficiency of the zone plate configuration.

11.4   Efficiency

The overall efficiency of the antenna system contains not only the aperture efficiency mentioned above (relating to the feed and the aperture distribution) for the transmitting situation but also the diffraction efficiency of a zone plate lens compared to a true lens as well as the ohmic transmission loss in the lens. The diffraction efficiency is determined by comparing the maximum on-axis, or focused intensity (due to a plane-wave illumination of the FZP) to that of an ideal lens, obviously relating to the receiving situation. (This presumes that the plane-wave illumination is propagating normal to the plane of the zone plate.) Generally, the diffraction efficiency of the zone plate is less than for a true lens, but the ohmic loss of the zone plate is less and in many cases the combination of these two efficiency factors is better (higher) for the zone plate [11].

The diffraction efficiency of the FZP compared to a lens has already been analyzed by numerous investigators, based on simplified plane-wave analysis. [1] These results will be described first. Other recent analyses that give different efficiency values are discussed in reference [1], but as yet the analyses have no experimental verification.

From the simplified analysis, the diffraction efficiency is given by

$$\frac{\sin^2(\phi/2)}{(\phi/2)^2}$$

where ϕ is the increment of phase correction. A plot of this function is shown in Figure 11.8. For a zone plate with half-wave phase correction ($\phi = \pi$) the efficiency is only 40.5%; for a quarter-wave plate ($\phi = 90°$) it is 81%, and for an eighth-wave plate ($\phi = 45°$) it is 95% (0.22 dB loss). When eighth-wave or smaller correction was used, the FZP with its lower attenuation loss has been analyzed to be better than a standard lens for a number of cases considered between 35 and 200 GHz [11]. In two recent papers, the authors have reported on a planar quarter-wave zone plate utilizing four dielectrics (one of which is air), rather than a grooved dielectric, and have measured exceptionally good efficiencies, as high as 56.7% aperture efficiency [2,12]. This is comparable to efficiencies for excellent parabolic reflector antennas. The authors did not measure diffraction efficiencies for their cases.

A new zone plate structure has recently been proposed, which offers higher efficiency, lower phase errors, and better phase correction [13,14]. In the usual grooved FZP, the grooves are milled out so that the bottoms of the grooves are parallel to the back of the zone plate. However, in the so-called stepped conical zone plate (SCZP), the bottoms of the zones are tilted so that they approximate the curve of a lens. For circular symmetry the tilted subzones form a section of a cone. When rays are traced through the SCZP, the path lengths show smaller errors and are better-corrected. For this configuration, the simplified analysis shows an improvement for the half-wave correction from 40.5% to 95.2%. For the quarter-wave

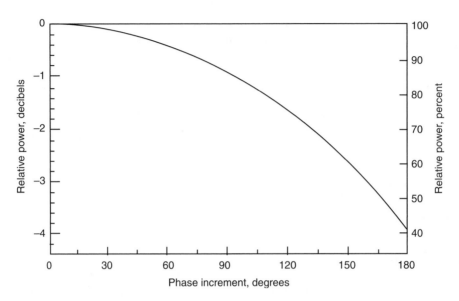

FIGURE 11.8 Relative power of subzoned plates. (From Wiltse, J.C., *Proceedings of SPIE*, Vol. 5104, Orlando, pp. 45–56, April 22, 2003.)

case the efficiency is improved from 81% to 99%. The tilted or tapered conical subzones also offer broader bandwidth capability.

At microwave and millimeter-wave frequencies the wavelengths of interest may range from 10 cm (3 GHz) to 2 mm (150 GHz). Fractional values (half-wave, quarter-wave) of these dimensions can easily be machined into dielectrics like polystyrene by conventional cutters on vertical milling machines or lathes. However, diffraction optics or binary optics for wavelengths shorter than 1 mm (300 GHz) require different approaches. Possible dielectric materials are formed by semiconductor growth techniques or laser machining. These optical diffractive elements employ phase correction to much smaller increments than quarter-wave or eighth-wave, and very complex analysis has been developed to explain the behavior of axially symmetric diffractive elements.

11.5 Loss Considerations

The overall efficiency of the FZP also depends on the attenuation or ohmic loss of the lens. Examples of such losses were calculated for the comparison between a true lens and a half-wave corrected FZP of 20 cm diameter and 20 cm focal length at frequencies from 35 to 200 GHz [11]. The extra attenuation of the true lens for polystyrene material ranges from about $1/4$ to 2.3 dB at these frequencies. Loss will also be reduced if the FZP center is open (has air dielectric), which is one possible design option [1,2,12]. A low dielectric constant is desirable to reduce lens or FZP surface and/or multiple internal reflection losses. An analysis carried out by van Houten and Herben showed that these reflection losses are low if the dielectric constant is below about 5, which is true for polystyrene, Teflon, polyethylene, polycarbonate, TPX, quartz and the foamed materials mentioned earlier [6].

11.6 Far-Field Patterns

The beamwidth of an FZP is essentially the same, perhaps 1–3% wider than that of a true lens, for the same illumination conditions. The usual way to obtain the antenna far-field or Fraunhofer pattern is to apply the Kirchoff–Fresnel integral knowing the phase and amplitude distribution across the aperture. When the distributions are not constant, the far-field integral (Fourier transform) usually cannot be solved in closed form, and computer numerical solutions are required [3,7,8]. However, we can see the improvement in the far-field pattern resulting from reduced phase errors by using the curves given by Baggen and Herben in Reference 8. Figure 11.5 of that reference [8] shows plots of the far-field patterns for three types of zone plates. One is the Soret type with alternate absorbing zones (not phase-correcting) and it has the lowest gain and the highest average sidelobes. The phase-correcting versions have improved gain and sidelobes. For half-wave correction ($P = 2$) the on-axis gain is about 6 dB higher and the average sidelobes are better. Going to quarter-wave correction ($P = 4$) gives 2 dB more on-axis gain and greatly reduced principal side lobes and average sidelobe level. This illustrates the improvement due to reducing the phase error.

11.7 Multiple Frequency Bands

Fresnel-zone plate antennas are normally designed for a particular center frequency and/or band of operation. The response of the zone plate is frequency dependent, however, and it is possible to design the FZP so that it will function well at two or three widely spaced bands [3,15]. This permits the development of high-gain, narrow-beamwidth patterns at 2 or more bands, or one can operate in a given band with narrow beamwidth and high gain, but suppress other bands to reduce interference or jamming, or for signal hiding purposes. It turns out, for example, that a quarter-wave corrected FZP designed for one frequency will normally operate as a half-wave corrected zone plate at twice the design frequency. Although the half-wave zone plate would have a 3 dB lower gain due to the poorer phase correction, doubling the frequency for a given aperture size would add a gain increment of about 6 dB (for a comparable amplitude distribution), so there would be an overall gain of about 3 dB at the doubled frequency.

Reference 15 gives the analysis and derives the conditions for such FZPs, while additional theory is given in the earlier references [3,4]. Examples include a design which behaves as a sixth-wave ($P = 6$) plate at 30 GHz and a half wave plate at 90 GHz, as well as a quarter-wave design at 30 GHz, and a half-wave plate at 60 GHz, all having high gain. A sample response curve is included, which shows relative gain as a function of frequency ratio for two zone plates with $F = 50 \lambda$; a quarter-period plate has 24 zones, while a half-period FZP has 12 zones.

11.8 Frequency Behavior

The response of the zone plate is frequency dependent, and the intensity at the focal point exhibits a periodic behavior with frequency. As frequency changes, the focal point moves along the axis of the zone plate. An increase in frequency moves the focus away from the lens, and a decrease does the reverse, with a decrease in signal level in either direction as the focus moves along the axis from the original position, reaching half-power levels at 1.5 or 0.5 times the design frequency. The effect may be used for "focal isolation" and frequency filtering.

Equation 11.1 may be rearranged to solve for F as a function of frequency

$$F = \frac{\mathrm{P}r_n^2}{2n\lambda} - \frac{n\lambda}{2P}$$

where $\lambda = c/f$ and $f =$ frequency. So

$$F = \frac{Pfr_n^2}{2nc} - \frac{nc}{2Pf} \tag{11.3}$$

When numerical examples are substituted, it turns out that the first term is usually larger than the second term. As frequency increases, the first term increases and the second (smaller) term decreases. Thus the response is not linear (but may be nearly linear in some cases). When the frequency increases, F also increases, and when frequency decreases, F decreases. For a frequency change of $\pm 10\%$, the focal point may move along the axis by a comparable percentage of the focal length (F). This means that an arriving signal having $\pm 10\%$ bandwidth about the center frequency will be focused at locations spread along the axis by a comparable percentage of the focal length. The zone plate thus acts as a frequency demodulator. Similarly, by reciprocity, narrow-band transmitter sources appropriately arranged along the axis about the focal point will be combined into a single collimated beam with wideband characteristics.

11.9 Summary Comments

Zone plate lens antennas offer several advantages over standard lenses and certain reflective antennas. Usually zone plates give comparable performance, while offering advantages of lower weight, volume, and cost, as well as simplicity of fabrication. Zone plate antennas have seen application in several fields, including satellite communications, radar, radiometry, point-to-point communications, and missile terminal guidance.

References

1. Wiltse, J.C., "Large-angle zone plate antennas," *Proceedings of SPIE*, Vol. 5104, Orlando, pp. 45–56, April 22, 2003.
2. Hristov, H.D. and Herben, M.H.A.J., "Millimeter-wave Fresnel zone plate lens and antenna," *IEEE Transactions on Microwave Theory and Techniques*, Vol. 43, pp. 2779–2785, December, 1995.
3. Black, D.N. and Wiltse, J.C., "Millimeter-wave characteristics of phase correcting Fresnel zone plates," *IEEE Transactions on Microwave Theory and Techniques*, Vol. MTT-35, pp. 1122–1129, December, 1987.
4. Hristov, H.D., *Fresnel Zones in Wireless Links, Zone Plate Lenses and Antennas*, Artech House, Boston, 2000.
5. Minin, I. and Minin, O., *Diffractional Optics of Millimetre Waves*, Institute of Physics, Bristol, UK, 2004.
6. van Houten, J.M. and Herben, M.H.A.J., "Analysis of a phase-correcting Fresnel-zone plate antenna with dielectric/transparent zones," *Journal of Electromagnetic Waves and Applications*, Vol. 8, No. 7, pp. 847–858, 1994.
7. Baggen, L.C. and Herben, M.H.A.J., "Calculating the radiation pattern of Fresnel-zone plate antenna: A comparison between UTD/GTD and PO," *Electromagnetics*, Vol. 15, pp. 321–345, 1995.
8. Baggen, L.C. and Herben, M.H.A.J., "Design procedure for a Fresnel-zone plate antenna," *International Journal of Infrared and Millimeter Waves*, Vol. 14, No. 6, pp. 1341–1352, 1993.
9. Wiltse, J.C., "Diffraction optics for terahertz waves," *Proceedings SPIE*, Vol. 5411, Orlando, April 12, 2004.
10. Goldsmith, P.F., "Radiation patterns of circular apertures with Gaussian illumination," *International Journal of Infrared and Millimeter Waves*, Vol. 8, pp. 771–781, July, 1987.
11. Wiltse, J.C., "High efficiency, high gain Fresnel zone plate antennas," *Proceedings of SPIE*, Vol. 3375, Orlando, pp. 286–290, April, 1998.
12. Hristov, H.D. and Herben, M.H.A.J., "Quarter-wave Fresnel zone planar lens and antenna," *IEEE Microwave and Guided Wave Letters*, Vol. 5, pp. 249–251, August, 1995.
13. Wiltse, J.C., "The stepped conical zone plate antenna," *Proceedings of SPIE*, Vol. 4386, Orlando, pp. 85–92, April 17, 2001.
14. Wiltse, J.C., "Bandwidth characteristics for the stepped conical zoned antenna," *Proceedings of SPIE*, Vol. 4732, Orlando, pp. 59–68, April 1, 2002.
15. Wiltse, J.C., "Dual-band Fresnel zone plate antennas," *Proceedings of SPIE*, Vol. 3062, Orlando, pp. 181–185, April 22, 1997.
16. Wiltse, J.C., "Recent developments in Fresnel zone plate antennas at microwave/millimeter wave," *Proceedings of SPIE*, Vol. 3464, San Diego, CA, pp. 146–154, July 22–23, 1998.

12

RF Package Design and Development

Successful RF and microwave package design involves adherence to a rigorous and systematic methodology in package development together with a multi-disciplined and comprehensive approach. This formal planning process and execution of the plan ultimately insures that the package and product will perform as expected, for the predicted lifetime duration in the customer's system, under the prescribed application conditions.

Probably the first concern is having a thorough and in-depth knowledge of the application and the system into which the microwave component or module will be placed. Once these are understood, then package design can begin. Elements that must be considered do not simply include proper electrical performance of the circuit within the proposed package. Mechanical aspects of the package design must be thoroughly analyzed to assure that the package will not come apart under the particular life conditions. Second, the substrates, components, or die within the package must not fracture or lose connection. Third, any solder, epoxy, or wire connections must be able to maintain their integrity throughout the thermal and mechanical excursions expected within the application. Once these elements are thoroughly investigated, the thermal aspects of the package must be simulated and analyzed to appropriately accommodate heat transfer to the system. Thermal management is probably one of the most critical aspects of the package design because it not only contributes to catastrophic circuit overload and failure in out-of-control conditions, but it could also contribute to reduced life of the product and fatigue failures over time. Thermal interactions with the various materials used for the package itself and within the package may augment mechanical stress of the entire package system, ultimately resulting in failure.

Once proper simulation and analysis have been completed from a mechanical and thermal point of view, the actual package design can be finalized. Material and electrical properties and parameters then become the primary concern. Circuit isolation and electromagnetic propagation paths within the package need to be thoroughly understood. In addition, impedance levels must be defined and designed for input and output to and from the package. New developments in package design systems have paved the way for rapid package prototyping through computer integrated manufacturing systems by tying the design itself to the machining equipment that will form the package. These systems can prototype a part in plastic for further study or can actually build the prototypes in metal for delivery of prototype samples.

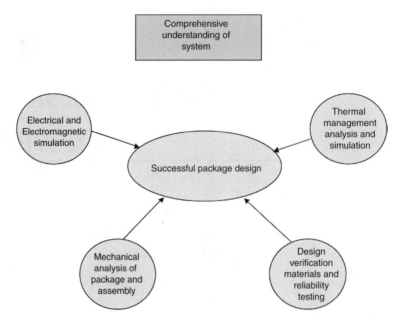

FIGURE 12.1   Elements of successful package design.

Finally, design verification must take place. The verification process typically includes the various long-term reliability tests that gives the designer, as well as his or her customer confidence that the package and its contents will live through the predicted lifetime and application conditions. Other testing may be more specific, such as fracture testing, material properties tests, or precise design tolerance testing. Much of the final testing may also include system-level integration tests. Usually specific power levels are defined and the packages, fully integrated into the system, are tested to these levels at particular environmental conditions.

These are some of the key elements in RF and microwave package development. Although this is not an all-encompassing list, these elements are critical to success in design implementation. These will be explored in the following discussion, hopefully defining a clear path to follow for RF package design and development. Figure 12.1 depicts these key elements leading to successful package design.

12.1   Thermal Management

From an MTBF (mean time before failure) point of view, the thermal aspects of the circuit/package interaction are one of the most important aspects of the package design itself. This can be specifically due to actual heat up of the circuit, reducing lifetime. It may also be due to thermal effects that degrade performance of the materials over time. A third effect may be a materials/heat interaction that causes severe thermal cycling of the materials resulting in stress concentrations and degradation over time.

It is clear that the package designer must have a fully encompassing knowledge of the performance objectives, duty cycles, and environmental conditions that the part will experience in the system environment. The engineer must also understand the thermal material properties within the entire thermal path. This includes the die, the solder or epoxy attachment of that die, the package or carrier base, package system attachment, and material connection to the chassis of the system. There are relatively good databases in the industry that provide the engineer with that information right at his or her fingertips. Among the many are the CINDAS [1] database and the materials' database developed at Georgia Institute of Technology. Other information may be gleaned from supplier datasheets or testing.

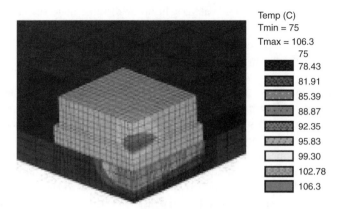

FIGURE 12.2   Ansys output showing thermal gradient across silicon die.

Thermal density within the package, and in particular, at the die level becomes an all-important consideration in the thermal management equation. To insure proper heat transfer and to eliminate any potential thermal failure modes (such as materials breakdown or diffusion and migration), analysis of heat transfer within the die must be completed at the die layout level. This thermal analysis will ultimately be parametrically incorporated into an analysis at the next level up, which may be at the circuit substrate or at the package level itself. The analysis is usually completed with standard finite element simulation techniques present in various software packages available in the industry. Ansys, MSC Nastran, Mechanica, Flowtherm, and Computational Fluid Dynamics (CFD) are some of those available. Material properties that are critical to input into the model would be thermal conductivity and the change in conductivity with temperature. Figure 12.2 shows a typical output of one of these software tools. The FEM simulation uses $1/4$ model symmetry. In this particular figure, the analysis demonstrates the thermal gradient across a silicon die, which is an 8 W power amplifier transistor, solder attached to a via structure, with 75°C applied to the bottom of the heat sink. The die junction is at 106.3°C.

It is through such simulated analysis that the entire heat transfer methodology of component to system can be developed. Assuming that there is good correlation between simulated and verified results, the engineer can then gain confidence that the product will have a reasonable lifetime within the specific application. The correlation is typically achieved through the use of infrared microscopy techniques. A number of these infrared microscopes are available in the industry. Usually, the component or module is fixtured on a test station under the infrared camera. The camera is focused on the top surface of the die, which is the heat-generating element. As power is applied to the component or module, the die begins to heat up to a steady-state level. The heat can be measured under RF or DC power conditions. A measurement is done of the die surface temperature. At the same time, a thermocouple impinges on the bottom of the case or package and makes a temperature measurement there. With the maximum die junction temperature (Tjmax in °C), the case temperature (Tc in °C) and the dissipated power in watts, the packaged device junction to case thermal resistance (in °C/W) can be calculated from the following equation:

$$\theta jc = \left(Tjmax - Tc\right)\big/ Pdis$$

In a fully correlated system, the agreement between simulated and measured results is usually within a few percentage points.

12.2   Mechanical Design

The mechanical design usually occurs concurrently with thermal analysis and heat transfer management. In order to adequately assess the robustness of the package system and its elements, an in-depth understanding of all the material properties must be achieved. In addition to the mechanical properties such as Young's modulus and stress and strain curves for materials, behavior of those materials under thermal loading conditions must be well understood. Once again, these material properties can be found in standard databases in the industry as mentioned above. Typically, the engineer will first insure that the packaging materials under consideration will not cause fracture of the semiconductor devices or of the substrates, solder joints, or other interconnects within the package. This involves knowing the Coefficient of Thermal Expansion (CTE) for each of these materials, and understanding the processing temperatures and the subsequent temperature ramp up that will be experienced under loading conditions in the application. The interaction of the CTEs of various materials may create a mismatched situation and create residual stresses that could result in fracture of any of the elements within the package. The engineer also assesses the structural requirements of the application and weight requirements in order to form appropriate decisions on what materials to use. For instance, a large microwave module may be housed within an iron-nickel (FeNi) package that sufficiently addresses all of the CTE concerns of the internal packaged elements. However, this large, heavy material might be inappropriate for an airborne application where a lightweight material such as AlSiC (aluminum silicon carbide) would be more suitable.

It is not sufficient to treat the packaged component or module as a closed structure without understanding and accounting for how this component or module will be mounted, attached, or enclosed within the actual system application. A number of different scenarios come to mind. For instance, in one situation a packaged component may be soldered onto a printed circuit board (PCB) of a wireless phone. Power levels would not be of concern in this situation, but the mechanical designer must develop confidence through simulation that the packaged RF component can be reliably attached to the PCB. He or she also must insure that the solder joints will not fracture over time due to the expansion coefficient of the PCB compared to the expansion of the leads of the package. Finally, the engineer must comprehend the expected lifetime in years of the product. Cost is obviously a major issue in this commercial application. A solution may be found that is perfectly acceptable from a thermo-mechanical perspective, but it may be cost prohibitive for a phone expected to live for three to five years and then be replaced.

Another scenario on the flip side of the same application is the power device or module that must be mounted into a base station. Here, obviously, the thermal aspects of the packaged device become all important. And great pain must be taken to insure that an acceptable heat transfer path is clearly delineated. With the additional heat from the power device and within the base station itself, heat degradation mechanisms are thoroughly investigated both with simulation techqniques and with rigorous testing. It is common for the RF power chains within base station circuits to dissipate 100 to 200 watts each. Since the expected lifetime of base stations may be over fifteen years, it would be a great temptation for a mechanical engineer to utilize optimum heat transfer materials for the package base, such as diamond for instance, with a thermal conductivity of 40.6 W/in°C. A high power device attached to diamond would operate much cooler than a device attached to FeNi or attached to ceramic. Since, over time, it is the heat degradation mechanisms that eventually cause failure of semiconductor devices, a high power die mounted over a diamond heat sink would be expected to have a much longer lifetime than one mounted over iron nickel or over ceramic. However, once again, the cost implications must enter into the equation. Within the multifunctioned team developing the package and the product, a cost trade-off analysis must be done to examine cost comparisons of materials vs. expected lifetimes. The mechanical package designer may develop several simulations with various materials to input into the cost-reliability matrix. It is necessary that such material substitutions can be done easily and effectively in the parametric model that was initially developed.

The mechanical analysis must encompass attachment of the RF or microwave component or module to the customer system. As we have discussed, in a base station, the thermal path is all important. In order to provide the best heat transfer path, engineers may inadvertently shortcut mechanical stress concerns, which then compromise package integrity. An example was a system mounting condition initially created for the eight-watt power device shown in Figure 12.2. This semiconductor die was packaged on a copper lead frame to which plastic encapsulation was applied. The lead frame was exposed on the bottom side of the device to insure that there would be a good thermal path to the customer chassis. The copper leads were solder attached (using the typical lead-tin, PbSn, solder) to the printed circuit board. At the same time, the bottom of the device was solder attached to a brass heat sink, as shown in Figure 12.3, which was then screw mounted to the aluminum chassis to provide thermal transfer to the chassis.

The CTE mismatch of materials resulting in residual stresses during thermal excursions, caused the plastic to rip away from the copper leads. It was the expansion of the aluminum chassis impacting the brass heat sink that created both tensile and shear forces on the leads of the device. The brass heat sink, in effect, became a piston pushing up at the center of the component. The stress levels in the plastic mold compound, which resulted in the failure of the mold/copper interface, can be seen in Figure 12.4.

Through subsequent simulation, a solution was found that provided the proper heat transfer for the eight-watt device as well as mechanical stability over time and temperature. This was verified through thousands of hours of temperature cycle testing and device power conditioning over temperature excursions.

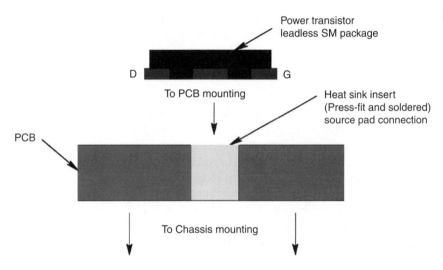

FIGURE 12.3   Eight watt power device attached to brass heat sink.

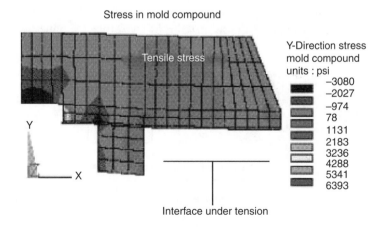

FIGURE 12.4   Modeled stresses in plastic mold compound resulting in failure.

12.3   Package Electrical and Electromagnetic Modeling

Quite obviously, the electrical design cannot stop at the circuit model for the silicon or GaAs die itself. Particularly at higher frequencies, such as those in the RF or microwave arena, the electromagnetic propagation due to all circuit elements create interactions, interference, and possibly circuit oscillations if these electrical effects are not accounted for and managed. Of course the customer's initial requirement will be that a packaged device, component, or module have a specific impedance into and out of their system. Typically, this has been 50 ohms for many microwave systems. It can be achieved through properly dimensioned microstrip input and output leads, through coaxial feeds, or through stripline to microstrip connections that feed into the customer system. These are modeled using standard industry software such as that provided by Hewlett Packard or Ansoft. The next consideration for the package designer is that all of the circuit functions that require isolation are provided that isolation. This can be accomplished through the use of actual metal wall structures within the package. It can also be done by burying those circuit elements in cavities surrounded by ground planes or through the use of solid vias all around the functional elements. These are only some of the predictive means of providing isolation. The need for isolating circuit elements and functions is ascertained by using full wave electromagnetic solvers such as HFSS, Sonnet, or other full wave tools. The EM analysis of the packaged structure will output an S parameter block. From this block, an electrical equivalent circuit can then be extracted with circuit optimization software such as Libra, MDS, ADS, etc. An example of an equivalent circuit representation can be seen in Figure 12.5.

After proper circuit isolation is achieved within the package, the designer must insure that there will not be inductive or capacitive effects due to such things as wire bonds, leads, or cavities. Wire bonds, if not controlled with respect to length in particular, could have serious inductive effects that result in poor RF performance with respect to things such as gain, efficiency, and intermodulation distortion, etc. In the worst case, uncontrolled wire bonds could result in circuit oscillation. In the same way, RF and microwave performance could be severely compromised if the capacitive effects of the leads and other capacitive elements are not accounted for. These are modeled with standard RF and microwave software tools, and then the materials or processes are controlled to maintain product performance within specifications. Most software tools have some type of "Monte Carlo" analysis capability in which one can alter the material or process conditions and predict the resulting circuit performance. This is especially useful if the processes have been fully characterized and process windows are fully defined and understood. The Monte Carlo analysis then can develop expected RF performance parameters for the characterized process within the defined process windows.

12.4   Design Verification, Materials, and Reliability Testing

After all of the required simulation and package design has been completed, the time has come to begin to build the first prototypes to verify the integrity of the design. During the simulation phase, various

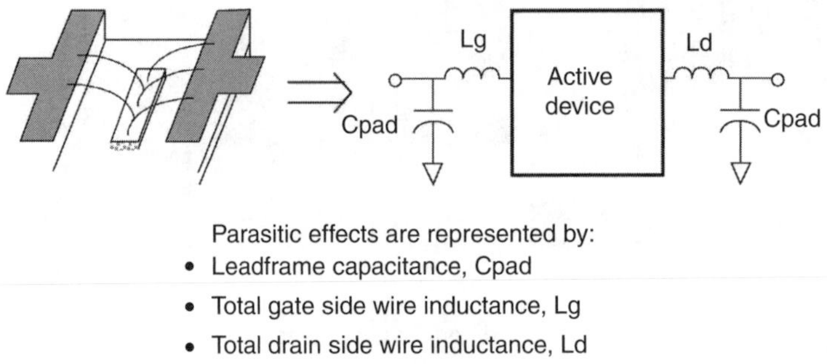

Parasitic effects are represented by:
- Leadframe capacitance, Cpad
- Total gate side wire inductance, Lg
- Total drain side wire inductance, Ld

FIGURE 12.5   Equivalent circuit representation of simple package.

material property studies may have been undertaken in order to insure that the correct properties are input into the various models. These may be studies of dielectric constant or loss on a new material, fracture studies to determine when fracture will occur on a uniquely manufactured die or on a substrate, or thermal studies, such as laser flash, to determine the precise thermal conductivity of a material. Figure 12.6 shows one technique used to measure the fracture strength of a GaAs or silicon die. A load is applied to a fixtured sphere, which then impacts the die at a precise force level. From the test, the critical value of the force to break the die is recorded. Then this force is converted to the maximum die stress via the following well-known [5] equation:

$$\sigma_t = \frac{3W}{2\pi m t^2}\left[(m+1)\ln\left(\frac{a}{r_o}\right)+(m-1)\left(1-\left(\frac{r_o}{a}\right)^2\right)\right]$$

After various material properties tests and all simulations and models have been completed, initial prototypes are built and tested. This next phase of tests typically assess long-term reliability of the product through thermal cycling, mechanical shock, variable random vibration, long-term storage, and high temperature and high humidity under biasing conditions. These, as well as other such tests, are the mainstay of common qualification programs. The levels of testing and cycles or hours experienced by the packaged device are often defined by the particular final application or system. For instance, a space-qualified product will require considerably more qualification assessment than a component or module going into a wireless handset that is expected to live 3 to 5 years. The temperature range of assessment for the space-qualified product may span from cryogenic temperatures to +150°C. The RF component for the wireless phone, on the other hand, may simply be tested from 0°C to 90°C.

In high power applications, often part of the reliability assessment involves powering up the device or module after it is mounted to a simulated customer board. The device is powered up and down at a

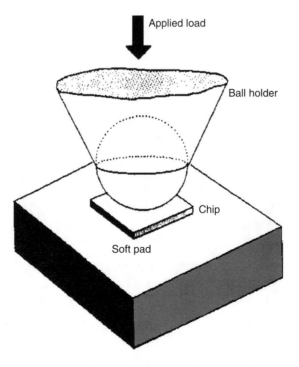

FIGURE 12.6 Technique used to measure fracture strength of semiconductor die.

specific duty cycle, through a number of cycles often as the ambient progresses through a series of thermal excursions. This represents what the RF packaged component would experience in the customer system, though usually at an accelerated power and/or temperature condition. Lifetime behavior can then be predicted, using standard prediction algorithms such as Black's equation, depending on the test results.

The RF product and packaging team submit a series of prototype lots through final, standard qualification/verification testing. If all results are positive, then samples are usually given to the customer at this time. These will undergo system accelerated life testing. The behavior of the system through this series of tests will be used to predict expected life cycle.

12.5   Computer-Integrated Manufacturing

As mentioned above, there are tools available in the industry that can be used to rapidly develop prototypes directly from the package design files. These prototypes may be constructed of plastic or of various metals for examination and further assessment. Parametric Technologies offers such design and assembly software modules, although they are not by any means the only company with this type of software. The package design is done parameterically in Pro-E so that elements of the design can easily be changed and/or uploaded to form the next higher assembly. The package design elements then go through a series of algorithms to which processing conditions can be attached. These algorithms translate the information into CNC machine code which is used to operate equipment such as a wire EDM for the cutting of metals. Thus, a lead frame is fashioned automatically, in a construction that is a perfect match to the requirements of the die to be assembled. A process flow chart for this rapid prototyping scenario is shown in Figure 12.7.

Computer-integrated manufacturing is also a highly effective tool utilized on the production floor, once the designed package has been accepted by the customer and is ready for production implementation. Here it is utilized for automated equipment operation, for statistical process control (SPC), for equipment shut down in out-of-control situations, etc. Coupled with neural networks, computer-integrated manufacturing can also be used for advanced automated process optimization techniques.

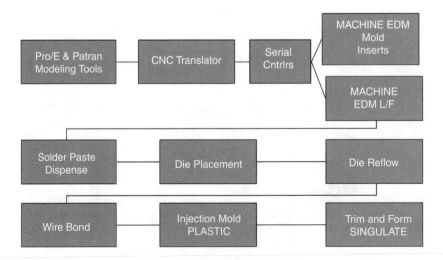

FIGURE 12.7   Rapid prototyping system.

12.6   Conclusions

The development and design of packages for RF and microwave applications must involve a rigorous and systematic application of the proper tools and methodology to create a design that "works the first time" and every time for the predicted lifetime of the product. This encompasses an in-depth knowledge of the system requirements, the environmental conditions, and the mounting method and materials to be used for package assembly into the customer's system. Then modeling and simulation can take place. Often, in order to understand material properties and to use these more effectively in the models, material studies are done on specific parameters. These are then inserted into electrical, mechanical, and thermal models, which must be completed for effective package design. After a full set of models is completed, verification testing of the design can be done on the first prototypes. Rapid prototyping is made simple through techniques that automatically convert design parameters into machine code for operation of machining equipment. Computer-integrated manufacturing is a highly effective technique that can be utilized at various levels of the product introduction. In package design, it is often used for rapid prototyping and as a tool for better understanding the design. At the production level, it is often used for automated equipment operation and for statistical process control.

Verification testing of the prototypes may include IR scanning to assess thermal transfer. It may include instron testing to test the integrity of a solder interface or of a package construction. It may include power cycling under DC or RF conditions to insure that the packaged design will work in the customer application.

The final phase of assessment is the full qualification of the RF packaged device or module. This certifies to the engineer, and ultimately to the customer that the packaged product can live through a series of thermal cycles, through high temperature and high humidity conditions. It certifies that there will be no degradation under high temperature storage conditions. And it certifies that the product will still perform after appropriate mechanical shock or vibration have been applied. Typically, predictive lifetime assessment can be made using performance to accelerated test conditions during qualification and applying these results to standard reliability equations.

These package design elements, when integrated in a multidisciplined approach, provide the basis for successful package development at RF and microwave frequencies.

References

1. CINDAS = Center for Information and Data Analysis; Operated by Purdue University; Package Materials Database created under SRC (Semiconductor Research Corporation) funding.
2. G. Hawkins, H. Berg, M. Mahalingam, G. Lewis, and L. Lofgran "Measurement of silicon strength as affected by wafer back processing," International Reliability Physics Symposium, 1987.
3. T. Liang, J. Pla, and M. Mahalingam, Electrical Package Modeling for High Power RF Semiconductor Devices, Radio and Wireless Conference, IEEE, Aug. 9-12, 1998.
4. R.J. Roark, *Formulas for Stress and Strain*, 4th Edition, McGraw-Hill, New York, 219.

II

Active Device Technologies

13

Varactors

Jan Stake
Chalmers University of Technology

13.1 Introduction

A varactor is a nonlinear reactive device used for harmonic generation, parametric amplification, mixing, detection, and voltage-variable tuning [1]. However, present applications of varactors are mostly harmonic generation at millimeter and submillimeter wave frequencies and as tuning elements in various RF & microwave applications. Varactors normally exhibit a voltage dependent capacitance and can be fabricated from a variety of semiconductor materials [2]. A common varactor is the reverse-biased Schottky diode. Advantages of varactors are low loss and low noise. The maximum frequency of operation is mainly limited by a parasitic series resistance; see Figure 13.1.

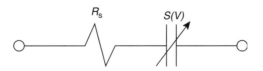

FIGURE 13.1 Equivalent circuit of a pure varactor (From Uhlir, A., *Proceedings IRE* 46, 1099–1115, 1958.)

13.2 Basic Concepts

Many frequencies may interact in a varactor, and of those, some may be useful inputs or outputs, while the others are *idlers* that, although they are necessary for the operation of the device, are not part of any input or output. For instance, to generate high harmonics in a frequency multiplier it is more or less necessary to allow current at intermediate harmonics (idlers) to flow. Such idler circuits are usually realized as short-circuit resonators, which maximizes the current at idler frequencies.

13.3 Manley–Rowe Formulas

The Manley–Rowe formulas [3] for *lossless* nonlinear reactances are useful for intuitive understanding of multipliers, frequency converters, and dividers. Consider a varactor excited at two frequencies f_p and f_s; the corresponding general Manley–Rowe formulas are

$$\sum_{m=1}^{\infty} \sum_{n=-\infty}^{\infty} \frac{mP_{m,n}}{nf_p + mf_s} = 0$$

$$\sum_{m=-\infty}^{\infty} \sum_{n=1}^{\infty} \frac{nP_{m,n}}{nf_p + mf_s} = 0$$

(13.1)

where m and n are integers representing different harmonics and $P_{m,n}$ is the average power flowing into the nonlinear reactance at the frequencies nf_p and mf_s.

- Frequency multiplier ($m = 0$): if the circuit is designed so that only real power can flow at the input frequency, f_p, and at the output frequency, nf_p, the above Equation 13.1 predicts a theoretical efficiency of 100%. The Manley–Rowe equation is $P_1 + P_n = 0$.
- Parametric amplifier and frequency converter: assume that the RF-signal at the frequency f_s is small compared to the pump signal at the frequency f_p. Then, the powers exchanged at sidebands of the frequencies nf_p and mf_s for m different from 1 and 0 are negligible. Furthermore, one of the Manley–Rowe formulas only involves the small-signal power as

$$\sum_{n=-\infty}^{\infty} \frac{P_{1,n}}{nf_p + f_s} = 0$$

(13.2)

Hence, the nonlinear reactance can act as an amplifying up-converter for the input signal at frequency f_s and output signal extracted at $f_u = f_s + f_p$ with a gain of

$$\frac{P_u}{P_s} = \frac{P_{1,1}}{P_{1,0}} = -\left(1 + \frac{f_p}{f_s}\right) = -\frac{f_u}{f_s}$$

(13.3)

13.4 Varactor Model

The intrinsic varactor model in Figure 13.1 has a constant series resistance, R_s, and a nonlinear differential elastance, $S(V) = dV/dQ = 1/C(V)$, where V is the voltage across the diode junction. This simple model is used to describe the basic properties of a varactor and is adequate as long as the displacement

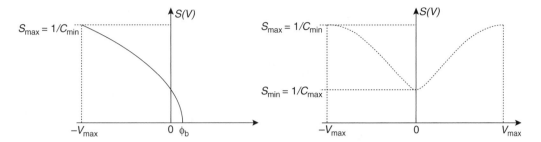

FIGURE 13.2 (a) Elastance as a function of voltage for one side junction diodes and (b) symmetric diodes.

current is much larger than any conduction current across the junction. A rigorous analysis should also include the effect of a frequency and voltage dependent series resistance, and the equivalent circuit of parasitic elements due to packaging and contacting. The differential elastance is the slope of the voltage–charge relation of the diode and the reciprocal of the differential capacitance, $C(V)$. Since the standard varactor model consists of a resistance in series with a nonlinear capacitance, the elastance is used rather than the capacitance. This simplifies the analysis and gives, generally, a better understanding of the varactor. The differential elastance can be measured directly and is used rather than the ratio of voltage to charge.

The elastance versus voltage for a conventional varactor and a symmetric varactor are shown in Figure 13.2. In both cases, the maximum available elastance is achieved at the breakdown voltage. The conventional varactor is reverse biased in order to allow maximum elastance swing and avoid any forward conduction current. A symmetric varactor will only produce odd harmonics when a sinusoidal signal is applied. This means that a varactor frequency tripler can be realized without any second harmonic idler circuit or DC-bias. This simplifies the design of such circuits and, hence, many novel symmetric varactors have been proposed. Among these, the Heterostructure Barrier Varactor (HBV) [5] and the Antiserial Schottky Varactor (ASV) [6] have so far shown the best performance.

13.5 Pumping

The pumping of a varactor is the process to pass a large current at frequency f_p through the varactor. The nonlinear varactor then behaves as a time-varying elastance, $S(t)$, and the series resistance dissipates power due to the large-signal current. The allowable swing in elastance is limited by the maximum elastance of the device used. Hence, the time-domain equation describing the varactor model in Figure 13.1 is given by

$$V(t) = R_s i(t) + \int S(t) i(t) \, dt \tag{13.4}$$

The above equation, which describes the varactor in the time-domain, must be solved together with equations describing the termination of the varactor. How the varactor is terminated at the input-, output- and idler-frequencies has a strong effect on the performance. The network has to terminate the varactor at some frequencies and couple the varactor to sources and loads at other frequencies. Since this embedding circuit is best described in the frequency domain, the above time-domain equation is converted to the frequency domain. Moreover, the varactor is usually pumped strongly at one frequency, f_p, by a local oscillator. If there are, in addition, other small signals present, the varactor can be analyzed in two steps: (1) a large-signal simulation of the pumped varactor at the frequency f_p, and (2) the varactor behaves like a time-varying linear elastance at the signal frequency, f_s. For the large signal analysis, the voltage, current, and differential elastance can be written in the form

$$i(t) = \sum_{k=-\infty}^{\infty} I_k e^{jk\omega_p t}, \, I_{-k} = I_k^*$$

$$V(t) = \sum_{k=-\infty}^{\infty} V_k e^{jk\omega_p t}, \, V_{-k} = V_k^* \qquad (13.5)$$

$$S(t) = \sum_{k=-\infty}^{\infty} S_k e^{jk\omega_p t}, \, S_{-k} = S_k^*$$

Hence, the time-domain equation that governs the above varactor model can be converted to the frequency domain and the relation between the Fourier coefficients, I_k, V_k, S_k, reads

$$V_k = R_s I_k + \frac{1}{jk\omega_p} \sum_{l=-\infty}^{\infty} I_l S_{k-l} \qquad (13.6)$$

The above Equation 13.6 is the general starting point for analyzing varactors. Since there is a relation between the Fourier coefficients S_k and I_k, the above equation is nonlinear and hard to solve for the general case. Today, the large-signal response is usually calculated with a technique called harmonic balance [7]. This type of nonlinear circuit solver is available from most commercial microwave CAD tools.

Assume that the varactor is fully pumped and terminated so that the voltage across the diode junction becomes sinusoidal. The corresponding elastance waveforms for the conventional and symmetrical varactor in Figure 13.2 are shown in Figure 13.3. It is important to note that the fundamental frequency of the symmetric elastance device is twice the pump frequency.

The nonlinear part of the elastance, $S(t)$-S_{min}, creates harmonics and its impedance should be large compared to the series resistance, R_s, for a good varactor. The impedance ratio of the series resistance and

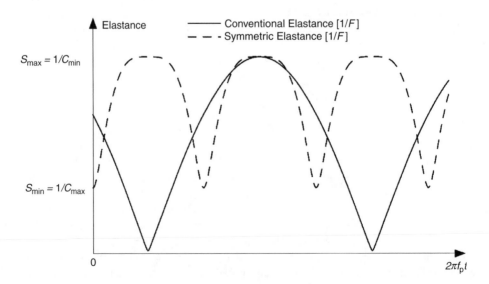

FIGURE 13.3   Elastance waveform $S(t)$ during full pumping with a sinusoidal voltage across the diode junction.

the nonlinear part of the elastance at the fundamental frequency can be written as

$$\frac{R_s}{\frac{S_{max} - S_{min}}{2\pi f_p}} = \frac{f_p}{f_c} \tag{13.7}$$

and the dynamic cut-off frequency is introduced as

$$f_c = \frac{S_{max} - S_{min}}{2\pi R_s} \tag{13.8}$$

The dynamic cut-off frequency is an important figure-of-merit for many varactor applications and a typical value for a state-of-the-art varactor is more than 1 THz. The starting point for a varactor design is, hence, to maximize the elastance swing, $S_{max} - S_{min}$, and minimize any losses, R_s. For semiconductor varactors, the maximum elastance swing is limited by at least one of the following conditions:

- Depletion layer punch-through
- Large electron conduction from impact ionization or forward conduction current
- Current saturation. The saturated electron velocity in the material determines the maximum length an electron can travel during a quarter of a pump-cycle [8]
- Maximum junction temperature. The dissipated energy due to large-signal pumping results in self-heating effects [9]

Increasing the pump power beyond any of the above conditions will result in reduced performance and probably introduce extra noise.

13.6 Varactor Applications

13.6.1 Frequency Multipliers

Varactor frequency multipliers are extensively used to provide LO power to sensitive millimeter- and submillimeter wavelength receivers [10]. Solid-state multipliers are relatively inexpensive, compact, light weight, and reliable compared to vacuum tube technology, which makes them suitable for space applications at these frequencies [11]. State-of-the-art balanced Schottky doublers can deliver approximately [12]: 24 dBm at 100 GHz, 17 dBm at 200 GHz (e.g., Rizzi et al. [13]), and 12 dBm at 300 GHz. State-of-the-art symmetric varactor triplers have so far been demonstrated to deliver: 23 dBm at 113 GHz (HBV [14]), 12 dBm at 228 GHz (ASV [6]), and 10 dBm at 300 GHz (HBV [15]).

Frequency multiplication or harmonic generation in devices occur due to their nonlinearity. Depending on if the multiplication is due to a nonlinear resistance or a nonlinear reactance, one can differentiate between the varistor and varactor type of multipliers. Varactor type multipliers have a high potential conversion efficiency, but exhibit a narrow bandwidth and a high sensitivity to operating conditions. According to the Page–Pantell inequality, multipliers that depend upon a nonlinear resistance have at most an efficiency of $1/n^2$, where n is the order of multiplication [16,17]. The absence of reactive energy storage in varistor frequency multipliers ensures a large bandwidth. For the ideal varactor multiplier, that is, a lossless nonlinear reactance, the theoretical limit is a conversion efficiency of 100% according to the Manley–Rowe formula. However, real devices exhibit properties and parameters that are a mixture of the ideal varistor and the ideal varactor multiplier. The following set of parameters are used to describe and compare properties of frequency multipliers:

- Conversion loss, L_n, is defined as the ratio of the available source power, P_{AVS}, to the output harmonic power, P_n, delivered to the load resistance. It is usually expressed in decibels. The inverted value of L_n that is, the conversion efficiency, η_n, is often expressed in percent.

- In order to minimize the conversion loss, the optimum source and load embedding impedances, Z_S and Z_L, should be provided to the diode. Optimum source and load impedances are found from maximizing, for example, the conversion efficiency, and they depend on each other and on the input signal level [18]. In a nonlinear circuit, such as a multiplier, it is not possible to define a true impedance. However, a "quasi-impedance," Z_n, can be defined for periodic signals as

$$Z_n = \frac{V_n}{I_n} \qquad (13.9)$$

where V_n and I_n are the voltage and the current, respectively, at the nth harmonic.

13.6.2 Noise in Varactor Multipliers

Being a mainly reactive device, a varactor generates very little noise. Thermal (white) noise is generated from the various contributions to the series resistance. However, these effects are almost negligible and the main concern is the phase noise, which is inevitably degraded by n^2, where n is the multiplication factor [18,19]. The strong capacitive nonlinearity can cause instabilities in practical circuits. The generation of such spurious side-band oscillations is chaotic and hence sometimes mistaken as a noise process.

13.6.3 Basic Principles of Single Diode Frequency Multipliers

Single diode frequency multipliers can either be shunt mounted or series mounted. In both cases the input and the output filter should provide optimum embedding impedances at the input and output frequencies, respectively. The output filter should also provide an open circuit for the shunt mounted varactor and a short circuit for the series-mounted varactor at the pump frequency. The same arguments apply to the input filter at the output frequency. Analysis and design of conventional doublers, and high-order varactor multipliers are well described in the book by Penfield et al. [1] and in the reference by Burckhardt [20].

Besides the above conditions, the correct impedances must be provided at the idler frequencies for a high-order multiplier. In general, it is hard to achieve optimum impedances at the different harmonics simultaneously. Therefore, a compromise has to be found (Figure 13.4).

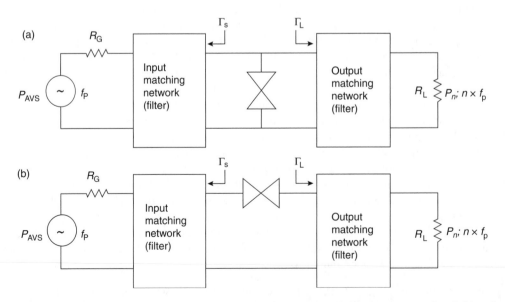

FIGURE 13.4 Block scheme of nth-order frequency multiplier circuit with (a) shunt-mounted and (b) series-mounted diodes.

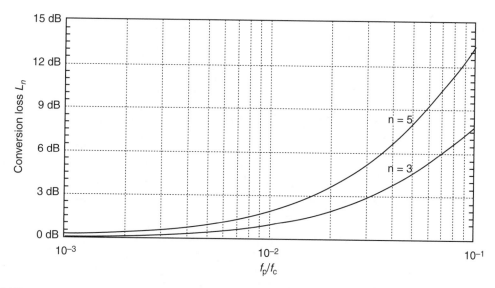

FIGURE 13.5 The minimum conversion loss for a tripler and a quintupler for the symmetric S–V characteristic shown in the pump frequency is normalized to the dynamic cut-off frequency. For the quintupler case, the idler circuit is an inductance in resonance with the diode capacitance.

13.6.4 Performance of Symmetric Varactor Frequency Multipliers

In Figure 13.5 a calculation of the minimum conversion loss for a tripler and a quintupler is shown. To systematically investigate how the tripler and quintupler performance depends on the shape of the S–V characteristic, a fifth degree polynomial model was employed by Dillner et al. [21]. The best efficiency is obtained for a S–V characteristic with a large nonlinearity at zero volt or a large average elastance during a pump cycle. The optimum idler circuit for the quintupler is an inductance in resonance with the diode capacitance (i.e., maximized third harmonic current).

Load-pull simulations are very useful for investigating impedance levels and their dependence on various parameters. Harmonic balance simulations are used to determine the performance as a function of the embedding impedance. In Figure 13.6, the conversion loss contours from simulations of a 500 GHz HBV quintupler are presented. When Z_1 is swept over the entire Smith chart, Z_3 and Z_5 are fixed at their respective optimum levels, and so forth. It is obvious that the input matching is very crucial for high-Q varactor multipliers.

13.6.5 Practical Multipliers

Since frequency multipliers find applications mostly as sources at higher millimeter and submillimeter wave frequencies, they are often realized in waveguide mounts, [22,23] see Figure 13.7. A classic design is the arrangement of crossed rectangular waveguides of widths specific for the input and output frequency bands. The advantages are

- The input signal does not excite the output waveguide, which is cut-off at the input frequency.
- Low losses.
- The height of the waveguide in the diode mounting plane may be chosen to provide the electrical matching conditions. Assuming a thin planar probe, the output embedding impedance is given by analytical expressions [24].
- Movable short circuits provide input/output tunability. However, modern EM software now allows for fixed tuned design of the back-short position.

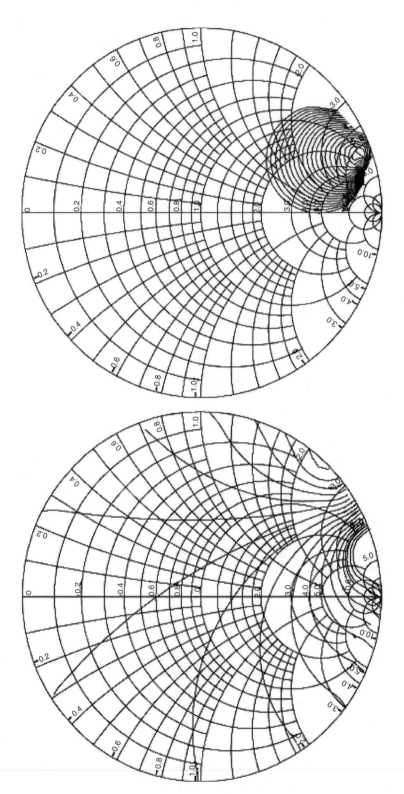

FIGURE 13.6 Conversion loss contours versus (top) input and (bottom) idler impedance. The results are obtained from harmonic balance simulations of a 500 GHz quintupler. The device assumed is a 2×3-barrier InP-based planar HBV with an area of 37 μm^2. The minimum conversion loss is 5.1 dB and the contours correspond to an increase in conversion loss of 1 dB.

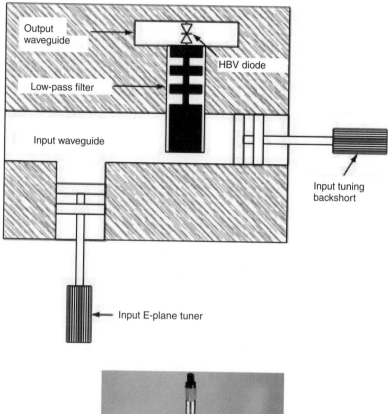

FIGURE 13.7 (a) cross-section of a crossed waveguide frequency multiplier and (b) photo of a 100-GHz HBV quintupler with two input tuners and one output tuner.

Today, whole waveguide mounts can be analyzed and designed using commercially available high frequency electromagnetic CAD tools.

The inherently limited bandwidth of varactors can be improved by employing a transmission line periodically loaded with varactors [25,26]. These Nonlinear Transmission Lines (NLTLs) can currently provide the largest bandwidth and still achieve a reasonable conversion efficiency as a frequency multiplier. Simultaneous effects of nonlinearity and dispersion may also be used for pulse compression (soliton propagation).

13.6.6 Frequency Converters

The varactor is useful as a frequency converter because of its good noise properties, and because gain can be achieved. The nonlinear reactance is pumped at a frequency f_p, and a small signal is introduced at a frequency f_s. Power will exchange at frequencies of the form $nf_p + f_s$ (n can be negative). If the output frequency is higher than the input frequency, the varactor acts as an up-converter, otherwise it is a down-converter. Furthermore, one differs between lower ($n = -1$) and upper sideband ($n = 1$) converters, and according to whether or not power is dissipated at any other sidebands (idlers).

Assume that the elastance is pumped sinusoidally (i.e., $S_k = 0$ for $|k| > 1$), the varactor is open-circuited at all frequencies except $f_s, f_p, f_u = f_p + f_s$, and that the varactor termination tunes out the average elastance. The source resistance is then adjusted to give optimum gain or optimum noise temperature. For the upper-sideband up converter, the minimal noise temperature is

$$T_{\min} = T_d \frac{2f_s}{m_1 f_c} \left[\frac{f_s}{m_1 f_c} + \sqrt{1 + \left(\frac{f_s}{m_1 f_c} \right)^2} \right] \tag{13.10}$$

when the source resistance is

$$R_0^T = R_s \sqrt{1 + \left(\frac{m_1 f_c}{f_s} \right)^2} \tag{13.11}$$

where T_d is the diode temperature, f_c is the dynamic cut-off frequency Equation 13.8, and m_1 is the modulation ratio defined as

$$m_1 = \frac{|S_1|}{S_{\max} - S_{\min}} \tag{13.12}$$

It can be shown that there is gain under optimum noise termination conditions only for signal frequencies smaller than $0.455 \, m_1 f_c$. [1] A different source resistance will result in maximum gain

$$R_0^G = R_s \sqrt{1 + \frac{m_1^2 f_c^2}{f_s f_u}} \tag{13.13}$$

The corresponding optimum gain is

$$G_{\text{MAX}} = \left(\frac{(m_1 f_c / f_s)}{1 + \sqrt{1 + (m_1^2 f_c^2 / f_s f_u)}} \right)^2 \tag{13.14}$$

As predicted by the Manley–Rowe formula for a lossless varactor, the gain increases as the output frequency, f_u, increases. The effect of an idler termination at $f_i = f_p - f_s$ can further increase the gain and reduce the noise temperature.

The above expressions for optimum noise and the corresponding source impedance are valid for the lower sideband up converter as well. However, the lower sideband up converter may have negative input and output resistances and an infinite gain causing stability problems and spurious oscillations. All pumped varactors may have such problems. By a proper choice of source impedance and pump frequency, it is possible to simultaneously minimize the noise and make the exchangeable gain infinite. This occurs for an "optimum" pump frequency of $f_p = \sqrt{m_1^2 f_c^2 + f_s^2}$ or approximately $m_1 f_c$ if the signal frequency is small. Further information on how to analyze, design, and optimize frequency converters can be found in the book by Penfield et al. [1].

13.6.7 Parametric Amplifiers

The parametric amplifier is a varactor pumped strongly at frequency f_p, with a signal introduced at frequency f_s. If the generated sidebands are terminated properly, the varactor can behave as a negative resistance at f_s. Especially the termination of the idler frequency, $f_p - f_s$, determines the real part of the impedance at the signal frequency. Hence, the varactor can operate as a negative resistance amplifier at the signal frequency, f_s. The series resistance limits the frequencies f_p and f_c for which amplification can be achieved and it also introduces noise.

The explanation of the effective negative resistance can be described as follows: The application of signal plus pump power to the nonlinear capacitance causes frequency mixing to occur. When current is allowed to flow at the idler frequency $f_p - f_s$, further frequency mixing occur at the pump and idler frequencies. This latter mixing creates harmonics of f_p and $f_p - f_s$, and power at f_s is generated. When the power generated through mixing exceeds that being supplied at the signal frequency f_s, the varactor appears to have a negative resistance. If idler current is not allowed to flow, the negative resistance vanishes. Assuming that the elastance is pumped sinusoidally (i.e., $S_k = 0$ for $|k| > 1$), and the varactor is open-circuited at all frequencies except $f_s, f_p, f_i = f_p - f_s$, and that the varactor termination tunes out the average elastance, gain can only be achieved if

$$f_s f_i (R_s + R_i) < R_s m_1^2 f_c^2 \tag{13.15}$$

where R_i is the idler resistance. By terminating the varactor reactively at the idler frequency, it can be shown that a parametric amplifier attains a minimum noise temperature when pumped at the optimum pump frequency, which is exactly the same as for the simple frequency converter. This is true for nondegenerated amplifiers where the frequencies are well separated. The degenerate parametric amplifier operates with f_i close to f_s, and can use the same physical circuit for idler and signal frequencies. The degenerate amplifier is easier to build but ordinary concepts of noise figure, noise temperature, and noise measure do not apply.

13.6.8 Voltage Tuning

One important application of varactors is voltage tuning. The variable capacitance is used to tune a resonant circuit with an externally applied voltage. This can be used to implement a voltage-controlled oscillator (VCO), since changing the varactor capacitance changes the frequency of oscillation within a certain range. As the bias is increased, the resonant frequency f_0 increases from $f_{0,\min}$ to $f_{0,\max}$ as the elastance changes from $S_{\min}$ to $S_{\max}$. If the present RF power is low, the main limitations are the finite tuning range implied by the minimum and maximum elastance and that the series resistance degrades the quality factor, Q, of the tuned circuit. The ratio of the maximum and minimum resonant frequency gives a good indication of the tunability

$$\frac{f_{0,\max}}{f_{0,\min}} \leq \sqrt{\frac{S_{\max}}{S_{\min}}} \tag{13.16}$$

However, if the present RF power level is large, the average elastance which determines the resonant frequency depends upon drive level as well as bias. Second, the allowed variation of voltage is reduced for large RF power levels.

Since the varactor elastance is nonlinear, quite steep at low voltages and almost flat at high voltages, the VCO tuning range is not naturally linear. However, an external bias circuit can improve the linearity of the VCO tuning range. It is also possible to optimize the doping profile of the varactor in terms of linearity, Q-value, or elastance ratio.

13.7 Varactor Devices

13.7.1 Conventional Diodes

Common conventional varactors at lower frequencies are reverse biased semiconductor abrupt p^+-n junction diodes made from GaAs or silicon [2]. The hyperabrupt p^+-n junction varactor diode has a nonuniform n-doping profile and is often used for voltage tuning. The n-doping concentration is very high close to the junction and the doping profile is tailored to improve elastance ratio and sensitivity. Such doping profiles can be achieved with epitaxial growth or by ion implantation. However, metal-semiconductor junction diodes (Schottky diodes) are superior at high frequencies since the carrier transport only rely on electrons (unipolar device). The effective mass is lower and the mobility is higher for electrons compared to holes. Furthermore, the metal-semiconductor junction can be made very precisely even at a submicron level. A reverse biased Schottky diode exhibits a nonlinear capacitance with a very low leakage current. High frequency diodes are made from GaAs since the electron mobility is much higher than for silicon.

13.7.2 The Heterostructure Barrier Varactor Diode

The HBV, [27] first introduced in 1989 by Kollberg et al. [5] is a symmetric varactor. The main advantage compared to the Schottky diode is that several barriers can be stacked epitaxially. Hence, an HBV diode can be tailored for a certain application in terms of both frequency and power handling capability. Moreover, since the HBV operates unbiased and is a symmetric device, it generates only odd harmonics. This greatly simplifies the design of high-order and broadband multipliers.

The HBV diode is an unipolar device and consists of a symmetric layer structure. An undoped high band gap material (barrier) is sandwiched between two moderately n-doped low band gap materials. The barrier prevents electron transport through the structure. Hence, the barrier should be undoped (no carriers), high and thick enough to minimize thermionic emission and tunneling of carriers. When the diode is biased a depleted region builds up (Figure 13.8), causing a nonlinear CV curve.

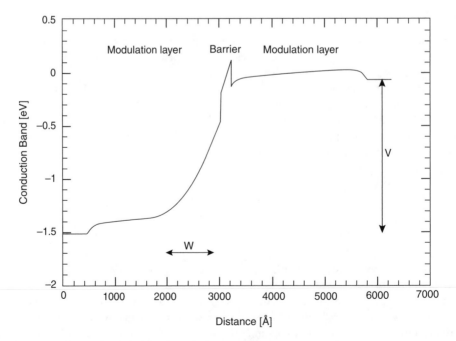

FIGURE 13.8 Conduction band of a biased GaAs/Al$_{0.7}$ GaAs HBV.

TABLE 13.1 Generic Layer Structure of an HBV. For N Epitaxially Stacked Barriers, the Layer Sequence 2–5 is Repeated N Times

Layer		Thickness (nm)	Doping level (cm^{-3})
7	Contact	~ 300	n^{++}
6	Modulation	$l \sim 300$	$N_d \sim 10^{17}$
5	Spacer	$s \sim 5$	Undoped
4	Barrier	$b \sim 15$	Undoped
3	Spacer	$s \sim 5$	Undoped
2	Modulation	$l \sim 300$	$N_d \sim 10^{17}$
1	Buffer	—	n^{++}
0	Substrate		n^{++} or SI

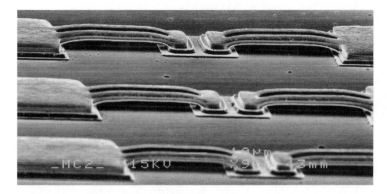

FIGURE 13.9 Planar four-barrier HBV (From Bryllert, T., Olsen, A. O., Vukusic, J., Emadi, T. A., Ingvarson, M., Stake, J., and Lippens, D., *Electronics Letters* 41(3), 30, 2005.)

Contrary to the Schottky diode, where the barrier is formed at the interface between a metallic contact and a semiconductor, the HBV uses a heterojunction as the blocking element. A heterojunction, that is, two adjacent epitaxial semiconductor layers with different band gaps, exhibits band discontinuities both in the valence and in the conduction band. Since the distance between the barriers (>1000 Å) is large compared to the de Broglie wavelength of the electron, it is possible to understand the stacked barrier structure as a series connection of *N* individual barriers. A generic layer structure of an HBV is shown in Table 13.1.

The substrate is either highly doped or semi-insulating (SI), depending on how the device is intended to be mounted. The contact layers (No. 1 and 7) should be optimized for low losses. Therefore, the buffer layer (No. 1) must be relatively thick ($\delta \sim 3 \mu m$) and highly doped for planar HBVs, see Figure 13.9. The barrier itself can consist of different layers to further improve the blocking characteristic. The spacer prevents diffusion of dopants into the barrier layer and increases the effective barrier height. The thickness of the barrier layer will not influence the cut-off frequency but it has some influence on the optimum embedding impedances. Hence, the thickness is chosen to be thick enough to avoid tunneling of carriers. Several III–V semiconductor material systems have been employed for HBVs. The best choices to date for HBVs are the lattice-matched $In_{0.53}Ga_{0.47}As/In_{0.52}Al_{0.48}As$ system grown on InP substrate and the lattice matched GaAs/AlGaAs system grown on GaAs substrate. High dynamic cut-off frequencies are achieved in both systems. However, the GaAs/AlGaAs system is well characterized and relatively easy to process, which increases the probability of reproducible results. The $In_{0.53}GaAs/In_{0.52}AlAs$ system exhibits a higher effective electron barrier and is therefore advantageous from a leakage current point of view. The thickness and doping concentration of the modulation layers should be optimized for maximal dynamic cut-off frequency [28] (Equation 13.8).

13.7.3 The HBV Capacitance

The parallel plate capacitor model, where the plate separation should be replaced with the sum of the barrier thickness, b, the spacer layer thickness, s, and the length of the depleted region, w, is normally an adequate description of the (differential) capacitance. The depletion length is bias dependent and the layer structure is symmetric, therefore the elastance is an even function of applied voltage and is given by

$$S = \frac{1}{C} = \frac{N}{A}\left(\frac{b}{\varepsilon_b} + \frac{s}{\varepsilon_d} + \frac{w}{\varepsilon_d}\right)$$

$$w = \sqrt{\frac{2\varepsilon_d |V_d|}{qN_d}}$$

(13.17)

where V_d is the voltage across the depleted region, N_d is the doping concentration in the modulation layers, b is the barrier thickness, s is the undoped spacer layer thickness, A is the device area, ε_b and ε_d are the dielectric constants in the barrier material and modulation layers, respectively. The maximum capacitance or the minimum elastance, $S_{\min}$, occurs at zero bias. However, due to screening effects, the minimum elastance, $S_{\min}$, must include the extrinsic Debye length, L_D, as

$$S_{\min} = \frac{1}{C_{\max}} = \frac{N}{A}\left(\frac{b}{\varepsilon_b} + \frac{2s}{\varepsilon_d} + \frac{2L_D}{\varepsilon_d}\right)$$

$$L_D \equiv \sqrt{\frac{\varepsilon_d kT}{q^2 N_d}}$$

(13.18)

To achieve a high $C_{\max}/C_{\min}$ ratio, the screening length can be minimized with a sheet doping, N_s, at the spacer/depletion layer interface. The minimum capacitance, $C_{\min}$, is normally obtained for punch-through condition, that is, $w = l$, or when the breakdown voltage, $V_{\max}$, is reached.

An accurate quasi-empirical expression for the $C-V$ characteristic of homogeneously doped HBVs has been derived by Dillner et al. [30]. The voltage across the nonlinear capacitor is expressed as a function of its charge as

$$V(Q) = N\left(\frac{bQ}{\varepsilon_b A} + 2\frac{sQ}{\varepsilon_d A} + \text{Sign}(Q)\left(\frac{Q^2}{2qN_d\varepsilon_d A^2} + \frac{4kT}{q}\left(1 - e^{-\frac{|Q|}{2L_D AqN_d}}\right)\right)\right)$$

(13.19)

where T is the device temperature, q is the elementary charge, and Q is the charge stored in the HBV. A complete electro-thermal model (Chalmers HBV Model) can be found in reference by Ingvarson et al. [31].

13.7.4 The Antiserial Schottky Varactor Diode

The antiserial Schottky varactor (ASV) consists of two Schottky diodes fabricated in an antiseries configuration. The symmetric structure of the overall device results in an antisymmetric $I-V$ and a symmetric $C-V$ characteristic. The dynamic elastance modulation relies on careful selection of a proper doping profile of the individual diodes. Frequency conversion will not occur for homogenously doped ASV diodes due to self-biasing caused by rectification of the individual diodes [32]. In the case of Schottky diodes with nonuniform doping profile, the space-charge region variation of the forward and reverse diode, respectively, differ from each other leading to a variable total elastance [6]. The nonuniform doping profile can simply be realized with two modulation regions with different thicknesses and doping concentrations (uniform). Experimental results with ASV triplers show an output power of 15 mW at 228 GHz and a corresponding flange-to-flange conversion efficiency of 22% [33].

13.7.5 The Si/SiO$_2$/Si Varactor Diode

By bonding two thin silicon wafers, [34,35] each with a thin layer of silicon dioxide, it is possible to form a structure similar to HBV diodes from III–V compounds. The SiO$_2$ layer blocks the conduction current very efficiently but the drawback is the relatively low mobility of silicon. If a method to stack several barriers can be developed, this material system may be interesting for lower frequencies where the series resistance is less critical and for voltage tuning circuits.

13.7.6 The Ferroelectric Varactor Diode

Ferroelectrics are dielectric materials characterized by an electric field and temperature-dependent dielectric constant [36]. Parallel plate capacitors made from such films can provide low loss and high tunability and hence can be used in varactor applications. Ferroelectric material, for example, (Ba, Sr)TiO$_3$ (BSTO) is a pseudo-cubic perovskite, which is naturally nonlinear. The nonlinearity originates from the polarization of the positive ions versus the negative within the crystal. In a ferroelectric film there are some intrinsic resonance frequencies at the transverse optical frequencies, for SrTiO$_3$ (STO) at room temperature, the first resonance peak is at 3 THz. The capacitance is frequency independent up to this resonance. Similar to the HBV, the $C(V)$ dependence for a ferroelectric varactor is symmetrical and can thereby be approximated with a series of even terms like $C(V) = C_0 + C_2 V^2 + C_4 V^4 + \ldots$.

References

1. Penfield, P. and Rafuse, R. P., *Varactor Applications*, M.I.T. Press, Cambridge, 1962.
2. Sze, S. M., *Physics of Semiconductor Devices*, 2nd ed., John Wiley & Sons, Singapore, 1981.
3. Manley, J. M. and Rowe, H. E., Some general properties of nonlinear elements, *IRE Proceedings* 44 (78), 904–913, 1956.
4. Uhlir, A., The potential of semiconductor diodes in high frequency communications, *Proceedings IRE* 46, 1099–1115, 1958.
5. Kollberg, E. L. and Rydberg, A., Quantum-barrier-varactor diode for high efficiency millimeter-wave multipliers, *Electronics Letters* 25, 1696–1697, 1989.
6. Krach, M., Freyer, J., and Claassen, M., Power generation at millimetre-wave frequencies using GaAs/GaAlAs triplers, *Physica Status Solidi C: Conferences* 1(8), 2160–2182, 2004.
7. Maas, S. A., Harmonic balance and large-signal-small-signal analysis, in *Nonlinear Microwave and RF Circuits*, 2nd ed., Artech House, Norwood MA, 2003, pp. 119–214.
8. Kollberg, E. L., Tolmunen, T. J., Frerking, M. A., and East, J. R., Current saturation in submillimeter wave varactors, *IEEE Trans. Microwave Theory Tech.* 40(5), 831–838, 1992.
9. Ingvarson, M., Alderman, B., Olsen, A. Ø., Vukusic, J., and Stake, J., Thermal constraint for heterostructure barrier varactor diodes, *IEEE Electron Device Letters* 25(11), 713–715, 2004.
10. Erickson, N., High efficiency submillimeter frequency multipliers, in *Microwave Theory and Techniques Society International Microwave Symposium*, Dallas, TX, 1990, pp. 1301–1304.
11. Chattopadhyay, G., Schlecht, E., Ward, J., Gill, J., Javadi, H., Maiwald, F., and Mehdi, I., An all-solid-state broad-band frequency multiplier chain at 1500 GHz, *IEEE Transaction on Microwave Theory and Techniques* 52(5), 1538–1547, 2004.
12. Porterfield, D., Hesler, J., Crowe, T., Bishop, W., and Woolard, D., Integrated terahertz transmit/receive modules, in *Proceedings of the 33rd European Microwave Conference*, Munich, Germany, 2003, pp. 1319–1322.
13. Rizzi, B. J., Crowe, T. W., and Erickson, N. R., A high-power millimeter-wave frequency doubler using a planar diode array, *IEEE Microwave and Guided Wave Letters* 3(6), 188–190, 1993.
14. Vukusic, J., Bryllert, T., Emadi, T. A., Sadeghi, M., and Stake, J., A 0.2-W heterostructure barrier varactor frequency tripler at 113 GHz, *IEEE Electron Device Letters* 28(5), 340–342, 2007.

15. Xiao, Q., Hesler, J. L., Crowe, T. W., Robert M. Weikle, I., and Duan, Y., High Efficiency Heterostructure-Barrier-Varactor Frequency Triplers Using AlN Substrates, in *IMS 2005*, Long Beach, CA, 2005.

16. Pantell, R. H., General power relationship for positive and negative nonlinear resistive elements, *Proceedings IRE* 46(December), 1910–1913, 1958.

17. Page, C. H., Harmonic generation with ideal rectifiers, *Proceedings IRE* 46(October), 1738–1740, 1958.

18. Faber, M. T., Chramiec, J., and Adamski, M. E., *Microwave and Millimeter-Wave Diode Frequency Multipliers* Artech House Publishers, Boston, 1995.

19. Maas, S. A., *Nonlinear Microwave and RF Circuits*, 2nd ed., Artech House, Norwood MA, 2003.

20. Burckhardt, C. B., Analysis of varactor frequency multipliers for arbitrary capacitance variation and drive level, *The Bell System Technical Journal* (April), 675–692, 1965.

21. Dillner, L., Stake, J., and Kollberg, E. L., Analysis of symmetric varactor frequency multipliers, *Microwave and Optical Technology Letters* 15(1), 26–29, 1997.

22. Archer, J. W., Millimeter wavelength frequency multipliers, *IEEE Transaction on Microwave Theory and Techniques* 29(6), 552–557, 1981.

23. Thornton, J., Mann, C. M., and Maagt, P. D., Optimization of a 250-GHz Schottky tripler using novel fabrication and design techniques, *IEEE Transaction on Microwave Theory and Techniques* 46(8), 1055–1061, 1998.

24. Eisenhart, E. L. and Khan, P. J., Theoretical and experimental analysis of a waveguide mounting structure, *IEEE Transaction on Microwave Theory and Techniques* 19(8), 706–719, 1971.

25. Hollung, S., Stake, J., Dillner, L., Ingvarson, M., and Kollberg, E. L., A Distributed heterostructure barrier varactor frequency tripler, *IEEE Microwave and Guided Wave Letters* 10(1), 24–26, 2000.

26. Carman, E., Case, M., Kamegawa, M., Yu, R., Giboney, K., and Rodwell, M. J., V-band and W-band broad-band, monolithic distributed frequency multipliers, *IEEE Microwave and Guided Wave Letters* 2(6), 253–254, 1992.

27. Dillner, L., Ingvarson, M., Kollberg, E. L., and Stake, J., High Efficiency HBV Multipliers for Millimetre Wave Generation, in *Terahertz Sources and Systems*, Miles, R. E., Harrison, P., and Lippens, D. Eds., Kluver Academic Publishers, Dordrecht, 2001, pp. 27–52.

28. Stake, J., Jones, S. H., Dillner, L., Hollung, S., and Kollberg, E. L., Heterostructure barrier varactor design, *IEEE Transaction on Microwave Theory and Techniques* 48(4, Part 2), 677–682, 2000.

29. Bryllert, T., Olsen, A. Ø., Vukusic, J., Emadi, T. A., Ingvarson, M., Stake, J., and Lippens, D., 11% efficiency 100 GHz InP-based heterostructure barrier varactor quintupler, *Electronics Letters* 41(3), 30, 2005.

30. Dillner, L., Stake, J., and Kollberg, E. L., Modeling of the Heterostructure Barrier Varactor Diode, in *International Semiconductor Device Research Symposium (ISDRS)*, Charlottesville, 1997, pp. 179–182.

31. Ingvarson, M., Vukusic, J., Olsen, A. Ø., Emadi, T. A., and Stake, J., An Electro-Thermal HBV Model, in *IMS 2005*, Long Beach, CA, 2005, pp. 1151–1153.

32. Bradley, R. F., *The Application of Planar Monolithic Technology to Schottky Varactor Millimeter-Wave Frequency Multipliers*, PhD thesis, University of Virginia, 1992.

33. Krach, M., Freyer, J., and Claassen, M., An Integrated ASV Frequency Tripler for Millimeter-Wave Applications, in *33rd European Microwave Conference*, Munich, Germany, 2003, pp. 1279–1281.

34. Fu, Y., Mamor, M., Willander, M., Bengtsson, S., and Dillner, L., n-Si/SiO$_2$/Si heterostructure barrier varactor diode design, *Applied Physics Letters* 77(1), 103–105, 2000.

35. Mamor, M., Fu, Y., Nur, O., Willander, M., and Bengtsson, S., Leakage current and capacitance characteristics of Si/SiO$_2$/Si single-barrier varactor, *Applied Physics A (Materials Science Processing)* A72(5), 633–637, 2001.

36. Tagantsev, A. K., Sherman, V. O., Astafiev, K. F., Venkatesh, J., and Setter, N., Ferroelectric materials for microwave tunable applications, *Journal of Electroceramics* 11(1–2), 5–66, 2003.

Further Reading

1. Faber, M. T., Chramiec, J., and Adamski, M. E., *Microwave and Millimeter-Wave Diode Frequency Multipliers*. Boston: Artech House Publishers, 1995.
2. Maas, S. A., *Nonlinear Microwave and RF Circuits*, 2nd ed., Artech House, Norwood MA, 2003.
3. Penfield, P., and Rafuse, R. P., *Varactor Applications*. Cambridge: M.I.T. Press, 1962.
4. Yngvesson, S., *Microwave Semiconductor Devices*. Boston: Kluwer Academic Publishers, 1991.

14

Schottky Diode Frequency Multipliers

Jack East
The University of Michigan

Imran Mehdi
Jet Propulsion Laboratory
California Institute of Technology

14.1   Introduction

Heterodyne receivers are an important component of most high-frequency communications systems and other receivers. In its simplest form, a receiver consists of a mixer being pumped by a local oscillator with an associated intermediate frequency (IF) circuit. At lower frequencies, a variety of local oscillator sources are available, but as the desired frequency of operation increases the local oscillator source options become more limited. The "lower frequencies" limit has increased with time. Early transistor oscillators were available in the MHz and low GHz range. Two terminal transit time devices such as impact ionization avalanche transit time (IMPATT) and Gunn diodes were developed for operation in X and Ka band in the early 1970s. However, higher frequency heterodyne receivers were needed for a variety of communications and science applications, so alternative local oscillator sources were needed. One option was vacuum tubes. A variety of vacuum tubes such as klystrons and backward wave oscillators grew out of the radar effort during World War II. These devices were able to produce large amounts of power over most of the desired frequency range. However they were large, bulky, and expensive, and suffered from modest lifetimes. They were also difficult to use in small science packages for space applications.

An alternative solid-state source was needed and the technology of the diode frequency multiplier was developed beginning in the 1950s. These devices use the nonlinear reactance or resistance of a simple semiconductor diode to produce high-frequency signals by frequency multiplication. These multipliers have been a part of many high-frequency communications and science applications since that time. As time has passed, the operating frequencies of both transistors and two terminal devices have increased. Silicon and GaAs transistors have been replaced by much higher frequency heterojunction field effect transistors

(HFETs) and heterojunction bipolar transistors (HBTs) with f_{max} values of hundreds of GHz. Two terminal IMPATT and Gunn diodes can produce more than 100 MW at frequencies above 100 GHz. However, the desired operating frequencies of communications and scientific applications have also increased. The most pressing needs are for a variety of science applications in the frequency range between several hundred GHz and several THz. Applications include space-based remote sensing of the earth's upper atmosphere to better understand the chemistry of ozone depletion, basic astrophysics to investigate the early history of the universe and ground-based remote sensing for chemical and biological detection applications. These and other applications will require heterodyne receivers with near THz local oscillators. Size, weight, and prime power will be important parameters. Alternative approaches include mixing of infrared lasers to produce the desired local oscillator frequency from higher frequencies, and a multiplier chain to produce the desired frequency from lower frequencies. Laser-based systems with the desired output frequencies and powers are available, but not with the desired size and weight. Recent advances in semiconductor diode-based frequency multipliers can now provide the required power for these applications even at THz frequencies along with the modest size and weight needed.

The goal of this chapter is to briefly describe the operation of diode frequency multipliers to better understand their performance and limitations. The chapter is organized as follows. Section 14.2 will describe the properties of Schottky barrier diodes, the most useful form of a varactor multiplier along with the analytic tools developed to predict multiplier operation. Two limitations, the reactive multiplier described by Manley and Rowe and the resistive multiplier discussed by Page, will be discussed. The results of these two descriptions can be used to understand the basic limits of multiplier operation. However, these analytic results do not provide enough information to design operating circuits. The nonlinear simulation tools needed to design multiplier embedding circuits will be described. A more realistic computer-based design approach is needed. This will be discussed in Section 14.3. With increasing frequency and power, the details of the device physics degrades the performance compared with the simple approach of Sections 14.2 and 14.3. A better understanding of device operation is needed. This will be discussed in Section 14.4. The discussion so far has assumed an ideal diode. In fact, parasitics, power limitations, and fabrication issues dominate the operation of realistic devices at THz frequencies. Section 14.5 will discuss the solution of some of these problems for state-of-the-art THz multipliers. Finally, Section 14.6 will give a brief overview of the performance of THz multipliers that are part of a space-based heterodyne receiver. A brief summary will be given in Section 14.7.

14.2 Analytic Description of Diode Multipliers

14.2.1 Schottky Diode Characteristics

Multiplier operation depends on the nonlinear properties of Schottky diodes. The diode has both a capacitive and a resistive nonlinearity. Consider a uniformly doped semiconductor structure with an ohmic contact on one end and a metal–semiconductor contact on the other. The semiconductor will have a depletion layer with a width w, with the remaining portion of the structure undepleted. The depletion layer can be represented as a capacitor and the undepleted portion can be represented as a resistor. The depletion layer will act as a parallel plate capacitor with a capacitance of

$$C = \frac{\varepsilon A}{w}, \tag{14.1}$$

where C is the capacitance, ε is the dielectric constant, A is the area, and w is the depletion width. The depletion width versus applied reverse voltage is

$$w = \sqrt{\frac{2\varepsilon(\phi_{bi} + V_{bias})}{qN_d}}, \tag{14.2}$$

where ϕ_{bi} is the built-in potential of the metal–semiconductor junction, V_{bias} is the applied reverse bias, q is the electronic charge, and N_d is the uniform semiconductor doping. This width versus applied bias will result in a capacitance versus applied voltage of the form

$$C(V) = \frac{C_{j0}}{\sqrt{\phi_{bi} + V_{bias}}}, \qquad (14.3)$$

where C_{j0} is the capacitance at zero applied bias. This inverse square root-dependent capacitance is the starting point for analytic varactor multiplier analysis. Other effects are also present. Under realistic pumping conditions, the diode can also be forward biased allowing forward conduction current flow of the form

$$I(V) = I_0 \left(\exp^{(v/nkT)} - 1 \right), \qquad (14.4)$$

where n is an ideality factor that is a measure of the metal–semiconductor junction quality and kT is the thermal voltage. This current versus voltage characteristic corresponds to a nonlinear resistance that is the starting point for analytic varistor multiplier analysis.

There are also parasitic elements. The undepleted region of the device and various contact and package resistances will appear in series with the nonlinear junction. Although the undepleted region width is voltage dependent, this series resistance is usually modeled with a constant value. However, at very high frequencies, the current through the undepleted region can crowd to the outside edge of the material due to the skin effect, increasing the resistance. This frequency-dependent resistance is sometimes included in multiplier simulations [1]. The varactor diode must be connected to the external circuit. The resulting physical connection usually results in a parasitic shunt capacitance associated with the connection and a parasitic series inductance associated with the current flow through the wire connection. It will be seen in Section 14.5 that these parasitics can dominate multiplier performance and must be considered to obtain realistic performance predictions. However, as a first approximation, a simple device model that includes only a nonlinear capacitor or resistor can be used to explore the limits of multiplier operation. The diode circuit elements can be combined to form the equivalent circuit shown in Figure 14.1. The equivalent capacitance and the diode current depend on the internal voltage across the diode depletion layer and the parasitic resistance depends on the current flow through the bulk undepleted portion of the device as well as loss due to ohmic contacts.

14.2.2 Analytic Descriptions of Diode Multipliers

The nonlinear capacitance versus voltage characteristic of a diode can be used as a frequency multiplier. Consider the charge Q in the nonlinear capacitance as a function of voltage written in terms of a power series in voltage

$$Q(V(t)) = a_0 + a_1 V(t) + a_2 V(t)^2 + a_3 V(t)^3 + \cdots . \qquad (14.5)$$

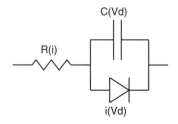

FIGURE 14.1 Diode equivalent circuit.

The current $I(t)$ is the time derivative of the charge

$$I(t) = \frac{dQ(t)}{dt} = \left[a_1 + 2a_2 V(t) + 3a_3 V(t)^2 + \cdots \right] \frac{dV(t)}{dt}. \qquad (14.6)$$

If $V(t)$ is sinusoidal, then the mixing products of various orders of $V(t)$ will produce harmonic currents. A similar expression will show harmonic generation in a nonlinear resistor. The earliest research on frequency multipliers were based on closed form descriptions of multiplier operation using these types of series expressions.

The Manley–Rowe relations describe the operation of ideal nonlinear reactance multipliers. The relations are a general description of power and frequency conversion in nonlinear reactive elements [2,3]. The earliest work on these devices sometimes used nonlinear inductances, but all present work involves the nonlinear capacitance versus voltage characteristic of a reverse biased Schottky barrier diode. Although the Manley–Rowe equations describe frequency multiplication, mixer operation, and parametric amplification, they are also useful as an upper limit on multiplier operation. If an ideal nonlinear capacitance is pumped with a local oscillator at frequency f_0, and an embedding circuit allows power flow at harmonic frequencies, then the sum of the powers into and out of the capacitor is zero,

$$\sum_{m=0}^{\infty} P_m = 0. \qquad (14.7)$$

This expression shows that an ideal frequency multiplier can convert input power to higher frequency output power with 100% efficiency. Nonlinear resistors can also be used as frequency multipliers [4,5]. For a nonlinear resistor pumped with a local oscillator at frequency f_0 the sum the powers is

$$\sum_{m=0}^{\infty} m^2 P_m \geq 0. \qquad (14.8)$$

For an mth order resistive harmonic generator, the efficiency is at best $1/m^2$ 25% for a doubler and 11% for a tripler.

Although Equations 14.7 and 14.8 give upper limits on the efficiency to be expected from a multiplier, they provide little design information for real multipliers. The next step in multiplier development was the development of closed form expressions for design based on varactor characteristics [6,7]. Burckardt [7] gives design tables for linear and abrupt junction multipliers based on closed form expressions for the charge in the diode. These references provide embedding impedances and information about the effects of parasitic resistances on multiplier performance. An example of multiplier design based on this approach is given by Maas [8]. However, this analytic approach makes simplifying assumptions that do not accurately describe multiplier operation. The next level of complexity involves numerical simulation of nonlinear device operation.

14.3 Computer-Based Design Approaches

The figures and much of the analysis in the next two sections were first presented by East [9]. The analytic tools discussed in the last section are useful to predict the ideal performance of various frequency multipliers and provide first-order estimates for efficiency and embedding impedances. However, more exact techniques are needed for useful designs. Important information such as input and output impedances, the effects of series resistance, and the effect of harmonic terminations at other harmonic frequencies are needed. This information requires detailed knowledge of the current and voltage in formation at the nonlinear device. Computer-based simulations are needed to provide this information. The general problem can be described with the help of Figure 14.2. The multiplier consists of a nonlinear microwave diode,

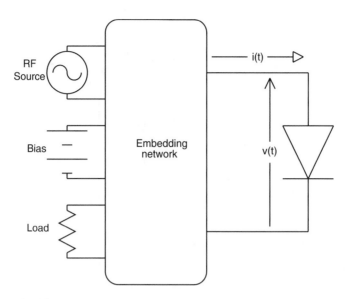

FIGURE 14.2   Generalized frequency multiplier.

an embedding network that provides coupling between the local oscillator source and the output load, provisions for dc bias and terminations for all other frequencies. The device sees embedding impedances at the fundamental frequency and each of the harmonic frequencies. The local oscillator power available is usually specified. The goal is to obtain the operating conditions, the output power and efficiency, and the input and output impedances of the overall circuit. The nonlinear nature of the problem makes the solution more difficult. Additional information about the Schottky diodes will be discussed in Section 14.4, but within the context of Figure 14.2, the current as a function of time is a nonlinear function of the voltage. Device impedances can be obtained from ratios of the Fourier components of the voltage and current at each frequency. The impedances of the diode must match the embedding impedances at the corresponding frequency. A "harmonic balance" is required for the correct solution. Many commercial software tools use nonlinear optimization techniques to solve this "harmonic balance" optimization problem. An alternative approach is the multiple reflection algorithm. This solution has the advantage of a very physical representation of the actual transient response of the circuit. This technique is discussed in References 1, 10, and 11, and will be described here. The basic problem is to calculate the diode current and voltage waveforms when the device is embedded in an external linear circuit. The diode waveforms are best represented in the time domain. However, the embedding circuit consists of linear elements and is best represented in the frequency or impedance domain. This technique splits the simulation into two parts, a nonlinear time domain description of the diode multiplier and a linear frequency domain description of the embedding network. The solution goal is to match the frequency domain impedances of the embedding circuit with the time domain impedances of the nonlinear device. A numerical transmission line with a length equal to an integral number of pump frequency wavelengths and an arbitrary impedance Z_0 can be added to circuit in Figure 14.2 here to give the circuit in Figure 14.3. Waves will propagate in both directions on this transmission line depending on the conditions on the ends. When steady-state conditions are reached, the waveforms at the circuit and the diode will be the same with or without the transmission line. This transmission line allows us to determine the waveforms across the diode as a series of reflections from the circuit. The signals on the transmission line are composed of left and right traveling current and voltage waves. The voltage at the diode is the sum of the right and left traveling wave voltages,

$$V(x) = V_r(x) + V_l(x),$$ (14.9)

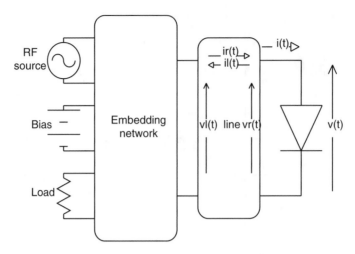

FIGURE 14.3   Multiple reflection circuit.

and the current at the diode is the difference of the left and right traveling current waves

$$I(x) = I_r(x) - I_l(x) = \frac{V_r(x) - V_l(x)}{Z_0}. \tag{14.10}$$

Since the transmission line is an integral number of pump frequency wavelengths long, the conditions are the same at each end under steady-state conditions,

$$V(x = 0) = V(x = 1) \tag{14.11}$$

$$I(x = 0) = I(x = 1). \tag{14.12}$$

At the start of the simulation, we assume that there is a right traveling wave associated with the dc bias along with a component at the local oscillator frequency. This wave will arrive at the diode at $x = 0$ as $V_r(t)$. The resulting voltage across the diode will produce a diode current. This current driving the transmission line will produce a first reflected voltage,

$$V_{\text{reflected}}^1 = V_1^1 = \left(V_{\text{diode}}^1 - I_{\text{diode}}^1(t)Z_0 \right) / 2. \tag{14.13}$$

This time domain reflected or left traveling wave will then propagate to the embedding network. Here it can be converted into the frequency domain with a Fourier transform. The resulting signal will contain harmonics of the local oscillator signal due to the nonlinear nature of the diode current versus voltage and charge versus voltage characteristic. The resulting frequency domain information can then be used to construct a new reflected voltage wave from the embedding network. This process of reflections from the diode and the circuit or "multiple reflections" continues until a steady-state solution is reached.

This computer-based solution has several advantages over the simpler analytic-based solutions. It can handle combinations of resistive and reactive nonlinearities. Most high-efficiency multipliers are pumped into forward conduction during a portion of the radio frequency (RF) cycle, so this is an important advantage for accurate performance predictions. In practice, the varactor diode will also have series resistances, parasitic capacitances, and series inductances. These additional elements are easily included in the multiplier simulation. At very high frequencies, the series resistance can be frequency dependent, due to current crowding in the semiconductor associated with the skin effect. This frequency dependence

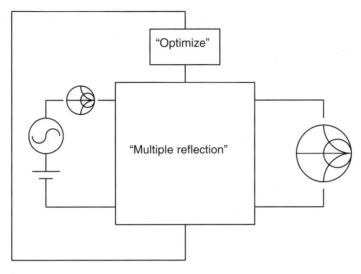

FIGURE 14.4 Performance optimization.

loss can also be included in simulations. Computer programs with these features are widely used to design high performance frequency multipliers.

An alternative solution technique is the fixed point method [12]. This technique uses a circuit similar to Figure 14.3, with a nonlinear device and an embedding network connected with a transmission line. However, this approach uses a fixed point iteration to arrive at a converged solution. The solution starts with arbitrary voltages at the local oscillator frequency at the harmonics. With these starting conditions, a new voltage is obtained from the existing conditions using

$$V_{n,k+1} = \frac{2Z_0}{Z_n^L + Z_0} \frac{Z_{n,k}^{NL}}{Z_{n,k}^{NL} - Z_0} V_n^S + \frac{Z_n^L - Z_0}{Z_n^L + Z_0} \frac{Z_{n,k}^{NL}}{Z_{n,k}^{NL} + Z_0} \left(V_{nk} - I_{n,k} Z_0 \right) \tag{14.14}$$

for the driven local oscillator frequency and

$$V_{n,k+1} = \frac{Z_0}{Z_n^L + Z_0} \left(V_{n,k} - I_{n,k} Z_0 \right) \tag{14.15}$$

at other frequencies, where Z_n^L are the embedding impedances at frequency n, Z_0 is the same line characteristic impedance used in the multiple reflection simulation, $Z_{n,k}^{NL}$ is the nonlinear device impedance at frequency n and iteration number k, $V_{n,k}$, and $I_{n,k}$ are the frequency domain voltage and current at iteration k, and V_n^S is the RF voltage at the pump source. These two equations provide an iterative solution of the nonlinear problem. They are particularly useful when a simple equivalent circuit for the nonlinear device is not available.

The solution described in the previous paragraph provides the current and voltage waveforms in the device for a given set of embedding impedances. However, the circuit designer usually wants to optimize the design, the set of impedance that provide the best output power, or efficiency, for example. The simulator of Figure 14.3 can be included in an optimization loop and the circuit tuned as shown in Figure 14.4. Since the circuit is nonlinear, the device impedance at each frequency will depend on the embedding impedance at all of the other frequencies. Consider the goal of finding the set of impedances that maximize the efficiency of the circuit in Figure 14.4 for a specified set of diode characteristics, and for a specified frequency and available local oscillator power. The "optimize" box in the figure will have several tasks. The outer loop of the optimization will be a search on the output frequency reflection coefficient plane for the best output load condition. However, the choice of the output will effect the power transfer from the pump to the output frequency and thus the value of the device input impedance. The combination of the device input impedance and the source embedding impedance will determine the fraction of the power that is available from the source that is adsorbed in the device. The optimal input embedding source impedance is a conjugate match to the device, so the next optimization is an adjustment of the source impedance to obtain an input conjugate match. The device impedances all also depend on the signal level. The power available from the source depends on the RF voltage magnitude at the pump voltage source at the real part of the source impedance. The inner optimization loop is an adjustment of the local oscillator voltage to obtain the specified local oscillator specified power.

14.4   Device Limitations

The simulation tools and simple device model described in the last section do a good job of predicting the performance of low-frequency diode multipliers. However, many multiplier applications require very high-frequency output signals with reasonable output power levels. Under these conditions, the output powers and efficiencies predicted are always higher than the experimental results. There are several possible reasons. Circuit loss increases with frequency, so the loss between the diode and the external connection should be higher. Measurements are less accurate at these frequencies, so the differences between the desired designed circuit embedding impedances and the actual values may be larger. Parasitic effects are also more important, degrading the performance. However, even when all these effects are taken into account, the experimental powers and efficiencies are still lower than expected. A more detailed description of the device physics is needed.

A first-order view of the problem can be described by rewriting Equation 14.2 for the depletion layer width with a time-dependent applied voltage

$$w(t) = \sqrt{\frac{2\varepsilon V(t)}{qN_d}}.$$ (14.16)

The time derivative of this equation is the velocity of the edge of the depletion layer. When this velocity, which depends on a combination of the frequency, V_{RF} and V_{dc} and the doping N_d, is larger than the saturated velocity imposed by the semiconductor physics, the voltage-dependent description of the capacitance is no longer correct [13]. The onset of saturation effects depends on a combination of the pump power, the frequency, and the device structure. Figure 14.5 is a plot of the frequency at which the peak velocity in a uniformly doped varactor is greater than 10^7 cm/s, a typical saturated velocity in silicon or GaAs for a range of epitaxial layer dopings and RF voltages. The voltage across the device is assumed to be $V(t) = V_{dc} + V_{RF} \sin(\omega t)$ with $V_{dc} = V_{RF}$. The region in the upper left portion of the figure is limited by saturation effects. Saturation effects can occur at lower millimeter wave frequencies for lightly doped structures. The effects can be avoided by increasing the doping. However, this will reduce breakdown voltage and thus limit the RF voltage swing and the power.

A simple device description that can be included in a nonlinear simulator can be used to investigate these saturation effects. The model can be understood with the help of Figure 14.6 [14]. The figure shows the electric field versus distance in a uniformly doped Schottky barrier diode. The structure has a metal contact and depletion layer on the right and an undepleted epitaxial layer on the left. The electric field in the depletion region has a constant slope that depends on doping via Poisson's equation over the width w_d and a constant field in the neutral undepleted region on the left. The area under the field curve is the negative of the total voltage across the structure. The current in the two regions is the same,

$$J_{device} = qnv(E_c) = \varepsilon \frac{dE}{dt}.$$ (14.17)

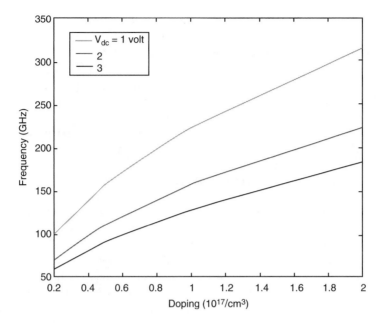

FIGURE 14.5 Velocity saturation in multipliers.

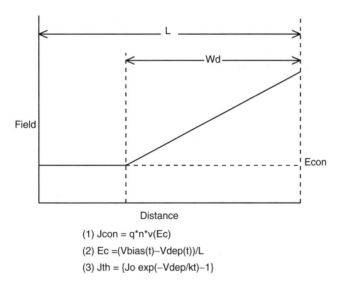

(1) Jcon = q*n*v(Ec)
(2) Ec =(Vbias(t)–Vdep(t))/L
(3) Jth = {Jo exp(–Vdep/kt)–1}

FIGURE 14.6 Diode velocity saturation model.

The depletion voltage associated with the area above the dashed line in the figure is

$$V_{\text{dep}} = \frac{qN_{\text{d}}w_{\text{d}}^2}{2\varepsilon}. \tag{14.18}$$

The thermionic current over the depletion layer is

$$J_{\text{thermionic}} = J_0 \left(\exp\left(\frac{V_{\text{dep}}}{kT}\right) - 1 \right). \tag{14.19}$$

These equations can be used to investigate velocity saturation in varactors.

When no current is flowing in the structure, the current in Equation 14.17 is zero and the corresponding depletion layer voltage in Equation 14.18 is equal to the terminal voltage. The device capacitance depends on the depletion layer width w_d, which is a nonlinear function of the voltage. The capacitance versus voltage for this structure is the same as the one in Equation 14.2 if the field in the undepleted region is neglected. However, if a realistic velocity versus electric field is used in Equation 14.17 then saturation effects occur. The velocity of the edge of the depletion layer is just the velocity of the electrons in the undepleted portion of the structure. The time rate of change of the nonlinear capacitance is limited by the saturated velocity of the semiconductor material. Driving the device beyond this saturation point will increase the voltage drop and resistance of the undepleted region and reduced the conversion efficiency of the multiplier. A sample plot of diode impedance as a function of frequency and RF voltage is shown in Figure 14.7. The example device has an epitaxial layer doping of $5 \times 10^{16}/\text{cm}^3$ and is operating between 75 and 225 GHz. The four curves correspond to RF voltages of 1, 2, 3, and 6 V, with the dc voltage equal to the RF voltage. The lowest RF voltage case is in the upper left corner of the figure. Over the entire frequency range, the current through the device is modest with no saturation. The resistance is small and constant and the reactance is $\propto 1/\omega C$. Note that the reactance is decreasing toward zero on the vertical axis. Increasing the dc voltage for the second curve decreases the capacitance and increases the magnitude of the reactance. At low frequencies, the resistance is similar to the value in the first curve. However, increasing the frequency increases the resistance. The final curve shows saturation effects even at the lowest frequency, with a resistance that has increased to the point of greatly degrading the diode performance at 225 GHz. This velocity saturation description for the nonlinear device can be used in the nonlinear simulator described in Section 14.3. Careful design optimization is needed to tradeoff the device and embedding circuit with the power and frequency requirements to obtain optimal performance at submillimeter wave and THz frequencies.

These first four sections have described the idealized operation of diode multipliers. In fact, these devices have a number of practical limitations. The next two sections briefly describe some of these limitations and then present state-of-the-art results for space-based THz source applications.

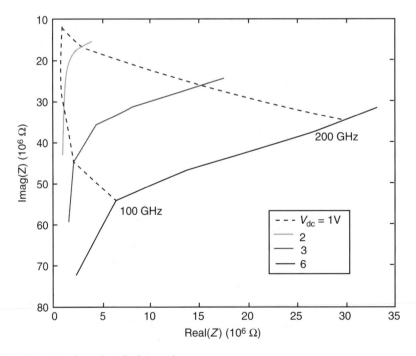

FIGURE 14.7   Saturation-dependent diode impedance.

14.5 Practical Submillimeter Wave Multipliers

This section will describe some of the practical design tradeoffs for state-of-the-art multipliers operating near THz frequencies. The figures and data in this section and in Section 14.6 were obtained from Mehdi [15]. The nonlinear modeling described in the earlier section is just the staring point in realizing a useful source. Parasitic resistances, losses, and shunt capacitance associated with semiconductor substrates, heat sinking, and power handling along with fabrication and assembly are all critical near THz frequencies. Research and development effort over the past decade has overcome many of these problems, resulting in practical multiplier chains producing useful power above 1 THz. Before discussing recent advances, it is useful to describe earlier multiplier configurations. Figure 14.8 shows a whisker-contacted diode structure. The figure shows a pointed metal whisker contacting a small Schottky barrier diode that is part of a "honeycomb" array. This configuration has several advantages. The semiconductor structure is relatively easy to fabricate and very small anode contacts are possible. The metal whisker contact has a well-understood impedance with low-parasitic capacitance. Perhaps the most important advantage was the years of experience invested in the design. However, there are disadvantages too. The semiconductor chip thickness does not easily scale down with decreasing waveguide dimensions at higher frequencies. The parasitic resistance depends on the spreading resistance from the circular anode. Assembly of the whisker can be an art form rather than a science, and although excellent performance is possible, reproducible results are sometimes difficult to obtain. Finally, the use of multiple junctions for additional power or for balanced configurations is not possible. Over the last decade, a planar diode technology has been developed to overcome these problems and to extend multipliers to above THz frequencies.

Planar diode technology uses fabrication tools and techniques that have been developed by the semiconductor industry to realize improved components at THz frequencies. Planar technology can be used to optimize the size and shape of the anode to reduce the parasitic capacitance and resistance, allow multiple anodes for higher power and improved heat distribution and enables balanced designs that reduce circuit complexity. An example die is shown in Figure 14.9. The circular anode of the whisker structure has been replaced with a long narrow anode derived from field effect transistor (FET) gate technology.

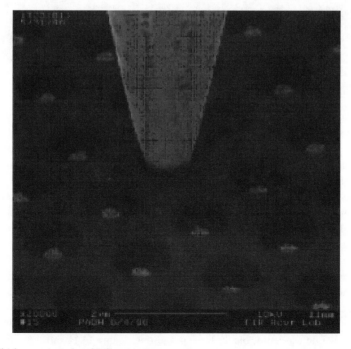

FIGURE 14.8 Whisker-contacted multiplier structure.

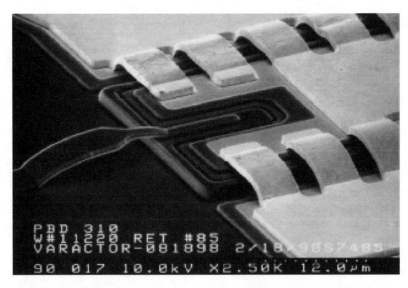

FIGURE 14.9   Planar THz diode.

FIGURE 14.10   Six-anode planar diode array.

This geometry has a reduced spreading resistance. The whisker has been replaced with the planar finger on the left of the photograph. Not clearly shown in this figure is the reduced volume of semiconductor in the planar structure. The substrate has been removed except for the region near the anode. This reduces the capacitance associated with the removed material and results in a truly "planar" structure. These structures can be designed to fit into very small waveguide blocks that will not accommodate thicker substrates. The planar technology can also be used to fabricate more complex structures. A six-anode array is shown in Figure 14.10. This configuration can only be realized in planar form and has important design advantages. The series combination of two groups of three diodes improves the impedance. The impedance of the individual anodes adds, allowing larger anodes for the same overall impedance level. The modest conversion efficiency of high-frequency sources implies that significant amounts of heat will be adsorbed. Increasing the number of anodes increases the amount of heat and local oscillator power that can be used. The center feed configuration allows a balanced configuration that reduces the harmonic termination requirements. Finally, this configuration reduces the assembly complexity. These planar circuits can be fitted directly into waveguide blocks for rapid testing. Figure 14.11 shows an assembled 600 GHz tripler. The input waveguide and coupling probe are on the right and the output waveguide and coupling

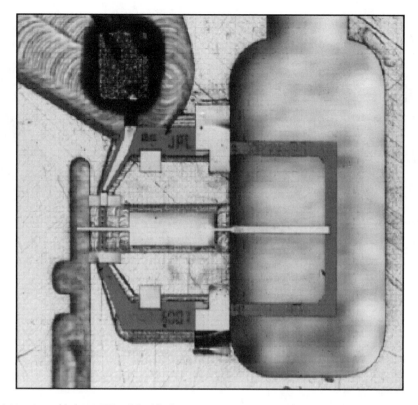

FIGURE 14.11   Assembled 600 GHz tripler block.

probe are on the left. The active planar circuit is located in the channel between the two waveguides. These individual doublers and triplers can be combined in chains with transistor-based power amplifiers in W band and synthesized sources to produce frequency sources with 10% bandwidth above 1 THz. Some example results will be described in the next section.

14.6 Multiplier-Based THz Sources

Although there are a variety of possible application for THz sources, a major focus has been space-based remote sensing and astrophysics applications. An ideal measurement requirement is the most sensitive possible performance over a wide bandwidth up to THz frequencies. A more realistic goal is a set of measurement bands with 10% bandwidth centered on interesting atomic and molecular lines between 100 GHz and 1.9 THz. There have been important advances in mixer and receiver technology to match the advances in multiplier technology. In particular, a new generation of nonlinear mixer elements based on cryogenic hot electron bolometers have extended the frequency range available from more conventional superconductor–insulator–superconductor (SIS) elements and greatly reduced the required local oscillator power [16–19]. Microwatt power levels are sufficient to pump these devices, greatly reducing the local oscillator requirements. A second advance is the development of wideband high-power amplifiers in W and D band. [20,21]. The individual amplifier chips can be cascaded to provide approximately 200 mW over an 87–105 GHz band. Power combining can be used to nearly double this power level to 400 mW over selected portions of this band and 300 mW over the entire frequency band. A synthesized source driving these amplifiers provides the pump source for the multiplier chain. Multiplier doublers and triplers can then be cascaded to produce the required THz signal. An assembled 1.2 THz multiplier chain is shown in Figure 14.12. The power amplifiers are on the right and the multipliers and a horn antenna are on the

FIGURE 14.12   1.2 THz multiplier chain.

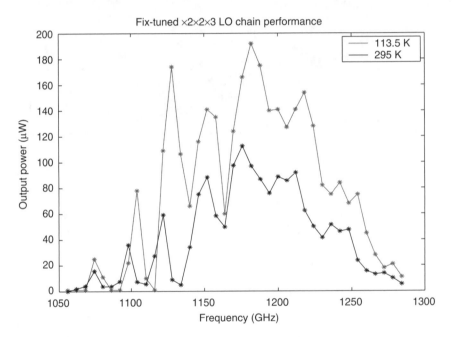

FIGURE 14.13   1.2 THz multiplier performance.

right. The ultimate realization of multiplier technology is all solid-state sources producing useful power at above 1 THz. Figure 14.13 shows the state-of-the-art performance of a 1.2 THz chain.

14.7   Summary and Conclusions

This chapter has presented a brief description of Schottky diode-based frequency multipliers. Although varactor technology was first used in the early 1950s and was considered mature by the 1960s, dramatic advances have occurred in the last decade. A better understanding of device physics allowed better device designs. Advances in fabrication technology have produced robust planar structures with low parasitics, multiple anodes, and balanced configurations. Monolithic microwave-integrated circuit (MMIC) amplifiers provide abundant pump power. The development of heterojunction effect transistor (HET) mixers has greatly reduced the required power at THz frequencies. All these advances have combined to make high performance all solid-state THz sources possible.

With the realization of these solid-state THz sources it is interesting to look at possible future directions for the field. One direction is the extension of the work to even higher frequencies. Here the competing technology is the quantum cascade laser (QCL) [22,23]. The QCL uses a cascade of heterostructure energy wells in an optical process to generate radiation. This technology was first developed for infrared wavelengths and has been extended to lower THz frequencies. It is an interesting alternative to more conventional electronic sources. Electronic sources work well at lower frequencies and are limited by capacitance at higher frequencies. QCL devices work well at optical wavelengths and are limited by the small energy level differences required for THz frequencies. Schottky devices have a very simple semiconductor structure but require a complex circuit. QCL devices can have hundreds of semiconductor layers that must be grown with precise molecular beam epitaxy (MBE) technology, but the external circuit requirements are modest. Frequency control is part of the operation of multipliers since the pump frequency is obtained from a synthesized source, but the bandwidth is limited by the requirements of several multipliers in cascade. The frequency control of the QCL is more complex, but the bandwidth can be very wide. These tradeoffs lead to clear advantages for electronic sources below 1 THz and for QCLs above 10 THz. Future research will determine the source options for the gap in the middle.

A second direction is the use of multiplier technology for a variety of ground-based sensing applications. Here the important requirement is modest cost rather than the very best performance. The sources described above were designed for a single application. The waveguide blocks require very precise and thus expensive machining. An alternative is the use of silicon fabrication technology to realize waveguide blocks. An example of WR10 waveguide is shown in Figure 14.14 [24]. The figure shows a waveguide etched with deep reactive ion etching (DRIE), two microchannels containing waveguide to finite ground coplanar transitions and alignment markers to assemble the complete structure. There are several advantages with these silicon structures. First, since they are batch fabricated, the potential cost is very low compared to more conventional machined metal blocks. Second, the structures are easily scaled to higher frequencies.

FIGURE 14.14   Silicon WR10 waveguide and probes.

Machining the very small dimensions required for THz structures is difficult. The silicon structures are defined by lithography and etching, so the structures become easier to fabricate with increasing frequency. Finally, the potential for lower cost will change the design approach. Instead of several blocks with associated waveguide feeds and flanges, a complete multiplier chain can be designed and fabricated in a single silicon structure. The lower cost approach will allow THz sources and systems to expand into a variety of new applications.

Acknowledgment

Part of the work described in this publication was supported by ARO under the MURI program on Biological and Chemical Sensing on Terahertz Frequencies (Contract DAAD-19-01-1-0622), Jack East. Part of the research described in this publication was carried out at the Jet Propulsion Laboratory, California Institute of Technology, under a contract with the National Aeronautics and Space Administration, Imran Mehdi.

References

1. P. Siegel, A. Kerr, and W. Hwang, *Topics in the Optimization of Millimeter Wave Mixers*, NASA Technical Paper 2287, March 1984.
2. J.M. Manley and H.E. Rowe, Some general properties of nonlinear elements, I: General energy relations, *Proceedings of the IRE*, Vol. 44, 904, 1956.
3. H.A. Watson, *Microwave Semiconductor Devices and Their Circuit Applications*, McGraw-Hill Book Company, New York, 1969, Chapter 8.
4. C.H. Page, Frequency conversion with positive nonlinear resistors, *Journal of Research of the National Bureau of Standards*, Vol. 56, 179–182, April, 1956.
5. R.H. Pantell, General power relationships for positive and negative nonlinear resistive elements, *Proceedings of the IRE*, Vol. 46, 1910–1913, December, 1958.
6. J.A. Morrison, Maximization of the fundamental power in nonlinear capacitance diodes, *Bell System Technical Journal*, Vol. 41, 677–721, 1962.
7. C.B. Burckardt, Analysis of varactor frequency multipliers for arbitrary capacitance variation and drive level, *Bell System Technical Journal*, Vol. 44, 675–692, 1965.
8. S.A. Maas, *Nonlinear Microwave Circuits*, Artech House, Boston, MA, 1988, Chapter 7.
9. Jack East, *Millimeter Wave Device Modeling*, Presented at The International Microwave Symposium Workshop on Millimeter Wave Devices and Circuits, June, 2005.
10. D.N. Held and A.R. Kerr, Conversion loss and noise of microwave and millimeter wave mixers: Part 1—theory; Part 2—experiment, *IEEE Transactions on Microwave Theory and Techniques*, Vol. MTT-26, 55–61, 1978.
11. S.A. Maas, *Microwave Mixers*, Artech House, 1986.
12. G.B. Tait, Efficient solution method unified nonlinear microwave circuit and numerical solid-state device simulation, *IEEE Microwave and Guided Wave Letters*, Vol. 4, 420–422, December, 1994.
13. E.L. Kollberg, T.J. Tolmunen, M.A. Frerking, and J.R. East, Current saturation in submillimeter wave varactors, *IEEE Transactions on Microwave Theory and Techniques*, Vol. MTT-40, 831–838, May, 1992.
14. J.R. East, E.L. Kollberg, and M.A. Frerking, Performance limitations of varactor multipliers, *Fourth International Conference on Space Terahertz Technology*, Ann Arbor, MI, March, 1993.
15. Imran Mehdi, Generation of THz signals: From chips to (space) ships, *High Frequency Microelectronics Symposium*, University of Michigan, Ann Arbor, MI, September 30, 2005.
16. A. Skalare, R.W. McGrath, A frequency-domain mixer model for diffusion-cooled hot-electron bolometers, *IEEE Transactions on Applied Superconductivity*, Vol. 9, 4444–4447, June, 1999.
17. K. Fiegle, D. Diehi, and K. Jacobs, Diffusion cooled superconducting hot electron bolometer heterodyne mixer between 630 and 820 GHz, *IEEE Transactions on Applied Superconductivity*, Vol. 7, 3552–3555, June, 1997.
18. A.M. Datesman, J.C. Schultz, A.W. Lichtenberger, D. Golish, C.K. Walker, and J. Kooi, Fabrication and characterization of niobium diffusion-cooled hot-electron bolometers on silicon nitride membranes, *IEEE Transactions on Applied Superconductivity*, Vol. 15, 928–931, June, 2005.

19. J.A. Stern, B. Bumble, J. Kawamura, and A. Skalare, Fabrication of terahertz frequency phonon cooled HEB mixer, *IEEE Transactions on Applied Superconductivity*, Vol. 15, 499–502, June, 2005.
20. C.W. Pobanz, M. Matloubian, M. Lui, H.-C. Sun, M. Case, C.M. Ngo, P. Janke, T. Gaierand, and L. Samoska, A high-gain monolithic D-band InP HEMT amplifier, *IEEE Journal of Solid State Circuits*, Vol. 34, 1219–1224, September, 1999.
21. W. Huei, L. Samoska, T. Gaier, A. Peralla, L. Hsin-Hsing, Y.C. Leong, S. Weinreb, Y.C. Chen, M. Nishimoto, and R. Lai, Power-amplifier modules covering 70–113 GHz using MMICs, *IEEE Transactions on Microwave Theory and Techniques*, Vol. 49, 9–16, January, 2001.
22. B.S. Williams, S. Kumar, Q. Hu, and J.L. Reno, High-power terahertz quantum-cascade lasers, *Electronics Letters*, Vol. 42, 8991, January 19, 2006.
23. H.W. Hubers, S.G. Pavlov, A.D. Semenov, R. Kohler, L. Mahler, A. Tredicucci. H. Beere, D. Ritchie, and E. Linfield, Characterization of a quantum cascade laser as local oscillator in a heterodyne receiver at 2.5 THz, *The Joint 30th International Conference on Infrared and Millimeter Waves and 13th International Conference on Terahertz Electronics*, Vol. 1, 82–83, September 2006.
24. Yongshik Lee, *Fully Micromachined Power Combining Module for Millimeter Wave Applications*, Ph.D. Thesis, The University of Michigan, Ann Arbor, MI, January, 2004.

15

Transit Time Microwave Devices

Robert J. Trew
North Carolina State University

There are several types of active two-terminal diodes that can oscillate or supply gain at microwave and millimeter-wave frequencies. These devices can be fabricated from a variety of semiconductor materials, but Si, GaAs, and InP are generally used. The most common types of active diodes are the IMPATT (an acronym for IMPact Avalanche Transit-Time) diode, and the Transferred Electron Device (generally called a Gunn diode). Tunnel diodes are also capable of producing active characteristics at microwave and millimeter-wave frequencies, but have been replaced in most applications by three-terminal transistors (such as GaAs MESFETs and AlGaAs/GaAs HEMTs), which have superior RF and noise performance, and are also much easier to use in systems applications. The IMPATT and Gunn diodes make use of a combination of internal feedback mechanisms and transit-time effects to create a phase delay between the RF current and voltage that is more than 90°, thereby generating active characteristics. These devices have high frequency capability since the saturated velocity of an electron in a semiconductor is high (generally on the order of $\sim 10^7$ cm/sec) and the transit time is short since the length of the region over which the electron transits can be made on the order of a micron (i.e., 10^{-4} cm) or less. The ability to fabricate devices with layer thicknesses on this scale permits these devices to operate at frequencies well into the millimeter-wave region. Oscillation frequency on the order of 400 GHz has been achieved with IMPATT diodes, and Gunn devices have produced oscillations up to about 150 GHz. These devices have been in practical use since the 1960s and their availability enabled a wide variety of solid-state system components to be designed and fabricated.

15.1 Semiconductor Material Properties

Active device operation is strongly dependent upon the charge transport, thermal, electronic breakdown, and mechanical characteristics of the semiconductor material from which the device is fabricated. The charge transport properties describe the ease with which free charge can flow through the material. This is described by the charge velocity-electric field characteristic, as shown in Figure 15.1 for several commonly used semiconductors. At low values of electric field, the charge transport is ohmic and the charge velocity is directly proportional to the magnitude of the electric field. The proportionality constant is called the mobility and has units of cm^2/V-sec. Above a critical value for the electric field, the charge velocity (units of cm/sec) saturates and either becomes constant (e.g., Si) or decreases with increasing

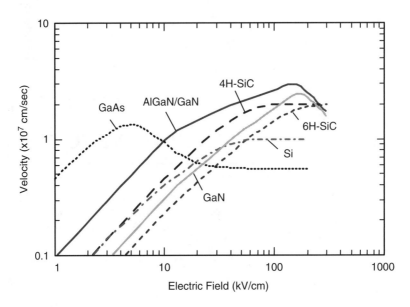

FIGURE 15.1   Electron velocity versus electric field characteristics for several semiconductors.

field (e.g., GaAs). Both of these behaviors have implications for device fabrication, especially for devices intended for high frequency operation. Generally, for transit time devices, a high velocity is desired since current is directly proportional to velocity. The greatest saturated velocity is demonstrated for electrons in the wide bandgap semiconductors, SiC and GaN. Both of these materials have saturated electron velocities on the order of $v_s \sim 2 \times 10^7$ cm/sec. This is one of the main reasons these materials are being developed for high frequency electronic devices. Also, a low value for the magnitude of the electric field at which velocity saturation occurs is desirable since this implies high charge mobility. High mobility produces low resistivity, and therefore low values for parasitic and access resistances for semiconductor devices.

The decreasing electron velocity with electric field characteristic for compound semiconductors such as GaAs and InP makes active two-terminal devices called Transferred Electron Devices (TED's) or Gunn diodes possible. The negative slope of the velocity versus electric field characteristic implies a decreasing current with increasing voltage. That is, the device has a negative resistance. When a properly sized piece of these materials is biased in the region of decreasing current with voltage, and placed in a resonant cavity, the device will be unstable up to very high frequencies. By proper selection of embedding impedances, oscillators or amplifiers can be constructed.

Other semiconductor material parameters of interest include thermal conductivity, dielectric constant, energy bandgap, electric breakdown critical field, and minority carrier lifetime. The thermal conductivity of the material is important because it describes how easily heat can be extracted from the device. The thermal conductivity has units of W/cm-°K, and in general, high thermal conductivity is desirable. Compound semiconductors, such as GaAs and InP, have relatively poor thermal conductivity compared to elemental semiconductors such as Si. Materials such as SiC have excellent thermal conductivity and are used in high power electronic devices. The dielectric constant is important since it represents capacitive loading and, therefore, affects the size of the semiconductor device. Low values of dielectric constant are desirable since this permits larger device area, which in turn results in increased RF current and increased RF power that can be developed. Electric breakdown characteristics are important since electronic breakdown limits the magnitudes of the DC and RF voltages that can be applied to the device. A low magnitude for electric field breakdown limits the DC bias that can be applied to a device, and thereby limits the RF power that can be handled or generated by the device. The electric breakdown for the material is generally described by the critical value of electric field that produces avalanche ionization.

TABLE 15.1   Material Parameters for Several Semiconductors

Material	E_g(eV)	ε_r	κ(W/K-cm) @ 300°K	E_c(V/cm)	$\tau_{minority}$ (sec)
Si	1.12	11.9	1.5	3×10^5	2.5×10^{-3}
GaAs	1.42	12.5	0.54	4×10^5	$\sim10^{-8}$
InP	1.34	12.4	0.67	4.5×10^5	$\sim10^{-8}$
6H-SiC	2.86	10.0	4.9	3.8×10^6	$\sim(10–100) \times 10^{-9}$
4H-SiC	3.2	10.0	4.9	3.5×10^6	$\sim(10–100) \times 10^{-9}$
3C-SiC	2.2	9.7	3.3	$(1–5) \times 10^6$	$\sim(10–100) \times 10^{-9}$
GaN	3.4	9.5	1.3	2×10^6	$\sim(1–100) \times 10^{-9}$

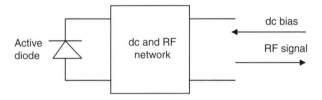

FIGURE 15.2   One-port network.

Minority carrier lifetime is important for bipolar devices, such as pn junction diodes, rectifiers, and bipolar junction transistors (BJTs). A low value for minority carrier lifetime is desirable for devices such as diode temperature sensors and switches where low reverse bias leakage current is desirable. A long minority carrier lifetime is desirable for devices such as bipolar transistors. For materials such as Si and SiC the minority carrier lifetime can be varied by controlled impurity doping. A comparison of some of the important material parameters for several common semiconductors is presented in Table 15.1. The large variation for minority lifetime shown in Table 15.1 for SiC and GaN is due to relatively immature materials growth technology for these wide bandgap semiconductors.

15.2   Two-Terminal Active Microwave Devices

The IMPATT diode, transferred electron device (often called a Gunn diode), and tunnel diode are the most commonly used two-terminal active devices. These devices can operate from the low microwave through high mm-wave frequencies, extending to several hundred GHz. They were the first semiconductor devices that could provide useful RF power levels at microwave and mm-wave frequencies and were extensively used in early systems as solid-state replacements for vacuum tubes. The three devices are similar in that they are fabricated from diode or diode-like semiconductor structures. DC bias is applied through two metal contacts that form the anode and cathode electrodes. The same electrodes are used for both the DC and RF ports and since only two electrodes are available, the devices must be operated as a one-port RF network, as shown in Figure 15.2. This causes little difficulty for oscillator circuits, but is problematic for amplifiers since a means of separating the input RF signal from the output RF signal must be devised. The use of a nonreciprocal device, such as a circulator can be used to accomplish the task. Circulators, however, are large, bulky, and their performance is sensitive to thermal variations. In general, circulators are difficult to use for integrated circuit applications. The one-port character of diodes has limited their use in modern microwave systems, particularly for amplifiers, since transistors, which have three terminals and are two-port networks, can be designed to operate with comparable RF performance, and are much easier to integrate. Diodes, however, are often used in oscillator circuits since these components are by nature one-port networks.

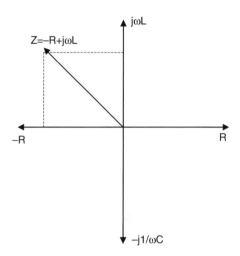

FIGURE 15.3   Complex impedance plane showing active characteristic.

IMPATT and Gunn diodes require a combination of charge injection and transit time effects to generate active characteristics and they operate as negative immittance components (the term "immittance" is a general reference that includes both "impedance" and "admittance"). When properly designed and biased, the active characteristics of the diodes can be described as either a negative resistance or a negative conductance. Which description to use is determined by the physical operating principles of the particular device, and the two descriptions are, in general, not interchangeable. Bias and RF circuits for the two active characteristics must satisfy different stability and impedance matching criteria.

Transit time effects alone cannot generate active characteristics. This is illustrated in Figure 15.3, which shows a general impedance plane. All passive circuits, no matter how complex or how many circuit elements are included, when arranged into a one-port network as shown in Figure 15.2, and viewed from an external vantage point, will have an input impedance that lies in the right-hand plane of Figure 15.3. The network resistance will be positive and real, and the reactance will be inductive or capacitive. This type of network is not capable of active performance and cannot add energy to a signal. Transit time effects can only produce terminal impedances with inductive or capacitive reactive effects, depending upon the magnitude of the delay relative to the RF period of the signal. In order to generate active characteristics it is necessary to develop an additional delay that will result in a phase delay between the terminal RF voltage and current that is greater than 90° and less than 270°. The additional delay can be generated by feedback that can be developed by physical phenomena internal to the device structure, or created by circuit design external to the device. The IMPATT and Gunn diodes make use of internal feedback resulting from electronic charge transfer within the semiconductor structure. The internal feedback generally produces a phase delay of ~90°, which when added to the transit time delay will produce a negative real component to the terminal immittance.

15.2.1.1   Tunnel Diodes

Tunnel diodes [1] generate active characteristics by an internal feedback mechanism involving the physical tunneling of electrons between energy bands in highly doped semiconductors, as illustrated in the energy band diagram shown in Figure 15.4. The illustration shows a p⁺n junction diode with heavily doped conduction and valence bands located in close proximity. When a bias is applied, charge carriers can tunnel through the electrostatic barrier separating the p-type and n-type regions, rather than be thermionically emitted over the barrier, as generally occurs in most diodes. When the diode is biased (either forward or reverse bias) current immediately flows and ohmic conduction characteristics are obtained. In the forward bias direction conduction occurs until the applied bias forces the conduction and valence

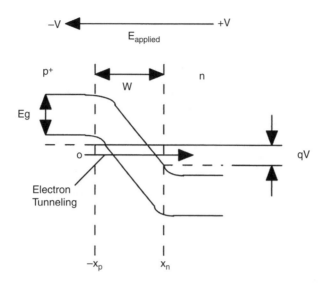

FIGURE 15.4   Energy-band diagram for a p⁺n semiconductor junction showing electron tunneling behavior.

bands to separate. The tunnel current then decreases and normal, thermionic junction conduction occurs. In the forward bias region where the tunnel current is decreasing with increasing bias voltage an N-type negative immittance characteristic is generated, as shown in Figure 15.5a. The immittance is called "N-type" because the I-V characteristic looks like the letter N. This type of active element is current driven and is short-circuit stable. It is described by a negative conductance in shunt with a capacitance, as shown in Figure 15.5b. Tunnel diodes are limited in operation frequency by the time it takes for charge carriers to tunnel through the junction. Since this time is very short (on the order of 10^{-12} s) operation frequency can be very high, approaching 1000 GHz.

Tunnel diodes have been operated at 100s of GHz, and are primarily limited in frequency response by practical packaging and parasitic impedance considerations. The RF power available from a tunnel diode is limited (~100s of mW level) since the maximum RF voltage swing that can be developed across the junction is limited by the forward turn-on characteristics of the device (typically 0.6 to 0.9 v). Increased RF power can only be obtained by increasing device area to increase RF current. However, increases in diode area will limit operation frequency due to increased diode capacitance. Tunnel diodes have moderate DC-to-RF conversion efficiency (<10%) and very low noise figures and have been used in low noise systems applications, such as microwave and mm-wave receivers used for radioastronomy.

15.2.1.2 Transferred Electron Devices

Transferred electron devices (i.e., Gunn diodes) [2] also have N-type active characteristics and can be modeled as a negative conductance in parallel with a capacitance, as shown in Figure 15.5b. Device operation, however, is based upon a fundamentally different principle. The negative conductance derives from the complex conduction band structure of certain compound semiconductors, such as GaAs and InP. In these direct bandgap materials the lower valley central (or Γ) conduction band is in close energy-momentum proximity to secondary, higher order conduction bands (i.e., the X and L) valleys (illustrated schematically as the L upper valley in Figure 15.6). The electron effective mass is determined by the shape of the conduction bands and the effective mass is "light" in the Γ valley, but "heavy" in the higher order X and L upper valleys. When the crystal is biased, current flow is initially due to electrons in the light effective mass Γ valley and conduction is ohmic. However, as the bias field is increased, an increasing proportion of the free electrons are transferred into the X and L valleys where the electrons have heavier

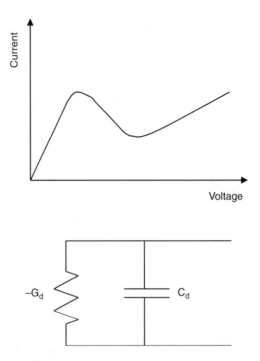

FIGURE 15.5   (a) Current-voltage characteristic for an N-type active device. (b) Small-signal RF equivalent circuit for an N-type active device.

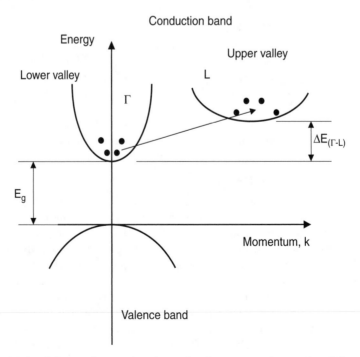

FIGURE 15.6   Energy band diagram for a semiconductor that demonstrates the transferred electron effect.

effective mass. The increased effective mass slows down the electrons, with a corresponding decrease in conduction current through the crystal. The net result is that the crystal displays a region of applied bias voltages where current decreases with increasing voltage. That is, a negative resistance is generated. A charge dipole domain is formed in the device, and this domain will travel through the device generating a transit time effect. The combination of the transferred electron effect and the transit time delay will produce a phase shift between the terminal RF current and voltage that is greater than 90°. The device is unstable and when placed in an RF circuit or resonant cavity, oscillators or amplifiers can be fabricated.

The Gunn device is not actually a diode since no pn or Schottky junction is used. The transferred electron phenomenon is a characteristic of the bulk material and the special structure of the conduction bands in certain compound semiconductors. In order to generate a transferred electron effect, a semiconductor must have Γ, X, and L valleys in the conduction bands in close proximity so that charge can be transferred from the lower Γ valley to the upper valleys at reasonable magnitude of applied electric field. It is desirable that the charge transfer occur at low values of applied bias voltage in order for the device to operate with good DC-to-RF conversion efficiency. Most semiconductors do not have the conduction band structure necessary for the transferred electron effect, and in practical use, Gunn diodes have only been fabricated from GaAs and InP. It should be noted that the name "Gunn diode" is actually a misnomer since the device is not actually a diode, but rather a piece of bulk semiconductor.

TEDs are widely used in oscillators from the microwave through high mm-wave frequency bands. They can be fabricated at low cost and provide an excellent price-to-performance ratio. They are, for example, the most common oscillator device used in police automotive radars. They have good RF output

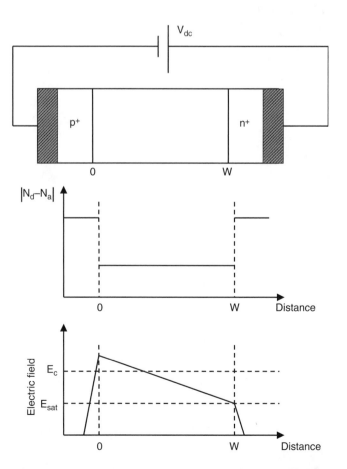

FIGURE 15.7 Diode structure, doping profile, and internal electric field for a p+nn+ IMPATT diode.

power capability (mW to W level), moderate efficiency (< 20%), and excellent noise and bandwidth capability. Octave or multi-octave band tunable oscillators are easily fabricated using devices such as YIG (yttrium iron garnet) resonators. High tuning speed can be achieved by using varactors as the tuning element. Many commercially available solid-state sources for 60 to 100 GHz operation (for example, automotive collision-avoidance radars) often use InP TEDs.

15.2.1.3 IMPATT Diodes

IMPATT (IMPact Avalanche Transit Time) diodes [3] are fabricated from pn or Schottky junctions. The doping profile in the device is generally tailored for optimum performance and a typical p^+nn^+ junction device structure as shown in Figure 15.7. The diode is designed so that when it is reverse biased, the n-region is depleted of free electrons. The electric field at the p^+n junction exceeds the critical magnitude for avalanche breakdown, and the electric field exceeds the magnitude required to maintain electron velocity saturation throughout the n-region. Saturated charge carrier velocity must be maintained throughout the RF cycle in order for the device to operate with maximum efficiency. In operation, the high electric field region at the p^+n interface will generate charge when the sum of the RF and DC bias voltage produces an electric field that exceeds the critical value. A pulse of free charge (electrons and holes) will be generated. The holes will be swept into the p^+ region and the electrons will be injected into the depleted n-region, where they will drift through the diode, inducing a current in the external circuit as shown in Figure 15.8. Due to the avalanche process, the RF current across the avalanche region lags the RF voltage by 90°. This inductive delay is not sufficient, by itself, to produce active characteristics. However, when the 90° phase shift is added to that arising from an additional inductive delay caused by the transit time of the carriers drifting through the remainder of the diode external to the avalanche

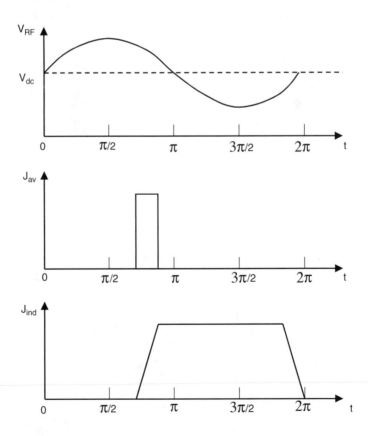

FIGURE 15.8 Terminal voltage, avalanche current, and induced current for an IMPATT oscillator.

region, a phase shift between the terminal RF voltage and current greater than 90° is obtained. A Fourier analysis of the resulting waveforms reveals a device impedance with a negative real part. That is, the device is active and can be used to generate or amplify RF signals. The device impedance has an "S-type" active i-v characteristic, as shown in Figure 15.9a, and the device equivalent circuit consists of a negative resistance in series with an inductor, as shown in Figure 15.9b. An S-Type active device is voltage driven and is open-circuit stable. For IMPATT diodes, the active characteristics only exist under RF conditions. That is, there is a lower frequency below which the diode does not generate a negative resistance. Also, the negative resistance is generally small in magnitude, and on the order of −1 Ω to −10 Ω Therefore, it is necessary to reduce all parasitic resistances in external circuits to the maximum extent possible since parasitic series resistance will degrade the device's negative resistance, thereby limiting device perfor-mance. An IMPATT diode has significant pn junction capacitance that must be considered and a complete

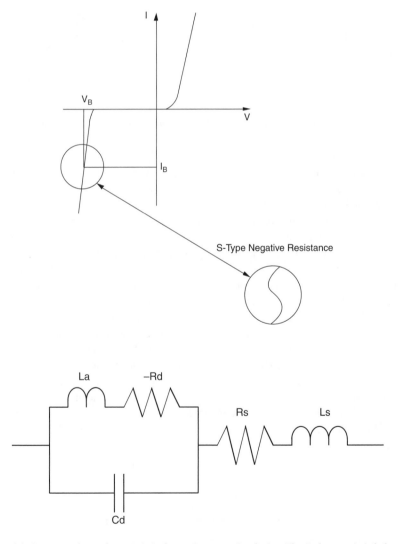

FIGURE 15.9 (a) Current-voltage characteristic for an S-type active device. The S-characteristic behavior does not exist under DC conditions for an IMPATT diode. The inset shows the dynamic i-v behavior that would exist about the DC operation point under RF conditions. (b) Small-signal RF equivalent circuit for an S-type active device.

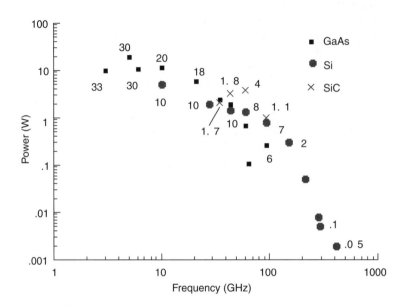

FIGURE 15.10   RF power-frequency performance for IMPATT oscillators (numbers associated with data points are DC-to-RF conversion efficiency).

equivalent circuit includes the device capacitance in parallel with the series negative resistance-inductance elements, as shown in Figure 15.9b.

For optimum performance the drift region is designed so that the electric field throughout the RF cycle is sufficiently high to produce velocity saturation for the charge carriers. In order to achieve this it is common to design complex structures consisting of alternating layers of highly doped and lightly doped semiconductor regions. These structures are called "high-low," "low-high-low," or "Read" diodes, after the man who first proposed their use [1]. They can also be fabricated in a back-to-back arrangement to form double-drift structures [4]. These devices are particularly attractive for mm-wave applications since the back-to-back arrangement permits the device to generate RF power from series-connected diodes, but each diode acts independently with regard to frequency response.

IMPATT diodes can be fabricated from most semiconductors, but are generally fabricated from Si or GaAs. The devices are capable of good RF output power (mW to tens of W) and good DC-to-RF conversion efficiency (~10 to 20%). They operate well into the mm-wave region and have been operated at frequencies in excess of 400 GHz. They have moderate bandwidth capability, but have relatively poor noise performance due to the impact ionization process associated with avalanche breakdown. The power-frequency performance of IMPATT diodes fabricated from Si, GaAs, and SiC is shown in Figure 15.10. The Si and GaAs data are experimental, and the SiC data are predicted from simulation. The numbers associated with the data points are the conversion efficiencies for each point.

Defining Terms

Immittance: A general term that refers to both impedance and admittance.

Active device: A device that can convert energy from a DC bias source to a signal at an RF frequency. Active devices are required in oscillators and amplifiers.

Two-terminal device: An electronic device, such as a diode, that has two contacts. The contacts are usually termed the cathode and anode.

One-port network: An electrical network that has only one RF port. This port must be used as both the input and output to the network. Two-terminal devices result in one-port networks.

Charge carriers: Units of electrical charge that produce current flow when moving. In a semiconductor two types of charge carriers exist: electrons and holes. Electrons carry unit negative charge and have an effective mass that is determined by the shape of the conduction band in energy-momentum space. The effective mass of an electron in a semiconductor is generally significantly less than an electron in free space. Holes have unit positive charge. Holes have an effective mass that is determined by the shape of the valence band in energy-momentum space. The effective mass of a hole is generally significantly larger than that for an electron. For this reason electrons generally move much faster than holes when an electric field is applied to the semiconductor.

References

1. Sze, S.M., 1981. *Physics of Semiconductor Devices*, 2nd Edition, Wiley-Interscience, New York.
2. Bosch, B.G., Engelmann, R.W. 1975. *Gunn-Effect Electronics*, Halsted Press, New York.
3. S.M. Sze, 1998. *Modern Semiconductor Device Physics*, Wiley-Interscience, New York.
4. P. Bhartia and I.J. Bahl, 1984. *Millimeter Wave Engineering and Applications*, Wiley-Interscience, New York.

16

Bipolar Junction Transistors (BJTs)

John C. Cowles
Analog Devices—Northwest Labs

16.1 Introduction

The topic of bipolar junction transistors (BJTs) is obviously quite broad and a full treatment would consume volumes. This work focuses on basic principles to develop an intuitive feel for the transistor behavior and its application in contemporary high-speed integrated circuits (ICs). The exponential growth in high bandwidth wired, wireless, and fiber communication systems coupled with advanced IC technologies has created an interesting convergence of two disparate worlds: the microwave and analog domains. The traditional microwave IC consists of a few transistors in discrete form or in low levels of integration surrounded by a sea of transmission lines and passive components. The modern high-speed analog IC usually involves tens to thousands of transistors with few passive components. The analog designer finds it a more cost-effective solution to use extra transistors rather than passive components to resolve performance issues. The microwave designer speaks in terms of noise figure (NF), IP3, power gain, stability factor, voltage standing wave ratio (VSWR), and s-parameters while the analog designer prefers noise voltages, harmonic distortion, voltage gain, phase margin, and impedance levels. Present BJT IC technologies fall exactly in this divide and thus force the two worlds together. Since parasitics within an IC are significantly lower than those associated with packages and external interconnects, analog techniques can be applied well into the microwave region and traditional microwave techniques such as impedance matching are only necessary when interfacing with the external world where reference impedances are the rule. This symbiosis has evolved into what is now termed radio frequency IC (RFIC) design. With this in mind, both analog and microwave aspects of bipolar transistors will be addressed. Throughout the discussion, the major differences between BJTs and field-effect transistors (FETs) will be mentioned.

16.2 A Brief History [1]

The origin of the BJT was more a fortuitous discovery rather than an invention. The team of Shockley, Brattain, and Bardeen at Bell Labs had been pursuing a field-effect device in the 1940s to replace vacuum tubes in telephony applications when they stumbled on bipolar transistor action in their experimental point-contact structure. The point-contact transistor consisted of a slab of n-type germanium sitting on a metal "base" with two metal contacts on either side. With proper applied bias, the "emitter" contact would inject current into the bulk, and then the "collector" contact would sweep it up. They announced their discovery in 1948 for which they later received the Nobel Prize in 1956. At that point, the device was described as a current-controlled voltage source. This misinterpretation of the device as a transresistor led to coining of the universal term transistor that today applies to both BJTs and their field-effect brethren.

Within a few years of the discovery of the transistor, its theory of operation was developed and refined. The decade of the 1950s was a race toward the development of a practical technique to fabricate them. Research into various aspects of the material sciences led to the development of the planar process with silicon as the cornerstone element. Substrate and epitaxial growth, dopant diffusion, ion-implantation, metalization, lithography, and oxidation had to be perfected and integrated into a manufacturable process flow. These efforts culminated with Robert Noyce's patent application for the planar silicon BJT IC concept that he invented while at Fairchild. In the 1960s, the first commercial ICs appeared on the market, launching the modern age of electronics. Since then, transistor performance and integration levels have skyrocketed driven primarily by advances in materials, metrology, and process technology as well as competition, cost, and opportunity.

16.3 Basic Operation [2,3]

The basic structure of the bipolar transistor given in Figure 16.1a consists of two back-to-back intimately coupled p–n junctions. While the npn transistor will be chosen as the example, note that the pnp structure operates identically albeit in complementary fashion. Bipolar transistor action is based on the injection of minority carrier electrons from the emitter into the base as dictated by the base–emitter voltage. These carriers diffuse across the thin base region and are swept by the collector, which is normally reverse biased.

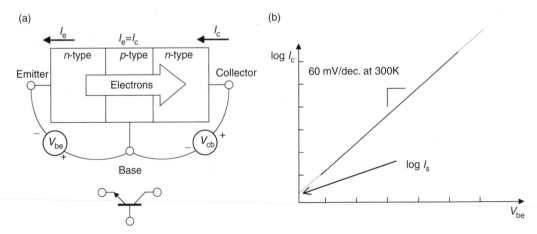

FIGURE 16.1 (a) BJT action involves carrier injection across the base from emitter to collector; (b) the ideal BJT exhibits a perfectly exponential input–output transfer function.

In high-speed ICs, BJTs are nearly always biased in this way that is known as the forward-active mode. The net effect is that an input-signal voltage presented across the base–emitter terminals causes a current to flow into the collector terminal. The relationship between bias current, I_c, and base–emitter voltage, V_{be}, originates from fundamental thermodynamic arguments via the statistical Boltzmann distribution of electrons and holes in solids. It is given by,

$$I_c = I_s \exp(V_{be}/V_t), \tag{16.1}$$

where I_s is called the saturation current and V_t is Boltzmann's thermal voltage kT/q, which is 26 mV at room temperature (RT). This expression plotted in Figure 16.1b is valid over many decades of current and its implications are far reaching in BJT IC design. Consider the small-signal transfer characteristics relating the output current to input voltage at a given I_c, that is the transconductance, g_m. It is given by the remarkably simple expression,

$$\delta I_c/\delta V_{be}|I_c = g_m = I_c/V_t. \tag{16.2}$$

The notion that transconductance is precisely and linearly proportional to I_c and inversely proportional to temperature through V_t is known as translinearity [4]. Notice that I_s, which has a strong dependence on temperature, base width, and other physical device parameters, has dropped out of the picture. This view of bipolar transistors as a voltage-controlled current source suggests that it should have been named the transductor rather than the transistor. Although bipolar transistors are usually considered hopelessly nonlinear because of the exponential law in Equation 16.1, extremely precise and robust linear and nonlinear functions can be realized at very high frequencies by proper circuit techniques that exploit the translinear property embodied in Equation 16.2.

The intrinsic bandwidth of high-frequency BJTs is ultimately limited by minority carrier storage primarily in the base region. Current is conveyed from the emitter to the collector by the relatively slow process of diffusion resulting from the gradient in the distribution of stored minority carriers in the base. In contrast, current in FETs is transported by the faster process of majority carrier drift in response to an electric field. By invoking the quasistatic approximation that assumes that signals of interest change on a much longer time scale than the device time constants, carrier storage can be modeled as a lumped diffusion capacitance, C_d given by

$$C_d = \delta Q_d/\delta V_{be} = \delta(I_c\tau_f)/\delta V_{be} = g_m\tau_f, \tag{16.3}$$

where Q_d is the stored charge and τ_f is the forward transit time. The parameter τ_f can be viewed as the average time an electron spends diffusing across the base and can be expressed as a function of basic material and device parameters,

$$\tau_f = W_B^2/2\eta D_n, \tag{16.4}$$

where W_B is the physical base width, D_n is the electron diffusion coefficient, and η is a dimensionless factor that accounts for any aiding fields in the base. Despite the limitations associated with carrier storage and diffusion, BJTs have demonstrated excellent high-frequency performance. This is partly due to their vertical nature where dimensions such as W_B can be significantly less than 100 nm. In contrast, FETs are laterally arranged devices where the critical dimension through which carriers drift is determined by lithography and is typically larger than 100 nm even in the most advanced RF technologies. Furthermore, the fundamental limit contained in τ_f is never achieved in practice since other factors, both device- and circuit-related conspire to reduce the actual bandwidth.

A first-order model of BJTs is illustrated in Figure 16.2 consisting of a diode at the base–emitter junction whose current are conveyed entirely to the collector according to Equation 16.1 and a diffusion capacitance

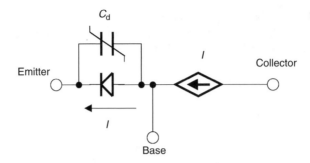

FIGURE 16.2   First order BJT model includes the nonlinear carrier injection and storage.

that models the carrier storage. From this simplistic model, the transistor f_t, defined as the frequency at which the common-emitter (CE) short-circuit current gain, h_{21}, reaches unity, can be calculated from

$$h_{21} = I_c/I_b = f_t/f. \tag{16.5a}$$

$$f_t = g_m/2\pi C_d = 1/2\pi \tau_f. \tag{16.5b}$$

Note that f_t is independent of current and is solely a function of material parameters and device design details. Pushing this figure of merit to higher frequencies has dominated industrial and academic research for years. Polysilicon emitter BJTs, Si/SiGe heterojunction BJTs (HBTs) and III-V-based HBTs are all attempts to craft the active layers in order to maximize f_t without significantly compromising other device aspects.

16.4 Parasitics and Refinements

The first-order model captures the essential operation of the BJT. However, actual devices deviate from that ideal in many aspects. As electrons diffuse across the base, some are lost to the process of recombination with holes as illustrated in Figure 16.3a. Excess holes are also parasitically injected into emitter. This overall loss of holes must originate at the base terminal and represents a finite base current, I_b, which is related to I_c via the parameter, β, known as the common-emitter current gain

$$\beta = I_c/I_b. \tag{16.6}$$

The effect of finite β is to modify the emitter–base diode current by a correction factor, α, known as the common-base (CB) current gain, which accounts for the extra base current. Since $I_e = I_b + I_c$, it is easy to show that

$$\alpha = \beta/\beta + 1 < 1 \tag{16.7}$$

so that the emitter–base diode current now becomes I_s/α. Many factors figure into determining β, including W_B and the highly variable recombination lifetime τ_r. Thus, β is a poorly controlled parameter with potential variations of $\pm 50\%$, making it inadvisable to design circuits that depend on its exact value. However, it is usually possible to design ICs that are insensitive to β as long as β is large, say >50. Figure 16.3b illustrates the well-known BJT Gummel plot, which shows the translinearity of I_c as well as the finite I_b in response to V_{be}. The separation between the two curves represents β. As shown in Figure 16.4, β represents a low-frequency asymptote of the generalized short-circuit ac current gain, h_{21}, which exhibits a single pole role-off towards f_t at a frequency, f_β given by f_t/β. Since RFICs most often operate well above

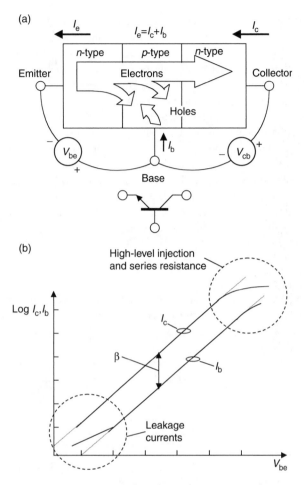

FIGURE 16.3 (a) Recombination of electrons and holes leads to finite current gain, β; (b) the Gummel plot illustrates the effects of finite β and other nonidealities.

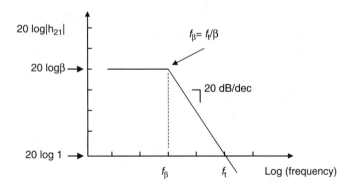

FIGURE 16.4 The frequency dependence of h_{21} captures finite β and f_t.

f_β, the useful measure of current gain is actually f_t/f rather than β, although high dc β is still important for low noise and ease of biasing.

Traditionally, BJTs have been characterized as current-controlled devices where a forced I_b drives an I_c into a load impedance, consistent with Shockley's transresistor description. Now if I_b is considered a parasitic nuisance rather than a fundamental aspect, it becomes even more appropriate to view the BJT as a voltage-controlled device that behaves as a transconductor albeit with exponential characteristics. In fact, contemporary analog IC design avoids operating the BJT as a current-controlled device due to the unpredictability of β. Instead, the designer takes advantage of device matching in an IC environment and translinearity to provide the appropriate voltage drive. This approach can be shown to be robust against device, process, temperature, and supply voltage variations.

Superimposed on the basic model are parasitic-ohmic resistances in series with each active terminal (R_b, R_e, R_c) and parasitic capacitances associated with all p–n junction depletion regions (C_{jc}, C_{je}, C_{js}), including the collector-substrate junction present to some extent in all technologies. Since parasitics degrade the idealized dc and RF characteristics captured in Equations 16.1 and 16.5, a major effort is focused on minimizing them through aggressive scaling and process engineering. Their values and bias dependencies can be estimated from the physical device structure and layout.

In particular, the base and emitter resistance, R_b and R_e, soften the elegant exponential characteristics in Equation 16.1 by essentially debiasing the junction since

$$I_c = I_s \exp([V_{be} - I_b R_b - I_e R_e]/V_t). \tag{16.8}$$

This effect is also illustrated in Figure 16.3b at high values of V_{be} where both curves appear to saturate. This departure from ideal translinearity can introduce unwelcome distortion into many otherwise linear ICs. Furthermore, these resistances add unwelcome noise and degeneration (voltage drops) as will be discussed later.

The idealized formulation for f_t given in Equation 16.5 also needs to be modified to account for parasitics. A more comprehensive expression for f_t based on a more complex equivalent circuit results in

$$f_t = (2\pi \times [\tau_f + (C_{je} + C_{jc})/g_m + C_{jc}(R_c + R_e)])^{-1}, \tag{16.9}$$

where now f_t is dependent on current through g_m charging of C_{je} and C_{jc}. To achieve peak f_t, high currents are required to overcome the capacitances. As the intrinsic device τ_f has been reduced, the parasitics have become a dominant part of the BJTs' high-frequency performance requiring even higher currents to reach the lofty peak values of f_t. It should also be kept in mind that f_t only captures a snapshot of the device's high-frequency performance. In circuits, transistors are rarely current driven and short circuited at the output. The base resistance and substrate capacitance that do not appear in Equation 16.9 can have significant impact on high frequency IC performance. While various other figures of merit such as f_{max} have been proposed, none can capture the complex effects of device interactions with source and load impedances in a compact form. The moral of the story is that ideal BJTs should have low inertia all around, that is, not only high peak f_t but also low-parasitic capacitances and resistances so that time constants in general can be minimized at the lowest possible currents.

At the high currents required for high f_t, second-order phenomena known in general as high-level injection begin to corrupt the dc and RF characteristics. Essentially, the electron concentration responsible for carrying the current becomes comparable to the background doping levels in the device causing deviations from the basic theory that assumes low-level injection. The dominant phenomenon known as the Kirk effect or base pushout manifests itself as a sudden widening of the base width at the expense of the collector. This translates into a departure from translinearity with dramatic drop in both β and f_t. A number of other effects have been identified over the years such as Webster effect, base conductivity modulation and current crowding. High-level injection sets a practical maximum current at which peak f_t can be realized. Figure 16.5 illustrates the typical behavior of f_t with I_c. To counteract high-level injection, doping levels throughout the device have increased at a cost of higher depletion capacitances and lower breakdown voltages. Since modeling of these effects is very complex and not necessarily included in many models, it is dangerous to design in this regime.

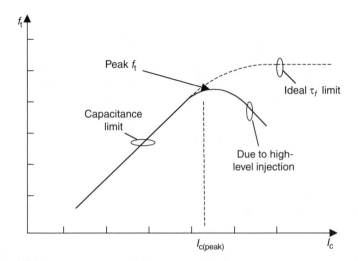

FIGURE 16.5 The f_t peaks at a particular current when high-level injection occurs.

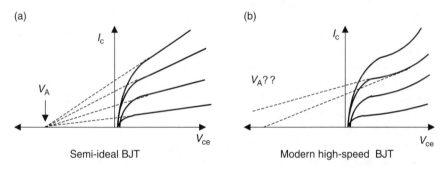

FIGURE 16.6 (a) The traditional interpretation of V_A is not unique in modern BJTs as shown in (b).

So far, the transistor output has been considered a perfect current source with infinite output resistance. In reality, as the output voltage swings, the base width is modulated causing I_s and thus I_c to vary for a fixed V_{be}. This is exactly the effect of an output resistance and is modeled by a parameter V_a, the early voltage,

$$r_o = \delta V_{ce}/\delta I_c = V_a/I_c. \tag{16.10}$$

As illustrated in the output CE characteristics, I_c versus V_{ce}, shown in Figure 16.6a, V_a represents the common extrapolation point where all tangents intersect the V_{ce}-axis. The effect of r_o is to set the maximum small-signal unloaded voltage gain since

$$A_v = \delta V_o/\delta V_i = (\delta I_c/\delta V_i)(\delta V_o/\delta I_c) = -g_m r_o = -V_a/V_t. \tag{16.11}$$

For typical values of $V_a = 50$ V, $A_v = 2000$ (66 dB) at room temperature. Note that this gain is independent of I_c and is quite high for a single device, representing one of the main advantages of BJTs over FETs where the maximum gain is usually limited to <40 dB for reasonable gate lengths. In modern devices, however, V_a is not constant due to complex interactions between flowing electrons and internal electric fields. The net effect illustrated in Figure 16.6b indicates a varying V_a depending on I_c and V_{ce}. This is often termed soft breakdown or weak avalanche in contrast to actual breakdown, which will be discussed shortly. The effect of a varying V_a is to introduce another form of distortion since the gain will

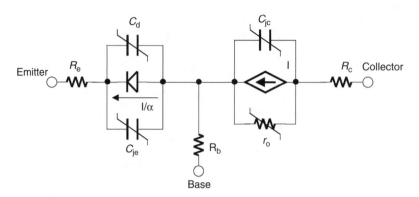

FIGURE 16.7   Augmented BJT model with parasitic resistances and finite base current.

vary according to bias point. Figure 16.7 shows a more complete model that includes the fundamental parameters as well as the parasitics and refinements considered so far. It resembles a simplified version of the popular Gummel–Poon model found in many simulators.

Another form of pseudobreakdown occurs in ultranarrow base BJTs operating at high voltages. If the early-effect or base-width modulation is taken to an extreme, the base eventually becomes completely depleted. After this point, a further change in collector voltage directly modulates the emitter–base junction leading to an exponential increase in current flow. This phenomenon known as "punchthrough" fortunately has been mitigated by the fact that as base widths have narrowed, the base doping has been forced to increase so as to maintain a reasonable base resistance. Furthermore, since higher collector doping levels have been necessary to fight high-level injection, true breakdown has become the voltage-limiting mechanism rather than punchthrough.

Just as high-level injection limits the maximum operating current, junction breakdown restricts the maximum operating voltage. When the collector voltage is raised, the collector base junction is reverse biased. The resulting electric field reaches a critical point where valence electrons are literally ripped out of their energy band and promoted to the conduction band while leaving holes in the valence band. The observed effect known as "avalanche breakdown" is a dramatic increase in current. The breakdown of the collector–base junction in isolation, that is, with the emitter open circuited, is termed the BV_{cbo} where the "o" refers to the emitter open. Another limiting case of interest is when the base is open while a collector-emitter voltage is applied. In this case, an initial avalanche current acts as base current that induces more current flow, which further drives the avalanche process. The resulting positive feedback process causes the BV_{ceo} to be significantly lower than BV_{cbo}. The relationship clearly depends on β and is empirically modeled as

$$BV_{ceo} = BV_{cbo}/\beta^{1/n}, \tag{16.12}$$

where n is a fit parameter. A third limit occurs when the base is ac shorted to ground via a low impedance. In this scenario, avalanche currents are shunted to ground before becoming base current and BV_{ces} ("s" means shorted) would be expected to be $BV_{cbo} + V_{be}$. Therefore, BV_{ces} represents an absolute maximum value for V_{ce}, while BV_{ceo} represents a pessimistic limit since the base is rarely open. Operating in the intermediate region requires care in setting the base impedance to ground and knowing its effect on breakdown. Figure 16.8 illustrates the transition from BV_{ceo} to BV_{ces}. The base–emitter junction is also sensitive to reverse bias. In this case, the breakdown is usually related to Zener tunneling and is represented by BV_{ebo}. Excessive excursions towards BV_{ebo} can cause injection of energetic carriers into surrounding dielectrics, leading to leakage currents, degraded reliability, and possibly device failure. Since these breakdown mechanisms in general are poorly modeled, it is recommended to operate well below BV_{ceo} and BV_{ebo}.

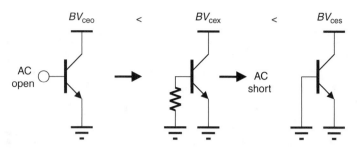

FIGURE 16.8 The AC base impedance helps determine the effective breakdown voltage.

Even when these guidelines are followed, there are still situations where peak voltages might exceed breakdown limits. For instance, modern ICs in portable products often require a disable mode in which the part consumes virtually no current so as to conserve battery life. When disabled, it is common for transistors that are normally in a safe operating condition to find themselves with reverse-biased junctions in excess of BV_{ceo} and BV_{ebo}. The resulting leakage currents can lead to expectedly high disable currents, accidental activation of certain circuit blocks and again to potential reliability failures. A second situation is when output stages drive variable reactive loads. It is possible (and likely) that certain conditions will cause 2–3 times the supply voltage to appear directly across a transistor leading to certain catastrophic breakdown.

As devices have been scaled down in size, isolated in dielectrics and driven with higher current densities, self-heating has become increasingly problematic. The thermal resistance, R_{th}, and capacitance, C_{th}, model the change in junction temperature with power and the time constant of its response. The dc and ac variations in the junction temperature induce global variations in nearly all model parameters causing bias shifts, distortion, and even potentially catastrophic thermal runaway. Furthermore, local heating in one part of a circuit, say an output power stage, can start affecting neighboring areas such as bias circuits creating havoc. It is essential to selectively model thermal hotspots and ensure that they are within maximum temperature limits and that their effect on other circuit blocks is minimized by proper layout.

16.5 Models

In modern high-speed silicon-based bipolars, the complex carrier distributions and storage in the device lead to further nonideal characteristics such as weak avalanche, quasi-saturation, current crowding, excess phase delay, and self-heating to name a few. The Gummel–Poon model developed earlier was based on the fundamental Ebers–Moll dc model of the bipolar as a pair of coupled junction augmented with a charge control approach to address dynamic performance [5]. Over the years, refinements to the models have accompanied the tremendous leaps in technology resulting in enhanced Gummel–Poon [6], VBIC [7], and Mextram [8] models, all of which are charge-control models that reduce back to the Gummel–Poon/Ebers–Moll models in their limits. Figures 16.9a through 16.9c illustrate the modeled Gummel Plot, CE, and h_{21} characteristics for a modern SiGe npn transistor, capturing high-level injection, excess base current, weak avalanche, and numerous other nonidealities. While the advanced models yield more accurate simulation results, the computation time can be greatly increased and convergence can become a nuisance, particularly in larger circuits.

16.6 Dynamic Range

The limits on signal integrity are bounded on the low end by noise and on the high end by distortion. It is not appropriate to claim a certain dynamic range (DR) for BJTs since the circuit topology has a large impact on it. However, the fundamental noise sources and nonlinearities in BJTs can be quantified

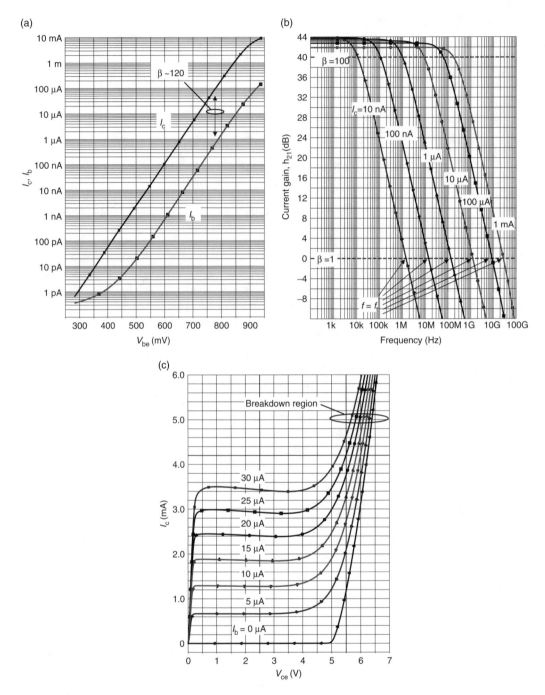

FIGURE 16.9 (a) The Gummel plot for SiGe BJT shows excellent junction characteristics. (b) The h_{21} curves as a function of frequency and current follow the theory and (c) The CE characteristics show the complex behavior in weak/strong avalanche.

and then manipulated by circuit design. It is customary to refer the effects of noise and nonlinearity to either the input or the output of the device or circuit to allow fair comparisons and to facilitate cascaded system-level analyses. Here, all effects will be referred to the input, noting that the output quantity is simply scaled up by the appropriate gain. Note that transistors fundamentally respond to and generate

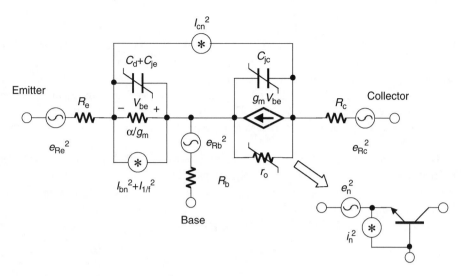

FIGURE 16.10 The fundamental noise sources can be transformed to equivalent input-referred sources.

voltages and currents; therefore, noise and distortion will be first presented in those terms rather than in terms more familiar to microwave engineers such as NF and 1 dB-compression power. Later, the more familiar microwave engineering expressions will be derived with the reference impedance level providing the link. The ability to comfortably migrate across both domains is key to RFIC design.

There are three fundamental noise mechanisms that are manifested in BJTs [9]. Figure 16.10 presents a linearized small-signal model that includes the noise sources. The object will be to refer all noise sources to equivalent voltage and current noise sources, e_n and i_n, respectively, at the input. Shot noise is associated with fluctuations in current caused by the discrete nature of electronic charge overcoming the junction potential. The presence of the forward biased base–emitter junction results in collector and base shot noise, i_{cn}, and i_{bn}, which are given by

$$i_{cn}^2 = 2qI_c; \quad i_{bn}^2 = 2qI_b, \tag{16.13}$$

where q is the electronic charge. This is a major disadvantage of BJTs with respect to FETs, which are free of shot noise due to the absence of potential barriers that determine current flow. Each physical resistance carries an associated Johnson noise represented as

$$e_{Rx}^2 = 4kTR_x, \tag{16.14}$$

where R_x is the appropriate resistance and k is Boltzmann's constant. Finally, a $1/f$ noise source, $i_{1/f}^2$ associated primarily with recombination at semiconductor surfaces appears at the input. The $1/f$ noise component decreases with increasing frequency, as expected. A corner frequency is often quoted as the point where the $1/f$ noise and the white thermal and shot noise crossover. It is an empirical value that varies from device to device and depends on bias conditions and source impedance. BJTs typically offer lower $1/f$ corner frequency than FETs since surfaces play a lesser role in BJTs. Except for broadband dc-coupled circuits and strongly nonlinear circuits such as oscillators and mixers that perform frequency conversion, $1/f$ noise is insignificant at high frequencies and will be ignored for simplicity. Note that treating noise in nonlinear systems is quite complex and remains an active research topic.

Once the noise sources are input referred, e_n and i_n are approximately given by

$$e_n^2 = 4kT(R_b + R_e + 1/2g_m) \tag{16.15a}$$

and

$$i_n^2 = 2q(I_b + I_c/|h_{21}(f)|^2).$$ (16.15b)

An arbitrarily higher I_c reduces e_n through the larger g_m but has mixed effects on i_n depending on how $h_{21}(f)$ varies with current. An arbitrarily larger device reduces e_n by lowering R_b and R_e but the commensurate increase in capacitance might degrade $h_{21}(f)$ and thus i_n. The moral here is that minimization of noise requires a trade-off in I_c and device size given an f_t equation and parasitic scaling rules with device geometry.

The typical measure of noise performance for communications is the NF defined loosely as the degradation in signal-to-noise ratio (SNR) through a system. Implicit in the definition of NF is an external input noise source associated with the source resistance, R_s (i.e., $n_{R_s} = (4\,KTR_s)^{1/2}$). If connected directly to a bipolar transistor or, for that matter, to any other circuit block with associated input-referred noise voltages and currents, e_n and i_n, the NF in dB is simply given by

$$NF(dB) = 10\log_{10}(1 + [e_n + i_n Rs]^2/n_{Rs}^2).$$ (16.16)

Note that for bipolar transistors, the expression within brackets cannot be separated into two terms as e_n^2 and $i_n^2 R_s^2$ because e_n and i_n are partly correlated via the collector current shot noise $2qI_c$. Equation 6.16 clearly illustrates that NF simply captures the amount of noise added by the target circuit relative to the noise available from R_s. Achieving minimum NF requires a very specific value of $R_{s,opt}$ that can be derived by minimizing the argument in brackets in Equation 6.16. Since e_n and i_n are in general complex, the optimal source impedance is complex as well. Converting the real, external source impedance to R_{opt} is an exercise known as impedance matching for noise. Because of the complicated equations that result, particularly when a more complete high-frequency model is used, microwave engineers rely on noise circles that illustrate on a Smith Chart the contours of constant NF as a function of complex source impedance [10]. Since NF represents a relative amount of noise, it is simple to invert in Equation 6.16 in order to express e_n and i_n in terms of NF and R_s.

RF communications system engineers make use of NF to determine noise power density levels in a signal chain. The noise power density available from any real source is KT or -174 dBm/Hz at 300 K, reflecting thermal equilibrium condition between the source impedance and the environment. Interestingly, this quantity is independent of the actual value of R_s. The total noise-power density referred to the input and the output, respectively, including that of the source impedance is given by

Total input noise density (dBm/Hz) $= -174$ dBm/Hz $+$ NF(dB) (16.17a)

Total output noise density (dBm/Hz) $= -174$dBm/Hz $+$ NF(dB) $+$ Gain(dB). (16.17b)

Note that for a noiseless device, the NF $= 0$ dB and the input noise is simply -174 dBm/Hz or the noise available from the source impedance. Note that these power densities are normalized to a $1 -$ Hz bandwidth, hence the peculiar dBc/Hz unit. In order to determine the total noise power, integration over a bandwidth is needed. Under conditions of a "white" noise spectrum (i.e., flat with frequency), a factor of $10\log_{10}[\text{bandwidth(Hz)}]$ with units dB Hz is added to Equation 16.17.

Distortion originates in nonlinearities within the device. For simplicity, distortion will be described in terms of a power series to derive expressions for harmonic distortion, intermodulation distortion and compression [11]. It is assumed that linearities only up to third order are considered, limiting the accuracy of the results as compression is approached. While this approach is valid at low frequencies, capacitive effects at higher frequencies greatly complicate nonlinear analysis and more generalized Volterra series must be invoked. However, the general idea is still valid. Note that not all systems can be mathematically modeled by power or Volterra series; for example, a hard-limiting amplifier that shows sudden changes in characteristics.

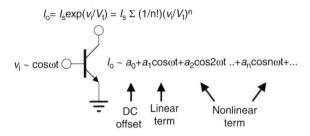

$$I_o = I_s \exp(v_i/V_t) = I_s \Sigma \, (1/n!)(v_i/V_t)^n$$

$v_i \sim \cos\omega t$ $I_o \sim a_0 + a_1\cos\omega t + a_2\cos2\omega t \,..+ a_n\cos n\omega t + ...$

DC offset   Linear term   Nonlinear term

FIGURE 16.11 A power series expansion of the exponential junction nonlinearity can be used to quantify distortion.

In BJTs, the fundamental transfer function given in Equation 16.1 represents one of strongest nonlinearities. As shown in Figure 16.11, by expanding Equation 16.1 in its power series up to third order, the following measures of input-referred figures of merit are derived in dB V_{pk} (mV),

$$P_{1db} = 0.933V_t = 24.3 \text{ mV} = -32.3\text{dB}V_{pk}, \tag{16.18a}$$

$$\text{IIP3} = 2.83V_t = 73.5 \text{ mV} = -22.7\text{dB}V_{pk}, \tag{16.18b}$$

$$\text{P3OI} = 4.89V_t = 127 \text{ mV} = -17.9\text{dB}V_{pk}, \tag{16.18c}$$

where P_{1db} is the $1 - \text{dB}$ gain compression point, IIP3 are the third-order intercept points and P3OI is the third-harmonic intercept point. Note that the conversion from dB V_{pk} to dBm referenced to 50 Ω is $+10$ dB. From the analysis, it can be inferred that P_{1db} occurs for peak input swings of about V_t while a third-order intermodulation level of 40 dBc is achieved for an input swing of only 0.0283 V_t or 0.73 mV at room temperature. These values are quite low, indicating that circuit linearization techniques must be invoked to exceed these limits. FETs on the other hand possess a more benign input nonlinearity that ranges from square law in long channel devices to nearly linear in short channels. While this is not the sole source of nonlinear behavior, FETs are expected to have superior linearity than BJTs from this perspective. It is interesting to observe that the commonly used rule-of-thumb stating that IIP3 is approximately 10 dB higher than P_{1db} is true in this case. As mentioned earlier, higher order terms, parasitics and reactive effects modify the results in Equation 16.18 but they do provide a useful order of magnitude.

A signal-processing chain is usually benchmarked by the range of signal levels that can be detected with fidelity. In digital systems "fidelity" is inevitably associated with a bit-error-rate (BER) which can be translated into aSNR specification more familiar to analog communication systems. There are several different measures of DR depending on the environment in which the system operates. In a situation where a communication system only has to deal with its own signal, the DR might be loosely specified as the ratio of P_{1dB} to input noise floor, representing the upper and lower limits, respectively:

$$\text{DR}_0(\text{dBHz}) = P_{1dB}(\text{dBm}) - (-174\text{dBm/Hz} + \text{NF(dB)}). \tag{16.19}$$

A slightly more sophisticated metric might relate the maximum level represented by P_{1dB} to a minimum detectable signal (MDS) defined according to an SNR and an assumed noise bandwidth. In this case the maximum signal might be an unwanted interferer that "blocks" or degrades the signal quality rather than a wanted signal. This is called the blocking dynamic range (BDR).

$$\text{BDR(dB)} = P_{1dB}(\text{dBm}) - [-174\text{dBm/Hz} + \text{NF(dB)} + \text{SNR(dB)} + \text{BW(dBHz)}], \tag{16.20}$$

where the quantity in brackets represents the MDS which sits above the integrated noise floor by the SNR. A third expression of DR is known as the spurious-free DR (SFDR). It defines the maximum signal level according to a two-tone scenario in which the third-order intermodulation products (or spurs) are just

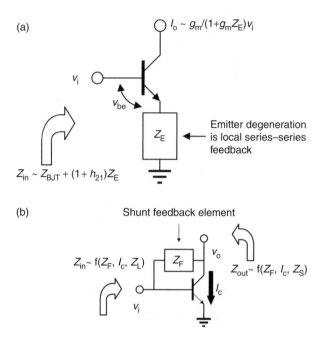

FIGURE 16.12 (a) Emitter degeneration can improve dynamic range and change impedance levels. (b) The shunt–shunt feedback amplifier can increase dynamic range and provide input/output impedance match.

equal to the noise floor. Noting that third-order products change by 3 dB for every 1 dB of signal level change, the SFDR can be shown to be

$$\text{SFDR(dB)} = 2/3(\text{IIP3(dBm)} - [-174\text{dBm/Hz} + \text{NF(dB)} + \text{BW(dBHz)}]) - \text{SNR}. \qquad (16.21)$$

A common technique for extending overall DR is emitter degeneration, which as illustrated in Figure 16.12a, essentially amounts to series–series feedback. The impedance in series with the emitter reduces the signal that appears directly across the nonlinear emitter–base leading to lower input-referred distortion. Note that the gain is also reduced by degeneration; however, the overall output-referred distortion is still improved. The potentially superior distortion performance of FETs is analogous to a strongly degenerated BJT with the accompanying lower gain. Since this degenerating impedance is directly in the signal path, it also affects noise. If the impedance is a resistor, it will contribute noise as if it were an extra R_e in Equation 16.15a. If R_e is not a dominant noise contributor, this approach can extend DR significantly. Another option is to use a degenerating inductor which at high frequencies provides adequate degeneration while only contributing noise associated with its finite quality factor, Q. Other feedback techniques using both active and passive elements such as the shunt–shunt feedback amplifier in Figure 16.12b are also possible to improve DR but care must be taken to ensure stability particularly around longer feedback loops. Additional benefits of using local feedback are desensitization of input impedance and gain to device parameters.

16.7 Complementary pnp

The availability of a pnp device to complement the npn offers numerous advantages, particularly at lower supply voltages. For example, pnp active loads can provide more voltage gain at a given supply voltage; signals can be "folded" up and down to avoid hitting the rails, and balanced complementary push–pull topologies can operate near rail-to-rail. In general, the *p*-type device is slower than the *n*-type device

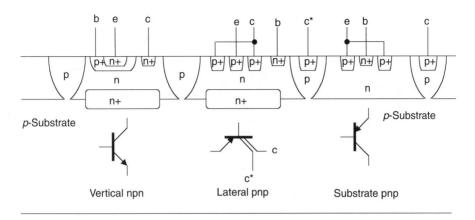

FIGURE 16.13   Both lateral and substrate pnp BJTs can be generated in a typical npn process.

across most technologies due to the fundamentally lower hole mobility with respect to electron mobility. Furthermore, in practice, the pnp transistor is often implemented as a parasitic device, that is, the fabrication process is optimized for the npn and the pnp shares the available layers. Figure 16.13 illustrates how two different pnps can be constructed in an npn junction-isolated process (*p*-type substrate). The so-called substrate pnp has the substrate double as its collector and thus only appears in a grounded collector arrangement. The lateral pnp frees the collector terminal but carries along a parasitic substrate pnp of its own, which injects large currents into the substrate and degrades f_t and β. In some cases, true complementary technologies are available in which both the npns and pnps are synthesized and optimized separately.

When only lateral and substrate pnps are available, they are usually used outside of the high-speed signal path due to their compromised RF performance. They appear prominently in bias generation and distribution as well as in low- and intermediate-frequency blocks. When using either pnp, care must be taken to account for current injection into the substrate that acts locally as a collector. This current can induce noise and signal coupling as well as unintentional voltage drops along the substrate. It is customary to add substrate contacts around the pnps to bring this current to the surface before it disperses throughout the IC. If true high-quality vertical pnps are available that are balanced in performance with respect to the npns, they can be used for high-frequency signal processing, enabling a number of circuit concepts commonly used at lower frequencies to be applied at RF.

16.8   Topologies

Several basic single-transistor and composite transistor configurations are commonly encountered in RFICs. Figure 16.14 shows the three basic single-transistor connections known as CE, CB and common-collector (CC). The name refers to the fact that the common terminal is ac grounded. Their properties are well covered in numerous texts and only a brief discussion on their general properties and applications will be discussed here.

The CE stage is the most commonly considered configuration since it provides a fairly high-input impedance, high-output impedance and both current gain, $h_{21}(f)$, and voltage gain, $g_m R_L$. The noise and distortion properties are the same as considered earlier for the BJT. It is commonly used in single-ended low-noise amplifiers (LNA) and power amplifiers (PA). A well-known drawback of the CE stage at high frequencies is the feedback C_{jc}, which degrades gain via Miller multiplication and couples input and output networks leading to detuning of matching networks and possible instability.

As noted earlier emitter degeneration can be used not only to improve DR but also to help set impedance levels. It is sometimes necessary for the IC to present a standard impedance level at its input and output

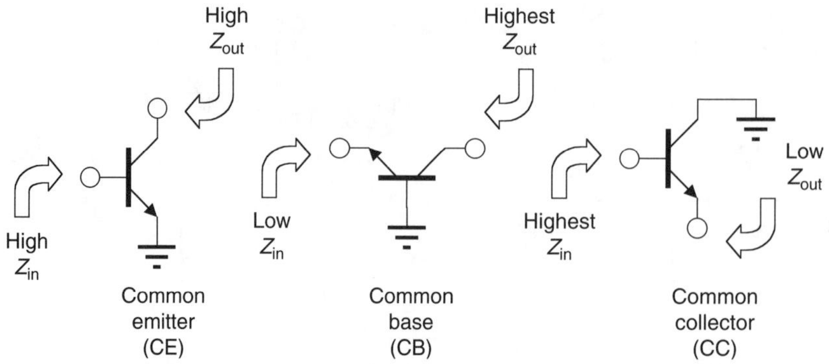

FIGURE 16.14   The three basic single BJT stages offer different features.

terminals to provide proper terminations. The use of emitter inductive degeneration has the fortuitous property of generating a synthetic resistive input impedance at high frequencies without contributing noise. As illustrated in Figure 16.12a earlier, the transformed value is given by

$$R_{\text{eff}} = h_{21}(f)Z_{\text{E}} = (f_{\text{t}}/f)(2\pi fL_{\text{E}}) = 2\pi f_{\text{t}}L_{\text{E}}, \tag{16.22}$$

which now appears in series with the input terminal.

The CB stage provides a well-predicted low-input impedance given by $1/g_m$, high-output impedance and voltage gain, $g_m R_{\text{L}}$, but has near unity current gain, α. It acts as an excellent current buffer and is often used to isolate circuit nodes. The CB appears most often in tight synergy with other transistors arranged in composite configurations. Although it lacks the capacitive feedback present in the CE that degrades gain and destabilizes operation, any series feedback at the grounded base node can lead to instability. The feedback element might be due to metalization, package parasitics, or even the actual device base resistance. Extreme care must be exercised when grounding the CB stage. Another difference is that with a grounded base, the larger BV_{cbo} sets the voltage limit. The noise and distortion properties are identical to the CE.

The CC stage is often known as an emitter follower. It is predominantly used as a level shifter and as a voltage buffer to isolate stages and provide extra drive. It offers high-input impedance, low-output imped-ance and near unity voltage gain. The impedance buffering factor is roughly $h_{21}(f)$ and at frequencies approaching f_{t}, its effectiveness is diminished as $h_{21}(f)$ nears unity. In this case it is common to see several stages of cascaded followers to provide adequate drive and buffering. The CC stage is a very wide-band stage since direct capacitive feedthrough via C_d and C_{je} cancels the dominant pole to first order. However, this same capacitive coupling from input to output can cause destabilization, particularly with capacitive loads. In fact this is the basis of operation of the Colpitts oscillator. From a noise point of view, the CC simply transfers noise at its output directly back to its input and then adds on its own noise; therefore it is not used where low noise is essential. From a distortion perspective, the CC can be fairly linear as long as the impedance that it is driving is significantly higher than its own output impedance $1/g_m$. For minimal added shot noise and low distortion, CC stages must run at high-bias currents.

The strength of IC technology is the ability to use transistors at will at practically no extra cost. Figure 16.15 illustrates five very common composite transistor configurations. Note that npn–pnp com-posite structures are also possible creating many permutations that might offer certain advantages. These are illustrated in Figure 16.16 in the form of the double up-down emitter follower, the folded cascade, the composite pnp and complementary mirror. The first two examples reduce the supply voltage requirements over their npn-only embodiment, the third transforms a poor quality pnp into a nearly ideal device and the fourth provides a voltage-controlled mirror ratio.

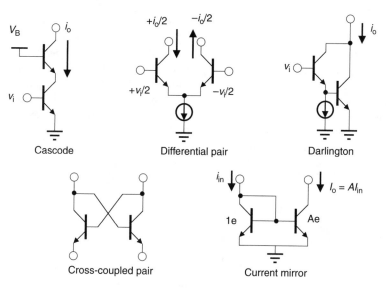

FIGURE 16.15 These common two-transistor configurations are used extensively in BJT ICs.

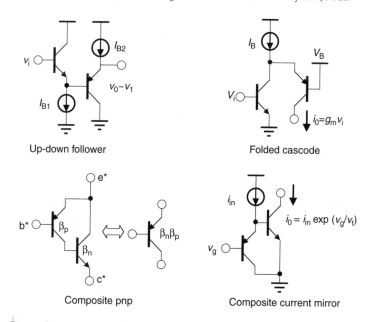

FIGURE 16.16 Composite npn–pnp combinations add functionality at lower supply voltages.

The cascade topology is a CE–CB connection. The CB provides a low impedance to the output of the CE stage, eliminating Miller multiplication of its C_{jc} and increasing the overall bandwidth. It also raises the output impedance of the single transistor by a factor of approximately β, which is useful in current sources. More importantly at RF, it improves input–output isolation, minimizing interactions that reduce the gain. The cascade configuration is common in LNAs since higher tuned power gain is achievable with minimal degradation in noise. Note that the CB transistor can be significantly smaller than the CE transistor, minimizing the parasitic capacitance at its output. The cascade requires a higher supply voltage than a stand-alone CE stage by at least a $V_{ce(sat)}$ to keep the lower BJT in its active region.

The differential pair (also known as a long-tailed pair) can be thought of as a CC–CB connection when driven in a single-ended fashion or as parallel CE stages when driven differentially. A tail current source or a current setting resistance is required to establish the operating point. This ubiquitous, canonical form has a long, distinguished history in operational amplifiers, mixers, IF/RF amplifiers, variable-gain amplifiers, digital ECL/CML gates, and even oscillators and latches when cross-coupled. The basic operation relies on the controlled steering of current from one branch to the other. Notice that the DR is modified by the fact that noise sources from two transistors contribute to the total noise while the signal is divided across two junctions. The noise associated with the tail current source appears as common mode noise and does not affect the input noise if differential signaling is used. The structure can be enhanced with the addition of emitter degeneration, cascading CB stages and buffering CC stages.

A variation on the differential pair that provides an extension of the DR is called the "doublet" in Figure 16.17 [12]. First consider a differential pair with tail current I_T and load R_L. The large signal gain can be expressed as $I_T R_L \text{sech}(v_{in}/V_t)$, which is plotted in Figure 16.18a. Within 20 mV, the gain has

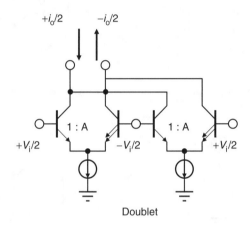

FIGURE 16.17   The doublet extends the input range with a modest increase in noise.

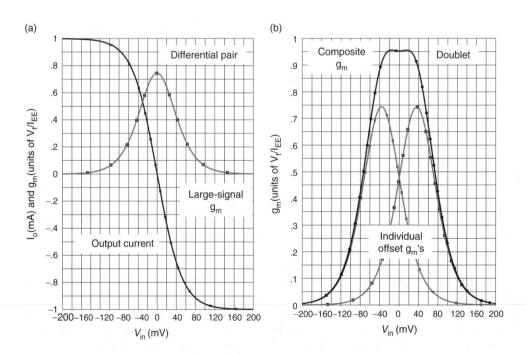

FIGURE 16.18   The superposition of two offset differential pairs extends the linear range.

dropped by 1 dB. Now consider a differential pair in which one of the transistors is scaled by a factor A larger than the other. The large signal gain is now shifted along the x-axis by V_t log A. By connecting two differential pairs with such area offsets in parallel, their superposition creates a flatter gain profile as shown in Figure 16.18b. Overall, the maximum signal capacity is increased to 45 mV for a 1 dB gain drop with a modest increase in noise.

The Darlington connection is a CE stage buffered at the input by CC stage. This configuration behaves essentially like a supertransistor with greater impedance buffering capability and higher $h_{21}(f)$. It typically appears at input and output interfaces where drive and buffering requirements are most severe. The Darlington- and the differential-pair are sometimes called f_t multipliers since the input signal appears across two transistor junctions in series while the output signal is a parallel combination of currents. For a fixed input bias current per device, the effective input capacitance is $C_d/2$ while the effective transconductance is still g_m giving from Equation 16.5 an $f_t^{\text{eff}} = 2f_{t\text{BJT}}$. This of course ignores additional higher order poles and requires twice the current of a single BJT. In analogy to the differential pair, noise and distortion occur across two junctions in series.

The cross-coupled connection is derived from the differential pair. Following the signals around the loops leads to a positive feedback condition that can be used to induce regeneration for Schmidt triggers, multivibrators, and latches or negative resistance for tuned VCOs and gain peaking. It has also been used to synthesize voltage-to-current converters, bias cells, multipliers, and other functions. Since positive feedback is inevitably invoked in cross-coupled devices, care must be taken to avoid instability when it is undesired.

16.9 Translinear Circuits

The concept of translinearity was observed in deriving the basic properties of BJTs and its usefulness was alluded to in general terms. In particular, if the signals of interest are conveyed as currents rather than voltages, a large class of linear and nonlinear functions can be synthesized. When current-mode signals are used, extremely large DRs are possible since the junction essentially compands the current signal logarithmically to a voltage; that is, a 60 dB (1000 times) signal current range corresponds to 180 mV of voltage swing at room temperature. The relatively small voltage swings represent a low impedance through which capacitances charge and discharge and thus the bandwidth of these circuits achieve broadband operation out to near the device limits represented by f_t. Furthermore, reduced internal voltage swings are consistent with lower supply voltages.

The simplest example of a translinear circuit is the well-known current mirror shown as the last example in Figure 16.15. The input current, I_{in}, is mirrored to the output, I_{out} according to a scaling factor associated with the ratio of device sizes. Inherent in this process is a nonlinear conversion from I_{in} to the common V_{be} and then a second related nonlinear conversion from V_{be} to I_{out}. The impedance at the common base node is nominally a parallel combination of $1/g_m$ and $(1 + A)C_d$, which results in a low time constant on the order of τ_f.

A large family of translinear circuits has been synthesized by applying the translinear principle stated as

> In a closed loop containing only junctions, the product of current densities flowing in the clockwise direction is equal to the product of current densities flowing in the counter-clockwise direction.

The principle has a corollary when voltage sources are inserted into the loop, namely that the products are equal to within a factor $\exp(V_a/V_t)$, where V_a is the applied voltage. It is merely a restatement of logarithmic multiplication and division, that is, addition and subtraction of V_{be}s is tantamount to products and quotients of currents. In this sense, the BJT can be thought of as the mathematical transistor. Note that since the signal is in current form, the distortion is not caused by the exponential characteristics

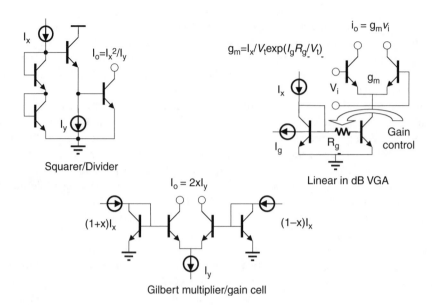

FIGURE 16.19 Numerous functions can be implemented using translinear principles.

but is actually due to *departure* from it (i.e., series resistances), high-level injection and Early voltage effects conspire to distort the signals.

Figure 16.19 illustrates three examples of translinear IC design. The first example is an analog squarer/divider with input signals I_x, I_y, I_z and output, I_o. By following the translinear principle, it can be shown that the mathematical operation

$$I_o = I_x I_y / I_z, \tag{16.23a}$$

is performed. The second example is the well-known Gilbert-cell multiplier with predistortion. Again, by following the rules, the differential output currents have the same form as Equation 16.23a where now the divisor is simply a dimensional scaling factor. The third example is a linear in dB variable gain amplifier. In this case a voltage V_g is inserted into the loop and Equation 16.23a is modified so that

$$I_o = I_x(I_y/I_z) \exp(V_g/V_t) = I_x G_o 10^{V_g/V_t}. \tag{16.23b}$$

Note that if V_g is engineered to be proportional to temperature, then the gain control becomes stable with temperature. In all cases, the operating frequency of these circuits approaches the technology f_t.

16.10   Biasing Techniques

Microwave IC designers have historically focused on the signal path with biasing treated as an after-thought. Analog IC designers, on the other hand, have placed a strong emphasis on biasing since it usually determines the sensitivity and robustness of a circuit. In particular, as was already seen in the variable gain amplifier, it is often necessary to generate a current that is proportional to absolute temperature (PTAT). Other times it is necessary to generate a complementary to absolute temperature (CTAT) current. Finally, sometimes a zero temperature coefficient current (ZTAT) is desired. The choice on temperature shaping depends on what is needed. In differential amplifiers where the gain is given by $g_m R_L = I_o R_L / 2V_t$, it is appropriate to choose a PTAT bias current for a stable gain; however, if the amplifier is actually a limiter,

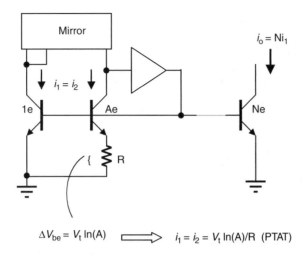

$$\Delta V_{be} = V_t \ln(A) \implies i_1 = i_2 = V_t \ln(A)/R \ \ (\text{PTAT})$$

FIGURE 16.20   The ΔV_{be} cell that generates a PTAT current is ubiquitous in BJT biasing.

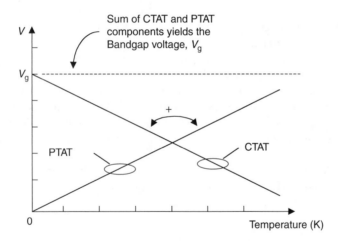

FIGURE 16.21 V_g is generated from properly proportioned PTAT and CTAT voltages.

then a stable limited voltage swing requires a ZTAT shaping. By far the most common and fundamental bias current shape for BJTs turns out to be PTAT. From the PTAT form other shapes can be derived.

The fundamental way to synthesize a bias sub-circuit is to start with a ΔV_{be} cell that generates a PTAT current. This cell illustrated in Figure 16.20, is a modified translinear loop. Two area-ratioed BJTs, each forced to carry the same current, generate a difference in V_{be} given by

$$\Delta V_{be} = V_t \ln(A) \tag{16.24}$$

where A is the area ratio. This voltage develops a current in R that is PTAT and can now be mirrored throughout the IC by adding a driver that buffers the ΔV_{be} cell.

In some cases, a stable voltage reference is desired for scaling, and it is beneficial to derive the various currents from this reference. To achieve a stable reference, the bandgap principle is invoked. As illustrated in Figure 16.21, the idea is to add a CTAT voltage to a PTAT voltage so that the sum is a constant with temperature. It can be shown that the desired reference voltage V_g nearly corresponds to the bandgap energy, E_g of silicon, a very fundamental property. Intuitively, this is expected since at the extreme

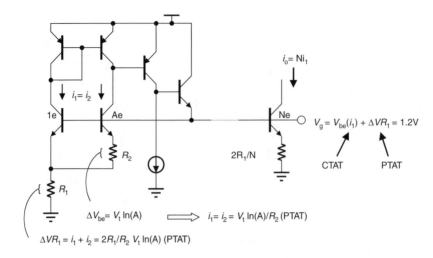

FIGURE 16.22   The Brokaw cell provides PTAT currents and a stable voltage reference.

temperature of 0 K, it takes an energy equivalent to E_g to promote an electron into conduction. It so happens that a transistor V_{be} at a fixed I_c can be shown to be CTAT while a PTAT voltage can be generated from the ΔV_{be} cell. With this in mind, the PTAT component should be scaled such that it adds to the CTAT component to synthesize V_g. A physical realization of this principle known as a bandgap reference is illustrated in its simplest form in Figure 16.22 [13]. It consists of a ΔV_{be} cell to generate a PTAT current and a resistor that converts it to the required PTAT voltage. Transistor Q1 plays a double role as part of the ΔV_{be} cell and as the V_{be} responsible for the CTAT voltage. The current mirror above ensures that both currents are the same, and the buffer provides drive to the reference output. This output can be used to generate PTAT currents elsewhere in the circuit as shown in Figure 16.22. The target transistor and resistor scaling are necessary to preserve the PTAT shape. This topology can be made insensitive to supply voltage, temperature, and process variations by the use of more sophisticated cells and by the prudent choice of component sizes and layout.

16.11   Fabrication Technology

The technology for fabricating BJTs has been performance and cost driven toward smaller transistor geometries, lower device and interconnect parasitics, higher yield/integration, and greater functionality. Silicon-based BJT technology has benefited greatly from the synergy with CMOS processes targeting VLSI/ULSI digital applications. Several variants of the simple npn process have evolved over the years that feature better npns and/or addition of other transistor types. Examples include BiCMOS that integrates BJTs with CMOS, fully complementary bipolar with true vertical npns and pnps, SiGe/Si HBT which offers higher f_t and lower R_b than traditional BJTs and silicon-on-insulator (SOI) processes that rely on an insulating substrate to isolate devices and reduce substrate parasitics. III–V based HBTs have developed in a different direction since raw device speed has been more than adequate but integration levels and process complexity have limited their availability. Process features can be grouped into two general categories: the active device that determines the transistor performance and the back-end process that defines the interconnect metalization, dielectric isolation, passive components, and through-vias. The back-end process is particularly critical in RFICs since they set limits to IC parasitics and are responsible for the quality of resistors, capacitors, and inductors.

The silicon BJT process has evolved from the junction-isolated buried–collector process shown in Figure 16.14 to the oxide trench isolated double-poly (emitter and base) process diagrammed in Figure 16.23. The double-poly structure with self-aligned spacers allows ultra small emitter–base structures

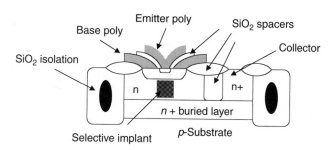

FIGURE 16.23 Modern processes feature double-poly, SiO_2 trenches and sub-μm lithography.

to be defined while simultaneously providing low-extrinsic base resistance and low-emitter charge storage. A selective collector implant under the active emitter helps delay the Kirk effect without significantly increasing the depletion capacitance. In general, the devices are defined by diffusions into the substrate with interconnects and passives constructed over dielectrics above the substrate. Standard aluminum metals have been silicided for improved reliability. Overall, this has led to significantly reduced device footprints and feature sizes with emitter geometries down to $<0.35\,\mu$m and f_t ranging from 30 GHz for standard BJTs to >100 GHz for SiGe HBTs. Refinements such as silicided base handles and optimized epitaxially deposited bases promise even further improvements in device performance. On the back-end side, thin-film resistors and metal–metal capacitors are replacing their polysilicon and diffused versions bringing lower temperature coefficients, reduced parasitics and coupling to the substrate, trimming capability and bias insensitivity. Furthermore, advanced multiple level interconnect modules and planarization techniques have been adopted from CMOS processes in the form of copper metalization, chemical–mechanical polishing, tungsten plugs, and low-dielectric insulators. For RFICs, advanced interconnects are desirable not for high-packing density but for defining high-quality passives away from the lossy substrate, particularly inductors that are essential in certain applications.

For the past few years, there has been growing interest in software-defined radios in which a single hardware platform can be reconfigured electronically to adapt in real time to changing signaling standards and channel conditions. While bipolar transistor technology offers excellent DR in terms of noise and distortion at a given power dissipation, low-power digital logic and high-quality switches are cumbersome to implement. In such cases, BiCMOS technologies that incorporates, for example, a SiGe bipolar technology with a 0.18–0.35 μm gate length CMOS, becomes an attractive option. While the SiGe bipolars are ideal for the signal path in high-performance systems, CMOS technology offers compact, low-power digital blocks and low-parasitic analog switches. It becomes possible to integrate programmable filters, data converters, and full synthesizers with fractional dividers and stepped-gain-control much more efficiently into the radio.

The fabrication technology for the III–V HBTs is significantly different from that of silicon BJTs. The main difference stems from the fact that the starting material must be grown by advanced epitaxial techniques such as molecular beam epitaxy (MBE) or metallo-organic chemical vapor deposition (MOCVD). Each layer in the stack is individually optimized as needed. As a result of the predefined layers as well as thermal limits that disallow high-temperature diffusion, the device is literally carved in to the substrate leading to the wedding-cake triple mesa structure depicted in Figure 16.24. The nature of the process limits minimum emitter geometries to about 1 μm although self-aligned structures are used to minimize extrinsic areas. Values of f_t range from 30 GHz for power devices to 200 GHz for the highest speed structures. Interestingly, the improvements being made to silicon BJTs such as epitaxial bases, polysilicon emitters and silicided bases, make use of deposited active layers which is reminiscent of III–V technologies based on epitaxy. The back-end process inevitably consists of thin film resistors and metal–metal capacitors. Since the substrate is semi-insulating by definition, high-Q monolithic inductors are already available. Some processes offer backside vias that literally provide a direct connection from the front-side to the back-side of the wafer enabling easy and effective grounding at any point on the circuit.

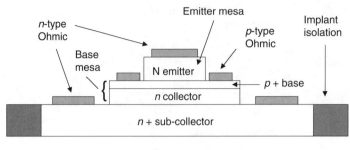

FIGURE 16.24   III–V HBTs are fabricated with mesa structures and relaxed lithography.

References

1. R.M. Warner and R. L. Grung, *Transistors: Fundamentals for the Integrated-Circuit Engineer*, John Wiley & Sons, 1983, Chapter 1.
2. S.M. Sze, *Physics of Semiconductor Devices*, John Wiley & Sons, 1981, Chapters 2–3.
3. R. Muller and T. Kamins, *Device Electronics for Integrated Circuits*, John Wiley & Sons, 2003, Chapters 4–7.
4. C. Toumazou et al, editor, *Analogue I.C. Design: The Current Mode Approach*, IEE, 1993, Chapter 1.
5. I. Getreu, *Modeling the Bipolar Transistor, Tektronics Inc.*, 1976
6. G. M. Kull, et. al., "A Unified Circuit Model for Bipolar Transistors Including Quasi-Saturation Effects," *IEEE Trans. Electron Devices*, Vol. ED-32, No. 6, June 1985, pp. 1103–1113.
7. C. McAndrew et. al., "VBIC95: An Improved Vertical, IC Bipolar Transistor Model," *Proceedings IEEE Bipolar Circuits and Technology Meeting*, 1995, pp. 170–177.
8. H. C. de Graaff, et. al., "Experience with the New Compact MEXTRAM Model for Bipolar Transistors," *IEEE 1989 Bipolar Circuits & Technology Meeting*, Paper 9.3.
9. P. Gray , R. Meyer et al., *Analysis and Design of Analog Integrated Circuits*, John Wiley & Sons, 2001, Chapters 11.
10. G. Gonzalez., *Microwave Transistor Amplifier: Analysis and Design*, Prentice-Hall, 1996, Chapter 4.
11. B. Razavi., *RF Microelectronics*, Prentice-Hall, 1997, Chapter 2.
12. B. Gilbert., "The Multi-tanh Principle: A Tutorial Overview," *IEEE JSSC*, Vol. 33, No. 1, Jan. 1998, pp. 2–17.
13. P. Brokaw., "A Simple Three-terminal IC Bandgap Reference," *IEEE JSSC*, Vol. 18, No. 6, Dec. 1974, pp. 388–393.

17

Heterostructure Bipolar Transistors (HBTs)

William Liu
Maxim Integrated Products

17.1 Basic Device Principle

Heterojunction bipolar transistors (HBTs) differ from conventional bipolar junction transistors (BJTs) in their use of hetero-structures, particularly in the base-emitter junction. In a uniform material, an electric field exerts the same amount of force on an electron and a hole, producing movements in opposite directions as shown in Figure 17.1a. With appropriate modifications in the semiconductor energy gap, the forces on an electron and a hole may differ, and at an extreme, drive the carriers along the same direction as shown in Figure 17.1b.[1] The ability to modify the material composition to independently control the movement of carriers is the key advantage of adopting hetero-structures for transistor design.

Figure 17.2 illustrates the band diagram of a *npn* BJT under normal operation, wherein the base-emitter bias (V_{BE}) is positive and the base-collector bias (V_{BC}) is negative. The bipolar transistor was

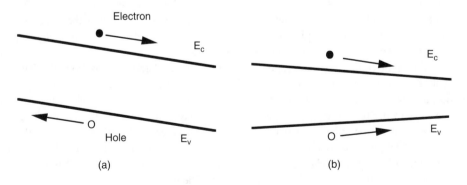

FIGURE 17.1 (a) An electric field exerts the same amount of force (but in opposite directions) on an electron and a hole. (b) Electron and hole can move in the same direction in a hetero-structure. (After Kromer, H, *Proc. IEEE*, 70, 13, 1982, with permission.)

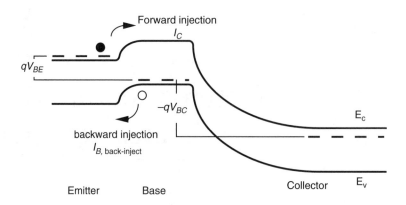

FIGURE 17.2   Band diagram of a *npn* bipolar junction transistor.

detailed in the previous section. Here we use Figure 17.2 to emphasize that the energy gap is the same throughout the entire transistor structure. The Fermi levels and the depletion regions of the band diagram reflect the BJT design constraint that the doping level is highest in the emitter, and lowest in the collector. The Fermi level in the emitter is above the conduction band, signifying that the emitter is degenerately doped.

A BJT has two key current components. The first is the forward-injection current conducted by the electrons. These electrons are emitted at the emitter, crossing the base-emitter potential barrier with the help of the V_{BE} bias, diffusing through the thin base layer as minority carriers, and then are swept by the large electric field in the reverse-biased base-collector junction. The carriers finally leave the collector, forming the collector current (I_C). In their journey, only a few electrons are lost through recombinations with the majority holes in the base. We therefore denote the electron current as I_C in Figure 17.2, even though the current magnitude in the emitter is slightly larger than that which finally exits the collector terminal. Of the three phases of the journey, the diffusion through the base layer is the rate-limiting step. Hence, a quantitative expression of I_C to be developed shortly, relates intimately to the base layer parameters.

By varying V_{BE}, we control the amount of the electron injection from the emitter to the base, and subsequently, the amount of electrons collected at the collector. This current flow, however, is independent of V_{BC}, as long as the base-collector junction is reverse biased. The property that the desired signal (I_C) is modified by the input bias (V_{BE}) but unaffected by the output bias (V_{BC}) fulfills the requirement of a sound three-terminal device. As the input and output ports are decoupled from one another, complicated circuits based on the transistors can be easily designed.

The second current component is composed of holes that are back-injected from the base to the emitter. This current does not get collected at the collector terminal and does not contribute to the desired output signal. However, the moment a V_{BE} is applied so that the desired electrons are emitted from the emitter to the base, these holes are back-injected from the base to the emitter. The bipolar transistor is so named to emphasize that both electron and hole play significant roles in the device operation. A good design of a bipolar transistor maximizes the electron current transfer while minimizing the hole current. As indicated on Figure 17.2, we refer the hole current as $I_{B,back-inject}$. It is a base current because the holes come from the base layer where they are the majority carriers.

We mentioned that I_C is limited by the diffusion process in the base, wherein the electrons are minority carriers. Likewise, $I_{B,back-inject}$ is limited by the diffusion in the emitter, wherein the holes are minority carriers. Fisk's law states that a diffusion current density across a layer is equal to the diffusion coefficient times the carrier concentration gradient.[2] (From the Einstein relationship, the diffusion coefficient can be taken to be kT/q times the carrier mobility.) The carrier concentrations at one end of the base layer (for the calculation of I_C) and the emitter layer (for the calculation of $I_{B,\text{back-inject}}$) are both proportional to $\exp(qV_{BE}/kT)$, and are 0 at the other end. An application of Fisk's Law leads to:[3,4]

$$I_C = \frac{qA_{\text{emit}}D_{n,\text{base}}}{X_{\text{base}}} \frac{n_{i,\text{base}}^2}{N_{\text{base}}} \exp\left(\frac{qV_{BE}}{kT}\right) \tag{17.1}$$

$$I_{B,\text{back-inject}} = \frac{qA_{\text{emit}}D_{p,\text{emit}}}{X_{\text{emit}}} \frac{n_{i,\text{emit}}^2}{N_{\text{emit}}} \exp\left(\frac{qV_{BE}}{kT}\right) \tag{17.2}$$

where, for example, A_{emit} is the emitter area; $D_{n,\text{base}}$ is the electron diffusion coefficient in the base layer; N_{base} is the base doping; X_{base} is the base thickness; and $n_{i,\text{base}}$ is the intrinsic carrier concentration in the base layer. I_C is proportional to the emitter area rather than the collector area because I_C is composed of the electrons injected from the emitter. For homojunction BJTs, $n_{i,\text{base}}$ is identical to $n_{i,\text{emit}}$. The ratio of the desired I_C to the undesired $I_{B,\text{back-inject}}$ is,

$$\frac{I_C}{I_{B,\text{back-inject}}} = \frac{X_{\text{emit}}}{X_{\text{base}}} \frac{D_{n,\text{base}}}{D_{p,\text{emit}}} \times \frac{N_{\text{emit}}}{N_{\text{base}}} \quad \left(\text{homojunction BJT}\right) \tag{17.3}$$

Because the diffusion coefficients and layer thicknesses are roughly equal on the first order, Equation 17.3 demonstrates that the emitter doping of a BJT must exceed the base doping in order for I_C to exceed $I_{B,\text{back-inject}}$. For most applications, we would actually prefer the base doping be high and the emitter doping, low. A high base doping results in a small base resistance and a low emitter doping reduces the base-emitter junction capacitance, both factors leading to improved high frequency performance. Unfortunately, these advantages must be compromised in the interest of minimizing $I_{B,\text{back-inject}}$ in homojunction transistors.

A heterojunction bipolar transistor (HBT), formed by a replacement of the homojunction emitter by a larger energy gap material, enables the design freedom of independently optimizing the $I_C/I_{B,\text{back-inject}}$ ratio and the doping levels. Figure 17.3 illustrates the band diagrams of *Npn* HBTs. The capital *N* in "*Npn*" rather than a small letter *n* emphasizes that the emitter is made of a larger energy gap material than the rest. It is implicit that the base-collector junction of an HBT is a homojunction. An HBT whose base-collector junction is also a heterojunction is called a double heterojunction bipolar transistor (DHBT), typically found in the InP/InGaAs material system.

The energy gap difference in the base-emitter heterojunction of an HBT burdens the holes from the base to experience a much larger energy barrier than the electrons from the emitter. With the same

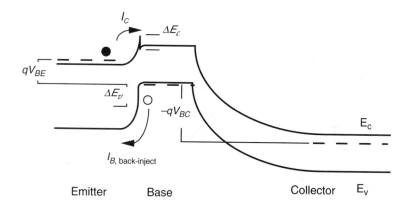

FIGURE 17.3 Band diagram of a *Npn* abrupt heterojunction bipolar transistor.

application of V_{BE}, the forces acting on the electrons and holes differ, favoring the electron injection from the emitter into the base to the hole back-injection from the base into the emitter. Figure 17.3 shows that the difference in the electron and hole barriers is ΔE_v, the valence band discontinuity. The $I_C/I_{B,\text{back-inject}}$ ratio of Equation 17.3 can be extended to the HBT as:[1]

$$\frac{I_C}{I_{B,\text{back-inject}}} = \frac{X_{\text{emit}}}{X_{\text{base}}} \frac{D_{n,\text{base}}}{D_{p,\text{emit}}} \times \frac{N_{\text{emit}}}{N_{\text{base}}} \exp\left(\frac{\Delta E_v}{kT}\right) \quad \left(\text{abrupt HBT}\right) \tag{17.4}$$

The exponential factor is the key to the fact that base doping can be made larger than the emitter doping without adversely affecting the HBT performance.

We qualify the expression in Equation 17.4: that is for an *abrupt* HBT. This means that an abrupt change of material composition exists between the emitter and the base. An example is an InP/In$_{0.53}$Ga$_{0.47}$As HBT. The indium and gallium compositions form a fixed ratio so that the resulting In$_{0.53}$Ga$_{0.47}$As layer has the same lattice constant as InP, the starting substrate material. Otherwise, the dislocations due to lattice mismatch prevent the device from being functional. In fact, although the idea of HBT is as old as the homojunction transistor itself,[5] HBTs have emerged as practical transistors only after molecular beam epitaxy (MBE) and metal-organic chemical vapor deposition (MOCVD) were developed to grow high-quality epitaxial layers in the 1980s.[6] Another example of an abrupt HBT is the Al$_{0.3}$Ga$_{0.7}$As/GaAs HBT, in which the entire emitter layer consists of Al$_{0.3}$Ga$_{0.7}$As, while the base and collector are GaAs. When the AlGaAs/GaAs material system is used, we can take advantage of the material property that AlAs and GaAs have nearly the same lattice constants. It is therefore possible to grow an Al$_x$Ga$_{1-x}$As layer with any composition x and still be lattice-matched to GaAs. When the aluminum composition of an intermediate AlGaAs layer is graded from 0 at the GaAs base to 30% at the Al$_{0.3}$Ga$_{0.7}$As emitter, then the resulting structure is called a *graded* HBT.

The band diagram of a graded HBT is shown in Figure 17.4. A graded HBT has the advantage that the hole barrier is larger than the electron barrier by ΔE_g, the energy-gap discontinuity between the emitter and the base materials. Previously, in an abrupt HBT, the difference in the hole and electron barriers was only ΔE_v. The additional barrier brings forth an even larger $I_C/I_{B,\text{back-inject}}$ ratio for the graded HBT:

$$\frac{I_C}{I_{B,\text{back-inject}}} = \frac{X_{\text{emit}}}{X_{\text{base}}} \frac{D_{n,\text{base}}}{D_{p,\text{emit}}} \times \frac{N_{\text{emit}}}{N_{\text{base}}} \exp\left(\frac{\Delta E_g}{kT}\right) \quad \left(\text{Graded HBT}\right) \tag{17.5}$$

Equation 17.5 is obtained from the inspection of the band diagram. It is also derivable from Equations 17.1 and 17.2 by noting that $n_{i,\text{base}}^2 = n_{i,\text{emit}}^2 \times \exp(\Delta E_g/kT)$.

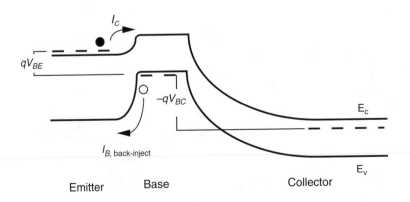

FIGURE 17.4   Band diagram of a *Npn* graded heterojunction bipolar transistor.

TABLE 17.1 Heterojunction Parameters at Room Temperature

	Al$_x$Ga$_{1-x}$As/GaAs ($x \leqslant 0.45$)	Ga$_{0.51}$In$_{0.49}$P/GaAs	InP/In$_{0.53}$Ga$_{0.47}$As	In$_{0.52}$Al$_{0.48}$As/In$_{0.53}$Ga$_{0.47}$As	Si/Si$_{0.8}$Ge$_{0.2}$
ΔE_g (eV)	$1.247 \cdot x$	Ordered: 0.43	0.60	0.71	0.165
		Disordered: 0.46			
ΔE_c (eV)	$0.697 \cdot x$	Ordered: 0.03	0.23	0.50	
		Disordered: 0.22			
ΔE_v (eV)	$0.55 \cdot x$	Ordered: 0.40	0.37	0.21	
		Disordered: 0.24			
E_g (eV)	GaAs: 1.424	Ordered: 1.424	InP: 1.35	InGaAs: 0.75	Si: 1.12

The amount of improvement from a homojunction to a heterojunction depends on the band alignment properties of the two hetero-materials. Table 17.1 lists the conduction band discontinuity (ΔE_c), valence band discontinuity (ΔE_v), and their sum which is the energy-gap discontinuity (ΔE_g), for several material systems used for HBTs. The energy gap of a semiconductor in the system is also given for convenience. The values shown in Table 17.1 are room temperature values.

The most popular III-V material system is Al$_x$Ga$_{1-x}$As/GaAs. The table lists its band alignment parameters for $x \leq 0.45$. At these aluminum mole fractions, the AlGaAs layer is a direct energy-gap material, in which the electron wave function's wave vectors **k** at both the conduction band minimum and the valence band maximum are along the same crystal direction.[2] Since nearly all semiconductor valence band maximums take place when **k** is [000], the conduction band minimum in a direct energy-gap material is also at **k** = [000]. The conduction band structure surrounding this **k** direction is called the Γ valley. There are two other **k** directions of interest because the energy band structures in such directions can have either the minimum or a local minimum in the conduction band. When **k** is in the [100] direction, the band structure surrounding it is called the X valley, and when **k** is in the [111] direction, the L valley. As the aluminum mole fraction of Al$_x$Ga$_{1-x}$As exceeds 0.45, it becomes an indirect energy-gap material, with its conduction band minimum residing in the X valley. The band alignment parameters at $x > 0.45$ are found elsewhere.[3] SiGe is an indirect energy-gap material; its conduction band minimum is also located in the X valley. Electron-hole generation or recombination in indirect energy-gap materials requires a change of momentum ($\mathbf{p} = \hbar\mathbf{k}$), a condition that generally precludes them from being useful for laser or light-emitting applications.

Particularly due to the perceived advantage in improved reliability, HBTs made with the GaInP/GaAs system have gained considerable interest.[7] The band alignment of GaInP/GaAs depends on whether the grown GaInP layer is ordered or disordered, as noted in Table 17.1. The crystalline structure in an *ordered* GaInP layer is such that sheets of pure Ga, P, In, and P atoms alternate on the (001) planes of the basic unit cell, without the intermixing of the Ga and In atoms on the same lattice plane.[8] When the Ga, In and P atoms randomly distribute themselves on a plane, the GaInP layer is termed *disordered*.

The processing of AlGaAs/GaAs and GaInP/GaAs HBTs is fairly simple. Both wet and dry etching are used in production environments, and ion implantation is an effective technique to isolate the active devices from the rest. The processing of InP/InGaAs and InAlAs/InGaAs materials, in contrast, is not straightforward. Because of the narrow energy-gap of the InGaAs layer, achieving an effective device isolation often requires a complete removal of the inactive area surrounding the device, literally digging out trenches to form islands of devices. Further, the dry-etching, and its associated advantages such as directionality of etching, is not readily/easily available for InGaAs.[9] However, the material advantages intrinsic to InP/InGaAs and InAlAs/InGaAs make them the choice for applications above 40 GHz. In addition, the turn-on voltage, the applied V_{BE} giving rise to a certain collector current, is smaller for HBTs formed with InGaAs base compared to GaAs base (due to the energy-gap difference). The turn-on characteristics of various HBTs are shown in Figure 17.5.

A calculation illustrates the advantage of an HBT. Consider an AlGaAs/GaAs HBT structure designed for power amplifier applications, as shown in Figure 17.6.[10] The emitter and the base layers of the transistor are: $N_{emit} = 5 \times 10^{17}$ cm^{-3}; $N_{base} = 4 \times 10^{19}$ cm^{-3}; $X_{emit} \approx 1300$ Å; and $X_{base} = 1000$ Å. We shall use the following diffusion coefficients for the calculation: $D_{n,base} = 20$ and $D_{p,emit} = 2.0$ cm^2/V-s. For a

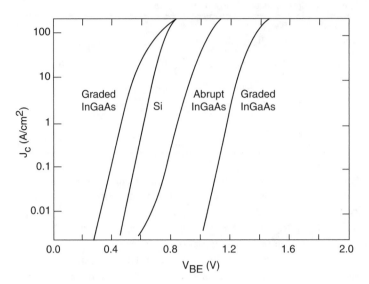

FIGURE 17.5 Turn-on characteristics of bipolar transistors based on various material systems. (After Asbeck et al., *IEEE Trans. Electron Devices*, 36, 2032–2041, 1989, with permission.)

Material	Thickness (Å)	Composition x	Doping (cm^{-3})	
n-In$_x$Ga$_{1-x}$As	800	$0 \rightarrow 0.6$	$> 3 \times 10^{19}$	
n-GaAs	2000		5×10^{18}	
N-Al$_x$Ga$_{1-x}$As	300	$0.3 \rightarrow 0$	1×10^{18}	
N-Al$_x$Ga$_{1-x}$As	1000	0.3	5×10^{17}	Emitter
N-Al$_x$Ga$_{1-x}$As	300	$0 \rightarrow 0.3$	5×10^{17}	
(The above grading layer is absent in abrupt HBT)				
p-GaAs	1000		4×10^{19}	Base
n-GaAs	7000		3×10^{16}	Collector
n-GaAs	6000		5×10^{18}	Subcollector
Semi-insulating GaAs Substrate				

FIGURE 17.6 Typical HBT epitaxial structure designed for power amplifier applications. (After Ali et al., *IEEE Microwave and Millimeter-Wave Monolithic Circuits Symposium*, 61–66, 1996, with permission.)

graded Al$_{0.3}$Ga$_{0.7}$As/GaAs heterojunction, ΔE_g is calculated from Table 17.1 to be $0.3 \times 1.247 = 0.374$ eV. The ratio for the graded HBT, according to Equation 17.5, is,

$$\frac{I_C}{I_{B,\text{back-inject}}} = \frac{1300}{1000} \frac{20}{2.0} \times \frac{5 \times 10^{17}}{4 \times 10^{19}} \exp\left(\frac{0.374}{0.0258}\right) = 2.5 \times 10^5$$

For an abrupt Al$_{0.3}$Ga$_{0.7}$As/GaAs HBT, ΔE_v is the parameter of interest. According to the table, ΔE_v at $x = 0.3$ is $0.55 \times 0.3 = 0.165$ eV. Therefore, Equation 17.4 leads to:

$$\frac{I_C}{I_{B,\text{back-inject}}} = \frac{1300}{1000} \frac{20}{2.0} \times \frac{5 \times 10^{17}}{4 \times 10^{19}} \exp\left(\frac{0.165}{0.0258}\right) = 97$$

Consider a Si BJT with identical doping levels, layer thicknesses, and diffusion coefficients, except that it is a homojunction transistor so that $\Delta E_g = 0$. Using Equation 17.1, we find the $I_C/I_{B,\text{back-inject}}$ ratio to be:

$$\frac{I_C}{I_{B,\text{back-inject}}} = \frac{1300}{1000}\frac{20}{2.0} \times \frac{5 \times 10^{17}}{4 \times 10^{19}} = 0.16.$$

The useful collector current in the homojunction transistor is only 1/6 of the undesirable back-injection current. This means the device is useless. In contrast, both the graded and the abrupt HBTs remain functional, despite the higher base doping in comparison to the emitter.

17.2 Base Current Components

$I_{B,\text{back-inject}}$ is only one of the five dominant base current components in a bipolar transistor. We have thus far considered only $I_{B,\text{back-inject}}$ because it is the distinguishing component between a HBT and a BJT. Once $I_{B,\text{back-inject}}$ is made small in a HBT through the use of a heterojunction, the remaining four components become noteworthy. All of these components are recombination currents; they differ only in the locations where the recombinations take place, as shown in Figure 17.7. They are: (1) extrinsic base surface recombination current, $I_{B,\text{surf}}$; (2) base contact surface recombination current, $I_{B,\text{cont}}$; (3) bulk recombination current in the base layer, $I_{B,\text{bulk}}$; and (4) space-charge recombination current in the base-emitter junction depletion region, $I_{B,\text{scr}}$. In the discussion of bipolar transistors, an easily measurable quantity of prime importance is the current gain (β), defined as the ratio of I_C to the total base current I_B:

$$\beta = \frac{I_C}{I_B} = \frac{I_C}{I_{B,\text{back-inject}} + I_{B,\text{surf}} + I_{B,\text{cont}} + I_{B,\text{bulk}} + I_{B,\text{scr}}} \tag{17.6}$$

Depending on the transistor geometrical layout, epitaxial layer design, and the processing details that shape each of the five base current components, the current gain can have various bias and temperature dependencies. In the following, the characteristics of each of the five base components are described, so that we can better interpret the current gain from measurement and establish some insight about the measured device.

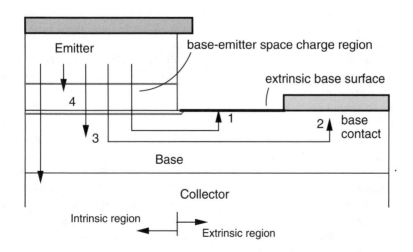

FIGURE 17.7 Locations of the four base recombination currents. The fifth base current component, not a recombination current, is $I_{B,\text{back-inject}}$ shown in Figure 17.2. (After Liu, W., Microwave and DC characterizations of Npn and Pnp HBTs, PhD. Dissertation, Stanford University, Stanford, CA, 1991.)

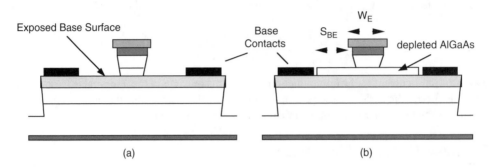

FIGURE 17.8   Schematic cross-sections of HBTs: (a) unpassivated HBTs, and (b) passivated HBTs. (After Liu, W., *Microwave and DC characterizations of Npn and Pnp HBTs*, PhD. Dissertation, Stanford University, Stanford, CA, 1991.)

Figure 17.8a illustrates a schematic cross-section of an HBT. Without special consideration, a conventional fabrication process results in an exposed base surface at the extrinsic base regions. (*Intrinsic region* is that underneath the emitter, and *extrinsic region* is outside the emitter mesa, as shown in Figure 17.7.) Because the exposed surface is near the emitter mesa where the minority carrier concentration is large, and because the surface recombination velocity in GaAs is high (on the order of 10^6 cm/s), $I_{B,surf}$ is significant in these unpassivated devices. Various surface passivation techniques have been tested. The most effective method is *ledge passivation*,[11,12] formed with, for example, an AlGaAs layer on top of the GaAs base. The AlGaAs ledge must be thin enough so that it is fully depleted by a combination of the free surface Fermi level pinning above and the base-emitter junction below. If the passivation ledge is not fully depleted, the active device area would be much larger than the designed emitter. The requirement for the AlGaAs layer to be fully depleted limits the AlGaAs thickness to the order of 1000 Å, and emitter doping to low to mid 10^{17} cm^{-3}. Although the ledge passivation was originally designed to minimize $I_{B,surf}$, it is also crucial to long-term reliability.[3,7]

Unlike I_C, $I_{B,surf}$ is proportional to the emitter periphery rather than the emitter area. For high frequency devices whose emitter is in a strip form (thus the perimeter-to-area ratio is large), $I_{B,surf}$ is a major component to the overall base current. The current gain is substantially reduced from that of a large squarish device whose perimeter-to-area ratio is small. The discrepancy in β due to emitter geometry is termed the *emitter-size effect*. Figure 17.9 displays β vs. I_C for both passivated and unpassivated devices with $A_{emit} = 4 \times 10$ μm^2. The emitter area is small enough to demonstrate the benefit of the surface passivation. A large device has negligible surface recombination current and the current gain does not depend on whether the surface is passivated or not. Because the two devices are fabricated simultaneously and physically adjacent to each other, the difference between the measured βs is attributed to the additional $I_{B,surf}$ of the unpassivated device.

The second base recombination current, $I_{B,cont}$, is in principle the same as $I_{B,surf}$. Both are surface recombination currents, except $I_{B,cont}$ takes place on the base contacts whereas $I_{B,surf}$, on the extrinsic base surfaces. Because the contacts are located further away from the intrinsic emitter than the extrinsic base surface, $I_{B,cont}$ is generally smaller than $I_{B,surf}$ when the surface is unpassivated. However, it may replace $I_{B,surf}$ in significance in passivated devices (or in Si BJTs whose silicon dioxide is famous in passivating silicon). There is a characteristic distance for the minority carrier concentration to decrease exponentially from the emitter edge toward the extrinsic base region.[3] As long as the base contact is placed at roughly 3 times this characteristic length away from the intrinsic emitter, $I_{B,cont}$ can be made small. The base contact cannot be placed too far from the emitter, however. An excessively wide separation increases the resistance in the extrinsic base region and degrades the transistor's high frequency performance.

The above two recombination currents occur in the extrinsic base region. Developing analytical expressions for them requires a solution of the two-dimensional carrier profile. Although this is possible without a full-blown numerical analysis,[3] the resulting analytical equations are quite complicated. The

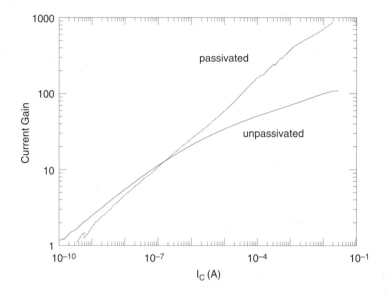

FIGURE 17.9 Measured current gain as a function of collector current for both passivated and unpassivated HBTs. (After Liu, W. et al., Diode ideality factor for surface recombination current in AlGaAs/GaAs heterojunction bipolar transistors, *IEEE Trans. Electron Devices*, 39, 2726–2732, 1992, with permission.)

base bulk recombination current, in contrast, can be accurately determined from a one-dimensional analysis since most of the base minority carriers reside in the intrinsic region. It is convenient to express $I_{B,\text{bulk}}$ through its ratio with I_C:

$$\frac{I_C}{I_{B,\text{bulk}}} = \frac{\tau_n}{\tau_b} \tag{17.7}$$

where τ_n is the minority carrier lifetime in the base, and τ_b, the minority carrier transit time across the base. In a typical Si BJT design, an electron spends about 10 ns diffusing through the base layer while in every 1 μs an electron is lost through recombination. The transistor then has a current gain of 1 μs/10 ns = 100. The recombination lifetime in the GaAs base material is significantly shorter than in Si, at about 1 ns. However, the transit time through the GaAs is also much shorter than in Si due to the higher carrier mobility in GaAs. A well-designed HBT has a τ_b of 0.01 ns; therefore, a $\beta = 100$ is also routinely obtainable in III-V HBTs.

Equation 17.7 indicates that $I_{B,\text{bulk}}$ is large if τ_n is small. The recombination lifetime of a semiconductor, a material property, is found to be inversely proportional to the doping level. When the base doping in an AlGaAs/GaAs (or GaInP/GaAs) HBT is 5×10^{18} cm^{-3}, the current gain easily exceeds 1000 when proper device passivation is made and the base contacts are placed far from the emitter. As the base doping increases to 10^{20} cm^{-3}, $I_{B,\text{bulk}}$ dominates all other base current components, and the current gain decreases to only about 10, independent of whether the extrinsic surface is passivated or not. The base doping in III-V HBTs for power applications, as shown in Figure 17.6, is around 3–5 × 10^{19} cm^{-3}. It is a compromise between achieving a reasonable current gain (between 40 and 200) and minimizing the intrinsic base resistance to boost the high frequency performance.

Equation 17.7 also reveals that $I_{B,\text{bulk}}$ is large when the base transit time is long. This is the time that a base minority carrier injected from the emitter takes to diffuse through the base. Unlike the carrier lifetime, which is mostly a material constant, the base transit time is a strong function of the base layer design:

$$\tau_b = \frac{X_{\text{base}}^2}{2D_{n,\text{base}}} \tag{17.8}$$

Because τ_b is proportional to the square of X_{base}, the base thickness is designed to be thin. Making the base too thin, however, degrades high frequency performance due to increased base resistance. A compromise between these two considerations results in a X_{base} at around 800–1000 Å, as shown in Figure 17.6.

The derivation of Equation 17.8 assumes that the minority carriers traverse through the base purely by diffusion. This is certainly the scenario in a bipolar transistor whose base layer is uniformly doped and of the same material composition. With energy-gap engineering, however, it is possible to shorten the base transit time (often by a factor of 3) by providing a drift field. In a Si/SiGe HBT for example, the Ge content can be linearly graded from 0 to 8% across a 300 Å base to result a *quasi-electric field* on the order of 30 to 50 kV/cm.[13] We used the term quasi-electric field to describe the electric field generated by grading the energy gap, or more specifically, the gradient of the electron affinity (χ_e). In a conventional bipolar transistor, an electric field can be established only in the presence of space charges (such as the depletion region in a *p-n* junction). In an HBT with a graded base, the overall electric field in the base layer is nonzero even though the entire base region remains charge neutral. A SiGe HBT band diagram, shown in Figure 17.10, illustrates how a minority carrier can speed up in the presence of the band grading. The figure shows that the energy gap becomes narrower as the Ge content increases. Because the base layer is heavily doped, the quasi-Fermi level in the base is pinned to be relatively flat with respect to position. The entire energy gap difference appears in the conduction band. The base quasi-electric field, being proportional to the slope of the band bending, propels the electrons to drift from the emitter toward the collector. As the carrier movement is enhanced by the drift motion, in addition to the diffusion, the base transit time decreases.

Figure 17.10 is characteristic of the SiGe HBT pioneered by a U.S. company,[14] in which the Ge content is placed nearly entirely in the base and graded in a way to create a base quasi-electric field. The base-emitter junction is practically a homojunction; therefore, it perhaps does not strictly fit the definition of being a heterojunction bipolar transistor and the base must be doped somewhat lighter than the emitter. This type of transistor resembles a drift homojunction transistor,[15] in which a base electric field is established by grading the base doping level. A drift transistor made with dopant grading suffers from the fact that part of the base must be lightly doped (at the collector side), thus bearing a large base resistance. An alternative school of SiGe HBT places a fixed Ge content in the base, mostly promoted by

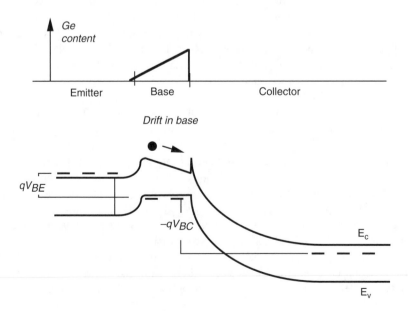

FIGURE 17.10   Band diagram of a SiGe HBT with a base quasi-electric field established by grading the germanium content.

European companies.[16] Transistors fabricated with the latter approach do not have a base quasi-electric field. However, the existence of the base-emitter heterojunction allows the emitter to be more lightly doped than the base, just as in III-V HBTs. In either type of SiGe HBTs, there can be a conduction band discontinuity between the base and collector layers. This base-collector junction spike (Figure 17.10), also notable in InP-based DHBT, has been known to cause current gain fall off.[3] The spike can be eliminated by grading the Ge content from the base to inside the collector layer.

Likely due to reliability concerns or for purely historical reasons, most commercial III-V HBTs have a uniformly doped base without a base quasi-electric field. If a base electric field is desired, in AlGaAs/"GaAs" HBTs in particular, the field can be established by grading of the aluminum concentration in the AlGaAs base layer.

The fourth recombination current is the space-charge recombination current in the base-emitter depletion region. $I_{B,scr}$ differs from the other base current components in its bias dependency. Equations 17.1 and 17.2 show that I_C and $I_{B,back-inject}$ are proportional to $\exp(qV_{BE}/nkT)$ with n, the *ideality factor*, being equal to 1. Equation 17.7 also shows that $I_{B,bulk}$ is directly proportional to I_C. Hence, $I_{B,bulk}$ has a unity ideality factor as well. Extensive measurement experiments and theoretical calculations indicate that $I_{B,surf}$ and hence, $I_{B,cont}$, have an ideality factor closer to 1 than 2.[3,17] The ideality factor of $I_{B,scr}$, in contrast, is nearly 2 because the electron and hole concentrations in the forward-biased base-emitter junction are both proportional to $\exp(qV_{BE}/2kT)$.[2] This means $I_{B,scr}$ is most significant when V_{BE} is small, at which operating region I_C is also small. A *Gummel plot*, I_B and I_C as a function of V_{BE} taken at $V_{BC} = 0$, illustrates dominance of $I_{B,scr}$ in low-current regions, as shown in Figure 17.11. There are times when $I_{B,scr}$ dominates other base current components even at high I_C levels, particularly in graded HBTs.[3]

The previous four base current components are all recombination current. The fifth component is the back-injection current, $I_{B,back-inject}$. This component is made small in a heterojunction transistor, at least at room temperature. However, as temperature increases, the extra energy barrier provided to the hole carriers becomes less effective in impeding the back injection. When the HBT is biased with a high I_C and a certain amount of V_{BC}, the power dissipated in the device can heat up the device itself (called *self-heating*). As the HBTs junction temperature rises, $I_{B,back-inject}$ increases and the current gain decreases. This β's temperature dependency is to be contrasted with silicon BJTs current gain, which increases with temperature.[18]

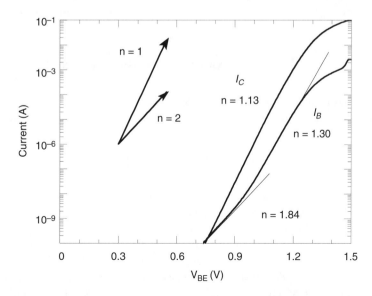

FIGURE 17.11 Measured Gummel plot of an abrupt AlGaAs/GaAs HBT. (After Liu, W., Experimental comparison of base recombination currents in abrupt and graded AlGaAs/GaAs heterojunction bipolar transistors, *Electronic Letters*, 27, 2115–2116, 1991, with permission.)

17.3   Kirk Effects

Understanding the properties of the various base current components facilitates the description of the measured device characteristics, such as the β vs. I_C curve of Figure 17.9. The previous analysis of the transistor currents implicitly assumes that the transistor operates in the normal bias condition, under which the transistor gain is seen to increase steadily with the collector current. However, I_C cannot increase indefinitely without adverse effects. Figure 17.9 reveals that, after a certain critical I_C is reached while V_{BC} is kept constant, the current gain plummets, rendering the device useless. For high-power applications, such as a power amplifier for wireless communications, HBTs are biased at a large current (with a current density on the order of 10^4 A/cm²), not only because the output power is directly proportional to I_C, but also because the high frequency performance is superior at large I_C (but before the current gain falls). Therefore, it is imperative to understand the factor setting the maximum I_C level, below which the current gain is maintained at some finite values.

The Poisson equation relates charges to the spatial variations of the electric field (ε).[2] It is a fundamental equation, applicable to any region at any time in a semiconductor:

$$\frac{d\varepsilon}{dx} = \frac{q}{\in_s}\left(p - n + N_d - N_a\right)$$

(17.9)

$\in_s$ is the dielectric constant of the semiconductor; N_d and N_a are the donor and acceptor doping levels, respectively; and n and p are the mobile electron and hole carrier concentrations, respectively. We apply this equation to the base-collector junction of HBTs, which is typically a homojunction. Since the base doping greatly exceeds the collector doping, most of the depletion region of the base-collector junction is at the collector side. In the depleted collector region where it is doped n-type, N_d in Equation 17.9 is the collector doping level, N_{coll}, and N_a is 0. The collector current flowing through the junction consists of electrons. If the field inside the depletion region was small, then these electrons would move at a speed equal to the product of the mobility and the electric field: $\mu_n \cdot \varepsilon$. It is a fundamental semiconductor property that, once the electric field exceeds a certain critical value ($\varepsilon_{crit} \sim 10^3$ V/cm for GaAs), the carrier travels at a constant velocity called the saturation velocity (v_{sat}). Because the electric field inside most of the depletion region exceeds 10^4 V/cm, practically all of these electrons travel at a constant speed of v_{sat}. The electron carrier concentration inside the collector can be related to the collector current density (J_C; equal to I_C/A_{emit}) as:

$$n(x) = \frac{J_C}{qv_{sat}} = \text{constant inside the base-collector junction}$$

(17.10)

Lastly, because there is no hole current, p in Equation 17.9 is zero.

Equation 17.9, when applied to the base-collector junction of a HBT, is simplified to:

$$\frac{d\varepsilon}{dx} = \frac{q}{\in_s}\left(-\frac{J_C}{qv_{sat}} + N_{coll}\right)$$

(17.11)

When J_C is small, the slope of the electric field is completely determined by the collector doping, N_{coll}. Because the doping is constant with position, solving Equation 17.11 at negligible J_C gives rise to a field profile that varies linearly with position, as shown in Figure 17.12a. As the current density increases, the mobile electron concentration constituting the current partially cancels the positive donor charge concentration N_{coll}. As the net charge concentration decreases, the slope of the field decreases, as shown in Figure 17.12b. While the current density increases, the base-collector bias V_{BC} remains unchanged. Therefore, the enclosed area of the electric field profile, which is basically the junction voltage, is the same before and after the current increase. The simultaneous requirements of having a decreasing field slope and a constant enclosed area imply that the depletion region extends toward the subcollector layer and the maximum electric field decreases. The depletion thickness continues to increase until the collector

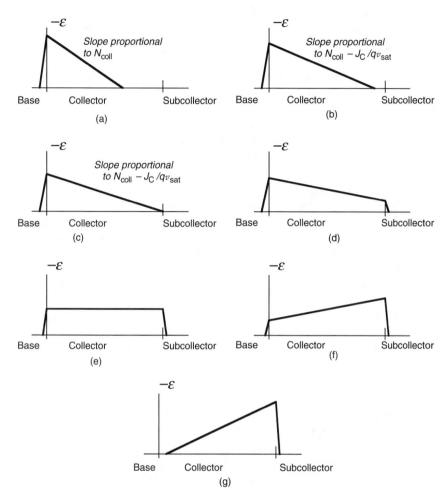

FIGURE 17.12 Electric field profile inside the base-collector junction of HBT during normal operation. (a) $J_C = 0$; (b) J_C is small; (c) J_C further increases so the entire collector is depleted; (d) further increase in J_C; (e) J_C reaches $qN_{coll} \cdot v_{sat}$; (f) $J_C > qN_{coll} \cdot v_{sat}$ and field reversal occurs; (g) further increase in J_C results in base push out as the electric field at the base-collector boundary decreases to zero.

is fully depleted, as shown in Figure 17.12c. The depletion thickness does not extend beyond the collector layer because the subcollector is a heavily doped layer. Afterwards, further increase of current results in a quadrangle field profile, as shown in Figure 17.12d, replacing the previous triangular profile. As the current density increases to a level such that $J_C = qN_{coll} \cdot v_{sat}$, the term inside the parentheses of Equation 17.11 becomes zero. A field gradient of zero means that the field profile stays constant with the position inside the junction (slope = 0). This situation, depicted in Figure 17.12e, marks the beginning the *field reversal*. When J_C increases further such that $J_C > qN_{coll} \cdot v_{sat}$, the mobile electrons brought about by the collector current more than compensates the fixed charges inside the collector. The net charge concentration for the first time becomes negative and the electric field takes on a negative slope (Figure 17.12f), with a smaller magnitude at the base side of the junction than at the subcollector side. As the trend progresses, the magnitude of the field base-collector junction eventually diminishes to zero (Figure 17.12g). When there is no more field to block the holes from "spilling" into the collector, the *base pushout* is said to occur and the current gain falls. The device characteristics as a result of the base pushout are referred to as *Kirk effects*.[19] The above description suggests that the threshold current due to Kirk effects increases if the collector doping increases. However, in many applications where the collector doping may not be increased arbitrarily (so the operating voltage is greater than a certain value), Kirk effects then become an important mechanism affecting the current gain falloff in HBTs. For an HBT with a collector doping of 3×10^{16} cm^{-3} (Figure 17.6), the threshold current density is roughly $J_C = qN_{coll} \cdot v_{sat} = 3.9 \times 10^4$ A/cm^2 (v_{sat} is ~ 8×10^6 cm/s). Clearly, the value of such threshold current density depends on the magnitude of the saturation velocity. Since the saturation velocity decreases with the ambient temperature, the threshold density due to Kirk effects is lower at higher temperatures.

Kirk effects confine the operating collector current to some values. Similarly, the collector-to-emitter bias (V_{CE}) has its limit, set by two physical phenomena. The first one, well analyzed in elementary device physics, is the avalanche breakdown in the base-collector junction. The base-collector junction is a reverse-biased junction with a finite electric field. The mobile electrons comprised of J_C, while moving through the junction, quickly accelerate and pick up energy from the field. When V_{CE} is small, the magnitude of the field is not too large. The energy the carriers acquired is small and is quickly dissipated in the lattice as the carriers impact upon the lattice atoms. The distance within which a carrier travels between successive impacts with the lattice atoms is called a *mean free path*. As V_{CE} increases such that the electric field approaches $10^5 - 10^6$ V/cm, the energy gained by a carrier within one mean free path can exceed the energy gap of the collector material. As the highly energetic carrier impacts the lattice atoms, the atoms are ionized. The act of a carrier impacting the lattice and thereby creating electron-hole pairs is called *impact ionization*. One single impact ionization creates an electron-hole pair, which leads to further impact ionization as the recently generated carriers also pick up enough energy to ionize the lattice. The net result of this positive feedback mechanism is a rapid rise of I_C, which puts the transistor out of useful (or controllable) range of operation. The V_{CE} corresponding the rapid rise in I_C is called the *breakdown voltage*.

17.4   Collapse of Current Gain

The breakdown voltage represents the absolute maximum bias that can be applied to a bipolar transistor. There is, in addition, one more physical phenomenon that further restricts V_{CE} to values smaller than the breakdown voltage. This phenomenon occurs in multi-finger HBTs, having roots in the thermal-electrical interaction in the device. It is termed the *collapse of current gain* (or *gain collapse*) to emphasize the abrupt decrease of current gain observed in measured I-V characteristics when V_{CE} increases past certain values. Figure 17.13 is one such example, measured from a 2-finger HBT. The figure reveals two distinct operating regions, separated by a dotted curve. When V_{CE} is small, I_C decreases gradually with V_{CE}, exhibiting a negative differential resistance (NDR). We briefly describe the cause of NDR, as it relates to the understanding of the gain collapse. The band diagrams in Figures 17.3 and 17.4 showed that the back-injected holes from the base into the emitter experience a larger energy barrier than the emitter electrons forward injected into the base. The ratio of the desirable I_C to the undesirable $I_{B,back-inject}$ is proportional to $\exp(\Delta E_g / kT)$ in a graded HBT, and $\exp(\Delta E_v / kT)$ in an abrupt HBT. At room temperature, this ratio is large in either HBT. However, as V_{CE} increases, the power dissipation in the HBT increases, gradually elevating the device temperature above the ambient temperature. $I_C / I_{B,back-inject}$, and hence the current gain, gradually decrease with increasing V_{CE}. Since Figure 17.13 is measured I_C for several constant I_B, the decreasing β directly translates to the gradual decrease of I_C.

As V_{CE} crosses the dotted curve, NDR gives in to the collapse phenomenon, as marked by a dramatic lowering of I_C. The *collapse locus*, the dotted curve, is the collection of I_C as a function of V_{CE} at which the gain collapse occurs. When several identical transistors are connected together to common emitter, base, and collector electrodes, we tend to expect each transistor to conduct the same amount of collector current for any biases. Contrary to our intuition, equal conduction takes place only when the power dissipation is low to moderate, such that the junction temperature rise above the ambient temperature is small. At high V_{CE} and/or high I_C operation where the transistor is operated at elevated temperatures, one transistor spontaneously starts to conduct more current than the others (even if all the transistors are ideal and identical). Eventually, one transistor conducts all the current while the others become

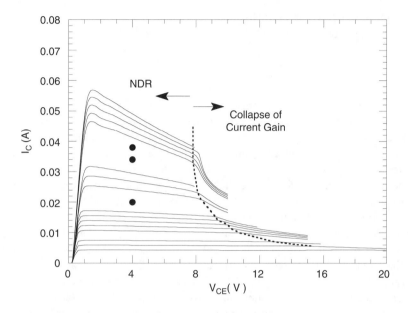

FIGURE 17.13   Measured I-V characteristics of a 2-finger AlGaAs/GaAs HBT showing two distinct regions of operation.

electrically inactive. This current imbalance originates from a universal bipolar transistor property that as the junction temperature increases, the bias required to turn on some arbitrary current level decreases. Quantitatively, this property is expressed by an empirical expression relating I_C, V_{BE}, and T:

$$I_C = I_{C,\text{sat}} \cdot \exp\left[\frac{q}{kT_0} \cdot \left(V_{BE,\text{junction}} - \phi \cdot \left(T - T_0\right)\right)\right] \qquad (17.12)$$

where $I_{C,\text{sat}}$ is the collector saturation current; $V_{BE,\text{junction}}$ is the bias across the base-emitter junction; T_0 is the ambient temperature; and T is the actual device temperature. The degree of the turn-on voltage change in response to a junction temperature change is characterized by ϕ, which is called the *thermal-electrical feedback coefficient*. ϕ decreases logarithmically with I_C^3, and can be approximately as 1.1 mV/°C. This means that when the junction temperature exceeds the ambient temperature by 1°C, turning on the same amount of I_C requires 1.1 mV less of $V_{BE,\text{junction}}$.

A multi-finger HBT can be viewed as consisting of several identical sub-HBTs, with their respective emitter, base, and collector leads connected together. If one finger (i.e., one sub-HBT) becomes slightly warmer than the others, its base-emitter junction turn-on voltage becomes slightly lower. Consequently, this particular finger conducts more current for a given fixed base-emitter voltage. The increased collector current, in turn, increases the power dissipation in the junction, raising the junction temperature even further. The gain collapse occurs when the junction temperature in one finger of the entire device becomes much hotter than the rest of the fingers, so that the feedback action of increased collector current with junction temperature quickly leads to the fact that just one particular finger conducts the entire device current. Since the transition from uniform current conduction to one finger domination occurs suddenly, the surge in junction temperature in the conducting finger quickly lowers the overall device current gain. The fundamental cause of both NDR and collapse is the current gain lowering at elevated temperatures. Their difference, however, lies in the degree of temperature increase as V_{CE} increases. In the NDR region, all fingers share relatively the same amount of current and the junction temperatures increase gradually with V_{CE}. In contrast, in the collapse region, as the device power is entirely dissipated in one finger and the junction temperature rises sharply, the current gain suddenly plummets.

The equation governing the collapse locus (per unit finger) is given by:[3,20]

$$I_{C,\text{collapse}} = \frac{kT_0}{q} \frac{1}{R_{th} \cdot \phi \cdot V_{CE} - R_E} \tag{17.13}$$

where R_{th} is the thermal resistance per finger and R_E is the emitter resistance per finger. When the individual finger current is below this critical current level (or when $I_{C,\text{collapse}}$ is negative), all fingers share the same amount of current. Above this critical current level, one finger conducts most of the current, whereas the rest of the fingers share the remaining current equally. Equation 17.13 shows that an effective method to increase $I_{C,\text{collapse}}$ is to increase R_E. The portion of the resistance that is intentionally introduced into device fabrication (such as by connecting a TaN thin-film resistor in series with the emitter electrode) is called the *ballasting resistance*. Alternatively, $I_{C,\text{collapse}}$ can be increased by reducing the thermal resistance, a goal often requiring more elaborate processes.[21]

Equation 17.13 neglects the contribution from the base resistance, R_B. For III-V HBTs, it is actually advantageous to use base ballasting; i.e., with the ballasting resistance placed in the base terminal.[22] The reason why the base ballasting approach is undesirable for Si BJT has been analyzed.[3]

The collapse of current gain occurring in III-V HBTs, closely relates to the thermal runaway in Si BJTs. HBTs suffering from gain collapse remain functional and can be biased out of the collapse region by reducing V_{CE}. Si BJTs suffering from thermal runaway, however, die instantly. The bias conditions triggering the thermal runaway in Si BJTs are identically given by the collapse locus equation [Equation 17.13]. The main cause of the difference between the collapse in HBTs and thermal runaway in Si BJTs is that the current gain increases with temperature in Si BJTs whereas it decreases with temperature in HBTs.[3]

17.5   High Frequency Performance

Current gain is the most important DC parameter characterizing a bipolar transistor. The high frequency properties are generally measured by two figures of merit: f_T, the cutoff frequency, and f_{max}, the maximum oscillation frequency. The cutoff frequency is the frequency at which the magnitude of the AC current gain (small-signal collector current divided by small-signal base current) decreases to unity. As far as analytical expression is concerned, it is easier to work with the related emitter-collector transit time (τ_{ec}), which is inversely proportional to f_T:

$$f_T = \frac{1}{2\pi \tau_{ec}}. \tag{17.14}$$

The emitter-collector transit time can be broken up into several components. The emitter charging time, τ_e, is the time required to change the base potential by charging up the capacitances through the differential base-emitter junction resistance:

$$\tau_e = \frac{kT}{q I_C} \cdot \left(C_{j,\text{BE}} + C_{j,\text{BC}}\right). \tag{17.15}$$

$C_{j,\text{BE}}$ and $C_{j,\text{BC}}$ denote the junction capacitances of the base-emitter and the base-collector junctions, respectively. The inverse dependence of τ_e on I_C is the primary reason why BJTs and HBTs are biased at high current levels. When the current density is below 10^4 A/cm^2, this term often dominates the overall collector-emitter transit time.

The second component is the base transit time, the time required for the forward-injected charges to diffuse/drift through base. It is given by,

$$\tau_b = \frac{X_{\text{base}}^2}{v \cdot D_{n,\text{base}}} \tag{17.16}$$

The value of v depends on the magnitude of the base quasi-electric field. In a uniform base without a base field, v is 2, as suggested by Equation 17.8. Depending on the amount of energy-gap grading, v can easily increase to 6 to 10.[3]

The space-charge transit time, τ_{sc}, is the time required for the electrons to drift through the depletion region of the base-collector junction. It is given by,

$$\tau_{sc} = \frac{X_{dep}}{2\,v_{sat}} \qquad (17.17)$$

where X_{dep} is the depletion thickness of the base-collector junction. The factor of 2 results from averaging the sinusoidal of carriers current over a time period.[3] It is assumed in the derivation that, because the electric field is large throughout the entire reverse-biased base-collector junction, the carriers travel at a constant saturation velocity, v_{sat}. With a p^- collector layer placed adjacent to the p^+ base of an otherwise conventional Npn HBT,[23] the electric field near the base can be made smaller than ε_{crit}. Consequently, the electrons travel at the velocity determined completely by the Γ valley. Without scattering to the L valley, the electrons continue to travel at a velocity that is much larger than v_{sat}, and τ_{sc} is significantly reduced. When carriers travel faster than v_{sat} under an off-equilibrium condition, *velocity overshoot* is said to occur.

The last term, the collector charging time, τ_c, is given by,

$$\tau_c = \left(R_E + R_C\right) \cdot C_{j,BC}, \qquad (17.18)$$

where R_E and R_C are the device emitter and collector resistances, respectively. The value of this charging time depends greatly on the parasitic resistances. This is the term that degrades the HBT's high frequency performance when the contacts are poorly fabricated.

The overall transit time is a sum of the four time constants:

$$\tau_{ec} = \frac{kT}{q\,I_C} \cdot \left(C_{j,BE} + C_{j,BC}\right) + \frac{X_{base}^2}{v \cdot D_{n,base}} + \frac{X_{dep}}{2\,v_{sat}} + \left(R_E + R_C\right) \cdot C_{j,BC} \qquad (17.19)$$

In most HBTs, R_E and R_C are dominated by the electrode resistances; the epitaxial resistances in the emitter and collector layers are insignificant. The cutoff frequency relates to τ_{ec} through Equation 17.14.

The maximum oscillation frequency is the frequency at which the unilateral power gain is equal to 1. The derivation is quite involved,[3] but the final result is elegant:

$$f_{max} = \sqrt{\frac{f_T}{8\pi\,R_B C_{j,BC}}}. \qquad (17.20)$$

The base resistance has three components that are roughly equal in magnitude. They are base electrode resistance ($R_{B,eltd}$); intrinsic base resistance ($R_{B,intrinsic}$); and extrinsic base resistance ($R_{B,extrinsic}$). $R_{B,eltd}$ is intimately related to processing, depending on the contact metal and the alloying recipe used to form the contact. The other two base resistance components, in contrast, depend purely on the designed base layer and the geometrical details. The HBT cross-sections illustrated in Figure 17.8 show two base contacts placed symmetrically beside the central emitter mesa. For this popular transistor geometry, the intrinsic base resistance is given by,

$$R_{B,intrinsic} = \frac{1}{12} \times R_{SH,base} \frac{W_E}{L_E} \qquad (17.21)$$

where W_E and L_E are the emitter width and length, respectively; and $R_{SH,base}$ is the base sheet resistance in Ω/square. Where does the 1/12 factor come from? If all of the base current that goes into one of the base contacts in Figure 17.8 leaves from the other contact, the intrinsic base resistance would follow our intuitive formula of simple resistance, equal to $R_{SH,base} \times W_E/L_E$. However, during the actual transistor operation, the base current enters through both contacts, but no current flows out at the other end. The holes constituting the base current gradually decrease in number as they are recombined at the extrinsic base surface, in the base bulk layer, and in the base-emitter space-charge region. Some base current carriers remain in the base layer for a longer portion of the width before getting recombined. Other carriers get recombined sooner. The factor 1/12 accounts for the distributed nature of the current conduction, as derived elsewhere.[3] Because of the way the current flows in the base layer, the intrinsic base resistance is called a *distributed resistance*. If instead there is only one base contact, the factor 1/12 is replaced by 1/3 (not 1/6!).[24] The distributed resistance also exists in the gate of MOS transistors,[25] or III-V field-effect transistors.[4]

The extrinsic base resistance is the resistance associated with the base epitaxial layer between the emitter and the base contacts. It is given by,

$$R_{B,extrinsic} = \frac{1}{2} \times R_{SH,base} \frac{S_{BE}}{L_E} \qquad (17.22)$$

where S_{BE} is the separation between the base and emitter contacts. The factor 1/2 appears in the transistor shown in Figure 17.8, which has two base contacts. The presence of $R_{B,extrinsic}$ is the reason why most transistors are fabricated with self-aligned base-emitter contacts, and that the base contacts are deposited right next to the emitter contacts, regardless of the finite alignment tolerance between the base and emitter contact photolithographical steps. (A detailed fabrication process will be described shortly.) With self-alignment, the distance S_{BE} is minimized, to around 3000 Å.

A detailed calculation of f_T and f_{max} has been performed for a $W_E \times L_E = 2 \times 30 \ \mu m^2$ HBT,[4] with a S_{BE} of 0.2 μm, a base thickness of 800 Å, a base sheet resistance of 280 Ω/square, and a base diffusion coefficient of 25.5 cm²/s. Although device parameters will have different values in other HBT structures and geometries, the exemplar calculation gives a good estimation of the relative magnitudes of various terms that determine a HBTs high frequency performance. We briefly list the key results here. The HBT has: $R_E = 0.45 \ \Omega$; $R_{B,eltd} = 7 \ \Omega$; $R_{B,extrinsic} = 0.94 \ \Omega$; $R_{B,intrinsic} = 1.56 \ \Omega$; $C_{j,BE} = 0.224$ pF; $C_{j,BC} = 0.026$ pF. It was further determined at the bias condition of $J_C = 10^4$ A/cm² and $V_{BC} = -4$ V, that the collector depletion thickness $X_{dep} = 0.59$ μm. Therefore,

$$\tau_e = \frac{kT}{q \, I_C} \cdot \left(C_{j,BE} + C_{j,BC} \right) = \frac{0.0258 \left(2.24 \times 10^{-13} + 2.59 \times 10^{-14} \right)}{1 \times 10^4 \cdot 2 \times 10^{-4} \cdot 3 \times 10^{-4}} = 1.08 \text{ ps.}$$

$$\tau_b = \frac{X^2 base}{v \cdot D_{n,base}} = \frac{\left(800 \times 10^{-8} \right)^2}{2 \cdot 25.5} = 1.25 \text{ ps.}$$

$$\tau_{sc} = \frac{X_{dep}}{2 \, v_{sat}} = \frac{0.591 \times 10^{-4}}{2 \cdot 8 \times 10^6} = 3.69 \text{ ps.}$$

$$\tau_c = \left(R_E + R_C \right) C_{j,BC} = \left(0.453 + 4.32 \right) \cdot 2.59 \times 10^{-14} = 0.124 \text{ ps.}$$

Summing up these four components, we find the emitter-collector transit time to be 6.14 ps. The cutoff frequency is therefore $1/(2\pi \cdot 6.14$ ps) = 26 GHz. As mentioned, although this calculation is specific to a particular power HBT design, the relative magnitudes of the four time constants are quite representative. Generally, the space-charge transit time is the major component of the overall transit time. This is unlike silicon BJTs, whose τ_b and τ_e usually dominate, because of the low base diffusion coefficient in the silicon material and the high base-emitter junction capacitance associated with high emitter doping. HBT design places a great emphasis on the collector design, while the Si BJT design concentrates on the base and emitter layers.

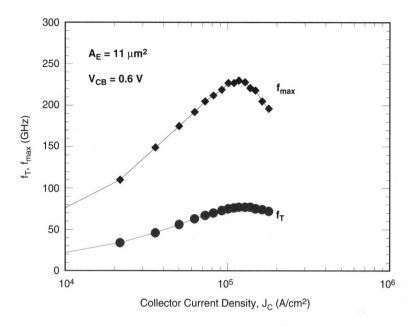

FIGURE 17.14 High-frequency performance of a state-of-the-art InAlAs/InGaAs HBT. (After Chau, H. and Kao, Y., High f_{max} InAlAs/InGaAs heterojunction bipolar transistors, IEEE IEDM, 783–786, 1993, with permission.)

The total base resistance is $R_B = 7 + 0.96 + 1.95 = 9.91\ \Omega$ The maximum oscillation frequency, calculated with Equation 17.20, is 65 GHz.

Figure 17.14 illustrates the cutoff and maximum oscillation frequencies of a state-of-art InAlAs/InGaAs HBT. The frequency responses are plotted as a function of the collector current to facilitate the circuit design at the proper bias condition. When the current is small, the cutoff frequency is small because the emitter charging time is large. As the current increases, f_T increases because the emitter charging time decreases. When the current density exceeds roughly 10^5 A/cm^2, Kirk effects take place and the transistor performance degrades rapidly. f_{max}'s current dependence follows f_T's, as governed by the relationship of Equation 17.13.

The high frequency transistor of the calculated example has a narrow width of 2 μm. What happens if W_E increases to a large value, such as 200 μm? A direct application of Equation 17.19 would still lead to a cutoff frequency in the GHz range. In practice, such a 200 μm wide device will have a cutoff frequency much smaller than 1 GHz. The reason for the discrepancy is simple; the time constants appearing in Equation 17.19, which all roughly scale with the emitter area, are based on the assumption that the current conduction is uniform within the entire emitter mesa. In HBTs, and more so in Si BJTs, the base resistance is significant. As the base current flows horizontally from the base contacts to the center of the emitter region, some finite voltage is developed, with a increasingly larger magnitude toward the center of the mesa. The effective base-emitter junction voltage is larger at the edge than at the center. Since the amount of carrier injection depends exponentially on the junction voltage at the given position, most of the injection takes place near the emitter edges, hence the term *emitter crowding*. Sometimes this phenomenon of nonuniform current conduction is referred to as the *base crowding*. Both terms are acceptable, depending on whether the focus is on the crowding of the emitter current or the base current. A figure of merit quantifying the severity of emitter crowding is the effective emitter width (W_{eff}). It is defined as the emitter width that would result in the same current level if current crowding were absent and the emitter current density were uniform at its edge value. Figure 17.15 illustrates the effective emitter width (normalized by the defined emitter width) as a function of base doping and collector current.

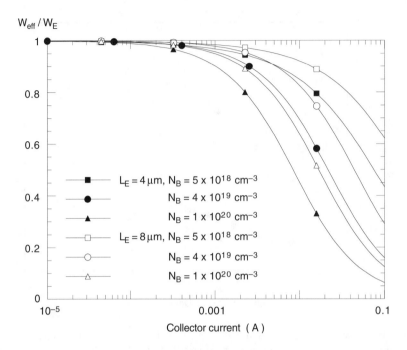

FIGURE 17.15   Calculated W_{eff}/W_E ratio. (After Liu, W. and Harris, J., Dependence of base crowding effect on base doping and thickness for Npn AlGaAs/GaAs heterojunction bipolar transistors, *Electronic Letters*, 27, 2048–2050, 1991, with permission.)

Because of the emitter crowding, all high-frequency III-V HBTs capable of GHz performance have a narrow emitter width on the order of 2 μm. For wireless applications in the 2 GHz range, 2 to 3 μm W_E is often used,[26] but a 6.4 μm W_E has also been reported.[27] The choice of the emitter width is a trade-off between the ease (cost) of device fabrication versus the transistor performance. We intuitively expected the W_{eff}/W_E ratio to increase as the doping increased (since the base resistance decreased). However, an increase in base doping was accompanied by shortened minority lifetime. The large increase in $I_{B,bulk}$ causes the emitter crowding to be more pronounced as N_{base} increases, as shown in Figure 17.15. Due to differences in material properties, the emitter width in Si BJT or SiGe HBTs tends to be on the order of 0.5 μm.

17.6   Device Fabrication

An HBT mask layout suitable for high-frequency applications is shown in Figure 17.16, and a corresponding fabrication process is delineated in Figure 17.17. The following discussion of processing steps assumes the AlGaAs/GaAs HBT epitaxial structure shown in Figure 17.6. The first step is to deposit ~7000 Å of silicon dioxide, onto which the first-mask pattern "ISOL" is defined, and 1.5 μm of aluminum, evaporated and lifted off. The aluminum protects the silicon dioxide underneath during the subsequent reactive ion etch (RIE), resulting in the profile shown in Figure 17.17a. Oxygen atoms or protons are then implanted everywhere on the wafer. This implantation is designed to make the region outside of "ISOL" pattern electrically inactive. A shallow etching is applied right after the implantation so that a trail mark of the active device area is preserved. This facilitates the alignment of future mask levels to the first mask. Afterward, both aluminum and oxide are removed with wet etching solutions, reexposing the fresh InGaAs top surface onto which a refractory metal (such as W) is sputtered. The "EMIT" mask level is used to define the emitter mesas, and Ti/Al, a conventional contact metal used in III-V technologies, is evaporated and lifted off. If a refractory metal is not inserted between the InGaAs semiconductor

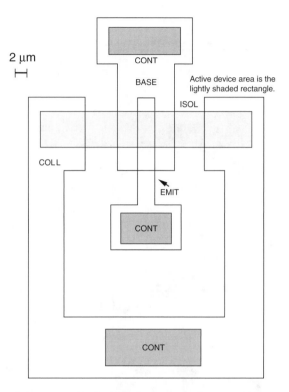

FIGURE 17.16 A HBT mask layout suitable for high-frequency applications. (After Liu, W., Microwave and DC Characterizations of Npn and Pnp HBTs, PhD. Dissertation, Stanford University, Stanford, CA, 1991.)

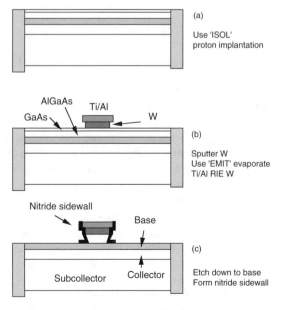

FIGURE 17.17 A high-frequency HBT process flow. (After Liu, W., Microwave and DC Characterizations of Npn and Pnp HBTs, PhD. Dissertation, Stanford University, Stanford, CA, 1991.)

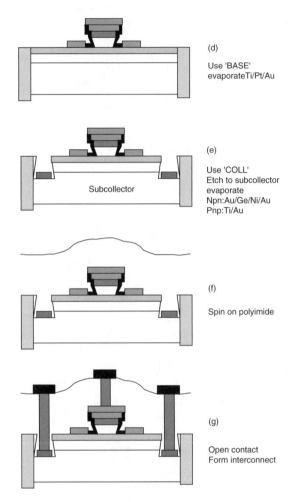

(d)

Use 'BASE'
evaporateTi/Pt/Au

(e)

Use 'COLL'
Etch to subcollector
evaporate
Npn:Au/Ge/Ni/Au
Pnp:Ti/Au

Subcollector

(f)

Spin on polyimide

(g)

Open contact
Form interconnect

FIGURE 17.17   (continued)

and the Ti/Al metal, long-term reliability problems can arise as titanium reacts with indium. During the ensuing RIE, the refractory metal not protected by the emitter metal is removed.

The resulting transistor profile is shown in Figure 17.17b. Wet or dry etching techniques are applied to remove the exposed GaAs cap and the AlGaAs active emitter layer, and eventually reach the top of the base layer. Silicon nitride is deposited by plasma enhanced chemical vapor deposition (PECVD). This deposition is conformal, forming a nitride layer everywhere, including the vertical edges. Immediately after the nitride deposition, the whole wafer is etched by RIE. Because a vertical electric field is set up in the RIE chamber to propel the chemical species in the vertical direction, the sidewall nitride covering the sides of the emitter mesas remains untouched while the nitride layer lying on a flat surface is etched away (Figure 17.17c). The nitride sidewall improves the device yield by cutting down the possible electrical short between the emitter contact and the soon-to-be-deposited base contacts.

The step after the nitride sidewall formation is the "BASE" lithography. As shown in the layout of Figure 17.16, there is no separation between the "BASE" and the "EMIT" levels. The base metal during the evaporation partly lands on the emitter mesa, and some of it lands on the base layer as desired (Figure 17.17d). The process has the desired feature that, even if the "BASE" level is somewhat misaligned with the "EMIT" level, the distance S_{BE} between the emitter and base contacts is unchanged, and is at the minimum value determined by the thickness of the nitride sidewall. In this manner, the base contact is said to be *self-aligned* to the emitter. Self-alignment reduces $R_{B\text{extrinsic}}$ [Equation 17.15] and $C_{j,BC}$, hence improving the high-frequency performance. Typically, Ti/Pt/Au is the choice for the base metal.

Following the base metal formation, the collector is defined and contact metal of Au/Ge/Ni/Au is deposited (Figure 17.17e). After the contact is alloyed at ~450°C for ~1 minute, a polyimide layer is spun to planarize the device as shown in Figure 17.17f. The contact holes are then defined and Ti/Au is evaporated to contact various electrodes. The final device cross-section after the entire process is shown in Figure 17.17g.

References

1. Kroemer, H, Heterostructure bipolar transistors and integrated circuits, *Proc. IEEE*, 70, 13, 1982.
2. Sah, C.T., *Fundamentals of Solid-State Electronics*, World Scientific, Singapore, 1991.
3. Liu, W., *Handbook of III-V Heterojunction Bipolar Transistors*, Wiley & Sons, New York, 1998. An in-depth discussion of several topics can be found in this handbook.
4. Liu, W., *Fundamentals of III-V Devices: HBTs, MESFETs, and HFETs/HEMTs*, Wiley & Sons, New York, 1999.
5. Kroemer, H., Theory of wide gap emitter for transistors, *Proc. IRE*, 45, 1535, 1957.
6. For pioneering HBT papers, see for example, Asbeck, P., Chang, M., Higgins, J., Sheng, N., Sullivan, G., and Wang, K., GaAlAs/GaAs heterojunction bipolar transistors: issues and prospects for application, *IEEE Trans. Electron Devices*, 36, 2032–2041, 1989. For introduction to MBE and MOCVD growth techniques, see References 3 and 4.
7. Henderson, T., Physics of degradation in GaAs-based heterojunction bipolar transistors, *Microelectronics Reliability*, 39, 1033–1042, 1999. See also, Low, T., et al., Migration from an AlGaAs to an InGaP emitter HBT IC process for improved reliability, *IEEE GaAs IC Symposium*, 153–156, 1998.
8. Liu, W. et al., Recent developments in GaInP/GaAs heterojunction bipolar transistors, in *Current Trends in Heterojunction Bipolar Transistors*, Chang, M.F., Ed., World Scientific, Singapore, 1996.
9. Chau, H. and Liu, W., Heterojunction bipolar transistors and circuit applications, in *InP-Based Material and Devices: Physics and Technology*, Wada, O. and Hasegawa, H., Eds., Wiley & Sons, New York, 1999.
10. Ali, F., Gupta, A. and Higgins, A., Advances in GaAs HBT power amplifiers for cellular phones and military applications, *IEEE Microwave and Millimeter-Wave Monolithic Circuits Symposium*, 61–66, 1996.
11. Lin, H. and Lee, S., Super-gain AlGaAs/GaAs heterojunction bipolar transistor using an emitter edge-thinning design, *Appl. Phys. Lett.*, 47, 839–841, 1985.
12. Lee, W., Ueda, D., Ma, T., Pao, Y. and Harris, J., Effect of emitter-base spacing on the current gain of AlGaAs/GaAs heterojunction bipolar transistors, *IEEE Electron Device Lett.*, 10, 200–202, 1989.
13. Patton, G., 75-GHz fr SiGe-base heterojunction bipolar transitors, *IEEE Electron Devices Lett.*, 11, 171–173, 1990.
14. Meyerson, B. et al., Silicon:Germanium heterojunction bipolar transistors; from experiment to technology, in *Current Trends in Heterojunction Bipolar Transistors*, Chang, M., Ed., World Scientific, Singapore, 1996.
15. Pritchard, R., *Electrical Characteristics of Transistors*, McGraw-Hill, New York, 1967.
16. Konig, U., SiGe & GaAs as competitive technologies for RF applications, *IEEE Bipolar Circuit Technology Meeting*, 87–92, 1998.
17. Tiwari, S., Frank, D. and Wright, S., Surface recombination current in GaAlAs/GaAs heterostructure bipolar transistors, *J. Appl. Phys.*, 64, 5009–5012, 1988.
18. Buhanan, D., Investigation of current-gain temperature dependence in silicon transistors, *IEEE Trans. Electron Devices*, 16, 117–124, 1969.
19. Kirk, C., A theory of transistor cutoff frequency falloff at high current densities, *IRE Trans. Electron Devices*, 9, 164–174, 1962.

20. Winkler, R., Thermal properties of high-power transistors, *IEEE Trans. Electron Devices*, 14, 1305–1306, 1958.

21. Hill, D., Katibzadeh, A., Liu, W., Kim, T. and Ikalainen, P., Novel HBT with reduced thermal impedance, *IEEE Microwave and Guided Wave Lett.*, 5, 373–375, 1995.

22. Khatibzadeh, A. and Liu, W., *Base Ballasting*, U.S. Patent Number 5,321,279, issued on June 14, 1994.

23. Maziar, C., Klausmeier-Brown, M. and Lundstrom, M., Proposed structure for collector transit time reduction in AlGaAs/GaAs bipolar transistors, *IEEE Electron Device Lett.*, 7, 483–385, 1986.

24. Hauser, J., The effects of distributed base potential on emitter-current injection density and effective base resistance for stripe transistor geometries, *IEEE Trans. Electron Devices*, 11, 238–242, 1964.

25. Liu, W., *MOSFET Models for SPICE Simulation, Including BSIM3v3 and BSIM4*, Wiley & Sons, New York, in press.

26. RF Microdevices Inc., A high efficiency HBT analog cellular power amplifier, *Microwave Journal*, 168–172, January 1996.

27. Yoshimasu, T., Tanba, N. and Hara, S., High-efficiency HBT MMIC linear power amplifier for L-band personal communication systems, *IEEE Microwave and Guided Wave Lett.*, 4, 65–67, 1994.

18

Metal-Oxide-Semiconductor Field-Effect Transistors (MOSFETs)

Julio Costa
Mike Carroll
G. Ali Rezvani
Tom McKay
RF Micro Devices

18.1   Introduction

The Metal-Oxide-Semiconductor Field-Effect Transistor (MOSFET) is today the most utilized electronic semiconductor device in the world, covering a vast range of electronic applications. Its usage ranges from simple single transistor designs to complex high-speed very large-scale integration (VLSI) circuits containing tens of millions of transistors, integrated on the same monolithic silicon die, covering a wide range of analog and digital applications. Because of its nearly ideal switching characteristics, low-fabrication costs and scaleable electronic behavior, MOSFET technologies have been and will continue to be the workhorse of the electronic industry for years to come; covering all ranges of speed, frequency, and voltage applications. With vast improvements in its switching speeds which are realized with modern semiconductor fabrication processes, MOSFET technology is now routinely used in high-speed and high-frequency analog and RF applications *in lieu* of other technologies traditionally used in this area, such as bipolar (BJT) transistors and III–V semiconductor technologies. It is often mentioned in the electronics design industry that "*if it can be done with MOSFET technology, then it should.*"

The MOSFET technology evolution itself is quite remarkable, not only in how the basic device technology has progressed over the last two decades, but also in the way it spurred the creation of a multi-hundred-billion dollar industry involving a large degree of synergy between process equipment suppliers, fabrication sites (commonly known as "fabs"), and a vast collection of accompanying computer-aided design (CAD) software tools for design, modeling and verification. Altogether, this technological synergy has allowed the product evolution of silicon MOSFET-integrated electronic designs of unparalleled complexity and unprecedented high levels of integration, implemented in a relatively short span of time.

In the following sections, we will cover the MOSFET technology from its basic device and fabrication to its modeling characteristics. The MOSFET technology itself has been successfully applied to RF applications since the early 1970s, although initially with a much stronger presence in high-voltage RF power transmitter (VHF/UHF) amplifiers rather than receiver applications. The reason for this disparity lies in the fact that until the MOSFET technology was scaled to submicron gate lengths ($<0.5\ \mu\text{m}$ or so), its noise and frequency response characteristics were significantly inferior to existing silicon bipolar device and III–V technologies of similar geometries such as GaAs Metal-Semiconductor Field-Effect Transistors (MESFETs). With the gate lengths of modern MOSFET transistors reduced to $0.5\ \mu\text{m}$ dimensions and below, the RF characteristics of MOSFET technologies improved to a point where they begin to rival existing traditional solutions. But because the cost structure of MOSFET technologies is significantly more favorable than the other alternative solutions and because of the integration factor it offers with digital complementary metal oxide semiconductor (CMOS) is so appealing, radio frequency CMOS (RFCMOS) technologies now are routinely employed in very complex single-chip RF systems. Today we speak in terms of optimizing RFCMOS technologies for mixed-mode (analog/digital) integration at the lowest possible system cost, involving multiple digital and analog previously developed IP (intellectual property) blocks, and integrating a vast array of RF passive and active components (such as inductors, capacitor, and varactors).

An example of the type of RF system integration in a RFCMOS technology is shown in Figure 18.1. The figure depicts a complete Bluetooth System-On-Chip (SOC) Version 1.2 system-integrated circuit, of dimensions 3.5×3.5 mm on a side, implemented in a $0.18\ \mu\text{m}$ RF CMOS technology with direct frequency conversion. The figure illustrates the unequaled integration power of the RFCMOS technology; where complex system blocks containing microprocessor, memory, millions of digital gates, analog, input/output cells, and RF blocks are monolithically integrated in a very small form factor, at a very economic scale. The semiconductor process complexity of the modern RFCMOS technologies is well illustrated in scanning electron microscope (SEM) cross-section of Figure 18.2, which depicts the multitude of metal, dielectric, and semiconductor layers necessary to create the mix of passive and active devices commonly required in modern mixed-mode (analog/digital) applications.

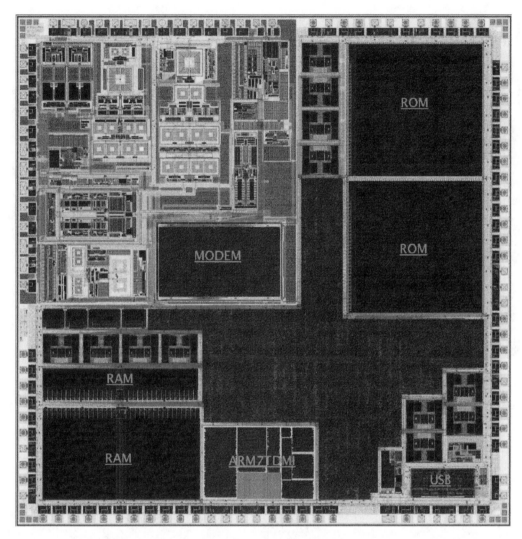

FIGURE 18.1 A complete SoC Bluetooth Version 1.2 RF CMOS chip built by RF Micro Devices. The chip was implemented in a 0.18 μm RFCMOS technology. Die dimensions are 3.5 × 3.5 mm².

18.2 A Brief History of the MOSFET

Figure 18.3 illustrates a cross section of a modern MOSFET device. In the next few sections, we will go over all of the basic terms and regions depicted in the illustration. Noticeably absent from the cross section of Figure 18.3 is the actual *METAL* gate region in the device (responsible for the letter "M" in the MOSFET abbreviation): metal gates have not been employed in the gate regions of MOSFETs for quite some time; instead polycrystalline-silicon (commonly referred as "poly") gate regions are routinely used. As it will be shown later in the fabrication section of this chapter, polysilicon gate regions allow for a much superior self-aligned process that is much more advantageous from an electrical performance point of view. Although the term insulated gate field-effect transistor (IGFET) was proposed earlier to describe this insulating gate device from a more generic point of view, this particular term is very rarely used and has for all practical purposes been discontinued from modern usage in favor of the MOSFET

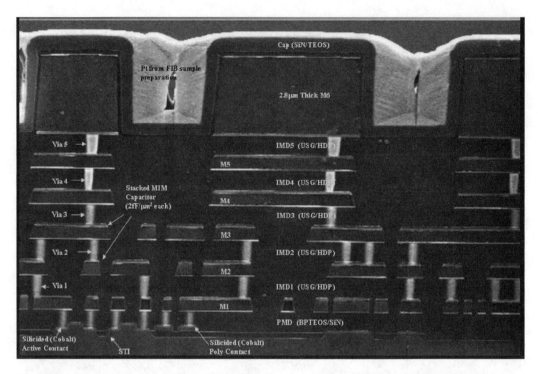

FIGURE 18.2   A SEM (scanning electron microscopy) cross section of a modern 0.18 μm RFCMOS process, illustrating the basic semiconductor, dielectric and metallization layers used in the process. Picture courtesy of JAZZ Semiconductor (Newport Beach, California).

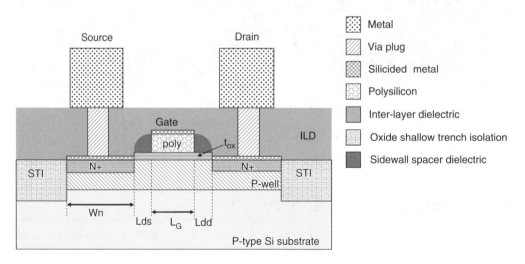

FIGURE 18.3   The basic modern MOSFET structure. An N-channel MOSFET device is depicted.

abbreviation. The term IGFET continues to be used however in the area of compound semiconductor field-effect transistors, such as the exploration of III–V oxide-gate FET technology.

The MOSFET device concept consists of a basic semiconductor "electronic valve," in the sense that the electrical current between the *drain* and *source* terminals, biased respectively with voltages V_D and V_S, can be modulated by the voltage applied at a third terminal at the gate terminal V_G. The gate region is defined by the gate length dimension L_G is separated from the actual channel by a thin oxide region t_{OX}, and the potential V_G modulates the actual conductance of the channel region, thereby controlling the magnitude of the electrical current I_{DS} that flows between source and drain terminals of the MOSFET. Because the gate is separated from the actual channel by a very high-quality insulating oxide region, no DC current flows from the gate region to the other electrodes under normal operation, allowing the MOSFET to operate essentially as a nearly ideal electronic valve or switch. The actual fabrication of a modern MOSFET and the physics involved in the electrical characteristics of the device are of course much more involved and complex than this simple concept would suggest and will be described in detail in later sections of this chapter. Nonetheless, the unmatched success of the MOSFET technology in modern electronics is due in great part to the fact that this remarkably simple concept of the MOSFET as a variable conductance element that can be switched ON, modulated, and then switched OFF via a third high-impedance node continues to be successfully applied, generation after generation, to higher performance MOSFET technologies.

Although there are a few authors who published works which could potentially be described as the precursor of the MOSFET concept (based mostly on intuitive concepts), [1,2] William Shockley can arguably be credited as being the inventor of the modern MOSFET device during his brilliant tenure at Bell Labs (USA) during 1936–1954. Shockley's main interests were in the area of conduction of electrons in solids, which at that time was an emerging field. Shockley developed at the time the required physical relationships between carrier concentration, charge and energy, which elegantly described the operation of many phenomena observed in the development of the bipolar device [3]. For this remarkable and revolutionary development, the Bell Labs team (Shockley, Bardeen, Brattain) received the 1956 Nobel Prize in Physics. Through the Solid Physics theoretical foundation developed for solid-state devices, Shockley envisioned that a field-effect transistor device could be potentially designed by the creation of an inverting conduction layer in the semiconductor region through a thin oxide region and a gate electrode [4].

Practically, however, the first attempts of fabricating this proposes MOSFET structures failed and a much stronger emphasis was placed on the development of bipolar semiconductor technology with respect to solid-state electronic devices. The leading cause of failure was readily identified as the quality of the dielectric region, which separated the gate region from the actual semiconductor channel [5]. Because the interface quality between the dielectric and the semiconductor of these initial attempts was very poor, changes in electrical potential applied at the gate of the device did not result in a reliable and efficient creation of an inversion channel thus resulting in very poor modulation of charge in the channel of the semiconductor device.

The practical proof of concept of the MOSFET device would occur later in 1960, when another research team at Bell Labs, composed of Dawon Khang and Martin Atalla, [6] implemented the MOSFET device concept using thermally grown thin silicon oxides on top of a semiconductor region. With the advent of a reliable oxide-semiconductor interface through thermal oxidation processes, the fabrication of the device became essentially straight forward, and a great number of researchers throughout the US, Europe, and Asia quickly developed in the following decade the theoretical and fabrication foundation of the modern MOSFET technology. Complementary N- and P-channel MOSFET operation was demonstrated in 1963 by Frank Wanlass at Fairchield Semiconductor, [7] and early in the 1970s, CMOS digital logic at large-scale integration (LSI) levels was integrated successfully, illustrating in detail the tremendous improvement in leakage currents, density, and efficiency when compared with bipolar digital technologies used at that time.

18.3 Moore's Law

Moore's Law is often mentioned in technical articles and discussion regarding MOSFET technology. As a colleague of Shockley, Gordon Moore (one of the founders of Intel) recognized at an early stage that the nearly ideal electrical and scaling characteristics of MOSFET devices that had been optimized in the early 60s, coupled with the successful demonstrations of increased integration levels, were leading to an unparalleled growth in electronic applications. In 1965, Moore theorized that such tremendous growth would

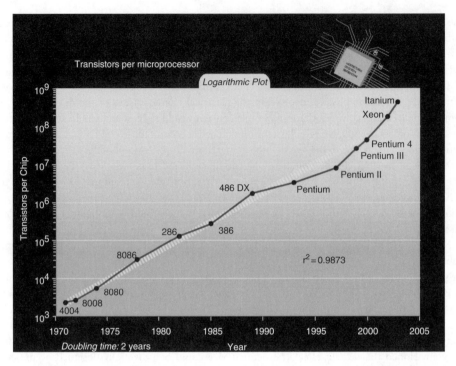

FIGURE 18.4 Moore's Law plotted from 1970 to 2005. The graph depicts the name of popular microprocessor generations. (From R. Kurzweil, The impact of 21st Century Technology on Human Health and Business, *Proceedings of the World Science Forum*, November 2006.)

create a constant demand for improved performance and miniaturization of the MOSFET technology. By observing recent developments and with an understanding of the scaling power of the device, he predicted that "*The number of transistors incorporated in a chip will double every 24 months.*" Figure 18.4 illustrates Moore's observation plotted over a span of 35 years. Moore's Law is obviously not a physical "law"; it was merely an extrapolation based on the observation of the progress of existing CMOS IC technology and applications. Despite that misnomer, Moore's law essentially has became a "self-fulfilling prophecy" in the digital CMOS industry, in the sense that it became a roadmap for technology development, providing the targets for device research, semiconductor manufacturing evolution, and product evolution throughout the MOSFET industry. Although Moore's Law is observed only for digital CMOS and has never been observed for RFCMOS applications, it is important for the RF engineer to understand its definition in view of the fact that modern integrated RFCMOS system-on-chip designs involve an ever-larger number of digital blocks.

Table 18.1 compares the evolution of MOSFET technology nodes over the last decade, showing some of the useful parameters that characterize the most popular CMOS technologies with respect to RF/analog applications. It should be noted that with respect to the utilization of MOSFET technology in RF or analog applications, there is a significant "lag" as compared with its usage in digital applications. The reason for this delay has mostly to do with the fact that during the initial release of a CMOS technology, a much stronger focus is placed on enabling the digital design environment (i.e., design kits).

18.4   The Metal-Oxide-Semiconductor Interface

Paramount to the understanding of the MOSFET structure is a basic knowledge of the Metal-Oxide-Semiconductor interface and some general concepts regarding the charge carriers responsible for electrical conduction in a semiconductor device. In this section, we will present the basic concepts of the MOS

TABLE 18.1 MOSFET Standard Technology RFCMOS Nodes*

Min Feature	Year of Large-Scale Utilization	Digital Density (gates/mm^2)	Supply Voltage (Volts)	t_{OX} (A)	IDSAT (uA/um) NFET	IDOFF (nA/um) NFET	FT/FMAX (GHz) NFET
0.5 μm	1990	10K	5	120	450	<0.01	12/20
0.35 μm	1993	20K	3.3	65	550	<0.01	20/35
0.25 μm	1996	40K	2.5	50	600	<0.01	32/45
0.18 μm	1999	100K	1.8	30	650	0.03	50/65
0.13 μm LL	2001	200K	1.2	20	550	0.5	75/120
HS					670	20	90/120
90 nm LL	2004	400K	1.0	18	600	2	150/200
HS					750	50	115/180
65 nm LL	2006	800K	1.0	18	620	2	180/200
HS					750	50	225/250

IDSAT: measured at VGS = VSUPPLY, VDS = VSUPPLY.
IDOFF: measured at VG = 0, VD = VSUPPLY.
FT/FMAX: peak measured values at VD = VSUPPLY.
* HS: High Speed device option ; LL: Low Leakage device option.

structure; for an in-depth analysis, dedicated textbooks are devoted to the complex physics that governs this important semiconductor stack [8].

18.4.1 Donors and Acceptors [9]

The semiconductor material commonly used to make MOSFETs is silicon. Pure or "intrinsic" silicon exists as an orderly three-dimensional array of atoms, arranged in a crystal lattice. The atoms of the lattice are bound together by covalent bonds containing silicon valence electrons. At absolute zero temperature, all valence electrons are locked into these covalent bonds and are unavailable for current conduction, but as the temperature is increased, it's possible for an electron to gain enough thermal energy so that it escapes from its covalent bond, and in the process leaves behind a covalent bond with a missing electron, or "hole." When that happens the electron that escaped is free to move about the crystal lattice. At the same time, another electron which is still trapped in nearby covalent bonds because of its lower energy state can move into the hole left by the escaping electron. The mechanism of current conduction in intrinsic silicon is therefore by hole–electron pair generation and the subsequent motion of free electrons and holes throughout the lattice.

At normal temperatures intrinsic silicon behaves as an insulator because the number of free electron–hole pairs available for conducting current is very low. The conductivity of silicon can be adjusted by adding foreign atoms to the silicon crystal. This process is called "doping," and a "doped" semiconductor is referred to as an "extrinsic" semiconductor. Depending on what type of material is added to the pure silicon, the resulting crystal structure can either have more electrons than the normal number needed for perfect bonding within the silicon structure, or less electrons than needed for perfect bonding. When the dopant material increases the number of free electrons in the silicon crystal, the dopant is called a "donor."

In a donor-doped semiconductor the number of free electrons is much larger than the number of holes, and so the free electrons are called the "majority carriers" and the holes are called the "minority carriers." Since electrons carry a negative charge and they are the majority carriers in a donor-doped silicon semiconductor, any semiconductor that is predominantly doped with donor impurities is known as "n-type." Semiconductors with extremely high donor doping concentrations are often denoted "n+ type." Dopant atoms that accept electrons from the silicon lattice are also used to alter the electrical characteristics of silicon semiconductors. These types of dopants are known as "acceptors." The number of holes in the lattice therefore increases. The holes are therefore the majority carriers and the electrons

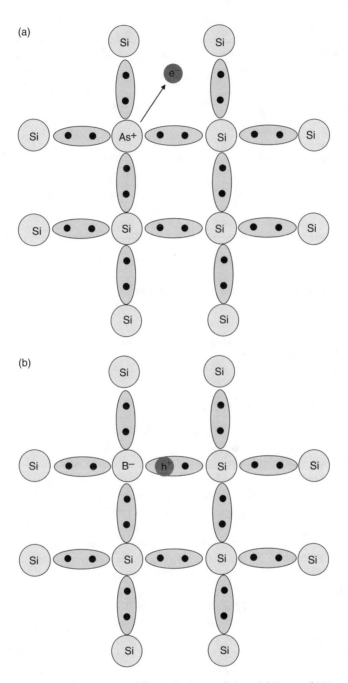

FIGURE 18.5   The silicon lattice depicting two different doping conditions: (a) P-type, (b) N-type.

are the minority carriers. Semiconductors doped with acceptor impurities are known as "p-type," since the majority carriers effectively carry a positive charge. Semiconductor regions with extremely high acceptor doping concentrations are called "p+ type." Figure 18.5 depicts the silicon lattice under these two different doping conditions, illustrating the case where the material is said to be of N-type (a), and P-type (b).

Under equilibrium (meaning without any external disturbance such as an applied voltage or incidence of light), the semiconductor material is strictly charge neutral. The charge neutrality equation

(Equation 18.1) can be written as

$$p + N_D^+ = n + N_A^-$$ (18.1)

Where n and p are the concentration of free electrons and free holes in the lattice (typically in units of concentration per cm^{-3}), and N_D^+ and N_A^- are the concentration of ionized donor and acceptor impurities. Typically the superscripts $^+$ and $^-$ are dropped in the nomenclature in the literature and it is assume that N_D and N_A are assumed to indicate the concentration of ionized impurities. If only acceptor impurities (like boron), the silicon lattice is said to be P-type, with a concentration approximated by Equation 18.2

$$p = N_A^-$$ (18.2)

On the other hand, if only donor impurities (like phosphorus and arsenic), the silicon lattice is said to be N-type, with a concentration approximated by Equation 18.3

$$n = N_D^+$$ (18.3)

18.4.2 The MOS Surface Charge Q_s and Surface Potential ψ_s

As mentioned previously, the concentration of carriers in the semiconductor/oxide interface can be manipulated by several orders of magnitude by an electrostatic potential applied between the metal and the semiconductor regions. It is this basic interface control property that allows us to create and to turnoff a conduction channel, which when properly biased allows for the operation of the MOSFET. In this section, we present the basic equation that controls the interface potential in an ideal MOS device under different external biases. For an in-depth understanding of this concept, the reader is referred to some of the references, where the concept of energy band theory and semiconductor charge statistics are discussed [10]. For the purposes of this chapter, let's define Q_s as the concentration of carriers at the surface of the oxide–semiconductor interface (in units of Coulombs/cm^2), and let's define ψ_s as the surface potential at the same oxide–semiconductor interface (in units of Volts).

Let's assume that the semiconductor material in this example is P-type and that an external applied voltage V_G is applied between the metal region and the semiconductor region. Figure 18.6 illustrates this MOS interface for a p-type semiconductor under two distinct conditions: *Inversion* and *Accumulation*. When an external applied voltage V_G (positive or negative) occurs across the MOS stack, the free carriers in the lattice are either attracted or repelled away from the oxide–semiconductor surface, and the surface potential ψ_s at the oxide–semiconductor interface adjusts itself according to Poisson's charge equation and Boltzmann carrier statistics equation. The exact solution of the system of equations, which governs the relationship between Q_s and ψ_s, is beyond the scope of this reference book. However, much can be learned from an investigation of the graphical solution of the surface charge equation, as depicted in Figure 18.7.

$$\text{ACCUMULATION}: \psi_s < 0(V_G < 0)$$

In this particular mode, electrons are repelled away from the semiconductor–oxide interface region, causing a condition of ACCUMULATION of holes. The accumulated charge in this case (P-type material) is composed of *holes* and the charge is positive. An equal charge exists (except in negative sign) at the metal–oxide interface.

$$\text{DEPLETION}: 2\psi_B > \psi_s > 0(V_G > 0)$$

A positive voltage on the gate causes electrons to be attracted to the semiconductor–oxide interface. These attracted negative charge carriers initially counterbalance the existing positive charge due

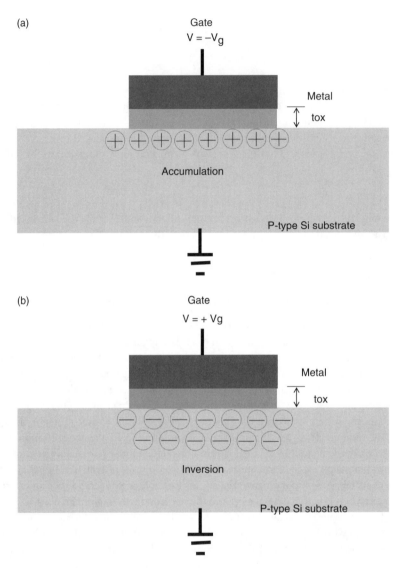

FIGURE 18.6 The basic MOS stack under (a) negative and (b) positive gate bias. A P-type semiconductor substrate structure is depicted.

to the substrate doping, causing what is referred to as a depletion region in the semiconductor near the oxide-semiconductor interface. When V_G is increased sufficiently to cause the surface potential to become positive, as depicted in Figure 18.7, we refer to this situation as causing a "depletion of holes." The quantity ψ_B used in the condition above is a reference potential used commonly in semiconductor band theory which can be approximated in most practical cases by the equation:

$$\psi_B = \frac{kT}{q} \ln\left(\frac{N_A}{n_i}\right) \tag{18.4}$$

for P-type semiconductors, where n_i is the intrinsic carrier concentration of silicon at a temperature T; n_i is the intrinsic carrier concentration of silicon and is approximately equal to 1.45×10^{10} cm^{-3} at room temperature. ψ_B is therefore logarithmically dependent on the amount of impurities added to dope the semiconductor region. Notice that as depicted in Figure 18.7, the amount of negative surface charge Q_s

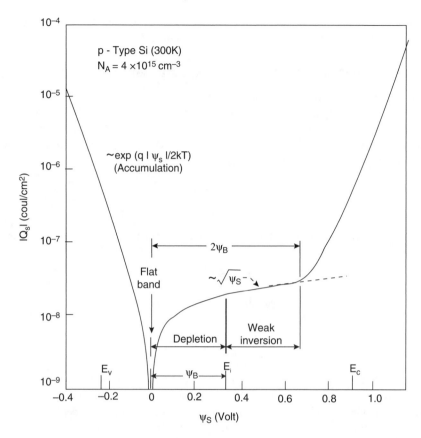

FIGURE 18.7 The Interface Surface Charge Q_S versus Surface Potential for a MOS stack at $T = 300$ K. The semiconductor is p-type, doped with NA $= 4 \times 10^{16}$ cm^{-3}. (From S.M. Sze, *Physics of Semiconductor Devices*, John Wiley & Sons, 1981.)

during the depletion regime ($\psi_s < 2\psi_B$) is only weakly dependent on the surface potential ψ_s. During this regime, one often invokes the "depletion approximation," which assumes that the electrons attracted to the semiconductor/oxide region have countered the doping charge in the semiconductor region creating a region depleted of any carriers to a depth of W_D from the interface with a uniform depletion charge of $Q_D = -qN_A$ where q is the electron charge ($q = 1.6 \times 10^{-19}$ C).

$$\text{INVERSION: } \psi_s > 2\psi_B (V_G > V_{TH})$$

As V_G is further increased beyond a certain threshold voltage V_{TH} (we'll define this quantity in a later section), the surface potential ψ_s eventually increases to a value greater than $2\psi_B$; as shown in Figure 18.7, this causes an extremely fast increase in the concentration of negative charge (electrons) Q_s at the semiconductor oxide interface. This condition is commonly referred as *inversion*. Further increases in gate voltage in the inversion regime cause the surface charge Q_S to increase at a very fast rate, as indicated in Figure 18.7.

The onset of the inversion regime for the MOS stack is roughly equivalent to the situation where the channel MOSFET device will be "*turned on*," so it is extremely important for the operation of the device. In a MOSFET device, very highly doped regions (the n$^+$ region in the case of an n-channel MOSFET (NFET) of Figure 18.3) are implanted immediately adjacent to the gate region. These implanted regions form the source and drain contacts regions of the MOSFET, which causes a drain-to-source electrical current to flow in the surface inversion channel with an externally applied drain-to-source bias V_{DS}. In the next sections, we examine the electrical characteristics of the MOSFET device in this active condition.

18.4.3 The Capacitance–Voltage Characteristics of the MOS Stack

Experimentally, the transition from accumulation to depletion and inversion of a MOS stack can be most easily observed by monitoring the capacitance–voltage characteristics between the gate and the body of a simple MOS diode structure. The external gate capacitance measured is therefore the series combination of the oxide capacitance C_{OX} and the capacitance of the interface region between the oxide and the semiconductor region C_S. This capacitance can be expressed as

$$C_G = \frac{C_{OX} \bullet C_S}{C_{OX} + C_S} \tag{18.5}$$

Where the oxide capacitance C_{OX} and the semiconductor interface capacitance C_S are defined by the relationships:

$$C_{OX} = \frac{\varepsilon \varepsilon_{OX} A}{t_{OX}} \tag{18.6}$$

And C_S is defined as the derivative of the interface charge equation:

$$C_S (V_G) = \frac{\partial Q_S (V_G)}{\partial \psi_S (V_G)} \tag{18.7}$$

Where ε is the dielectric constant of silicon and ε_o is permittivity constant in vacuum. Figure 18.8 depicts the theoretical gate-to-body capacitance C_G of a MOS stack as a function of the gate voltage V_G. As illustrated in the figure, the C–V characteristics of the MOS diode clearly depict the conditions of accumulation, depletion and inversion of the interface charge Q_S, and are therefore commonly employed in the characterization and modeling of the semiconductor MOS device.

    It is also necessary to mention that in the case of an N-type semiconductor (as in a p-channel MOSFET), a very similar situation occurs regarding the formation of a channel at the interface, except the polarity of the inversion charge and sign of gate voltage V_G are opposites from the P-type case. In this case,

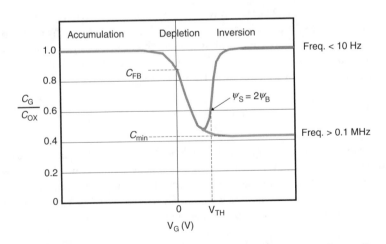

FIGURE 18.8 Gate Capacitance C_G versus External Gate Bias V_G for a p-type MOS stack. (From A.S. Grove, E.H. Snow, B.E. Deal, and C.T. Sah, Investigation of thermally oxidized silicon surfaces using metal-oxide-semiconductors, *Solid State Electron*, Vol. 8, p. 145, 1964. At high C–V characterization frequencies, the p-type semiconductor is not capable of generating electrons fast enough for the inversion channel, and thus the measured capacitance does not rise about C_{mim}.)

the threshold voltage will be negative ($V_{TH} < 0$), and holes (as opposed to electrons in the P-type semiconductor case) cause the inversion channel at the interface of the semiconductor oxide interface.

18.5 The 1-Dimensional Long-Channel MOSFET I–V Model

The previous section dealt with the characteristics of the surface charge in a MOS stack under an applied gate bias. In this section, we will examine an idealized simple case of a long-channel MOSFET to gain insight into the current–voltage relationship of the device. Whereas this simple model has many deficiencies when applied to modern deep-submicron gate MOSFET technologies, it still serves as a very useful stepping stone in the understanding of the modern MOSFET structure.

Consider again the MOSFET structure depicted in Figure 18.6. As shown in the previous section, under a sufficiently large gate voltage V_G applied between the gate and the body B contact, an inversion channel of negative charge (electrons) appears at the interface of the semiconductor oxide interface. Under an external applied bias V_{DS}, this channel provides a conduction path for a current I_{DS} composed of electrons injected from the source and collected by the drain of the device. A detailed derivation of the current–voltage characteristics of the device is readily found in many dedicated textbooks on MOSFET technology [5–8]. In this chapter, we'll cover the main concepts of this simple model, which are necessary for the understanding of more complex short-channel effects that occur in the most commonly used MOSFET devices.

18.5.1 The MOSFET Threshold Voltage V_{TH}

The simple analysis of surface charge Q_S versus applied gate voltage of the simple MOS stack provides the basis for the definition of a threshold voltage V_{TH} in a MOSFET device. It is necessary to point out to the reader that the concept of a threshold voltage V_{TH} is strictly *nonphysical*: an analysis of Figure 18.7 shows that there are no abrupt and discrete points in the charge function Q_S versus gate voltage, merely a region of fast increase in surface charge corresponding to a condition where the gate voltage V_G causes a surface charge ψ_S (as defined in the previous section) that is roughly equivalent to twice the quantity ψ_B. Based on this condition, the threshold voltage V_{TH} is commonly defined in the MOSFET literature [11] by the relationship:

$$V_{TH} = V_{TO} + \gamma \left(\sqrt{|-2\phi_B + V_{SB}|} - \sqrt{2\phi_B} \right) \tag{18.8}$$

where V_{TO} is the threshold voltage of the MOSFET device when the source and the body potential are shorted together ($V_{SB} = 0$). Under our simple model, the quantity V_{TO} is a function of the characteristics of the oxide and the semiconductor region: the oxide thickness and doping in the semiconductor region, the work-function difference between the oxide, semiconductor and the metal regions, and it is also a function of any stored charges that may exist in the oxide region and the oxide-interface regions due to the characteristics of the process utilized. The quantity γ is called the body-effect coefficient, and it is a relationship added to the current equation to account for the effect of the source-body voltage on the threshold voltage V_{TH} of the MOSFET device; it is defined for an uniformly doped channel by the relationship:

$$\gamma = \frac{1}{C_{OX}} \sqrt{2q\varepsilon_{Si}N_A} \tag{18.9}$$

For modern submicron MOSFET devices, the actual threshold is of course a much more complex function than described by this simple model, being dependent on the actual characteristics of the materials used in the polysilicon gate region, the implanted species in the channel, and also a function of the actual gate length of the device. The threshold voltage can also vary across the channel of the MOSFET transistor due to the complexity of the doping profiles under the gate region. The more complex models provided later in this chapter include additional modeling parameters to account for these more complex characteristics of the V_{TH}. Nonetheless, this simple relationship of Equation 18.9 does capture the first order dependencies and relationships of this important MOSFET parameter.

18.6 The Long-Channel MOSFET Current–Voltage Model

The simple long-channel MOSFET model presented in this section serves to establish a foundation for the reader, which will be fundamental in order to understand later sections of this chapter where more advance electrical models are used to describe modern deep-submicron MOSFET devices in which short-channel effects become very important. It is also instructive to note that in most analog designs, there will likely be a combination of long- and narrow-channel devices, so the long-channel MOSFET theory provides more than just a good foundation to the reader.

Let us again consider the example of the NFET transistor depicted in Figure 18.6, biased in the active region ($V_{GS} > V_{TH}$, and $V_{DS} > 0$). When V_{GS} exceeds the threshold voltage V_{TH} of the long-channel MOSFET device, an inversion charge region occurs at the interface between the semiconductor and the oxide region, allowing electrons (and hence an electrical current) to flow between the source and the drain terminals, creating a drain-to-source current I_{DS}. Depending on the value of the drain-to-source voltage V_{DS}, the long-channel approximation gives us two distinct regions of operation:

Linear Region: At low V_{DS} values ($V_{DS} < V_{GS} - V_{TH}$), the current I_{DS} can be derived to yield the following equation:

$$I_{DS} = k \frac{W}{L_G} \left[(V_{GS} - V_{TH}) - \frac{1}{2} V_{DS}^2 \right] \tag{18.10}$$

where the process transconductance parameter k is defined by:

$$k = \mu_e \frac{\varepsilon_O \varepsilon_{OX}}{t_{OX}} \tag{18.11}$$

where μ_e is the mobility of electrons in the inverted surface channel of the device (typically defined in cm^2/V-s), and ε_O is the permittivity in vacuum (8.854×10^{-14} F/cm) and ε_{OX} is the dielectric constant of the oxide (typically equal to 3.9). It is important to note that the mobility of electrons in the inversion channel is significantly lower than the mobility of electrons measured in a bulk silicon crystal; typical values of μ_e for an MOS inversion channel are in the range of 500–700 cm^2/V-s, typically 2X to 3X lower than its respective bulk value.

Saturation Region: If V_{DS} is further increase beyond ($V_{GS} - V_{TH}$), the current–voltage characteristics are said to "saturate," yielding a region of nearly flat I_{DS} characteristics versus V_{DS}. Physically, the model predicts a region near the gate drain region where the electric field created by the different potentials causes the channel to be "*pinched-off*," meaning that the concentration of electrons has dropped to a value equal to zero in that region. (The reader should keep in mind that this mechanism for current saturation is typically *not applicable* to MOSFET devices typically used modern design, where actually velocity saturation of carriers is responsible for the saturation characteristics of I_{DS})

The long-channel I_{DS} current in the regime of $V_{DS} > V_{GS} - V_{TH}$ is expressed under these simplifications by the following equation:

$$I_{DS} = \frac{k}{2} \frac{W}{L_G} (V_{GS} - V_{TH})^2 (1 + \lambda V_{DS}) \tag{18.12}$$

where the nonphysical term $(1 + \lambda V_{DS})$ was arbitrarily added to the current equation to account for the fact that the current I_{DS} still has a weak dependence on the drain voltage V_{DS} even in the saturation region. The parameter λ is called as the drain- or channel-length modulation parameter and its unit is in V^{-1}.

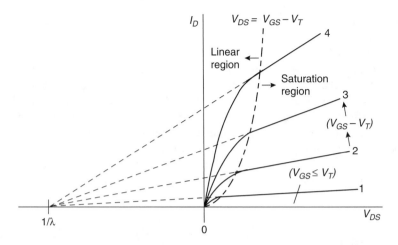

FIGURE 18.9 IDS versus VDS characteristics of a MOSFET device using the Long-Channel model. This model is more commonly referred to as the MOSFET Level 1 model.

Figure 18.9 illustrates the current I_{DS} characteristics predicted by the simple long-channel model, depicting the transition between the linear and the saturation regions of the model. The reader should also realize by now that the simple model does not contain any upper bounds on the current I_{DS} as a function of either V_{DS} or V_{GS}; more advanced models presented in a later section of this chapter will in fact correct this deficiency.

An important quantity in RF design is the transconductance parameter G_M, which is defined by the relationship:

$$G_M = \frac{\partial I_{DS}}{\partial V_{GS}} \tag{18.13}$$

This important parameter for the RF engineer essentially corresponds to the current gain of the MOSFET device as a function of gate voltage V_G for a given drain bias V_{DS}. For the simple long-channel model presented, the transconductance G_M in the saturation region ($V_{DS} > V_{GS} - V_{TH}$) becomes:

$$G_M = k\frac{W}{L}(V_{GS} - V_{TH}) \tag{18.14}$$

Equation 1.14 predicts therefore a perfectly linear dependence of G_M on the value of V_{GS} in the case of the long-channel MOSFET; as show in later sections, this relationship does not occur in modern MOSFET devices.

18.6.1 The Subthreshold Region

It is often necessary to model the region where the gate voltage is below the threshold voltage V_{TH} ($V_{GS} < V_{TH}$) of the MOSFET device, rather than just assume that the current is equal to zero below threshold. The so-called subthreshold region is of particular importance in the design of low power RF and digital circuits which operate at or near the threshold voltage of the device. As we learned during the analysis of the MOS stack (see Figure 18.7), when V_{GS} is just below V_{TH}, the channel is said to be in the depletion/weak inversion regime. It turns out that, in this regime, the device operates very similarly to a bipolar device, with the drain region emulating a collector region, the weakly-inverted channel operating as a base and the source region serving as the collector region of the device.

A detailed derivation of the resulting diffusion current which accounts for I_{DS} during the subthreshold regime is beyond the scope of this chapter but can be readily found in the literature [12]. Because of the diffusion nature of the current in this regime, the current can be shown to have an exponential behavior with a $(1 - e^{-\beta V}\text{DS})$ dependence on V_{DS}. A later section dealing with a more advanced electrical model will present a more complete treatment of the current in this regime. A very important characteristic of the MOSFET device is the so-called subthreshold-swing parameter S, which is a measure of the steepness of the turn-off behavior of the MOSFET as a function of the drain voltage, and hence a measure of ideality when used as a current switch. The parameter S is defined in mV/decade, and it should be thought as the required change in drain voltage V_{DS} in the subthreshold regime necessary to reduce the current I_{DS} by one order of magnitude. The subthreshold swing parameter S is a function of the particular MOSFET technology being employed as well as the geometry of the particular device, and it is typically of the order of 80–100 mV/decade for modern MOSFET technologies used for RF design. Figure 18.10 shows the measured subthreshold characteristics of a typical modern MOSFET device for different drain-to-source voltages V_{DS} .

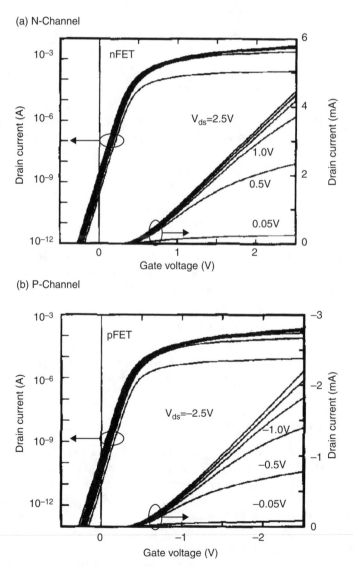

FIGURE 18.10   Measured subthreshold current characteristics of an n-channel MOSFET. $L = 2.5\ \mu$m, $W = 10\ \mu$m, $T = 300$ K. [20]

18.6.2 Short-Channel Effects in Modern MOSFETs

As pointed out in the previous sections of this chapter, the electrical models presented for threshold voltage V_{TH} and current–voltage relationships were derived for the case of a long one-dimensional channel approximation of a MOSFET device. Whereas the previous model presents some very useful first order relationships that the MOSFET designer needs to be aware of, there are several effects which occur on modern submicron devices. We typically refer to these effects as "short-channel effects," in the sense that they are responsible for the departure from the idealized long-channel device. The more advanced models presented in a later section of this chapter deal with many of these effects and how they scale with device geometry, so designers generally do not need to be directly concerned with regard to how these effects need to be incorporated in their design methodology. Some of the most important effects regarding modern submicron MOSFET devices are listed in the sections below:

18.6.2.1 Mobility Degradation

When charge carriers are subjected to a small applied external electric field, the charge carriers will move with an average drift velocity v_D. It is this movement of charge of course that gives rise to the electrical current I_{DS} in the MOSFET device. In this low-electric field regime, the relationship between carrier drift velocity v_D and electric field E is given by

$$\vec{v}_D = \mu_e \times \vec{E} = \mu_e \times \frac{\partial V(x)}{\partial x} \tag{18.15}$$

It is this simple relationship that yields the simple model presented in the previous section. In reality, however, the mobility of charge carriers in an inversion channel is a rather strong function of the applied electric field. This effect is very important in modeling the degradation of drain current and transconductance at high fields, which occur when large gate and drain voltages are applied. Advanced current–voltage models such as the one presented in a later section of this chapter contain more elaborate mobility-field relationships which capture this relationship.

18.6.2.2 Velocity Saturation

The transition between the linear and saturation regime in modern submicron devices is determined in large part due to the saturation velocity of charge carriers in an applied electric field. This effect is ignored in the long-channel model presented earlier. As the carriers in an inversion channel are further accelerated by the effective electric field E, their average drift velocity actually begins to saturate. Further increases in electric field after this regime cannot enhance the velocity beyond this saturation velocity v_{SAT}. A relationship commonly used in MOSFET modeling that captures this particular transition between carrier velocity and applied electric field is given in the following equation:

For $E < E_{SAT}$

$$v = \frac{\mu(E) \cdot E}{1 + \frac{E}{E_{SAT}}} \tag{18.16}$$

And for $E > E_{SAT}$

$$v = v_{SAT} \tag{18.17}$$

where v_{SAT} is typically around 1×10^7 cm/s; rather than being measured directly, the saturation velocity v_{SAT} is typically a fitting parameter which the modeling engineer will manipulate to provide better overall agreement with an electrical model.

Once the velocity of the charge carriers in the inversion region reach v_{SAT}, the drain current I_{DS} of the MOSFET device becomes saturated as well. Velocity saturation is the dominant mechanism responsible for the saturation characteristics for modern submicron MOSFET devices. In the velocity-saturation regime, the current I_{DS} of the MOSFET can be shown in a simple model to be

$$I_{DS} = v_{SAT} \times W \times C_{OX} \times (V_{GS} - V_{TH} - VDS_{SAT}) \tag{18.18}$$

where VDS_{SAT} is the value of the drain voltage VDS that causes the electric field near the gate drain edge to be large enough to cause velocity saturation of the charge carriers in the inversion channel. It is very important to note that under this saturation mechanism, the drain current I_{DS} of the MOSFET device has a *linear* dependence on the value of $(V_{GS} - V_{TH})$, rather than a quadratic dependence as predicted by the simple long-channel model of the previous section (see Equation 18.12). Another important characteristic of the MOSFET in the velocity-saturation regime is that its transconductance G_M is now *independent* of the actual value of $(V_{GS} - V_{TH})$, whereas the long-channel model of Equation (18.14) predicted a linear dependence.

18.6.2.3 Hot-Carrier Effects

The simple models presented earlier do not contain any upper bounds in the gate and drain operating voltages of the MOSFET device. Of course, such is not the case in a real device and indeed upper limits on the gate and drain biases must be established for a given technology in order to guarantee a reliable operating region for the device. For modern MOSFET technology, hot-carrier effects are extremely important and essentially determine the safe operating bias region for the MOSFET [13,14]. If the devices are allowed to operate beyond these safe operating limits, significant reliability degradation will likely result.

We apply the term "hot-carrier" to describe a charge carrier such as an electron or a hole that is not in thermal equilibrium with the lattice [15]. In the specific case of the MOSFET device, hot-carriers are created in the regions of the semiconductor device where the carriers are subjected to very high-electric fields, which occur near the gate drain edge region of the MOSFET device under an applied drain bias V_D. A very high-electric field in the gate drain region creates the generation of electron–hole pairs by impact ionization; electrons generated by this mechanism may have enough energy to actually penetrate the gate-oxide region, where they typically cause permanent degradation to the MOSFET device [16]. Hot-carriers create severe reliability issues because once these hot-carriers are created by the high-electric fields, they may have enough energy to migrate into the oxide region of the MOS stack, creating permanent changes in threshold voltages and current drive characteristics of the MOSFET. Hot-carriers provide a cumulative effect: the infusion of additional hot-carriers continue to degrade the characteristics of the transistor. MOSFET technologists generally define the maximum operating voltage of the MOSFET to allow for a 10-year operating lifetime for the device with no more than a 10% reduction in operating current due to hot-carrier effects.

There are well-established methods to characterize the effects of hot-carrier degradation on MOSFET device reliability [17], generally relying on the characterization of substrate and device currents at stressing biases under accelerated temperature conditions.

18.6.2.4 Drain-Induced Barrier Lowering

Drain-induced barrier lowering (DIBL) is particularly important in minimum short-channel devices used commonly in RFCMOS applications. Just as in hot-carrier phenomena, DIBL is caused by large voltages which exist at the drain terminal. The large electric field caused by this potential actually creates a condition where the surface potential of the channel is significantly degraded by attracting electrons to the interface region. The end effect of this condition is a reduction of the actual threshold voltage of the device with a large drain voltage VD, creating thus an increase in the drain current of the device in the region where DIBL exists.

DIBL effects can be minimized to a great degree by engineering of lightly doped drain regions next to the channel of the device, which dramatically reduce the electric fields at the gate drain edge region. DIBL effects are incorporated in the large signal models commonly employed for modern CMOS technologies.

18.6.2.5 Other Short-Channel Effects

The MOSFET theory which was presented earlier essentially was a one-dimensional model solution of the Poisson equation of the channel region under a potential applied between the drain and the source terminals. In reality, the electrical characteristics of short-channel modern MOSFET devices are greatly affected by both lateral and longitudinal potentials, and in fact, a more rigorous solution of the system requires that two-dimensional physics be applied.

Another major effect consists of the fact that so far constant doping has been assumed in the different regions of the MOSFET device. Modern devices are of course obtained with ion-implantation of N- and P-type species, and the actual profiles obtained create a situation of nonuniform and nonabrupt channel doping in the devices.

As with other short-channel effects, all of these effects should be adequately captured by the large signal MOSFET modeling strategy developed for a particular RFCMOS technology.

18.7 MOSFET Fabrication Section

18.7.1 Introduction to MOSFET Fabrication

This section is intended to give the reader an overview of modern RFCMOS processes and the common fabrication process flow employed in modern fabrication facilities (commonly referred to as "fabs").

18.7.2 Example CMOS Device Cross Section

A cross-section showing N-channel and P-channel MOSFETs in a typical CMOS process is shown in Figure 18.11. This figure shows the final devices after completion of the process. Sections 18.7.4 and 18.7.5 presents a series of cross-sections, showing the major steps involved in fabrication of the devices. The N-channel MOSFET (NFET) is most commonly formed in a P-type doped silicon region (P-well). The P-well serves as the body of the NFET. The gate dielectric is formed on the surface of the silicon, over the P-well. The gate electrode is typically formed from polycrystalline-silicon (polysilicon), which is heavily doped N-type. The source and drain contact regions of the NFET consist of heavily doped N-type silicon on either side of the gate electrode. The P-channel MOSFET (PFET) is formed in an N-type doped silicon region (N-well). The N-well serves as the body of the PFET. The gate dielectric is formed on the surface of the silicon, over the N-well. The gate electrode is typically formed from doped P-type polysilicon. The source and drain contact regions of the PFET consist of heavily doped P-type silicon on either side of the gate electrode.

18.7.3 Example CMOS Layout

An important step in the design process of a new integrated circuit is physical design, or layout. Features are drawn using CAD software to represent the layers to be transferred to the silicon wafer during CMOS processing. An example CMOS layout with NFET and PFET devices is shown in Figure 18.12. The typical

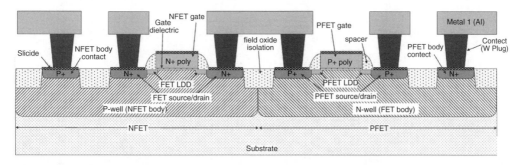

FIGURE 18.11 Cross-section of NFET and PFET devices in a CMOS process.

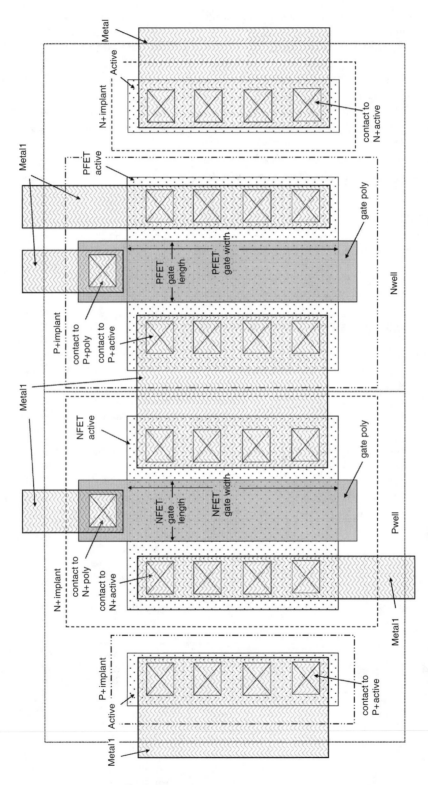

FIGURE 18.12 Layout of NFET and PFET devices in a CMOS process.

layers drawn during circuit layout are

Active – This layer designates areas where active devices are formed or contacts to the silicon wafer are made.

N-well and P-well Implant – These layers designate areas which will receive N-well and P-well implants. These implants create the body of a MOSFET.

Poly – This layer designates areas where the gate polysilicon will remain on the wafer. When poly and active features overlap, the channel region of a MOSFET is created.

N-LDD and P-LDD Implant – These layers designate areas which will receive N-type and P-type lightly doped drain (LDD) implants.

N+ and P+ Implant – These layers designate areas which will receive high-dose N-type and P-type implants used for MOSFET source/drain contact areas.

Contact – This layer designates where contacts to silicon or gate polysilicon will be made.

Metal1, Metal2, Metal3, and so forth – These layers designate where metal features will remain on the wafer. The number of metal layers in a CMOS process may range from 1 to 10 (or more).

Via1, Via2, Via3, and so forth – These layers designate where vias will be created to connect between different metal layers. Via1 will provide an electrical connection between Metal1 and Metal2. Via2 will provide an electrical connection between Metal2 and Metal3. The number of via layers will depend on the number of metal layers in the process.

18.7.4   Modern RFCMOS Front-End Process Flow

The example process flow in this chapter describes a modern CMOS process containing submicron MOSFETs, with both low-voltage and input/output (I/O) devices integrated into the same process. The process is planarized, containing multiple metal levels. This is the type of CMOS process which is currently used for microprocessors, analog integrated circuits, and also becoming widely used for radio frequency integrated circuits (RFICs). In this section, the front-end portion of the process flow is described, where MOSFETs and other active devices are formed.

18.7.4.1   Field Oxide Formation

The first steps in a modern CMOS process typically involve formation of the field oxide regions, which separate active device regions as shown in Figure 1.11. The active device regions are where MOSFETs or other active devices will be built, and where contacts to the silicon wafer are made. The field oxide regions electrically isolate the doped regions at the surface of the wafer, such as the N+ and P+ contact regions shown in Figure 18.11. The field oxide is typically thick enough to block the implant of the doped regions, establishing a well controlled separation of these regions. There are several methods for forming the field oxide. Most modern CMOS processes use shallow trench isolation (STI) for the field oxide.

The STI process sequence is shown in Figure 18.13a through Figure 18.13i. The first step in the STI process sequence is to form SiO_2 and Si_3N_4 layers on the silicon surface, typically through thermal oxidation and chemical vapor deposition (CVD). These layers serve as the hard mask during the shallow trench etch. Next, the Active photolithography step is done, leaving photoresist covering the active device areas on the wafer. The STI SiO_2/Si_3N_4 hard mask is patterned using plasma etching, followed by removal of the photoresist. Next, a trench is etched into the silicon wafer using plasma etching. After trench etch, a thermal oxidation is done to form a high-quality SiO_2/silicon interface at the surface of the shallow trench. Next, the trench is filled with SiO_2 through a CVD process. After the shallow trench is filled, the thick SiO_2 layer covering the active device areas must be removed. Chemical mechanical polishing (CMP) is used to planarize the surface of the wafer. A Reverse Active photolithography step may be done prior to CMP to allow more effective planarization of the wafer. After Reverse Active photolithography, the photoresist will be removed over large active areas. The SiO_2 over these active areas is then etched away using a plasma etch, leaving only the sidewalls of SiO_2 in most areas of the wafer. For processes using a

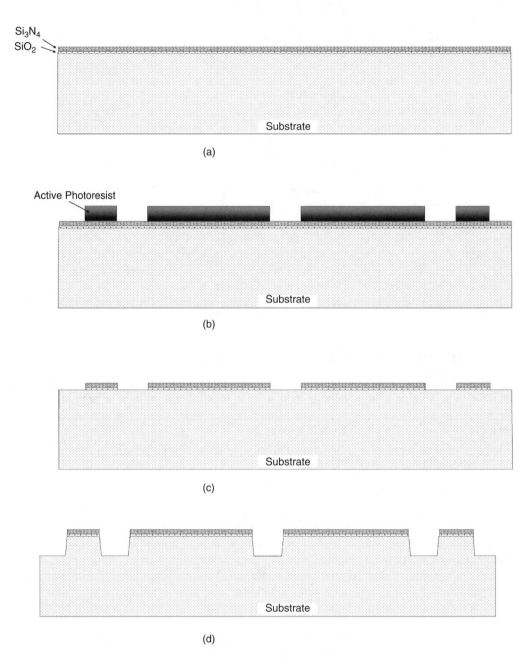

FIGURE 18.13   Shallow trench isolation (STI) CMOS process flow.

Reverse Active step, the STI CMP process mainly serves to remove the remaining sidewalls of SiO_2. After STI CMP, the SiO_2/Si_3N_4 hard mask is removed, completing the STI process sequence.

18.7.4.2 Well Formation

Following field oxide formation, the next steps in the CMOS process form the N-well and P-well. The well serves as the body of the MOSFET and determines the MOSFET channel characteristics. The well formation sequence is show in Figure 18.13j through Figure 18.13m. The first step in well formation is to form a thin SiO_2 layer on the surface of the wafer, which serves as the screen layer during the well implant.

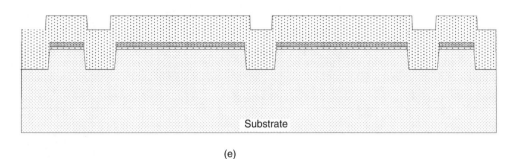

(e)

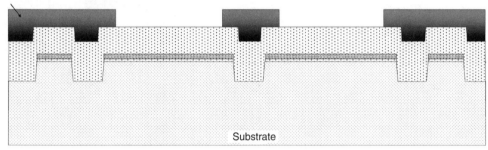

(f)

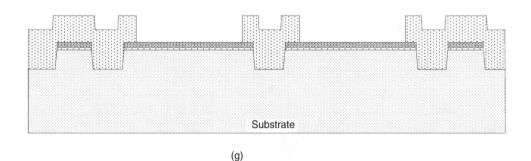

(g)

FIGURE 18.13 Continued.

This SiO_2 layer, often called the sacrificial oxide layer, is usually formed through thermal oxidation or CVD. Next, the N-well photolithography step is done, opening areas in the photoresist where the N-well will be formed. The N-well is then formed through one or multiple ion implants of n-type atoms (generally As or P). This creates a well with relatively uniform doping concentration from the surface of the wafer to a depth of around 0.5–1 μm. The doping concentration at the surface of the N-well controls the threshold voltage of the PFET. After removal of the N-well photoresist, the P-well photolithography step is done and the P-well implant (generally B or BF_2) follows. The doping concentration at the surface of the P-well controls the threshold voltage of the NFET. For technologies with both low-voltage and high-votlage MOSFETs, separate well formation steps may be used for the four different MOSFETs. In some cases, some of the well-formation steps may be able to be combined to reduce the number of process steps. It is normally not important whether the N-well or P-well is done first.

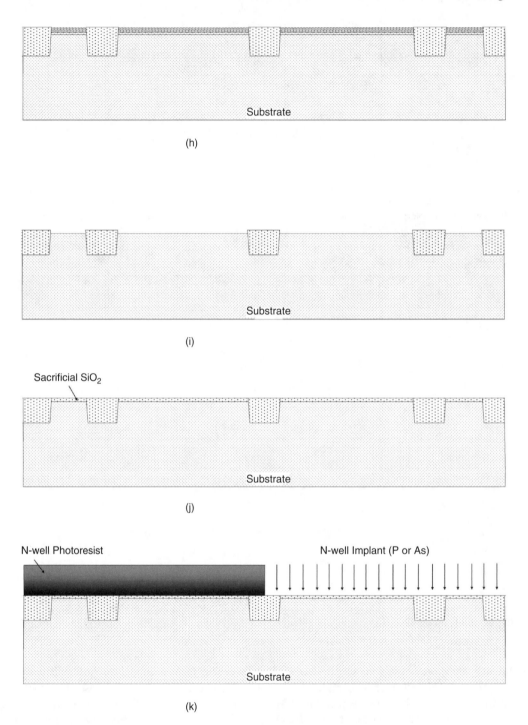

FIGURE 18.13 Continued.

18.7.4.3 Gate Formation

The next part of the CMOS process involves formation of the MOSFET gate, as shown in Figure 18.13n through Figure 18.13r. First, the sacrificial oxide is removed using a wet etch. Next, the gate dielectric (normally SiO_2) is formed. The thickness of the gate oxide depends on the voltage rating of the MOSFET.

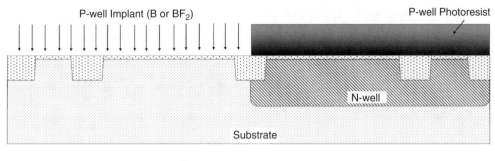

(l)

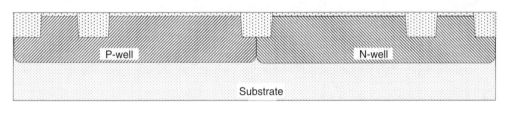

(m)

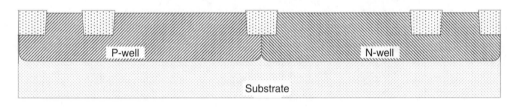

(n)

FIGURE 18.13 Continued.

For processes with both low-voltage and high-voltage MOSFETs, two different gate oxide thicknesses are normally used in the process. In this case, the thick-gate oxide is grown first, followed by a photolithography step. Photoresist remains in the thick-gate oxide regions to protect these areas, while the thick-gate oxide is removed by wet etching in the remaining areas. After removing the photoresist, the thin-gate oxide is grown.

Following the gate dielectric formation, the gate electrode layer is deposited. For most CMOS processes, polysilicon is used as the gate electrode. The polysilicon may be doped through ion implant at this point in the process, or it may be left undoped until later in the process. In NFET areas the gate polysilicon will normally be doped n-type. In PFET areas the gate polysilicon will normally be doped P-type. For some processes, the gate polysilicon may be doped N-type everywhere on the wafer. In this case, PFETs will have N-type gate polysilicon, and the MOSFET will be a buried channel device.

Next, the gate photolithography step is done to define the MOSFET gate electrodes. This is the most critical photolithography step in a CMOS process, since the gate length determines many of the electrical

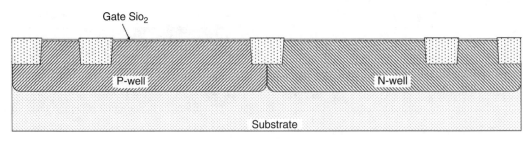

(o)

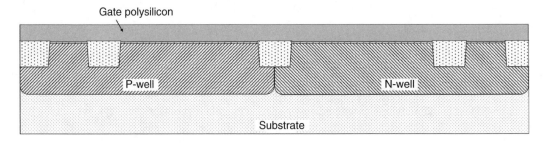

(p)

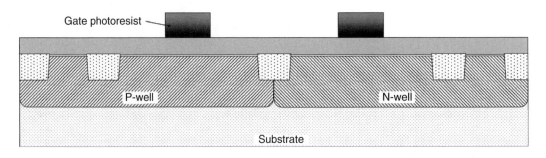

(q)

FIGURE 18.13   Continued.

properties of the device. After the gate photolithography, the gate polysilicon is patterned using plasma etching. The polysilicon must be completely etched across the entire wafer, while not breaking through the underlying gate oxide and potentially damaging the silicon surface.

18.7.4.4   Source/Drain Extension (LDD) Formation

After the gate polysilicon is defined, the next step in the process is to form the source/drain extension, which is often called the lightly doped drain (LDD) region. The sequence for LDD formation is shown in Figure 18.13s through Figure 18.13y. The purpose of the LDD is to improve the MOSFET hot-carrier reliability by reducing the doping concentration in the drain contact region near the edge of the gate electrode. Since MOSFETs in most CMOS processes are symmetrical (source and drain may be interchanged), the LDD is formed on both the source and drain side of the device. The N-type LDD photolithography step is done, and photoresist is patterned so that it is removed only in the NFET regions that will receive the N-LDD implant. After removing the N-LDD photoresist, the P-LDD photolithography step is done, and

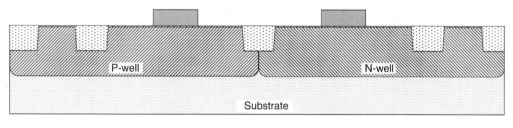

(r)

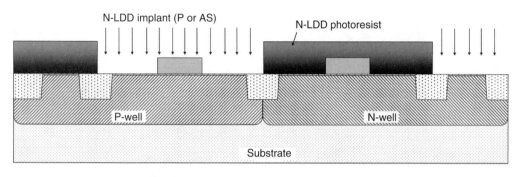

(s)

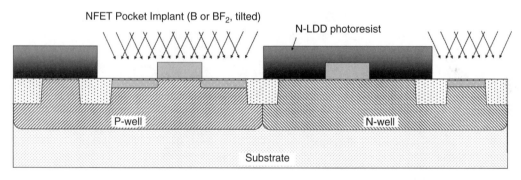

(t)

FIGURE 18.13 Continued.

photoresist is patterned so that it is removed only in the PFET regions that will receive the P-LDD implant. If there are both low-voltage and high-voltage MOSFETs in the process, there may be four separate LDD implant steps so that each MOSFET has an optimized LDD region.

In addition to the LDD implants, pockets (or halo) implants may also be done at the same photolithography steps for both NFETs and PFETs. The purpose of the pocket implant is to increase the body doping concentration in the channel region of deep-submicron MOSFETs to minimize short-channel effects.

After LDD and pocket implants, the gate spacer layer is deposited. The spacer layer is generally SiO_2 or a SiO_2/Si_3N_4 stack. The thickness of the spacer layer is around 500–2000A. The bottom layer of the spacer is typically SiO_2, to minimize the interface trap density at the silicon surface. After deposition, the spacer layer is etched back as shown in Figure 18.13y. The purpose of the gate spacer is to block the source/drain

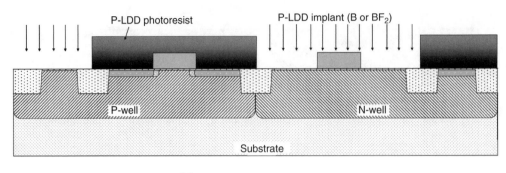

(u)

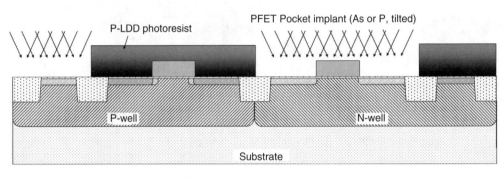

(v)

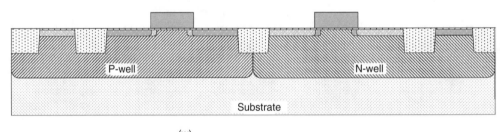

(w)

FIGURE 18.13   Continued.

implant in the LDD region of the MOSFET, creating a precisely controlled separation of the MOSFET channel from the heavily doped source and drain contact regions.

18.7.4.5   Source/Drain Formation

After the gate spacer is defined, the source and drain region of the MOSFET are created through ion implants, as shown in Figure 18.13z through 18.13bb. The N+ implant photolithography step is done, and photoresist is patterned so that it is removed only in the regions which will receive the N+ implant. These regions include the NFET source and drain and contacts to N-wells. The N+ implant is usually As or P. After removing the N+ implant photoresist, the P+ implant photolithography step is done to define area to receive the P+ implant. These regions include the PFET source and drain and contacts to P-wells. The P+ implant is usually B or BF$_2$. The N+ and P+ doping profiles are normally kept as shallow as possible. After both N+ and P+ implants, an implant activation anneal is done to make the source/drain dopants electrically active.

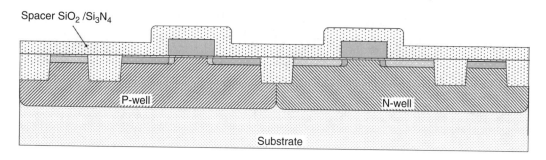

(x)

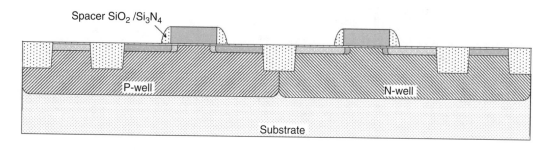

(y)

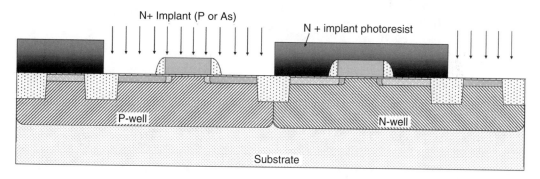

(z)

FIGURE 18.13 Continued.

18.7.4.6 Silicide Formation

After source/drain formation, the final step in the front end of the CMOS process is the formation of silicide, as shown in Figure 18.13cc through 18.13ee. The purpose of the silicide is to reduce the sheet resistance of the silicon source/drain and gate polysilicon regions, as well as reduce the resistance of contacts formed in these areas. Silicide will only form in areas of the wafer where silicon or polysilicon is exposed, therefore a self-aligned silicide (salicide) process is typically used. In the salicide process, the first step is a short wet etch on the wafer surface to remove remaining SiO_2 in active areas of the wafer. Next, the silicide metal is deposited. After the metal is annealed to react with the silicon, an etch step is used to remove the unreacted silicide metal.

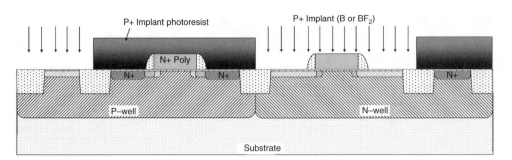

(aa)

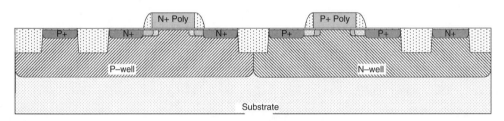

(bb)

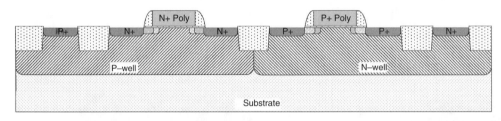

(cc)

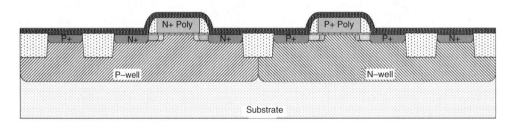

(dd)

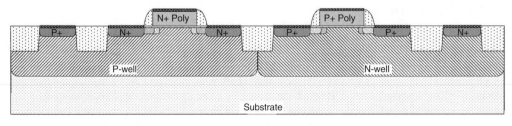

(ee)

FIGURE 18.13   Continued.

18.7.5   CMOS Back-End Process Flow

In the section, the back-end portion of a typical CMOS process flow is described, where multiple-level metallization are formed to connect MOSFETs, resistors and other devices together to form an IC. In modern CMOS processes, the wafer is planarized at every metal level to allow stacking of many metal levels.

18.7.5.1   Al Metal/W-plug Process

The Al metal/W-plug process is used in most CMOS processes from with feature sizes from 0.35 to 0.13 μm. The advantage of a W-plug process over older unplanarized processes is that many metal layers may be stacked to allow reduced die sizes for complex ICs. The process sequence for an Al metal/W-plug process is shown in Figure 18.14a through 18.4r. The first step is to deposit the dielectric layer between the silicon and first layer of metal. This first dielectric layer typically consists of one or more layers of SiO_2. After depositing the first dielectric layer, this layer is planarized using CMP. Then, the contact photolithography step is done, and photoresist is patterned so that it is removed in areas where contacts to silicon or gate polysilicon are to be made. Next, the first dielectric is etched using anisotropic plasma etching until reaching the silicon or polysilicon. After removal of the photoresist, a thin layer (100–300Å) of TiN is deposited, followed by a thicker layer of W. Next, the W layer remaining above the first dielectric is removed using CMP. After W-CMP, the first Al metal layer (Metal 1) is deposited by sputtering. The Metal 1 layer is usually a stack consisting of Al with thin barrier layers of Ti or TiN on top and bottom. The Ti or TiN barrier layers are used to improve the reliability of the Al conductor by reducing voids or hillocks that may occur in the Al during processing. Next, the Metal 1 photolithography step is done, and photoresist is patterned so that it remains in areas where the metal will be left after etching. Then, the Metal 1 layer is etched using plasma etching.

After Metal 1 is patterned, the second dielectric—called the intermetal dielectric (IMD)—is deposited. The IMD may be undoped SiO_2, or it may be doped with fluorine. Fluorine silicate glass (FSG) has a dielectric constant of 3.5 as compared to 3.9 for SiO_2. The lower dielectric constant reduces the capacitance between metal interconnects, resulting in reduced switching times and allowing higher frequency operation of the IC. After the IMD is deposited, it is planarized using CMP. Then, the Via 1 photolithography step is done, and photoresist is patterned so that it is removed in areas where vias between Metal 1 and Metal 2 are to be made. Next, the IMD is etched using anisotropic plama etching until reaching Metal 1. After removal of the photoresist, the TiN liner and W layer are deposited, as was done during the Contact process. CMP is used to remove the W layer remaining above the IMD. Next, the second Al metal layer (Metal 2) is deposited. The Metal 2 photolithography step is done, and photoresist is patterned so that it remains in areas where the metal will be left after etching. Then, the Metal 2 layer is etched.

Additional IMD layers,vias, and metal layers may be added using the same process sequence as described for forming the Via 1 and Metal 2 layers. After the final metal level is patterned, a passivation layer is deposited on the surface of the wafer. The passivation is typically two layer stack of SiO_2 and Si_3N_4. Polyimide may also be added to the passivation layer for additional protection of the wafer. After the passivation layer is deposited, an opening is etching in the passivation layer over areas of the wafer where the IC will connect to the package (called bond pads or bump pads). The wafer then receives a final anneal in forming gas (N_2 and H_2) at around 400°C. The hydrogen in the forming gas is absorbed into the wafer, and has been shown to improve the quality of Si/SiO_2 interfaces and improve transistor reliability.

For technologies beyond the 0.13 μm node, CMOS processes typically utilize an alternative metallization process from the one described in this chapter, involving dual copper damascene plating and low-K dielectric materials.

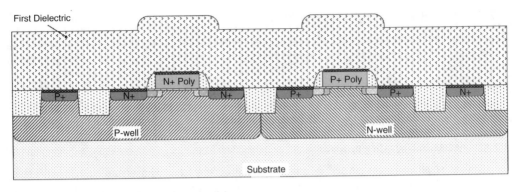

(a)

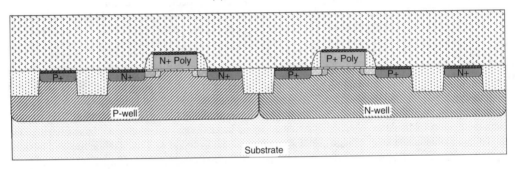

(b)

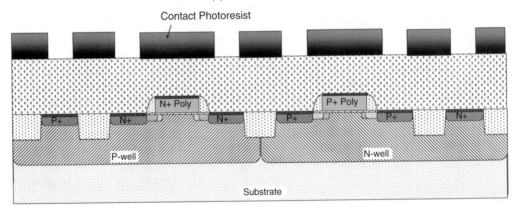

(c)

(d)

FIGURE 18.14   Modern CMOS back-end process flow.

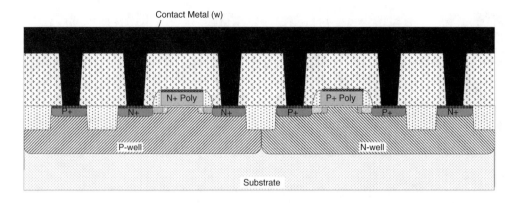

(e)

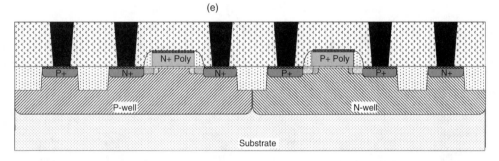

(f)

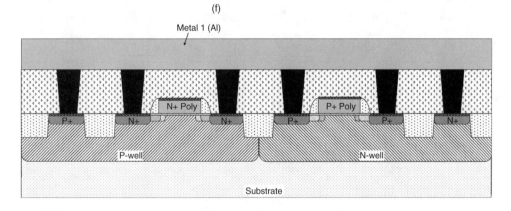

(g)

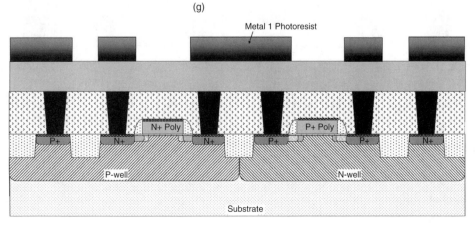

(h)

FIGURE 18.14 Continued.

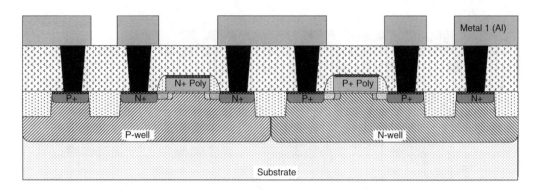

(i)

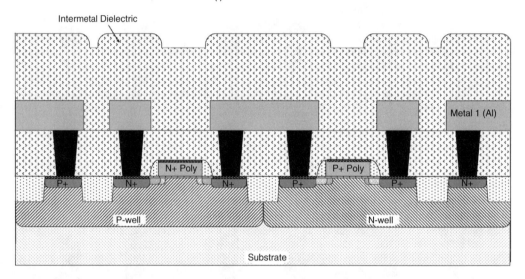

(j)

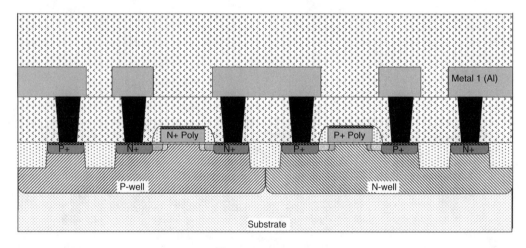

(k)

FIGURE 18.14   Continued.

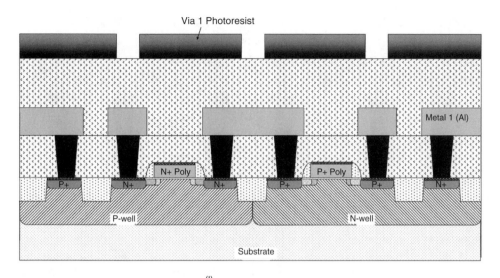

(l)

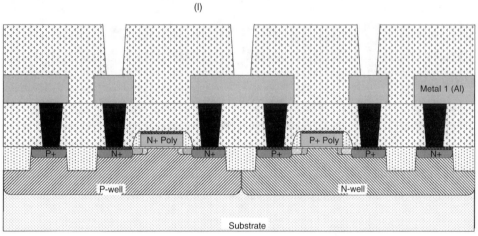

(m)

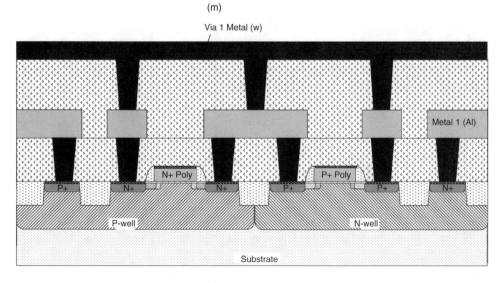

(n)

FIGURE 18.14   Continued.

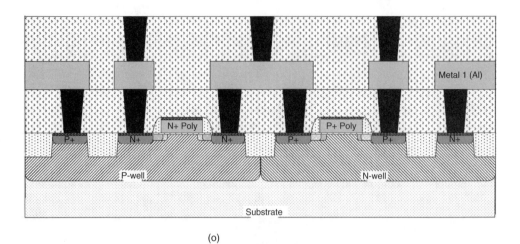

(o)

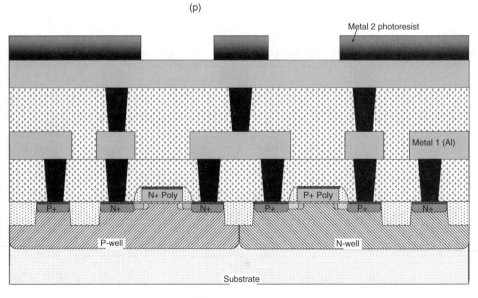

(p)

(q)

FIGURE 18.14 Continued.

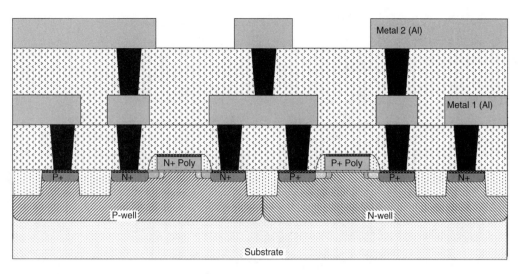

(r)

FIGURE 18.14 Continued.

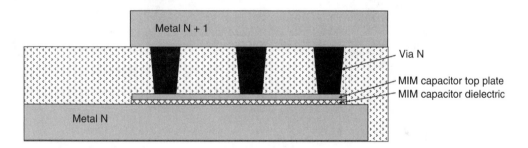

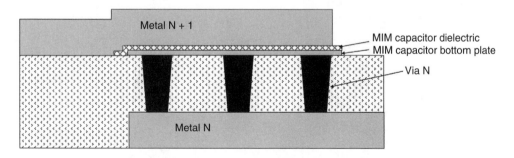

FIGURE 18.15 Example MIM capacitor structures.

18.7.6 MIM Capacitors

Metal insulator metal (MIM) capacitors may be integrated into the back end of the CMOS process to provide a high-density capacitor with excellent linearity. There are several methods for forming a MIM capacitor, and two common examples of MIM capacitor structures are shown in Figure 18.15. In both cases the MIM dielectric and an additional metal layer are introduced between two existing metal layers in the CMOS process to create the capacitor. The MIM dielectric is typically SiO_2 or Si_3N_4. Si_3N_4 has a

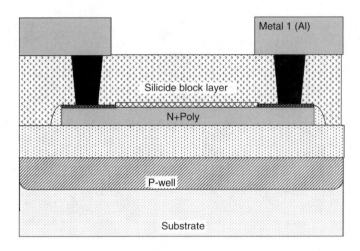

(a) Polysilicon resistor

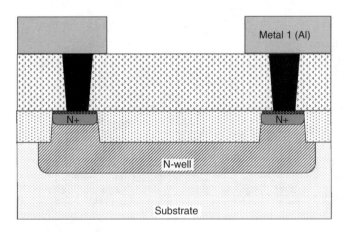

(b) N-well resistor

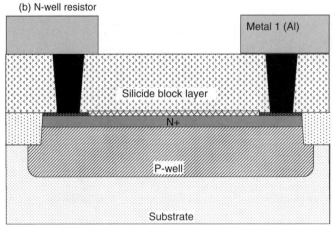

(c) N+ source/drain resistor

FIGURE 18.16   Resistors processes in RFCMOS.

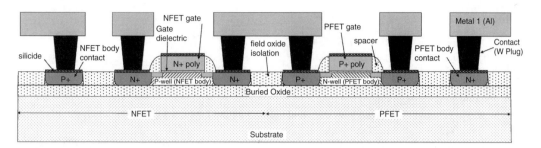

FIGURE 18.17   SOI CMOS cross section.

dielectric constant of around 7.0 as compared to 3.9 for SiO_2. This results in improved capacitor density. Tantalum pentoxide (Ta_2O_5), with a dielectric contact of around 25, may also be used for very high-density MIM capacitors. In the top example in Figure 18.15, the MIM dielectric and MIM top metal are deposited on top of an existing metal layer (Metal N) in the process. The MIM top metal and dielectric are then patterned, using photolithography and plasma etching. The vias between Metal N and Metal N + 1 are then used to connect the MIM top plate to Metal N + 1, creating a capacitor between Metal N and Metal N + 1. In the bottom example in Figure 18.15, the MIM bottom metal and dielectric are deposited on top of the vias between Metal N and Metal N+1. After the MIM bottom plate and dielectric are patterned, the Metal N + 1 layer is formed completing the capacitor.

18.7.7 Resistors

Resistors may be created in a CMOS process using the existing layers. The most common types of resistors are poly resistors, well resistors, and source/drain resistors, as shown in Figure 18.16a through 18.16c. For poly resistors and source/drain resistors, the silicide in the body of the resistor must be blocked or removed to create a reasonably high-sheet resistance. Typically, an additional layer of SiO_2 or Si_3N_4 is added, along with a photolithography step to create the ability to block the silicide formation in areas of the wafer. To create even higher sheet resistance, some of the poly gate implant may be blocked in the body of the resistor, using another added photolithography step. For well resistors, as shown in Figure 18.16b, the standard N-well or P-well layers are used as a resistor. Typically, well resistors have sheet resistance of around 500 Ω/sq to 1 kΩ/sq. Poly resistors have sheet resistance of 100 Ω/sq to 1 kΩ/sq. Source/drain resistors have sheet resistance of 100–200 Ω/sq.

18.7.8 Silicon on Insulator Processes

Silicon on insulator (SOI) CMOS processes are becoming more widely used in manufacturing, as the cost and availability of SOI substrates improves. For SOI processes, the bulk silicon substrate is replaced with a substrate containing a buried SiO_2 layer, as shown in Figure 18.17. Most production SOI substrates are created using wafer bonding, where two wafers with thermally grown layers on their surfaces are fused together at high temperature. The silicon is thinned on one of the two wafers, creating a thin device layer above the buried oxide layer. For most advanced CMOS SOI processes, the device layer is 0.2–0.3 μ m thick. In this case, the STI will extend down to the buried oxide. The advantage of SOI processes over bulk silicon CMOS processes is lower junction capacitance for MOSFETs, since the N+ and P+ source/drain regions are fully surrounded by SiO_2. The dielectric constant of SiO_2 is 3.9 as compared to 11.7 for silicon. The lower junction capacitance improves the performance of the IC and significantly improves the isolation of the digital and analog blocks.

References

1. J.E. Lillienfeld, US Patents 1,745,175 (1930)/1,877,140 (1932)/1,900,018 (1933).
2. O. Heil, British Patent 439,457 (1939).
3. W. Shockley, *Electrons and Holes in Semiconductors*, Van Nostrand, New York, 1950.
4. W. Shockley, The Path to the Conception of the Junction Transistor, *IEEE Transaction on Electron Devices*, 23, p. 597, 1976.
5. R. Warner, and B. Grung, *Transistor Fundamentals for the Integrated-Circuit Engineer*, John Wiley & Sons, New York, p. 25, 1983.
6. D. Khang and M. Atalla, Silicon-Silicon Dioxide Field Induced Surface Devices, *Proceedings of the IRE Solid State Research Conference*, Pittsburg, PA, June 1960.
7. F. Wanlass, Low Stand-By Power Complementary Field Effect Circuitry, United States Patent # 3,356,858.
8. E.H. Nicollian and J.R. Brews, *MOS (Metal Oxide Semiconductor) Physics and Technology*, John Wiley & Sons, Inc. New York, 1982.
9. L. MacEachern and T. Manku, MOSFET Chapter, CRC Reference RF Guide, 2005.
10. S.M. Sze, *Physics of Semiconductor Devices*, John Wiley & Sons, 1981.
11. D. Hodges and H. Jackson, Analysis and Design of Digital Integrated Circuits, McGraw Hill, New York, 1983.
12. Z. Liu, C. Hu, J. Huang, T. Chan, M. Jeng, P. Ko, and Y. Cheng, "Threshold voltage model for deep submicrometer MOSFETs," *IEEE Transaction on Electron Devices*, Vol. 40, No. 1, p. 86, 1993.
13. H. Hara, Y. Okamoto, and H. Ohnuma, "A new instability in MOS transistors caused by hot electron and hole injection from drain avalanche plasma into gate oxide," *Japanese Journal of Applied Physics*, Vol. 9, pp. 1103–1112, September 1970.
14. S. Tam, P. Ko and C. Hu, "Lucky electron model of channel hot-electron injection in MOSFETs," *IEEE Transactions on Electron Devices*, Vol. 31, No. 9, pp. 1116–1125, September 1984.
15. E. Takeda, A. Shimizu, and T. Hagiwara, "Device performance degradation due to hot-carrier injection at energies below the Si-SiO2 barrier," *IEDM Technical Digest*, p. 386, 1987.
16. S. Naseh and J. Deen, "RFCMOS Reliability," in CMOS RF Modeling and Characterization and Applications, p.363, World Scientific, 2002.
17. C. Hu, S.C. Tam, F. Hsu, P. Ko, T. Chan, and K. Terrill, "Hot-electron-induced MOSFET degradation—model, monitor, and improvement," *IEEE Transactions electron Devices*, Vol. 32, Issue 2, pp. 375–385, February 1985.
18. R. Kurzweil, "The impact of 21st Century Technology on Human Health and Business," *Proceedings of the World Science Forum*, November 2006.
19. A.S. Grove, E.H. Snow, B.E. Deal, and C.T. Sah, "Investigation of thermally oxidized silicon surfaces using metal-oxide-semiconductors," *Solid State Electron*, Vol. 8, p. 145, 1964.
20. W. Chang, B. Davari, M. Wordeman, Y. Taur, C. Hsu, and M. Rodriguez "A high performance 0.25 μm CMOS Technology: I – Design and Characterization," *IEEE Transaction on Electron Devices*, Vol. 39, No. 4, April 1992.

19

RFCMOS Modeling and Circuit Applications

Julio Costa
Mike Carroll
G. Ali Rezvani
Tom McKay
RF Micro Devices

19.1 The Need for RFCMOS Modeling

Before 0.35 μm process technology node (mid-1990), the RFCMOS terminology was not widely recognized in the technical community as it is today. CMOS processing was driven mostly by the needs of digital design and solid-state memories. Something crucial to the popularization of RFCMOS happened around 1997: 0.25 μm CMOS process technology was suddenly accessible to a wide range of designers with the popularization of silicon foundries. The cut-off frequency of NMOS devices at this node exceeded 25 GHz: this was an important factor as one of the main applications driving integrated RF design was Bluetooth radios which operate at 2.4 GHz. The rule of thumb among many RFCMOS designers is that the cut-off frequency of transistors (f_T) in a given technology should be at least ten times the application frequency; it is important to note that there are many exceptions to this guideline, especially in the area of III–V microwave and millimeter-wave applications.

At the same time as the technology nodes progressed towards deeper submicron geometry nodes, the necessity for more compact digital blocks with increased RF and digital functionality drove the need for an increased number of metallization layers in the process. As the number of metal layers increased, it also brought with it further distance between the topmost metal layer and the substrate. This feature helped spark the development of integrated passive components such as on-chip inductors, with larger quality factor due to the greater separation of metal and semiconductor. With the advent of 0.18 μm CMOS technology nodes with five or more layers of metallization, CMOS foundries also started providing an additional thick top layer metal, ranging in thickness from 2 to 6 μm thick. It is interesting to note that there is a limit to the advantage taken from the increase in the thickness of the top metal in the design of on-chip inductors. This is due to the fact that skin effect does not allow the full cross section of the metal to participate in current conduction at higher frequencies and therefore, the series resistance would not decrease as a function of metal thickness with the same rate as in lower frequencies. Consequently, beyond certain metal thickness the degree of improvement of quality factor of inductor may not be so much to justify the adoption of the more expensive ultra thick metal processes.

Another feature that came with use of CMOS in RF design was the concept of dc- and ac-isolation. High-frequency operation means that the nodes separated from each other are still capacitively coupled and thus can suffer from cross talk. This leads the technologists to think of new concepts like triple well processes (sometimes called a "deep n-well") which allowed separation between different blocks through extra junction capacitances across the well junction to substrate. Another feature that came with introduction of high-frequency operation was the advent of RF characterization for CMOS. Before the era of RFCMOS, the characterization of MOS devices consisted of dc measurements and modeling of devices in the MHz regime. The high frequency effects mostly meant the proper characterization and modeling of the oxide capacitance, junction capacitances, and MOS intrinsic capacitances. In the digital world this meant that the models should have been good for up to a few hundred MHz of frequency. In RF applications, the characterization procedure typically employs network analyzers with frequency sweeps up to 50 GHz and beyond. As a result, the parasitics and proper removal of them (commonly referred as "de-embedding") became increasingly important. Calibration and de-embeding procedures are discussed in detail in Volume 3: *Circuits, Measurements and Modeling* of this handbook. These techniques essentially are RF concepts which were commonly employed in the high frequency microwave field being applied in the rapidly evolving CMOS world.

It is important to note that modeling of RFCMOS devices requires significantly more complex modeling strategies than commonly employed for digital CMOS circuit simulation. Whereas the simulation of digital behavior generally requires the modeling of the initial and the final I–V characteristics to model the transition between digital states, RF models typically have to provide very accurate dc and ac characteristics for several different biases. In addition, the modeling of nonlinear and large-signal behavior typically also requires the prediction of second, third, and higher order derivatives of the I–V and capacitance equations.

In this section we will provide an overview of modeling for RFCMOS design. The earlier models for CMOS devices are presented in some detail as these are the type of models that can be easily used for quick calculations. The more advanced models are not presented in much detail as they are too complex to be covered in an overview and they are not very suitable for quick calculations and they must be used in a circuit simulator environment. In this overview, the goal is to report the important points and direct the reader to the current status of RFCMOS modeling and point out the areas of concern. There are numerous books and papers that have to be studied in order to obtain the details of the more advanced models.

19.2 DC Modeling

The first model for MOS transistors that was used in SPICE2 was the model that was proposed by Shichman and Hodges [1]. The authors of this model used the equations that were previously developed by Wallmark and Johnson [2] and created a simple model to represent a metal gated NMOS and PMOS device. They neglected the subthreshold conduction completely, but did take into account the body effect and channel length modulation. These authors were more concerned about the simulation of switching circuits and therefore were able to neglect some of the more advanced effects in MOSFET device but still get simulation results that were very close to measurement results. This is the model that is referred to as MOSFET Level 1 Model in most circuit simulators such as HSPICE, Spectre, and so forth. The model equations for the Level 1 Model as implemented in most simulators can be summarized as below:

$$I_{DS} = \mu_0 C_{ox} \left(\frac{W}{L} \right) \left[(V_{GS} - V_{th}) V_{DS} - \frac{V_{DS}^2}{2} \right] \quad \text{if } V_{GS} > V_{th} \text{ in linear region} \tag{19.1}$$

$$I_{DS} = \mu_0 C_{ox} \left(\frac{W}{2L} \right) (V_{GS} - V_{th})^2 \quad \text{if } V_{GS} > V_{th} \text{ in saturation region} \tag{19.2}$$

$$I_{DS} = 0 \quad \text{if } V_{GS} < V_{th} \tag{19.3}$$

And the threshold voltage is modeled using the familiar form below, in which φ is the surface potential and γ is the body factor.

$$V_{th} = V_{th0} + \gamma \left(\sqrt{V_{BS} + \varphi} - \sqrt{\varphi} \right) \tag{19.4}$$

Later on another model was created that was referred to as MOSFET Level 2 Model. This model did improve on an important approximation made in the Level 1 Model. In this approximation, the voltage difference between the channel and substrate or body across the channel is neglected. In this approximation the charge induced in the channel was considered to be dependent on the voltage difference between the source and body only and therefore the local potential in the channel was neglected. This was improved by Meyer [3] and resulted in the following equation for drain current:

$$I_{DS} = \mu_0 C_{ox} \frac{W}{L} \left\{ \left(V_{GS} - V_{FB} - 2\varphi - \frac{V_{DS}}{2} \right) V_{DS} - \frac{2}{3}\gamma \left[(V_{DS} - V_{BS} + 2\varphi)^{\frac{3}{2}} - (-V_{BS} + 2\varphi)^{\frac{3}{2}} \right] \right\} \tag{19.5}$$

$V_{D,sat}$ can be shown to be

$$V_{D,sat} = V_{GS} - V_{FB} - 2\varphi + \frac{\gamma^2}{2} \left[1 - \sqrt{1 + \frac{4}{\gamma^2} (V_{GS} - V_{FB} - V_{BS})} \right] \tag{19.6}$$

And $I_{D,sat}$ can be found by setting V_{DS} equal to $V_{D,sat}$ in the above equation for V_{DS}. The mobility in the above equation was assumed to be constant, independent of the bias, but with further application of these models and further advances in processing it was observed that the mobility is dependent on the bias. In particular the vertical field mobility reduction became a limitation at thinner oxides when the vertical field resulted in further interaction between the electrons in the channel and the interface. As a result several mobility models were developed to take into account such effects, with the first one suggested by Frohman-Bentchowsky and Grove [4]. For this reason, sometimes the MOSFET Level 2 Model is referred to as Frohman–Grove model. One example of such models for mobility uses the following form:

$$\mu = \mu_0 \left(\frac{U_c t_{ox}}{V_{GS} - V_{TH} - U_t V_{DS}} \right)^{U_e} \tag{19.7}$$

Where the parameter U_c represents the critical field in the vertical direction above which the mobility starts to decrease. In further advancements of the MOSFET Level 2 Model, the mobility model was improved.

The next enhancement to MOS model that was introduced in the Level 2 Model came from an improvement of current conduction model in the subthreshold regime. Contrary to the assumptions of the Level 1 Model, which assumed drain current to be zero when $V_{GS} < V_{th}$, in the MOSFET Level 2 Model the current in subthreshold regime was modeled on the basis of the works of Swanson and Meindel [5]. In addition, there were additional enhancements such as channel length modulation, short channel effects on threshold voltage, the narrow width effects on threshold voltage, and velocity saturation effects, which were added to the MOSFET Level 2 Model.

With the evolution of silicon processes, the need for better models for shorter channel devices became apparent. The MOSFET Level 3 Model was an initial response to this need and it was first formulated by Dang [6]. In the MOSFET Level 3 Model, the equations of the device in the linear region were significantly simplified in order to make the model more efficient and faster for the computation of larger circuits. The mobility model that models the vertical field mobility reduction was simplified as well for this same purpose. In addition, the impact of lateral field on mobility was taken into account in the MOSFET Level 3 model, which had not been previously employed in the earlier models; the effect of width on threshold voltage and effect of velocity saturation of saturation regime of device were also taken into account in the MOSFET Level 3 model.

The BSIM (*B*erkeley *S*hort channel *I*GFET *M*odel) model was introduced in 1985 to address some of the device physics of the evolving short channel devices in more advanced CMOS technologies and to improve the agreement of the models with MOSFET characterization results [7]. The generations of BSIM1, BSIM2, BSIM3, and BSIM4 models were an effort in this direction. Better mobility models, nonuniform channel doping, geometrical effects, velocity saturation, better subthreshold regime models, and better high-frequency noise models were among the improvements in BSIM models.

In practice it was found that fitting the model to the I–V characteristics across all regions of operation and across all geometries (scaling) with small relative error was very challenging. As a result, the binning method was introduced early on to make the model parameters length and width dependent. As a result, the agreement of the models across all regions of operation and across all channel lengths becomes increasingly better. The binning approach was adopted by many of the modeling groups in the industry in the 1990s, and even today most of the models available from major foundries in both BSIM3 and BSIM4 are binned models. There are two approaches to binning that are referred to in the literature. In both approaches device model parameters are extracted for certain devices selected across the allowed space of width and length. The selection of these devices is based on the behavior of the critical characteristics of the device as a function of the width and length. For example, the threshold voltage of a transistor may show reverse short-channel effect. Therefore, it would be important to select a number of devices with different gate length that capture the reverse short channel effect. The same consideration applies to selection of devices that capture width effect in a given technology.

In the first approach, a general functional form for width and length dependence is used in order to model the width and length dependence of the model parameters. The fitting parameters in this functional form are referred to as bin combining parameters. In this approach the final model parameter used in the circuit simulation is going to be different than the model parameter extracted for that device even for the device sizes used for primary model parameter extraction. This approach is the scalable approach.

In the second approach a different functional form is used in order to model the width and length dependence. In this functional form the width and length dependence is modeled in such a way that at the device sizes used for primary model extraction the final model parameter becomes equal to the primary model parameter extracted at that device size. In this approach, there are many different sets of model

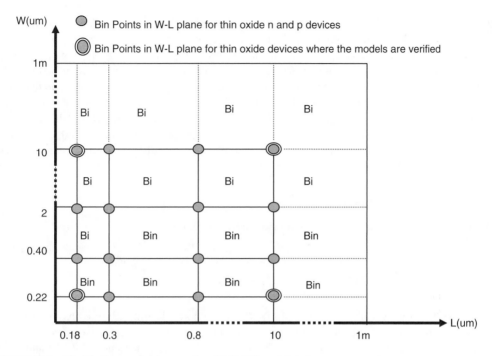

FIGURE 19.1 The binning strategy in a modern RF CMOS model deck.

parameters, each set corresponding to a separate bin. For device sizes falling in between the sizes used for model extraction, the model parameters are calculated as some kind of interpolation. This method is referred to as binned approach. Consequently in binned approach the model files available from the foundries are very long and involved, consisting of thousands of model parameters that span the whole binning space for a given technology.

In such cases, where the models provided are binned models, model file consists of several models that are developed at various bins across W (width) and L (length) dimension. Each bin is specified by its boundary defined by L_{min}, L_{max}, W_{min}, and W_{max}. For any given device the simulator chooses the proper bin according to the following condition:

$$L_{min} \leq L_{drawn} < L_{max}$$
$$W_{min} \leq W_{drawn} < W_{max}$$

(19.8)

where W_{drawn} and L_{drawn} are the drawn geometries of the device. The bin space is shown for some example devices in Figure 19.1. In this methodology the model parameters are extracted for several devices in the $W-L$ space. The devices at which the model parameters are extracted are shown as dots in the figure.

19.2.1 Scalable Approach

The relationship between the final parameter P and its associated bin combining parameters are indicated below:

$$P = P_0 + \frac{LP}{L_{eff}} + \frac{WP}{W_{eff}} + \frac{PP}{L_{eff} W_{eff}}$$

(19.9)

In the extraction methodology for scalable models of the BSIM model, the parameter $P0$ is first extracted for several transistors by fitting the IV characteristics of the model to measured data. For example, for the set of geometries shown in Figure 19.1, we will end up with P01, P02, ..., and P016 for the devices with W and L corresponding to the lower left corner of each bin shown. These define 16 points in the $P-W-L$ space, which are the optimum points corresponding to each device geometry. On the other hand, the above relation between $P-L_{eff}-W_{eff}$ defines a surface in $P-W-L$ space. Then parameters P_0, LP, WP, and PP are found by performing a least square fit such that the defined surface is the best fit to the original 16 points. This approach will result in a scalable model, which may not be a very good fit at each device size, but is reasonably good fit for all geometries in the respective bin. A potential problem with this approach is that it forces a W and L dependence that may not match the physics based dependence of that parameter on W and L. Another problem with the above approach is that the final value of $P0$ does not correspond to any of the parameters P01, P02,…, P016. It is, instead, some kind of average of these values resulting from least square fitting.

19.2.2 Binned Approach

The approach that allows the extraction of binned model is as follows:

$$P = P_0 + \text{LP}(1/L_{eff} - 1/L_{eff,ref}) + \text{WP}(1/W_{eff} - 1/W_{eff,ref})$$
$$+ \text{PP}(1/L_{eff,ref})(1/W_{eff} - 1/W_{eff,ref}) \tag{19.10}$$

The above approach will allow the value of LP, WP, and PP to be determined for each bin separately. First the value of $P0$ is determined by fitting the modeled IV characteristics to the measured results for selected device sizes corresponding to the dots in the Figure 19.1. Next, the value of LP, WP, and PP in each bin can be determined by equating the device parameter across the bin boundaries. This strategy will provide a method for calculation of parameters LP, WP, and PP for each bin separately. Note that in scalable approach one finds one set of LP, WP, and PP parameters across all geometries, but in binned approach one finds different sets of LP, WP, and PP parameters for each bin. As a result, binned approach will provide a more accurate model across geometry space, but it makes the model files very involved and cumbersome. The scalable approach is less accurate at each bin but will provide a uniform model across geometry since it will end up with only one model for all points in L-W space. In each bin the reference point is always the lower left corner. This is a matter of convention and is not essential.

19.3 AC Modeling

Once the dc models are established, the ac characteristics of the device need to be adjusted. At this step, the junction capacitances and channel intrinsic capacitances are adjusted based on the measurements of test structures that contain large area junction caps and large periphery junction caps. The reason for having two separate structures, one with large area but relatively small periphery, such as square, and one with lots of edges but not so much area is to be able to extract the area and periphery components of the junction capacitances. In the more advanced processes, one needs to also distinguish between the edge of the junction adjacent to the gate and edge of the junction adjacent to the field oxide (FOX) or shallow trench isolation (STI) regions. For this reason the modeling engineers will provide two types of periphery dominated components, one with gate edge and one with FOX or STI edge.

The intrinsic components are much more elaborate. Most models use the intrinsic capacitance models developed by Ward and Dutton [8]. The validation of intrinsic capacitances of the MOS device is not trivial at all. These are very small capacitances and they cannot be measured very easily because they typically are much smaller than the parasitic capacitances that exist in the test structure. There has been methods proposed for characterization of these devices, but most modeling groups do not perform such characterization and do not perform model validation for the intrinsic capacitances. These models are typically assumed to be correct and are verified by characterization at the ac level, for example by checking the gate delay in a delay chain or ring oscillator and/or by checking the S-parameters of the device versus frequency and the derived device characteristics such as Cgd, Cgs, and so forth.

Once the junction caps and intrinsic capacitances are characterized and the corresponding junction cap models are created, the next step will be to adjust the parameters that correspond to the gate to drain overlap capacitance and the gate side wall to drain and source capacitance. These parameters corresponding to these components of capacitance are adjusted by fitting the simulated to measured gate delay.

This method was used heavily in the era of modeling work for digital design and low frequency analog applications. In the era of RFCMOS, the ac components can be adjusted by looking at the RF characteristics. In the era of RFCMOS, the transistors are characterized using vector network analyzers to frequencies up to several tens of GHz.

19.4 RF Modeling

As the frequency of operation of the device increases, the device operation will get into the regime that the channel does not respond to the changes in the potential of the gate. Once the characteristics time for the carriers to travel across the channel from source to drain becomes comparable to the characteristic time of the change in gate voltage, for example, the period of the signal at the input, then the dc and low-frequency models alone are not sufficient for description of the device behavior. The gate resistance, substrate distributed resistance, the capacitive and inductive parasitics of the metallization connecting the source, drain, gate, and substrate of the device to the outside circuits then become very important.

In addition, the test structure design and the method of measurement and characterization needs to be significantly more advanced in RFCMOS as opposed to the digital and low-frequency analog CMOS applications. Even in the new generation of digital circuits in which the clock frequency reaches GHz range and above, the same concerns as in the RF designs become important. It is increasingly important that device modeling engineers learn the RF and microwave concepts such as scattering parameters, insertion loss, maximum available gain, and so forth.

The ac characterization and modeling that was traditionally done with classical C–V meters and ring oscillators, now has to be replaced by network analyzers, spectrum analyzers, noise figure analyzers, and so forth. In addition, the efficiency of test structures, that is, the amount of information or characterization that can be done per area of silicon consumed started decreasing. For example, in Figures 19.2 and 19.3, a typical ground-signal-ground (or as is referred to it in industry GSG) structure is shown in which the characteristics, such as reflection and transmission scattering parameters, of one NMOS device is measured. The size of this structure is of the order of about 250×250 μm. In order to develop models for such a device, one has to have several of these structures on silicon. Realizing that the current generation of MOS technologies, for example, 90 nm RFCMOS, has several flavors of NMOS and PMOS devices with various oxide thicknesses, one can imagine that the cost of model development just from the point of view of wafer cost by itself is tremendously high. In particular in very deep submicron generations in which the cost of silicon real state is very high, the development of the models can become very expensive.

In RF transistors design several concerns will impact the device layout. One is that in order to increase the f_{MAX} of the device one may use multiple finger poly gate and contact the gate at both ends to reduce the effective gate resistance (by a factor of 12 for poly contact at both ends). The other concern is the substrate resistance. In order to minimize substrate resistance, one may use diffusion islands and insert substrate contact around these islands in order to reduce substrate resistance. While the area of the source and drain is kept at its minimum to reduce the capacitance of the junctions, the current conduction through the device both during dc and ac operation must be considered from electro-migration reliability point of view.

As can be seen from the size of the pads and interconnect around these devices, the impact of these extra structures on the measured characteristics of the device becomes significant. As a result, the concept of de-embedding of the parasitics has come to the discussions related to the RF characterization. The subject of calibration and de-embedding is an important subject of research for on wafer characterization at high frequency. We do not cover this subject here as it is beyond the scope of the current discussion.

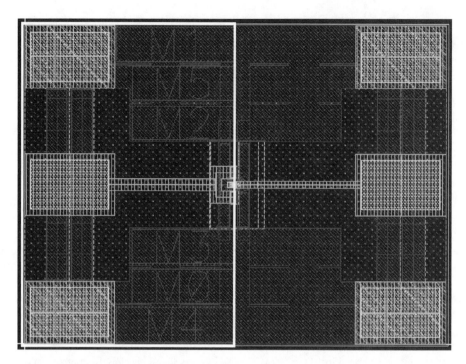

FIGURE 19.2 An RFCMOS characterization structure.

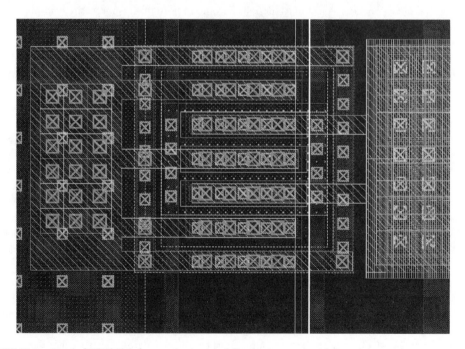

FIGURE 19.3 A typical RFCMOS characterization/modeling layout.

In BSIM3, the model equations of the MOSFET do not take into account the distributed effects in the substrate and the effect of the gate resistance. As a result the modeling groups working with BSIM3 provide an RF extension model, which consists of the intrinsic device with some external elements to model these effects. The following steps are typically taken in order to create such RF extension to the supplied BSIM models:

> *The BSIM3 MOS diode model is disabled by setting AS, AD, PS, and PD to 0.*
> *NQS=0*
> R_{sub} *is added to model substrate resistance*
> R_{gate} *is added to model gate resistance*
> *The Drain/Source to bulk junction diodes are added separately in order to have access to the substrate node.*
> C_{ds} *and* R_{ds} *elements are added to model proximity capacitance and associated high-frequency resistance*

In order to extract the device parameters for the RF model extension, we assume the transistor schematic as shown in the Figures 19.4 and 19.5, which represents a PI structure. Based on this schematic, we can solve for Y_{ij} and obtain

$$Y_{gd} = -Y12$$

$$Y_{gs} = Y11 + Y12$$

$$Y_{ds} = Y22 + Y12$$

$$Y_{gm} = Y21 - Y12 = g_m * \exp(-j\omega T)$$

$$C_{gs} = -1/(2 * \pi * \text{freq} * \text{IMAG}(1/Y_{gs}))$$

$$R_{gs} = \text{REAL}(1/Y_{gs})$$

$$C_{ds} = \text{IMAG}(Y_{ds})/(2 * \pi * \text{freq})$$

$$R_{ds} = 1/\text{REAL}(Y_{ds})$$

$$C_{gd} = -1/(2 * \pi * \text{freq} * \text{IMAG}(1/Y_{gd}))$$

$$R_{gd} = \text{REAL}(1/Y_{gd}) \tag{19.11}$$

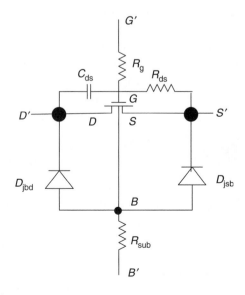

FIGURE 19.4   RF extension for RFCMOS.

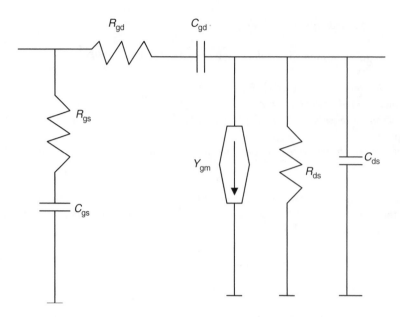

FIGURE 19.5   RFCMOS equivalent scheme.

This allows us to define the following important RF relationships for the MOSFET device:

$$\text{Current gain } |H21| = |Y21|/|Y11| \tag{19.12}$$

$$\text{Power gain} = \frac{|Y21 - Y12|^2}{4 * (REAL(Y11) * REAL(Y22) - REAL(Y21) * REAL(Y12))} \tag{19.13}$$

where f_T and f_{MAX} are defined as frequencies at which current gain and power gain equal 1, respectively. Therefore, they can be extracted by plotting current gain (in dB) and power gain (in dB) versus frequency and the frequency corresponding to 0 dB can be found. In practice in deep submicron technologies where f_T and f_{MAX} can be as high as 100 GHz or more, and most of the available instruments can measure the S-parameters only up to several tens of GHz, one may have to extrapolate in order to be able to find f_T and f_{MAX}. These extrapolations may not be very clear in some cases due to the nature of data, test structure limitations, shortcomings in de-embedding techniques, etc.

The following equations are typically used to extract gate resistance and substrate resistance

$$R_g \cong \frac{\text{Re}\{Y_{11}\}}{\text{Im}\{Y_{11}\}^2}$$

$$R_{sub} \cong \frac{\text{Re}\{Y_{22}\}}{(\text{Im}\{Y_{22}\})^2} \tag{19.14}$$

The Y-parameters used in the calculations above come from transformation of measured S-Parameters to Y-Parameters. As an example, Figure 19.6 demonstrates the BSIM3 model versus measured dc and RF characteristics for a typical 0.18 μm NMOS device with 32 fingers and 2.5 μm active width in a conventional RFCMOS technology.

19.5   Flicker Noise Modeling

Low-frequency noise in CMOS devices is generally attributed to the trapping and release of carriers at the interface between silicon and silicon dioxide. Various theories about the nature of flicker noise

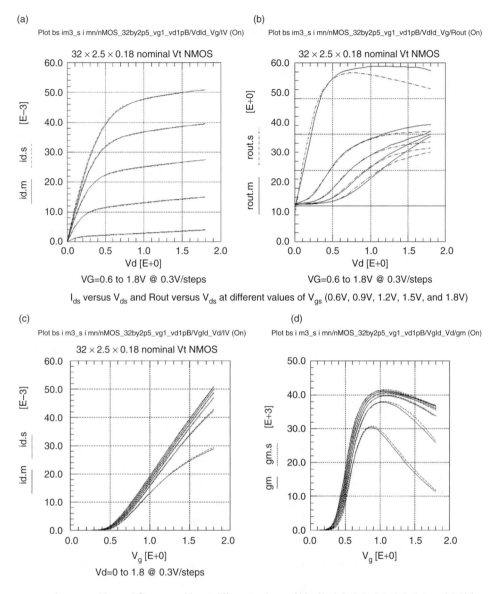

FIGURE 19.6   Modeled versus Measured characteristics of a MOSFET device using the BSIM RFCMOS model (W/L = 32 × 2.5/0.18 NMOS). In this figure the dotted line represents modeled (or simulated) and solid line represents the measured results.

have been discussed in the literature referring to number fluctuation versus mobility fluctuation models. Correspondingly, there have been several models for flicker noise implemented in circuit simulators. For example in BSIM3, the models based on G_m or I_{ds} are available with proper parameters that will select the preferred models. The model equations for these flicker noise models are as below:

$$S_{I_D}(f) = \frac{K_F I_D^{AF}}{f^{EF} C_{ox} L_{eff}^2}$$

$$S_{I_D}(f) = \frac{K_F I_D^{AF}}{f^{EF} C_{ox} W_{eff} L_{eff}}$$

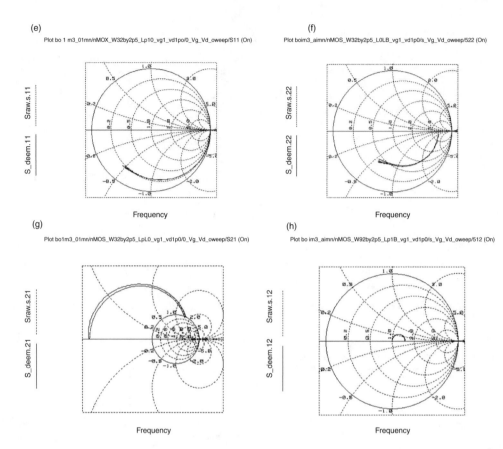

(e)
Plot bo 1 m3_01mn/nMOX_W32by2p5_Lp10_vg1_vd1po/0_Vg_Vd_oweep/S11 (On)

(f)
Plot boim3_aimn/nMOS_W32by2p5_L0LB_vg1_vd1p0/s_Vg_Vd_oweep/522 (On)

(g)
Plot bo1m3_01mn/nMOS_W32by2p5_LpL0_vg1_vd1p0/0_Vg_Vd_oweep/S21 (On)

(h)
Plot bo im3_aimn/nMOS_W92by2p5_Lp1B_vg1_vd1p0/s_Vg_Vd_oweep/512 (On)

FIGURE 19.6   Continued. In this figure the measured scattering parameters, referred to as S_deem, are compared with simulated or modeled scattering parameters, referred to as Sraw.s.

$$S_{I_D}(f) = \frac{K_F g_m^2}{f^{EF} C_{ox} W_{eff} L_{eff}}$$

In addition to the above models, in BSIM3 there is another more advanced model available in which the correlation between mobility fluctuation and number fluctuation is taken into account. This model is as follows:

If $V_{gs} > V_{th} + 0.1$

$$S_{I_D}(f) = \frac{q^2 kT \mu_{eff} I_{DS}}{C_{ox} L_{eff}^2 f^{ef} \times 10^8} \left\{ Noia \times \log \left(\frac{N_0 + 2 \times 10^{14}}{N_I + 2 \times 10^{14}} \right) + Noib \times (N_0 - N_I) + \frac{Noic}{2} \left(N_0^2 - N_I^2 \right) \right\}$$

$$+ \frac{V_{tm} I_{DS} \Delta L_{clm}}{W_{eff} L_{eff}^2 f^{ef} \times 10^{14}} \times \frac{Noia + Noib \times N_I + Noic \times N_I^2}{(N_I + 2 \times 10^{14})^2}$$

where V_{tm} is the thermal voltage, μ_{eff} is the effective mobility calculated at a given bias condition, ΔL_{clm} is the channel length reduction due to channel length modulation, and N_0 is the charge density at the source side given by

$$N_0 = \frac{C_{ox}(V_{gs} - V_{th})}{q}$$

and the parameter N_I is the charge density at the drain end given by

$$N_I = \frac{C_{ox} \left(V_{gs} - V_{th} - \min(V_{DS}, V_{DSAT})\right)}{q}$$

Otherwise (i.e., if $V_{gs} \leq V_{th} + 0.1$),

$$S_{I_D}(f) = \frac{S_{\lim it} \cdot S_{wi}}{S_{\lim it} + S_{wi}}$$

where, $S_{\lim it}$ is the flicker noise calculated at $V_{gs} = V_{th} + 0.1$ and S_{wi} is given by

$$S_{wi}(f) = \frac{Noia \cdot V_{tm} \cdot I_{DS}^2}{W_{eff} L_{eff} f^{ef} \cdot 4 \cdot 10^{36}}$$

Flicker noise in certain blocks of RF transceivers is very important. In particular, in the voltage-controlled oscillators, the low-frequency noise in NMOS devices gets folded into the high frequency and will impact the phase noise of the oscillator quite drastically. PMOS devices tend to have lower flicker noise. One concern about flicker noise in CMOS devices and their modeling is to decide on how to deal with the variation of flicker noise across a wafer and across a lot. Process corner modeling for basic devices characteristics like threshold voltage, gate length, and oxide thickness variation do not reproduce the observed variation of flicker noise. Methods for proper corner modeling to take into account the variation of flicker noise across the process is a subject of research at this time. Owing to the fact that the measurement of flicker noise is quite time consuming and tedious, the foundries do not have proper monitor of flicker noise on every lot, the way they monitor the parametric data. Consequently, the variation of flicker noise across a lot is a new concept among device modeling engineers and it has to be addressed properly. As RFCMOS becomes more and more abundant, and with the pressure for keeping the yield high and variation of oscillator performance tight, these type of concerns about the proper modeling and statistical fluctuation of flicker noise will become more and more important.

19.6 High-Frequency Noise Modeling

High frequency noise in CMOS devices is an important aspect of the device that has to be modeled properly in order to predict the performance of critical circuit blocks such as LNA in an RF system. In the study of high-frequency noise in CMOS devices we treat the device as a two-port network. Typically the high frequency noise characteristics of a two-port network are expressed in terms of the four noise parameters of the network. The four noise parameters are NF_{min}, the minimum noise figure, Γ_{opt}, the optimum source reflection coefficient, and r_n, the equivalent normalized noise resistance. Γ_{opt} is a complex number with real and imaginary components, and so is counted as two noise parameters. Therefore, we have four noise parameters that characterize a noisy two-port system. The relationship between the noise figure of the system and source impedance is given by the following equation:

$$NF = NF_{min} + \frac{4r_n|\Gamma_s - \Gamma_{opt}|^2}{(1 - |\Gamma_s|^2)|1 + \Gamma_{opt}|^2}$$

in which NF is the noise figure of the system for arbitrary source reflection coefficient Γ_s. When the source reflection coefficient matches Γ_{opt}, system noise figure NF will be equal to minimum noise figure NF_{min}.

In order to model the high frequency noise of CMOS devices, we should extract the noise parameters of the device from measurements done on the device and compare them with the parameters extracted from the device in the simulation environment. Extraction of device noise parameters from the measurement is not a trivial task. Typically these measurements are done by connecting known mismatched impedance at the input of the device and measuring the noise at the output of the device. The mismatched impedance at the input serves as Γ_s at the input and the measured noise figure at the output will be NF for that particular Γ_s. Theoretically if one measures NF for four different mismatched impedances, then it should be possible to determine all noise parameters of the device by solving four equations with four unknowns. In practice however, because of the error in measurements, relying on four measurements will not produce stable results. Typically noise figure is measured at several impedances at the input and the four noise parameters are determined by fitting the noise equation to the measured data.

The most important source of high-frequency noise in CMOS devices is due to the scattering of the carriers in the channel during their transport from source to drain. The next source of high frequency noise in CMOS devices is the resistance of the gate material that results in small fluctuation of gate voltage over the channel, which in turn results in high-frequency noise in the channel. The third source of high-frequency noise is referred to as gate-induced noise, which is due to variation of gate voltage due to the fluctuation of potential in the channel. The local fluctuation of potential in the channel, as a result of carrier scatterings, will induce a variation in gate potential, which will in turn result in random changes (or noise) in channel current due to transconductance of the device. In deep submicron technologies the thickness of gate oxide has decreased quite drastically, which results in more tunneling current across the gate oxide in these devices. The tunneling current through the thin oxide is another source of noise in deep submicron devices and has to be taken into account for proper modeling of noise in CMOS devices.

The noise models available in industry standard models usually model different noise mechanisms in CMOS device. Today BSIM3v3 is the most popular model that is used in technologies as advanced as 0.13 μm technology node. In BSIM3v3 there are two types of high-frequency noise models available. One is based on the SPICE2 model, which relates the channel thermal noise to various conductances in the channel as below:

$$S_{id} = \frac{8kT}{3}(G_m + G_{mbs} + G_{ds})$$

And the other, which is referred to as BSIM3v3 model, is as below:

$$S_{id} = \frac{4kT\mu_{eff}}{L_{eff}^2 + \mu_{eff}|Q_{inv}| \cdot R_{DS}}|Q_{inv}|$$

Where Q_{inv} is the inversion charge in the channel based on long channel theory, or the Van der Zeil model, with a modification to include short-channel effect. In this model the induced gate noise is not taken into account. One can use the above models in the circuit simulation tools in order to simulate the noise parameters of the device and then compare with the measured noise parameters of the device as was mentioned earlier. In the case of the BSIM3v3 it was found by many modeling groups in the field that the measured noise parameters of the devices were not matching the modeled noise parameters. In BSIM3 there are no extra parameters for high-frequency noise adjustment. Basically one models dc characteristics and S-parameters, and beyond that there are no other parameters to adjust upon mismatch between measured and modeled noise parameters. As a result, it is required to add extra noise sources in the channel and on the gate in order to match the measured and modeled noise levels.

In recent years BSIM4 models have been developed and released and are used by many in the industry, in particular in 90 nm node. In BSIM4 the high-frequency noise of the device is improved through inclusion of many of the noise sources such as channel thermal noise, gate-induced noise, thermal noise due to physical resistances such as source/drain resistance, gate material, substrate resistance, and shot noise resulting from the leakage due to tunneling through gate dielectric. The correlation between the channel thermal noise and gate-induced noise is accounted for in BSIM4, and more adjustable parameters are provided in order to fit the model against measured noise parameters. In 2005 the Compact Modeling Council (CMC) adopted PSP, a surface potential-based model, as the next standard model for MOS devices. In this model the high-frequency noise is modeled even more thoroughly through inclusion of velocity saturation effects. One of the main characteristics of the PSP model is that there is no extra parameter to adjust for high-frequency noise, and the developers of the model have shown that the measured and modeled noise parameters match fairly well across frequency and bias conditions.

19.7 Large-Signal Modeling

RFCMOS devices so far have been used mainly in small-signal RF circuit blocks. In the case of voltage-controlled oscillators (VCOs), there is some level of large signal present but is not as significant as in the case of high power devices such as power amplifiers that experience a really large signal at the gate on MOS device. Concerns about the nonlinearity of the CMOS devices under large-signal operation and validity of models like BSIM3 models in predicting the operation of the device under large-signal regime have become prevalent only recently. It is believed that BSIM3 and BSIM4 models can adequately predict device behavior under large-signal operation in common source configuration. In common gate configuration, it has been shown by various authors that the BSIM models suffer from discontinuity of derivatives of transconductances at around $V_{ds} = 0$. This resulted in much concern about the use of BSIM models for simulation of switches and passive mixers. Newer generation of models that are formulated based on surface potential (e.g., PSP models) have fixed this problem. Some of the companies in the industry have moved to these newer models while major foundries still have not adopted the newer models. Furthermore, the newer models will address some of the features of the devices in very deep submicron nodes such as 45 nm and beyond. Also the importance of nonlinear and large-signal modeling has implications about the requirement of a different set of characterization tools such as load-pull system, two-tone tests, and large-signal network analyzers, which like the network analyzers and spectrum analyzers of the late 1990s are finding their ways into the CMOS device modeling groups across the industry.

19.8 Introduction to RF and Microwave Design for Highly Integrated Systems

CMOS process technology evolution targeted for digital application represents one of the greatest continuing manufacturing achievements in the history of electronics. Around the year 2000, the 180 nm node became available in production from merchant foundries, and was found to provide peak f_Ts around 60 GHz, low power dissipation for 20 GHz operating f_T, minimum noise figures of 1 dB and below [9]. While maximum voltage for the core device was limited to around 2 V, an ID_{SAT} of around 0.6 A/mm was reached with knee voltages around perhaps 0.5 V. As these technologies were following Moore's law, by the time they achieved the high-density digital circuit manufacturing capability, they had implicitly passed the test for radio-frequency and microwave function application. Because of these improvements in the basic characteristics of modern CMOS technologies, wireless personal area network (WPAN), wireless local area network (WLAN), and cellular transceivers operating through about 6 GHz have been implemented and delivered in production using CMOS technologies at 180 nm and below.

While many innovations at the system and subsystem level have been driven by this reality, dramatically lowering the cost of wireless transceivers, in this section we will explore the benefits of radio-frequency and microwave concepts for developing world-class radio transceivers. Recognizing the tremendous contribution of pioneering efforts [10,11] and continued innovation of these and many other researchers, we take a step back to focus modern high-frequency radio techniques on the highly integrated systems now commonplace in CMOS.

The emphasis here will be on useful concepts, where they apply and short-reference lists to probe further. Given the nature of this section, there is no possibility of being complete or exhaustive, but rather to connect topics that may not be brought together in undergraduate course work. The hope is that this section is useful both to recent graduates and practicing engineers interested in wireless interface design in standard CMOS.

19.9 Basic Building Blocks

19.9.1 RF and Microwave Circuit Design Principles

Circuit design principles are built upon a solid analytic foundation. In this section, electrical network classification energy conservation in network analysis, the application of scattering parameters, and stability analysis methods for highly integrated transceivers are discussed.

19.9.1.1 Electrical Network Classifications

In applying various analysis and design methods, it is important to classify electrical networks for which a useful set of general characteristics apply. Presumably these classifications also have some practical benefit.

In a wireless receiver, energy impinging on an antenna is typically grossly filtered, amplified somewhat, and down-converted with a mixer to a lower frequency for further amplification and filtering. For many purposes the filter and amplifier can be considered linear, time-invariant (LTI) networks, and the mixer a linear, but time-varying system. It should be stated that linearity is a very valuable idealization, allowing considerable analytic and creative insight.

Often the broad filter is fabricated out of good conductors and dielectrics whether lumped-element or distributed, typically external to a CMOS IC. Such a network is linear over an extremely wide input signal range. The amplifier may be considered linear over an input signal range below a certain, application-specific value, but nonlinear above that range.

19.9.1.1.1 Linear Time-Invariant

An LTI system network response to a stimulus satisfies the properties of scaling and superposition. For example, given a scalar a, if we have an output y, due an input x, the output due to an input $a\,x$ is $a\,y$. Suppose we take the response to two inputs x_1 and later x_2 and find the response to be y_1 and y_2 respectively. If we input $x_1 + x_2$, superposition implies that the output will be $y_1 + y_2$.

19.9.1.1.2 Linear Time-Varying

In Figure 19.7, two examples of relevant linear, time-varying networks are shown. The first is a switch with finite on- and off-resistance switching at a rate of f_{LO}. This network is linear in that, over the complete operating cycle, we can say that scaling and superposition apply. Another linear time-varying network is the mixer and local oscillator (LO) combination shown in Figure 19.7b. The mixer driven on the LO terminals by a very large signal is typically a nonlinear system, since for example, the gate capacitances are a function of the signal swing. If we choose to evaluate the mixer and LO in combination, however, we can consider only small-signal stimulus and this network is linear.

19.9.1.1.3 Nonlinear

Networks which do not satisfy superposition and scaling are certainly not linear. The class of circuits which is not linear is very broad. A useful subclass of nonlinear network often taken is along the idea of autonomous versus nonautonomous networks. One working definition of an autonomous network which separates it from a linear time-varying network is one which produces output for no external input. In Figure 19.8, an oscillator and its energy source are included in the definition of this system.

19.9.1.2 Energy Conservation and Finite Impedance

In pure analog integrated circuit design, input and output signals of interest are typically taken as either voltage or current. At lower frequencies and for many designs this convention is very useful, since, for example, the MOS transistor can represent very nearly an open circuit input at 10 MHz and below. The output impedance of a preceding stage may be quite low relative to that high input impedance, so that for all practical purposes the ac input current is zero. In that convention, the signal itself has zero energy; it is only a voltage and the associated current is vanishingly small.

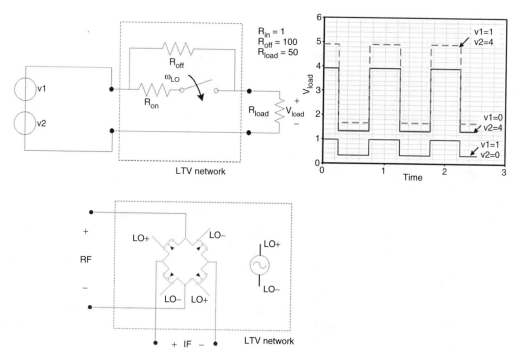

FIGURE 19.7   The periodically closing switch with finite on and off resistance (1a) is a linear, time-varying (LTV) network. Superposition is illustrated by considering the load voltage for an input of v1+v2 and recognizing that it is the sum of the outputs for v1 and v2 alone. The MOSFET ring mixer with local oscillator (1b) is LTV as observed through the RF and IF ports for sufficiently small signals.

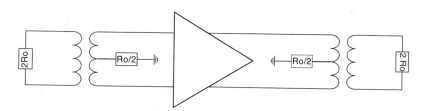

FIGURE 19.8   Mixed-mode S-parameter concept. The differential reference impedance is twice the single-ended reference impedance, Ro. The common-mode reference impedance is Ro/2. The ideal center-tapped transformer can be implemented in most simulators using controlled-source equivalent circuits.

   In processing a signal incident on a receiving antenna, care must be taken to utilize all the energy possible at the antenna output. In radio-frequency circuit design and analysis, this predisposition for energy conservation is captured in finite source and load impedances. For example, at the antenna–LNA interface, the antenna radiation resistance [12] forms source impedance of the receive electronics. By careful small-signal and low-noise impedance matching, the available energy at the antenna port is utilized, which comprises both a voltage and current. The most energy is transferred from the antenna when it is loaded with a resistance equal to its own resistance. If there is a reactive component to the antenna impedance, this can be cancelled with an opposite reactance as seen looking into the receive electronics.
   Since power is energy per unit time, we alter the terminology to the more common power signals, such as power available from the source, power delivered to a load.

19.9.1.3   Role of S-Parameters

There are many ways to describe LTI networks. If one considers voltage response to a current input, the impedance parameters of the network apply. If the current response to voltage input is considered, the admittance parameters apply. Scattering parameters are another way to capture network behavior, relating reflected waves output to incident wave input. The wave approach normalizes total voltage and current to a reference impedance. This reference impedance is often taken as 50 Ω in many systems, both physically and analytically, but this value is somewhat arbitrary. The key is that the reference impedance is finite.

S-parameters are useful in CMOS radio front-end design; however, it is common for an amplifier for example to be operating at different impedance levels on the input and output. This is in contrast to nonintegrated systems where the source and load impedances might be kept at 50 Ω in an effort to have a predetermined or controlled impedance level independent of varying phase lengths present between circuit blocks.

Beneficial attributes of a scattering-parameter approach include accommodating with nonunilateral behavior (finite isolation), a common way to think about impedance matching and trade-offs, and as an element of stability investigation.

Impedances with real part greater than zero, when mapped to reflection coefficient, lie within the unit circle. This includes the open, which has infinite impedance. A very useful concept available in two-port S-parameter network theory is the idea that a network *maps* a load reflection coefficient to an input reflection coefficient. Load reflection coefficients falling on a circle are mapped to a circle looking into the combination, generally scaled, offset and rotated. The mapping may or may not turn the load reflection region inside the circle to the outside of the circle.

19.9.1.4   Stability

Those skilled in microwave network theory applied to circuit block design will be familiar with the two-port stability measures that can be applied. These measures are incomplete, but are very useful taken in conjunction with other methods and have advantages over other methods.

A network comprised of lumped elements can be analyzed for the presence of right-half plane poles, which imply exponentially growing solutions and instability. Using the mapping concept of two-ports, it is simple to analyze the case where the input, for example, is loaded with all possible passive impedances. If the mapping to the output stays within the unit circle, then stability is assured. This analysis would be somewhat tedious to perform with pole-zero analysis, since all possible reactive terminations would need to be applied and a pole-zero analysis performed over those reactive terminations.

However, it has been shown that the mapping-related analyses do not guarantee stability, but cover the case of arbitrary loads if the unloaded network has no right-half plane poles [13]. The network determinant function (NDF) can be used to determine the presence of right-half-plane poles. It is applied by performing a Nyquist analysis on each active device, one at a time.

Alternatively, methods relying on transient simulation can be very practical. A common approach to stability verification on analog circuits is to step the supply voltage in a transient simulation, in addition to gain and phase margin analysis. An approximate, convenient, time-domain version of NDF has been applied with some success to radio-frequency circuits. The method seeks to create a broadband excitation at each active device, as is done in NDF, by providing a time-domain current impulse. The current source can be added in parallel to existing circuitry without loading. A circuit detected as unstable would not return to a quiescent state after impulse excitation.

19.10   Radio-Frequency Circuit Blocks and Interfaces

19.10.1   Antenna Interface

Transmitted and received electromagnetic energy is converted to a localized electrical format by an antenna. The format can be differential or single-ended and has a particular nominal impedance level, typically in the range of 50–300 Ω. A low loss filter operating at this same impedance level is often required to eliminate out of band blocker signals at the receiver input or to prevent noise and spurious output from the transmitter from reaching the antenna. Since this filter is a low-loss passive and reciprocal device, any impedance changes appearing at the antenna terminals are passed along to the LNA input.

The input impedance of the antenna and reflection of signals back into the antenna can change the impedance of the antenna, change the amount of LO signal at the input and affect the amount of signal transferred to the antenna by the transmitter. In addition, manufacturing tolerances of antenna and filter can also cause variation from the nominal impedance level. These effects are usually considered somewhat random, and specified by a magnitude of reflection coefficient or equivalently standing-wave ratio at arbitrary phase.

While the antenna interface must accommodate significant variation, interfaces on-chip are subject to much random impedance variation since the distances are extremely short. One of the key differences between monolithically integrated transceivers and those comprised of separate ICs is the control of the impedance levels. This allows moving from a nominal antenna impedance level to typically higher impedance levels on-chip.

19.10.2 LNA/Mixer Design

NMOS minimum noise figure below about 1 dB, equivalent noise resistance of about 50 ohms, and 2 GHz associated gain of 17 dB at an Idd of 2 mA were demonstrated in the 180 nm generation of digital CMOS. One of the challenges to exploiting the device capability lies in the relatively high quality factor (Q) required by the source impedance. To lower the Q to manageable levels, source inductive feedback is usually employed. LNA design analysis of Shaeffer and Lee [14] outline some basic considerations for noise and linearity performance for CMOS LNAs with inductive source feedback to lower the input Q while maintaining good noise figure performance. A common approach is to lower the Q with some inductive feedback and then use a reactive element impedance or noise match to complete the design. It is possible to adjust the inductive feedback to make the source impedance for small-signal match and noise match align. More aggressive design techniques can result in improved performance, such as transformer-coupled feedback [15].

It is interesting to note that minimum noise figure is not an invariant to lossless feedback and minimum noise figure can actually be reduced with lossless feedback. Of course, the gain is reduced with feedback also. Vendelin [16] observed that minimum noise figure reduced when adding parasitic inductive source feedback and pointed out that a different metric, minimum noise measure, which accounts for gain, cannot be reduced by the use of lossless feedback. In practice, losses in the reactive feedback elements compensate the potential reduction in minimum noise figure.

The LNA output network often consists of a differential inductor, center tapped for providing bias current. The differential inductance forms part of the impedance match to the mixer.

It is important to ac couple the LNA to the mixer to reject any second-order distortion produced in the LNA. A common technique in bipolar LNAs to improve third-order distortion involves optimizing the input circuit at the difference frequency. This technique does not apply to CMOS LNAs. Since the drain current–gate voltage characteristic changes from exponential in weak inversion to a power law in strong inversion, Aparin [17] has shown that third order cancellation is possible. If the technology is capable of withstanding large voltage swings, it has been shown larger IP3 can be obtained with simpler approaches [18].

Mixer designs are classified into passive and active. The passive mixer has zero average drain current in the switching devices and has the advantage of very low flicker noise and low-thermal noise driving a capacitive IF load. The low flicker noise of the passive mixer is based on the notion that since there is no average current flowing there is no mechanism for the external circuit to sense the slowly varying channel carrier mobility or number fluctuation, the mechanisms thought responsible for flicker noise. It has been pointed out [19] that while flicker noise output is zero for zero input, flicker noise can be detected for nonzero input even though there is zero average current. In this reference, the possibility of significant output flicker noise in the presence of a large unwanted blocker signal is raised as a consideration.

Active CMOS mixers do show large $1/f$ noise contribution even with zero input. Various methods are being considered to mitigate these effects.

19.10.3 Integration Considerations

The LNA is required to process very small signals and one of the challenges in highly integrated design, particularly when significant digital functionality is cointegrated, is avoiding self-interference from harmonics of the digital clock and high speed and strength digital drivers. Several techniques can be employed to circumvent this interference from the LNA point of view.

> Pin-out planning
> Differential LNA topology
> Substrate isolation techniques

Pin function assignment for highly integrated systems is best driven from the application board level through the module or package and into the die level. Driving the IC layout floor plan from the reference design level has historically been challenging, since the habit from lower level integration RFIC development is often driven from the IC to the board. Historically, the application circuit is designed after a significant amount of IC work is completed. However, it is the board level that restricts layout significantly due to much larger line, space and via design rules but which must play a part in maintaining signal isolation. IC layout flexibility is fundamentally much higher, given several orders of magnitude finer design rules and higher metal layer count of perhaps six or so, compared to two to four layers on module laminates or system PC boards.

The desired signal level at the input to the LNA may be on the order of a microvolt, if interfering signals appear at the LNA input on the same order of magnitude, within a channel bandwidth, the desired signal would be lost and the system noise figure severely degraded. If a logic voltage signal on the order 1 V is present on the chip, a voltage mode isolation requirement of more than 120 dB is implied. This was a source of concern with many practitioners in the late 1990s, who felt it was not possible to integrate highly sensitive radio-receive circuitry with digital functionality. There are several considerations which this simple argument overlooks, however.

- Relatively low clock frequency resulting in smaller, high-order harmonics at the receive frequency
- Delay spread in the digital circuit resulting in spectral spread of the energy
- Use of relatively wide on-chip RF-digital interfaces reducing required clock rates for high throughput
- The relatively low impedance level of radio frequency circuitry as compared to digital gate functionality
- Benefits of differential topologies for radio-frequency signal processing
- The fortuitous upward trend in substrate resistivity foundry CMOS processes
- Straightforward addition of a deep n-type implant to decouple the body of the NMOS device from the p-type substrate
- Benefit of extreme miniaturization relative to cross-coupling caused by package, wire bond and board level geometries

Low power requirements of digital systems tend to push clock frequencies down whenever possible. A clock frequency on the order of 20 MHz for a highly-integrated WPAN solution might be typical. Digital circuitry is sized according to timing requirements to minimize current consumption which has the effect of lowering the energy available for interference, and often for causing some spread in the spectrum, since longer RC delays are allowed automatically resulting in a spread in gate transition times.

When integrating the radio–baseband interface functionality on-chip, wider digital interfaces are possible and can have lower clock rates and slower edges than interfaces with fewer signals and the same information bandwidth requirements.

RF node impedances are somewhat moderate, achieving levels on the order of a few kilo-ohms at most at 3 dB bandwidths of 10%, and this level only rarely. The impedance level of typical antennas sets the source impedance of the receiver to 50–100 Ω, and on-chip inductor quality factors of around 10 or 20 limits the impedance transformation to a similar number. It is difficult to generalize on the source impedance of a digital aggressor, but it is overly pessimistic to assume the victim circuitry is operating at an open circuit, capturing all the voltage drop of the aggressor.

Within a small region on-chip, digital noise signals can be considered common-mode, so differential LNA topologies, which can provide perhaps 20–30 dB of common-mode rejection, are sometimes chosen. Rejection of power supply, ground, ESD network and substrate noise are required in the signal band and differential techniques can help.

When analyzing linear differential systems, mixed-mode S-parameters [20] are useful. To capture common-mode to differential-mode conversion effects, care must be taken to terminate the ports properly. An ideal balun (i.e., center-tapped ideal transformer) terminated on its differential port by twice the single-ended impedance and by half the single-ended impedance on the common-mode port is a means for considering networks self-consistently.

Fortuitously, the epitaxial substrates having a bulk resistivity on the order of mΩ-cm have been replaced by substrates having a bulk resistivity on the order of 10 Ω-cm in more recent CMOS foundry processes. This 10,000-fold increase in substrate resistivity increases greatly the impedance achievable between regions on-chip, at the expense of more complex latch-up rules and a higher dependence on precharacterized I/O libraries. The addition of a deep n-well implant provides a low cost decoupling of the NMOS transistor body from the substrate, which provides a significant component of isolation at frequencies below 1 GHz.

While integration places sensitive circuitry near noisy circuitry, one thing to keep in mind is that the alternative technology that involves separate packaged ICs with large metal areas and lengths, which can create capacitive, inductive, and radiation coupling paths.

19.11 Polyphase Concept

In working with low-IF or direct conversion systems, it is helpful to consider the vector nature of modulation riding on the carrier frequency. Conceptually, the signal is considered a vector, decomposed into in-phase and quadrature components. The frequency of this vector is the rate of change of the phase angle. Without loss of generality, one can consider observing the vector from the viewpoint of a reference already rotating at the carrier frequency. It is in this sense that positive frequency (clockwise rotation) and negative frequency (counter-clockwise rotation) can be identified. In a zero IF system, the output vector will rotate with negative frequency if the input is below the LO frequency and will rotate with positive frequency if the input is above the LO frequency.

Networks that operate on multiple phases simultaneously are called polyphase networks. The polyphase RC network discussion of Gingell [21] and the active polyphase work of Minnis [22] are useful concepts in the generation of quadrature signals and in IF signal processing.

19.12 Transmit Subsystem

The transmit subsystem operates under very different constraints than the receive side. Modulation quality, spurious output, and efficiency are key requirements.

Transmit architectures choices are heavily dependent on the modulation scheme. For constant envelope systems, such as Gaussian minimum shift-keying (GMSK), saturated output stages can be used. For nonconstant envelope systems, such as eight-phase-shift keying (8-PSK) with relatively low modulation bandwidths in the 200 kHz range, polar modulation can be used [23]. In polar modulation schemes, the amplitude and phase information are split into separate paths, the phase of the desired signal being passed in through the phase-locked loop controlling the voltage-controlled oscillator. The amplitude is passed to a control signal that modulates the supply voltage of a power amplifier or driver amplifier.

A more generally applicable scheme of generating a transmit signal is by utilizing in-phase and quadrature mixers, driven by digital-to-analog converters (DAC) typically operating at the modulation rate. Calibration is usually needed to remove IF dc offsets since any dc present at the DAC outputs results in carrier feed through degrading modulation quality. Phase and amplitude imbalance result in undesired sidebands. With calibration, carrier and sideband suppression better than 30 dB can be readily achieved.

19.13   Local Oscillator Subsystem

The LO is required to drive transmit and receive sections. Key requirements on this subsystem are phase noise, spurious output, and quadrature balance. Very low VCO phase noise eliminates the need for external transmit noise filtering in many applications. Phase noise on the receive side will cause the mixer to convert larger undesired signals to the same frequency as smaller desired signals, degrading frequency selectivity.

19.13.1   Voltage-Controlled Oscillator

The best starting point for VCO design is to consider the cross-coupled PMOS pair, which provides a negative conductance over a broad frequency range. Connecting a parallel resonant network with high-quality factor fixes the frequency of oscillation. It is common to use either a p+/n-well varactor or an "NMOS in n-well" varactor [24] to realize a high-quality factor voltage-tunable capacitance.

Switched capacitance, using some form of fixed capacitors, such as metal-insulator-metal (MIM) capacitors and NMOS switches, is used to band switch or calibrate for manufacturing offsets.

19.14   Transmit and Receive Antenna Interfaces

Ideally, the transmitter and receiver are interfaced to the antenna with separate filtering and matching networks. This sometimes may not be necessary, and some implementations have been put forward that simply connect, transmit, and receive together, which is workable for small output powers and moderate sensitivities.

It is more common to incorporate some form of transmit receive switching, which for small signals is readily achieved in bulk CMOS up through 2.5 GHz [25].

19.15   Design Flow Considerations

When designing integrated RF circuitry, it is important to think through the hierarchy of the design to maintain manageable simulation times throughout the design flow.

Following the discipline of modern custom layout tools can greatly assist in the layout refinement process. Parasitic extraction including both capacitance and resistance is required to determine approximate circuit performance. Hierarchy should break circuitry down into manageable pieces, with consideration given to which circuits and subsystems are critical to simulate together. For example, in a multiband receiver, it might be advantageous to simulate each receive band signal path with LO drive rather than all receive bands together.

19.15.1 Simulation Technique Features and Benefits

Time-domain simulation remains the most reliable and widest range of applicability. For many functional radio-frequency circuits, however, techniques which take advantage of the knowledge of the periodicity of the solution are exploited for greatly increased speed and improved dynamic range.

Periodic steady-state or shooting methods operate in the time domain but target the response after the transients have died away. Often a short transient simulation is run to provide a starting point for the steady-state solution. These simulations can be very robust, and are noted for having no trouble with rapidly changing signals over the simulation period. Once the steady-state solution is found, AC-like and S-parameter-like analyses can be performed following the LTV concept.

Harmonic balance methods solve the linear portion of the circuit in the frequency domain and the nonlinear portion in the time domain. For situations with stimulus frequencies very close to each other, such as in low-IF receivers, this method can have advantage over the shooting methods. Also, distributed linear components are less of an issue with harmonic balance, whereas with shooting methods these elements need to be converted into Laplace transform models so they can be analyzed in the time domain.

References

1. Shichman and Hodges, "Modeling and simulation of insulate gate field effect transistor switching circuits," *IEEE J. Sol. Stat. Cir.*, SC-3, 285–289, 1968.
2. Wallmark, J.T. and Johnson, H., *Field Effect Transistors*, Edgewood Cliffs, NJ, Prentice Hall, pp. 113–115, 1966.
3. Meyer, J.E., *MOS Models and Circuit Simulation*, RCA Rev. 32, 1971.
4. Frohman-Bentshkowsky, D. and Grove, A.S., "On the effect of Mobility Variation on MOS Characteristics," *Proc. IEEE*, 56, 217–218, 1968.
5. Swanson, R.M. and Meindl, J.D., "Ion implanted complementary MOS transistor in low voltage circuits," *IEEE J. Sol. Stat. Cir.*, 14, 1979.
6. Dang, L.M., "A simple current model for short channel IGFET and Its application in circuit simulation" *IEEE J. Sol. Stat. Cir.* 14, 1979.
7. Sheu, B.J., "MOS transistor modeling and characterization for circuit simulation," Electronics Research Laboratory, Report Number ERL-M85/85, University of California, Berkeley, 1985.
8. Ward, D.E. and Dutton, R.W., "A charge-oriented model for MOS transistor capacitances," *IEEE J. Sol. Stat. Cir.*, 13, 703–708, 1978.
9. McKay, T., "Toward Highly Integrated RFCMOS Radios," *IEEE MTT-S WSFA Workshop*, Anaheim, California, June 13, 1999.
10. Shaeffer, D.K., Shahani, A.R., Mohan, S.S., Samavati, H., Rategh, H.R., del Mar Hershenson, M., Min Xu Yue, C.P., Eddleman, D.J., and Lee, T.H, " A 115-mW, 0.5-μm CMOS GPS receiver with wide dynamic-range active filters," *IEEE J. Sol. Stat. Circ.*, 33(12), 2219–2231, December 1998.
11. Cho, S., Dukatz, T., Mack, E., Macnally, M., Marringa, D., Mehta, M., Nilson, S., Plouvier, C., Rabii, L., "A single-chip CMOS direct-conversion transceiver for 900 MHz spread-spectrum digital cordless phones," 1999 IEEE Solid-State Circuits Conference, Digest of Technical Papers, February 15–17, 1999, pp. 228–229.
12. Harrington, R. *Time Harmonic Electromagnetic Fields*, New York, NY, Wiley-IEEE Press, 2nd Edition, 2001, Section 2-10.
13. Platzker, A. and Struble, W., "Rigorous determination of the stability of linear n-node circuits from network determinants and the appropriate role of the stability factor K of their reduced two-ports" *Integrated Nonlinear Microwave and Millimeterwave Circuits, 1994. Third International Workshop*, October 5–7, 1994, pp. 93–107.
14. Shaeffer, D.K. and Lee, T.H., "A 1.5-V, 1.5-GHz CMOS low noise amplifier," *IEEE J. Sol. Stat. Circ.*, 32(5), 745–759, May 1997.
15. Cassan, D.J. and Long, J.R., "A 1-V transformer-feedback low-noise amplifier for 5-GHz wireless LAN in 0.18 um CMOS," *IEEE J. Sol. Stat. Circ.*, 38(3), 427–435, March 2003.
16. Vendelin, G.D., "Feedback effects on the noise performance of GaAs MESFETs," *MTT-S International Microwave Symposium Digest*, 75(1), 324–326, May 1975.

17. Aparin, V. and Larson, L.E., "Modified derivative superposition method for linearizing FET low-noise amplifiers," *IEEE Trans. Micro. Theory Tech.*, 53(2), 571–581, February 2005.
18. Griffith, D., "A +7.9dBm IIP3 LNA for CDMA2000 in a 90 nm digital CMOS process," *Radio Frequency Integrated Circuits (RFIC) Symposium, 2006 IEEE*, June 11–13, 2006, 4 pp.
19. Chehrazi, S., Bagheri, R., and Abidi, A.A., "Noise in passive FET mixers: a simple physical model," *Custom Integrated Circuits Conference*, 2004. Proceedings of the IEEE 2004, October 3–6, 2004, pp. 375–378.
20. Bockelman, D.E. and Eisenstadt, W.R., "Combined differential and common-mode scattering parameters: theory and simulation," *IEEE Trans. Micro. Theory Tech.*, 43(7), Part 1–2, pp. 1530–1539, July 1995.
21. Gingell, M.J., "Single sideband modulation using sequence asymmetric polyphase networks," *Elect. Comm.*, 48 (1 and 2), 1973.
22. Minnis, B.J., Moore, P.A., Payne, A.W., Caswell, A.C., and Barnard, M.E., "A low-IF polyphase receiver for GSM using log-domain signal processing," *Radio Frequency Integrated Circuits (RFIC) Symposium*, 2000. Digest of Papers. 2000 IEEE, June 11–13, 2000, pp. 83–86.
23. Hietala, A.W., "A quad-band 8PSK/GMSK polar transceiver," *IEEE J. Sol. Stat. Circ.*, 41(5), 1133–1141, May 2006.
24. Bunch, R.L. and Raman, S., "Large-signal analysis of MOS varactors in CMOS-Gm LC VCOs," *IEEE J. Sol. Stat. Circ.*, 38(8), pp. 1325–1332, August 2003.
25. Gan, H. and Simon Wong, S., "Integrated transformer baluns for RF low noise and power amplifiers," *Radio Frequency Integrated Circuits (RFIC) Symposium*, 2006 IEEE, June 11–13, 2006, 4 pp.

20

Metal Semiconductor Field Effect Transistors

Michael S. Shur
Rensselaer Polytechnic Institute

20.1 Introduction

Silicon Metal Oxide Semiconductor Field Effect Transistors (MOSFETs) dominate modern microelectronics. Gallium Arsenide Metal Semiconductor Field Effect Transistors (GaAs MESFETs) are "runners-up," and they find many important niche applications in high-speed or high-frequency circuits. After the first successful fabrication of GaAs MESFETs by Mead in 1966[1] and after the demonstration of their performance at microwave frequencies in 1967 by Hooper and Lehrer,[2] these devices emerged as contenders to silicon MOSFETs, bipolar transistors, and High-Electron Mobility Transistors. In the late 1970s and early 1980s, high-quality semiinsulating substrates and ion-implantation processing techniques made it possible to fabricate GaAs MESFET VLSI, such as 16×16 multiplier with a multiplication time of 10.5 ns and less than 1 W power dissipation.[3] More recently, CMOS and SiGe displaced GaAs MESFETs from digital electronics applications. However, these devices are still being widely used in analog RF electronics, because they have a relatively low noise and a high product of the cutoff frequency times the breakdown voltage. For example, Fujitsu has a 240 W output device in mass production operating at 2.14 GHz with the power-added efficiency of 54%.

The microwave performance of GaAs MESFETs approaches that of Heterostructure Field Effect Transistors (HFETs).[4] As discussed below, the record maximum frequency of oscillations and the record cutoff frequency, f_T, for GaAs MESFETs reached 190 and 168 GHz, respectively. Even though the record numbers of f_{max} and f_T for GaAs-and InP-based HFETs reach over 500 and 300 GHz, respectively, their more typical f_{max} and f_T are well within the reach of GaAs MESFET technology.

Emerging materials for MESFET applications are SiC and GaN—wide bandgap semiconductors that have a much higher breakdown voltage, a higher thermal conductivity, and a higher electron velocity than GaAs. SiC MESFETs are predicted to reach breakdown voltages above 100 V.[5]

However, SiGe and even advanced, deep submicron Si technologies emerge as a serious competitor to GaAs MESFET technology at relatively low frequencies (below 40 GHz or so). This trend toward SiGe and Si might be alleviated by a shift toward 150-mm GaAs substrates, which are now used by the leading GaAs IC manufacturers, such as Vitesse, Anadigics, Infineon, Motorola, Tektronix, and RFMD.

In this chapter, we first discuss the MESFET principle of operation. Then, we will review the material properties of semiconductors competing for applications in MESFETs and the properties of Schottky barrier contacts followed by a brief review of MESFET fabrication and MESFET modeling. We also consider wide bandgap semiconductor MESFETs, new emerging heterodimensional MESFETs, and discuss applications of the MESFET technology.

20.2   Principle of Operation

Figure 20.1 shows a schematic MESFET structure. In n-channel MESFETs, an n-type channel connects n^+ drain and source regions. The depletion layer under the Schottky barrier gate contact constricts the current flow across the channel between the source and drain.

The gate bias changes the depletion region thickness, and hence modulates the channel conductivity.

This device is very different from a silicon MOSFET, where a silicon dioxide layer separates the gate from the channel. MOSFETs are mainstream devices in silicon technology, and silicon MESFETs are not common. Compound semiconductors, such as GaAs, do not have a stable oxide, and a Schottky gate allows one to avoid problems related to traps in the gate insulator, such as hot electron trapping in the gate insulator, threshold voltage shift due to charge trapped in the gate insulator, and so on.

In the *normally off* (*enhancement mode*) MESFETs, the channel is totally depleted by the gate built-in potential even at zero gate voltage (Figure 20.2). The threshold voltage of normally-off devices is positive. In *normally-on* (*depletion mode*) MESFETs, the conducting channel has a finite cross-section at zero gate voltage. The drawback of normally-off MESFET technology is a limited gate voltage swing due to the low turn-on voltage of the Schottky gate. This limitation is much less important in depletion mode FETs with a negative threshold voltage. Also, this limitation is less important in low-power digital circuits operating with a low supply voltage.

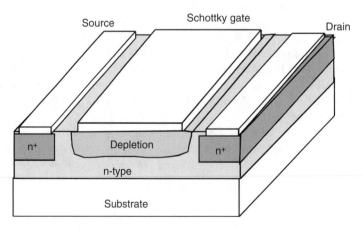

FIGURE 20.1   Schematic MESFET structure.

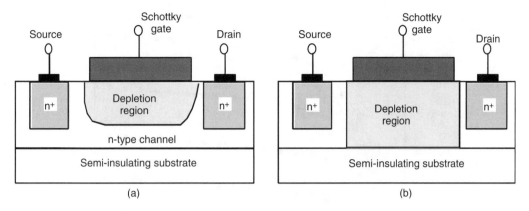

FIGURE 20.2 Normally-on (a) and normally-off (b) MESFETs at zero gate voltage.

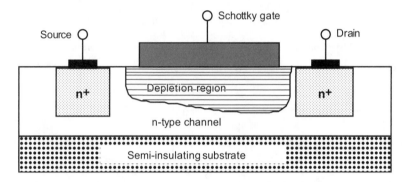

FIGURE 20.3 Depletion region in MESFET with positive drain bias.

Usually, the source is grounded, and the drain is biased positively. A schematic diagram of the depletion region under the gate of a MESFET for a finite drain-to-source voltage is shown in Figure 20.3.

The depletion region is wider and closer to the drain because the positive drain voltage provides an additional reverse bias across the channel-to-gate junction. With an increase in the drain-to-source bias, the channel at the drain side of the gate becomes more and more constricted. Finally, the velocity of electrons saturates leading to the current saturation (see Figure 20.4 that shows typical MESFET current–voltage characteristics).

MESFETs have been fabricated using many different semiconductor materials. However, GaAs MESFETs are mainstream MESFET devices. In many cases, GaAs MESFETs are fabricated by direct ion implantation into a GaAs semiinsulating substrate, making GaAs IC fabrication less complicated than silicon CMOS fabrication.

20.3 Properties of Semiconductor Materials Used in MESFET Technology

The effective mass of electrons in GaAs is very small (0.067 m_e in GaAs compared to 0.98 m_e longitudinal effective mass and 0.19 m_e transverse effective mass in Si, where m_e is the free electron mass). This leads to much higher electron mobility in GaAs—approximately 8500 cm^2/Vs in pure GaAs at room temperature compared to 1500 cm^2/Vs in Si. As shown in Figure 20.5, the electron velocity in GaAs exceeds that for the electrons in Si. This is an important advantage for modern day short-channel devices, where the electric fields are higher than the peak velocity field under normal operating conditions. The

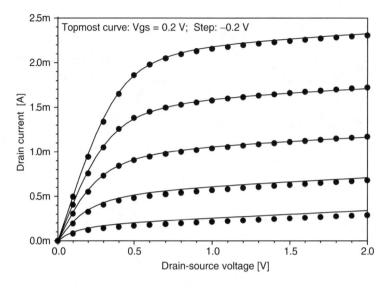

FIGURE 20.4 Measured (symbols) and simulated (using AIM-Spice, lines) drain current characteristics of MESFET operating at room temperature. (After Ytterdal et al., *IEEE Trans. Elect. Dev.*, Vol. 42, No. 10, pp. 1724–1734, 1995.)

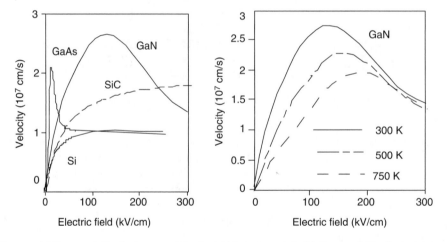

FIGURE 20.5 (a) Electron drift velocity at 300 K in GaN, SiC, GaAs, and Si. (b) Electron drift velocity in GaN at 300, 500, and 750 K. (Foutz et al., *Appl. Phys. Lett.*, Vol. 70, No. 21, pp. 2849–2851, 1997.)

light electrons in GaAs also experience so-called overshoot or even ballistic transport in short-channel devices,[7–11] where the electron transit time becomes comparable to or even smaller than the electron energy or even momentum relaxation time. This boosts the electron velocity well above the expected steady-state values (Figure 20.6). GaN also has a high-electron velocity and pronounced overshoot effects (see Figures 20.5 and 20.6).

Another important advantage of GaAs and related compound semiconductors is the availability of semiinsulating material that could serve as a substrate for device and circuit fabrication. A typical resistivity of semiinsulating GaAs is $10^7 \Omega$-cm or larger, compared to 2.5×10^5 Ω-cm for intrinsic silicon at room temperature. The semiinsulating GaAs is used as a substrate for fabricating GaAs MESFETs and other devices. Passive elements can be also fabricated on the same substrate, which is a big advantage for fabricating Monolithic Microwave Integrated Circuits (MMICs). As mentioned above, an important

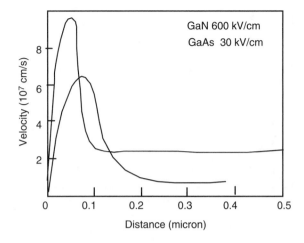

FIGURE 20.6 Computed velocity of electrons injected with low velocities into a constant electric field region into GaN and GaAs (After[13]).

TABLE 20.1 Material Properties of Si, GaAs, α-SiC, and GaN

Property	Si	GaAs	α-SiC (6H)	GaN
Energy gap (eV)	1.12	1.42	2.9	3.4
Lattice constant(a) Å	5.43107	5.6533	3.081	3.189
Lattice constant (c), Å	—	—	15.117	5.185
Density (g/cm^3)	2.329	5.3176	3.211	6.1
Dielectric constant	11.7	12.9	9.66($\perp$) 10.03($\parallel$)	9.5 (8.9)
Electron mobility (cm^2/Vs)	1,500	8,500	330	1,200
Hole mobility (cm^2/Vs)	500	400	60	< 30
Saturation velocity (m/s)	10^5	1.2×10^5	2–2.5×10^5	2–2.5×10^5
Electron effective mass ratio	0.92/0.19	0.067	0.25/1.5	0.22
Light hole mass ratio	0.16	0.076	0.33	0.7
OPTICAL phonon energy (eV)	0.063	0.035	0.104	0.092
Thermal conductivity (W/cm°C)	1.31	0.46	4.9	1.5

advantage of the GaAs MESFET is a possibility to fabricate these devices and integrated circuits using a direct implantation into the semiinsulating GaAs substrate.

Since GaAs, InP, and related semiconducting materials are direct gap materials, they are widely used in optoelectronic applications. Hence, electronic and photonic devices can be integrated on the same chip for use in optical interconnects or in optoelectronic circuits.

The direct band gap leads to a high-recombination rate, which improves radiation hardness. GaAs-based devices can survive over 100 megarads of ionizing radiation.[14]

GaAs and, especially SiC and GaN MESFETs are suitable for use in high-temperature electronics and as power devices (because of a small intrinsic carrier concentration and a high-breakdown field).

Table 20.1 summarizes important material properties of semiconductors used in MESFET technology.

20.4 Schottky Barrier Contacts

Figure 20.7 shows the energy band diagram of a GaAs Schottky barrier metal–semiconductor contact at zero bias. The Fermi level will be constant throughout the entire metal–semiconductor system, and the energy band diagram in the semiconductor is similar to that for an *n*-type semiconductor in a p^+-*n* junction.

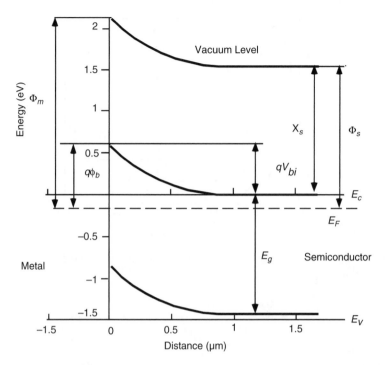

FIGURE 20.7 Simplified energy diagram of GaAs metal–semiconductor barrier. $q\phi_b$ is the barrier height (0.75 eV), X_s is the electron affinity in the semiconductor, Φ_s and Φ_m are the semiconductor and the metal work functions, and V_{bi} (0.59 V) is the built-in voltage. Donor concentration in GaAs is 10^{15} cm^{-3} (From Shur, *Introduction to Electronic Devices*, 1996.)

Energies Φ_m and Φ_s shown in Figure 20.7 are called the metal and the semiconductor *work functions*. The *work function* is equal to the difference between the *vacuum level* (which is defined as a free electron energy in vacuum) and the Fermi level. The *electron affinity* of the semiconductor, X_s (also shown in Figure 20.7), corresponds to the energy separation between the vacuum level and the conduction band edge of the semiconductor.

In the idealized picture of the Schottky junction shown in Figure 20.7, the energy barrier between the semiconductor and the metal is

$$q\phi_b = \Phi_m - X_s. \tag{20.1}$$

Here, ϕ_b is the Schottky barrier height, q is the electronic charge (in Equation 20.1 and in Figure 20.7, ϕ_b is measured in V and Φ_s and Φ_m are measured in Joules). Since $\Phi_m > \Phi_s$, the metal is charged negatively. The positive net space charge in the semiconductor leads to a band bending

$$qV_{bi} = \Phi_m - \Phi_s, \tag{20.2}$$

where V_{bi} is called the *built-in voltage*, in analogy with the corresponding quantity in a *p-n* junction.

Equation 20.1 and Figure 20.7 are not quite correct. In reality, a change in the metal work function, Φ_m, is not equal to the corresponding change in the barrier height, ϕ_b, as predicted by Equation 20.1. In actual Schottky diodes, ϕ_b increases with an increase in Φ_m but only by 0.1–0.3 eV when Φ_m increases by 1–2 eV. This difference is caused by interface states and is determined by the properties of a thin interfacial layer. However, even though a detailed and accurate understanding of Schottky barrier formation remains a challenge, many properties of Schottky barriers may be understood independently of the exact mechanism determining the barrier height. In other words, we can simply determine the effective barrier height from experimental data.

A forward bias decreases the potential barrier for electrons moving from the semiconductor into the metal and leads to an exponential rise in current. At high forward biases (approaching the built-in voltage), the voltage drop across the series resistance (composed of the contact resistance and the resistance of the neutral region between the ohmic contact and the depletion region) becomes important, and the overall current–voltage characteristic of a Schottky diode can be described by the following *diode equation*

$$I = I_s \left[\exp\left(\frac{V - IR_s}{\eta V_{th}}\right) - 1 \right],$$
(20.3)

where I_s is the saturation current, R_s is the series resistance, $V_{th} = k_B T / q$ is the thermal voltage, η is the ideality factor (η typically varies from 1.02 to 1.6), q is the electronic charge, k_B is the Boltzmann constant, and T is temperature.

The diode saturation current, I_s, is typically much larger for Schottky barrier diodes than in *p–n* junction diodes since the Schottky barrier height is smaller than the barrier height in *p-n* junction diodes. For a *p-n* junction, the height of the barrier separating electrons in the conduction band of the *n*-type region from the bottom of the conduction band in the *p*-region is on the order of the energy gap. For a Schottky diode, the barrier height could be very roughly estimated as 2/3 of the energy gap or less.

The current flow mechanism in Schottky diodes depends on the doping level. In a relatively low-doped semiconductor, the depletion region between the semiconductor and the metal is very wide, and electrons can only reach the metal going over the barrier. In higher doped samples, the barrier near the top is narrow enough for the electrons to tunnel through. Finally, in very highly doped structures, the barrier is thin enough for tunneling at the Fermi level. Figure 20.8 shows the band diagrams illustrating these three conduction mechanisms.

For low-doped devices, the saturation current density, j_{ss}, in a Schottky diode is given by

$$j_{ss} = A^* T^2 \exp\left(-\frac{q\phi_b}{k_B T}\right),$$
(20.4)

where A^* is called the Richardson constant, T is temperature (in degrees Kelvin), and k_B is the Boltzmann constant. For a conduction band minimum with the spherical surface of equal energy (such as the Γ minimum in GaAs),

$$A^* = \alpha \frac{m_n q k_B^2}{2\pi^2 \hbar^3} \approx 120\alpha \frac{m_n}{m_e} \left(\frac{A}{cm^2 K^2}\right),$$
(20.5)

where A^* is called the Richardson constant, T is temperature (in degrees Kelvin), and k_B is the Boltzmann constant. For a conduction band minimum with the spherical surface of equal energy (such as the Γ minimum in GaAs),

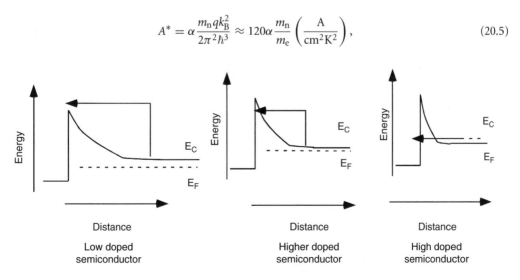

FIGURE 20.8 Current paths in low doped, higher doped, and high-doped Schottky diodes.

where m_n is the effective mass, m_e is the free electron mass, $\hbar$ is the Planck constant, and α is an empirical factor on the order of unity. The Schottky diode model described by Equations 20.4 and 20.5 is called the *thermionic emission model*. For Schottky barrier diodes fabricated on the {111} surfaces of Si, $A^* = 96$ A/(cm^{2}K^2). For GaAs, $A^* = 4.4$ A/(cm^{2}K^2).

As stated above, in higher doped semiconductors, the depletion region becomes so narrow that electrons can tunnel through the barrier near the top. This conduction mechanism is called thermionic-field emission.

The current–voltage characteristic of a Schottky diode in the case of thermionic-field emission (i.e., for higher doped semiconductors) under forward bias is given by

$$j = j_{stf} \exp\left(\frac{qV}{E_0}\right), \tag{20.6}$$

where

$$E_0 = E_{oo} \coth\left(\frac{E_{oo}}{k_B T}\right) \tag{20.7}$$

$$E_{oo} = \frac{qh}{4\pi}\sqrt{\frac{N_d}{m_n \varepsilon_s}} = 1.85 \times 10^{-11}\left[\frac{N_d(\text{cm}^{-3})}{(m_n/m_e)(\varepsilon_s/\varepsilon_o)}\right]^{1/2} \text{(eV)} \tag{20.8}$$

$$j_{stf} = \frac{A^* T\sqrt{\pi E_{oo}(\phi_b - qV - E_c + E_{Fn})}}{k_B \cosh(E_{oo}/k_B T)} \exp\left[-\frac{E_c - E_{Fn}}{k_B T} - \frac{(\phi_b - E_c + E_{Fn})}{E_0}\right] \tag{20.9}$$

Here E_c is the bottom of the conduction band in a semiconductor (outside of the depletion region, E_{Fn} is the electron quasi-Fermi level and ε_o is the dielectric permittivity of vacuum. In GaAs Schottky diodes, the thermionic-field emission becomes important for $N_d > 10^{17}$ cm^{-3} at 300 K and for $N_d > 10^{16}$ cm^{-3} at 77 K. In silicon, the corresponding values of N_d are several times larger.

In degenerative semiconductors, especially in semiconductors with a small electron effective mass, such as GaAs, electrons can tunnel through the barrier near or at the Fermi level, and the tunneling current is dominant. This mechanism is called *field emission*. The resistance of the Schottky barrier in the field emission regime is quite low. Metal-n^+ contacts operated in this regime are used as ohmic contacts.

Figure 20.9 shows a small signal equivalent circuit of a Schottky diode, which includes a parallel combination of the differential resistance of the Schottky barrier

$$R_d = \frac{dV}{dI} \tag{20.10}$$

and the differential capacitance of the space charge region:

$$C_{dep} = S\sqrt{\frac{qN_d\varepsilon_s}{2(V_{bi} - V)}}. \tag{20.11}$$

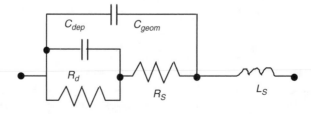

FIGURE 20.9   The small signal equivalent circuit of a Schottky diode.

Here V and I are the voltage drop across the Schottky diode and the current flowing through the Schottky diode, respectively, V_{bi} is the built-in voltage of the Schottky barrier, and N_d is the ionized donor concentration in the semiconductor. The equivalent circuit also includes the series resistance, R_s, which accounts for the contact resistance and the resistance of the neutral semiconductor region between the ohmic contact and the depletion region, the equivalent series inductance, L_s, and the device geometric capacitance:

$$C_{geom} = \varepsilon_S S/L, \qquad (20.12)$$

where L is the device length and S is the device cross-section.

20.5 MESFET Technology

The most popular MESFETs are GaAs MESFETs that find applications in both analog microwave circuits (including applications in Microwave Monolithic Integrated Circuits) and in digital integrated circuits. Ion-implanted GaAs MESFETs represent the dominant technology for applications in digital integrated circuits. They are also found in microwave applications. Figure 20.10 shows a typical process sequence for ion-implanted GaAs MESFETs (developed in late 70s[16,17]).

In a typical fabrication process, a GaAs semiinsulating substrate is coated with a thin silicon nitride (Si$_3$N$_4$) film. Implantations steps shown in Figure 20.10 are carried out through this layer. As shown in Figure 20.10, the first implant defines the active layer including the MESFET channel. A deeper and a

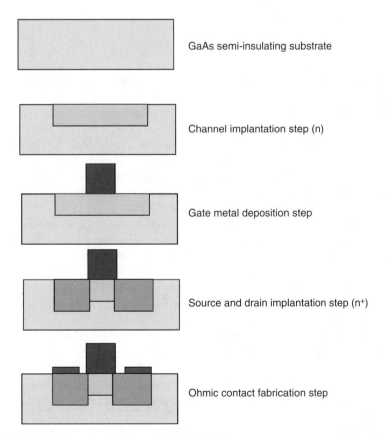

FIGURE 20.10   Fabrication steps for self-aligned GaAs MESFET.

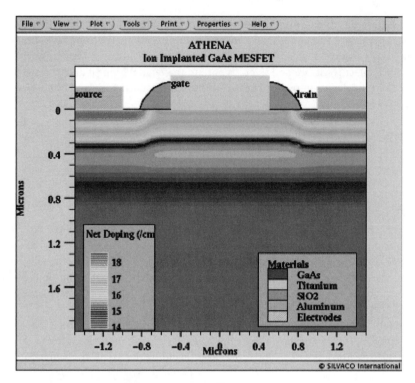

FIGURE 20.11   Simulated GaAs MESFET device structure.[19]

higher dose implant is used for ohmic contacts. This implant is often done as a self-aligned implant. In this case, a temperature stable refractory metal-silicide gate (typically tungsten silicide) is used as a mask for implanting the n^+ source and drain contacts. This technique reduces parasitic resistances. Also, this fabrication process is planar. However, the n^+ implant straggle under the gate might increase gate leakage current and also cause carrier injection into the substrate.[18]

After the implants, an additional insulator is deposited in order to cap the GaAs surface for the subsequent annealing step. This annealing (at 800°C or above) activates the implants.

Figure 20.11 (from[19]) shows simulated GaAs MESFET device structure. The doping in the gate region was formed by low dose 100 keV Be and Si implants and the source and drain regions were fabricated by higher dose 50 keV Si implants. The structure was annealed at 850°C.

For microwave applications, the devices are often grown by Molecular Beam Epitaxy. In this design, a top of n^+ layer doping extending from the source and drain contacts helps minimize the series resistances.

Figure 20.12 shows the recessed gate MESFET structure, where the thickness of the active layer under the gate is reduced. A thick n-doped layer between the gate, the source, and the drain contacts leads to a relatively low series source and drain resistances. The position and the shape of the recess are very important, since they strongly affect the electric field distribution and the device breakdown voltage.

In power devices, the gate contact in the recess is usually closer to the source than to the drain (Figure 20.13). Such placement reduces the source parasitic resistance and enhances the drain-source breakdown voltage by allowing additional expansion space for the high-field region at the drain side of the gate.

Another important issue is the reduction of the gate series resistance. This can be achieved by using a T-shape gate or a so-called mushroom gate (which might be obtained by side etching the gate), see Figure 20.14.

In this design, the gate series resistance is reduced without an increase in the gate length, which determines the device cut-off frequency.

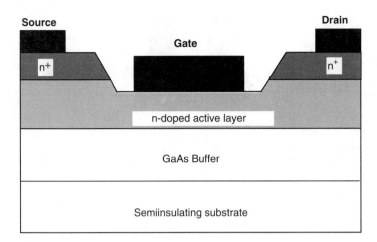

FIGURE 20.12   Recessed gate structure.

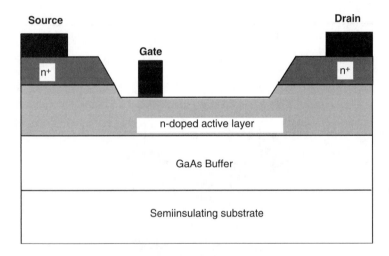

FIGURE 20.13   Recessed structure with offset gate for power devices.

MESFETs are usually passivated by a Si_3N_4 layer. This passivation affects the surface states and the surface depletion layer, stress-related and piezoelectric effects can lead to shifts in the threshold voltage.[20]

A more detailed discussion of GaAs MESFET fabrication can be found in references [21,22]

Wide bandgap semiconductors, such as SiC, GaN, and related materials, might potentially compete with GaAs for applications in MESFETs and other solid-state devices (Figure 20.15).

SiC exists in more than 170 polytypes. The three most important polytypes are hexagonal 6H (α-SiC) and 4H, and cubic 3C (β-SiC). As stated in Table 20.1, SiC has the electron saturation drift velocity of 2×10^7 cm/s (approximately twice that of silicon), a breakdown field larger than 2,500–5,000 kV/cm (compared to 300 kV/cm for silicon), and a high-thermal conductivity of 4.9 W/cm°C (compared to 1.3 W/cm°C for silicon and 0.5 W/cm°C for GaAs).

These properties make SiC important for potential applications in high-power, high-frequency devices as well as in devices operating at high temperatures and/or in harsh environments.

Palmour et al.[24] reported operation of α-SiC MESFETs at a temperature of 773 K. In a 6H-SiC MES-FET fabricated by CREE (gate length 24 μm, channel depth 600 nm, doping 6.5×10^{16} cm^{-3}), the room temperature transconductance was approximately 4 mS/mm. At elevated temperatures, the device transconductance decreases owing to the decrease in mobility. MESFETs did not exhibit breakdown even

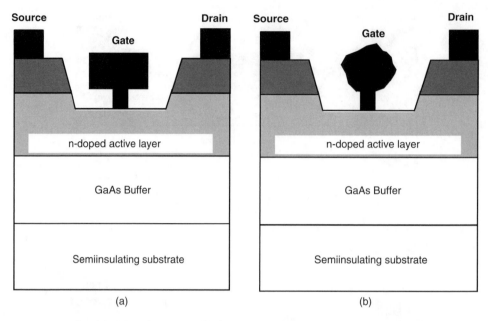

FIGURE 20.14   T-gate (a) and mushroom gate (b) for gate series resistance reduction.

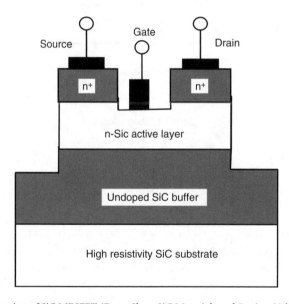

FIGURE 20.15   Cross-section of SiC MESFET (From Shur, *SiC Materials and Devices*, Vol. 52, pp. 161–193 1998.

with drain voltages up to 100 V. Using the square law MESFET model (described below), Kelner and Shur[25] estimated that the field effect mobility in these MESFETs was approximately 300 cm^2/Vs. β-SiC MESFETs have also been fabricated[26] but α-SiC MESFETs exhibit a better performance because of a better material quality.

α-SiC MESFETs have achieved microwave operation.[27] A cutoff frequency of 5 GHz, 12 dB gain at 2 GHz and a breakdown voltage of 200 V was demonstrated in an α-SiC MESFET with a 0.4 μm gate length.[28]

GaN is another material that is potentially important for MESFET applications. For GaN at room temperature and with an n-type doping density of 10^{17} cm^{-3}, Monte Carlo simulations predict a high-peak velocity (2.7×10^5 m/s), a high-saturation velocity (1.5×10^5 m/s), and a high-electron mobility (1000 cm^2/Vs).[29–32] Khan et al.[33] and Binari et al.[34] reported on microwave performance of GaN MESFETs. However, most of the research on GaN-based FETs has concentrated on GaN-based Heterostructure Field Effect Transistors.[35–38]

20.6 MESFET Modeling

MESFET modeling has been done at several different levels. Most advanced numerical simulation techniques rely on self-consistent simulation based on the Monte Carlo approach. In this approach, random number generators are used to simulate random electron-scattering processes. The motion of these electrons is simulated in the electric field that is calculated self-consistently by solving the Poisson equation iteratively. The particle movements between scattering events are described by the laws of classical mechanics, while the probabilities of the various scattering processes and the associated transition rates are derived from quantum mechanical calculations. Some of the results obtained by using this approach were reviewed in reference[39]. Table 20.2 from this reference[39] describes the Monte Carlo algorithm in more detail.

Self-consistent Monte Carlo simulations are very useful for revealing the device physics and verifying novel device concepts and ideas.[23,40–44]

A less rigorous but also a less numerically demanding approach relies on solving the balance equations. These partial differential equations describe conservation laws derived from the Boltzmann transport equation.[45,46] Two-dimensional device simulators based on the balance equations and on the drift-diffusion model can be used to optimize device design and link the device characteristics to the device fabrication process.[47–48]

A more simplistic but also an easier approach is to use conventional drift-diffusion equations implemented in commercial two-dimensional and three-dimensional device simulators such as ATLAS, MEDICI, or DESSIS. However, even this approach might be too complicated and too numerically involved for the simulation of MESFET-based digital VLSI and/or for the simulation of MESFET-based analog circuits.

The simplest model that relates the MESFET current–voltage characteristics to the electron mobility, the electron saturation velocity, the device dimensions, and applied voltages is called the square-law model. This model predicts the following equation for the drain saturation current:

$$I_{\text{sat}} = \beta (V_{\text{GS}} - V_{\text{T}})^2, \qquad (20.13)$$

TABLE 20.2 Monte Carlo Algorithm[39]

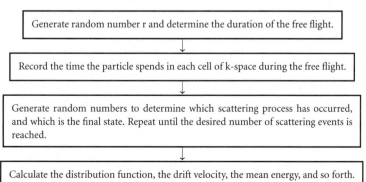

where[50,51]

$$\beta = \frac{2\varepsilon_s \mu_n v_s W}{A \left(\mu_n V_{po} + 3 v_s L\right)} \tag{20.14}$$

is the transconductance parameter,

$$V_T = V_{bi} - V_{po} \tag{20.15}$$

is the threshold voltage, V_{GS} is the intrinsic gate-to-source voltage, and

$$V_{po} = \frac{q N_d A^2}{2\varepsilon_s} \tag{20.16}$$

is the pinch-off voltage. Here A is the channel thickness, μ_n is a low-field mobility, and v_s is the electron saturation velocity.

This "square law" model is fairly accurate for devices with relatively low pinch-off voltages ($V_{po} = V_{bi} - V_T \le 1.5 \sim 2$ V). For devices with higher pinch-off voltages, the model called the *Raytheon model* (which is implemented in many versions of SPICE) yields a better agreement with experimental data:

$$I_{sat} = \frac{\beta (V_{GS} - V_T)^2}{1 + t_c (V_{GS} - V_T)} \tag{20.17}$$

Here t_c is an empirical parameter that depends on the doping profile in the MESFET channel. Another empirical model (called the *Sakurai–Newton model*) is also quite useful for MESFET modeling:

$$I_{sat} = \beta_{sn} (V_{GS} - V_T)^{m_{sn}}. \tag{20.18}$$

The advantage of this model is simplicity. The disadvantage is that the empirical parameters β_{sn} and m_{sn} cannot be directly related to the device and material parameters. (The Sakurai–Newton model is implemented in several versions of SPICE. In AIM-Spice [49,50,], this model is implemented as Level 6 MOSFET model.)

The source and drain series resistances, R_s and R_d, may play an important role in determining the current–voltage characteristics of GaAs MESFETs. The intrinsic gate-to-source voltage, V_{GS}, is given by

$$V_{GS} = V_{gs} - I_{ds} R_s, \tag{20.19}$$

where V_{gs} is the applied (extrinsic) gate-to-source voltage. Substituting Equation 20.19 into Equation 20.16 and solving for I_{sat} we obtain

$$I_{sat} = \frac{2\beta V_{gt}^2}{1 + 2\beta V_{gt} R_s + \sqrt{1 + 4\beta V_{gt} R_s}}. \tag{20.20}$$

In device modeling suitable for computer-aided design, one has to model the current–voltage characteristics in the entire range of drain-to-source voltages, not only in the saturation regime. In 1980, Curtice proposed the use of a hyperbolic tangent function for the interpolation of MESFET current–voltage characteristics

$$I_d = I_{sat} (1 + \lambda V_{ds}) \tanh\left(\frac{g_{ch}}{I_{sat}}\right), \tag{20.21}$$

where

$$g_{ch} = \frac{g_i}{1 + g_i(R_s + R_d)} \tag{20.22}$$

is the MESFET conductance at low drain-to-source voltages, and

$$g_i = g_{cho}\left(1 - \sqrt{\frac{V_{bi} - V_{GS}}{V_{po}}}\right) \tag{20.23}$$

is the intrinsic channel conductance at low drain-to-source voltages predicted by the Shockley model.

The constant λ in Equation 20.21 is an empirical constant that accounts for the output conductance in the saturation regime. This output conductance may be related to short-channel effects and also to parasitic leakage currents in the substrate. Hence, output conductance may be reduced by using a heterojunction buffer layer between the device channel and the substrate or by using a p-type buffer layer. Such a layer creates an additional barrier, which prevents carrier injection into the substrate.[18]

The *Curtice model* is implemented in PSpice[tm]. The Curtice model and the Raytheon model [see Equation (20.21)] have become the most popular models used for MESFET circuit modeling. A more sophisticated model, which describes both subthreshold and above-threshold regimes of MESFET operation, is implemented in AIM-Spice. This model accurately reproduces current–voltage characteristics over several decades of currents and is suitable for both analog and digital circuit simulations.[54] One of the simulation results obtained using this model is depicted in Figure 20.16.

In order to simulate MESFET circuits, one also needs to have a model describing the MESFET capacitances. Meyer[56] proposed a simple charge-control model, in which device capacitances ($C_{ij} = C_{ji}$) were obtained as derivatives of the gate charge with respect to the various terminal voltages. Fjeldly et al.[52,54]

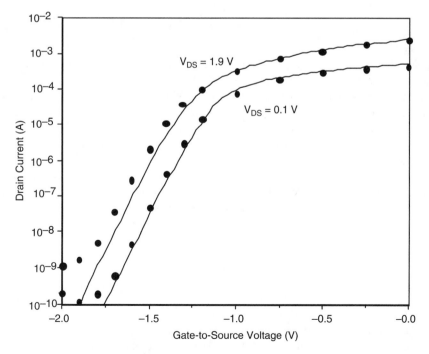

FIGURE 20.16 Subthreshold experimental (symbols) and calculated (solid lines) I–V characteristics for ion-implanted MESFET with nominal gate length $L = 1$ μm.[55]

approximated a unified gate-channel capacitance, C_{gc}, of a MESFET at zero drain-source bias by the following combination of the above-threshold capacitance, C_a, and the below-threshold capacitance, C_b,:

$$C_{gc} = \frac{C_a C_b}{C_a + C_b}. \tag{20.24}$$

This approach in conjunction with Meyer's model leads to the following expressions for the gate-source (C_{gs}) and gate-drain capacitance (C_{gd}) valid for the subthreshold and the above-threshold regimes:

$$C_{gs} = \frac{2}{3} C_{gc} \left[1 - \left(\frac{V_{sat} - V_{dse}}{2V_{sat} - V_{dse}} \right)^2 \right], \tag{20.25}$$

$$C_{gs} = \frac{2}{3} C_{gc} \left[1 - \left(\frac{V_{sat}}{2V_{sat} - V_{dse}} \right)^2 \right]. \tag{20.26}$$

Here, V_{sat} is the extrinsic saturation voltage and V_{dse} is an effective extrinsic drain-source voltage that equals V_{ds} for $V_{ds} < V_{sat}$ and V_{sat} for $V_{ds} > V_{sat}$.

A more accurate model of the intrinsic capacitances requires an analysis of the variation of the charge distribution in the channel versus terminal bias voltages. For the MESFET, the depletion charge under the gate has to be partitioned between the source and drain terminals.[57]

Finally, the gate leakage current has to be modeled in order to accurately reproduce MESFET current voltage characteristics in the entire range of bias voltages including positive gate biases. To a first order approximation, the gate leakage current can be described in terms of simple diode equations assuming that each "diode" represents half of the gate area:

$$I_g = J_{ss} \frac{LW}{2} \left[\exp \left(\frac{V_{gs}}{m_{gs} V_{th}} \right) + \exp \left(\frac{V_{gd}}{m_{gd} V_{th}} \right) - 2 \right]. \tag{20.27}$$

Here L and W are the gate length and width, respectively, J_{ss} is the saturation current, V_{gs} and V_{gd} are gate-to-source and gate-to-drain voltages, V_{th} is the thermal voltage, and m_{gs} and m_{gd} are the ideality factors. Figure 20.17 shows a more accurate equivalent circuit, which accounts for the effect of the leakage current on the drain current. In the equivalent circuit shown in Figure 20.17, this effect is accounted for by the current controlled current source, I_{corr}.

Here, J_{ss} is the reverse saturation current density, and m_{gs} and m_{gd} are the gate-source and gate-drain ideality factors, respectively.

A more accurate description proposed by Berroth et al.[58] introduced effective electron temperatures at the source side and the drain side of the channel. The electron temperature at the source side of the channel, T_s, is taken to be close to the lattice temperature, and the drain side electron temperature, T_d, is assumed to increase with the drain-source voltage to reflect the heating of the electrons in this part of the channel. The resulting gate leakage current can be written as

$$I_g = J_{gs} \frac{LW}{2} \left[\exp \left(\frac{V_{gs}}{m_{gs} V_{ths}} \right) - 1 \right] + \frac{LW}{2} \left[J_{gd} \exp \left(\frac{V_{gd}}{m_{gd} V_{thd}} \right) - J_{gs} \right], \tag{20.28}$$

where J_{gs} and J_{gd} are the reverse saturation current densities for the gate-source and the gate-drain diodes, respectively, and $V_{ths} = k_B T_s / q$ and $V_{thd} = k_B T_d / q$. The second term in Equation 20.28 accounts for the gate-drain leakage current and for the fact that the effective temperature of the electrons in the metal is maintained at the ambient temperature.

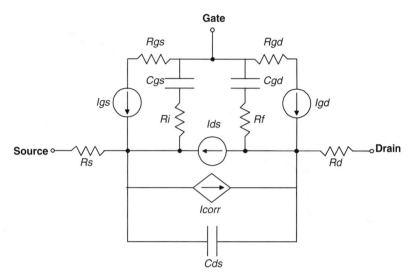

FIGURE 20.17   MESFET equivalent circuit (After[52]).

In GaAs MESFETs, the reverse gate current is usually also dependent on the reverse bias.[59] The following expression accounts for this dependence[52]:

$$I_g = J_{gs} \frac{LW}{2} \left[\exp\left(\frac{V_{gs}}{m_{gs} V_{ths}} \right) - 1 \right] + \frac{LW}{2} g_{gs} V_{gs} \exp\left(-\frac{q V_{gs} \delta_g}{k_B T_s} \right)$$
$$+ \frac{LW}{2} \left[J_{gd} \exp\left(\frac{V_{gd}}{m_{gd} V_{thd}} \right) - J_{gs} \right] + \frac{LW}{2} g_{gd} V_{gd} \exp\left(-\frac{q V_{gd} \delta_g}{k_B T_s} \right), \qquad (20.29)$$

where g_{gs} and g_{gd} are the reverse diode conductances and δ_g is the reverse bias conductance parameter. However, using the above expressions directly will cause a kink in the gate current and a discontinuity in its derivatives at zero applied voltage. Equation 20.29 is valid for both negative and positive values of V_{gs} and V_{gd}.

Figure 20.18 compares the measured gate leakage current with the model implemented in AIM-Spice and described above.[6]

Figure 20.19 shows that the GaAs MESFET model implemented in AIM-Spice accurately reproduces the differential characteristics of the devices. Therefore, this model is suitable for the simulations of analog, microwave, and mixed-mode circuits.

20.7   Heterodimensional (2D MESFETs)

Heterodimensional MESFET technology utilizes the Schottky contact between a 3D metal and a 2D electron gas in a semiconductor. This technology holds promise for the fabrication of high-speed devices with low-power consumption.[60–65] However, this is still a very immature technology that has not found its way into production.

Figure 20.20 shows the 3D–2D Schottky barrier junction.

The depletion width d_{dep} of the semiconductor 2-D electron gas for a reverse biased 3D–2D Schottky barrier shows a linear instead of a square root dependence of voltage[58]:

$$d_{dep} = \frac{\varepsilon}{q n_s} (V_{bi} - V). \qquad (20.30)$$

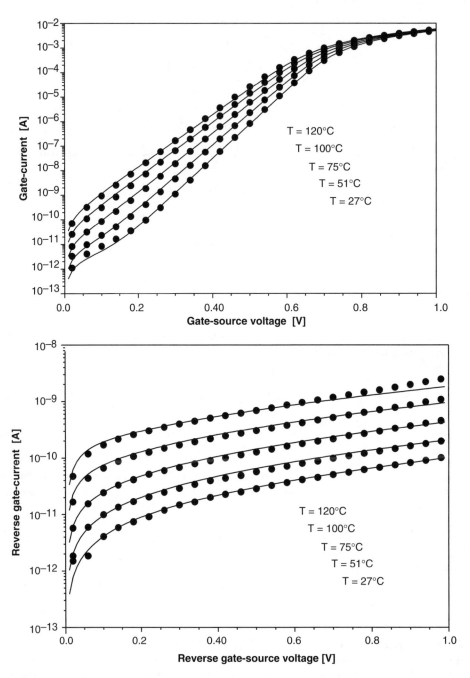

FIGURE 20.18   Measured (symbols) and simulated (lines) gate current versus gate bias for (a) positive and (b) negative gate-source voltages at different temperatures. Temperature parameters: $\Phi_{b1} = 0.96$ meV/K, $\xi = 0.033$ K^{-1} (After Ytterdal et al., *IEEE Trans. Elect. Dev.*, Vol. 42, No. 10, pp. 1724–1734, 1995, © IEEE.)

Here n_s is the sheet density in the 2-D electron gas (2-DEG), V_{bi} is the built-in voltage of the junction, and V is the voltage applied to the junction.

Figure 20.21 shows the two-dimensional metal–semiconductor field-effect transistor (2-D MESFET). This transistor utilizes Schottky gates on both sides of a degenerative 2-DEG channel to laterally modulate the current between the drain and source.[67-69]

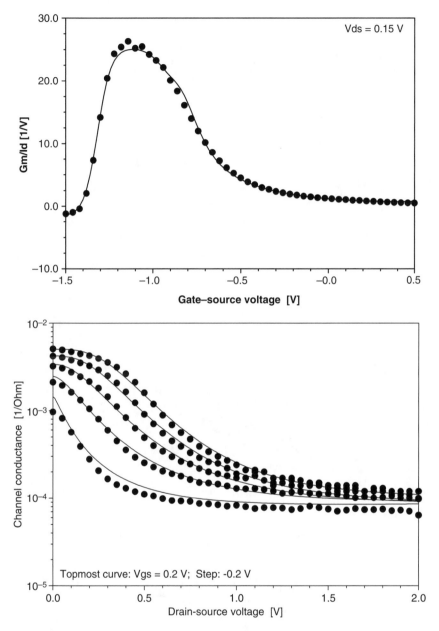

FIGURE 20.19 Measured (symbols) and simulated using AIM-Spice (lines) drain current characteristics of device operating at room temperature; (a) ratio of transconductance and drain current, (b) channel conductance (After Ytterdal et al., *IEEE Trans. Elect. Dev.*, Vol. 42, No. 10, pp. 1724–1734, 1995, © IEEE.)

The novel geometry of this 2-D MESFET eliminates or reduces parasitic effects associated with top planar contacts of conventional FETs, such as narrow-channel and short-channel effects. The output conductance in the saturation regime is quite small, and the junction capacitance of the 3D-2D Schottky diode is also small. This results in a low power-delay product.

The functionality of the 2-D MESFET can be further enhanced by using multiple gates on both sides of the channel, as shown in Figure 20.22. Two- and three-gate 2-D MESFETs with excellent electrical performance have been fabricated.[71] The 2-D MESFET also holds promise for microwave analog applications,

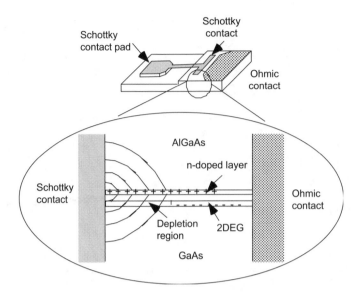

FIGURE 20.20   Schematic structure of a 3D-2D Schottky diode (After Peatman et al., *IEEE Elect. Dev. Lett.*, Vol. 13, No. 1, 1992, © IEEE.)

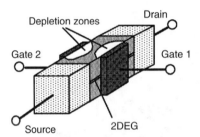

FIGURE 20.21   Schematic structure of a 2-D MESFET (After Peatman et al., *Proc. 1993 IEEE/Cornell Conf*, p. 314, 1993.)

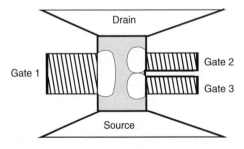

FIGURE 20.22   Structure of a three-gate 2-D MESFET (After Iniguez et al., *IEEE Trans. Elect. Dev.*, Vol. 46, No. 8, 1999.)

where the small capacitance of the 3D-2D junction should lead to a low channel and amplifier noise. Also, the high transconductance, and the fact that the transconductance and the output conductance do not vary much across the broad range of gate biases, are advantageous factors for linear amplification.

The 2-D MESFET prototype devices were fabricated using a pseudomorphic $Al_{0.25}Ga_{0.75}As/In_{0.2}Ga_{0.8}As$ heterostructure grown on a semi-insulating GaAs substrate.[68–70,72–74] Ni/Ge/Au ohmic contacts were formed using standard contact UV lithography and evaporation/lift-off techniques. The gate pattern was

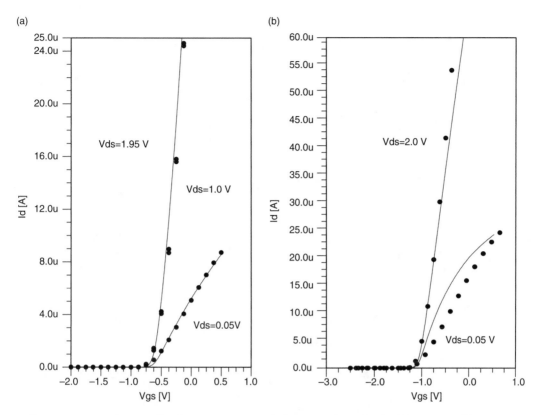

FIGURE 20.23 I–V characteristics for 2-D MESFETs with the gates tied together for (a) $L = 3~\mu m$, (b) $L = 0.5~\mu m$ (From Iniguez et al., *IEEE Trans. Elect. Dev.*, Vol. 46, No. 8, 1999.)

defined using electron beam lithography. The Pt/Au gates were deposited into the gate trench using capacitor discharge electroplating. Cr/Au ohmic contact pads were evaporated on the wafer and a wet etch was used to isolate the ohmic and Schottky pads (see reference[64] for more details.)

A unified 2-D MESFETs is described in Ref.[72] Figures 20.23 and 20.24 show the comparison between the measured and calculated 2D-MESFET I-V characteristics.

20.8 Applications

GaAs MESFETs play an important role in both analog and digital applications, such as for satellite and fiber-optic communication systems, in the cellular phones and other wireless equipment, in automatic IC test equipment, and for other diverse civilian and military uses. GaAs MESFETs have been used in highly efficient microwave power amplifiers, since they combine the low ON resistance and the high cutoff frequency. GaAs semi-insulating substrates also present a major advantage for microwave applications, since they decrease parasitic capacitance and allow for fabrication of passive elements with low parasitics for microwave monolithic integration. GaAs MESFETs have also found applications in linear low-noise amplifiers.

Figure 20.25 compares f_T and f_{max} for different high-frequency technologies.

As can be seen from the figure, GaAs MESFETs exhibit quite respectable microwave performance and, given a lower cost of GaAs MESFETs compared to more advanced heterostructure devices, they could capture a sizeable portion of the microwave market.

As an example of GaAs MESFET performance, we can point out power GaAs MESFET developed by Oki Electric Industry Co., Ltd. This device (Figures 20.26 and 20.27) reached 10 W output power in

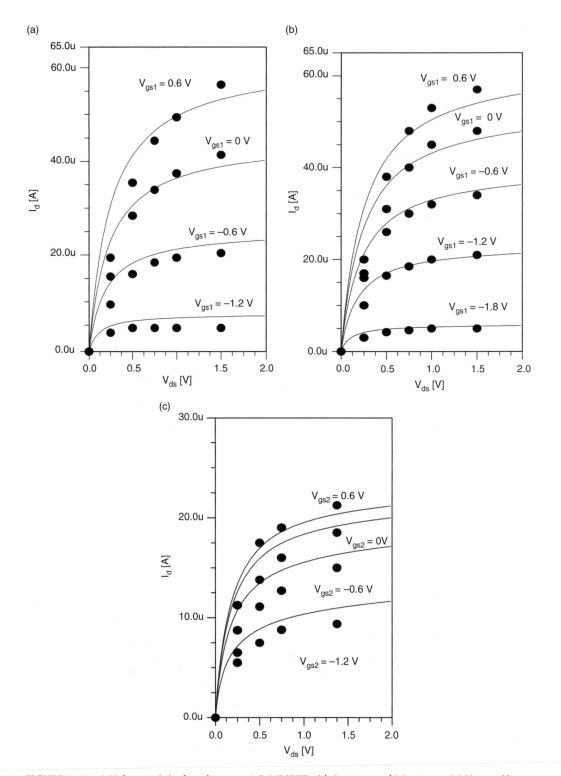

FIGURE 20.24 I–V characteristics for a three-gate 2-D MESFET with $L = 2\,\mu m$ and $W = 1\,\mu m$; (a) $V_{GS1} = V_{GS2} = V_{GS2} = V_{GS3}$, (b) $V_{GS2} = V_{GS3} = 0.6$ V. (c) $V_{GS1} = -1.2$ V, $V_{GS3} = 0.6$ V. Symbols: measurements. Solid lines: AIM-Spice simulations.

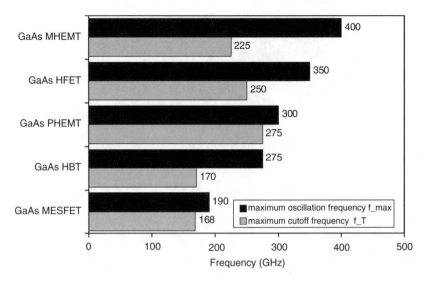

FIGURE 20.25 Maximum frequency of oscillation f_{max} and maximum cutoff frequency f_T of different GaAs technologies (After Werner and Shur, GaAs Microwave Transistors, unpublished (2000), http://www.oki.com/en/press/2005/z05006e.html).

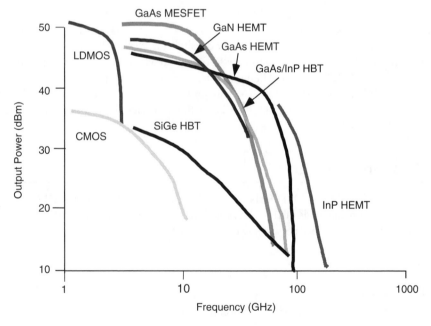

FIGURE 20.26 Maximum output power for different technologies.[74]

the 6 GHz range with high efficiency (over 55%) and high voltage (over 20 Volts) operation.[75] This was achieved by using gate recess structure and an optimized layer structure of the channel region.

GaAs MESFET technology has also been used in efficient DC-to-DC converters that demonstrated a high-switching speed.[76] These devices are capable of operating at higher switching speeds and require a less complex circuitry.

Vitesse Semiconductor Corporation is one of the leaders in digital GaAs MESFET technology. VSC8141 and the VSC8144 SONET/SDH OC-48 multirate transceivers include multiplexer and demultiplexer with integrated clock generation capabilities for the physical layer.[77] Both ICs dissipate the lowest power

FIGURE 20.27 Oki Electric Industry Co., Ltd MESFET (After http://www.oki.com/en/press/2005/z05006e.html).

available in the industry today—1.2 W typical. These integrated circuits are suitable for transmission systems, optical networking equipment, networking and digital cross connect systems, and they have 20% lower power dissipation than competing products.

CREE inc. developed an impressive SiC MESFET 0.45 μm gate length MMIC process that allowed for 50-volt drain operation. The circuits include NiCr resistors and high-voltage MIM capacitors. Typical RF MESFET performance at 3.5 GHz is 4 W/mm power density with 2 dB small signal gain 60% PAE.[78]

Toshiba Electronics, Europe has a new industry-leading product—a small, low-profile GaAs MESFET single-pole/dual-throw (SPDT) switch. The new switch can be used in multiband/multimode cellular antenna switch modules, Bluetooth modules and wireless LAN applications. The device has low insertion loss and high isolation and operates from DC to 3GHz. Chip scale package (CSP) mass production line is used to manufacture this device at low cost.[79]

20.9 To Probe Further

Several books dealing with GaAs MESFETs include "Microwave MESFET's and HEMT's Edited by J. Michael Golio,[80] "Fundamentals of III-V Devices: HBTs, MESFETs, and HFETs/HEMTs" by W. Liu Wiley,[81] and "Electrical and Thermal Characterization of MESFETs, HEMTs, and HBTs" by Robert Anholt.[82]

Acknowledgment

The author is grateful to Tobias Werner for useful comments.

References

1. C. A. Mead, "Schottky barrier gate field effect transistor," *Proc. IEEE*, Vol. 54, No. 2, pp. 307–308, February 1966.
2. W. W. Hooper and W. I. Lehrer, "An epitaxial GaAs field-effect transistor," *Proc. IEEE*, Vol. 55, No. 7, pp. 1237–1238, July 1967.
3. Y. Nakayama, K. Suyama, H. Shimizu, N. Yokoyama, H. Ohnishi, A. Shibatomi, and H. Ishikawa, "A 16x16 bit parallel multiplier," *IEEE J. Sol. Stat. Circ.*, Vol. SC-18, pp. 599–603, 1983.
4. M. Feng, C. L. Lau, V. Eu, and C. Ito, "Does the two-dimensional electron gas effect contribute to high-frequency and high speed performance of field-effect transistors?" *Appl. Phys. Lett.* Vol. 57, p. 1233, 1990.

5. A. Mills, "The GaAs IC business never so healthy! GaAs IC report," *III-Vs Review*, Vol. 13, No. 1, pp. 35–39, January 2000.

6. T. Ytterdal, B-J. Moon, T. A. Fjeldly, and M. S. Shur, "Enhanced GaAs MESFET CAD model for a wide range of temperatures," *IEEE Trans. Elect. Dev.*, Vol. 42, No. 10, pp. 1724–1734, 1995.

7. M. S. Shur and L. F. Eastman, "Ballistic transport in semiconductors at low-temperatures for low power high speed logic," *IEEE Trans. Elect. Dev.*, Vol. 26, No. 11, pp. 1677–1683, November 1979.

8. M. Heiblum, M. I. Nathan, D. C. Thomas, and C. M. Knoedler, "Direct observation of ballistic transport in GaAs," *Phys. Rev. Lett.*, Vol. 55, p. 2200, 1985.

9. A. F. J. Levi, J. R. Hayes, P. M. Platzman, and W. Wiegmann, "Injected hot electron transport in GaAs," *Phys. Rev. Lett.*, Vol. 55, pp. 2071–2073, 1985.

10. G. Ruch, "Electronics dynamics in short channel field-effect transistors," *IEEE Trans. Elect. Dev.*, Vol. ED-19, pp. 652–654, 1972.

11. A. Cappy, B. Carnes, R. Fauquembergue, G. Salmer, and E. Constant, "Comparative potential performance of Si, GaAs, GaInAs, InAs submicrometer-gate," *IEEE Trans. Elect. Dev.*, Vol. ED-27, pp. 2158–2168, 1980.

12. M. S. Shur and M. Asif Khan, Electronic and Optoelectronic AlGaN/GaN Heterostructure Field Effect Transistors, in Proceeding of the Symposium on Wide Band Gap Semiconductors and the Twenty-Third State-of-the-Art Program on Compound Semiconductors (SOTAPOCS XXIII), F. Ren, D. N. Buckley, S. J. Pearton, G. Van Daele, G. C. Chi, T. Kamijoh, and F. Schuermeyer, Editors, Proceedings Vol. 95-21, pp. 128–135, The Electrochemical Society, inc., New Jersey, 1995.

13. B. E. Foutz, L. F. Eastman, U. V. Bhapkar, and M. S. Shur, "Comparison of high electron transport in GaN and GaAs," *Appl. Phys. Lett.*, Vol. 70, No. 21, pp. 2849–2851, 1997.

14. S. Roosild, in *Microprocessor Design for GaAs Technology*, V. Milutinovic, Editor, Prentice Hall Advanced Reference Series, Engineering, New Jersey, 1990.

15. M. S. Shur, *Introduction to Electronic Devices*, John Wiley and Sons, New York, 1996.

16. J. A. Higgins, R. L. Kuvaas, F. H. Eisen, and D. R. Chen, "Low-noise GaAs FET's prepared by ion implantation," *IEEE Trans. Elect. Dev.*, Vol. ED-25, pp. 587–596, 1978.

17. B. M. Welch and R. C. Eden, "Planar GaAs integrated circuits fabricated by ion implantation," *Int. Solid State Circuits Conf. Tech. Digest*, pp. 205–208, 1977.

18. L. F. Eastman and M. S. "Shur, Substrate Current in GaAs MESFET's," *IEEE Trans. Elect. Dev.*, Vol. ED-26, No. 9, pp. 1359–61, September 1979.

19. http://www.silvaco.com/products/vwf/athena/ss4/ss4_br.html

20. C. H. Chen, A. Peczalski, M. S. Shur, and H. K. "Chung, Orientation and Ion-implanted transverse effects in self-aligned GaAs MESFETs," *IEEE Trans. Elect. Dev.*, Vol. ED-34, No. 7, pp. 1470–1481, July 1987.

21. M. S. Shur, *GaAs Devices and Circuits*, Plenum Publishing Corporation, New York, 1987.

22. I. Brodie and J. J. Murray, *The Physics of Microfabrication*, Plenum Press, New York and London, 1982.

23. M. S. Shur, "SiC Transistors," in *SiC Materials and Devices*, Y. S. Park, Editor, Academic Press, Semiconductors and Semimetals, Vol. 52, pp. 161–193, 1998.

24. J. W. Palmour, H. S. Kong, D. G. Waltz, J. A. Edmond, C. H. Carter, Jr., "6H-silicon carbide transistors for high temperature operation," *Proc. of First Intern. High Temperature Electronics Conference*, Albuquerque, NM, pp. 511–518, 1991.

25. G. Kelner and M. Shur, "SiC Devices," in *Properties of Silicon Carbide*, G. Harris, Editor, M. Faraday House, IEE, England, 1995.

26. G. Kelner, S. Binari, K. Sleger, and H. Kong, "β-SiC MESFET's and buried-gate JFET's," *IEEE Electr. Dev. Lett.*, Vol. 8, p. 428, 1987.

27. J. W. Palmour and J. A. Edmond, "Epitaxial thin film growth and device development in monocrystalline alpha and beta silicon carbide," *Proc. 14th IEEE Cornell Conf.*, Ithaca, NY, 1991.

28. S. Sriram, R. C. Clarke, M. H. Hanes, P. G. McMullin, C. D. Brandt, T. J. Smith, A. A. Burk, Jr., H. M. Hobgood, D. L. Barrett, and R. H. Hopkins, "SiC microwave power MESFETS," *Inst. Phys. Conf. Ser.*, Vol. 137, pp. 491–494, 1993.

29. R. F. Davis, G. Kelner, M. Shur, J. W. Palmour, and J. A. Edmond, "Thin film deposition and microelectronic and optoelectronic device fabrication and characterization in monocrystalline alpha and beta silicon carbide," *Proc. of IEEE*, Vol. 79, No. 5, pp. 677–701, May 1991.

30. H. Morkoc, S. Strite, G. B. Gao,. M. E. Lin, B. Sverdlov, and M. Burns, "Large-band-gap SiC, III-V Nitride, and II-VI ZnSe-based semiconductor device technologies," *J. Appl. Phys.*, Vol. 76, No. 3, pp. 1363–1398, August 1994.

31. M. A. Littlejohn, J. R. Hauser, and T. H. Glisson, "Monte Carlo calculation of the velocity-field relationship for gallium nitride," *Appl. Phys. Lett.* Vol. 26, p. 625, 1975.

32. B. Gelmont, K. S. Kim, and M. Shur, "Monte Carlo simulation of electron transport in gallium nitride," *J. Appl. Phys.*, Vol. 74, p. 1818, 1993.

33. M. A. Khan, J. N. Kuznia, D. T. Olson, W. Schaff, G. Burm, M. S. Shur, and C. Eppers presented at *Device Research Conference*, Boulder, Colorado, 1994.

34. S. Binari, L. B. Rowland, W. Kruppa, G. Kelner, K. Doverspike, and D. K. Gatskill, "Microwave performance of GaN MESFETs," *Elect. Lett.*, Vol. 30, No 15, p. 1248, July 1994.

35. G. Gaska, M. S. Shur, and A. Khan, "AlGaN/GaN high electron mobility transistors," in *Optoelectronic Properties of Semiconductors and Superlattices*, Vol. 16, Taylor and Francis Books, Inc., ISBN 1-56032-974-2, 2003, E. T. Yu and M. O. Manasreh, Editors, pp. 193–269.

36. M. S. Shur and M. A. Khan, "GaN and AlGaN devices: field effect transistors and photodetectors," in *GaN and Related Materials II*, S. Pearton, Editor, Gordon and Breach Science Publishers, Series Optoelectronic Properties of Semiconductors and Superlattices, Vol. 7, pp. 47–86, 1999.

37. S. Karmalkar, M. S. Shur, and R. Gaska, "GaN-based power high electron mobility transistors," in *Wide Energy Bandgap Electronic Devices*, Fan Ren and John Zolper, Editors, World Scientific, 2003, ISBN 981-238-246-1, pp. 173–216.

38. M. S. Shur and R. Davis, Preface, from *GaN-based Materials and Devices: Growth, Fabrication, Characterization and Performance*, M. S. Shur and R. Davis, Editors, World Scientific, ISBN 981-238-844-3, 2004.

39. G. U. Jensen, B. Lund, M. S. Shur, and T. A. Fjeldly, "Monte carlo simulation of semiconductor devices," *Com. Phys. Comm.*, Vol. 67, No. 1, pp. 1–61, 1991.

40. K. Hess and C. Kizilyalli, "Scaling and transport properties of high electron mobility transistors," *IEDM Technical Digest*, Los Angeles, pp. 556–558, 1986.

41. A. Afzalikushaa and G. Haddad, "High-frequency characteristics of MESFETs," *Sol. Stat. Elect.*, Vol. 38, No. 2, pp. 401–406, February 1995.

42. C. Jacoboni and P. Lugli, *The Monte Carlo Method for Semiconductor Simulation*, Springer Verlag, Vienna, 1989.

43. C. Moglestue, *Monte Carlo Simulation of Semiconductor Devices*, Chapman & Hall, London, 1993.

44. K. Tomizawa, *Numerical Simulation of Submicron Semiconductor Devices*, Artech House, Boston, 1993.

45. M. Lundstrom, *Fundamentals of Carrier Transport*, Vol. X of the Modular Series on Solid State Devices, Addison-Wesley, Reading, MA, 1990.

46. K. Bløtekjær, "Transport equations for two-valley semiconductors," *IEEE Trans. Elect. Dev.*, Vol. 17, pp. 38–47, 1970.

47. J. Jyegal and T. A. Demassa, "New nonstationary velocity overshoot phenomenon in submicron Gallium-Arsenide field-effect transistors," *J. Appl. Phys.*, Vol. 75, No. 6, pp. 3169–3175, March 1994.

48. S. H. Lo and C. P. Lee, "Analysis of surface-state effect on gate lag phenomena in GaAs-MESFETs," *IEEE Trans. Elect. Dev.*, Vol. 41, No. 9, pp. 1504–1512, September 1994.

49. BLAZE, *Atlas II User's Manual*, Silvaco, 1993.

50. M. S. Shur, "Analytical models of GaAs FETs," *IEEE Trans. Elect. Dev.*, Vol. ED-32, No. 1, pp. 70–72, January 1985.

51. M. S. Shur, *Introduction to Electronic Devices*, John Wiley and Sons, New York, ISBN 0-471-10348-9, 1996.

52. K. Lee, M. S. Shur, T. A. Fjeldly, and T. Ytterdal, *Semiconductor Device Modeling for VLSI*, Prentice Hall, Englewood Cliffs, NJ, 1993.

53. T. Fjeldly, T. Ytterdal, and M. S. Shur, *Introduction to Device and Circuit Modeling for VLSI*, John Wiley and Sons, New York, ISBN 0-471-15778-3, 1998.

54. B. Iñiguez, T. A. Fjeldly, M. S. Shur, T. Ytterdal, Spice Modeling of Compound Semiconductor Devices, in "Silicon and beyond. Advanced device models and circuit simulators, M. S. Shur and T. A. Fjeldly, Editors, World Scientific, 2000, pp. 55–112.

55. M. S. Shur, T. Fjeldly, Y. Ytterdal, and K. Lee, "Unified GaAs MESFET model for CIRCUIT Simulations," *Int. J. High Speed Elect.*, Vol. 3, No. 2, pp. 201–233, June 1992.

56. J. E. Meyer, "MOS models and circuit simulation," *RCA Review*, Vol. 32, pp. 42–63, 1971.

57. M. Nawaz and T. A. Fjeldly, "A new charge conserving capacitance model for GaAs MESFETs," *IEEE Trans. Elect. Dev.*, Vol. 44, No. 11, pp. 1813–1821, 1997.

58. M. Berroth, M. Shur, and W. Haydl, "Experimental studies of hot electron effects in GaAs MESFETs," in Extended Abstracts of the 20th International Conf. on Solid State Devices and Materials (SSDM-88), Tokyo, pp. 255–258, 1988.

59. C. Dunn, *Microwave Semiconductor Devices and Their Applications*, H. A. Watson, Editor, McGraw Hill, New York, 1969.

60. B. Gelmont, M. Shur and C. Moglestue, "Theory of junction between two-dimensional electron gas and p-type semiconductor," *IEEE Trans. Elect. Dev.*, Vol. 39, No. 5, pp. 1216–1222, 1992.

61. S. G. Petrosyan and Y. Shik, "Contact phenomena in a two dimensional electron gas," *Sov. Phys.-Semicond.*, Vol. 23, No. 6, pp. 696–697, 1989.

62. W. C. B. Peatman, T. W. Crowe, and M. S. Shur, "A novel Schottky/2-DEG diode for millimeter and submillimeter wave multiplier applications," *IEEE Elect. Dev. Lett.*, Vol. 13, pp. 11–13, 1992.

63. W. C. B. Peatman, H. Park, B. Gelmont, M. S. Shur, P. Maki, E. R. Brown, and M. J. Rooks, "Novel Metal/2-DEG junction transistors," *Proc. 1993 IEEE/Cornell Conference*, Ithaca, NY, pp. 314–319, 1993.

64. W. C. B. Peatman, H. Park, and M. Shur, "Two-dimensional metal-semiconductor field effect transistor for ultra low power circuit applications," *IEEE Elect. Dev. Lett.*, Vol. 15, No. 7, pp. 245–247, 1994.

65. M. S. Shur, W. C. B. Peatman, H. Park, W. Grimm, and M. Hurt, "Novel heterodimensional diodes and transistors," *Sol. Stat. Elect.*, Vol. 38, No. 9, pp. 1727–1730, 1995.

66. W. C. B. Peatman, T. W. Crowe, and M. S. Shur, "A novel Schottky/2-DEG diode for millimeter and submillimeter wave multiplier applications," *IEEE Elect. Dev. Lett.*, Vol. 13, No. 1, 1992.

67. W. C. B. Peatman, M. J. Hurt, H. Park, T. Ytterdal, R. Tsai, and M. Shur, "Narrow channel 2-D MESFET for low power electronics," *IEEE Trans. Elect. Devi.*, Vol. 42, No. 9, pp. 1569–1573, 1995.

68. W. C. B. Peatman, R. Tsai, T. Ytterdal, M. Hurt, H. Park, J. Gonzales, and M. Shur, "Sub-half-micrometer width 2-D MESFET," *IEEE Elect. Dev. Lett.*, Vol. 17, No. 2, pp. 40–42, 1996.

69. M. Hurt, W. C. B. Peatman, R. Tsai, T. Ytterdal, M. Shur, and B. J. Moon, "An ion-implanted 0.4 μm wide 2-D MESFET for low-power electronics," *Elect. Lett.*, Vol. 32, No. 8, pp. 772–773, 1996.

70. W. C. B. Peatman, H. Park, B. Gelmont, M. S. Shur, P. Maki, E. R. Brown, and M. J. Rooks, "Novel Metal/2-DEG junction transistors," in *Proc. 1993 IEEE/Cornell Conf.*, Ithaca, NY, p. 314, 1993.

71. J. Robertson, T. Ytterdal, W. C. B. Peatman, R. Tsai, E. Brown, and M. Shur, "RTD/2-D MESFET/RTD logic elements for compact, ultra low-power electronics," *IEEE Trans. Elect. Dev.*, Vol. 44, No. 7, pp. 1033–1039, 1997.

72. B. Iñiguez, J.-Q. Lü, M. Hurt, W. C. B. Peatman, and M. S. Shur, "Modeling and simulation of single and multiple gate 2-D MESFETs," *IEEE Trans. Elect. Dev.*, Vol. 46, No. 8, 1999.

73. T. Werner and M. S. Shur, *GaAs Microwave Transistors*, unpublished (2000).

74. Data from J. A del Alamo, see http://www-mtl.mit.edu/~alamo/pdf/2005-2004/2005/RC-108.pdf.

75. http://www.oki.com/en/press/2005/z05006e.html

76. http://www.anadigics.com/GaAsline/mesfets.html

77. http://www.vitesse.com/news/101199.htm

78. http://www.cree.com/Products/mmic.htm

79. http://www.electronicstalk.com/news/tos/tos249.html

80. J. Michael Golio, Editor, *Microwave MESFET's and HEMT's*, Artech House Publishers, March 1991.

81. W. Liu, "Fundamentals of III-V Devices: HBTs, MESFETs, and HFETs/HEMTs", Wiley, 1999.

82. R. Anholt, *Electrical and Thermal Characterization of MESFETs, HEMTs, and HBTs*, Artech House Publishers, December, 1994.

21

High Electron Mobility Transistors (HEMTs)

Prashant Chavarkar
Infinera

Umesh Mishra
University of California

The concept of modulation doping was first introduced in 1978.[1] In this technique electrons from remote donors in a higher bandgap material transfer to an adjacent lower gap material. The electrostatics of the heterojunction results in the formation of a triangular well at the interface, which confines the electrons in a two-dimensional (2D) electron gas (2DEG). The separation of the 2DEG from the ionized donors significantly reduces ionized impurity scattering resulting in high electron mobility and saturation velocity.

Modulation-doped field effect transistors (MODFETs) or high electron mobility transistors (HEMTs), which use the 2DEG as the current conducting channel have proved to be excellent candidates for micro-wave and millimeter-wave analog applications and high-speed digital applications. This progress has been enabled by advances in crystal growth techniques such as molecular beam epitaxy (MBE) and metal-organic chemical vapor deposition (MOCVD) and advances in device processing techniques, most notably electron beam lithography, which has enabled the fabrication of HEMTs with gate lengths down to 0.05 μm.

However, using a high electron mobility channel alone does not guarantee superior high-frequency performance. It is crucial to understand the principles of device operation and to take into consideration the effect of scaling to design a microwave or millimeter-wave HEMT device. The advantages and limitations of the material system used to implement the device also need to be considered. This section therefore begins with a discussion on the device operation of a HEMT. This is followed by a discussion of scaling issues in HEMT, which are of prime importance, as the reduction of gate length is required to increase the operating frequency of the device.

The first HEMT was demonstrated in the AlGaAs/GaAs material system in 1981. It demonstrated significant performance improvements over the GaAs MESFET at microwave frequencies. However, the high-frequency performance was not sufficient for operation at millimeter-wave frequencies. In the past twenty years, the AlGaAs/InGaAs psuedomorphic HEMT on GaAs substrate (referred to as GaAs pHEMT) and the AlInAs/GaInAs HEMT on InP substrate (referred to as InP HEMT) have emerged as premier devices for microwave and millimeter-wave circuit applications. This highlights the importance of choosing the appropriate material system for device implementation. This will be discussed in the section on Material Systems for HEMT Devices.

The next two sections will discuss the major advances in the development of the GaAs pHEMT and InP HEMT. Traditionally these devices have been used in low-volume, high-performance and high-cost military and space-based electronic systems. Recently the phenomenal growth of commercial wireless and optical fiber-based communication systems has opened up new applications for these devices. This also means that new issues like manufacturability and operation at low bias voltage have to be addressed.

21.1 HEMT Device Operation and Design

21.1.1 Linear Charge Control Model

The current control mechanism in the HEMT is control of the 2DEG density at the heterojunction interface by the gate voltage. Figure 21.1 shows the band diagram along the direction perpendicular to the heterojunction interface using the AlGaAs/GaAs interface as an example.

The first HEMT charge control model was proposed by Delagebeaudeuf and Linh in 1982.[2] The potential well at the AlGaAs/GaAs interface is approximated by a triangular well. The energy levels in this triangular well and the maximum 2DEG density, n_{sm} can be calculated by solving the Schrödinger equation in the triangular well and Poisson equation in AlGaAs donor layers.[3] For $0 < n_s < n_{sm}$, the sheet charge density n_s as a function of gate voltage V_g can be expressed as

$$qn_s = C_s\left(V_g - V_{th}\right) \tag{21.1}$$

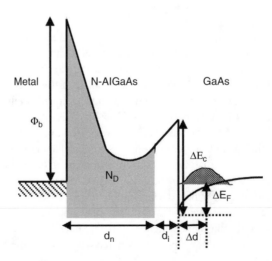

FIGURE 21.1   Schematic of conduction band diagram at the AlGaAs/GaAs interface.

where C_s is the 2DEG capacitance per unit area and is given by the following expression:

$$C_s = \frac{\varepsilon}{d_n + d_i + \Delta d} \tag{21.2}$$

Here Δd is the distance of the centroid of the 2DEG distribution from the AlGaAs/GaAs interface and is typically or the order of 80 Å for $n_s \sim 10^{12}/cm^2$. Here V_{th} is the threshold voltage or pinch-off voltage and is given by,

$$V_{th} = \phi_b - \frac{qN_D}{2\varepsilon}d_n^2 - \Delta E_c + \Delta E_F \tag{21.3}$$

where ϕ_b, N_D, and d_n are the Schottky barrier height on the donor layer, doping density, and doped layer thickness as illustrated in Figure 21.1. Here, ΔE_F is the Fermi potential of the 2DEG with respect to the bottom of the conduction band. It can be expressed as a function of 2DEG density as follows

$$\Delta E_F = \Delta E_{FO}\left(T\right) + a n_s \tag{21.4}$$

where $\Delta E_{FO}(T) = 0$ at 300 K, $a = 0.125 \times 10^{-16}$ V/m^2.

This simplified version of charge control is accurate only at low temperature. At room temperature, apart from the 2DEG charge density n_s, the gate voltage also modulates the bound carrier density, n_{bound} in the donor layer and the free electrons, n_{free} in the donor layers. This results in premature saturation of the sheet charge and degradation of device performance. A more accurate model for charge control, which solves Poisson's and Schrödinger's equations in a self-consistent manner was proposed by Vinter.[4]

21.1.2 Modulation Efficiency

The parasitic modulation of charge in the higher bandgap donor layer reduces the efficiency of the gate voltage to modulate the drain current, as the carriers in the donor layers do not contribute the drain current. The modulation efficiency (η) of the FET is proportional to ratio between the change in drain current (δI_{ds}) and the change in total charge (δQ_{tot}) required to cause this change.[5] This ratio is defined as follows,

$$\eta \propto \frac{\delta I_{ds}}{\delta Q_{tot}} = \frac{\delta\left(qv_{sat}n_s\right)}{\delta q\left(n_s + n_{bound} + n_{free}\right)} \tag{21.5}$$

Dividing the numerator and denominator by the change in gate voltage, δV_g that is required to cause this change, the following expression is obtained,

$$\frac{\delta I_{ds}}{\delta Q_{tot}} = v_{sat}\frac{\delta\left(n_s\right)\big/\delta V_g}{\delta\left(n_s + n_{bound} + n_{free}\right)\big/\delta V_g} \tag{21.6}$$

The modulation efficiency is defined as the ratio of the rate of change of the useful charge, i.e., the 2DEG over that of the total charge,

$$\eta = \frac{\delta\left(n_s\right)\big/\delta V_g}{\delta\left(n_s + n_{bound} + n_{free}\right)\big/\delta V_g} = \frac{\delta\left(n_s\right)\big/\delta V_g}{C_{TOT}} = \frac{C_s}{C_{TOT}} \tag{21.7}$$

The relation between the modulation efficiency and high frequency performance of the FET is evident in the expressions for transconductance (g_m) and current gain cutoff frequency (f_T).

$$g_m = \frac{\delta I_{ds}}{\delta V_g} = \frac{\delta\left(qv_{sat}n_s\right)}{\delta V_g} = qv_{sat}\left(\delta n_s/\delta V_g\right);$$

$$C_{gs} = C_{TOT}L_g;$$ (21.8)

$$f_T = \frac{g_m}{2\pi C_{gs}} = \frac{qv_{sat}\left(\delta n_s/\delta V_g\right)}{2\pi L_g C_{TOT}} = \frac{v_{sat}}{2\pi L_g}\eta$$

Hence, to improve the high frequency performance it is essential to improve the modulation efficiency.

Equation 21.8 must be used with caution in case of short gate length HEMTs. The saturation velocity v_{sat} may be replaced by the effective velocity v_{eff}. Usually v_{eff} is higher than v_{sat} due to high field and velocity overshoot effects. Using v_{sat} in this case may lead to values of modulation efficiency that are greater than 100%.

21.1.3 Current-Voltage (I-V) Models for HEMTs

By assuming linear charge control, gradual channel approximation, and a 2-piece linear velocity-field model, the expression for the saturated drain current I_{DSS} in a HEMT is given by,[2]

$$I_{DSS} = C_s v_{sat}\left(\sqrt{\left(E_c L_g\right)^2 + \left(V_g - V_c\left(0\right) - V_{th}\right)^2} - E_c L_g\right)$$ (21.9)

Here E_c is defined as the critical electric field at which the electrons reach their saturation velocity v_{sat} and $V_c(0)$ is the channel potential at the source end of the gate. For a long gate length HEMT, Equation 21.9 is valid until the onset of donor charge modulation, that is, $0 < n_s < n_{sm}$. The intrinsic transconductance of the device obtained by differentiating this expression with respect to the gate voltage and is expressed as follows:

$$g_{mo} = \frac{\delta I_{ds}}{\delta V_g} = C_s v_{sat}\frac{V_g - V_c\left(0\right) - V_{th}}{\sqrt{\left(V_g - V_c\left(0\right) - V_{th}\right)^2 + \left(E_c L_g\right)^2}}$$ (21.10)

For a short gate length HEMT, the electric field in the channel is much greater in magnitude than the critical electric field E_c. Assuming that the entire channel of the FET operates in saturated velocity mode, we can make the following assumption, that is, $V_g - V_c(0) - V_{th} \gg E_c L_g$. Then using Equations 21.1, (21.9), and (21.10) are reduced to the following:

$$I_{DSS} = qn_s v_{sat}$$ (21.11)

$$g_m = C_s v_{sat}$$ (21.12)

More insight can be obtained in terms of device parameters if the equation for charge control [Equation 21.1] is substituted in the expressions for I_{ds} and V_g as follows:[6]

$$I_{ds} = qv_{sat}n_s\left[\sqrt{1 + \left(\frac{n_c}{n_s}\right)^2} - \frac{n_c}{n_s}\right]$$ (21.13)

$$g_{mo} = C_s v_{sat}\frac{1}{\sqrt{1 + \left(n_c/n_s\right)^2}}$$ (21.14)

where $n_c = E_c C_s L_g / q$ and $0 < n_s < n_{sm}$. Dividing both sides of Equation 21.14 by $C_s v_{sat}$ the following expression for modulation efficiency is obtained:

$$\eta = \frac{1}{\sqrt{1 + \left(n_c / n_s\right)^2}} \tag{21.15}$$

Hence it is necessary to maximize the 2DEG density n_s to maximize the current drive, transconductance, and modulation efficiency of the HEMT. Although this is in contrast with the saturated-velocity model, it agrees with the experimental results. The foregoing results can also be used to select the appropriate material system and layer structure for the fabrication of high-performance microwave and millimeter-wave HEMTs.

Although the analytical model of device operation as was described here provides great insight into the principles of device operation and performance optimization, it fails to predict some of the nonlinear phenomena such as reduction of g_m at high current levels (g_m compression) and soft pinch-off characteristics. A model has been developed to explain these phenomena.[5] The total charge in the HEMT is divided into three components. The first, Q_{SVM}, is the charge required to support a given I_{ds} under the saturated velocity model (SVM). This charge is uniformly distributed under the gate. In reality this is not the case as the electron velocity under the gate varies. To maintain the current continuity under the gradual channel approximation (GCA), extra charge under the channel has to be introduced. This is defined as Q_{GCA} and is maximum at the source end of the gate and minimum at the drain end.

The excess charge in the wide bandgap electron supply layer is denoted by Q_{SL}. Figure 21.2 shows the location and distribution of these charges in the HEMT. Only Q_{SVM} supports current density and thus contributes to the transconductance of the HEMT. The other two components contribute only to the total capacitance of the device. Hence the modulation efficiency (ME) of the HEMT in terms of these charges is expressed as

$$\eta = \frac{\delta Q_{SVM}}{\delta \left(Q_{SVM} + Q_{GCA} + Q_{SL}\right)} \tag{21.16}$$

and the transconductance can be expressed as $g_m = C_s v_{sat} \eta$.

Figure 21.3 shows the variation of ME as a function of drain current density for an AlGaAs/GaAs HEMT and an AlGaAs/InGaAs pHEMT. At low current density, ME is low as most of the charge in the 2DEG channel has to satisfy the gradual channel approximation. This low value of ME results in low transconductance and soft pinch-off characteristics at low drain current densities. In the high current regime, modulation of Q_{SL} reduces the ME, resulting in gain compression. In the intermediate current regime the ME is maximum. However, if there exists a bias condition where both Q_{GCA} and Q_{SL} are modulated (as in the low band offset AlGaAs/GaAs system), it severely affects the ME.

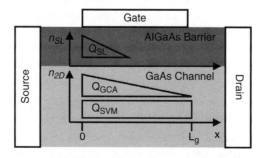

FIGURE 21.2 Schematic diagram showing the location and distribution of Q_{SVM}, Q_{GCA}, and Q_{SL} in a HEMT (Foisy *IEEE Transactions of Electron Devices*, 35, 87 1988).

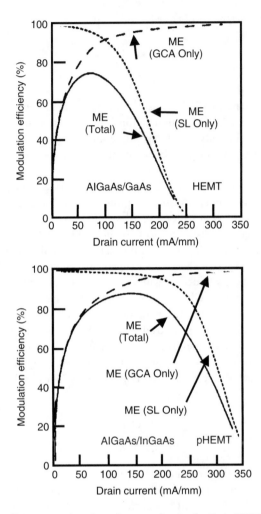

FIGURE 21.3   Modulation efficiency as a function of current density for GaAs HEMT and GaAs pHEMT.

For optimal high-power and high-frequency performance, it is necessary to maximize the range of current densities in which ME is high. The drop off in ME due to parasitic charge modulation in the donor layers can be pushed to higher current density by increasing the maximum 2DEG density n_{sm}. The 2DEG density can be maximized by using planar doping in the donor layer and by increasing the conduction band discontinuity at the barrier/2DEG interface. The drop off in ME due to operation in gradual channel mode can be pushed to lower current densities by reducing the saturation voltage V_{Dsat}. This is achieved by increasing the mobility of the electrons in the 2DEG channel and by reducing the gate length. As seen from Figure 21.3 higher modulation efficiency is achieved over a larger range of current density for the AlGaAs/InGaAs pHEMT, which has higher sheet charge density, mobility, and band discontinuity at the interface than the AlGaAs/GaAs HEMT.

21.1.4   Small Signal Equivalent Circuit Model of HEMT

The small signal equivalent circuit model of the HEMT is essential for designing HEMT-based amplifiers. The model can also provide insights into the role of various parameters in the high-frequency performance of the device. Figure 21.4 shows the small signal equivalent circuit for a HEMT. The grey box highlights the intrinsic device. The circuit elements in the preceding model are determined using microwave S-parameter

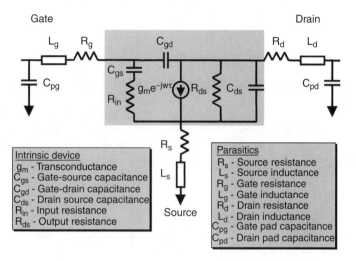

FIGURE 21.4 Small signal equivalent circuit of a HEMT.

measurements.[7,8] The intrinsic circuit elements are a function of the DC bias, whereas the extrinsic circuit elements or parasitics are independent of it. The two measures of the high frequency performance of a FET can now be defined in terms of the small signal model of the device as follows. The current gain cutoff frequency, f_T can be defined as

$$f_T = \frac{g_m}{2\pi\left(C_{gs} + C_{gd}\right)} \tag{21.17}$$

Hence, to increase the current gain cutoff frequency it is essential to increase the g_m and reduce C_{gs} and C_{gd}. Referring to Equation 21.8, it is clear that this can be achieved by increasing electron velocity in the channel and reducing gate length. The current gain cutoff frequency is mainly a physical measure of device performance. A more practical measure of high-frequency device performance is f_{max}, the power gain cutoff frequency. This is the frequency at which the power gain of the FET is unity. It is defined as follows,[9]

$$f_{max} = \frac{f_T}{\sqrt{4g_{ds}\left(R_{in} + \dfrac{R_s + R_g}{1 + g_m R_s}\right) + \dfrac{4}{5}\dfrac{C_{gd}}{C_{gs}}\left(1 + \dfrac{2.5C_{gd}}{C_{gs}}\right)\left(1 + g_m R_s\right)^2}} \tag{21.18}$$

A simple form of Equation 21.18 is:

$$f_{max} = f_T\sqrt{\frac{R_{ds}}{4R_{in}}} = \frac{f_T}{\sqrt{4g_{ds}R_{in}}} \tag{21.19}$$

To improve the f_{max} of the device it is necessary to minimize the quantities in the denominator of Equation 21.18. The crucial parameters here are the output conductance of the device g_{ds}, the source and gate parasitic resistances R_s and R_g, and the gate-drain feedback capacitance C_{gd} that need to be minimized. Reduction of g_{ds} can be achieved by appropriate vertical scaling (to be discussed in the next section). Reduction of R_s and R_g depends mainly on the process technology. Reduction of C_{gd} can be achieved by proper design of the gate-drain region of the FET. The crucial parameter in the design of the gate drain depletion region is the gate-drain separation L_{gd}.[10] Increasing L_{gd} reduces C_{gd} but also increases the effective gate length of the device, reducing the short channel effects. The optimum value of L_{gd} is 2.3 times that of the gate length L_g. Thus it is clear that f_{max} is a better measure of the high-frequency performance of a FET as it is determined not only by the material system used but also by the process technology and device design parameters. Large signal models of HEMTs are essential for designing power amplifiers and are similar to those of MESFETs.

21.2 Scaling Issues in Ultra-High-Speed HEMTs

The frequency at which a HEMT operates is limited by the electron transit time from the source to the drain. Therefore to increase the frequency of operation it is necessary to reduce the gate length. However, as the gate length approaches 0.1 μm it is necessary to reduce the other parasitic delays in the device and take into account short channel effects to maintain the high-frequency performance of the HEMT.

21.2.1 Delay Time Analysis

The reduction of parasitic delays in a FET is essential to improve the high frequency performance as these delays can be as high as 45% of the intrinsic delay.[11] Considering the small-signal model of a FET, the total delay t_T in a FET can be expressed as follows:[12]

$$t_T = t_{pad} + t_{fringe} + t_{channel} + t_{transit} + t_{drain} = 1/\left(2\pi f_T\right) \tag{21.20}$$

Here t_{pad} is the charging time for the parasitic pad capacitance and is given by

$$t_{pad} = C_{pad}/g_m \cdot W \tag{21.21}$$

where C_{pad} is the pad capacitance and is typically 10 fF per 50 μm × 50 μm bonding pad, g_m is the extrinsic transconductance per unit gate width, and W is the width of the device. To minimize t_{pad} it is necessary to have a high gate width, high transconductance HEMT.

The gate fringe capacitance charging time (t_{fringe}) is given by

$$t_{fringe} = C_{fringe}/g_{mo} \tag{21.22}$$

where g_{mo} is the intrinsic transconductance of the HEMT and is related to the extrinsic transconductance (g_m) and source resistance R_s by the following expression:

$$g_m = g_{mo}/\left(1 + g_{mo} \cdot R_s\right) \tag{21.23}$$

The gate fringe capacitance C_{fringe} is typically 0.18 pF/mm, hence for a HEMT with an intrinsic transconductance of 1000 to 1500 mS/mm, t_{fringe} is approximately 0.1 to 0.2 ps.

Channel charging delay $t_{channel}$ is associated with RC delays and is proportional to channel resistance. The channel charging delay is minimum at high current densities. The channel charging delay can be considered as a measure of the effectiveness of a FET operating in the saturated velocity mode.

The transit delay of the FET, $t_{transit}$, can be expressed as the time required to traverse under the gate and is given by

$$t_{transit} = L_g/v_{sat} \tag{21.24}$$

The drain delay (t_{drain}) is the time required by the electron to traverse the depletion region between the gate and the drain and is a function of bias conditions.[13] The drain delay increases with drain bias as the length of the depletion region beyond the gate increases. Drain delay is an important parameter for millimeter-wave power HEMTs. To increase the breakdown voltage of the device, the gate-to-drain spacing has to be increased. When the device is biased at a high drain voltage to maximize the power output, it creates a drain depletion region that is on the order of gate length of the device. Thus the drain delay becomes a major component of the total delay in the device, and can limit the maximum f_T and f_{max}.

21.2.2 Vertical Scaling

Aspect ratio (the ratio between the gate length L_g and the gate-to-channel separation $d_{Barrier}$) needs to be maintained when gate length is reduced. Aspect ratio is a critical factor affecting the operation of the field effect transistor and should be maintained above five. As the gate length is reduced, the distance between the gate and 2DEG (the distance $d_n + d_i$ as seen in Figure 21.1) has to be reduced so that the aspect ratio of the device is maintained.

However, maintaining the aspect ratio alone does not guarantee improvement in device performance. This is clear if the variation of threshold voltage with the reduction in $d_{Barrier}$ is examined. It is clear that d_i cannot be reduced, as it will result in degradation of mobility in the 2DEG channel due to scattering from the donors in the barrier layers. Therefore, to maintain aspect ratio, the thickness of the doped barrier layer d_n has to be reduced. By examining Equation 21.3 for threshold voltage, it is clear that this makes the threshold voltage more positive.

At first glance, this does not seem to affect device performance. The effect of the more positive threshold voltage is clear if the access regions of the device are considered. A more positive threshold voltage results in reduction of sheet charge in the access region of the device. This increases the source and drain resistance of the device, which reduces the extrinsic transconductance [see Equation 21.23] and also increases the channel charging time (due to increased RC delays). Thus the increased parasitic resistances nullify the improvements in speed in the intrinsic device.

The threshold voltage of the device must be kept constant with the reduction in d_n. From Equation 21.3 it can be seen that the doping density in the high bandgap donor has to be increased. Since the threshold voltage varies as a square of the doped barrier thickness, a reduction in its thickness by a factor of 2 requires that the doping density be increased by a factor of 4. High doping densities can be difficult to achieve in wide bandgap materials such as AlGaAs due to the presence of DX centers. Increased doping also results in higher gate leakage current, higher output conductance, and a lower breakdown voltage. Utilizing planar or delta doping wherein all the dopants are located in a single plane can alleviate these problems. This leaves most of the higher bandgap layer undoped and enables reduction of its thickness.

The threshold voltage of a planar-doped HEMT is given as follows[14]

$$V_T = \phi_B - \frac{qN_{2D}d_n}{\varepsilon} - \Delta E_c + \Delta E_F \tag{21.25}$$

where N_{2D} is the per unit area concentration of donors in the doping plane and d_n is the distance of the doping plane from the gate. In this case, the 2D doping density has to increase linearly with the reduction in barrier thickness. The transfer efficiency of electrons from the donors to the 2DEG channel also is increased, as all the dopant atoms are close to the 2DEG channel. Hence higher 2DEG sheet densities can be achieved in the channel and thus planar doping enables efficient vertical scaling of devices with reduction in gate length.[15] From a materials point of view, efficient vertical scaling of a HEMT requires a high bandgap donor/barrier semiconductor that can be doped efficiently.

The voltage gain of the device (g_m/g_{ds}) can be considered as a measure of short channel effects in the device. The reduction of gate length and the gate-to-channel separation results in an increase in the transconductance of the device. However, to reduce the output conductance g_{ds} of the device, it is also necessary to reduce the channel thickness, which then increases the carrier confinement in the channel. Enoki et al. have investigated the effect of the donor/barrier and channel layer thickness on the voltage gain of the device.[16] The gate-to-channel separation ($d_{Barrier}$) and the channel thickness ($d_{channel}$) were varied for a 0.08-μm gate length AlInAs/GaInAs HEMT. For a $d_{Barrier}$ of 170 Å and a $d_{channel}$ of 300 Å, the g_m was 790 mS/mm and g_{ds} was 99 mS/mm, resulting in a voltage gain of 8. When $d_{Barrier}$ was reduced to 100 Å and $d_{channel}$ was reduced to 150 Å, the g_m increased to 1100 mS/mm and g_{ds} reduced to 69 mS/mm; this doubled the voltage gain to 16. This illustrates the necessity to reduce the channel thickness to improve charge control in ultra-short gate length devices.

Subthreshold slope is an important parameter to evaluate short channel effects for digital devices. A high value of subthreshold slope is necessary to minimize the off-state power dissipation and to increase the device speed. Two-dimensional simulations performed by Enoki et al. indicate that reduction in channel thickness is more effective than the reduction in barrier thickness, for maintaining the subthreshold slope with reduction in gate length.[16]

The high-frequency performance of a device is a function of the electrical gate length $L_{g,eff}$ of the device, which larger than the metallurgical gate length L_g due to lateral depletion effects near the gate. The relation between $L_{g,eff}$ and L_g is given by,[14]

$$L_{g,eff} = L_g + \beta\left(d_{Barrier} + \Delta d\right)$$ (21.26)

where $d_{Barrier}$ is the total thickness of the barrier layers, Δd is the distance of the centroid of the 2DEG from the channel barrier interface and is on the order of 80 Å. The value of parameter β is 2.

Consider a long gate length HEMT ($L_g = 1$ μm) with a barrier thickness of 300 Å. Using Equation 21.26, the value of 1.076 μm is obtained for $L_{g,eff}$. Thus the effective gate length is only 7.6% higher than the metallurgical gate length. Now consider an ultra-short gate length HEMT ($L_g = 0.05$ μm) with an optimally scaled barrier thickness of 100 Å. Using the same analysis, a value of 0.086 μm is obtained for $L_{g,eff}$. In this case the effective gate length is 43% higher than the metallurgical gate length. Hence, to improve the high-frequency performance of a ultra-short gate length HEMT, effective gate length reduction along with vertical scaling is required.

21.2.3   Horizontal Scaling

Reduced gate length is required for the best high-frequency performance. However, it should be kept in mind that the gate series resistance increases with the reduction in gate length. This problem can be solved with a T-shaped gate. This configuration lowers the gate series resistance while maintaining a small footprint. Another advantage of the T-shaped gate is reduced susceptibility to electromigration under large signal RF drive as the large gate cross-section reduces current density. For a 0.1-μm gate length using a T-gate instead of a straight gate, reduces the gate resistance from 2000 Ω/mm to 200 Ω/mm.

The simplified expression for f_T as expressed in Equation 21.17 does not include the effect of parasitics on the delay time in a FET. A more rigorous expression for f_T, which includes the effects of parasitics on f_T was derived by Tasker and Hughes and is given here,[17]

$$f_T = \frac{g_m/2\pi}{\left[C_{gs}+C_{gd}\right]\left[1+\left(R_s+R_d\right)/R_{ds}\right]+C_{gd}g_m\left(R_s+R_d\right)}$$ (21.27)

It is clear from Equation 21.27 that it is necessary to reduce source and drain resistances R_s and R_d, respectively, to increase the f_T of a FET. Mishra et al. demonstrated a record f_T of 250 GHz for a 0.15-μm device with a self-aligned gate, which reduces the gate-source and gate-drain spacing and results in the reduction of R_s and R_d.[18] Equation 21.26 can be rearranged as follows,[17]

$$\frac{1}{2\pi f_T} = \frac{\left(C_{gs}+C_{gd}\right)}{g_m} + \frac{\left(C_{gs}+C_{gd}\right)\left(R_s+R_d\right)}{g_m R_{ds}} + C_{gd}\left(R_s+R_d\right)$$ (21.28)

where the first term on the right-hand side is the intrinsic delay of the device (τ_{int}) and the rest of the terms contribute to parasitic delay (τ_p). From this equation the ratio of parasitic delay to the total delay ($\tau_t = \tau_p + \tau_{int}$) is given as

$$\frac{\tau_p}{\tau_t} = g_m\left(R_s+R_d\right)\left[\frac{G_{ds}}{g_m} + \frac{1}{\left[1+C_{gs}/C_{gd}\right]}\right]$$ (21.29)

Hence to improve the f_T of the device, the parasitic source and drain resistance have to be reduced as the gate length of the device is reduced. This minimizes the contribution of the parasitic delays to the total delay of the device.

21.3 Material Systems for HEMT Devices

The previous portions of this section discussed the various device parameters crucial to high-frequency performance of HEMTs. In this section the relationship between material and device parameters will be discussed. This will enable the selection of the appropriate material system for a particular device application. Table 21.1 illustrates the relationship between the device parameters and material parameters for the various constituent layers of the HEMT, namely the high bandgap donor and buffer layers, and the 2DEG channel. Figure 21.5 shows a schematic diagram of a HEMT, illustrating the material requirements from each component layer.

The first HEMT was implemented in the lattice-matched AlGaAs/GaAs system in 1981.[19] The AlGaAs/GaAs HEMT demonstrated significant improvement in low noise and power performance over GaAs MESFET due to superior electronic transport properties of the 2DEG at the AlGaAs/GaAs interface and better scaling properties. However, the limited band discontinuity at the AlGaAs/GaAs interface limits the 2DEG sheet charge density. Other undesirable effects, such as formation of a parasitic MESFET in the donor layer and real space transfer of electrons from the channel to donor, are prevalent, however.

One way to increase band discontinuity is to increase the Al composition in AlGaAs. However, the presence of deep level centers (DX centers) associated with Si donors in AlGaAs prevents the use of high Al composition AlGaAs donor layers to increase the band discontinuity and also limits doping efficiency. Problems relating to low band discontinuity can also be solved by reducing the bandgap of the channel, and by using a material that has higher electron mobility and electron saturation velocity. The first step in this direction was taken by the implementation of an AlGaAs/InGaAs pseudomorphic HEMT (GaAs pHEMT).[20] In an AlGaAs/InGaAs pHEMT the electron channel consists of a thin layer of narrow bandgap InGaAs that is lattice mismatched to GaAs by 1 to 2%. The thickness of the InGaAs channel is thin enough (~200 Å) so that the mismatch strain is accommodated coherently in the quantum well, resulting in a dislocation free "pseudomorphic" material. However the indium content in the InGaAs channel can be increased only up to 25%. Beyond this limit the introduction of dislocations due to high lattice mismatch degrades the electronic properties of the channel. The maximum Al composition that can be used in the barrier is 25% and the maximum indium composition that can be used in the channel is 25%.

TABLE 21.1 Relationship between Device and Material Parameters

Device Type	Device Parameters	Material Parameters	
		2DEG Channel Layer	Barrier/Buffer Layer
Short Gate Length Devices	High Electron Velocity	High Electron Velocity High Electron Mobility	
Power Devices	High Aspect Ratio High Current Density Low Gate Leakage High Breakdown Voltage Low Output Conductance Good Charge Control Low Frequency Dispersion	High 2DEG Density High Breakdown Field High Modulation Efficiency	High Doping Efficiency High Schottky Barrier High Breakdown Field High Quality Buffer Low Trap Density
Low Noise Devices	Low Rs High Electron Velocity	High 2DEG density High Electron Velocity High Electron Mobility	
Digital Devices	Low Gate Leakage Current High Current Drive	High 2DEG Density	High Schottky Barrier

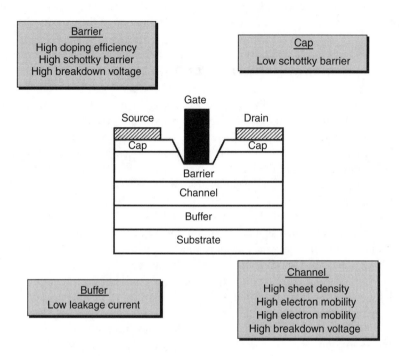

FIGURE 21.5   Material requirements for HEMT devices.

Using the $Al_{0.48}In_{0.52}As/Ga_{0.47}In_{0.53}As$ material system lattice matched to InP can simultaneously solve the limitations of the high bandgap barrier material and the lower bandgap channel material. The AlInAs/GaInAs HEMT (InP HEMT) has demonstrated excellent low-noise and power performance that extends well into the millimeter-wave range; they currently hold all the high-frequency performance records for FETs. The GaInAs channel has high electron mobility (>10,000 cm^2/Vs at room temperature), high electron saturation velocity (2.6 × 10^7 cm/s) and higher intervalley (Γ-L) energy separation. The higher conduction band offset at the AlInAs/GaInAs interface (ΔE_c = 0.5 eV) and the higher doping efficiency of AlInAs (compared to AlGaAs) results in a sheet charge density that is twice that of the AlGaAs/InGaAs material system. Higher doping efficiency of AlInAs also enables efficient vertical scaling of short gate length HEMTs. The combination of high sheet charge and electron mobility in the channel results in low source resistance, which is necessary to achieve high transconductance. However, the low bandgap of the InGaAs channel results in low breakdown voltage due to high impact ionization rates.

Table 21.2 summarizes the material properties of the three main material systems used for the fabrication of HEMTs. The emergence of growth techniques like Metal Organic Chemical Vapor Deposition (MOCVD) and Gas Source Molecular Beam Epitaxy (GSMBE) and continuing improvement in the existing growth techniques like molecular beam epitaxy (MBE) have enabled a new class of phosphorus-based material systems for fabrication of HEMTs. On the GaAs substrate, the GaInP/InGaAs has emerged as an alternative to the AlGaAs/InGaAs material system. GaInP has a higher bandgap than AlGaAs and hence enables high 2DEG densities due to the increased conduction band discontinuity (ΔE_c) at the GaInP/InGaAs interface. As GaInP has no aluminum it is less susceptible to environmental oxidation. The availability of high selectivity etchants for GaAs and GaInP simplifies device processing. However, the high conduction band discontinuity is achieved only for disordered GaInP, which has a bandgap of 1.9 eV. Using graded GaInP barrier layers and an $In_{0.22}Ga_{0.78}As$ channel, 2DEG density as high as 5 × 10^{12}/cm^2 and a mobility of 6000 cm^2/Vs was demonstrated.[21]

On InP substrates, the InP/InGaAs material system can be used in place of the AlInAs/GaInAs material system. The presence of deep levels and traps in AlInAs degrades the low frequency noise performance of AlInAs/GaInAs HEMT. Replacing the AlInAs barrier by InP or pseudomorphic InGaP can solve this

TABLE 21.2 Material Parameters of AlGaAs/GaAs, AlGaAs/InGaAs, and AlInAs/GaInAs Material Systems

Material Parameter	AlGaAs/GaAs	AlGaAs/InGaAs	AlInAs/GaInAs
ΔE_c	0.22 eV	0.42 eV	0.51 eV
Maximum donor doping	$5 \times 10^{18}/cm^3$	$5 \times 10^{18}/cm^3$	$1 \times 10^{19}/cm^3$
Sheet charge density	$1 \times 10^{12}/cm^2$	$1.5 \times 10^{12}/cm^2$	$3 \times 10^{12}/cm^2$
Mobility	8000 cm²/Vs	6000 cm²/Vs	12,000 cm²/Vs
Peak electron velocity	2×10^7 cm/s		2.7×10^7 cm/s
Γ-L valley separation	0.33 eV		0.5 eV
Schottky Barrier	1.0 eV	1.0 eV	0.45 eV

problem. One disadvantage of using the InP-based barrier is the reduced band discontinuity (0.25 eV compared to 0.5 eV for AlInAs/GaInAs) at the InP/InGaAs interface. This reduces 2DEG density at the interface and modulation efficiency. Increasing the indium content up to 75% in the InGaAs channel can increase the band discontinuity at the InP/InGaAs interface. The poor Schottky characteristics on InP necessitate the use of higher bandgap InGaP barrier layers or depleted p-type InP layers. A sheet density of $3.5 \times 10^{12}/cm^2$ and mobility of 11,400 cm²/Vs was demonstrated in an InP/In$_{0.75}$Ga$_{0.25}$As/InP double heterostructure.[22]

Despite the large number of material systems available for fabrication of HEMTs, the GaAs pHEMT implemented in the Al$_x$Ga$_{1-x}$As/In$_y$Ga$_{1-y}$As (x ~ 0.25; y ~ 0.22) material system and the InP HEMT implemented in the Al$_{0.48}$In$_{0.52}$As/Ga$_{0.47}$In$_{0.53}$As material system have emerged as industry vehicles for implementation of millimeter-wave analog and ultra high-speed digital circuits. The next two portions of this section will discuss the various performance aspects of GaAs pHEMT and InP HEMT.

21.4 AlGaAs/InGaAs/GaAs Pseudomorphic HEMT (GaAs pHEMT)

The first AlGaAs/InGaAs pseudomorphic HEMT was demonstrated in 1985.[23] Significant performance improvement over AlGaAs/GaAs HEMT was observed. Devices with a 1-µm gate length had peak transconductance of 270 mS/mm and maximum drain current density of 290 mA/mm.[20] The current gain cutoff frequency (f_T) was 24.5 GHz and the power gain cutoff frequency (f_{max}) was 40 GHz. An f_T of 120 GHz was reported for 0.2-µm gate length devices with In$_{0.25}$Ga$_{0.75}$As channel.[24] Devices with a 0.1-µm gate length with an f_{max} of 270 GHz were demonstrated in 1989.[14]

21.4.1 Millimeter-Wave Power GaAs pHEMT

In the past few years, the GaAs pHEMT has emerged as a device of choice for implementing microwave and millimeter-wave power amplifiers. To achieve a high output power density, device structures with higher current density and consequently higher sheet charge are required. As the sheet charge density in a single heterojunction AlGaAs/InGaAs pHEMT is limited to $2.3 \times 10^{12}/cm^2$, a double heterojunction (DH) device structure must be used to increase the sheet charge. In a DH GaAs pHEMT, carriers are introduced in the InGaAs channel by doping the AlGaAs barriers on both sides of the InGaAs channel. The AlGaAs barriers are doped with silicon using atomic planar doping to increase the electron transfer efficiency. A typical charge density of $3.5 \times 10^{12}/cm^2$ and a mobility of 5000 cm²/Vs is obtained for a double heterojunction GaAs pHEMT structure. The high sheet charge thus obtained enables higher current drive and power handling capability. Figure 21.6 shows the layer structure of a typical millimeter-wave power GaAs pHEMT. In some cases a doped InGaAs channel is also used to increase sheet charge density.[25,26]

Breakdown voltage is an important parameter for power devices. A device with high breakdown voltage can be biased at high drain voltages, which increases the drain efficiency, voltage gain, and power added efficiency (PAE). Typical breakdown voltages of GaAs pHEMTs range from 8 to 15 V. The breakdown

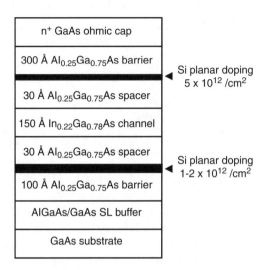

n⁺ GaAs ohmic cap	

FIGURE 21.6　Layer structure of a GaAs power pHEMT.

mechanism of a GaAs pHEMT can be either at the surface in the gate-drain of the device or in the channel (due to impact ionization).

There are several approaches used to increase the breakdown voltage of a GaAs pHEMT. The planar doping of AlGaAs barriers (as already described) helps in maintaining a high breakdown voltage, as most of the AlGaAs barrier is undoped. Another approach to increase the breakdown voltage uses a low-temperature grown (LTG) GaAs buffer below the channel. Using this approach, a 45% increase in channel breakdown voltage with a 12% increase in output power was demonstrated.[27] Using a double recessed gate structure to tailor the electric field in the gate drain depletion region can also increase breakdown voltage. The increase in breakdown voltage is mainly due to reduction in the electric field at the gate edge by surface states in the exposed recess region.[28]

The output power obtained from a HEMT also depends on the biasing conditions. To achieve high efficiency devices (as in Class B operation), the device is biased near pinch off, and therefore, high gain is required near pinch off. The mode of operation is ideally suited for pHEMTs, which typically have high transconductance near pinch off due to their superior charge-control properties. The effect for gate bias on the power performance of HEMT has been investigated.[29] Higher gain is achieved under Class A conditions. Biasing the device at higher drain voltages can increase the output power. Table 21.3 presents a summary of power performance of GaAs pHEMTs at various microwave and millimeter-wave frequencies.

From Table 21.3 it can be seen that at a given frequency, the device power output and gain increases with a decrease in gate length due to better high-frequency operation. Reduction in power gain is also observed for wider devices. This is due to the increase in source inductance, which increases with gate width and frequency and is due to the increase in gate resistance as a square of gate width. Low inductance via hole source grounding and proper gate layout is required to reduce these parasitics.

Reliability is important for space applications, typically a mean-time-to-failure (MTTF) of 10^7 h (1142 yr) is required for space applications. GaAs pHEMTs have demonstrated a MTTF of 1×10^7 h at a channel temperature of 125°C. The main failure mechanism is the atmospheric oxidation of the exposed AlGaAs barrier layers and interdiffusion of the gate metallization with the AlGaAs barrier layers (gate sinking).[30] Using dielectric passivation layers to reduce the oxidation of AlGaAs can solve these problems. Using refractory metal for gate contacts will minimize their interaction with the AlGaAs barrier layers. A MTTF of 1.5×10^7 hours at a channel temperature of 150°C was achieved using Molybdenum based gate contacts.[31]

Traveling wave tubes (TWT) have been traditionally used as multiwatt power sources for microwave applications up to K-band (20 GHz). Using GaAs pHEMT in place of TWT for these applications has many advantages, including lower cost, smaller size, smaller weight, and higher reliability. However, the

TABLE 21.3 Summary of Power Performance of GaAs pHEMTs

Frequency	Gate Length (µm)	Gate Width	Power Density (W/mm)	Output Power[a]	Gain[b] (dB)	PAE[b] (%)	Device, Drain Bias (Reference)
12 GHz (Ku-Band)	0.45	1.05 mm	0.77	0.81 W	10.0	60	Double HJ[c] V_{ds} = 7 V[32]
	0.25	1.6 mm	1.37	2.2 W	14.0	39	Double HJ[33]
20 GHz (K-Band)	0.25	600 µm	0.51	306 mW	7.4	45	Prematched, V_{ds} = 7 V [34]
	0.15	400 µm	1.04	416 mW	10.5	63	LTG Buffer, V_{ds} = 5.9 V[27]
	0.15	600 µm	0.12	72 mW	8.6	68	V_{ds} = 2 V[35]
	0.15	600 µm	0.84	501 mW	11	60	V_{ds} = 8 V[35]
35 GHz (Ku-Band)	0.25	500 µm	0.62	310 mW	6.8	40	Double HJ, V_{ds} = 5 V[36]
	0.15	150 µm	0.63	95 mW	9.0	51	Double HJ[37]
			0.91	137 mW	7.6	40	
44 GHz (Q-Band)	0.25		0.79		5.1	41	Doped Channel[d] [38]
	0.15	400 µm	0.5	200 mW	9.0	41	Double HJ, V_{ds} = 5 V[39]
	0.2	600 µm	0.53	318 mW	5.0	30	Double HJ, V_{ds} = 5 V[40]
	0.15	1.8 mm	0.44	800 mW	5.8	25	Double HJ, V_{ds} = 5 V[41]
55 GHz (V-Band)	0.25	400 µm	0.46	184 mW	4.6	25	Doped Channel, V_{ds} = 5.5 V[42]
	0.2	50 µm	0.85	42 mW	3.3	22	Doped Channel, V_{ds} = 4.3[43]
60 GHz (V-Band)	0.15	150 µm	0.83	125 mW	4.5	32	Double HJ[37]
			0.55	82 mW	4.7	38	
	0.15	400 µm	0.55	225 mW	4.5	25	Double HJ, V_{ds} = 5 V[44]
			0.44	174 mW	4.4	28	
94 GHz W-Band	0.25	75 µm	0.43	32 mW	3.0	15	Doped Channel, V_{ds} = 4.3 V[25]
				25 mW	4.0	14	
	0.15	150 µm	0.38	57 mW	2.0	16	Double HJ[37]
			0.30	45 mW	3.0	16	
	0.1	40 µm	0.31	13 mW	6.0	13	Doped Channel V_{ds} = 3.4 V[26]
	0.1	160 µm	0.39	63 mW	4.0	13	Doped Channel V_{ds} = 4.3 V[26]

[a] *Output Power* in italics indicates device biased for maximum power output.
[b] *PAE/Gain* in italics indicates device biased for maximum PAE/Gain.
[c] Double HJ — Double heterojunction GaAs pHEMT.
[d] Doped channel — Doped channel GaAs pHEMT.

TABLE 21.4 Summary of Power Performance of Multiwatt pHEMT Power Modules

Width (mm)	Frequency (GHz)	Power (W)	Gain (dB)	PAE (%)	Ref.
16.8	2.45	10.0	13.5	63.0	V_{ds} = 7 V[46]
24	2.45	11.7	14.0	58.2	V_{ds} = 8 V[47]
32.4	8.5–10.5	12.0	7.2	40.0	V_{ds} = 7 V[48]
16.8	12	12.0	10.1	48.0	[32]
25.2	12	15.8	9.6	36.0	V_{ds} = 9 V[49]
8	12	6	10.8	53.0	V_{ds} = 9 V[50]
9.72	18–21.2	4.7	7.5	41.4	V_{ds} = 5.5 V[45]

Note: V_{ds} = Drain bias for power measurements.

typical power density of a GaAs pHEMT at 20 GHz is on the order of 1 W/mm. Hence, the output power from a large number of devices has to be combined. To minimize combining losses, it is desirable to maximize power output of a single device. When large devices (gate width on the order of mm) are used to increase the total power output, other factors such as device layout, input signal distribution, output power combining networks, and substrate thickness are of critical importance. Several multiwatt GaAs pHEMT power modules have been demonstrated recently. A power module with 9.72-mm wide GaAs pHEMTs that delivered an output power of 4.7 W in the 18 to 21.2 GHz band with a PAE of 38% was demonstrated.[45] Table 21.4 summarizes the recent results of multiwatt GaAs pHEMT power modules.

21.4.2 Low Noise GaAs pHEMTs

Figure 21.7 shows the structure of a generic low-noise GaAs pHEMT. As the drain current requirement for a low-noise bias is low, single-side doped heterojunctions are sufficient for low-noise devices. The emphasis here is on achieving higher mobility to reduce the parasitic source resistance. As already discussed, the AlInAs/GaInAs material system is the ideal choice for fabrication of low-noise microwave and millimeter-wave devices. However, GaAs pHEMTs also find significant use in millimeter-wave low noise applications due to wafer size, cost, and process maturity related advantages.

Henderson et al. first reported on the low noise performance of GaAs pHEMTs in 1986. Devices with 0.25-μm gate length had a noise figure of 2.4 dB and an associated gain of 4.4 dB at 62 GHz.[51] A 0.15-μm gate length $Al_{0.25}Ga_{0.75}As/In_{0.28}Ga_{0.72}As$ pHEMT with a noise figure of 1.5 dB and an associated gain of 6.1 dB at 61.5 GHz was demonstrated in 1991.[52] The reduction in noise figure was a direct result of reducing the gate length, which increased the f_T of the device. Low noise operation of a GaAs pHEMT was also demonstrated at 94 GHz.[53] For a 0.1-μm gate length device a noise figure of 3.0 dB and associated gain of 5.1 dB was achieved. A noise figure of 2.1 dB and an associated gain of 6.3 dB was reported for a 0.1 μm gate length GaAs pHEMT at 94 GHz.[54] The improvement in noise figure is attributed to the use of a T-shaped gate with end-to-end resistance of 160 Ω/mm by Tan et al.,[52] compared to a trapezoidal gate with end-to-end resistance of 1700 Ω/mm used by Chao et al.[53] This further emphasizes the need to reduce parasitic resistances in low-noise devices.

One of the main system applications of low-noise GaAs pHEMTs is satellite direct broadcasting receiver systems (DBS) that are in increasing demand worldwide. Low-noise amplifiers operating at 12 GHz are a critical component in these systems. The low-noise performance of GaAs pHEMTs is more than adequate for these applications. A 0.25-μm gate length GaAs pHEMT with a noise figure of 0.6 dB and an associated gain of 11.3 dB at 12 GHz was reported by Tokue et al.[55] Performance coupled with low cost packaging is one of the crucial factors in the high volume DBS market. Hwang et al. have demonstrated a 0.2-μm gate length GaAs pHEMT in plastic packaging with a 1.0 dB noise figure and 9.9 dB associated gain at 12 GHz.[56] A plastic packaged GaAs pHEMT device with a gate length of 0.17 μm demonstrated a noise figure of 0.35 dB, and 12.5 dB associated gain at 12 GHz.[57] Table 21.5 summarizes the low-noise performance of GaAs pHEMTs at various microwave and millimeter-wave frequencies.

21.4.3 GaAs pHEMT for Wireless Applications

The explosive growth of the wireless communication industry has opened up a new area of application for GaAs pHEMTs. Unlike the millimeter-wave military and space applications, the frequencies of operation of these applications are much lower. The frequencies used in typical cellular phones range from 850 MHz for the American Mobile Phone System (AMPS) to 1.9 GHz for the Japanese Personal Handy Phone System (PHS) and the Digital European Cordless Telephone (DECT). The device parameters of

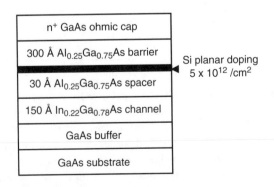

FIGURE 21.7 Layer structure of a low noise GaAs pHEMT.

TABLE 21.5 Summary of Low Noise Performance of GaAs pHEMTs

Frequency (GHz)	Gate Length (μm)	F_{min} (dB)	Ga (dB)	Ref.	Comments
12	0.25	0.6	11.3	55	
12	0.17	0.35	12.5	57	Packaged Device
18	0.25	0.9	10.4	51	
62	0.25	2.4	4.4	51	Improvement by
60	0.15	1.5	6.1	52	Reduction in L_g
94	0.10	3.0	5.1	53	Improvement by
94	0.10	2.1	6.3	54	Reduction in R_g

interest when considering device technologies for wireless applications are operating voltages, which must be positive, power density, output match and linearity requirements, and gate leakage current.[58] Using enhancement mode devices (threshold/pinch-off voltage > 0) eliminates the negative supply voltage generator and power-cutoff switch. The high gate turn-on voltage of an enhancement mode GaAs pHEMT, as compared to the enhancement mode GaAs MESFET, enables higher input voltage swing.

The power performance of MESFET and GaAs pHEMT for wireless applications has been compared.[59] For the same saturation drain current density, at a frequency of 950 MHz, the saturated power output from the pHEMT is 2.5 W, whereas it is 1.8 W from the MESFET. The power-added efficiency of the pHEMT is 68%, which is 8% higher than that of the MESFET. This difference is due to the transfer characteristics of the two devices. The pHEMT performs as a better power amplifier than the MESFET because the input power is effectively amplified with higher g_m near the pinch-off voltage. This is a direct consequence of better charge control properties of the HEMT when compared to the MESFET. The pHEMT also has a lower gate leakage current than the MESFET due to a higher Schottky barrier on AlGaAs.

The power performance of enhancement mode GaAs pHEMT with a threshold voltage of 0.05 V for wireless applications has also been investigated.[60] A device with a gate width of 3.2 mm delivered an output power of 22 dBm with power-added-efficiency of 41.7%. The standby current at a gate bias of 0 V was 150 μA. Enhancement mode GaAs HFET with a higher threshold voltage of 0.5 V has also been demonstrated.[61] A 1-μm gate length device with a gate width of 12 mm delivered an output power of 31.5 dBm with a PAE of 75% at 850 MHz and at a drain bias of 3.5 V. The standby current at a gate bias of 0 V was 1 μA. This eliminates the need for a switch in the drain current of the power amplifier. The device was manufactured using Motorola's CGaAs™ process, which is cost effective as it uses processes that are similar to standard silicon MOS and bipolar processes. Table 21.6 shows a summary of power performance of GaAs pHEMTs for cellular phones.

21.5 AlInAs/GaInAs/InP (InP HEMT)

Future military and commercial electronic applications will require high-performance microwave and millimeter-wave devices. Important applications include low-noise amplifiers for receiver front ends, power amplifiers for phased-array radars, ultra high-speed digital circuits for prescalers, and MUX/DEMUX electronics for high-speed (>40 Gb/s) optical links.

A HEMT device capable of operating at millimeter-wave frequency requires a channel with high electron velocity, high current density, and minimal parasitics. As discussed, the $Al_{0.48}In_{0.52}As/Ga_{0.47}In_{0.53}As$ material system lattice matched to InP satisfies these criteria. A 1-μm gate length AlInAs/GaInAs HEMT with extrinsic transconductance as high as 400 mS/mm was demonstrated.[77] The microwave performance of 1-μm gate-length devices showed an improvement of 20 to 30% over the AlGaAs/GaAs HEMT.[78] In 1988 Mishra et al. demonstrated a 0.1 μm InP HEMT with a f_T of 170 GHz.[79] Using a T-gate to self-align the source and drain contacts results in reduction of source-gate and source-drain spacing. This not only reduces the parasitic source and drain resistances, but also the drain delay. Using the preceding technique an f_T of 250 GHz was achieved in a 0.13-μm gate length self-aligned HEMT.[18] Recently, a 0.07-μm AlInAs/GaInAs HEMT with an f_T of 300 GHz and an f_{max} of 400 GHz was reported.[80]

TABLE 21.6 Summary of Power Performance of GaAs pHEMTs for Cellular Phones

Frequency	Device Width (μm)	Drain Bias (V)	Power Output	PAE (%)	ACPL	Device/Ref.
850 MHz	5	1.2	19.6 dBm	65.2		AlGaInP/InGaAs pHEMT[62]
	10	1.3	21.5 dBm	57.4		InGaP/InGaAs pHEMT[63]
	12	3.5	31.5 dBm	75		Enhancement Mode HFET[61]
	12	3.5	33.1 dBm	84.8		GaAs pHEMT[64]
	30	3.7	31.0 dBm	59.0	−30.1 dBc @ 30 kHz	1 μm GaAs MESFET[65]
900 MHz	12	3.5	31.5 dBm	75		Enhancement Mode CGaAs™[66]
	21	2.3	31.3 dBm	68		0.8 μm MESFET[67]
	14	3.0	32.3 dBm	71		pHEMT[68]
	40	1.5	31.5 dBm	65		MESFET[69]
950 MHz	28	1.2	1.1 W	54		pHEMT[70]
	21	2.2	32.7 dBm	62.8	−50.5 dBc	PHEMT[71]
	12	3.0	1.4 W	60.0	—	pHEMT[72]
	7	3.4	30.9 dBm	56.3	−51.5 dBc @ 50 kHz	pHEMT[73]
	16	3.4	1.42 W	60.0	−48.2 dBc @ 50 kHz	pHEMT[74]
	12	4.7	2.5 W	68.0		pHEMT[59]
	12	4.7	1.8 W	60.0		MESFET[59]
1.9 GHz	1	2.0	20.2 dBm	45.3	−55.2 dBc @ 600 kHz	pHEMT[75]
	2.4	2.0	21.1 dBm	54.4	−55 dBc @ 600 kHz	MESFET[76]
	3.2	3.0	22.0 dBm	41.7	−58.2 dBc @ 600 kHz	Enhancement Mode pHEMT[60]
	12	3.5	30 dBm	50	−30 dBc	Enhancement Mode pHEMT[61]
	5	2.0	25.0 dBm	53.0		AlGaInP/InGaAs pHEMT[62]

The high-frequency performance of the InP HEMT can be further improved by using a pseudomorphic InGaAs channel with an indium content as high as 80%. The f_T of a 0.1-μm InP HEMT increased from 175 to 205 GHz when the indium content in the channel was increased from 53 to 62%.[81] Although devices with high indium content channels have low breakdown voltages, they are ideal for low noise applications and ultra high-speed digital applications. An f_T of 340 GHz was achieved in a 0.05-μm gate length psuedomorphic InP HEMT with a composite $In_{0.8}Ga_{0.2}As/In_{0.53}Ga_{0.47}As$ channel.[82] This is the highest reported f_T of any 3-terminal device.

Compared to the GaAs pHEMT, the AlInAs/GaInAs HEMT has a higher current density that makes it suitable for ultra high-speed digital applications. The high current gain cutoff frequency and low parasitics make the AlInAs/GaInAs HEMT the most suitable choice for low- noise applications extending well beyond 100 GHz. The high current density and superior high-frequency performance can be utilized for high-performance millimeter-wave power applications provided the breakdown voltage is improved. Some of the state-of-the-art millimeter-wave analog circuits and ultra high-speed digital circuits have been implemented using InP HEMTs. A low-noise amplifier with 12 dB gain at a frequency of 155 GHz using a 0.1-μm InP HEMT with a $In_{0.65}Ga_{0.35}As$ psuedomorphic channel was demonstrated.[83] Pobanz et al. demonstrated an amplifier with 5 dB gain at 184 GHz using a 0.1 μm gate $In_{0.8}Ga_{0.2}As/InP$ composite channel HEMT.[84]

21.5.1 Low-Noise AlInAs/GaInAs HEMT

The superior electronic properties of the GaInAs channel enable the fabrication of extremely high f_T and f_{max} devices. The superior carrier confinement at the AlInAs/GaInAs interface results in a highly linear transfer characteristic. High transconductance is also maintained very close to pinch off. This is essential because the noise contribution of the FET is minimized at low drain current levels. Hence high gain can be achieved at millimeter-wave frequencies under low-noise bias conditions. The high mobility at the AlInAs/GaInAs interface also results in reduced parasitic source resistance of the device. AlInAs/GaInAs HEMTs with 0.25-μm gate length exhibited a noise figure of 1.2 dB at 58 GHz.[85] At 95 GHz a noise figure of 1.4 dB with associated gain of 6.6 dB was achieved in a 0.15-μm gate length device.[86] Table 21.7 summarizes the low-noise performance of AlInAs/GaInAs HEMTs.

TABLE 21.7   Summary of Low-Noise Performance of AlInAs/GaInAs HEMTs

Frequency (GHz)	Gate Length (μm)	F_{min} (dB)	G_a (dB)	Comments/Ref.
12	0.15	0.39	16.5	$In_{0.7}Ga_{0.3}As$ channel[87]
18	0.25	0.5	15.2	[85]
18	0.15	0.3	17.2	[86]
26	0.18	0.43	8.5	Passivated device [88]
57	0.25	1.2	8.5	[85]
60	0.1	0.8	8.9	[89]
63	0.1	0.8	7.6	Passivated[90]
		0.7	8.6	Unpassivated
94	0.15	1.4	6.6	[86]
94	0.1	1.2	7.2	[89]

21.5.2   Millimeter-Wave AlInAs/GaInAs Power HEMT

The millimeter-wave power capability of single heterojunction AlInAs/GaInAs HEMTs has been demonstrated.[91,92] The requirements for power HEMT are high gain, high current density, high breakdown voltage, low access resistance, and low knee voltage to increase power output and power-added efficiency. The AlInAs/GaInAs HEMT satisfies all of these requirements with the exception of breakdown voltage. This limitation can be overcome by operating at a lower drain bias. In fact, the high gain and PAE characteristics of InP HEMTs at low drain bias voltages make them ideal candidates for battery-powered applications.[93] Another advantage is the use of InP substrate that has a 40% higher thermal conductivity than GaAs. This allows higher dissipated power per unit area of the device or lower operating temperature for the same power dissipation. As low breakdown voltage is a major factor that limits the power performance of InP HEMTs, this section will discuss in detail the various approaches used to increase breakdown voltage.

Breakdown in InP HEMT is a combination of electron injection from the gate contact and impact ionization in the channel.[94] The breakdown mechanism in the off-state (when the device is pinched off) is electron injection from the gate. The low Schottky barrier height of AlInAs results in increased electron injection from the gate and, consequently, higher gate leakage current compared to the GaAs pHEMT. These injected hot electrons cause impact ionization in the high-field drain end of the GaInAs channel. The high impact ionization rate in the low bandgap GaInAs channel is the main mechanism that determines the on-state breakdown. Some of the holes generated by impact ionization are collected by the negatively biased gate and result in increased gate leakage. The potential at the source end of the channel is modulated by holes collected by the source. This results in increased output conductance.

Lowering the electric field in the gate-drain region can reduce the impact ionization rate. This is achieved by using a double recess gate fabrication process that increases the breakdown voltage from 9 to 16 V.[95] A gate-drain breakdown voltage of 11.2 V was demonstrated for 0.15-μm gate length devices with a 0.6-μm recess width.[96] In addition, reduction in output conductance (g_{ds}) and gate-drain feedback capacitance (C_{gd}) was observed when compared to single recessed devices. The f_{max} of a double recessed device increased from 200 to 300 GHz.[97] Hence, it is desirable for power devices. Another approach to reduce electric field in the gate-drain region is to use an undoped GaInAs cap instead of a doped GaInAs cap.[98] The output conductance can be reduced from 50 to 20 mS/mm for a 0.15-μm gate length device by replacing the doped GaInAs cap by an undoped cap.[99] This also improved the breakdown voltage from 5 to 10 V. The reduction in C_{gd} and g_{ds} resulted in an f_{max} as high as 455 GHz. Redistribution of the dopants in the AlInAs barrier layers can also increase breakdown voltage. An increase in breakdown voltage from 4 to 9 V is achieved by reducing doping in the top AlInAs barrier layer and transferring it to the AlInAs barrier layers below the channel.[92]

The gate leakage current can be reduced and the breakdown voltage increased by using a higher bandgap strained AlInAs barrier.[100,101] By increasing the Al composition in the barrier layers from 48 to 70%, the gate-to-drain breakdown voltage was increased from 4 to 7 V. This also results in reduction of gate leakage as the Schottky barrier height increases from 0.5 to 0.8 eV. The use of $Al_{0.25}In_{0.75}P$ as an Schottky barrier improves the breakdown voltage from −6 to −12V.[102]

The on-state breakdown can be improved in two ways. The first is to reduce the gate leakage current by the impact ionization generated holes by increasing the barrier height for holes. This was achieved by increasing the valence band discontinuity at the channel-barrier interface. The use of a strained 25-Å $In_{0.5}Ga_{0.5}P$ spacer instead of AlInAs increases the valence band discontinuity at the interface from 0.2 to 0.37 eV. An on-state breakdown voltage of 8 V at a drain current density of 400 mA/mm for a 0.7-μm gate length InP HEMT was achieved by using a strained InGaP barrier.[103]

The various approaches to increase the breakdown voltage, as already discussed here, concentrate on reducing the electron injection from the Schottky gate and reducing the gate leakage current. These approaches also have an inherent disadvantage as Al-rich barriers result in high source resistance and are more susceptible to atmospheric oxidation. Additionally, these approaches do not address the problem of a high impact ionization rate in the GaInAs channel and carrier injection from contacts. In the recent past, various new approaches have been investigated to increase breakdown voltage without compromising the source resistance or atmospheric stability of the device. These include the junction-modulated AlInAs/GaInAs HEMT (JHEMT), the composite GaInAs/InP channel HEMT, and the use of regrown contacts. Table 21.8 summarizes the power performance of AlInAs/GaInAs HEMTs.

21.5.3　GaInAs/InP Composite Channel HEMT

The high speed and power performance of InP HEMT can be improved by the use of composite channels that are composed of two materials with complementing electronic properties. The high-speed perfor-

TABLE 21.8　Summary of Power Performance of InP HEMTs

Frequency (GHz)	Gate Length (μm)	Gate Width	Power Density (W/mm)	Power Output (mW)	Gain (dB)	PAE (%)	Device/Drain Bias (Ref.)
4	0.5	2 mm	0.13	269	18	66	V_{ds} = 2.5 V[93]
	0.15	0.8mm	0.4	320	18	57	V_{ds} = 3 V[93]
12	0.22	150 μm	0.78	117	8.4	47	V_{ds} = 4 V[92]
			0.47	70	11.3	59	V_{ds} = 3 V, Double HJ
18	0.15	600 μm	0.74	446	13	59	Double recessed V_{ds} = 7 V[104]
20	0.15	50 μm	0.78	39	10.2	44	V_{ds} = 4.9 V[91]
(K-Band)			0.41	21	10.5	52	V_{ds} = 2.5 V
							Single heterojunction
	0.15	50 μm	0.61	30	12.2	44	Single heterojunction
							$In_{0.69}Ga_{0.31}As$ channel
							V_{ds} = 4.1 V[91]
	0.15	800 μm	0.65	516	7.1	47	70% AlInAs, V_{ds} = 4 V[105]
44	0.15	450 μm	0.55	251	8.5	33	$Al_{0.6}In_{0.4}As$ barrier
(Q-Band)			0.88	398	6.7	30	Doped Channel[100]
	0.2	600 μm	0.37	225	5	39	$Al_{0.6}In_{0.4}As$ barrier
							Single heterojunction, V_{ds} = 4 V[106]
57	0.22	450 μm	0.33	150	3.6	20	$Al_{0.6}In_{0.4}As$ barrier
(V-Band)			0.44	200		17	Doped channel, V_{ds} = 3.5 V[101]
60	0.15	50 μm	0.35		7.2	41	V_{ds} = 2.6 V[91]
(V-Band)			0.52	26	5.9	33	V_{ds} = 3.6 V
							Single heterojunction
	0.1	50 μm	0.30	15	8.6	49	V_{ds} = 3.35 V[107]
			0.41	21	8.0	45	
	0.1	400 μm	0.48	192	4.0	30	V_{ds} = 4.12 V, 67% In[107]
94	0.15	50 μm	0.30	15	4.6	21	Single HJ, Passivated device
(W-Band)							V_{ds} = 2.6 V[108]
	0.15	640 μm	0.20	130	4.0	13	Double HJ, V_{ds} = 2.7[109]
	0.1	200 μm	0.29	58		33	$In_{0.68}Ga_{0.32}As$ channel[110]

mance of an InP HEMT can be improved by inserting InAs layers in the InGaAs channel. The current gain cutoff frequency of a 0.15-µm gate length device increased from 179 to 209 GHz due to improved electron properties.[111] An f_T as high as 264 GHz was achieved for a 0.08-µm gate length device.

The GaInAs channel has excellent electronic properties at a low electric field, but suffers from high impact ionization at high electric fields. On the other hand, InP has excellent electronic transport properties at high fields but has lower electron mobility. In a composite InGaAs/InP channel HEMT, the electrons are in the InGaAs channel at the low field source end of the channel and are in the InP channel at the high field drain end of the channel. This improves the device characteristics at high drain bias while still maintaining the advantages of the GaInAs channel at low bias voltages.[112] A typical submicron gatelength AlInAs/GaInAs HEMT has an off-state breakdown voltage (BV_{dsoff}) of 7 V, and on-state breakdown voltage (BV_{dson}) of 3.5 V. Using a composite channel, (30 Å GaInAs/50 Å InP/100 Å n^+ InP), Matloubian et al. demonstrated a BV_{dsoff} of 10 V and BV_{dson} of 8 V for a 0.15-µm gate length device.[113] A 0.25-µm GaInAs/InP composite channel HEMT with a two-terminal gate drain voltage of 18 V was also demonstrated.[114]

The increased breakdown voltage of a composite channel HEMT enables operation at a higher drain bias. This increases the drain efficiency and the PAE of the device. An output power of 0.9 W/mm with a PAE of 76% at 7 GHz was demonstrated for a 0.15-µm GaInAs/InP composite channel HEMT at a drain bias of 5 V.[115] At 20 GHz, an output power density of 0.62 W/mm (280 mW) and a PAE 46% was achieved for a 0.15 µm gate length device at a drain bias of 6 V.[113] At 60 GHz, a 0.15-µm GaInAs/InP composite channel HEMT demonstrated an output power of 0.35 W/mm, and a power gain of 6.2 dB with a PAE of 12% at a drain bias of 2.5 V.[116]

21.6 Technology Comparisons

21.6.1 Comparison between FETs and HBTs for RF/ Microwave Applications

In recent years hetero-structure bipolar transistors (HBTs) have emerged as strong contenders for wireless, microwave, and millimeter-wave applications. HBTs have similar high frequency performance with modest lateral dimensions as transport is along the vertical direction. This also offers higher current drivability per unit chip area. On the other hand, submicron gate lengths are required to achieve microwave and millimeter-wave operation in FETs. In addition, HBTs have high transconductance due to an exponential relationship between the input (base) voltage and output (collector) current. Device uniformity over a wafer can be easily achieved in a HBT as the turn-on voltage depends on the built-in voltage of a pn junction (and is governed by the uniformity of the epitaxial growth technique). The threshold voltage of a HEMT (which is a measure of device uniformity) is mainly determined by uniformity in gate recess etching. The FET is mainly a surface device and therefore has a higher 1/f noise due generation-recombination processes in the surface traps. On the other hand, current transport in a HBT is in the bulk, hence HBTs have lower 1/f noise which makes them ideal for low-phase-noise microwave and millimeter-wave oscillators.

Semiconductor processing of HBTs is more complex than FETs owing to the vertical structure. Another advantage of FETs when compared to HBTs is lower operating voltage. The turn-on voltage of GaAs HBTs is significantly higher (1.4 V) than that of HEMTs (0.5 V). This makes the HEMT an ideal candidate for low-voltage battery-operated applications. Owing to low parasitics, HEMTs have a lower noise figure than HBTs at any frequency and therefore are preferred for low-noise receivers. For microwave and millimeter-wave power applications gain, HEMTs are again the preferred choice. This is because at millimeter-wave frequency (> 50 GHz) the gain in a HBT is mainly limited by the high collector base feedback capacitance. Another disadvantage of large area HBTs is thermal runaway due to nonuniform junction temperatures. This can be solved by the use of emitter ballast resistors, but this reduces device gain and processing complexity. In the following sections, the various HEMT technologies are compared for specific applications.

21.6.2 Power Amplifiers for Wireless Phones

The two main aspects that govern the choice of technology for wireless power amplifiers are the operating voltage and the cost. As far as operating voltages are concerned, the use of GaAs HBTs is limited to 3.6 V supply voltages. This is due to the high turn-on voltage (1.4 V) for the emitter-base junction in the GaAs HBT. On the other hand, GaAs MESFETs and pHEMT can be used in 1.2 V systems due to their low knee voltages. As far as cost is concerned, the major contenders are ion-implanted GaAs MESFETs and GaAs HBTs. However, GaAs MESFETs require a negative gate voltage, which potentially increases circuit complexity and cost. The enhancement mode GaAs pHEMT, which has a low knee voltage (like a GaAs MESFET) and positive voltage operation (like a GaAs HBT) is therefore a viable technology option for high performance, low voltage wireless phones.

21.6.3 Microwave Power Amplifiers

The GaAs pHEMT and the InP HEMT are the two major competing technologies for microwave and millimeter-wave power amplifiers. The relation between output power density (P_{out}) and power-added efficiency (PAE) and device parameters is given by the following expressions,

$$P_{out} = \frac{1}{8}\left(I_{max}\right)\left(BV_{gd} - V_{knee}\right) \tag{21.30}$$

$$PAE = \alpha\left(\frac{V_{DD} - V_{knee}}{V_{DD}}\right)\left(1 - \frac{1}{G_a}\right) \tag{21.31}$$

In Equation 21.31, α is ½ for Class A operation and $\pi/4$ for Class B operation. Figure 21.8 compares the P_{out} and PAE of GaAs pHEMTs and InP HEMTs as a function of operating frequency. It can be seen that GaAs pHEMTs have a higher power density than InP HEMTs at all frequencies than InP HEMT. This is due to higher breakdown voltages (BV_{gd}) in GaAs pHEMTs that enables higher operating voltage (V_{DD}). The InP HEMT operating voltage is limited by the low breakdown voltage, hence the power output is low. However, InP HEMTs have comparable power output at millimeter-wave frequencies. This is enabled by the low knee voltage (V_{knee}) and high current drive (I_{max}). On the other hand, due to their low knee voltage and high gain (G_a), InP HEMTs have higher gain and PAE at frequencies exceeding 60 GHz. The potential of InP HEMTs as millimeter-wave power devices is evident in the fact that comparable power performance is achieved at drain biases 2 to 3 V lower than those for GaAs pHEMTs. Hence a high voltage InP HEMT technology should be able to outperform the GaAs pHEMT for power applications at all frequencies.

21.6.4 Low Noise Amplifiers

Figure 21.9 compares the gain and noise figure of GaAs pHEMTs and InP HEMTs as a function of frequency. At any given frequency, the InP HEMT has a noise figure about 1 dB lower than the GaAs pHEMT and gain 2 dB higher than the GaAs pHEMT. This is mainly due to the excellent transport properties of the AlInAs/GaInAs material system.

Digital applications — High current driving capability with low voltage operation is essential for high speed, low power digital circuits. High f_T devices are also required for increasing the frequency of operation. In this area, GaAs MESFETs and GaAs pHEMTs are sufficient for 10 Gb/s digital circuits. However, for digital circuits operating at bit rates exceeding 40 Gb/s, the InP HEMT is the preferred technology option.

Table 21.9 lists the various frequency bands and military and space applications in each band with the appropriate HEMT device technology for each application.

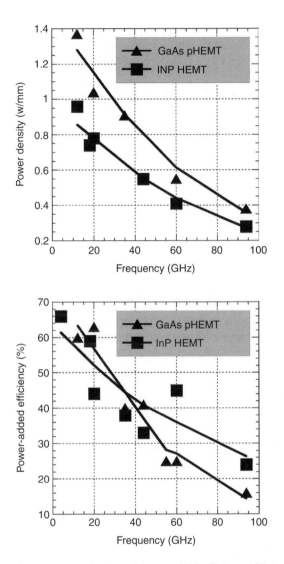

FIGURE 21.8 Comparison of output power density and power-added-efficiency of GaAs pHEMTs and InP HEMTs as a function of frequency.

21.7 Conclusion

GaAs- and InP-based high electron mobility transistors have emerged as premier devices for the implementation of millimeter-wave analog circuits and ultra high-speed digital circuits. In this section, principles of HEMT operation were discussed. The design aspects of HEMT for both low-noise and high-power applications were discussed. Reduction in gate length is essential for improved performance at high frequencies. Appropriate device scaling with gate length reduction is necessary to minimize the effect of parasitics on device performance.

Millimeter-wave power modules have been demonstrated using GaAs pHEMT devices. The superior device performance of GaAs pHEMTs is being used to improve the performance of power amplifiers for wireless phone systems. The superior material characteristics of the AlInAs/GaInAs material system have been used to achieve record low-noise performance at millimeter-wave frequencies using InP HEMTs. Despite their low breakdown voltage, InP HEMTs have demonstrated superior power performance at

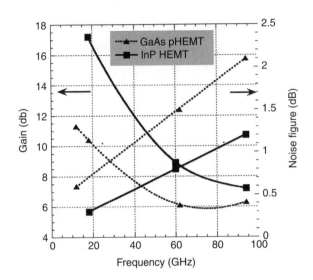

FIGURE 21.9   Comparison of noise figure and gain of GaAs pHEMTs and InP HEMTs as a function of frequency.

TABLE 21.9   Frequency Bands and Military and Commercial Applications

Frequency	Military/Space	Commercial	Device Technology
850 MHz–1.9 GHz		Wireless	Low noise — GaAs pHEMT Power — GaAs pHEMT
12 GHz (Ku-Band)	Phased array radar	Direct broadcast satellite	Low noise — GaAs pHEMT Power — GaAs pHEMT
20 GHz (K-Band)	Satellite downlinks		
27–35 GHz (Ka-Band)	Missile seekers	LMDS — Local multipoint distribution system	Low noise — GaAs pHEMT Power — GaAs pHEMT
44 GHz (Q-Band)	SATCOM ground terminals	MVDS — Multipoint video distribution system	Low noise — InP HEMT Power — GaAs pHEMT
60 GHz (V-Band)	Satellite crosslinks	Wireless LAN	Low noise — InP HEMT Power — GaAs pHEMT/InP HEMT
77 GHz		Collision avoidance radar	Low noise — InP HEMT Power — GaAs pHEMT/InP HEMT
94 GHz (W-Band)	FMCW radar		Low noise — InP HEMT Power — InP HEMT
100–140 GHz	Radio astronomy		Low noise — InP HEMT
Digital 10 Gb/s		Fiber-optic communication	GaAs pHEMT
Digital 40 Gb/s		Fiber-optic communication	InP HEMT

millimeter-wave frequency. Improving the breakdown voltage using approaches that include composite channel GaInAs/InP HEMT and junction-modulated HEMT will further improve power performance. The development of the GaAs pHEMT and InP HEMT technology was traditionally supported by low-volume, high-cost military and space applications. The recent emergence of high-volume commercial applications such as wireless and optical communications systems has new constraints that include manufacturability and low-voltage operation.

References

1. Dingle, R., Stromer, H. L., Gossard, A. C., and Wiemann, W., Electron mobility in modulation doped semiconductor superlattices, *Applied Physics Letters*, 33, 665, 1978.

2. Delagebeaudeuf, D. and Linh, N. T., Metal-(n) AlGaAs-GaAs two-dimensional electron gas FET, *IEEE Transactions on Electron Devices*, 29, 955, 1982.

3. Drummond, T. J., Masselink, W. T., and Morkoc, H., Modulation Doped GaAs/(Al,Ga)As hetero-junction field-effect transistors: MODFETs, *Proceedings of the IEEE*, 74, 773, 1986.

4. Vinter, B., Subbands and charge control in two-dimensional electron gas field effect transistor, *Applied Physics Letters*, 44, 307, 1984.

5. Foisy, M. C., Tasker, P. J., Hughes, B., and Eastman, L. F., The role of insufficient charge modulation in limiting the current gain cutoff frequency of the MODFET, *IEEE Transactions of Electron Devices*, 35, 871, 1988.

6. Nguyen, L. D., Larson, L. E., and Mishra, U. K., Ultra-high-speed modulation-doped field-effect transistors: A Tutorial Review, *Proceedings of the IEEE*, 80, 494, 1992.

7. Dambrine, G., Cappy, A., Heliodore, F., and Playez, E., A new method for determining the FET small-signal equivalent circuit, *IEEE Transactions on Microwave Theory and Techniques*, 36, 1151, 1988.

8. Berroth, M. and Bosch, R., Broad-band determination of the FET small-signal equivalent circuit, *IEEE Transactions on Microwave Theory and Techniques*, 38, 891, 1990.

9. Das, M. B., A high aspect ratio design approach to millimeter-wave HEMT structures, *IEEE Transactions on Electron Devices*, 32, 11, 1985.

10. Lester, L. F., Smith, P. M., Ho, P., Chao, P. C., Tiberio, R. C., Duh, K. H. G., and Wolf, E. D., 0.15 μm gate length double recess psuedomorphic HEMT with f_{max} of 350 GHz, *IEDM Technical Digest*, 172, 1988.

11. Nguyen, L. D., Tasker, P. J., Radulescu, D. C., and Eastman, L. F., Characterization of ultra-high speed pseudomorphic AlGaAs/InGaAs (on GaAs) MODFET's, *IEEE Transactions on Electron Devices*, 36, 2243, 1989.

12. Nguyen, L. D. and Tasker, P. J., Scaling issue of ultra-high speed HEMTs, *SPIE Conf. on High Speed Electronics and Device Scaling*, 1288, 251, 1990.

13. Moll, N., Hueschen, M. R., and Fischer-Colbrie, A., Pulse doped AlGaAs/InGaAs pseudomorphic MODFETs, *IEEE Transactions on Electron Devices*, 35, 879, 1988.

14. Chao, P. C., Shur, M. S., Tiberio, R. C., Duh, K. H. G., Smith, P. M., Ballingall, J. M., Ho, P., and Jabra, A. A., DC and microwave characteristics of Sub 0.1 μm gate length planar doped pseudo-morphic HEMTs, *IEEE Transactions on Electron Devices*, 36, 461, 1989.

15. Nguyen, L. D., Brown, A., Delaney, M., Mishra, U., Larson, L., Jelloian, L., Melendes, M., Hooper, C., and Thompson, M., Vertical scaling of ultra-high-speed AlInAs-GaInAs HEMTs, *IEDM Technical Digest*, 105, 1989.

16. Enoki, T., Tomizawa, M., Umeda, Y., and Ishii, Y., 0.05 μm Gate InAlAs/InGaAs High Electron mobility transistor and reduction of its short channel effects, *Japanese Journal of Applied Physics*, 33, 798, 1994.

17. Tasker, P. J. and Hughes, B., Importance of source and drain resistance to the maximum f_T of millimeter-wave MODFETs, *IEEE Electron Device Letters*, 10, 291, 1989.

18. Mishra, U. K., Brown, A. S., Jelloian, L. M., Thompson, M., Nguyen, L. D., and Rosenbaum, S. E., Novel high performance self aligned 0.15 μm long T-gate AlInAs-GaInAs HEMTs, *IEDM Technical Digest*, 101, 1989.

19. Mimura, T., Hiyamizu, S., and Hikosak, S., Enhancement mode high electron mobility transistors for logic applications, *Japanese Journal of Applied Physics*, L317, 1981.

20. Ketterson, A. A., Masselink, W. T., Gedymin, J. S., Klem, J., Peng, C.-K., Kopp, W. F., Morkoç, H., and Gleason, K. R., Characterization of InGaAs/AlGaAs pseudomorphic modulation-doped field-effect transistors, *IEEE Transactions on Electron Devices*, 33, 564, 1986.

21. Pereiaslavets, B., Bachem, K. H., Braunstein, J., and Eastman, L. F., GaInP/InGaAs/GaAs graded barrier MODFET grown by OMVPE: design, fabrication, and device results, *IEEE Transactions on Electron Devices*, 43, 1659, 1996.

22. Mesquida-Kusters, A. and Heime, K., Al-Free InP-based high electron mobility transistors: design, fabrication and performance, *Solid State Electronics*, 41, 1159, 1997.

23. Ketterson, A., Moloney, M., Masselink, W. T., Klem, J., Fischer, R., Kopp, W., and Morkoc, H., High transconductance InGaAs/AlGaAs pseudomorphic modulation doped field effect transistors, *IEEE Electron Device Letters*, 6, 628, 1985.

24. Nguyen, L. D., Radulescu, D. C., Tasker, P. J., Schaff, W. J., and Eastman, L. F., 0.2 μm gate-length atomic-planar doped pseudomorphic $Al_{0.3}Ga_{0.7}As/In_{0.25}Ga_{0.75}As$ MODFET's with f_t over 120 GHz, *IEEE Electron Device Letters*, 9, 374, 1988.

25. Smith, P. M., Lester, L. F., Chao, P.-C., Ho, P., Smith, R. P., Ballingall, J. M., and Kao, M.-Y., A 0.25 μm gate length pseudomorphic HFET with 32 mW Output power at 94 GHz, *IEEE Electron Device Letters*, 10, 437, 1989.

26. Streit, D. C., Tan, K. L., Dia, R. M., Liu, J. K., Han, A. C., Velebir, J. R., Wang, S. K., Trinh, T. Q., Chow, P. D., Liu, P. H., and Yen, H. C., High gain W-band pseudomorphic InGaAs power HEMTs, *IEEE Electron Device Letters*, 12, 149, 1991.

27. Actis, R., Nichols, K. B., Kopp, W. F., Rogers, T. J., and Smith, F. W., High performance 0.15 μm gate length pHEMTs enhanced with a low temperature grown GaAs buffer, *IEEE MTT-S Int. Microwave Symp. Dig.*, 445, 1995.

28. Huang, J. C., Boulais, W., Platzker, A., Kazior, T., Aucoin, L., Shanfield, S., Bertrand, A., Vafiades, M., and Niedzwiecki, M., The effect of channel dimensions on the millimeter wave performance of pseudomorphic HEMT, *Proceedings of GaAs IC Symposium*, 177, 1993.

29. Danzilio, D., White, P., Hanes, L. K., Lauterwasser, B., Ostrowski, B., and Rose, F., A high efficiency 0.25 μm pseudomorphic HEMT power process, *Proceedings of GaAs IC Symposium*, 255, 1992.

30. Chen, C. H., Saito, Y., Yen, H. C., Tan, K., Onak, G., and Mancini, J., Reliability study on pseudomorphic InGaAs power HEMT devices at 60 GHz, *IEEE MTT-S Int. Microwave Symp. Dig.*, 817, 1994.

31. Hori, Y., Onda, K., Funabashi, M., Mizutani, H., Maruhashi, K., Fujihara, A., Hosoya, K., Inoue, T., and Kuzuhara, M., Manufacturable and reliable millimeter wave HJFET MMIC technology using novel 0.15 μm MoTiPtAu gates, *IEEE MTT-S Int. Microwave Symp. Dig.*, 431, 1995.

32. Matsunaga, K., Okamoto, Y., and M. Kuzuhara, A 12-GHz, 12-W HJFET amplifier with 48% peak power added-efficiency, *IEEE Microwave and Guided Wave Letters*, 5, 402, 1995.

33. Helms, D., Komiak, J. J., Kopp, W. F., Ho, P., Smith, P. M., Smith, R. P., and Hogue, D., Ku band power amplifier using pseudomorphic HEMT devices for improved efficiency, *IEEE MTT-S Int. Microwave Symp. Dig.*, 819, 1991.

34. Yarborough, R., Heston, D., Saunier, P., Tserng, H. Q., Salzman, K., and B. Smith, Four watt, Kt-band MMIC amplifier, *IEEE MTT-S Int. Microwave Symp. Dig.*, 797, 1994.

35. Kao, M.-Y., Saunier, P., Ketterson, A. A., Yarborough, R., and Tserng, H. Q., 20 GHz power PHEMTs with power-added efficiency of 68% at 2 volts, *IEDM Technical Digest*, 931, 1996.

36. Dow, G. S., Tan, K., Ton, N., Abell, J., Siddiqui, M., Gorospe, B., Streit, D., Liu, P., and Sholley, M., Ka-band high efficiency 1 watt power amplifier, *IEEE MTT-S Int. Microwave Symp. Dig.*, 579, 1992.

37. Kao, M.-Y., Smith, P. M., Ho, P., Chao, P.-C., Duh, K. H. G., Abra, A. A. J., and Ballingall, J. M., Very high power-added efficiency and low noise 0.15 μm gate-length pseudomorphic HEMTs, *IEEE Electron Device Letters*, 10, 580, 1989.

38. Ferguson, D. W., Smith, P. M., Chao, P. C., Lester, L. F., Smith, R. P., Ho, P., Jabra, A., and Ballingall, J. M., 44 GHz hybrid HEMT power amplifiers, *IEEE MTT-S Int. Microwave Symp. Dig.*, 987, 1989.

39. Kasody, R., Wang, H., Biedenbender, M., Callejo, L., Dow, G. S., and Allen, B. R., Q Band high efficiency monolithic HEMT power prematch structures, *Electronics Letters*, 31, 505, 1995.

40. Boulais, W., Donahue, R. S., Platzker, A., Huang, J., Aucoin, L., Shanfield, S., and Vafiades, M., A high power Q-band GaAs pseudomorphic HEMT monolithic amplifier, *IEEE MTT-S Int. Microwave Symp. Dig.*, 649, 1994.

41. Smith, P. M., Creamer, C. T., Kopp, W. F., Ferguson, D. W., Ho, P., and Willhite, J. R., A high power Q-band HEMT for communication terminal applications, *IEEE MTT-S Int. Microwave Symp. Dig.*, 809, 1994.

42. Tan, K. L., Streit, D. C., Dia, R. M., Wang, S. K., Han, A. C., Chow, P. D., Trinh, T. Q., Liu, P. H., Velebir, J. R., and Yen, H. C., High power V-band pseudomorphic InGaAs HEMT, *IEEE Electron Device Letters*, 12, 213, 1991.

43. Saunier, P., Matyi, R. J., and Bradshaw, K., A double-heterojunction doped-channel pseudomorphic power HEMT with a power density of 0.85 W/mm at 55 GHz, *IEEE Electron Device Letters*, 9, 397, 1988.

44. Lai, R., Wojtowicz, M., Chen, C. H., Biedenbender, M., Yen, H. C., Streit, D. C., Tan, K. L., and Liu, P. H., High power 0.15 μm V-band pseudomorphic InGaAs-AlGaAs-GaAs HEMT, *IEEE Microwave and Guided Wave Letters*, 3, 363, 1993.

45. Kraemer, B., Basset, R., Baughman, C., Chye, P., Day, D., and Wei, J., Power PHEMT module delivers 4 Watts, 38% PAE over the 18.0 to 21.2 GHz band, *IEEE MTT-S Int. Microwave Symp. Dig.*, 801, 1994.

46. Aucoin, L., Bouthillette, S., Platzker, A., Shanfield, S., Bertrand, A., Hoke, W., and Lyman, P., Large periphery, high power pseudomorphic HEMTs, *Proceedings of GaAs IC Symposium*, 351, 1993.

47. Bouthillette, S., Platzker, A., and Aucoin, L., High efficiency 40 Watt PsHEMT S-band MIC power amplifiers, *IEEE MTT-S Int. Microwave Symp. Dig.*, 667, 1994.

48. Kraemer, B., Basset, R., Chye, P., Day, D., and Wei, J., Power PHEMT module delivers 12 watts, 40% PAE over the 8.5 to 10.5 GHz band, *IEEE MTT-S Int. Microwave Symp. Dig.*, 683, 1994.

49. Matsunaga, K., Okamoto, Y., Miura, I., and Kuzuhara, M., Ku-band 15 W single chip HJFET power amplifier, *IEEE MTT-S Int. Microwave Symp. Dig.*, 697, 1996.

50. Fu, S. T., Lester, L. F., and Rogers, T., Ku band high power high efficiency pseudomorphic HEMT, *IEEE MTT-S Int. Microwave Symp. Dig.*, 793, 1994.

51. Henderson, T., Aksun, M. I., Peng, C. K., Morkoc, H., Chao, P. C., Smith, P. M., Duh, K.-H. G., and Lester, L. F., Microwave performance of quarter micrometer gate low noise pseudomorphic InGaAs/AlGaAs modulation doped field effect transistor, *IEEE Electron Device Letters*, 7, 649, 1986.

52. Tan, K. L., Dia, R. M., Streit, D. C., Shaw, L. K., Han, A. C., Sholley, M. D., Liu, P. H., Trinh, T. Q., Lin, T., and Yen, H. C., 60 GHz pseudomorphic $Al_{0.25}Ga_{0.75}As/In_{0.28}Ga_{0.72}As$ low- noise HEMTs, *IEEE Electron Device Letters*, 12, 23, 1991.

53. Chao, P. C., Duh, K. H. G., Ho, P., Smith, P. M., Ballingall, J. M., Jabra, A. A., and Tiberio, R. C., 94 GHz low noise HEMT, *Electronics Letters*, 25, 504, 1989.

54. Tan, K. L., Dia, R. M., Streit, D. C., Lin, T., Trinh, T. Q., Han, A. C., Liu, P. H., Chow, P. D., and Yen, H. C., 94 GHz 0.1 μm T-gate low noise pseudomorphic InGaAs HEMTs, *IEEE Electron Device Letters*, 11, 585, 1990.

55. Tokue, T., Nashimoto, Y., Hirokawa, T., Mese, A., Ichikawa, S., Negishi, H., Toda, T., Kimura, T., Fujita, M., Nagasako, I., and Itoh, T., Ku band super low noise pseudomorphic heterojunction field effect transistor (HJFET) with high producibility and high reliability, *IEEE MTT-S Int. Microwave Symp. Dig.*, 705, 1991.

56. Hwang, T., Kao, T. M., Glajchen, D., and Chye, P., Pseudomorphic AlGaAs/InGaAs/GaAs HEMTs in low-cost plastic packaging for DBS application, *Electronics Letters*, 32, 141, 1996.

57. Hirokawa, T., Negishi, H., Nishimura, Y., Ichikawa, S., Tanaka, J., Kimura, T., Watanbe, K., and Nashimoto, Y., A Ku-band ultra super low-noise pseudomorphic heterojunction FET in a hollow plastic PKG, *IEEE MTT-S Int. Microwave Symp. Dig.*, 1603, 1996.

58. Halchin, D. and Golio, M., Trends for portable wireless applications, *Microwave Journal*, 40, 62, 1997.

59. Ota, Y., Adachi, C., Takehara, H., Yangihara, M., Fujimoto, H., Masato, H., and Inoue, K., Application of heterojunction FET to power amplifier for cellular telephone, *Electronics Letters*, 30, 906, 1994.

60. Kunihisa, T., Yokoyama, T., Nishijima, M., Yamamoto, S., Nishitsuji, M., K. Nishii, Nakayama, and Ishikawa, O., A high-efficiency normally-off MODFET power MMIC for PHS operating under 3.0V single-supply condition, *Proceedings of GaAs IC Symposium*, 37, 1997.

61. Glass, E., Huang, J.-H., Abrokwah, J., Bernhradt, B., Majerus, M., Spears, E., Droopad, R., and Ooms, B., A true enhancement mode single supply power HFET for portable applications, *IEEE MTT-S Int. Microwave Symp. Dig.*, 1399, 1997.

62. Wang, Y. C. and J.M. Kuo, J. R. L., F. Ren, H.S. Tsai, J.S. Weiner, J. Lin, A. Tate, Y.K. Chen, W.E. Mayo, An $In_{0.5}(Al_{0.3}Ga_{0.7})_{0.5}P/In_{0.2}Ga_{0.8}As$ power HEMT with 65.2% power-added efficiency under 1.2 V operation, *Electronics Letters*, 34, 594, 1998.

63. Ren, F., Lothian, J. R., Tsai, H. S., Kuo, J. M., Lin, J., Weiner, J. S., Ryan, R. W., Tate, A., and Chen, Y. K., High performance pseudomorphic InGaP/InGaAs power HEMTs, *Solid State Electronics*, 41, 1913, 1997.

64. Martinez, M. J., Schirmann, E., Durlam, M., Halchin, D., Burton, R., Huang, J.-H., Tehrani, S., Reyes, A., Green, D., and Cody, N., P-HEMTs for low-voltage portable applications using filled gate fabrication process, *Proceedings of GaAs IC Symposium*, 241, 1996.

65. Masato, H., Maeda, M., Fujimoto, H., Morimoto, S., Nakamura, M., Yoshikawa, Y., Ikeda, H., Kosugi, H., and Ota, Y., Analogue/digital dual power module using ion-implanted GaAs MESFETs, *IEEE MTT-S Int. Microwave Symp. Dig.*, 567, 1995.

66. Huang, J. H. and E. Glass, J. A., B. Bernhardt, M. Majerus, E. Spears, J.M. Parsey Jr., D. Scheitlin, R. Droopad, L.A. Mills, K. Hawthorne, J. Blaugh, Device and process optimization for a low voltage enhancement mode power heterojunction FET for portable applications, *Proceedings of GaAs IC Symposium*, 55, 1997.

67. Lee, J.-L., Mun, J. K., Kim, H., Lee, J.-J., and Park, H.-M., A 68% PAE, GaAs power MESFET operating at 2.3 V drain bias for low distortion power applications, *IEEE Transactions on Electron Devices*, 43, 519, 1996.

68. Inosako, K., Matsunaga, K., Okamoto, Y., and Kuzuhara, M., Highly efficient double doped heterojunction FET's for battery operated portable power applications, *IEEE Electron Device Letters*, 15, 248, 1994.

69. Tanaka, T., Furukawa, H., Takenaka, H., Ueda, T., Noma, A., Fukui, T., Tateoka, K., and Ueda, D., 1.5 V Operation GaAs spike-gate power FET with 65% power-added efficiency, *IEDM Technical Digest*, 181, 1995.

70. Inosako, K., Iwata, N., and Kuzuhara, M., 1.2 V operation 1.1 W heterojunction FET for portable radio applications, *IEEE Electron Devices Meeting*, 185, 1995.

71. Iwata, N., Inosako, K., and Kuzuhara, M., 2.2 V operation power heterojunction FET for personal digital cellular telephones, *Electronics Letters*, 31, 2213, 1995.

72. Iwata, N., Inosako, K., and Kuzuhara, M., 3V operation L-band power double-doped heterojunction FETs, *IEEE MTT-S Int. Microwave Symp. Dig.*, 1465, 1993.

73. Iwata, N., Tomita, M., Yamaguchi, K., Oikawa, H., and Kuzuhara, M., 7 mm gate width power heterojunction FETs for Li-ion battery operated personal digital cellular phones, *Proceedings of GaAs IC Symposium*, 119, 1996.

74. Bito, Y., Iwata, N., and Tomita, M., Single 3.4 V operation power heterojunction FET with 60% efficiency for personal digital cellular phones, *Electronics Letters*, 34, 600, 1998.

75. Lai, Y.-L., Chang, E. Y., Chang, C.-Y., Liu, T. H., Wang, S. P., and Hsu, H. T., 2-V-operation δ-doped power HEMT's for personal handy-phone systems, *IEEE Microwave and Guided Wave Letters*, 7, 219, 1997.

76. Choumei, K., Yamamoto, K., Kasai, N., Moriwaki, T., Y, Y., Fujii, T., Otsuji, J., Miyazaki, Y., Tanino, N., and Sato, K., A high efficiency, 2 V single-supply voltage operation RF front-end MMIC for 1.9 GHz personal hand phone systems, *Proceedings of GaAs IC Symposium*, 73, 1998.

77. Hirose, K., Ohata, K., Mizutani, T., Itoh, T., and Ogawa, M., 700 mS/mm 2DEGFETs fabricated from high electron mobility MBE-grown n-AlInAs/GaInAs heterostructures, *GaAs and Related Compounds*, 529, 1985.

78. Palamateer, L. F., Tasker, P. J., Itoh, T., Brown, A. S., Wicks, G. W., and Eastman, L. F., Microwave characterization of 1 μm gate $Al_{0.48}In_{0.52}As/Ga_{0.47}In_{0.53}As/InP$ MODFETs, *Electronics Letters*, 23, 53, 1987.

79. Mishra, U. K., Brown, A. S., Rosenbaum, S. E., Hooper, C. E., Pierce, M. W., Delaney, M. J., Vaughn, S., and White, K., Microwave performance of AlInAs-GaInAs HEMT's with 0.2 μm and 0.1 μm gate length, *IEEE Electron Device Letters*, 9, 647, 1988.

80. Suemitsu, T., Enoki, T., Yokoyama, H., Umeda, Y., and Ishii, Y., Impact of two-step-recessed gate structure on RF performance of InP-based HEMTs, *Electronics Letters*, 34, 220, 1998.

81. Mishra, U. K., Brown, A. S., and Rosenbaum, S. E., DC and RF performance of 0.1 μm gate length $Al_{0.48}In_{0.52}As$-$Ga_{0.38}In_{0.62}As$ pseudomorphic HEMTs, *IEDM Technical Digest*, 180, 1988.

82. Nguyen, L. D., Brown, A. S., Thompson, M. A., and Jelloian, L. M., 50-nm self-aligned-gate pseudomorphic AlInAs/GaInAs high electron mobility transistors, *IEEE Transactions on Electron Devices*, 39, 2007, 1992.

83. Lai, R., Wang, H., Chen, Y. C., Block, T., Liu, P. H., Streit, D. C., Tran, D., Barsky, M., Jones, W., Siegel, P., and Gaier, T., 155 GHz MMIC LNAs with 12 dB gain fabricated using a high yield InP HEMT MMIC process, *Microwave Journal*, 40, 166, 1997.

84. Pobanz, C., Matloubian, M., Lui, M., Sun, H.-C., Case, M., Ngo, C., Janke, P., Gaier, T., and Samoska, L., A high-gain monolithic D-band InP HEMT amplifier, *Proceedings of GaAs IC Symposium*, 41, 1998.

85. Ho, P., Chao, P. C., Du, K. H. G., Jabra, A. A., Ballingall, J. M., and Smith, P. M., Extremely high gain, low noise InAlAs/InGaAs HEMTs grown by molecular beam epitaxy, *IEDM Technical Digest*, 184, 1988.

86. Chao, P. C., Tessmer, A. J., Duh, K. H. G., Ho, P., Kao, M.-Y., Smith, P. M., Ballingall, J. M., Liu, S.-M. J., and Jabra, A. A., W-band low-noise InAlAs/InGaAs lattice matched HEMTs, *IEEE Electron Device Letters*, 11, 59, 1990.

87. Onda, K., Fujihara, A., Miyamoto, H., Nakayama, T., Mizuki, E., Samoto, N., and Kuzuhara, M., Low noise and high gain InAlAs/InGaAs heterojunction FETs with high indium composition channels, *GaAs and Related Compounds*, 139, 1993.

88. Umeda, Y., Enoki, T., Arai, K., and Ishii, Y., Silicon nitride passivated ultra low noise InAlAs/InGaAs HEMT's with n^+ InGaAs/n^+-InAlAs cap layer, *IEICE Transactions on Electronics*, E75-C, 649, 1992.

89. Duh, K. H. G., Chao, P. C., Liu, S. M. J., Ho, P., Kao, M. Y., and Ballingall, J. M., A super low noise 0.1 μm T-gate InAlAs-InGaAs-InP HEMT, *IEEE Microwave and Guided Wave Letters*, 1, 114, 1991.

90. Kao, M.-Y., Duh, K. H. G., Ho, P., and Chao, P.-C., An extremely low noise InP-based HEMT with silicon nitride passivation, *IEDM Technical Digest*, 907, 1994.

91. Kao, M. Y., Smith, P. M., Chao, P. C., and Ho, P., Millimeter wave power performance of InAlAs/InGaAs/InP HEMTs, *IEEE/Cornell Conference on Advanced Concepts in High Speed Semiconductor Devices and Circuits*, 469, 1991.

92. Matloubian, M., Nguyen, L. D., Brown, A. S., Larson, L. E., Melendes, M. A., and Thompson, M. A., High power and high efficiency AlInAs/GaInAs on InP HEMTs, *IEEE MTT-S Int. Microwave Symp. Dig.*, 721, 1991.

93. Larson, L. E., Matloubian, M., Brown, J. J., Brown, A. S., Rhodes, R., Crampton, D., and Thompson, M., AlInAs/GaInAs on InP HEMTs for low power supply voltage operation of high power-added efficiency microwave amplifiers, *Electronics Letters*, 29, 1324, 1993.

94. Bahl, S. R., Azzam, W. J., delAlamo, J. A., Dickmann, J., and Schildberg, S., Off-state breakdown in InAlAs/InGaAs MODFET's, *IEEE Transactions on Electron Devices*, 42, 15, 1995.

95. Boos, J. B. and Kruppa, W., InAlAs/InGaAs/InP HEMTs with High Breakdown voltages using double recess gate process, *Electronics Letters*, 27, 1909, 1991.

96. Hur, K. Y., McTaggart, R. A., LeBlanc, B. W., Hoke, W. E., Lemonias, P. J., Miller, A. B., Kazior, T. E., and Aucoin, L. M., Double recessed AlInAs/GaInAs/InP HEMTs with high breakdown voltages, *Proceedings of GaAs IC Symposium*, 101, 1995.

97. Hur, K. Y., Mctaggart, R. A., Miller, A. B., Hoke, W. E., Lemonias, P. J., and Aucoin, L. M., DC and RF characteristics of double recessed and double pulse doped AlInAs/GaInAs/InP HEMTs, *Electronics Letters*, 31, 135, 1995.

98. Pao, Y. C., Nishimoto, C. K., Majidi-Ahy, R., Archer, J., Betchel, N. G., and Harris, J. S., Characterization of surface undoped $In_{0.52}Al_{0.48}As/In_{0.53}Ga_{0.47}As$ high electron mobility transistors, *IEEE Transactions on Electron Devices*, 37, 2165, 1990.

99. Ho, P., Kao, M. Y., Chao, P. C., Duh, K. H. G., Ballingall, J. M., Allen, S. T., Tessmer, A. J., and Smith, P. M., Extremely high gain 0.15 μm gate-length InAlAs/InGaAs/InP HEMTs, *Electronics Letters*, 27, 325, 1991.

100. Matloubian, M., Larson, L., Brown, A., Jelloian, L., Nguyen, L., Lui, M., Liu, T., Brown, J., Thompson, M., Lam, W., Kurdoghlian, A., Rhodes, R., Delaney, M., and Pence, J., InP based HEMTs for the realization of ultra high efficiency millimeter wave power amplifiers, *IEEE/Cornell Conference on Advanced Concepts in High Speed Semiconductor Devices and Circuits*, 520, 1993.

101. Matloubian, M., Brown, A. S., Nguyen, L. D., Melendes, M. A., Larson, L. E., Delaney, M. J., Pence, J. E., Rhodes, R. A., Thompson, M. A., and Henige, J. A., High power V-band AlInAs/GaInAs on InP HEMT's, *IEEE Electron Device Letters*, 14, 188, 1993.

102. Brown, J. J., Matloubian, M., Liu, T. K., Jelloian, L. M., Schmitz, A. E., Wilson, R. G., Lui, M., Larson, L., Melendes, M. A., and Thompson, M. A., InP Based HEMTs with $Al_xIn_{1-x}P$ Schottky layers grown by gas source MBE, *Proceedings of International Conference on InP and Related Materials*, 419, 1994.

103. Scheffer, F., Heedt, C., Reuter, R., Lindner, A., Liu, Q., Prost, W., and Tegude, F. J., High breakdown voltage InGaAs/InAlAs HFET using $In_{0.5}Ga_{0.5}P$ spacer layer, *Electronics Letters*, 30, 169, 1994.

104. Hur, K. Y., McTaggart, R. A., Lemonias, P. J., and Hoke, W. E., Development of double recessed AlInAs/GaInAs/InP HEMTs for millimeter wave power applications, *Solid State Electronics*, 41, 1581, 1997.

105. Matloubian, M., Brown, A. S., Nguyen, L. D., Melendes, M. A., Larson, L. E., Delaney, M. J., Thompson, M. A., Rhodes, R. A., and Pence, J. A., 20 GHz high efficiency AlInAs-GaInAs on InP power HEMT, *IEEE Microwave and Guided Wave Letters*, 3, 142, 1993.

106. Hur, K. Y., McTaggart, R. A., Ventresca, M. P., Wohlert, R., Hoke, W. E., Lemonias, P. J., Kazior, T. E., and Aucoin, L. M., High efficiency single pulse doped $Al_{0.60}In_{0.40}As/GaInAs$ HEMTs for Q band power applications, *Electronics Letters*, 31, 585, 1995.

107. Ho, P., Smith, P. M., Hwang, K. C., Wang, S. C., Kao, M. Y., Chao, P. C., and Liu, S. M. J., 60 GHz power performance of 0.1 μm gate length InAlAs/GaInAs HEMTs, *Proceedings of International Conference on InP and Related Materials*, 411, 1994.

108. Hwang, K. C., Ho, P., Kao, M. Y., Fu, S. T., Liu, J., Chao, P. C., Smith, P. M., and Swanson, A. W., W Band high power passivated 0.15 μm InAlAs/InGaAs HEMT device, *Proceedings of International Conference on InP and Related Materials*, 18, 1994.

109. Chen, Y. C., Lai, R., Lin, E., Wang, H., Block, T., Yen, H. C., Streit, D., Jones, W., Liu, P. H., Dia, R. M., Huang, T.-W., Huang, P.-P., and Stamper, K., A 94-GHz 130-mW InGaAs/InAlAs/InP HEMT High-Power MMIC Amplifier, *IEEE Microwave and Guided Wave Letters*, 7, 133, 1997.

110. Smith, P. M., Liu, S. M. J., Kao, M. Y., Ho, P., Wang, S. C., Duh, K. H. G., Fu, S. T., and Chao, P. C., W-Band high efficiency InP-based power HEMT with 600 GHz f_{max}, *IEEE Microwave and Guided Wave Letters*, 5, 230, 1995.

111. Akazaki, T., Enoki, T., Arai, K., Umeda, Y., and Ishii, Y., High frequency performance for sub 0.1 μm Gate InAs-inserted-channel InAlAs/InGaAs HEMT, *Electronics Letters*, 28, 1230, 1992.

112. Enoki, T., Arai, K., Kohzen, A., and Ishii, Y., InGaAs/InP Double channel HEMT on InP, *Proceedings of International Conference on InP and Related Materials*, 14, 1992.

113. Matloubian, M., Liu, T., Jelloian, L. M., Thompson, M. A., and Rhodes, R. A., K-Band GaInAs/InP channel power HEMTs, *Electronics Letters*, 31, 761, 1995.

114. Shealy, J. B., Matloubian, M., Liu, T. Y., Thompson, M. A., Hashemi, M. M., DenBaars, S. P., and Mishra, U. K., High-performance submicrometer gatelength GaInAs/InP composite channel HEMT's with regrown contacts, *IEEE Electron Device Letters*, 16, 540, 1996.

115. Shealy, J. B., Matloubian, M., Liu, T. Y., Lam, W., and Ngo, C., 0.9 W/mm, 76% PAE (7 GHz) GaInAs/InP composite channel HEMTs, *Proceedings of International Conference on InP and Related Materials*, 20, 1997.
116. Chevalier, P., Wallart, X., Bonte, B., and Fauquembergue, R., V-band high-power/low-voltage InGaAs/InP composite channel HEMTs, *Electronics Letters*, 34, 409, 1998.

22

Nitride Devices

Robert J. Trew
North Carolina State University

22.1 Introduction

In the past few years, a class of wide energy bandgap semiconductors composed of III-nitride materials, primarily AlN, GaN, InN and related substitutional alloys composed of these compounds, have become of significant interest owing to their application in optoelectronics and high power RF devices. These materials have energy bandgaps in the range of Eg ∼1.9 to 6.2 eV. The materials crystallize in two phases, the hexagonal wurtzite and cubic zincblende polytypes. The wurtzite phases of AlN, GaN, and InN have direct bandgap of 6.2, 3.4, and 1.9 eV, respectively, and the cubic zincblende phases have bandgaps of Eg = 4.9, 3.2, and 1.7 eV, respectively [1]. The hexagonal wurtzite phase is the thermodynamically most stable structure of all the III-nitride polytypes and therefore, this polytype is the focus of virtually all research and development. Virtually all nitride-based devices of interest for commercial development are fabricated from hexagonal wurtzite GaN and related heterostructures. The wurtzite crystal structure is polar due to the charge redistribution between different atoms. A variety of RF devices can be fabricated from these materials, and these devices have improved performance over devices fabricated from standard materials such as Si, GaAs, and related compounds. The various devices that can be fabricated and their RF performance are reviewed in this chapter.

22.2 Materials Background

From a microscopic perspective a strain parallel or perpendicular to the c-axis of a nitride semiconductor produces an internal displacement of the metal sublattice (i.e., gallium sublattice) with respect to the nitrogen atoms. The magnitude of the piezoelectric polarization is determined by the change of the macroscopic lattice constants, and the magnitude of the piezoelectric polarization increases with strain and for crystals or epitaxial layers under the same strain from GaN to InN and AlN [2]. Spontaneous polarization is the polarization that exists at zero strain and results from electron orbital asymmetry in the bonding structure, such that the crystal has a net negative charge on one face and a net positive charge on the opposite face, while maintaining overall electrical neutrality. For the commonly used heterostructures where AlGaN is grown on GaN epitaxial layers, the sign of the spontaneous polarization is found to be negative, while the piezoelectric polarization is negative for tensile strain and positive for compressive strain [2]. Therefore, the orientation of the piezoelectric and spontaneous polarization is parallel in the case of tensile strain, and antiparallel in the case of compressively strained layers. The AlGaN on GaN structure will be under tensile strain, and for this case, the spontaneous and piezoelectric polarization vectors point in the same direction. The magnitude of the total polarization is the sum of the two component polarizations:

$$P = P_{\text{PZ}} + P_{\text{SP}} \tag{22.1}$$

The total polarization of both the AlGaN and GaN layers is directed towards the c-Al_2O_3 substrate for the Ga-face and towards the surface for the N-face polarity crystals.

The polarization is related to a free charge density whenever a gradient in the polarization exists, according to Poisson's equation,

$$\rho_{\text{P}} = -\nabla P \tag{22.2}$$

That is, the polarization establishes a charge density that manifests itself as a free charge density at the heterointerface of the AlGaN and GaN epitaxial layers. At this location a quantum well exits in the conduction band discontinuity, as shown in Figure 22.1 [3]. The charge density that is induced is essentially two-dimensional since the quantum well is on the order of 25Å thickness, while the lateral dimensions are on the nanometer or micron scale. Therefore, the charge density is called a two-dimensional electron gas (2DEG) with units of cm^{-2} and has a very high magnitude since the polarization electric field is very high, and on the order of 10^6 V/cm [4]. The 2DEG sheet charge density is typically on the order

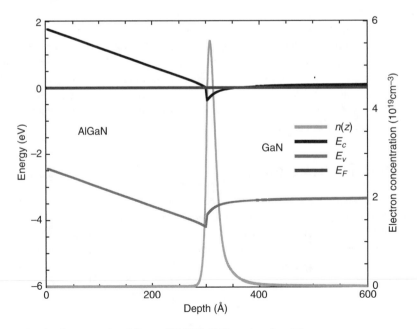

FIGURE 22.1   Conduction energy band for an AlGaN/GaN Heterointerface [3].

of $n_{ss} \sim 10^{13}$ cm^{-2} and varies with the strain at the AlGaN/GaN interface. Therefore, the sheet charge density is a function of the Al content in the AlGaN layer.

The first AlGaN/GaN heterostructures were fabricated by growing thin AlGaN layers on thicker GaN material. Originally, neither the AlGaN nor the GaN layers were doped. However, it was observed that despite the lack of electrons from intentionally doping, a 2DEG was, in fact, established at the heterointerface. As previously mentioned, the 2DEG has very high sheet charge density, on the order of $n_{ss} \sim 10^{13}$cm^{-2}, which is a factor of five greater than that produced in the AlGaAs/GaAs system. The fundamental question is: Since there are no intentionally introduced impurity atoms to supply electrons, what is the source of the electrons that form the 2DEG? Measured data for heterostructures fabricated with a variety of growth conditions always produce a high density 2DEG. It has been established that the density of the 2DEG varies with Al concentration in the AlGaN layer, with higher sheet charge density obtained for higher Al concentration [2].

An argument for the formation of the AlGaN/GaN 2DEG can by explained by reference to the model shown in the sketch in Figure 22.2 [4]. According to this model, the electrons that form the 2DEG result from the growth process. Since the AlGaN semiconductor layer is polar, with both spontaneous and piezoelectric polarizations, during growth the crystal atoms line up so that the positive side of the atomic layers are aligned towards the GaN layer. As the layer thickness increases during growth, the atomic layers continue to align, creating an electric field internal to the AlGaN layer, with the positive side of the dipole facing the GaN and the negative side of the dipole facing the growth surface. The magnitude of the electric field is very high, and is a function of the strain created by the Al concentration. The electric field can be expressed as

$$E(x) = (-9.5x - 2.1x^2)MV/cm \tag{22.3}$$

where x is the Al fraction. The magnitude of the electric field normal to the interface can be described as

$$\vec{E}_n = \frac{q\rho_+}{\varepsilon_r \varepsilon_o} \hat{n} \sim 10^6 V/cm \tag{22.4}$$

where q is the electronic charge, ρ_+ is the positive charge density, and ε_r and ε_o are the relative and free space dielectric constants. The magnitude of the electric field is sufficient to ionize some of the covalent

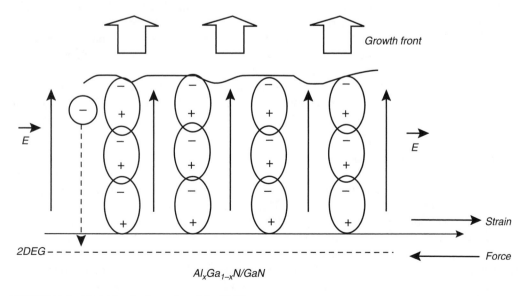

FIGURE 22.2   Model for the formation of the 2DEG.

electrons, as well as any impurities that happen to be present in the material. A large reservoir of electrons is also available from loosely bonded surface electrons. The electric field will ionize electrons and will cause them to drift towards the heterointerface, where they fall into the quantum well, thereby creating the 2DEG. As the electrons move from the AlGaN into the GaN the magnitude of the electric field is reduced, thereby acting as a feedback mechanism to quench the electron transfer process. An equilibrium condition is established when sufficient electrons are transferred into the quantum well to reduce the magnitude of the electric field in the AlGaN to the point where no further electrons are transferred. The Al concentration in the AlGaN becomes a control on the density of the 2DEG since it produces stress at the AlGaN/GaN interface that increases the polarization charge density, which defines the electric field in the AlGaN, according to Gauss' Law.

22.3   Material Growth

Epitaxial layers of GaN and AlGaN can be grown by a variety of techniques. The most popular approach is Organo-Metallic Chemical Vapor Deposition (OMCVD). Although there was significant research directed toward the growth of GaN epitaxial layers extending back at least into the 1970s [5] and 1980s [6,7], the successful demonstration of blue LEDs fabricated from GaN grown by OMCVD by Nakamura in 1994 [8] resulted in increased interest in this growth technique. Many other groups have developed OMCVD growth of GaN and AlGaN and progress has been rapid. Molecular Beam Epitaxy (MBE) is also used for the growth of GaN epitaxial layers [9], and MBE and OMCVD materials are of comparable quality, which has continually improved, particularly the reduction of defect density. Initially, GaN epitaxial growth was performed on sapphire (Al_2O_3) substrates due to lack of suitable GaN substrate availability. However, the lattice mismatch is poor (about 13.8%) and sapphire is a very poor thermal conductor. The large lattice mismatch results in a high density of surface cracks, although this can be relieved by the use of AlN interface layers. Also, the poor thermal conductivity of sapphire permits only small area devices to be fabricated. Sapphire was used primarily because the material is low cost, high purity, and readily available. For electronic device applications, sapphire is no longer considered a suitable substrate for GaN growth, and most research and development shifted to the use of SiC substrates. The lattice mismatch between GaN and SiC is about 3.5%, which is much improved over that between GaN and sapphire. Both 6H-SiC and 4H-SiC are used as substrates. Generally, 6H-SiC is used for highly conducting substrates and 4H-SiC is preferred for insulating highly resistive substrates. Lateral growth by OMCVD was shown to produce low-defect density GaN layers grown on both SiO_2 and 6H-SiC substrates when AlN buffer layers were employed [10]. The lateral growth technique is used to grow low-defect density GaN epitaxial layers over a variety of foreign substrates in a process now called "pendeo-epitaxy." Most notable, the technique is used to grow GaN layers on Si substrates [11]. In this work, the GaN growth is accomplished using a thin 3C-SiC transition layer between the GaN and Si substrate. Lateral growth and pendeo-epitaxy techniques are now being employed by several commercial firms for development of communications band power amplifiers using AlGaN/GaN HFETs fabricated on Si substrates.

22.4   Electronic Properties

Basically, a current is defined as the movement of charge and expressed as the product between the charge density and transport velocity. Therefore, the DC and RF currents that flow through a device are directly dependent upon the charge carrier velocity versus electric field transport characteristics of the semiconductor material. Generally, for high currents and high frequency, high charge carrier mobility and high saturation velocity are desirable. The v–E characteristic is described in terms of charge carrier mobility μ_n, (units of cm^2/V-sec) defined from the slope of the v–E characteristic at low-electric field, and the saturated velocity v_s (units of cm/sec), defined when the carrier velocity obtains a constant, field-independent magnitude, generally at high electric field. The nitride materials, GaN and the AlGaN/GaN heterojunction 2DEG, generally have good low-field electron mobility and excellent high field electron saturated velocity.

Various groups have developed electron transport characteristics using Monte Carlo simulation techniques [12–14]. The electron mobility decreases with temperature, with a room temperature magnitude of about $\mu_n \sim 1000$ cm^2/V-sec. At $T = 50$ K the mobility peaks with a value slightly over 5000 cm^2/V-sec. The calculated electron velocity for wurtzite GaN [13] for room temperature operation predicts that the mobility is about the same as for the AlGaN/GaN 2DEG, with a value $\mu_n \sim 1000$ cm^2/V-sec. However, as temperature is reduced the simulations show a reduced peak mobility of 2000 cm^2/V-sec compared to that for the AlGaN/GaN 2DEG at a temperature of $T = 150$ K.

The mobility increases with Al content in the AlGaN layer [15,16]. As the Al content is increased, the strain at the heterointerface increases, which increases the magnitude of the polarization electric field. As a result the 2DEG sheet charge density increases. The mobility peaks for an Al-mole fraction of about 34% and decreases for both lower and higher mole fractions. Measured 2DEG electron density and mobility as a function of temperature for an AlGaN/GaN heterostructure grown on sapphire is shown in Figure 22.3 [17]. As indicated, the sheet charge density is essentially independent of temperature, but the mobility decreases up to room temperature. Typically, room temperature mobility in the range of $\mu_n \sim 1000$– 1500 cm^2/V-sec are obtained for device applications. The mobility has a T$^{-2.3}$ dependence. Interface roughness is a dominant source of scattering and the interface scattering increases with Al-mole content.

In general, it is very difficult to directly measure the electron velocity versus electric field characteristic for a semiconductor. Time-of-flight techniques have been used, but these techniques generally require a depleted region so they are limited to lightly doped material. Time-resolved electroabsorption can be used to measure the transient electron velocity overshoot in GaN [18,19]. These measurements were performed on AlGaN/GaN pin diodes and the results provide evidence of the existence of electron velocity overshoot. A peak electron transient velocity as high as $v_s \sim 7.5 \times 10^7$ cm/sec was observed within the first 200 psec after photoexcitation occurred at an electric field of 320 kV/cm. The measurements also provide evidence for the existence of a negative differential resistivity region beyond 320 kV/cm, in agreement with theoretical Monte Carlo simulations.

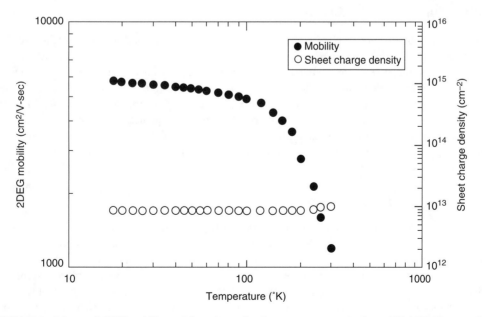

FIGURE 22.3 Measured 2DEG mobility and sheet charge density versus temperature for an AlGaN/GaN on sapphire structure [17].

Most measurements of the high electric field saturated electron velocity that have been reported use indirect techniques, where the current or f_T of a transistor is measured and the saturation velocity inferred [20–23]. These techniques yield values for the "effective" saturation velocity, which is defined as the saturation velocity required in models to produce the same measured channel current or device f_T. Using these indirect techniques, values for the electron saturation velocity range from $v_s \sim 1 \times 10^7$ cm/sec to $v_s \sim 2 \times 10^7$ cm/s. The value appears to be related to the Al-mole fraction and to the roughness of the interface between the AlGaN and GaN layers.

The combination of good mobility and high saturation velocity permit the AlGaN/GaN HFETs to produce high currents, which are significantly greater than available with Si- or GaAs-based devices.

22.5 Electronic Breakdown

Electronic breakdown is a fundamental phenomenon that limits the RF power capability of an electronic device. At a critical magnitude of electric field some of the covalent bonds are broken, creating free electrons and holes. The avalanche breakdown process produces a rapid increase in current through the device and the device impedance tends towards a short circuit, thereby degrading performance. The critical electric field defines the maximum dc and RF voltages that can be applied to the device, and therefore, the RF power capability of the device. The avalanche breakdown process is defined in terms of ionization rates for electrons and holes, and these rates can generally be calculated theoretically, or measured. The electron and hole ionization rates for zincblende and wurtzite GaN have been calculated using Monte Carlo simulation techniques by Farahmand and Brennan [24,25]. The theoretical simulations predict critical electric fields for avalanche breakdown on the order of $E_c \sim (3-4) \times 10^6$ V/cm. Monte Carlo simulations predict a "soft" breakdown for hexagonal GaN [26] and avalanche breakdown on p-π-n GaN diodes, as determined by electroluminescence measurements, are in qualitive agreement [27]. Avalanche breakdown ionization rates for bulk GaN have been measured and compared to those for GaAs by Kunihiro et al. [28]. These measurements indicate a critical electric field for avalanche breakdown on the order of $E_c \sim (2-4) \times 10^6$ V/cm, which is in good agreement with the theoretically generated values.

22.6 Electronic Devices

A variety of electronic devices, including both high power RF and microwave devices can be fabricated from nitride-based semiconductors. Microwave devices include MESFETs, Static-Induction Transistors (SITs), Heterojunction Bipolar Transistors (HBTs), and Heterojunction Field-Effect Transistors (HFETs). Devices for high power applications include diodes, MOSFETs, MOSHFETs, and bipolar transistors. Excellent performance has been demonstrated, although these devices have not yet found widespread insertion into commercial systems. A variety of reliability problems limit acceptance, and these problems need to be solved before these devices find use in practical applications. Microwave HFETs for communications band base station amplifiers are progressing rapidly and are commercially available. The performance of various devices is reviewed in this section.

22.6.1 High Voltage Devices

The key to achieving high voltage operation of standard HFETs is control of the electric field at the edge of the gate electrode on the drain side. Due to breakdown at this location standard power GaAs FET designs are generally limited to drain bias voltages in the range of 8–12 V, which limits the RF voltage and RF output power that can be developed [29,30]. It has been shown that the use of field-plate technology suppresses gate breakdown and permits significantly higher drain bias voltages to be applied [31]. Field-plate power GaAs FETs biased with drain voltage of 35 V have produced RF power density of 1.7 W/mm of gate periphery, and a 230 W amplifier when the FET was biased at $V_{ds} = 24$ V was reported [31]. Wide bandgap semiconductors such as those based upon the III-N materials system have much improved

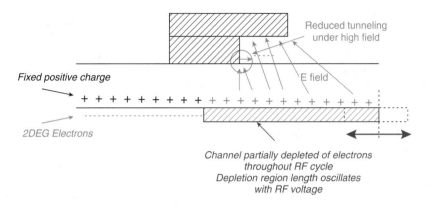

FIGURE 22.4 Field-plate AlGaN/GaN HFET design.

critical electric fields for breakdown compared to GaAs and HFETs fabricated from these materials can sustain significantly improved bias voltages, with $V_{ds} > 40$ V before breakdown is observed. Field-plate technology is also being widely used with nitride-based HFETs [32] to permit even greater drain voltage to be applied, and a field-plate HFET when biased at a $V_{ds} = 120$ V produced over 30 W/mm RF power density at S-band [33], and over 5 W/mm at 30 GHz with a drain bias of $V_{ds} = 30$ V [34].

The concept for a field-plate dates back to work on the development of guard rings for high voltage diodes. Basically, the principle is to provide a conducting plane near junctions and other locations where high electric fields exist. The conducting plane provides a means to terminate and smooth high electric fields, thereby reducing the high electric field peaks that result in electronic breakdown. In this manner, it is possible to engineer devices for increased breakdown voltage by spreading the electric field over a larger region.

The field-plate is applied to field-effect transistors (FETs) as shown in the sketch in Figure 22.4 for an AlGaN/GaN HFET. The field-plate consists of a conducting electrode placed over the gate and extending into the gate-drain region of the FET structure. The field-plate is generally designed so that it is electrically connected to the gate, but is held at a short distance above the surface of the semiconductor. The field-plate can also be connected to the source, or left free floating and electrically isolated. Normally, the field-plate metal is located on top of a surface passivation dielectric, which is designed for a specific thickness. Both the extension of the field-plate into the gate-drain region and the thickness of the dielectric are critical design parameters for optimum performance. The field-plate functions by reducing the electric field at the edge of the gate on the drain side. Owing to the geometry of the FET, the electric field at this position is very high (normally on the order of $E \sim 10^6$ V/cm), which is sufficient to produce a tunneling current from the gate metal to the semiconductor. It has been shown that the primary breakdown mechanism for FETs is tunnel leakage of electrons from the gate to the surface region of the semiconductor [35]. Under high-voltage operation, the electrons can tunnel into the semiconductor with sufficient energy to produce avalanche ionization. Field-plates produce sufficient reduction of the electric field at the gate edge to permit high voltages to be applied, and this in turn, permits high RF power to be generated.

Recently, a self-aligned "slant field-plate" design was introduced by Dora et al. [36]. The technology makes use of a tapered gate shape to create a field-plate extending over the gate-drain region. It was also found necessary to test the devices for high voltage capability in a Fluorinert atmosphere in order to suppress breakdown in the air outside the device. The off-state breakdown voltage was over 1.6 kV. When the same devices were tested in air without the Fluorinert, the breakdown voltage was significantly reduced and on the order of about 400–500 V. These results indicate that high voltage operation is possible with these devices, but that passivation and the operating environment must be carefully controlled.

Very high voltage HFETs have also been fabricated by Tipirneni et al. [37]. These AlGaN/GaN HFETs were grown on sapphire substrates. Measurements on these devices indicate that breakdown occurs by a "flashover" process, whereby the air directly over the device breaks down by dielectric breakdown, rather than by avalanche ionization in the conducting channel. An analysis of the breakdown mechanism revealed that breakdown occurred by gate metal evaporation. The high breakdown voltage was obtained using a Fluorinert solution. Fluorinert solution has a high dielectric strength of 1800 MV/cm, as compared to 300 MV/cm for air. Channel breakdown was not observed although the average electric field in the channel was about 400 MV/cm. This electric field magnitude is well above the critical breakdown field for GaN which is in the range of $E_c \sim$ (2–4) MV/cm. The reason for this discrepancy is not clear, but the authors speculate that depletion regions must exist to support the voltage. The device also demonstrated low on-resistance.

Nitride semiconductor devices that can switch very large voltages can be fabricated. Metal-Oxide-Semiconductor-Heterostructures (MOSH) have been fabricated for high power, low-loss fast RF switches [38]. These devices utilize the MOSH structure for the gate of an AlGaN/GaN HFET. The device switched RF power from 20 to 60 W/mm over a frequency range of 2–20 GHz with an insertion loss less than 2 dB.

22.6.2 Microwave HFETs

The AlGaN/GaN HFET also demonstrates excellent RF performance. High sheet charge density resulting from high Al incorporation in the AlGaN layer permits high channel current to be obtained [39]. Initial HFETs were fabricated on sapphire substrates, but recent work has focused upon the use of semi-insulating or p-type SiC substrates [40–42]. Excellent RF performance has been achieved at S-band through Ka-band frequencies, with the greatest RF power density obtained at S-band and up to X-band. Most AlGaN/GaN HFETs are fabricated with unintentionally doped AlGaN and GaN epitaxial layers. However, it is also possible to fabricate AlGaN/GaN HFETs with good RF performance using doped channel designs [43], and 1.73 W/mm RF output power with good gain was obtained at 8.4 GHz. The small signal performance of these devices demonstrated gain bandwidth products of $f_T = 39$ GHz and $f_{max} = 45$ GHz. Small-signal performance with intrinsic current gain-bandwidth products up to $f_T = 106$ GHz for a $L_g = 0.15$ μm device has been obtained [44]. These devices produced about 4 W/mm RF power and 41% PAE at 4 GHz. Very high RF power density has also been obtained and 9.8 W/mm RF power density with 47% PAE at 8 GHz has been obtained [45]. The devices had gate widths of $W = 2$ mm and the devices were flip-chip mounted to AlN substrates for improved thermal conductance. Other devices fabricated using SiC substrates produced RF power as high as 10.7 W/mm at 10 GHz with 40% PAE [46], with further improvements yielding slightly over 11 W/mm. Devices fabricated using AlN interfacial layers between the AlGaN and GaN produced RF output power of 8.4 W/mm with a power-added efficiency of 28% at 8 GHz [47]. The introduction of field-plate technology suppresses the electric field at the gate edge and permits larger drain bias to be applied. Higher RF output power results. A high RF output power density of > 30 W/mm was reported for a field-plate device biased at a drain voltage of 120 V [33].

High power-added efficiency has also been reported, and an AlGaN/GaN HFET grown by MBE on a 4H-SiC substrate produced 8.4 W/mm with 67% PAE with a drain bias of 30 V [48]. Silicon has emerged as a viable substrate material for AlGaN/GaN HFETs and excellent RF performance has been obtained. Johnson et al. [49] reported RF output power of 12 W/mm with 52.7% PAE and 15.3 dB gain for a 0.7 μm gate length device. The HFET was biased at 50 V and operated at 2.14 GHz. The transistor is intended for communications band applications. Dumka et al. report 7 W/mm with 38% PAE and 9.1 dB gain at 10 GHz from a AlGaN/GaN HFET fabricated on a Si(111) substrate [50]. The device was biased at a drain voltage of 40 V. Reduction of the drain bias to 20 V resulted in a decrease in RF power to 3.9 W/mm, but an improvement of the power-added efficiency to 52%.

High-frequency Ka-band performance has also been reported. At 18 GHz, Ducatteau et al. report an RF power density of 5.1 W/mm with 20% PAE and 9.1 dB gain from a nitride HFET fabricated on a Si substrate [51]. The device had a 0.25 μm gate length and a current gain bandwidth of $f_T = 50$ GHz. An AlGaN/GaN HFET fabricated on a SiC produced 5 W/mm with 30.1% PAE and 5.24 dB gain at 26 GHz [52]. Lee et al. [53] report 4.13 W/mm with 23% PAE and 7.54 dB gain at 35 GHz. The HFET was biased with a drain voltage of 30 V. The HFET was fabricated on a SiC substrate. At 40 GHz an RF power density of 2.8 W/mm, 10% PAE, and 5.1 dB gain was obtained from a device with a 0.18 μm gate length device [54]. The performance of the device was sensitive to frequency, and RF output power density increased to 3.4 W/mm by reduction of the operating frequency to 38 GHz. Using a recess gate design an RF output power density of 5.7 W/mm with 45% PAE was obtained with a drain bias of 20 V [55]. Increasing the drain bias to 28 V resulted in an increase in RF output power density to 6.9 W/mm. Palacios et al [56] report excellent RF performance at 40 GHz from an AlGaN/GaN HFET fabricated on a 4H-SiC(0001) substrate. Devices with similar structures were fabricated using both OMCVD and MBE. The device produced 8.6 W/mm with 29% PAE and gain of about 5 dB. The OMCVD grown device had improved performance, with 10.5 W/mm, 33% PAE, and about 6 dB gain.

Attempts to improve device performance include novel surface passivation and charge confinement. Lau and her colleagues [57] introduced a surface passivation technique involving a fluoride-based plasma treatment. The fluoride-based plasma treatment, along with a post-gate rapid thermal annealing step, was found to effectively incorporate negatively charged fluorine ions into the AlGaN barrier and positively shift the threshold voltage. The technique was used to fabricate an enhancement-mode (E-mode), HFET. Shen et al. [58] used the fluorine plasma process, along with a deeply recessed gate HFET design, to fabricate a device that produced 17.8 W/mm with 50% PAE and 15 dB gain at 4 GHz. The passivation process limited gate leakage and thereby permitted a drain voltage of $V_{ds} = 80$ V to be applied, without the use of a field-plate.

The strong polarization effects of the AlGaN/GaN structure may be a source of some of the reliability problems experienced with these devices. Attempts to investigate this include utilization of alternate barrier materials that are less polar. One such structure can be fabricated using InAlN, rather than the commonly employed AlGaN. An InAlN/GaN HFET with a gate length of 0.7 μm produced a gain-bandwidth product of $f_T = 13$ GHz and $f_{max} = 11$ GHz. The 2DEG was very high, with $n_{ss} = 4 \times 10^{13}$ cm^{-2} and an electron mobility of $\mu_n = 750$ cm^2/V-sec [59,60]. An InGaN layer was used as a back-barrier to improve confinement of the 2DEG electrons [61]. The confinement improved the output resistance, and a device with a gate length of 100 nm produced a gain-bandwidth product of $f_T = 153$ GHz and an $f_{max} = 198$ GHz. By adjusting the bias, the same device produced an $f_{max} = 230$ GHz. A double heterojunction device design using an InGaN notch fabricated on a sapphire substrate produced RF output power of 3.4 W/mm and 41% PAE at 2 GHz [62]. Good RF performance has also been obtained from a GaN FET, fabricated using a novel surface passivation consisting of a thin AlN layer located between the GaN channel and a SiN surface passivation [63]. The resulting structure is basically a metal-insulator–semiconductor (MIS) FET. A device with a gate length of 60 nm produced $f_T = 107$ GHz and $f_{max} = 171$ GHz.

Most of early results were for devices with very narrow gate widths to minimize device heating and thermal effects. More recent work has focused upon producing high power devices and amplifiers suitable for use in applications such as communications base station transmitters. Ando et al. [64] reported RF output power of 10.3 W with 47% power-added efficiency and 18 dB linear gain at 2 GHz. This result was for a device with a gate width of 1 mm. Linear gate width scaling with drain current and RF output power has also been demonstrated. A high power integrated circuit using 8 mm of gate periphery yielded 51 W RF output power at 6 GHz under pulse bias conditions [65]. A communications band amplifier, using AlGaN/GaN HFETs fabricated on SiC substrates and biased at 48 V produced CW RF output power of 100 W at 2.14 GHz [66]. A C-band amplifier using a 0.4 μm gate length and 50.4 mm gate width AlGaN/GaN HFET fabricated on a SiC substrate produced RF output power of 140 W with 25% PAE. The amplifier was operated with a pulse bias of 40 V [67]. A push-pull transmitter amplifier for 3G wireless base station applications was constructed using AlGaN/GaN HFETs fabricated on SiC substrates [68,69]. At a drain bias of 50 V the amplifier produced 250 W RF output power, and using digital predistortion linearization, an adjacent channel leakage power ration (ACLR) of less than −50 dBc for 4-carrier W-CDMA signals was obtained. Very high RF output power was obtained from wide gate width AlGaN/GaN HFETs fabricated on Si(111) substrates [70]. The individual HFETs had a gate width of 36 mm and when operated under CDMA modulation produced 20 W RF power with a drain efficiency of 27% when biased at a $V_{ds} = 28$ V. The amplifier was fabricated using two of the devices and produced a maximum RF output power of 156 W with 65% drain efficiency at 2.14 GHz and no modulation. The same authors report further improvements by employing a source-grounded field plate on the HFET, and when biased at $V_{ds} = 60$ V and under pulsed RF conditions, a saturated RF output power of 368 W with 70% PAE was obtained [71]. Wide gate width devices require effective means for grounding, and a laser-assisted processing procedure for fabricating via holes was reported [72]. The process permits wide gate width devices to be effectively grounded, and a 20 mm gate width device biased at $V_{ds} = 26$ V produced 41.6 W with 55% PAE at 2 GHz.

22.6.3 MISFETs, MOSFETs, and MOSHFETs

A variety of novel devices can be fabricated using various materials located under the gate region. Khan et al. have reported metal-oxide-semiconductor-based FET structures [73] (called MOSHFETs) and demonstrated their performance. The primary innovation is the insertion of a thin SiO_2 dielectric layer between the gate metal and the AlGaN epitaxial layer. The oxide produces an additional barrier and gate leakage is dramatically reduced. The MOSHFET capacitance is basically insensitive to gate voltage, which indicates that these devices should demonstrate excellent linearity. The MOSHFET can operate with significant forward gate bias, which results in large channel current, and about a factor of two greater than can be obtained with standard HFETs. The SiO_2/AlGaN heterointerface is of high quality and this results in very low reverse bias gate leakage current. The gate leakage current is about six orders of magnitude lower than typically obtained from standard HFETs. A large periphery device using the same technology has been reported [74]. The device had a gate width of 6 mm and supported a saturated current of 5.1 A, while maintaining a gate leakage current of about 1 $\mu A/cm^2$. The device had a cutoff frequency of $f_T = 8$ GHz and produced 1.4 W (about 2.9 W/mm) RF power with 40% power-added efficiency at 2 GHz. The large gate width device used a SiO_2 over-bridging process to connect the source fingers. The device is attractive for high-power RF switching applications, and a 1-mm gate width device demonstrated a 0.27 dB insertion loss and more than 40 dB isolation [75]. The device is capable of switching 40 W. The MOSHFET also has high-frequency potential, and a device grown on a 4H-SiC substrate and fabricated with a 1.1 μm gate produced 19 W/mm with about 50% PAE when biased at a drain voltage of 55 V [76]. The RF output power remained constant up to 100 h, and did not demonstrate the degradation with time, typically experienced by standard AlGaN/GaN HFETs.

Enhancement/Depletion (E/D) HFETs have been reported [77,78]. In these reports the E-mode was achieved by using a fluoride-based plasma treatment for the AlGaN surface under the gate electrode. The devices demonstrated excellent switching performance, and at a supply voltage of 3.3 V, the E/D-mode converters showed an output swing of 2.85 V, with the logic-low and logic-high noise margins at 0.34 and 1.47 V, respectively [77]. An RF device fabricated with the same technology produced a very high transconductance of $g_m > 400$ mS/mm. The cutoff frequency and f_{max} for the device are $f_T = 85$ GHz and $f_{max} = 150$ GHz, respectively.

MISHFETs have been fabricated by using thin layers of Si_3N_4 on the AlGaN surface under the gate [79]. The gate leakage current was very small, and at room temperature was only 90 pA/mm, increasing to 1000 pA/mm at 300°C. A MISHFET using a catalytic chemical vapor deposition process (called Cat-CVD) to form the surface insulator under the gate demonstrated a cutoff frequency of $f_T = 163$ GHz [80]. The RESURF technique has been applied to AlGaN/GaN HFETs to fabricate a device for high-voltage power switching applications [81]. The RESURF design distributes the electric field over the gate to drain region, thereby increasing the device breakdown voltage to over a kV.

22.7   Summary

The nitride-based devices produce RF output power about ten times that available from Si, GaAs, or InP-based devices of comparable size. Progress has been rapid in recent years on development of nitride-based transistors, primarily AlGaN/GaN heterojunction devices grown on either SiC or Si substrates. HFETs are rapidly approaching sufficient maturity for commercial application, and initial S- and C-band devices are available for communications base station transmitter applications.

Despite the rapid development progress a number of materials-related, as well as device-related problems still exist. These problems are generally termed "reliability" problems, although the term actually refers to a range of physical phenomena occurring within the device at the high internal electric fields at which the device operates. A class of materials-related problems is correlated to a high density of traps caused by defects, dislocations, and impurities present in the material. These traps produce a current collapse phenomenon where the drain current for AlGaN/GaN HFETs is observed to decrease upon application of a drain bias voltage. Gate leakage under RF operation is a major problem for the AlGaN/GaN HFETs. This leakage occurs by electron tunneling from the gate metal-to-surface states or traps in the semiconductor adjacent to the gate edge on the drain side. This phenomenon is the fundamental physical reason for the "reliability" problems experienced by these devices.

The thermal operation of GaN-based electronic devices is a major problem that needs to be addressed. As mentioned at the beginning of this chapter, the initial GaN-based devices were fabricated on sapphire due to the ready availability of high-quality and low-cost sapphire substrates. However, the thermal conductivity of sapphire is poor, and therefore the material is not suitable for use as a substrate for high power devices. The initial AlGaN/GaN HFETs addressed this problem by use of very narrow gate width devices in order to minimize thermal heating effects. This led to the wide spread use of normalized RF power in reports where the RF power obtained was quoted as an RF power density in the units of W/mm of gate periphery. This permitted the capability of various devices to be compared, but does not represent the true power capability of a device since RF power density typically significantly degrades with increasing gate width for a variety of thermal and impedance matching issues. The search for improved thermal conductivity substrates is a major thrust of device research and development, and the use of SiC and Si substrates is now commonly employed. Both these materials have much improved thermal conductivity compared to sapphire. The thermal conductivity of SiC is about a factor of 3–4 superior to that of Si, but the integration of AlGaN/GaN epitaxial layers on Si offers an attractive technology for potentially low-cost integration with the large Si-based technology. There is a continuing interest in the development of alternate substrate materials, including AlN and GaN. The use of GaN would be, of course, near ideal since the lattice match is perfect. Bulk growth of GaN for substrates has proved difficult and growth temperature greater than 1800°C has been required. However, a serendipitous discovery by the DiSalvo group [82] found that GaN single crystals could be grown at much more modest temperatures and pressures (700°C and 50 atm) in a sodium metal flux. Much work has been done to optimize the growth parameters, and the current limiting factor is the scale of the reaction. Bulk GaN growth may prove practical and homoepitaxy may find application. The thermal conductivity of bulk GaN has been found to be related to dislocation density, and reduction of the dislocation density to below mid 10^6 cm^{-3} results in a thermal conductivity on the order of 2.3 W/K-cm, and the thermal conductivity varies with temperature with a $T^{-1.43}$ dependence.

References

1. O. Ambacher, "Growth and Applications of Group III-Nitrides," *J. Phys. D: Appl. Phys.* vol. 31, pp. 2653–2710, Oct. 1998.
2. O. Ambacher, B. Foutz, J. Smart, J.R. Shealy, N.G. Weimann, K. Chu, M. Murphy, A.J. Sierakowski, W.J. Schaff, and L.F. Eastman, "Two Dimensional Electron Gases Induced by Spontaneous and Piezo-electric Polarization in Undoped and Doped AlGaN/GaN Heterostructures," *J. Appl. Phys.*, vol. 87, pp. 334–344, Jan. 2000.
3. B. Jogai, "Influence of Surface States on the Two-Dimensional Electron Gas in AlGaN/GaN Heterojunction Field-Effect Transistors," *J. Appl. Phys.* vol. 93, pp. 1631–1635, Feb. 2003.
4. R.J. Trew, G.L. Bilbro, W. Kuang, Y. Liu, and H. Yin, "Microwave AlGaN/GaN HFET's," *IEEE Microwave Magazine*, pp. 56–66, Mar. 2005.
5. H. Maruska, D. Stevenson, and J. Pankove, "Violet Luminescence of Mg-Doped GaN," *Appl. Phys. Lett.*, vol. 22, pp. 303–305, Mar. 1973.

6. Y. Ohki, Y. Toyoda, H. Kobayashi, and I. Akasaki, *Inst. Phys. Conf. Ser.*, vol. 63, p. 479, 1981.

7. H. Amano, N. Sawaki, I. Akasaki, and Y. Toyoda, "Metalorganic Vapor Phase Epitaxial Growth of a High Quality GaN Film Using an AlN Buffer Layer," *Appl. Phys. Lett.*, vol. 48, pp. 353–355, Feb. 1986.

8. S. Nakamura, T. Mukai, and M. Senoh, "Candela-class high-brightness InGaN/AlGaN Double-Heterostructure Blue-Light-Emitting Diodes," *Appl. Phys. Lett.*, vol. 64, pp. 1687–1689, Mar. 1994.

9. S. Yoahida, S. Misawa, and S. Gonda, "Improvements on the Electrical and Luminescent Properties of Reactive Molecular Beam Epitaxially Grown GaN Films by Using AlN-Coated Sapphire Substrates," *Appl. Phys. Lett.*, vol. 42, pp. 427–429, Mar. 1983.

10. O-H. Nam, M.D. Bremser, T.S. Tsvetanka, S. Zheleva, and R.F. Davis, "Lateral Epitaxy of Low Defect Density GaN Layers via Organometallic Vapor Phase Epitaxy," *Appl. Phys. Lett.*, vol. 71, pp. 2638–2640, Nov. 1997.

11. K.J. Linthicum, T. Gehrke, D. Thomson, C. Ronning, E.P. Carlson, C.A. Zorman, M. Mehregany, and R.F. Davis, "Pendeo-Epitaxial Growth of GaN on SiC and Silicon Substrates vis Metalorganic Chemical Vapor Depositon," *Mat. Res. Soc. Symp. Proc.*, vol. 572, pp. 307–313, 1999.

12. O. Katz, A. Horn, G. Bahir, and J. Salzman, "Electron Mobility in an AlGaN/GaN Two-Dimensional Electron Gas I-Carrier Concentration Dependent Mobility," *IEEE Trans. Electron Dev.*, vol. 50, pp. 2002–2008, Oct. 2003.

13. U.V. Bhapkar and M.S. Shur, "Monte Carlo Calculation of Velocity-Field Characteristics of Wurtzite GaN," *J. Appl. Phys.*, vol. 82, pp. 1649–1655, Aug. 1997.

14. Y. Zhang, and J. Singh, "Monte Carlo Studies of Two Dimensional Transport in GaN/AlGaN Transistors: Comparison with Transport in AlGaAs/GaAs Channels," *J. Appl. Phys.*, vol. 89, pp. 386–389, Jan. 2001.

15. S. Arulkumaran, T. Egawa, H. Ishikawa, and T. Jimbo, "Characterization of Different Al-Content Al_xGa_{1-x} N/GaN Heterostructures and High-Electron Mobility Transistors on Sapphire," *J. Vac. Sci. Technol. B*, vol. 21, pp. 888–894, Mar./Apr. 2003.

16. Y. Zhang, I.P. Smorchkova, C.R. Elsass, S. Keller, J.P. Ibbetson, S. Denbaars, U.K. Mishra, and J. Singh, "Charge Control and Mobility in AlGaN/GaN Transistors: Experimental and Theoretical Studies," *J. Appl. Phys.*, vol. 87, pp. 7981–7987, June 2000.

17. R.J. Trew, "SiC and GaN Transistors: Is There One Winner for Microwave Power Applications?" *Proc. IEEE, Special Issue on Wide Bandgap Semiconductors*, vol. 90, pp. 1032–1047, June 2002.

18. M. Wrabck, H. Shen, J.C. Carrano, C.J. Collins, J.C. Campbell, R.D. Dupuis, M.J. Schurman, and I.T. Ferguson, "Time-Resolved Electroabsorption Measurement of the Transient Electron Velocity Overshoot in GaN," *Appl. Phys. Lett.*, vol. 79, pp. 1303–1305, Aug. 2001.

19. M. Wraback, H. Shen, S. Rudin, E. Bellotti, M. Goano, J.C. Carrano, C.J. Collins, J.C. Campbell, and R.D. Dupuis, "Direction-Dependent Band Nonparabolicity Effects on High-Field Transient Electron Transport in GaN," *Appl. Phys. Lett.*, vol. 82, pp. 3674–3676, May 2003.

20. M. Akita, S. Kishimoto, K. Maezawa, and T. Mizutani, "Evaluation of Effective Electron Velocity in AlGaN/GaN HEMTs," *Electron. Lett.*, vol. 36, pp. 1736–1737, Sep. 2000.

21. C.R. Bolognesi, A.C. Kwan, and D.W. SiSanto, "Transistor Delay Analysis and Effective Channel Velocity Extraction in AlGaN/GaN HFETs," *IEDM Tech. Digest*, pp. 685–688, 2002.

22. L. Ardaravicius, A. Matulionis, J. Liberis, O. Kiptrjanovic, M. Ramonas, L.F. Eastman, J.R. Shealy, and A. Vertiatchikh, "Electron Drift Velocity in AlGaN/GaN Channel at High Electric Fields," *Appl. Phys. Lett.*, vol. 83, pp. 4038–4040, Nov. 2003.

23. C.H. Oxley, and M.J. Uren, "Measurements of Unity Gain Cutoff Frequency and Saturation Velocity of a GaN HEMT Transistor," *IEEE Trans. Electron Dev.*, vol. 52, pp. 165–169, Feb. 2005.

24. M. Farahmand, and K.F. Brennan, "Full Band Monte Carlo Simulation of Zincblende GaN MESFETs Including Realistic Impact Ionization Rates," *IEEE Trans. Electron Dev.*, vol. 46, pp. 1319–1325, July 1999.

25. M. Farahmand, and K.F. Brennan, "Comparison between Wurtzite Phase and Zincblende Phase GaN MESFETs Using a Full Band Monte Carlo Simulation," *IEEE Trans. Electron Dev.*, vol. 47, pp. 493–497, Mar. 2000.

26. J. Kolnik, I.H. Oguzman, K.F. Brennan, R. Wang, and P.P. Ruden, "Monte-Carlo Calculation of Electron Initiated Impact Ionization in Bulk Zinc-Blende and Wurtzite GaN," *J. Appl. Phys.*, vol. 81, pp. 726–733, 1997.

27. A. Osinsky, M.S. Shur, R. Gaska, and Q. Chen, "Avalanche Breakdown and Breakdown Luminescence in p-π-n GaN Diodes," *Electron. Lett.*, vol. 34, pp. 691–692, Apr. 1998.
28. K. Kunihiro, K. Kasahara, Y. Takahashi, and Y. Ohno, "Experiemental Evaluation of Impact Ionization Coefficients in GaN," *IEEE Electron Dev. Lett.*, pp. 608–610, Dec. 1999.
29. R.J. Trew, Y. Liu, W. Kuang, H. Yin, G.L. Bilbro, J.B. Shealy, R. Vetury, P.M. Garber, M.J. Poulton, "RF Breakdown and Large-Signal Modeling of AlGaN/GaN HFET's," 2006 *IEEE MTT-S International Microwave Symposium Digest*, 11–16 June 2006, San Francisco, CA, pp. 643–646.
30. T.A. Winslow, D. Fan, and R.J. Trew, "Gate-Drain Breakdown Effects Upon the Large Signal RF Performance of GaAs MESFETs," 1990 *IEEE MTT-S International Microwave Symposium Digest*, 7–10 May 1990, Dallas, TX, pp. 315–317.
31. A. Asano, Y. Miyoshi, K. Ishikura, Y. Nashimoto, M. Kuzuhara, and M. Mizuta, "Novel High Power AlGaAs/GaAs HFET with a Field-Modulating Plate Operated at 35v Drain Voltage," 1998 *IEDM Digest*, 6–9 Dec. 1998, Washington, DC, pp. 59–62.
32. S. Karmalkar, and U.K. Mishra, "Enhancement of Breakdown Voltage in AlGaN/GaN High Electron Mobility Transistors Using a Field Plate," *IEEE Trans. Electron Dev.*, vol. 48, pp. 1515–1521, Aug. 2001.
33. Y-F. Wu, A. Saxler, M. Moore, R.P. Smith, S. Sheppard, P.M. Chavarkar, T. Wisleder, U.K. Mishra, and P. Parikh,"30-W/mm GaN HEMTs by Field Plate Optimization," *IEEE Electron Dev. Lett.*, vol. 25, pp. 117–119, Nov. 2004.
34. C. Lee, P. Saunier, J. Yang, and M.A. Khan, "AlGaN-GaN HFEMT's or SiC with CW Power Performance >4 W/mm and 23% PAE at 35 GHz," *IEEE Electron Dev. Lett.*, vol. 24, pp. 616–618, Oct. 2003.
35. R.J. Trew and U.K. Mishra, "Gate Breakdown in MESFETs and HEMTs," *IEEE Electron Dev. Lett.*, vol. 12, pp. 524–526, Oct. 1991.
36. Y. Dora, A. Chakraborty, L. McCarthy, S. Keller, S.P. DenBaars, and U.K. Mishra, "High Breakdown Voltage Achieved on AlGaN/GaN HEMTs with Integrated Slant Field Plates," *IEEE Electron Dev. Lett.*, vol. 27, pp. 713–715, Sep. 2006.
37. N. Tipirneni, A. Koudymov, V. Adivarahan, J. Yang, G. Simin, and M.A. Khan, "The 1.6-kV AlGaN/GaN HFETs," *IEEE Electron Dev. Lett.*, vol. 27, pp. 716–718, Sep. 2006.
38. G. Simin, M.A. Khan, M.S. Shur, and R. Gaska, "High-Power Switching Using III-Nitride Metal-Oxide-Semiconductor Heterostructures," *Int. J. High Speed Electronics and Systems*, vol. 16, pp. 455–468, 2006.
39. Y.F. Wu, B.P. Keller, P. Fini, S. Keller, T.J. Jenkins, L.T. Kehias, S.P. Denbaars, and U.K. Mishra, "High Al-Content AlGaN/GaN MODFET's for Ultrahigh Performance," *IEEE Electron Dev. Lett.*, pp. 50–53, Feb. 1998.
40. A.T. Ping, Q. Chen, J.W. Yang, M.A. Khan, and I. Adesida, "DC and Microwave Performance of High-Current AlGaN/GaN Heterostructure Field Effect Transistors Grown on p-Type SiC Substrates," *IEEE Electron Dev. Lett.*, pp. 54–56, Feb. 1998.
41. G.J. Sullivan, M.Y. Chen, J.A. Higgins, J.W. Yang, Q. Chen, R.L. Pierson, and B.T. McDermott, "High Power 10 GHz Operation of AlGaN HFET's on Insulating SiC," *IEEE Electron Dev. Lett.*, pp. 198–200, June 1998.
42. S.T. Sheppard, K. Doverspike, W.L. Pribble, S.T. Allen, J.W. Palmour, L.T. Kehia, and T.J. Jenkins, "High-Power Microwave GaN/AlGaN HEMTs on Semi-Insulating Silicon Carbide Substrates," *IEEE Electron Dev. Lett.*, vol. 20, pp. 161–163, Apr. 1999.
43. Q. Chen, J.W. Yang, R. Gaska, M.A. Khan, M.S. Shur, G.J. Sullivan, A.L. Sailor, J.A. Higgings, A.T. Ping, and I. Adesida, "High-Power 0.25-μm Gate Doped-Channel GaN/AlGaN Heterostructure Field Effect Transistor," *IEEE Electron Dev. Lett.*, pp. 44–46, Feb. 1998.
44. L.F. Eastman, et al., "Undoped AlGaN/GaN HEMT's for Microwave Power Amplification," *IEEE Trans. Electron Dev.*, vol. 48, pp. 479–485, Mar. 2001.
45. Y.F. Wu, D. Kapolnek, J.P. Ibbetson, P. Parikh, B. Keller, and U.K. Mishra, "Very-High Power Density AlGaN/GaN HEMT's," *IEEE Trans. Electron Dev.*, vol. 48, pp. 586–590, Mar. 2001.
46. V. Tilak, B. Green, V. Kaper, H. Kim, T. Prunty, J. Smart, J. Shealy, and L. Eastman, "Influence of Barrier Thickness on the High-Power Performance of AlGaN/GaN HEMTs," *IEEE Electron Dev. Lett.*, vol. 22, pp. 504–506, Nov. 2001.
47. L. Shen, S. Heikman, B. Moran, R. Coffie, N.Q. Zhang, D. Buttari, I.P. Smorchkova, S. Keller, S.P. DenBaars, and U.K. Mishra, "AlGaN/AlN/GaN High-Power Microwave HEMT," *IEEE Electron Dev. Lett.*, vol. 22, pp. 457–459, Oct. 2001.

48. A. Corrion, C. Poblenz, P. Waltereit, T. Palacios, S. Rajan, U.K. Mishra, and J.S. Speck, "Review of Recent Developments in Growth of AlGaN/GaN High-Electron Mobility Transistors on 4H-SiC by Plasma-Assisted Molecular Beam Epitaxy," *IEICE Trans. Electronics*, vol. E89-C, No. 7, pp. 906–912, 2006.

49. J.W. Johnson, E.L. Piner, A. Vescan, R. Therrien, P. Rajagopal, J.C. Roberts, J.D. Brown, S. Singhal, and K.J. Linthicum, "12 W/mm AlGaN-GaN HFETs on Silicon Substrates," *IEEE Electron Dev. Lett.*, vol. 25, pp. 459–461, July 2004.

50. D.C. Dumka, C. Lee, H.Q. Tserng, P. Saunier, and M. Kumar, "AlGaN/GaN HEMTs on Si Substrates with 7 W/mm Output Power Density at 10 GHz," *Electron. Lett.*, vol. 40, No. 16, Aug. 2004.

51. D. Ducatteau, A. Minko, V. Hoel, E. Morvan, E. Delos, B. Grimbert, H. Lahreche, P. Bove, C. Gaquiere, J.C. De Jaeger, and S. Delage, "Output Power Density of 5.1 W/mm at 18 GHz with an AlGaN/GaN HEMT os Si Substrate," *IEEE Electron Dev. Lett.*, vol. 27, pp. 7–9, Jan. 2006.

52. C. Lee, H. Wang, J. Yang, L. Witkowski, M. Muir, M.A. Khan, and P. Saunier, "State-of-Art CW Power Density Achieved at 26 GHz by AlGaN/GaN HEMTs," *Electron. Lett.*, vol. 38, pp. 924–925, Aug. 2002.

53. C. Lee, P. Saunier, J. Yang, and M.A. Khan, "AlGaN-GaN HEMTs on SiC with CW Power Performance >4 W/mm and 23% PAE at 35%," *IEEE Electron Dev.* Lett., vol. 24, pp. 616–618, Oct. 2003.

54. K. Boutros, M. Regan, P. Rowell, D. Gotthold, R. Birkhahn, and B. Brar, "High Performance GaN HEMTs at 40 GHz with Power Density of 2.8 W/mm," *IEDM Tech. Digest*, pp. 981–982, 2003.

55. J.S. Moon, S. Wu, D. Wong, I. Milosavljevic, A. Conway, P. Hashimoto, M. Hu, M. Antcliffe, and M. Micovic, "Gate-Recessed AlGaN-GaN HEMTs for High Performance Millimeter-Wave Applications," *IEEE Electron Dev. Lett.*, vol. 26, pp. 348–350, June 2005.

56. T. Palacios, A. Chakroborty, S. Rajan, C. Poblenz, S. Keller, S.P. DenBaars, J.S. Speck, and U.K. Mishra, "High-Power AlGaN/GaN HEMTs for Ka-Band Applications," *IEEE Electron Dev. Lett.*, vol. 26, pp. 781–783, Nov. 2005.

57. Y. Cai, Y. Zhou, K.J. Chen, and K.M. Lau, "High-Performance Enhancement-Mode AlGaN/GaN HEMTs Using Fluoride-Based Plasma Treatment," *IEEE Electron Dev. Lett.*, vol. 26, pp. 435–437, July 2005.

58. L. Shen, T. Palacios, C. Poblenz, A. Corrion, A. Chakraborty, N. Fichtenbaum, S. Keller, S.P. DenBaars, J.S. Speck, and U.K. Mishra, "Unpassivated High Power Deeply Recessed GaN HEMTs with Fluorine-Plasma Surface Treatment," *IEEE Electron Dev. Lett.*, vol. 27, pp. 214–216, Apr. 2006.

59. O. Katz, D. Mistele, B. Meyler, G. Bahir, and J. Salzman, "Polarization Engineering of InAlN/GaN HFET and the Effect on DC and RF Performance," *IEDM Tech. Digest*, pp. 1035–1038, 2004.

60. O. Katz, D. Mistele, B. Meyler, G. Bahir, and J. Salzman, "Characteristics of InAlN-GaN High-Electron Mobility Field-Effect Transistor," *IEEE Trans. Electron Dev.*, vol. 52, pp. 146–150, Feb. 2005.

61. T. Palacios, A. Chakraborty, S. Heikman, S. Keller, S.P. DenBaars, and U.K. Mishra, "AlGaN/GaN High Electron Mobility Transistors with InGaN Back-Barriers," *IEEE Electron Dev. Lett.*, vol. 27, pp. 13–15, Jan. 2006.

62. J. Liu, Y. Zhou, J. Zhu, K.M. Lau, and K.J. Chen, "AlGaN/GaN/InGaN/GaN DH-HEMTs with an InGaN Notch for Enhanced Carrier Confinement," *IEEE Electron Dev. Lett.*, vol. 27, pp. 10–12, Jan. 2006.

63. M. Higashiwaki, T. Mimura, and T. Matsui, "AlN/GaN Insulated-Gate HFETs Using Cat-CVD SiN," *IEEE Electron Dev. Lett.*, vol. 27, pp. 719–721, Sep. 2006.

64. Y. Ando, Y. Okamoto, H. Miyamoto, T. Nakamura, T. Inoue, and M. Kuzuhara, "10-W/mm AlGaN-GaN HFET with a Field Modulating Plate," *IEEE Electron Dev. Lett.*, vol. 24, pp. 289–291, May 2003.

65. Y.F. Wu, P.M. Chavarkar, M. Moore, P. Parikh, B.P. Keller, and U.K. Mishra, "A 50W AlGaN/GaN HEMT Amplifier," *2000 International Electron Devices Meeting Digest*, pp. 375–376.

66. R. Vetury, Y. Wei, D.S. Green, S.R. Gibb, T.W. Mercier, K. Leverich, P.M. Garber, M.J. Poulton, and J.B. Shealy, "High Power, High Efficiency, AlGaN/GaN HEMT Technology for Wireless Base Station Applications," *2005 IMS Digest*, pp. 487–490.

67. Y. Kamo, et al., "A C-Band AlGaN/GaN HEMT with Cat-CVD SiN Passivation Developed for an Over 100 W Operation," *IEEE IMS Tech. Digest*, pp. 495–498, 2005.

68. T. Kikkawa, et al., "An Over 200 W Output Power GaN HEMT Push-Pull Amplifier with High Reliability," *IEEE IMS Tech. Digest*, pp. 1347–1350, 2004.

69. T. Kikkawa, and K. Joshin, "High Power GaN-HEMT for Wireless Base Station Applications," *IEICE Trans. on Electronics*, vol. E89-C, No. 5, pp. 608–615, 2006.

70. W. Nagy, et al., "150 W GaN-on-Si RF Power Transistor," *IEEE IMS Tech. Digest*, pp. 483–486, 2005.

71. R. Therrien, et al., "A 36mm GaN-on-Si HFET Producing 368 W at 60 V with 70% Drain Efficiency," *IEEE IEDM Tech. Digest*, 2005.

72. O. Kruger, G. Schone, T. Wernicke, R. Lossy, A. Liero, F. Schnieder, J. Wurfl, and G. Trankle, "Laser-Assisted Processing of VIAs for AlGaN/GaN HEMTs on SiC Substrates," *IEEE Electron Dev. Lett.*, vol. 27, pp. 425–427, June 2006.

73. M.A. Khan, X. Hu, A. Tarakji, G. Simin, and J. Yang, "AlGaN/GaN Metal-Oxide-Semiconductor Heterostructure Field-Effect Transistors on SiC Substrates," *Appl. Phys. Lett.*, vol. 77, pp. 1339–1341, Aug. 2000.

74. G. Simin, X. Hu, N. Ilinskaya, J. Zhang, A. Tarakji, A. Kumar, J. Yang, M.A. Khan, R. Gaska, and M.S. Shur, "Large Periphery High-Power AlGaN/GaN Metal-Oxide-Semiconductor Heterostructure Field Effect Transistors on SiC with Oxide-Bridging," *IEEE Electron Dev. Lett.*, vol. 22, pp. 53–55, Feb. 2001.

75. A. Koudymov, X. Hu, K. Simin, G. Simin, M. Ali, J. Yang, and M.A. Khan, "Low-Loss High Power RF Switching Using Multifinger AlGaN/GaN MOSHFETs," *IEEE Electron Dev. Lett.*, vol. 23, pp. 449–451, Aug. 2002.

76. V. Adivarahan, J. Yang, A. Koudymov, G. Simin, and M.A. Khan, "Stable CW Operation of Field-Plated GaN-AlGaN MOSHFETs at 19 W/mm," *IEEE Electron Dev. Lett.*, vol. 26, pp. 535–537, Aug. 2005.

77. R. Wang, Y. Cai, W. Tang, K.M. Lau, and K.J. Chen, "Planar Integration of E/D-Mode AlGaN/GaN HEMTs Using Fluoride-Based Plasma Treatment," *IEEE Electron Dev. Lett.*, vol. 27, pp. 633–635, Aug. 2006.

78. T. Palacios, C-S. Suh, A. Chakraborty, S. Keller, S.P. DenBaars, and U.K. Mishra, "High-Performance E-Mode AlGaN/GaN HEMTs," *IEEE Electron Dev. Lett.*, vol. 27, pp. 428–430, June 2006.

79. X. Hu, A. Koudymov, G. Simin, J. Yang, and M.A. Khan, "Si_3N_4/AlGaN/GaN-Metal-Insulator-Semiconductor Heterostructure Field-Effect Transistors," *Appl. Phys. Lett.*, vol. 79, pp. 2832–2834, Oct. 2001.

80. M. Higashiwaki, T. Matsui, and T. Mimura, "AlGaN/GaN MIS-HFETs with f_T of 163 GHz Using Cat-CVD SiN Gate-Insulating and Passivation Layers," *IEEE Electron Dev. Lett.*, vol. 27, pp. 16–18, Jan. 2006.

81. S. Karmalkar, J. Deng, M.S. Shur, and R. Gaska, "RESURF AlGaN/GaN HEMT for High Voltage Power Switching," *IEEE Electron Dev. Lett.*, vol. 22, pp. 373–375, Aug. 2001.

82. H. Yamane, M. Shimada, S.J. Clarke, and F.J. DiSalvo, "Preparation of GaN Single Crystals Using a Na Flux," *Chem. Mater.* vol. 9, pp. 413–416, 1997.

23

Microwave Vacuum Tubes

Jerry C. Whitaker
Advance Television Systems Committee

Robert R. Weirather
Harris Corporation

Thomas K. Ishii
Marquette University

23.1 Microwave Power Tubes

Jerry C. Whitaker

23.1.1 Introduction

Microwave power tubes span a wide range of applications, operating at frequencies from 300 MHz to 300 GHz with output powers from a few hundred watts to more than 10 MW. Applications range from the familiar to the exotic. The following devices are included under the general description of microwave power tubes:

- Klystron, including the *reflex* and *multicavity* klystron
- *Multistage depressed collector* (MSDC) klystron
- *Klystrode* (IOT) tube

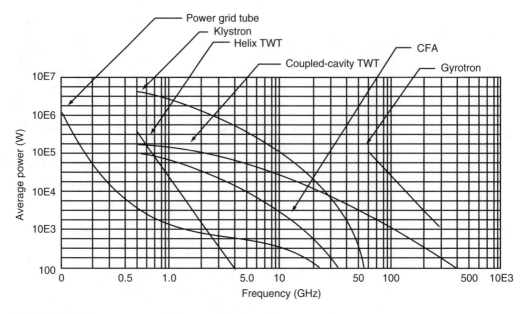

FIGURE 23.1   Microwave power tube type as a function of frequency and output power.

- Traveling-wave tube (TWT)
- **Crossed-field tube**
- *Coaxial magnetron*
- *Gyrotron*
- **Planar triode**
- High-frequency tetrode

This wide variety of microwave devices has been developed to meet a wide range of applications. Some common uses include

- UHF-TV transmission
- Shipboard and ground-based radar
- Weapons guidance systems
- Electronic countermeasure (ECM) systems
- Satellite communications
- Tropospheric scatter communications
- Fusion research

As new applications are identified, improved devices are designed to meet the need. Microwave power tubes manufacturers continue to push the limits of frequency, operating power, and efficiency. Microwave technology is an evolving science. Figure 23.1 charts device type as a function of operating frequency and power output.

Two principal classes of microwave vacuum devices are in common use today:

- **Linear beam** tubes
- Crossed-field tubes

Each class serves a specific range of applications. In addition to these primary classes, some power grid tubes are also used at microwave frequencies.

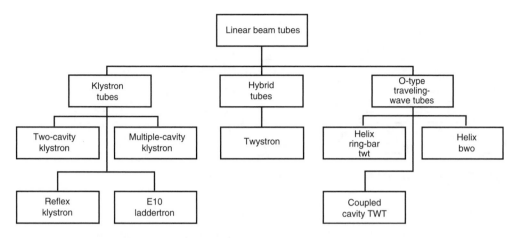

FIGURE 23.2   Types of linear beam microwave tubes.

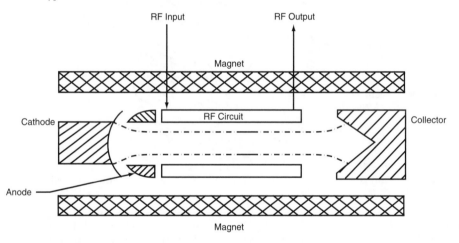

FIGURE 23.3   Schematic diagram of a linear beam tube.

Linear Beam Tubes

As the name implies, in a linear beam tube the electron beam and the circuit elements with which it interacts are arranged linearly. The major classifications of linear beam tubes are shown in Figure 23.2. In such a device, a voltage applied to an anode accelerates electrons drawn from a cathode, creating a beam of kinetic energy. Power supply potential energy is converted to kinetic energy in the electron beam as it travels toward the microwave circuit. A portion of this kinetic energy is transferred to microwave energy as RF waves slow down the electrons. The remaining beam energy is either dissipated as heat or returned to the power supply at the collector. Because electrons will respell one another, there usually is an applied magnetic focusing field to maintain the beam during the interaction process. The magnetic field is supplied either by a solenoid or permanent magnets. Figure 23.3 shows a simplified schematic of a linear beam tube.

Crossed-Field Tubes

The magnetron is the pioneering device of the family of crossed-field tubes. The family tree of this class of devices is shown in Figure 23.4. Although the physical appearance differs from that of linear beam tubes, which are usually circular in format, the major difference is in the interaction physics that requires a magnetic field at right angles to the applied electric field. Whereas the linear beam tube sometimes requires a magnetic field to maintain the beam, the crossed-field tube always requires a magnetic focusing field.

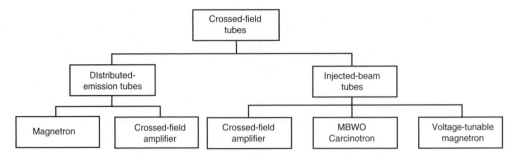

FIGURE 23.4   Types of crossed-field microwave tubes.

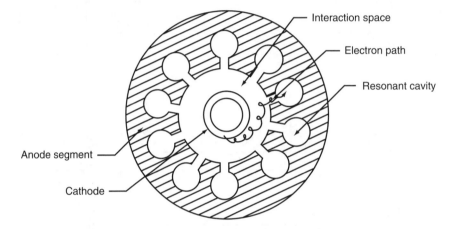

FIGURE 23.5   Magnetron electron path looking down into the cavity with the magnetic field applied.

Figure 23.5 shows a cross section of the magnetron, including the magnetic field applied perpendicular to the cathode–anode plane. The device is basically a diode with the anode composed of a plurality of resonant cavities. The interaction between the electrons emitted from the cathode and the crossed electric and magnetic fields produces a series of space charge spokes that travel around the anode–cathode space in a manner that transfers energy to the RF signal supported by the multicavity circuit. The mechanism is highly efficient.

Crossed-Field Amplifiers
Figure 23.6 shows the family tree of the crossed-field amplifier (CFA). The configuration of a typical present day distributed emission amplifier is similar to that of the magnetron except that the device has an input for the introduction of RF into the circuit. Current is obtained primarily by secondary emission from the negative electrode that serves as a cathode throughout all or most of the interaction space. The earliest versions of this tube type were called **amplitrons**.

The CFA is deployed in radar systems operating from the UHF to the Ku band and at power levels up to several megawatts. In general, bandwidth ranges from a few percent to as much as 25% of the center frequency.

23.1.2 Grid Vacuum Tubes

The physical construction of a vacuum tube causes the output power and available gain to decrease with increasing frequency. The principal limitations faced by grid-based devices include the following:

- Physical size. Ideally, the RF voltages between electrodes should be uniform; however, this condition cannot be realized unless the major electrode dimensions are significantly less than $\frac{1}{4}$-wavelength

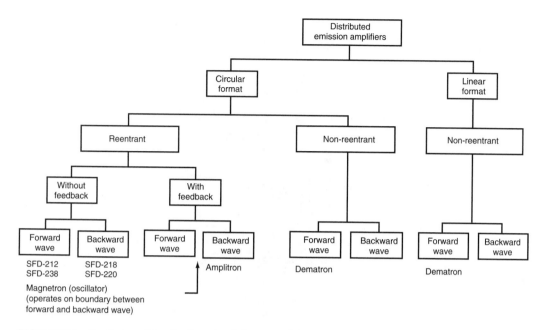

FIGURE 23.6 Family tree of the distributed emission crossed-field amplifier (CFA).

at the operating frequency. This restriction presents no problems at VHF frequencies; however, as the operating frequency increases into the microwave range, severe restrictions are placed on the physical size of individual tube elements.

- Electron transit-time. Interelectrode spacing, principally between the grid and the cathode, must be scaled inversely with frequency to avoid problems associated with electron transit time. Possible adverse conditions include: (1) excessive loading of the drive source, (2) reduction in power gain, (3) back heating of the cathode as a result of electron bombardment, and (4) reduced conversion efficiency.

- Voltage standoff. High-power tubes operate at high voltages. This presents significant problems for microwave vacuum tubes. For example, at 1 GHz the grid-cathode spacing must not exceed a few mils. This places restrictions on the operating voltages that may be applied to individual elements.

- Circulating currents. Substantial RF currents may develop as a result of the inherent interelectrode capacitances and stray inductances/capacitances of the device. Significant heating of the grid, connecting leads, and vacuum seals may result.

- Heat dissipation. Because the elements of a microwave grid tube must be kept small, power dissipation is limited.

Still, some grid vacuum tubes find applications at high frequencies. Planar triodes are available that operate at several gigahertz, with output powers of 1–2 kW in pulsed service. Efficiency (again for pulsed applications) ranges from 30 to 60%, depending on the frequency.

Planar Triode

A cross-sectional diagram of a planar triode is shown in Figure 23.7. The envelope is made of ceramic, with metal members penetrating the ceramic to provide for connection points. The metal members are shaped either as disks or as disks with cylindrical projections.

The cathode is typically oxide coated and indirectly heated. The key design objective for a cathode is high-emission density and long tube life. Low-temperature emitters are preferred because high cathode temperatures typically result in more evaporation and shorter life.

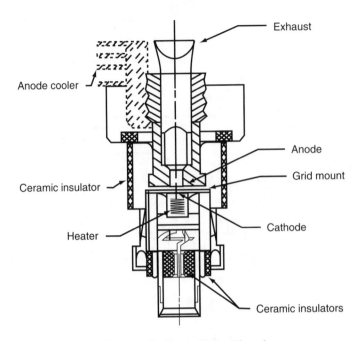

FIGURE 23.7 Cross section of a 7289 planar triode. (*Source:* Varian/Eimac.)

The grid of the planar triode represents perhaps the greatest design challenge for tube manufacturers. Close spacing of small-sized elements is needed, at tight tolerances. Good thermal stability is also required, because the grid is subjected to heating from currents in the element itself, plus heating from the cathode and bombardment of electrons from the cathode.

The anode, usually made of copper, conducts the heat of electron bombardment to an external heat sink. Most planar triodes are air cooled.

Planar triodes designed for operation at 1 GHz and above are used in a variety of circuits. The grounded grid configuration is most common. The plate resonant circuit is cavity based, using waveguide, coaxial line, or stripline. Electrically, the operation of planar triode is much more complicated at microwave frequencies than at low frequencies. Figure 23.8 compares the elements at work for a grounded-grid amplifier operating at (a) low frequencies and (b) at microwave frequencies. The equivalent circuit is made more complicated by

- Stray inductance and capacitance of the tube elements
- Effects of the tube contact rings and socket elements
- Distributed reactance of cavity resonators and the device itself
- Electron transit-time effects, which result in resistive loading and phase shifts

Reasonable gains of 5–10 dB may be achieved with a planar triode. Increased gain is available by cascading stages. Interstage coupling may consist of waveguide or coaxial-line elements. Tuning is accomplished by varying the cavity inductance or capacitance. Additional bandwidth is possible by stagger tuning cascaded stages.

High-Power UHF Tetrode

New advancements in vacuum tube technology have permitted the construction of high-power UHF transmitters based on tetrodes. Such devices are attractive because they inherently operate in an efficient class A–B mode. UHF tetrodes operating at high-power levels provide essentially the same specifications, gain, and efficiency as tubes operating at lower powers. The anode power supply is much lower in voltage

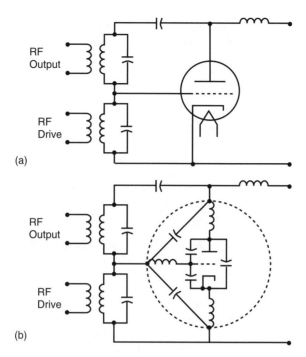

FIGURE 23.8 Grounded-grid equivalent circuits: (a) low-frequency operation, (b) microwave frequency operation. The cathode-heating and grid-bias circuits are not shown.

than the collector potential of a klystron-based system (8 kV is common). The tetrode also does not require focusing magnets.

Efficient removal of heat is the key to making a tetrode practical at high-power levels. Such devices typically use water or vapor-phase cooling. Air cooling at such levels is impractical because of the fin size that would be required. Also, the blower for the tube would have to be quite large, reducing the overall transmitter AC-to-RF efficiency.

The expected lifetime of a tetrode in UHF service is usually shorter than a klystron of the same power level. Typical lifetimes of 8000–15,000 h have been reported.

Defining Terms

Amplitron: A classic crossed-field amplifier in which output current is obtained primarily by secondary emission from the negative electrode that serves as a cathode throughout all or most of the interaction space.

Crossed-field tube: A class of microwave devices where the electron beam and the circuit elements with which it interact are arranged in a nonlinear, usually circular, format. The interaction physics of a crossed-field tube requires a magnetic field at right angles to the applied electric field.

Linear beam tube: A class of microwave devices where the electron beam and the circuit elements with which it interacts are arranged linearly.

Planar triode: A grid power tube whose physical design make it practical for use at microwave frequencies.

References

Badger, G. 1986. The Klystrode: A new high-efficiency UHF-TV power amplifier. In *Proceedings of the NAB Engineering Conference.* National Association of Broadcasters, Washington, DC.

Clayworth, G.T., Bohlen, H.P., and Heppinstall, R. 1992. Some exciting adventures in the IOT business. In *NAB 1992 Broadcast Engineering Conference Proceedings,* pp. 200–208. National Association of Broadcasters, Washington, DC.

Collins, G.B. 1947. *Radar System Engineering.* McGraw-Hill, New York.

Crutchfield, E.B. 1992. *NAB Engineering Handbook,* 8th ed. National Association of Broadcasters, Washington, DC.

Fink, D. and Christensen, D., eds. 1989. *Electronics Engineer's Handbook,* 3rd ed. McGraw-Hill, New York.

IEEE. 1984. *IEEE Standard Dictionary of Electrical and Electronics Terms.* Institute of Electrical and Electronics Engineers, Inc., New York.

McCune, E. 1988. Final report: The multi-stage depressed collector project. In *Proceedings of the NAB Engineering Conference.* National Association of Broadcasters, Washington, DC.

Ostroff, N., Kiesel, A., Whiteside, A., and See, A. 1989. Klystrode-equipped UHF-TV transmitters: Report on the initial full service station installations. In *Proceedings of the NAB Engineering Conference.* National Association of Broadcasters, Washington, DC.

Pierce, J.A. 1945. Reflex oscillators. *Proc. IRE* 33(Feb.):112.

Pierce, J.R. 1947. Theory of the beam-type traveling wave tube. *Proc. IRE* 35(Feb.):111.

Pond, N.H. and Lob, C.G. 1988. Fifty years ago today or on choosing a microwave tube. *Microwave Journal* (Sept.):226–238.

Spangenberg, K. 1947. *Vacuum Tubes.* McGraw-Hill, New York.

Terman, F.E. 1947. *Radio Engineering.* McGraw-Hill, New York.

Varian, R. and Varian, S. 1939. A high-frequency oscillator and amplifier. *J. Applied Phys.* 10(May):321.

Whitaker, J.C. 1991. *Radio Frequency Transmission Systems: Design and Operation.* McGraw-Hill, New York.

Whitaker, J.C. *Power Vacuum Tubes Handbook.* 2nd ed. CRC PRESS, New York.

Further Information

Specific information on the application of microwave tubes can be obtained from the manufacturers of those devices. More general application information can be found in the following publications:

Radio Frequency Transmission Systems: Design and Operation, written by Jerry C. Whitaker, McGraw-Hill, New York, 1991.
Power Vacuum Tubes Handbook 2nd ed., written by Jerry C. Whitaker, CRC Press, New York, 1994.

The following classic texts are also recommended:

Radio Engineering, 2nd ed., written by Frederick E. Terman, McGraw-Hill, New York, 1947.
Vacuum Tubes, written by Karl R. Spangenberg, McGraw-Hill, New York, 1948.

23.2 Klystron

Robert R. Weirather

23.2.1 Introduction

Energy in a microwave system may be viewed as with the passage of traveling electromagnetic waves along a waveguide axis. The generation of these waves can be resolved to one of energy conversion: the source potential energy (DC) is converted to the energy of the electromagnetic fields. At lower radio frequencies one use of the familiar multielectrode vacuum tube is as a valve controlling the flow of electrons in the anode conductor. (RF sources can use semiconductor devices for microwave power but these are generally limited to less than a kilowatt per device.) To generate considerable power at higher frequencies, the vacuum tube is less desirable than other means such as klystrons.

Conventional vacuum tubes are not efficient sources of currents at microwave frequencies. The assumption upon which their efficient design is based, namely, that the electrons pass from **cathode** to **anode** in a time short compared with the period of the anode potential, no longer holds true at 10^9–10^{10} Hz. The transit time of the order of magnitude of 10^{-8} s may be required for an electron to travel from cathode to anode of a conventional vacuum tube at higher frequencies. The anode and grid potentials may alternate from 10 to 100 times during the electron transit. The anode potential during the negative part of its cycle thus removes energy that was given to the electrons during the phase of positive anode potential. The overall result of transit-time effects is that of a reduction in the operating efficiency of the vacuum tube becoming more marked with increasing frequency. Reducing the transit time by decreasing cathode-to-anode spacing and increasing anode potential creates excessive interelectrode capacitance and overheating of the anode. Thus, the necessity for the design of tubes that are specifically suited for operation in the higher frequency range. Consider a tube the electron beam of which interacts with the time-varying fields of that part of the microwave system that is included within the tube. This interaction proceeds in such a way that the velocity of the electrons is lower after the interaction than before, while the kinetic energy removed from the electrons can reappear as energy of the electromagnetic fields. Practical tubes in use differ among themselves in the details of the mode of interaction between the electron beam and the rapidly varying fields. this chapter reviews one such tube: The klystron and some of its derivatives.

23.2.2 The Klystron

The three major functions of the high-power linear-beam tube (beam generation, interaction, and dissipation) are accomplished in three separate and well-defined sections of the tube, each of which can be optimized almost independently of the others. A relatively large high-convergence **electron gun** may be used to provide a small beam diameter (compared to wavelength) for interaction with a high-frequency circuit. Beam focusing can be made sufficiently good so that interception of current can often be neglected, and only RF circuit losses need be considered. The beam is usually allowed to expand into a large collector when it leaves the interaction region. Electron guns, beam focusing, **collectors**, and output windows are independent of the type of interaction structure used and are common to all tubes.

23.2.3 Typical Klystron Amplifier

Figure 23.9 shows a cutaway representation of a typical klystron amplifier. Schematically it is very similar to a triode tube in that it includes an electron source, resonant circuits, and a collector (which is roughly equivalent to the plate of a triode). In fact, the klystron amplifier consists of three separate sections—the electron gun for beam generation, the RF section for beam interaction, and the collector section for electron energy dissipation.

The Electron Gun

Let us consider, first, the electron gun structure: As in the triode, the electron gun consists of heater and cathode, a control grid (sometimes), and an anode. Electrons are emitted by the hot cathode surface and are drawn toward the anode that is operated at a positive potential with respect to the cathode. The electrons are formed into a small, dense beam by either electrostatic or magnetic focusing techniques, similar to the techniques used for beam formation in a cathode ray tube. In some klystron amplifiers, a control grid is used to permit adjustment of the number of electrons that reach the anode region; this control grid may be used to turn the tube completely on or completely off in certain pulsed-amplifier applications. The electron beam is well formed by the time it reaches the anode. It passes through a hole in the anode, passes on to the RF section of the tube, and eventually the electrons are intercepted by the collector. The electrons are returned to the cathode through an external power supply. It is evident that the collector in the klystron acts much like the plate of a triode insofar as collecting of the electrons is concerned, however, there is one important difference; the plate of a triode is normally connected in some fashion to the output RF circuit, whereas, in a klystron amplifier, the collector has no connection to the RF circuitry at all. From the preceding discussion, it is apparent that the klystron amplifier, insofar as the electron flow is concerned, is quite analogous to a stretched-out triode tube in that electrons are emitted by the cathode, controlled in number by the control grid, and collected eventually by the collector.

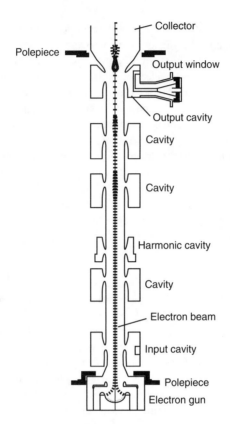

FIGURE 23.9   Sectional view of a klystron showing the principal elements.

RF Section

Now let us consider the RF section of a klystron amplifier. This part of the tube is physically quite different from a gridded tube amplifier. One of the major differences is in the physical configuration of the resonant circuit used in a klystron amplifier. The resonant circuit used with a triode circuit, at lower frequencies, is generally composed of an inductance and a capacitor, whereas the resonant circuit used in a microwave tube is almost invariably a metal-enclosed chamber known as a cavity resonator. A very crude analogy can be made between the resonant cavity and a conventional L-C resonant circuit. The gap in the cavity is roughly analogous to the capacitor in a conventional low-frequency resonant circuit in that alternating voltages, at the RF frequencies, can be made to appear across the cavity gap. Circulating currents will flow between the two sides of the gap through the metal walls of the cavity, roughly analogous to the flow of RF current in the inductance of an L-C resonant circuit. Since RF voltages appear across the sides of the cavity gap, it is apparent that an electric field will be present, oscillating at the radio frequency, between the two surfaces of the cavity gap. When a cavity is the correct size, it will resonate. To bring to resonance, a cavity can be tuned by adjusting its size by some mechanical means.

As shown in Figure 23.9, electrons pass through the cavity in each of the resonators, and pass through cylindrical metal tubes between the various gaps. These metal tubes are called *drift tubes*. In a klystron amplifier the low-level RF input signal is coupled to the first or input resonator, which is called the *buncher* cavity. The signal may be coupled in through either a waveguide or a coaxial connection. The RF input signal will excite oscillating currents in the cavity walls. These oscillating currents will cause the alternate sides of the buncher gap to become first positive and then negative in potential at a frequency equal to the frequency of the RF input signal. Therefore, an electric field will appear across the buncher gap, alternating at the RF frequency. This electric field will, for half a cycle, be in a direction that will tend to speed up the electrons flowing through the gap; on the other half of the cycle the electric field will be in a direction that will tend to slow down the electrons as they cross the buncher gap. This effect is called velocity modulation, and it is the mechanism that permits the klystron amplifier to operate at frequencies higher than the triode.

After leaving the buncher gap the electrons proceed toward the collector in the drift tube region. Ignore for the moment the intermediate resonators shown in Figure 23.9, and let us consider the simple case of a two-cavity klystron amplifier. In the drift tube region the electrons that have been speeded up by the electric field in the buncher gap will tend to overtake those electrons that have previously been slowed down (by the preceding half of the RF wave across the buncher gap). It is apparent that, since some electrons are tending to overtake other electrons, clumps or bunches of electrons will be formed in the drift tube region. If the average velocity of the electron stream is correct, as determined by the original voltage between anode and cathode, and if the length of the drift tube is proper, these bunches of electrons will be quite completely formed by the time they reach the catcher gap of the last or output cavity (which is called the catcher). This results in bunches of electrons flowing through the catcher gap periodically, and during the time between these bunches relatively fewer electrons flow through the catcher gap. The time between arrival of bunches of electrons is equal to the time of one cycle of the RF input signal.

These bunches of electrons will induce alternating current flow in the metal walls of the catcher cavity as they pass through the catcher gap. The output cavity will have large oscillating currents generated in its walls. These currents cause electric fields to exist, at the radio frequency, within the output cavity. These electric fields can be coupled from the cavity (to output waveguide, or coaxial transmission lines) resulting in the RF output power from the tube.

Collector

The electron beam energy is coupled in the last or output cavity to provide the desired RF power. The residual electron beam energy is absorbed in the collector. Collector design is based on the shape and energy of the electron beam. Electrons may have considerable energy upon impact on the collector and thus create undesired secondary emissions. Collector design includes coating and geometry considerations to minimize secondary emission. The collector may have to dissipate the full energy of the beam and is most frequently liquid cooled at higher powers.

23.2.4 High-Efficiency Klystron Design

The early analysis of ballistic bunching in a two-cavity klystron led to a maximum predicted efficiency of 58%. Early researchers recognized very early that tuning the penultimate cavity to the high-frequency side of the carrier could increase the efficiency over that which had been predicted. This occurs because it presents an inductive impedance to the beam and creates fields that squeeze the bunch formed at previous cavities. Locating a cavity tuned to the second harmonic of the operating frequency at the point where the second-harmonic content of the beam had increased enough to interact with that cavity, one could form a bunch that contained a higher fundamental component of current than by using conventional fundamental bunching only. The combination of fundamental bunching plus properly phased second harmonic bunching at a subsequent cavity is an approximation to a sawtooth voltage at a single gap. In theory, 100% efficiency would be possible. As we will see, further improvements have led to very efficient DC to RF power conversion.

External and Internal Cavities

The high-power, multicavity klystron can be manufactured with the cavities either integral with the body or as external to the body. These two types are generally referred to klystron constructions as either with external or internal cavities. The internal (integral) cavities are manufactured in place by the OEM and thus field replacement of a failed tube requires shipment back to the vendor. The advantage with internal cavities is that details of installation are completely under the control of the OEM and less subject to installation error. Additionally, the cooling and heat flow is controlled better. With external cavity models, the cavities can be removed and the tubes replaced. This is a feature that lowers cost and hastens repair.

Multistage Depressed Collector Klystron

Development efforts at NASA to achieve highly efficient microwave transmitters for space applications resulted in achieving multistage depressed collector (MSDC) technology with great potential for efficiency enhancement. The UHF television community recognized the potential of this technology for reducing operating costs, and preliminary analysis indicated that applying MSDC technology to the power klystron used in UHF-TV transmitters could reduce amplifier power consumption by half.

A development program was organized and achieved an MSDC klystron design. A four-stage design was selected that would operate with equal voltage steps per stage (see Figure 23.10). These MSDC designs

FIGURE 23.10   Multistage depressed collector (MSDC) for a klystron. (*Source:* Courtesy of Varian Associates, San Carlos, CA.)

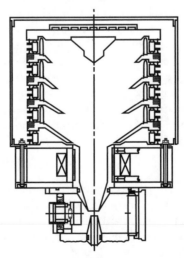

FIGURE 23.11   Cross-sectional view of a klystrode multistage depressed collector with four stages. (*Source:* Courtesy of Varian Associates, San Carlos, CA.)

are identical to the ordinary klystron except a different collector is used. The collector is segmented by different voltage stages. These different stages collect the various electron energies at lower voltages and thus reduce the overall dissipation resulting in increased efficiency (see Figure 23.11). Power efficiency varies from 74% with no RF output to 57% at saturated output. The overall efficiency is 71% with no RF output to 57% at saturation output. The overall efficiency is 71% at this point, compared with 55% without the MSDC. For television operation, the input power is reduced from 118 to 51.8 kW by use of the MSDC, consequently, reducing the power costs to less than one-half.

23.2.5 Broadband Klystron Operation

For television service in the United States, at least a 5.25-MHz bandwidth is required. Some European systems require as much as 8 MHz. At the power levels used in television service, it is necessary to stagger tune the cavities of a klystron to obtain this bandwidth. This is particularly true at the low end of the UHF band, but less so at the high end where the percentage bandwidth is lower.

The stagger tuning of a klystron in order to improve bandwidth is much more complicated, from a theoretical point of view, than is, for example, the stagger tuning of an IF amplifier. In an IF amplifier, the gain functions of the various stages may simply be multiplied together. In the klystron, however, the velocity modulation applied to the beam at a cavity gap will contribute a component to the exciting current at each and every succeeding gap. Thus, the interaction of tuning and the beam shaping for optimum power, bandwidth, efficiency, and gain is a tradeoff. To achieve the desired operation, the OEM must do studies and tests to determine the cavity tuning.

23.2.6 Frequency of Operation

Practical considerations limit klystron designs at the lower frequencies. The length of a klystron would be several meters at 100 MHz, but only 1 m long at 1 GHz. Triodes and tetrodes are more practical tubes at lower frequencies, whereas the klystron becomes practical at 300 MHz and above. At frequencies above 20 GHz, the design of klystrons is limited by the transit time.

23.2.7 Power and Linearity

The power of a klystron is limited to a theoretical 100% of that in the electron beam ($P_v = I_b V_b$). In some areas, efficiencies of 70–80% have been achieved. Normally, efficiency of 40–60% can be expected. RF power bandwidth, of course, is dependent on the number of cavities and their tuning. Typically, klystrons with multiple cavities can be stagger tuned to provide a flat RF passband at some reduction in efficiency and gain. For example, in four cavity tubes, staggered tuning results in a-1-dB bandwidth of 1% with a gain of 40 dB.

Klystron linearity (both amplitude and phase) is considered fair to good. For RF power, several decibels below saturation, the amplitude linearity exhibits a curvature as shown in Figure 23.12. As drive is increased, the RF output reaches saturation and further increases in drive reduces power output. The linearity curvature and saturation characteristic will generate harmonic and intermodulation products, which may be filtered or precorrected.

The phase shift versus drive is also nonlinear. As the output power reaches saturation, the phase shift increases and may reach 10° or more. This nonlinearity creates AM to PM conversion. Compensation circuits can be used in the RF drive to correct this condition.

23.2.8 Klystron Summary

When the need for high frequency and high power is indicated, the klystron is one of the first candidates for service. The klystron continues to be workhorse with long life and dependable performance. Klystrons are a powerful, high-frequency answer to many RF problems. Klystrons come in a variety of powers from a 1 kW to several megawatts. Modern designs have years of life, high gains, and improved efficiency. Found

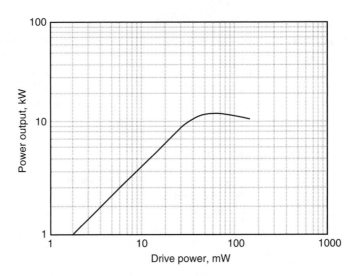

FIGURE 23.12   Typical gain and saturation characteristics of a klystron.

in radars, accelerators, television transmitters, and communication systems, these tubes operate in outer space to fixed base locations.

23.2.9   Inductive Output Tube A Derivative

The inductive output tube (IOT), also known as the **klystrode**, is a hybrid device with some attributes of a gridded tube and a klystron. The distinct advantage of a klystron over a gridded tube is its ability to produce high powers at high frequencies. One shortcoming of a klystron is its cost. The IOT attempts to lower the cost gap by using a grid to do some bunching and utilizes a drift region to complete this process. Another shortcoming of a klystron is its class of operation (class A). Operation of a klystron conduction cycle is class A, which delivers superior linearity but comes up short in efficiency. On the other hand, the IOT grid can be biased to completely cut off **beam current** for class B or be biased with some beam current for class AB, or even used in class A mode. Thus, the application of the klystrode is best suited for high-power, high-efficiency operation, whereas the klystron is best for high linearity.

History and Names for the IOT Klystrode

The klystrode was invented by Varian in the 1980s. Its name was chosen for its hybrid nature and describes the device quite well. Other manufacturers began work on similar devices and use other descriptions. English Electric Valve, Ltd., of the United Kingdom, chose the name inductive output tube (IOT) (see Figure 23.13). This name was chosen due to a reference in the literature to this device possibility in the 1930s. Thus, it was 50 years before the klystrode (or IOT) was made commercially practical.

IOT Description

The klystron relies on the velocity bunching of the electrons from a heated cathode over a body length of 1 m or more. This is necessary to permit orderly physical spacing of the buncher cavities and the individual electron velocities to form properly. The superiority of the klystron over the gridded tube (typically a tetrode) is its output coupling, a cavity. Contrasted, it is quite difficult to design circuits to couple from tetrodes at higher power due to the increasing capacitance, which lowers the impedance. The superiority of the gridded tube is in its compact size and the general possibility of lower cost.

The IOT takes advantage of two superior characteristics of the gridded tube and the klystron for high frequencies. A grid is used to provide simple control of the electron beam while a cavity is used to couple the RF energy from the beam. This combination makes possible a smaller, lower cost, high-frequency,

FIGURE 23.13 60-kW IOT (Klystrode). (*Source:* Courtesy of EEV.)

high-power tube. Still required, however, are focus magnets to maintain the beam, liquid cooling at higher powers, high beam supply voltages to accelerate the electrons, and the necessity for an external circuit called a **crowbar**. A crowbar is required for an IOT due to the possibility of the tube developing an internal arc. This internal arc creates a surge of current from the stored energy of the power supply, which would damage the grid and thus must be diverted. A crowbar is a device such as a thyratron used in shunt that diverts the powers supply stored energy based on the onset of an internal arc.

Gun, Grid, and Gain

The IOT electron gun is identical to that of the klystron except a grid is inserted between the cathode and the anode to control the electrons created by the cathode. The grid bunches the gun-generated electrons, and the beam drifts through the field-free region and interacts with the RF field in an output cavity to produce the RF output. The gun resembles a klystron gun but also has a grid much like a tetrode. Thus, control of the beam is like a gridded tube and operation is similar. The grid can be used to turn on or off the beam, modulate with RF, or both. Various classes of operation are possible. The most desired is class B, which is the most efficient and linear. Frequently used is class AB, which permits some beam current with no RF drive thus minimizing the class B crossover distortions.

The IOT gain is not as high as that of the klystron. Gain is typically 20–23 dB whereas the klystron is 35–40 dB. These gains are compared to those of a tetrode, which are typically 15 dB at higher frequencies. The lack of gain is not a severe limit for the IOT since solid-state drivers at 100th the output power (20 dB less) usually can be economical. The even higher gain of the klystron simplifies the driver stages even more.

Typical Operation

The IOT and klystron can be compared in operational parameters by reviewing Table 23.1. As this data clearly shows the klystron and IOT are close cousins in the high-frequency tube family. Not shown with this data is the other important performance parameters of amplitude linearity, phase distortion, bandwidth, size, or expected life. Depending on the application, these parameters may be more important than efficiency or gain.

TABLE 23.1　Comparison of Klystron and IOT Operating Parameters

Parameter	Klystron	IOT	Units
Frequency	UHF	UHF	—
Peak power	60	60	kW
Beam voltage	28	32	kV max
Beam current	7	4	A max
Drive power	8	300	W
Efficiency	42	60	%
Collector dis.	150	50	kW max

Acknowledgments

We kindly acknowledge the assistance of Earl McCune of Varian, Robert Symons of Litton, and Michael Kirk of EEV.

Defining Terms

Anode: A positive portion of the tube that creates the field to accelerate the electrons produced by the cathode for passage down the body of the tube.

Beam current: The current that flows from anode to cathode.

Cathode: The most negative terminal of a vacuum tube; the source of electrons.

Collector: The portion of the tube that collects the electrons. The most electrically positive portion of the tube and dissipates residual the beam energy.

Crowbar: A triggered, shunt device that diverts the stored energy of the beam power supply.

Electron gun (gun): An electron emission material that is heated to produce the electron beam. Usually physically shaped and electrically controlled to form a pencil lead sized beam.

Focus coil: Coils of wire encircling the tube forming an electromagnet to focus the beam electrons. The focus coils have DC applied during operation.

Grid or modulating anode: A terminal in the tube used to control the electron beam usually placed near the cathode.

Klystrode: A hybrid tube that uses elements from the klystron and the tetrode.

References

Badger, G. 1985. The klystrode, a new high-efficiency UHF TV power amplifier. Varian, *NAB Proceedings of the Engineering Conference.*

Heppinstall, R. 1986. Klystron operating efficiencies: Is 100 percent realistic? *NAB Engineering Conference*, EEV.

McCune, E.W. 1988. UHF-TV klystron multistage depressed collector development program. NASA Report.

Nelson, B. 1963. *Introduction to Klystron Amplifiers.* Varian Associates, San Carlos, CA, April.

Staprans, A., McCune, E.W., and Ruetz, J.A. High-Power Linear-Beam Tubes. *Proceedings IEEE*, vol. 61.

Symons, R. 1982. Klystrons for UHF television. *Proceedings IEEE* (Nov.).

Symons, R.S. 1986. Scaling laws and power limits for klystrons. *IEDM Digest.*

Symons, R.S. and Vaughan, J.R.M. 1994. The linear theory of the clustered-cavity^{TM} klystron. *IEEE Transactions* (Oct.).

Varian. 1972. Notes on Klystron Operation. Varian Associates, San Carlos, CA, April.

Varian. Klystrons for Ku-band satellite ground transmitters. Varian Associates, San Carlos, CA, Pub. No. 4084.

Further Information

For further information refer to:

Klystrons and Microwave Triodes. Hamilton, Knipp, Kuper, Radiation Laboratory Series.

Practical Theory and Operation of UHF-TV Klystrons. Reginal Perkins, R&L Technical Publishers, Boston, MA.

Very High Frequency Techniques. Radio Research Laboratory, Harvard Univ. Press, Boston, MA.

23.3 Traveling Wave Tubes

Thomas K. Ishii

23.3.1 Definitions (Ishii, 1989, 1990)

A **traveling wave tube** (TWT) is a vacuum device in which traveling microwave electromagnetic fields and an electron beam interact with each other. In most **traveling wave** tubes (TWTs), the direction of traveling microwave propagation inside the tube is made parallel to and in the same direction with the **electron beam**. Microwaves propagating in the same direction as the electron beam are termed **forward waves**. The TWTs based on the interaction with forward waves and the electron beam are termed the *forward wave tubes.* Most forward wave tubes are microwave amplifiers.

To make interaction between the traveling microwave electromagnetic fields and the electron beam efficient, in a TWT, the phase velocity or propagation speed of microwaves in the tube is slowed down to approximately equal speed with the speed of electrons in the electron beam. In most practical TWTs, the speed of electrons in the beam is approximately 1/10 of the speed of light. In most TWTs, there is a transmission line in the device. Microwave energy to be amplified is fed to one end of the transmission line to which the electron beam is coupled. The transmission line is designed so that the speed of microwave propagation is also slowed down to approximately 1/10 of speed of light. For the best result, the speed of microwave propagation on the transmission line and the speed of electrons in the coupled electron beam must be approximately equal. Because the speed of microwave propagation is slowed down in the transmission line, the line is termed a **slow-wave structure**.

Thus in a TWT, microwaves and electrons travel together with approximately the same speed and interact with each other. During the interaction, the kinetic energy of the electrons is transferred or converted into microwave energy. As both microwaves and electrons are traveling together, the electrons lose their kinetic energy and the microwaves gain the energy of electromagnetic fields. Thus microwave electromagnetic fields are amplified while traveling inside the traveling wave tube. Amplified microwave energy is extracted at the other end of the transmission line.

Since a traveling wave tube does not have a resonator, if the input circuit and the output circuit are well impedance matched, a wide operating frequency bandwidth can be expected. In practice, the operating frequency bandwidth usually exceeds an octave (EEV, 1991, 1995).

The gain of a traveling wave tube is, among other things, dependent on the length of the tube; the longer the tube, the higher is the gain. Therefore, the physical shape of a traveling wave tube is long to accommodate the transmission line. The transmission line must have an input coupling circuit to couple the input microwave signal that is to be amplified into the transmission line. At the end of the transmission line, there must be another coupling circuit to take the amplified microwave out to the output circuit.

In a traveling wave tube, the electron beam is generated by an **electron gun** that consists of a hot cathode and electrostatic focusing electrodes. Electrons are generated from the heated cathode by **thermionic emission**. The launched electron beam to couple with the transmission line is focused by longitudinally applied DC magnetic fields. These **focusing** DC **magnetic fields** are provided by either a stack of permanent magnet rings or a solenoid. Beyond the end of the transmission line, there is a positive electrode to collect the used electron beam. This electrode is termed the **collector**.

A generic configuration of a traveling wave tube is shown in Figure 23.14. The electron beam is coupled to a slow-wave structure transmission line to which the microwave signal to be amplified is fed through an input waveguide and a coupling antenna. The amplified microwaves are coupled out to the output waveguide through the output coupling antenna and the coupling waveguide circuit. As seen from Figure 23.14, the electron gun is biased negatively by the use of the DC power supply and radio frequency

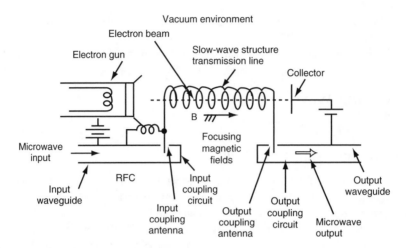

FIGURE 23.14   Generic configuration of a traveling wave tube (TWT).

choke (RFC) against the waveguide, the transmission line, and the collector. During the interaction between the electron beam and microwaves, the electron beam loses kinetic energy, and the lost energy is transferred into the electromagnetic field energy of the microwaves traveling in the transmission line.

Usually a traveling wave tube means a forward wave tube. In the vacuum tube, the electron beam and forward waves interact with each other. There is a vacuum tube in which the electron beam interacts with **backward waves**. This type of tube is termed the **backward wave tube**. The backward wave tube is inherently highly regenerative (built-in positive feedback); therefore, it is usually an oscillator. Such an oscillator is termed a **backward wave oscillator (BWO)**. The BWO has a **traveling wave** structure inside, but usually it is not called a traveling wave tube. Details of a BWO are presented in Section 23.4.3. The oscillation frequency of a BWO is dominated by the electron speed, which is determined by the anode voltage. Therefore, a BWO is a **microwave frequency voltage-controlled oscillator** (VCO).

23.3.2   Design Principles (Ishii, 1990)

By design, in a traveling wave tube, the speed of electrons in the electron beam u (m/s) and the **longitudinal phase velocity of** microwave electromagnetic field in the slow-wave structure transmission line v (m/s) are made approximately equal. When an electron comes out of the electron gun, the speed of an electron is

$$u_0 = \sqrt{\frac{2qV_a}{m}} \tag{23.1}$$

where q is the electric charge of an electron, 1.602×10^{-19} coulombs; m is the mass of an electron, 9.1085×10^{-31} kg; and V_a is the acceleration anode voltage.

But the speed of electrons changes as it travels within the beam due to interaction with traveling electromagnetic waves in the transmission line slow-wave structure. In the beginning, the electrons are velocity modulated. Some are accelerated and some are decelerated by the microwave electromagnetic field in the slow-wave transmission line. The result of **velocity modulation** is **electron bunching**. In a traveling wave tube, bunched electrons are traveling with the traveling electromagnetic waves. The speed of bunched electrons is given by

$$u = u_0 + u_m e^{-\alpha_e z} \cdot e^{-j(\beta_e z - \omega t)} \tag{23.2}$$

In this equation, a one-dimensional coordinate system z is laid along the direction of traveling waves and the electron beam. It is assumed that the interaction of electrons and traveling electromagnetic waves

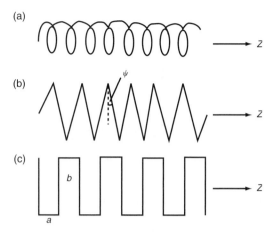

FIGURE 23.15 Simple slow-wave transmission line structures: (a) helical transmission line, (b) zig-zag transmission line, (c) interdigital transmission line.

started at $z = 0$ and the time $t = 0$. In Eq. (23.2), u_m is the magnitude of velocity modulation at the initial location, α_e the attenuation constant of the velocity modulation, β_e the phase constant of the electronic velocity modulation waves in the beam, and ω the operating microwave frequency.

Both α_e and β_e are a function of the magnitude of interacting microwave electric fields and time, which change as the electrons travel in the beam. When the TWT is amplifying, the traveling wave electric field grows as it travels. Therefore, the magnitude of the velocity modulation also grows as the electronic waves travel in the beam.

The speed of the electromagnetic waves propagating on the slow-wave structure transmission line depends on the structure of the line itself. For example, if the structure is a simple helical device as shown in Figure 23.15(a) is

$$v_z = c \sin \psi \tag{23.3}$$

where c is the speed of light in the vacuum and ψ the pitch angle of the helix. If the slow-wave structure is a simple zig-zag transmission with the angle of the zig-zag at 2ψ (as shown in Figure 23.2(b)), the longitudinal velocity of electromagnetic wave propagation in the z direction is

$$v_z = c \sin \psi \tag{23.4}$$

If the slow-wave structure is a simple interdigital circuit with the pitch dimensions a and b, as shown in Figure 23.2(c),

$$v_z = \frac{a}{a + b} c \tag{23.5}$$

These equations assume that the coupling between the electron beam and the transmission line is light. It is further assumed that the waves propagate along the conducting wire slow-wave transmission line with the speed c. In reality, the electron beam will load these transmission lines. Then the speed of the electrons in the beam will be slightly different from these equations. In an electron-beam-coupled generic transmission line

$$v_z = \frac{\omega}{\beta_c - \dfrac{1}{2}b} \tag{23.6}$$

where ω is the operating angular frequency, β_c is the phase constant of the slow-wave transmission line with the series impedance $\dot{Z}$ (Ω/m) and the shunt admittance of $\dot{Y}$ (S/m), which is

$$\beta_c = I_m \sqrt{\dot{Z}\dot{Y}} \tag{23.7}$$

and

$$b = (\dot{\gamma} - \dot{\gamma}_c)e^{j\frac{5\pi}{6}} \tag{23.8}$$

where

$$\dot{\gamma}_c = \sqrt{\dot{Z}\dot{Y}} \tag{23.9}$$

and

$$\dot{\gamma} = \left(\alpha_c - \frac{\sqrt{3}}{2}b\right) + j\left(\beta_c - \frac{1}{2}b\right) \tag{23.10}$$

$$\alpha_c = R_e\sqrt{\dot{Z}\dot{Y}} \tag{23.11}$$

and β_c is given in Eq. (23.7).

The **voltage gain** per meter of propagation in **neper** per meter and **decibel** per meter is

$$\alpha = \alpha_c - \frac{\sqrt{3}}{2}b \tag{23.12}$$

$$\alpha = 8.686\left[\alpha_c - \frac{\sqrt{3}}{2}b\right] \tag{23.13}$$

If the interaction transmission line length of the TWT is ℓ m long, then the voltage gain is

$$G = e^{\alpha\ell} \tag{23.14}$$

The **frequency bandwidth** is calculated by the following procedure. At the midband frequency ω_0

$$e^{\alpha(\omega_0)} = \exp\left\{\alpha_c(\omega_0) - \frac{\sqrt{3}}{2}b(\omega_0)\right\} \tag{23.15}$$

At the lower band edge ω^-

$$\exp\left\{\alpha_c(\omega^-) - \frac{\sqrt{3}}{2}b(\omega)\right\} = \frac{1}{\sqrt{2}}e^{\alpha(\omega_0)} \tag{23.16}$$

At the upper band edge ω^+

$$\exp\left\{\alpha_c(\omega^+) - \frac{\sqrt{3}}{2}b(\omega^+)\right\} = \frac{1}{\sqrt{2}}e^{\alpha(\omega_0)} \tag{23.17}$$

Solving Eq. (23.16) ω^- is found. Solving Eq. (23.17) ω^+ is found. Then the frequency bandwidth is

$$\Delta\omega = \omega^+ - \omega^- \tag{23.18}$$

$$\Delta f = \frac{\omega^+ - \omega^-}{2\pi} \tag{23.19}$$

The **noise figure** (NF) of a TWT ℓ-m-long interaction transmission line can be calculated by the following procedure:

$$\text{NF} = \frac{(kT\Delta fG_p + \int_0^\ell (P_s + P_p + P_f + kT_N\Delta f)dz + kT\Delta f)}{kT\Delta fG_p} \tag{23.20}$$

where T is the TWT circuit temperature of the input port and the output port, Δf the frequency bandwidth, G_p the **power gain** and

$$G_p = e^{2\alpha\ell} \tag{23.21}$$

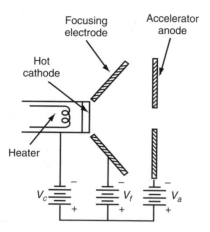

FIGURE 23.16 A schematic diagram of a generic electron gun.

and k is Boltzmann constant, 1.38054×10^{-23} J/K; P_s is the **shot noise** power per meter; P_p is the **partition noise** power per meter; P_f is the **flicker noise** per meter of propagation in the interaction transmission line section; and T_N is the **noise temperature** per meter of the interaction transmission line.

23.3.3 Electron Gun

The electron gun of a traveling wave tube is a device that supplies the electron beam to the tube. A schematic diagram of a generic electron gun is shown in Figure 23.16. A generic electron gun consists of a hot cathode heated by an electric heater, a negatively biased **focusing electrode** or **focuser** and positively biased accelerating anode. Figure 23.16 is a cross-sectional view, and the structure can be a two-dimensional or three-dimensional coaxial structure (Liao, 1988; Sims and Stephenson, 1953). If it is two dimensional, it is for a ribbon-shaped electron beam and if it is coaxial, it is for a cylindrical electron beam. The focusing of the electron beam and the amount of the beam current can be adjusted by adjusting the heater voltage V_H and the bias voltage for the cathode V_c, for the focusing electrode or focuser V_f and the accelerator anode voltage V_a.

Electron emission current density (A/m^2) can be calculated by the following **Richardson–Dushman equation** (Chodorow and Susskind, 1964):

$$J = 6.02 \times 10\, T^2 e^{-b/T} \tag{23.22}$$

where T is the absolute temperature of the cathode in kelvin, and b also in kelvin is (Chodorow and Susskind, 1964)

$$b = 1.1605 \times 10^4 \phi \tag{23.23}$$

where ϕ is the **work function** of the cathode material. If the cathode material is tungsten, $\phi = 4.5$ V. If it is tantalum, $\phi = 4.1$ V. If it is an oxide coated cathode by B_aO, S_rO or C_aO, $\phi = 1\text{–}2$ V. If it is a combination of B_aO and S_rO, then $\phi = 1.0$ V. If it is thoriated tungsten, $\phi = 2.6$ V. The operating temperature of the cathode is about $1300\sim1600$ K. Once electrons are thermally emitted from the cathode, electrons are pushed toward the center axis of the structure by the **coulomb force**

$$F_r = qE_r aa \tag{23.24}$$

where q is the electric charge of an electron 1.602×10^{-19} C and E_r is the radial electric field intensity (V/m). E_r is determined by the geometry of various electrodes and the bias voltages V_c, V_f, and V_a. However, V_f has the most direct influence on it. This negative electrode pushes the electron toward the center axis. The electrons are also pulled along the center axis by the positive **accelerator** anode. The longitudinal coulomb force acting on the electron is

$$F_z = qE_z \tag{23.25}$$

where E_z is the longitudinal electric field intensity in volt per meter.

E_z is determined by the geometry of the electrodes and the bias voltages V_c, V_f, and V_a. However, V_a has the most direct effect on pulling electrons out of the hot cathode.

The procedure to obtain the electron trajectory in the electron gun is to solve **Newton's equations of motion** for the electron. In a generic electron motion, a tangential force may be involved, if it is not balanced. It is

$$F_\phi = qE_\phi \tag{23.26}$$

where E_ϕ is a tangential electric field. If the electron gun is structured axially circularly symmetrical, then E_ϕ does not exist. Otherwise, E_ϕ must be considered. Newton's equations of motion of an electron in the electron gun are

$$m\frac{d^2r}{dt^2} = qE_r \tag{23.27}$$

$$m\frac{d^2z}{dt^2} = qE_z \tag{23.28}$$

$$m\frac{d^2(r\phi)}{dt^2} = qE_\phi \tag{23.29}$$

E_r, E_z, and E_ϕ are obtainable from the electric potential V within the electron gun

$$E_r = -\frac{\partial V}{\partial r} \tag{23.30}$$

$$E_z = -\frac{\partial V}{\partial z} \tag{23.31}$$

$$E_\phi = -\frac{1}{r}\frac{\partial V}{\partial \phi} \tag{23.32}$$

The electric potential V within the electron gun is obtainable by solving a Laplace equation with specifically given boundary conditions,

$$\frac{1}{r}\frac{\partial}{\partial r}\left(r\frac{\partial V}{\partial r}\right) + \frac{1}{r^2}\frac{\partial^2 V}{\partial \phi^2} + \frac{\partial^2 V}{\partial z^2} = 0 \tag{23.33}$$

An axially symmetrical solid cylindrical electron beam is produced by the gun structure illustrated in Figure 23.16, if the structure is axially cylindrically symmetrical. If the middle of the hot cathode is made nonemitting and only the edge of the cathode is emitting, then the cathode becomes an **annular cathode**. The annular cathode produces a **hollow beam**. The annular electron beam can save the electron beam current for equal microwave output.

If the gun structure shown in Figure 23.16 is two dimensional, then it produces a ribbon-shaped electron beam. A ribbon-shaped beam is used for a TWT of a two-dimensional structure.

If the angle of the focusing electrode against the axis of the electron beam is 67.5° and the anode is also tilted forward to produce a rectilinear flow, which is an electron flow parallel to the z axis in Figure 23.16, then, such electron gun is termed the **Pierce gun**.

In practice the hot cathode surface is curved as shown in Figure 23.17 to increase the electron-emitting surface and to obtain a high-density electron beam.

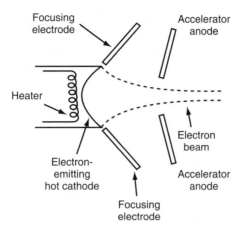

FIGURE 23.17 Cross-sectional view of an electron gun with curved hot cathode.

23.3.4 Electron Beam Focusing

Electrons in an electron beam mutually repel each other by the electron's own coulomb force due to their own negative charge. In addition, the electron beam usually exists in proximity to the positively biased slow-wave structure as seen in Figure 23.14. Therefore, the electron beam tends to diverge. The process of confining the electron beam within the designed trajectory against the mutual repulsion and diverging force from the slow-wave structure is termed *electron beam focusing*.

The electron beam in a TWT is usually focused by a DC magnetic flux applied parallel to the direction of the electron beam, which is coaxial to the slow-wave transmission line as illustrated in Figure 23.14. If there is no focusing magnetic flux only at the output of the electron gun, then such electron flow in the beam is termed **Brillouin flow** (Hutter, 1960). If the electron gun itself is exposed to and unshielded from the focusing para-axial longitudinal magnetic flux, then the electron flow produced is termed the **immersed flow** (Hutter, 1960). If the electron gun is not shielded from focusing magnetic flux, which is not para-axial, then the electron flow produced in such a focusing system is neither Brillouin flow nor immersed flow, but it is a *generic flow*.

In a generic flow device, the focusing force or equation of motion of an electron in the electron beam is

$$m\frac{\partial^2 r}{\partial t^2} = mr\left(\frac{\partial\phi}{\partial t}\right)^2 + qr\frac{\partial\phi}{\partial t}B_z + \frac{qr_0^2\rho_0}{2\epsilon_0 r} + qE_r \qquad (23.34)$$

where r_0 is the radius of a part of the electron beam that is repelling the electron in question, (r, ϕ, z) is the cylindrical coordinate of the electron in question, q is the charge of the electron, which is 1.602×10^{-19} C. B_z is the z component of applied magnetic flux density, m is the mass of the electron, which is 9.109×10^{-31} kg. Also, ρ_0 is the electronic charge density of the electron beam, and ϵ_0 is the **permittivity** of the vacuum, which is 8.854×10^{-12} F/m.

The angular velocity of the electron trajectory $\partial\phi/\partial t$ may be calculated from the following equation of **Busch's theorem** (Hutter, 1960; Chodorow and Susskind, 1964; Harman, 1953):

$$\frac{\partial\phi}{\partial t} = \frac{\eta}{2\pi r^2}(\psi - \psi_c) \qquad (23.35)$$

where η is a constant q/m, which is 1.759×10^{11} C/kg,

$$\psi = \pi r^2 B_z \qquad (23.36)$$

$$\psi_c = \pi r_c^2 B_z \qquad (23.37)$$

where r_c is the radial position of the electron in question where it first encountered the focusing magnetic flux.

If the electron beam is the immersed flow or Brillouin flow, then

$$m\frac{\partial^2 r}{\partial t^2} = mr\left(\frac{\partial \phi}{\partial t}\right)^2 + qr\frac{\partial \phi}{\partial t}B_z + \frac{qr_0^2\rho_0}{2\epsilon_0 r} + qE_r \tag{23.38}$$

Focusing conditions are

$$\left.\frac{\partial r}{\partial t}\right|_{r=a} = 0 \tag{23.39}$$

$$\left.\frac{\partial^2 r}{\partial t^2}\right|_{r=a} = 0 \tag{23.40}$$

where a is the radius of the electron beam.

In practice, the focusing magnetic flux is created by either a solenoid or a structure of permanent magnetic rings.

23.3.5 Electron Velocity Modulation (Ishii, 1990)

If there are no microwaves on the slow-wave transmission line of a TWT, the velocity of an electron in the electron beam emitted from the electron gun is constant and uniform. As seen from Figure 23.1, there is practically no longitudinal DC electric field in the transmission line section. Therefore, the electrons are just drifting with the speed given by the electron gun, which is

$$u_0 = \sqrt{\frac{2qV_a}{m}} \tag{23.41}$$

where q is the charge of the electron, m is the mass of the electron, and V_a is the accelerating electric potential of the electron gun. Now, if microwaves are launched on the slow-wave transmission line from the input as illustrated in Figure 23.14 then there will be interaction between electrons in the electron beam and microwave electromagnetic fields on the slow-wave transmission line. In a TWT, it is designed that the electron velocity u_0 and the longitudinal propagation velocity of microwaves on the slow wave structure v_z are made approximately equal to each other.

$$u_0 \approx v_z \tag{23.42}$$

Therefore, electrons under the acceleration field due to traveling microwaves are kept accelerating and other electrons under the decelerating electric field of the traveling microwaves are kept decelerating. The longitudinal electric field component of traveling microwaves on the slow-wave transmission line is expressed by

$$E_z = E_0 e^{-\alpha z} e^{-j(\beta z - \omega t)} \tag{23.43}$$

where z is a one-dimensional coordinate set along the slow-wave transmission line and $z = 0$ is the point where microwaves are launched. E_0 is the magnitude of the longitudinal component of the microwave electric field at $z = 0$, α is the **attenuation constant** of the line, and β is the phase constant of the line. Depending on the phase angle of E_z at specific time t and location z, which is $(\beta z - \omega t)$, the electron may be kept accelerating or may be kept decelerating. This acceleration and deceleration of electrons due to applied microwave electric fields is termed the velocity modulation of an electron beam. The modulated electron velocity is

$$u = u_0 + u_m \exp\{-\alpha_e z - j(\beta_e z - \omega t)\} \tag{23.44}$$

where u_m is the magnitude of velocity modulation of the electron velocity. Here, α_e is the attenuation constant and β_e is the phase constant of the propagation of the velocity modulation on the electron beam.

23.3.6 Electron Bunching (Ishii, 1989)

If the speed of electrons in a focused electron beam is velocity modulated, as seen in Eq. (23.44), then the electron flow is no longer uniform and smooth. Some electrons in the beam are faster or slower than others, and as a result electrons tend to *bunch* together as they travel from the electron gun in the beginning to the electron **collector** at the end. The traveling electron charge density in the velocity modulated electron beam is expressed by

$$\rho = \rho_0 + \rho_m e^{-\alpha_e z} e^{-j(\beta_e z - \omega t)} \tag{23.45}$$

where ρ_0 is the electron charge density of an unmodulated electron beam, ρ_m is the magnitude of charge density modulation as the result of the velocity modulation of the electrons, α_e is the attenuation constant of the electron charge wave, and β_e is the phase constant of the wave.

The effect of electron bunching or propagation of electron charge waves in the electron beam is observed in a form of an electron beam current. If the cross-sectional area of the electron beam is S, (in m^2), then the electron beam current that contains bunched electrons is

$$I = I_0 + I_1 e^{-\alpha_e z} \cdot e^{-j(\beta_e z - \omega t)} + I_2 e^{-2\alpha_e z} \cdot e^{-2j(\beta_e z - \omega t)} \tag{23.46}$$

where I_0 is the DC beam current or unmodulated electron beam current, I_1 is the magnitude of beam current modulation at the frequency of microwaves ω, and I_2 is the magnitude of beam current modulation at the second harmonic frequency of operating frequency,

$$I_0 = S\rho_0 u_0 \tag{23.47}$$

$$I_1 = S(\rho_0 u_m + \rho_m u_0) \tag{23.48}$$

and

$$I_2 = S\rho_m u_m \tag{23.49}$$

23.3.7 Electron Beam Coupling

As seen from Figure 23.14, to give electrons velocity modulation, microwave energy at or in proximity to the input must couple to the electron beam. To extract microwave energy from the bunched electrons at or in proximity to the output, the energy of the bounched electrons must be coupled to the slow-wave transmission line. This coupling of energy between the electron beam and the slow-wave transmission line is termed the **electron beam coupling**.

The strength of the beam coupling is represented by the beam coupling coefficient, which is defined as

$$k = \frac{I_c}{I_1} \tag{23.50}$$

where I_c is the microwave current in the slow-wave transmission line at operating frequency and I_1 is the magnitude of microwave beam current in the electron beam at the operating frequency ω. The beam coupling coefficient for a helical slow-wave line is obtained by

$$k = \frac{\pi u_0^2}{\omega \left(u_m \dfrac{\rho_0}{\rho_m} + u_0 \right) \sin \psi} \tag{23.51}$$

If the slow-wave transmission line is not a helix line, then the slow-wave ratio (less than 1) must be used instead of $\sin \psi$ where ψ is the **pitch angle of the helix**.

23.3.8 Power Gain

As microwaves propagate along the slow wave transmission line, they receive energy from the bunched electron beam. The loss of kinetic energy due to the microwave deceleration is the gain of microwave energy. Microwave power increase per meter of propagation is given by

$$\frac{dP}{dz} = \frac{d}{dz}\left(\frac{k^2}{2}mu^3NS\right) = 2\alpha P \tag{23.52}$$

where P is microwave power flow on the transmission line, z is a one-dimensional coordinate set up along the slow-wave transmission line, k the beam coupling coefficient, m the mass of an electron, u the velocity of the electron, N the density of electrons in the beam, S the cross-sectional area of the beam, and α the gain parameter in neper/m that represents the neper power increase on the transmission line per meter of propagation.

The **gain parameter** α in neper per meter is given by (Hutter, 1960)

$$\alpha = \frac{\sqrt{3}}{2}\beta_e\left\{Z_0\frac{I_0}{4V_a}\right\}^{\frac{1}{3}} \tag{23.53}$$

where Z_0 is the **characteristic impedance** of the slow-wave transmission line, I_0 the DC beam current, V_a the DC acceleration anode voltage, and β_e the phase constant of the electron charge wave.

The output power P_0 of a traveling wave tube that is the length of the interaction of the slow-wave transmission line ℓ m long is

$$P_0 = P_ie^{2\alpha\ell} \tag{23.54}$$

where P_i is the input power to the slow wave transmission line. Therefore, the power gain in number is

$$G = \frac{P_0}{P_i} = e^{2\alpha\ell} \tag{23.55}$$

The power gain in neper is

$$G = \ell n\frac{P_0}{P_i} = 2\alpha\ell \tag{23.56}$$

The power gain in decibel is

$$\begin{aligned} G = 10\log_{10}\frac{P_0}{P_i} &= 10\log_{10}e^{2\alpha\ell} \\ &= 8.686\,\alpha\ell \end{aligned} \tag{23.57}$$

23.3.9 Frequency Bandwidth

The gain parameter of a traveling wave tube is frequency independent. As seen in Eq. (23.53)

$$\beta_e = \frac{\omega}{u_0} \tag{23.58}$$

where u_0 is the DC electron velocity, and

$$Z_0 = \sqrt{\frac{R + j\omega L}{G + j\omega C}} = \sqrt{\frac{Z}{Y}}e^{j\frac{\phi-\psi}{2}} \tag{23.59}$$

where R, L, G, and C are resistance, inductance, conductance, and capacitance per meter of the slow-wave transmission line and

$$Z = \sqrt{R^2 + (\omega L)^2} \tag{23.60}$$
$$Y = \sqrt{G^2 + (\omega C)^2}$$

$$\phi = \tan^{-1} \frac{\omega L}{R} \tag{23.61}$$

$$\psi = \tan^{-1} \frac{\omega C}{G}$$

and

$$\alpha = \frac{\sqrt{3}}{2} \frac{\omega}{u_0} \left\{ \frac{Z(\omega)}{Y(\omega)} \frac{I_0}{4V_a} \right\}^{\frac{1}{3}} \cos \frac{\phi(\omega) - \psi(\omega)}{6} \tag{23.62}$$

If the maximum gain occurs at $\omega = \omega_0$, then the maximum power gain is

$$G(\omega_0) = e^{2\alpha(\omega_0)\ell} \tag{23.63}$$

If the power falls to $\frac{1}{2} \epsilon^{2\alpha(\omega_0)\ell}$ at $\omega = \omega_+$ and $\omega = \omega_-$, then

$$G(\omega_\pm) = \frac{1}{2} e^{2\alpha(\omega_0)\ell} \tag{23.64}$$

Solving Eq. (23.64) for both ω_+ and ω_-, the frequency bandwidth is obtained as

$$\Delta f = \frac{\omega_+ - \omega_-}{2\pi} \tag{23.65}$$

Equation (23.64) is usually written as transcendental equation:

$$\exp\left[\frac{\sqrt{3}\omega_\pm}{u_0} \left\{ \frac{Z(\omega_\pm)}{Y(\omega_\pm)} \frac{I_0}{4V_a} \right\}^{\frac{1}{3}} \cos \frac{\phi(\omega_\pm) - \psi(\omega_\pm)}{6} \right]$$
$$= \frac{1}{2} \exp\left[\frac{\sqrt{3}\omega_0}{u_0} \left\{ \frac{Z(\omega_0)}{Y(\omega_0)} \frac{I_0}{4V_a} \right\}^{\frac{1}{3}} \cos \frac{\phi(\omega_0) - \psi(\omega_0)}{6} \right] \tag{23.66}$$

23.3.10 Noise

As seen from Eq. (23.20), the **noise** output of a TWT with the interaction length of the electron beam and the slow-wave transmission line ℓ, power gain G_p, frequency bandwidth Δf, and temperature of both the input and the output circuits $T(\text{K})$ is

$$P_N = kT\Delta f G_p + \int_0^\ell (P_s + P_p + P_F + kT_N\Delta f)dz + kT\Delta f \tag{23.67}$$

Other symbols are as defined after Eq. (23.20). Here, $kT\Delta f$ is the thermal noise input power, also, $kT\Delta f$ is the **thermal noise** generated at the output circuit. Boltzmann's constant k is 1.38×10^{-23} J/K.

The shot noise power in the beam coupled out to the transmission line per meter is

$$P_s = 2\xi q I_0 \Delta f \tag{23.68}$$

where q is the electron charge, I_0 the DC beam current, and ξ the power coupling coefficient between the beam and the transmission line.

The flicker noise P_F is due to random variation of **emissivity** of the cathode and

$$P_F = \eta \frac{1}{f} \tag{23.69}$$

where f is the operating frequency and η a proportionality constant.

The partition noise P_p is generated by electron hitting the slow-wave transmission line and is proportional to the speed of the electron u and the beam current I_0,

$$P_p = \zeta I_0 u \tag{23.70}$$

The noise figure of a TWT is presented with Eq. (23.20).

23.3.11 Sensitivity

The amount of the input power that makes the signal-to-noise ratio at the output of a TWT unity is termed the **tangential sensitivity** of the TWT. The output power of the TWT is

$$P_0 = P_i G_p \tag{23.71}$$

where P_i is the input power and G_p is the power gain of the TWT.

For the tangential sensitivity

$$P_0 = P_N \tag{23.72}$$

where P_N is the noise output power presented by Eq. (23.67). For the tangential sensitivity P_s, $P_i = P_s$ and

$$P_i = P_s \quad \text{and} \quad P_s G_p = P_N \tag{23.73}$$

Combining Eq. (23.73) with Eq. (23.67),

$$P_S = \frac{P_N}{G_P} = kT\Delta f + \frac{1}{G_p}\left[\int_0^\ell (P_s + P_p + P_F + kT_N \Delta f)dz + kT\Delta f\right] \tag{23.74}$$

For a generic **sensitivity**

$$P_0 = nP_N \tag{23.75}$$

where n is an arbitrary specified number.

Since

$$P_0 = nP_N \tag{23.76}$$

$$P_S = \frac{P_N}{G_P} = kT\Delta f + \frac{1}{G_p}\left[\int_0^\ell (P_s + P_p + P_F + kT_N \Delta f)dz + kT\Delta f\right] \tag{23.77}$$

The value of n can be

$$n \gtreqless 1 \tag{23.78}$$

23.3.12 Collector

As seen from Figure 23.14, the electron beam emitted from the electron gun goes through the interaction region with the microwave slow-wave transmission line and hits the electron collector at the end. Various configurations of the electron collector are shown schematically in Figure 23.18. Collectors are hit by high-energy electrons of several thousand electron volts or more (EEV 1991, 1995). Collectors, therefore, often need forced air or water cooling. In practical TWTs, collectors and transmission lines are biased to the ground potential or near ground potential and the electron gun or cathode are biased to a high negative potential for operational security reasons.

For low-power TWTs, a flat collector as shown in Figure 23.18(a) is adequate. To prevent electrons from returning to the transmission line, the collector is sometimes biased slightly positively. If electrons have sufficient kinetic energy and no chance of going back to the interaction region, then the collector may be biased negatively to make the electrons soft landing on the collector to prevent overheating the collector.

Regardless, the flat corrector is not adequate for a high-power TWT. The cone-shaped collector as shown in Figure 23.18(b) has more electron collection area than the flat collector shown in Figure 23.18(a). Therefore, it reduces collector current density on the collector. This reduces the heating problem of the collector. A larger electron collection area can be obtained by making the collector configuration a curved cone shape, shown in Figure 23.18(c), or a cylindrical shape, shown in Figure 23.18(d) to reduce the heating of the collector by electron bombardment.

For a high-power TWT, the electron energy may be on the order of tens of thousand electron volts (EEV 1991, 1995). Typically, electrons in the collector region are gradually slowed down for soft landing to prevent excess heating of the collector structure. In Figure 23.18(e) (Liao, 1988) the cylindrical collector is negatively biased and electrons are pushed back and collected by a grounded plate at the end of the interaction region. The electron collection region may be expanded further as illustrated in Figure 23.18(f) (Liao, 1988). The electron collection is accomplished in two stages, as seen from the figure. For an extremely high-power operation a three-stage collector, as shown in Figure 23.18(g) (Liao, 1988), is employed. Each stage is progressively negatively biased for soft landing of electrons to prevent excess heating of the collector structure.

The energy of an electron at the collector is approximately,

$$w = qV_a \tag{23.79}$$

where q is the charge of an electron and V_a is the acceleration voltage. Therefore, for complete soft landing, the collector must provide

$$w_c = -qV_a = qV_c \tag{23.80}$$

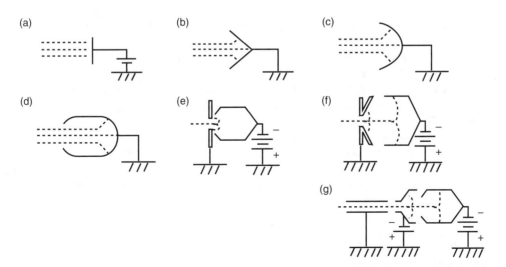

FIGURE 23.18 Schematic cross-sectional view of various collector configurations for a TWT: (a) plate, (b) cone, (c) curved cone, (d) cylinder, (e) depressed potential cylinder, (f) two stage, and (g) three stage.

where V_c is the collector voltage

$$V_c = -V_a \tag{23.81}$$

For any specified soft landing

$$V_c = -bV_a \tag{23.82}$$

where b is a specified constant that is less than unity.

23.3.13 Practical Examples

Traveling wave tubes are amplifiers. They are used for microwave communications, radar, industrial and biomedical sensing, and scientific research, as exemplified by the radio telescope. TWT are used wherever wide frequency band microwave amplification is needed. They are widely used for multichannel telephone and television channel transmission. They are also useful for short pulsed radar amplification and high data rate data transmission. A TWT can be used as transmitter amplifier of high power, or it can be used as a receiver amplifier of high sensitivity. A TWT may be for continuous wave operation, or for pulsed operation. Depending on the application, there are a great variety of TWTs in terms of power, voltage, current, mode of operation, size and weight.

Defining Terms

Accelerator: A positive electrode in a vacuum tube to accelerate emitted electrons from its cathode by coulomb force in a desired direction.

Annular cathode: A cathode of a vacuum tube with the shape of the emitting surface of the cathode is annular. The annular cathode can produce a hollow electron beam.

Attenuation constant: A constant representing attenuation of magnitude of propagating quantity such as the electromagnetic waves and acoustic waves.

Backward wave: Electromagnetic wave that travels in the direction opposite to the direction of electrons in the coupled electron beam.

Backward wave oscillator (BWO): In which the electron beam and microwaves travel in opposite direction to each other.

Backward wave tube: A backward wave TWT; the microwave input is located in proximity to the collector and the microwave output is located in proximity to the electron gun. The direction of traveling microwave and direction of electron motion in the beam are opposite to one another.

Brillouin flow: A stream of electron beam emitted from an electron gun that is not exposed to a focusing magnetic field.

Busch's theorem: Angular velocity of an electron in a magnetically focused electron beam is proportional to the difference of magnetic flux at the electron beam cross section, where the electron first encounters the focusing magnetic field, and at the cross section that includes the observation point. It is inversely proportional to the beam cross-sectional area at the observation point. The proportionality constant is 0.879×10^{11} C/kg.

Characteristic impedance: Ratio of a transmission line voltage of a forward voltage traveling wave to the accompanying transmission line current of a forward current traveling wave.

Collector: An electron collector; a positive electrode against the cathode. It is located at the final termination of the electron beam in a vacuum tube to collect electrons finished by the electronic process of the vacuum tube.

Coulomb force: Electric force exerted on an electrically charged body, which is proportional to the amount of the charge and the electric field strength in which the charged body is placed.

Decibel (dB): Ten times the common logarithm of the power ratio or 20 times the common logarithm of voltage ratio.

Duty cycle: In a periodical on-off system, ratio of the on time duration to the entire period of time of repetition.

Electron beam: Collimated and focused stream of electrons.

Electron beam coupling: Electromagnetic coupling between an electron beam and an electric circuit or a transmission line.

Electron bunching: Gathering of electrons due to both the velocity modulation and the drifting of electrons.

Electron gun: An electronic structure that contains a thermionic cathode, an accelerator, and a focusing electrode to emit an electron beam.

Electronic countermeasure: Electronic equipment, system, or procedure to sense, detect, disturb, or destroy a transmitter by detecting the transmitter signals.

Emissitivity: Ratio of emission current to the anode voltage under the given temperature of the cathode.

Focuser: A focusing electrode for an electron steam in a vacuum tube.

Focusing electrode: A special purpose electrode or an assembly of electrodes negatively charged to focus the electron beam by pushing electrons in desirable directions.

Focusing magnetic fields: Moving electrons are pushed by applied magnetic force or Lortentz force, when the direction and strength of magnetic fields are properly adjusted. It is possible to focus an electron beam in a desired direction. Such magnetic fields are termed the focusing magnetic fields.

Forward waves: In a TWT, microwaves propagate in the same direction with electrons in the electron beam. Electromagnetic waves travel in the positive direction of the coordinate system.

Frequency bandwidth: A range of frequencies within which the variation of the gain of the device is less than 3 dB.

Flicker noise: Electrical noise current in an electron stream due to stochastic variation of a heated cathode surface; it is inversely proportional to operating frequency of interest, the $1/f$ noise.

Gain parameter: Neper gain constant per meter of propagation.

Hollow beam: An electron beam with hollow core.

Immersed flow: A flow of electrons emitted from an electron gun exposed to the focusing magnetic fields.

Longitudinal phase velocity: Phase velocity of microwave propagation in the axial direction of a slow-wave structure of a TWT.

Microwave voltage controlled oscillator (MVCO): A backward wave oscillator (BWO) is an example.

Neper: A natural logarithm of a ratio.

Newton's equation of motion: The force is equal to a product of the mass and the acceleration.

Noise: Electrical noise that is unwanted and/or undesirable electrical variations.

Noise temperature: For a passive device, it is the ratio of noise power to the product of Boltzmann's constant and operating frequency bandwidth. For an amplifier, it is the ratio of output noise power to a product of Boltzmann's constant, power gain, and operating frequency bandwidth.

Noise figure: Ratio of actual noise output of an actual two-port network to fictitious ideal noise output of an ideal network that does not produce noise within the circuit but has transferable noise from the input. The ratio of the input S/N ratio to the output S/N ratio.

Permittivity: A ratio of the electric flux density to the electric field strength.

Partition noise: Electrical noise generated within a vacuum tube when the electron stream hits obstacles and is divided.

Pierce gun: An electron gun with which the focusing electrode is tilted 67.5° from the axis of the electron beam and the accelerator anode is also tilted forward to produce a rectilinear flow of an electron beam.

Pitch angle of a helix: An angle between a tangent to the helix and another tangent to a cylinder that contains the helix and is perpendicular to the cylinder axis at a common tangential point on the helix.

Power gain: A ratio of the output power to the input power. The ratio can be expressed by a number, in decibel, or in neper.

Richardson–Dushman equation: An equation of electron saturation emission current density from a heated cathode. It states that the emission electron current density is proportional to the square of the cathode temperature and an exponential of negative constant times a reciprocal of the absolute temperature of the cathode. The proportionality constant is 6.02×10^{-3} and the negative constant depends on the cathode material.

Sensitivity: Amount of input signal power to a two-port network that produces a designed signal-to-noise ratio at the output.

Shot noise: Electrical noise presents an electron current due to stochastic modulation of electronic emission from the source.

Slow-wave structure: A short microwave transmission line in a TWT in which the longitudinal phase velocity of traveling microwave is slowed down to almost equal speed of electrons in the interacting electron beam of the tube.

Tangential sensitivity: Amount of input signal power to a two-port network to produce the output signal-to-noise power ratio unity.

Thermal noise: Electric noise power that is due to heat. It is equal to the product of Boltzmann's constant, the absolute temperature of the subject, and the operating frequency bandwidth.

Thermionic emission: Electron emission from a heated cathode.

Traveling wave: Electromagnetic waves propagating into a specified direction coherently.

Traveling wave tube (TWT): A microwave amplifier vacuum tube that is based on interaction between the traveling microwave and the electron beam.

Velocity modulation: A process that induces variation of velocities to electrons in an electron beam with otherwise uniform velocity.

Voltage gain: Ratio of the output voltage of an amplifier to the input voltage to the amplifier. The voltage gain is expressed by a number, in decibel, or in Neper.

Work function: Amount of energy necessary to take out an electron from a material.

References

Chodorow, M. and Susskind, C. 1964. *Fundamentals of Microwave Electronics.* McGraw-Hill, New York.
EEV. 1991. *Microwave Products.* Chelmsford, England.
EEV. 1995. *Traveling Wave Tubes and Amplifiers.* Chelmsford, England.
Harman, W.W. 1953. *Fundamentals of Electronic Motion.* McGraw-Hill, New York.
Hutter, R.G.E. 1960. *Beam and Wave Electronics in Microwave Tubes.* D. Van Nostrand, Princeton, NJ.
Ishii, T.K. 1989. *Microwave Engineering.* Harcourt Brace Jovanovich, San Diego, CA.
Ishii, T.K. 1990. *Practical Microwave Electron Devices.* Academic Press, San Diego, CA.
Liao, S.Y. 1988. *Microwave Electron Tube Devices.* Prentice-Hall, Englewood Cliffs, NJ.
Sims, G.D. and Stephenson, I.M. 1963. *Microwave Tubes and Semiconductor Devices.* Interscience, London.

Further Information

Additional information on the topic of TWTs is available from the following sources.

IEEE Electron Device Meeting Digest.

IEEE Transactions on Electron Devices, published by the Institute of Electrical and Electronic Engineers Inc., 345 East 47th St., New York, NY 10017.

Manufacturers' catalogs such as from EEV, Hughes, Thomson Tubes Electroniques, Varian, and NEC.

23.4   Other Microwave Vacuum Devices

Thomas K. Ishii

23.4.1   Introduction

Thus far in Sections 23.1–23.3 planar microwave tubes, **klystrons**, and TWTs in microwave vacuum devices have been covered. There are a great number of other microwave devices available, and it is almost impossible to cover every available microwave vacuum device in detail. Therefore, in this chapter, some selected microwave vacuum devices are presented in some detail. These microwave devices include **magnetrons, backward wave oscillators (BWO), carcinotrons, twystrons, laddertrons, ubitrons, gyrotrons, helical beam tubes, amplitrons,** and **platinotrons** (Ishii, 1989).

23.4.2 Magnetrons

Magnetrons are used for high-power applications. The device is widely used for radar transmitters and as a power source for microwave heating processes. Though one of the most common types of magnetron is a radial magnetron, the linear magnetrons and inverted magnetrons may also be used depending on the applications. Both magnetrons and inverted magnetrons are oscillators, and the linear magnetron is an amplifier.

Schematic diagrams of cross-sectional views of linear magnetrons are shown in Figure 23.19(a) and Figure 23.19(b). In the **O-type linear magnetron** in Figure 23.19(a), the **electron beam** emitted from the **electron gun** is focused by a longitudinally applied DC magnetic flux density B as is the case in the TWT. In Figure 23.19(a), both the anode and the **collector** are electropositive against the electron gun and the **sole**. The sole is a nonemitting cathode. In practice, for safety reasons, the anode block is grounded and the cathode and the sole are biased to a high negative voltage. In this biasing arrangement, the elementary trajectory of an electron in an O-type linear magnetron is a helix. But as a whole, it is considered to be a uniform flow electron beam if there is no microwave excitation. As seen from Figure 23.19(a), the anode block has a number of slots. These slots are cut one-quarter wavelength deep. The slots are quarter-wavelength cavity resonators. The structure is a series of microwave **cavity resonators** coupling to an electron beam. This is similar to the multicavity klystron presented in Chap. 31. Therefore, the principles of a **linear magnetron** are similar to the principles of a multicavity klystron. The input microwaves impart velocity modulation to the electron beam and distribute microwave energy to the first cavity and following cavities. The **velocity-modulated** electrons are bunched, and the tightly bunched electrons induce a large amount of amplified microwave energy at the output cavity. The energy in the

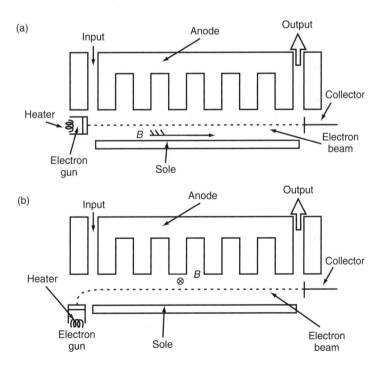

FIGURE 23.19 Cross-sectional view of linear magnetrons: (a) O-type linear magnetron, (b) M-type linear magnetron.

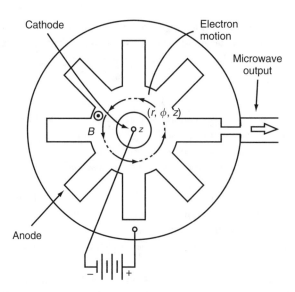

FIGURE 23.20 Cross-sectional view of a radial linear magnetron.

output is coupled to the output circuit. Usually a linear magnetron is an amplifier of high gain, but it is also an amplifier of a narrow frequency bandwidth.

When the orientation of the electron gun and the direction of focusing magnetic flux density B are changed, as shown in Figure 23.19(b), the magnetron is termed as **M-type linear magnetron.** In this arrangement, the elementary trajectory of an electron in the beam is a cycloid.

If the slots in the anode are not resonating, the tube is no longer a linear magnetron. If many nonresonating slots are cut in the anode block, then the device is a TWT. This type of TWT is termed a **crossed field amplifier (CFA)** or **crossed field device (CFD).**

If the anode block of M-type linear magnetron is bent and made re-entrant, in which the end of the structure is connected to the beginning of the structure as shown in Figure 23.20, then the magnetron is termed a **radial magnetron.** In the radial magnetron, there is no nonemitting sole. The cathode is an emitting hot cathode. Electrons emitted from the hot cathode are pulled toward the anode by the radial electric field. Therefore, the radial force is expressed in a cylindrical coordinate coaxially set up with the radial magnetron as shown in Figure 23.20. Ishii (1989) is

$$m\left\{\frac{\partial^2 r}{\partial t^2} - r\left(\frac{\partial \phi}{\partial t}\right)^2\right\} = q\left\{\frac{\partial V}{\partial y} - Br\frac{\partial \phi}{\partial t}\right\} \tag{23.83}$$

where (r, ϕ, z) is a cylindrical coordinate of an electron in a radial magnetron, m is an electronic mass 9.1091×10^{-31} kg, q is the electronic charge 1.6021×10^{-19} C, V is the electric potential in volts at (r, ϕ, z), and B is the DC magnetic flux density in tesla at (r, ϕ, z). The first term of Eq. (23.83) is the radial force toward the anode, which inwardly counteracted the centripetal force in the second term of Eq. (23.83). The third term is the electric radial force pushing the electron toward the anode and the fourth term of Eq. (23.83) is the magnetic force counteracting inwardly against the electric force.

The tangential force on the electron at (r, ϕ, z) is (Ishii, 1989)

$$r\left\{q\frac{1}{r}\frac{\partial V}{\partial \phi} + qB\frac{\partial r}{\partial t}\right\} = \frac{\partial}{\partial t}\left(rmr\frac{\partial \phi}{\partial t}\right) \tag{23.84}$$

The first term on the left-hand side of Eq. (23.84) is the torque due to the tangential electric force of an electron at (r, ϕ, z) against the center axis. The second term on the left-hand side of Eq. (23.84) is the torque against the center axis due to the tangential magnetic force acting on the electron moving radially. The right-hand side of Eq. (23.84) is the time rate of change of the torque of the electron at (r, ϕ, z) against the center axis due to tangential motion of the electron.

The energy of the electron at (r, ϕ, z) is (Ishii, 1989)

$$\frac{1}{2}m\{\dot{r}^2 + (r\dot{\phi})^2\} = qV \tag{23.85}$$

The first term on the left-hand side is the radial **kinetic energy** and the second term is the tangential kinetic energy. The right-hand side of Eq. (23.85) is the electric potential energy of the electron at the specific location (r, ϕ, z). Therefore, the electric potential V in Eqs. (23.83–23.85) should contain a solution of **Poisson's equation** and a solution of **Helmhortz's wave equation**. Since the work is very involved, however, most designers use a solution of **Laplace's** equation only for the first approximation. Solving Eqs. (23.83–23.85), the trajectory of an electron in a radial magnetron will be known. As a result, electrons emitted from the cathode surface in all directions circle around the cathode forming an **electron cloud**. The relationship between the electron cloud and the cavities in the anode of a radial magnetron is similar to the electron beam in the M-type linear magnetron to its cavity resonators. A small microwave voltage induced at the opening of the cavity resonator generated from background electrical noise, such as **thermal noise**, **shot noise**, **partition noise**, and **flicker noise**, produces velocity modulation to the circulating electron cloud. Because of this velocity modulation, **electron bunching** is produced. The bunched electrons in a radial magnetron are termed the **electron pole**. In most magnetrons, the number of electron poles formed is

$$n = \frac{N}{2} \tag{23.86}$$

where N is the number of cavity resonators in the anode block of the radial magnetron.

As the electron poles rotate around the cathode, the electron bunching gets tighter. The high-density electron pole produces a high microwave voltage across the opening or the gap of the cavity resonator. Thus, if an output transmission line such as a coaxial line or a waveguide is connected to one of the cavity resonators as shown in Figure 23.20, the induced microwave voltage is coupled to the outside of the magnetron. Thus, a radial magnetron is inherently an oscillator. On the other hand, a linear magnetron is inherently an amplifier.

A radial magnetron is, if the axial magnetic field is not applied, just a vacuum diode. If an anode voltage is applied, the anode current flows. If the axial magnetic field is applied, the anode current keeps flowing approximately the same amount as initially, as long as the anode voltage is kept constant. Then, at a certain value of magnetic flux density B_c, the anode current will drop to extremely small value. This indicates that the magnetic force has become so large that the **electron orbit** has been bent excessively. Electrons cannot reach the anode with this amount of magnetic force. The magnetic flux density that causes the anode current to drop sharply is termed the **cutoff magnetic flux density**. Magnetron oscillation is frequently found near the cutoff magnetic flux density. Oscillation under extremely small anode current is one of the features of a radial magnetron that provides high efficiency. The relationship among the anode voltage V_a, the cutoff magnetic flux density B_c, the radius of the cathode r_c, and the radius of the anode V_a is given by (Ishii, 1989)

$$B_c = \sqrt{45.5V_a}\frac{r_a}{r_a^2 - r_c^2} \tag{23.87}$$

The high efficiency of radial magnetrons stems from this low operating anode current as well as the **re-entrant use** of electron poles. In practice, whenever the high efficiency and high-power generation of microwave power is considered, magnetrons are the first to be considered. Radar and industrial and domestic microwave ovens are examples of applications of radial magnetrons.

In the radial magnetron shown in Figure 23.20, the output waveguide is connected to only one of the cavity resonators in the anode block. The output transmission line can be connected to all of the cavity resonators if the cavity resonators form a coaxial configuration with the outer jacket to which the output

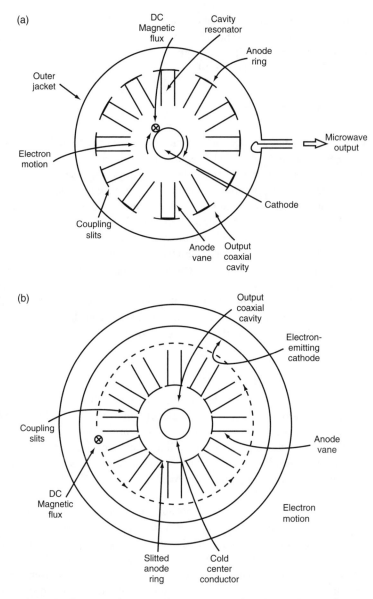

FIGURE 23.21 Schematic diagrams of a co-axial magnetron and an inverted coaxial magnetron: (a) coaxial magnetron, (b) inverted coaxial magnetron.

transmission line is connected, as shown in Figure 23.21(a). This type of magnetron is termed the **coaxial magnetron**. Microwaves generated by the rotating electron poles at the anode vane cavity resonators are coupled to the output coaxial cavity through intervane tip gaps and slits cut on the anode ring between the vane cavities, as shown in Figure 23.21(a). All vane cavity resonators are thus connected to a coaxial cavity consisting of the conducting outer jacket and the anode ring. Thus, if a transmission line is coupled to this coaxial cavity, generated microwaves are coupled to the output circuit through the output transmission line in a side way as shown in Figure 23.21(a). An alternate coupling configuration to the output is in the direction perpendicular to the plane of the drawing. The outer jacket is tapered to an outer conductor of the output coaxial line and the slitted anode ring is tapered into the center conductor of the output coaxial line. The side opposite to the output is terminated by a conducting endplate at an impedance matching location, which is approximately one-quarter wavelength from the active part of the coaxial magnetron.

To obtain a large emission current for higher power, the electron-emitting hot cathode can be the inside wall of the outermost cylinder as shown in Figure 23.21(b). This type of coaxial magnetron is termed the **inverted coaxial magnetron**. Microwave cavity resonators formed from a number of anode vanes are tied together with a cylindrical slitted anode ring. Through the slits in the slitted anode ring, microwave power induced in the vane cavity resonators is coupled to the central coaxial cavity, which is formed from the slitted anode ring and the center conductor (which is not emitting electrons). Therefore, if the center conductor is tapered to a center conductor of an output coaxial line and the slitted anode ring is tapered into an outer conductor of the output coaxial line, then the output will be taken from the coaxial line in the direction perpendicular to the plane of the drawing. The side opposite to the output coaxial line of the inverted coaxial magnetron is terminated by a conducting endplate at an impedance-matching location, which is approximately one-quarter wavelength away from the active part of the inverted coaxial magnetron.

23.4.3 Backward Wave Oscillators

In a TWT, if the microwave signal to be amplified is propagating in the slow-wave structure backwardly to the direction of the electron beam, as shown in Figure 23.22(a), the device is termed a **backward wave oscillator** (BWO). When the electron beam is focused through the use of longitudinal magnetic flux density as shown in Figure 23.22(a), the device is *O-type BWO*. Regardless the direction of microwave propagation in the slow-wave structure, electrons are velocity modulated in the neighborhood of the electron gun and bunched as the electrons travel toward the collector, and the bunched electrons induce a large microwave voltage in the slow-wave structure at a location near the collector. Microwaves traveling backward carry positive feedback energy toward the electron gun for stronger velocity modulation and bunching. Thus, the system is inherently an oscillator rather than a stable amplifier. It can be used as a high-gain amplifier, but usually it is used as a microwave voltage-controlled oscillator (MVCO). The input is typically terminated by an impedance-matched reflectionless termination, as shown in Figure 23.22. The oscillation frequency is determined by the speed of electrons and the time constant of the feedback. The speed of electrons is controlled by the anode voltage.

The BWO can take a form of an *M-type radial BWO* as shown in Figure 23.22(b). The direction of electron pole motion and the direction of microwave propagation along the annular reentrant type slow-wave structure are opposite to each other. It should be noted that the depth of the slits cut in the inner surface of the anode is very shallow and it is much less than one-quarter wavelength deep. In other words, the slits are not in resonance or the slits are not cavity resonators as in the case of magnetrons; rather the slits are not resonating, as is in the case of TWTs. In the M-type radial BWO, the electron beam is focused by a magnetic flux density applied perpendicular to the electron beam, as seen from Figure 23.22(b). Again the input terminal of a M-type radial BWO is usually terminated by an impedance-matched reflectionless termination. An M-type radial BWO is sometime termed the carcinotron by a trade name. A key feature of the carcinotron is its wide voltage tunability over a broad frequency range.

23.4.4 Strap-Fed Devices

In a radial magnetron, sometimes every other pole of the anode resonators is conductively tied for microwave potential equalization, as shown in Figure 23.23(a). These conducting tie rings are termed **straps**; this technique of using strap rings is termed *strapping*. Strapping ensures good synchronization of microwaves in the magnetron's resonators with the rotation of electron poles.

The technique of strapping is extended and modified for an M-type radial BWO, as shown in Figure 23.23(b). Strapping rings tie every other pole of the radial slow wave structure, as in the case of a strapped radial magnetron, but the strapping rings are no longer reentrant. Microwave energy to be amplified is fed to the strap at one end and the amplified output can be taken out from the other end. This type of electron device is termed the strap-fed device.

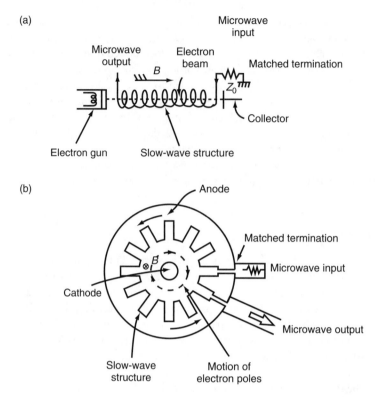

FIGURE 23.22 Schematic diagrams of backward wave oscillators: (a) O-type linear BWO, (b) M-type radial BWO.

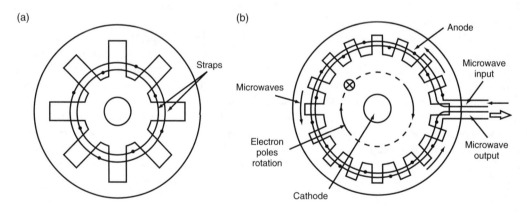

FIGURE 23.23 Strapped devices: (a) strapped radial magnetron, (b) nonreentrant strapping of a BWO.

If an M-type radial BWO is strapped, usually it does not start oscillation by itself. But if microwaves are fed through the strap from the outside using an external microwave power source to the microwave input, then the oscillation starts and even if the exciter source is turned off, the oscillation continues. This type of M-type radial BWO is termed the **platinotron** (Ishii, 1989).

In a platinotron, if the output of the tube is fed back to the input through a high-Q cavity resonator, it becomes a self-starting oscillator. The oscillation frequency is stabilized by the high-Q cavity resonator. This type of high-power frequency stabilized strapped radial BWO is termed the *stabilotron* (Ishii, 1989). The operating power levels are at kilowatt and megawatt levels.

Performance of a platinotron depends on, among other things, the design of the slow-wave structure. For example, the interdigital slow-wave structure as shown in Figure 23.23 has a limited power handling capability and frequency bandwidth. Design of a slow-wave structure with greater power handling capacity and stable with broader frequency bandwidth is possible. For example, instead of an anode with an interdigital slow-wave structure, the anode could be made of an annular open conducting duct loading with number of pairs of conducting posts across the open duct. Strapping is done at every other tip of the pairs of conducting posts. This type of strapping loads the slow-wave structure, stabilizing it and preventing oscillation. The structure of the anode with an annular duct and pairs of posts increases the power handling capability. This type of loaded radial BWO is termed the *amplitron* (Ishii, 1989; Liao, 1988). The amplitron is capable of amplifying high-power microwave signals with pulses and **continuous waves**. It is used for long-range pulsed radar transmitter amplifiers and industrial microwave heating generators. The operating power levels are at kilowatt and megawatt levels.

23.4.5 Modified Klystrons

Instead of using a multiple number of cavity resonators as in a multicavity klystron, the cavities may be combined to a single cavity, as shown in Figure 23.24(a). Multiple cavities are hard to fabricate at millimeter wavelengths. A multimode single cavity as shown in Figure 23.24(a) is more convenient. Beam coupling is accomplished through the use of ladder-shaped grids, as shown by the dashed lines with the single cavity in Figure 23.24(a). This type of klystron is termed the laddertron (Ishii, 1989). There will be a standing wave on the ladder line. Velocity modulation is performed in the ladder structure near the electron gun and bunching begins near the middle of the electron stream. Catching is done in the ladder structure near the collector. Since this is only a one-cavity resonator, there is a built-in positive feedback, and the oscillation starts like a reflex klystron, which has a single cavity. At millimeter wavelengths, a dominant mode cavity resonator becomes very small and loses its power handling capability. In a laddertron, the cavity resonator is a multiple higher mode resonator. Therefore, the quality factor Q of the cavity is high and the size of the cavity resonator is reasonably large. It can handle tens of watts of power at frequency as high as 50 GHz.

The catcher of a two-cavity klystron amplifier can be replaced by the slow-wave structure of a TWT, as shown in Figure 23.24(b). This type of tube is termed the *twystron* (Ishii, 1989). The slow-wave structure provides a broader frequency bandwidth than a regular two-cavity klystron. A microwave input fed into the buncher cavity produces velocity modulation to the electron beam. The electron beam is bunched while drifting, and the bunched electrons induce a microwave voltage in the microwave slow-wave structure. The electron beam is focused by the use of longitudinally applied magnetic flux density B.

Commercial twystrons operate in the **S-band** (2.6–3.95 GHz) and **C-band** (3.95–5.85 GHz), and are used for pulsed radar transmitter power amplifiers. The pulsed peak output ranges from 1 to 7 MW with a

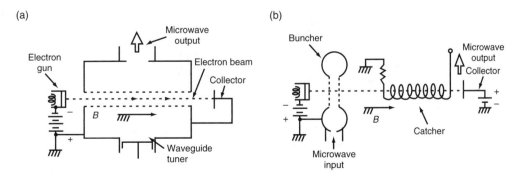

FIGURE 23.24 Schematic diagrams of modified klystrons: (a) laddertron, (b) twystron.

pulse width of 10–50 μs. This results in an average output power in the range of 1–28 kW with efficiency of 30–35%. The tube requires a beam voltage of 80–117 kV and a beam current of 45–150 A. The microwave input power to drive the amplifier twystron is in a range of 0.3–3.0 kW (Liao, 1988).

23.4.6 Electron Gyration Devices

As seen in an O-type traveling wave tube, electron trajectory of an electron in an electron beam focused by a longitudinally applied magnetic field is a helix. If the electron velocity, **electron injection angle** and applied longitudinal magnetic flux density are varied, then an electron beam of helical form with different size and pitch will be formed. A coil-shaped electron beam will be produced by adjusting the acceleration voltage, applied magnetic flux density, and the electron injection angle to the focusing magnetic field. The coil of the electron beam can be a simple single coils; or, depending on the adjustment of the aforementioned three parameters, it can be an electron beam of a double coil, or a large coil made of thin small coils. In the case of the double-coil trajectory, the large coil-shaped trajectory is termed the **major orbit** and smaller coil trajectory is termed the **minor orbit**.

If a single coil-shaped electron beam is launched in a waveguide as shown in Figure 23.25, then microwaves in the waveguide interact with the helical beam. This type of vacuum tube is termed the helical beam tube (Ishii, 1989). A single-coil helical beam is launched into a TE_{10}-mode (Transverse Electric one-zero mode) rectangular waveguide. Inside the waveguide, microwaves travel from right to left and the helical beam travels in an opposite direction. Therefore, the microwaves–electron beam interaction is the **backward wave interaction**. If the microwave frequency, the focusing magnetic flux density B, and the acceleration voltage V_a are properly adjusted, this device will function as a backward wave amplifier. Electrons in the helical beam interact with the transverse microwave electric fields and velocity modulated at the left-hand side of the waveguide as the beam enters into the wave-guide. The velocity-modulated electrons in the helical beam are bunched as they travel toward the right. If the alternating microwave electric field synchronizes its period and phase with the helical motion of bunched electrons so that the electrons always receive retardation from microwave transverse electric fields, then the electrons loose their kinetic energy and microwave gains in the electric field energy by the **kinetic energy conservation principles**. Thus, the amplified microwave power emerges at the waveguide output at the left, since the microwaves travel backward. The amplified microwaves at the left give strong velocity modulation to the helical electron beam at the left.

In Figure 23.25, if the microwave input port and the output port are interchanged with each other, then the system becomes a forward wave amplifier. Such a forward wave amplifier is termed a peniotron (Ishii, 1990).

If the electron gun is modified to a side emitting cathode and the waveguide is changed to TE_{11}**-mode** (Transverse Electric one-one mode) oversized circular waveguide as shown in Figure 23.26(a), this type of tube is termed the gyrotron. In this gyrotron, both ends of the waveguide are open and there are

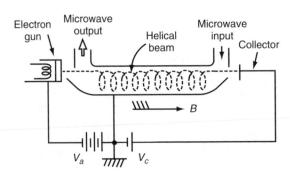

FIGURE 23.25 Schematic diagram of a helical beam tube.

sufficient reflections in the waveguide for positive feedback. The gyrotron is the **forward wave oscillator**. It is capable of generating extremely high power at extremely high frequency. The principles of operation are identical to the helical beam tubes.

If the anode voltage and the focusing flux density are readjusted, the electron beam is made into a double helical coil as shown in Figure 23.26(b), and the oversized circular waveguide is operated in TE_{01} mode (Transverse Electric zero-one mode), then a *double-coil helical beam gyrotron* is formed. In the TE_{01} mode, microwave transverse electric fields are concentric circles. Therefore, the tangential electric fields interact with electrons in the small coil trajectory. The alternating tangential microwave electric fields are made to synchronize with the tangential motion of electrons in the minor coil-shaped trajectory. Thus, electron velocity modulation takes place near the cathode and bunching takes place in the middle of the tube. Microwave kinetic energy transfer takes place as beam approaches the right. The focusing magnetic flux density B is applied only in the **interaction region**. Therefore, if electron beam comes out of the interaction region, it is defocused and collected by the anode waveguide as depicted in Figure 23.26(b). If the circular waveguide is operated in an oversized TE_{11} mode, with the double-coil helical beam, then it is termed the **tornadotron** (Ishii, 1989). Microwave–electron interaction occurs between the parallel component of tangential motion of the small helical trajectory and the TE_{11}-mode microwave electric field. If the phase of microwave electric field is decelerating bunched electrons, then the lost kinetic energy of the bunched electrons is transferred to the microwave signal and the oscillation starts.

In a single-coiled helical beam gyrotron as shown in Figure 23.26(a), frequency of the helical motion of electrons is the **cyclotron frequency** and it is given by (Ishii, 1990)

$$f_c = \frac{qB}{2\pi m} \tag{23.88}$$

where q is the charge of the electron, 1.6021×10^{-19} C; B is the longitudinal magnetic flux density in tesla; and m is the mass of an electron, 9.1091×10^{-31} kg. The amount of necessary magnetic flux density

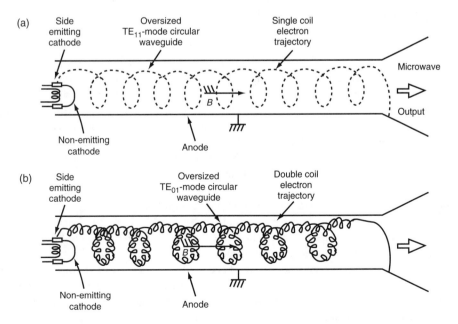

FIGURE 23.26 Schematic diagram of gyrotrons: (a) single coiled helical beam gyrotron, (b) electron trajectory of double-coil helical beam gyrotron.

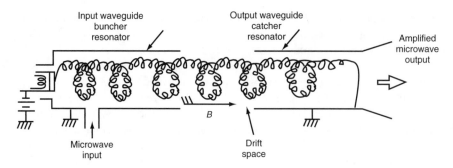

FIGURE 23.27   The gyroklystron amplifier.

to produce desired cyclotron frequency f_c is, then (Ishii, 1990)

$$B = \frac{2\pi\, m f_c}{q} \tag{23.89}$$

The radius of curvature r of the helical motion is given by (Ishii, 1990)

$$r = \frac{u_\phi}{2\pi f_c} \tag{23.90}$$

where u_ϕ is the tangential velocity of the electron and f_c is the cyclotron-frequency of the electron.

In the double-coiled helical beam gyrotron, the cycloidal frequency f of the double-coiled trajectory is calculated to be

$$f = \frac{qB}{2\pi\, mc} \tag{23.91}$$

This cycloidal frequency must be made equal to the operating microwave frequency.

Gyrotrons are capable of producing high microwave power on the order of 100 kW at microwave frequency on the order of 100 GHz.

There are a family of gyrotrons. Gyrotrons with a single waveguide, as shown in Figure 23.26 are termed **gyromonotron**. There are many other variations in gyrotrons (Coleman, 1982; McCune, 1985).

When the circular waveguide is split as shown in Figure 23.27, the tube is termed the **gyroklystron** amplifier. Both waveguides resonate to the input frequency. There are strong standing waves in both waveguide resonators. The input microwave signal to be amplified is fed through a side opening to the input waveguide resonator. This is the buncher resonator as in the case of a klystron. This buncher resonator gives velocity modulation to gyrating electrons in the double helical coil-shaped electron beam. There is a **drift space** between the buncher resonator and the **catcher** resonator at the output. While drifting electrons bunch and bunched electrons enter into the output waveguide catcher resonator, electron speed is adjusted in such a way that electrons are decelerated by the resonating microwave electric field. The lost kinetic energy in bunched electrons is transformed into microwave energy and microwaves in catcher resonator are amplified. The amplified power appears at the output of the tube.

If the waveguide is an unsplit one-piece waveguide that is impedance matched and not resonating, as shown in Figure 23.28, the tube is termed the **gyrotron traveling wave tube amplifier**. In this tube the input microwaves are fed through an opening in the waveguide near the electron gun, as shown in Figure 23.28. Microwaves in the waveguide are amplified gradually as they travel toward the output port by interacting with the double-coiled helical electron beam, which is velocity modulated and bunched. There are no significant standing waves in the waveguide. Microwaves grow gradually in the waveguide as they travel toward the output port by interaction with electrons.

If the electron gun is moved to the side of the waveguide and the microwave power is extracted from the wave-guide opening in proximity to the electron gun, as shown in Figure 23.29, then it is termed

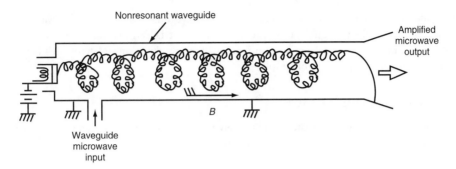

FIGURE 23.28   Gyrotron traveling wave tube amplifier.

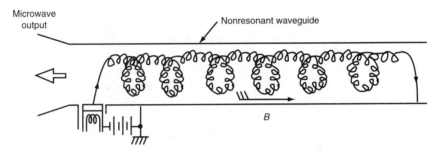

FIGURE 23.29   Gyrotron backward oscillator.

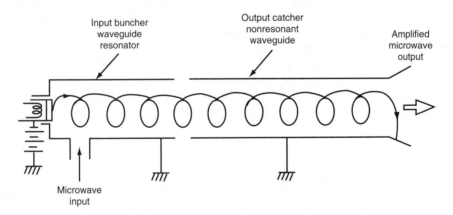

FIGURE 23.30   Gyrotwystron amplifier.

a **gyrotron backward oscillator**. The principle is similar to the backward wave oscillator; the process of velocity modulation, drifting, bunching, and catching are similar to the klystron. Microwave energy induced in the waveguide travels in both directions. The circuit is adjusted to emphasize the waves traveling in a backward direction. The backward waves become the output of the tube and, at the same time, they carry the positive feedback energy to the electrons just emitted and to be velocity modulated. Thus, the system goes into oscillation. This is a backward wave oscillator in a gyrotron form.

If the waveguide is split into two again, but this time the input side waveguide is short and the output side waveguide is long, as shown in Figure 23.30, then the tube is termed a **gyrotwystron** amplifier. The input side waveguide resonator is the same as the input resonator wave-guide of a gyroklystron, as shown in Figure 23.27. There are strong standing waves in the input bunched **waveguide** resonator. There is no drift space between the two waveguides. The output side waveguide is a long impedance-matched waveguide and there is no microwave standing wave in the waveguide. This is a traveling wave waveguide. As microwaves travel in the waveguide, they interact with bunched electrons and the microwaves grow as they travel toward the output port. This is a combination of the gyroklystron and gyrotron TWT amplifier. Therefore it is termed the gyrotwystron amplifier.

23.4.7 Quasiquantum Devices

In a gyrotron, if electrons are accelerated by extremely high voltage, electrons become *relativistic*, or the mass of an electron becomes a function of the velocity as

$$m = \frac{m_0}{\sqrt{1 - \left(\frac{c}{v}\right)^2}} \tag{23.92}$$

where m is the relativistic mass of an electron, m_0 the static mass of an electron, c the speed of light in vacuum, and v the speed of the electron in question.

The energy states of electrons are specified by a set of **quantum numbers**, and the transition of energy state occurs only between the energy state described by a set of quantum numbers. The set of quantum numbers specify a high-energy orbit of an electron and a low-energy orbit of an electron. With the interaction with the waveguide or waveguide resonator at the cyclotron frequency, if the electrons are stimulated to induce a downward transition of high-energy orbit electrons to the low-energy orbit, then microwave radiation results. This is a *cyclotron frequency maser* (Ishii, 1989).

Another **quasiquantum** vacuum **device** is the *free electron laser*, or the *ubitron*. A schematic diagram of a ubitron is shown in Figure 23.31. A high-speed relativistic electron beam is emitted from an electron gun focused by a longitudinally applied DC magnetic flux density B and periodically deflected by deflection magnetics. These repetitive deflections create among relativistic electrons high-energy states and low-energy states. If the high-energy electrons in the deflection waveguides are stimulated by the resonance frequency of the waveguide resonator, then the downward transition (which is the transition of an energy state of electrons in the high-energy state to the low-energy state) occurs. Then, there will be microwave emissions due to the energy transition at the stimulation frequency. The emission frequency is

$$f = \frac{\Delta E}{h} \tag{23.93}$$

where ΔE is the energy difference between the high-energy state and the low-energy state in Joule and h is Planck's constant, which is 6.6256×10^{-34} J-s. The operating frequency of ubitrons are in millimeter wave or submillimeter wave frequencies and operate at a high-power level.

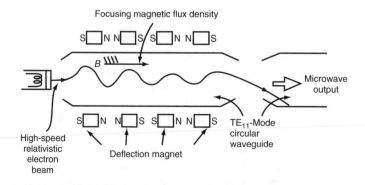

FIGURE 23.31 Schematic diagram of a ubitron.

23.4.8 Features of other Microwave Vacuum Devices

It is interesting to note that no matter what kind of microwave vacuum device, most electron devices function based on the principles of velocity modulation, electron bunching, and electrodynamic induction for microwave generation and amplification. The only microwave vacuum devices not based on velocity modulation, electron bunching, and electrodynamic induction are monolithically developed solid-state vacuum diode or triode. These devices are microscopically small and incorporate cathodes. The operating electric fields are strong enough for cold emission from the cathode.

Defining Terms

Amplitron: A strap-fed radial crossed-field microwave amplifier.

Backward wave interaction: Interaction between backward propagating microwave electric fields against an electron stream and the electron in the electron beam. The direction of propagating microwaves and the direction of motion of electrons in the beam are opposite each other.

Backward wave oscillators (BWOs): A microwave oscillator tube that is based on a backward wave interaction.

Carcinotron: A forward radial traveling wave amplifier in which microwave signals are fed to the radial slow-wave structure.

Catcher: A cavity resonator of a multicavity klystron proximite to the collector to catch microwave energy from the bunched electrons.

Cavity resonator: A resonating space surrounded by conducting walls. An electromagnetic wave energy storage device.

C-band: Microwave frequency range, 3.95–5.85 GHz.

Coaxial magnetron: A radial magnetron where the anode and cathode are gradually transformed into a coaxial line.

Continuous wave: A wave of a constant and uninterrupted amplitude, unmodulated waves, carrier waves.

Collector: An electrode at the end of the electron beam in a microwave vacuum tube to collect electrons finished with the electronic processes.

Crossed-field amplifier (CFA): Radial or linear traveling wave amplifier where radial or transverse DC accelerating electric fields are perpendicular to axial or longitudinal DC magnetic fields, respectively.

Crossed-field devices (CFD): Radial or linear forward traveling wave amplifier or backward wave oscillator where radial or transverse DC accelerating electric fields are perpendicular to axial or longitudinal DC magnetic fields, respectively.

Cutoff magnetic flux density: In magnetrons operated under a constant anode voltage, a value of magnetic flux density at which the anode current decreases abruptly.

Cyclotron frequency: A frequency of electron oscillation in a cyclotron. A frequency of circular motion of an electron under magnetic fields applied perpendicularly to the plane of the circular motion.

Drift space: Space where electrons move only due to their inertia.

Duty cycles: A ratio of activated time duration to the whole period in a periodically activated system.

Electron beam: A focused stream of electrons.

Electron bunching: A gathering of electrons as a result of velocity modulation and drifting.

Electron cloud: A gathering of electrons due to thermionic emission, space charge effect, Coulomb force, and Lorentz force by applied electric and magnetic fields, respectively.

Electron gun: A source of electron beam that shoots out electron beam in a specified direction in a focused and collimated fashion. It contains a hot cathode, accelerating anode, and focusing electrodes.

Electron injection angle: An angle between injected electron beam and focusing magnetic fields.

Electron orbit: Trajectory of an electron motion.

Electron pole: In a radial magnetron, gathering of electrons formed by tangential velocity modulation and the electromagnetic fields. It rotates around the cathode.

Flicker noise: An electrical noise current contained in a thermal emission current due to stochastic change on conditions of cathode surface. For example in case of the $1/f$ noise, the noise power is inversely proportional to the operating frequency of interest.

Forward wave oscillator: A traveling wave amplifier with a positive feedback.

Gyroklystron: A gyrotron with a catcher waveguide and a buncher waveguide.

Gyromonotron: A gyrotron with a single resonating waveguide.

Gyrotron: A high-power millimeter wave generator based on transverse interaction on helically gyrating electrons against the millimeter wave electric field that is traveling longitudinally in a waveguide.

Gyrotron backward wave oscillator: A gyrotron in which the gyrating helical electron beam and traveling millimeter waves in a waveguide are in opposite directions.

Gyrotron traveling wave amplifier: A gyrotron with a matched waveguide. The waveguide carries only forward waves. The gyrating electrons interact with forward millimeter waves only.

Gyrotwystron: A gyrotron with a resonating buncher waveguide and a nonresonating traveling wave catcher waveguide.

Helical beam tube: A backward wave amplifier based on interaction between a forward helical electron beam launched into a waveguide and backward traveling microwave in the waveguide.

Helmhortz' wave equation: A second-order linear differential equation of electric potential, magnetic potential, electric field, magnetic field, or any other field quantities associated with propagating waves and/or standing waves.

Interaction region: A region where an electron beam and microwaves transfer energy to each other.

Inverted magnetron: A radial magnetron in which the anode structure is at inside and the cathode structure is at outside.

Kinetic energy: Energy of motion.

Kinetic energy conservation principle: Any change of kinetic energy of an electron is transformed into a change in electric potential energy with which the electron is interacting, and vice versa.

Klystron: A microwave vacuum tube amplifier or oscillator that contains buncher cavity or cavities, drift space, and a catcher cavity, among other components.

Laddertron: A microwave vacuum tube oscillator with a slow-wave structure coupled to a single-cavity resonator.

Laplace's equation: A second-order linear homogeneous differential equation of electric potential or any other function.

Linear magnetron: A magnetron in which the cavity resonator loaded anode, a sole (nonemitting cathode), and the electron beam are arranged in rectilinear fashion.

Magnetron: A microwave vacuum tube oscillator in which microwaves in a number of cavity resonators interact with bunched electrons in motion rotating or rectilinearly. The basic electron motion is formed by the applied accelerating electric fields and the DC magnetic fields.

Major orbit: A larger helical orbit of an electron beam in a gyrotron.

Minor orbit: A smaller helical orbit of electron beam in a double-coil electron beam gyrotron.

M-type linear magnetron: A magnetron in which the cavity-loaded anode, the sole (nonemitting cathode), and the electron beam are arranged in a rectilinear fashion. Focusing DC magnetic fields are applied perpendicularly to both the electron beam and the DC electric fields so that the individual electron orbit is a cycloid.

O-type linear magnetron: A magnetron in which the cavity-loaded anode, the sole (nonemitting cathode), and the electron beam are arranged in a rectilinear fashion. Focusing DC magnetic fields are applied parallel to the electron beam and perpendicularly to the DC electric fields so that the individual electron orbit is a helix.

Partition noise: Electrical noise current in an electron stream in an electron device due to separation of electron stream owing to the electron stream striking an obstacle or obstacles in the device.

Platinotron: A strap-fed starting radial M-type backward wave oscillator.

Poisson's equation: Second-order linear inhomogeneous differential equation of electric potential or other electrical or magnetic functions.

Quantum numbers: A group of integrating constants in a solution of Schroedinger's equation.

Quasiquantum devices: Electron devices in which the principle of the device function is more easily explainable by quantum mechanics than by Newtonian classical dynamics.

Radial magnetron: A magnetron in which the anode and the cathode are arranged in a coaxial radial fashion.

Re-entrant use: Operation of a device in which a majority of the output of the device re-enters the input of the device.

S-band: A range of microwave frequency band, 2.6–3.95 GHz.

Shot noise: Electrical noise current in an electron stream in an electron device due to stochastic emission of electrons from the cathode.

Sole: A nonemitting cathode.

Strap: A conducting ring that ties tips of poles of magnetron or magnetronlike devices in a specified fashion for microwave potential and phase equalization.

Strap fed devices: Strapped magnetronlike devices that operate by microwaves fed through the strapping such as amplitron amplifies and platinotron oscillators.

TE$_{01}$ mode: A mode of circular or rectangular waveguide excitation in which all microwave electric fields are transverse to the direction of microwave propagation and coaxially concentric.

TE$_{11}$ mode: A mode of a circular or rectangular waveguide excitation in which all microwave electric fields are transverse to the direction of microwave propagation and approximately parallel to each other ending at the waveguide wall.

Thermal noise: Electric noise power of a device or component that is due to thermal vibration of atoms, molecules, and electrons. It is proportional to a product of the frequency bandwidth of interest and the absolute temperature of the device or component. The proportionality constant is Boltzmann's constant.

Tornadotron: A gyrotron with a simple helical beam.

Ubitron: A millimeter wave high-power quasiquantum generator with relativistically high-speed electron beam. Millimeter waves are generated due to quantum transition between two energy states of electrons and amplified due to the velocity modulation and kinetic energy transfer principles.

Velocity modulation: Production of variation of electron speed in a stream of electrons.

Waveguide: A transmission line that guides electromagnetic waves.

References

Coleman, J.T. 1982. *Microwave Devices*. Reston Pub., Reston, VA.

EEV Ltd. 1991. *EEV Microwave Products*. EEV Ltd., Chelmsford, England.

Ishii, T.K. 1989. *Microwave Engineering*. Harcourt Brace Jovanovich, San Diego, CA.

Ishii, T.K. 1990. *Practical Microwave Electron Devices*. Academic Press, San Diego, CA.

Liao, S.Y. 1988. *Microwave Electron-Tube Devices*. Prentice-Hall, Englewood Cliffs, NJ.

McCune, E.W. 1985. Fusion plasma heating with high power microwave and millimeter wave tubes. *Journal of Microwave Power*, vol. 20 (April) pp. 131–136.

Richardson Electronics Ltd. 1994. *Industrial Microwave Catalog*, June, LaFox, IL.

Veley, V.F. *Modern Microwave Technology*. Prentice-Hall, Englewood Cliffs, NJ.

Further Information

Additional information on the topic of other microwave vacuum devices may be obtained from the following sources:

IEEE Electron Device Meeting Digest.

IEEE Transactions on Electron Devices, published by the Institute of Electrical and Electronics Engineers Inc., 345 East 47th St., New York, NY 10017.

Manufacturers' catalog, application notes, and other technical publications such as from EEV, Hughes, Thomson Tubes Electroniques, Varian, and NEC.

23.5   Operational Considerations for Microwave Tubes

Jerry C. Whitaker

23.5.1   Introduction

Long-term reliability of a microwave power tube requires detailed attention to the operating environment of the device, including supply voltages, load point, and cooling. Optimum performance of the system can only be achieved when all elements are functioning within specified parameters.

23.5.2   Microwave Tube Life

Any analysis of microwave tube life must first identify the parameters that define *life*. The primary *wear out* mechanism in a microwave power tube is the electron gun at the cathode. In principle, the cathode will eventually evaporate the activating material and cease to produce the required output power. Tubes, however, rarely fail because of low emission, but for a variety of other reasons that are usually external to the device.

Power tubes designed for microwave applications provide long life when operated within their designed parameters. The point at which the device fails to produce the required output power can be predicted with some accuracy, based on design data and in-service experience. Most power tubes, however, fail because of mechanisms other than predictable chemical reactions inside the device itself. External forces, such as transient overvoltages caused by lightning, cooling system faults, and improper tuning more often than not lead to the failure of a microwave tube.

23.5.3   Life-Support System

Transmitter control logic is usually configured for two states of operation:

- An *operational level*, which requires all of the life-support systems to be present before the high-voltage (HV) command is enabled
- An *overload level*, which removes HV when one or more fault conditions occur

The cooling system is the primary life-support element in most RF generators. The cooling system should be fully operational before the application of voltages to the tube. Likewise, a cool-down period is usually recommended between the removal of beam and filament voltages and shut-down of the cooling system.

Most microwave power tubes require a high-voltage removal time of less than 100 ms from the occurrence of an overload. If the trip time is longer, damage may result to the device. **Arc detectors** are often installed in the cavities of high-power tubes to sense fault conditions, and shut down the high-voltage power supply before damage can be done to the tube. Depending on the circuit parameters, arcs can be sustaining, requiring removal of high voltage to squelch the arc. A number of factors can cause RF arcing, including

- Overdrive condition
- Mistuning of one or more cavities
- Poor cavity fit (applies to external types only)
- Under coupling of the output to the load
- Lightning strike at the antenna
- High voltage standing wave ratio (VSWR)

Regardless of the cause, arcing can destroy internal elements or the vacuum seal if drive and/or high voltage are not removed quickly. A lamp is usually included with each arc detector photocell for test purposes.

Protection Measures

A microwave power tube must be protected by control devices in the amplifier system. Such devices offer either visual indications, aural alarm warnings, or actuate interlocks within the system. Figure 23.32 shows a klystron amplifier and the basic components associated with its operation, including metering for each of the power supplies. Other types of microwave power devices use similar protection schemes. Sections of coaxial transmission line, representing essential components, are shown in the figure attached to the RF input and RF output ports of the tube. A single magnet coil is shown to represent any coil configuration that may exist; its position in the drawing is for convenience only and does not represent the true position in the system.

Heater Supply

The heater power supply can be either AC or DC. If it is DC, the positive terminal must be connected to the common heater-cathode terminal and the negative terminal to the heater terminal. The amount of power supplied to the heater is important because it establishes the cathode operating temperature. The temperature must be high enough to provide ample electron emission but not so high that emission life is jeopardized. The test performance sheet accompanying each tube lists the proper operating values of heater voltage and current for that tube. Heater voltage should be measured at the terminals of the device using a true-rms-reading voltmeter.

Because the cathode and heater are connected to the negative side of the beam supply, they must be insulated to withstand the full beam potential.

Beam Supply

The high-voltage **beam supply** furnishes the DC input power to the klystron. The positive side of the beam supply is connected to the body and collector of the klystron. The negative terminal is connected to the common heater-cathode terminal. Never connect the negative terminal of the beam supply to the heater-only terminal because the beam current will then flow through the heater to the cathode and cause

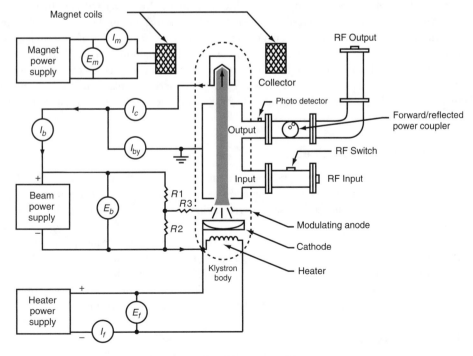

FIGURE 23.32 Protection and metering system for a klystron amplifier.

premature heater failure. The voltmeter, E_b in Figure 23.32, measures the beam voltage applied between the cathode and the body of the klystron.

Current meter I_c measures collector current, typically 95% or more of the total device current. Current meter I_{by} measures **body current**. An interlock should interrupt the beam supply if the body current exceeds a specified maximum value.

The sum of the body current I_{by} and collector current I_c is equal to the beam current I_b, which should stay constant as long as the beam voltage and modulating anode voltage are held constant.

Magnet Supply

Electrical connections to the DC magnet include two meters, one for measuring current through the circuit (I_m in Figure 23.32) and one for measuring voltage (E_m). When the device is first installed in its magnet assembly, both parameters should be measured and recorded for future reference. If excessive body current or other unusual symptoms occur, these data will be valuable for system analysis.

Undercurrent protection should be provided to remove beam voltage if the magnetic circuit current falls below a preset value. The interlock should also prevent the beam voltage from being applied if the magnetic circuit is not energized. This scheme will not, however, provide protection if the coils are shorted. Shorted conditions can be determined by measuring the normal values of voltage and current and recording them for future reference.

The body-current overload protection should actuate if the magnetic field is reduced for any reason.

RF Circuits

In Figure 23.32, monitoring devices are shown on the RF input and output of the device. These monitors protect the tube should a failure occur in the RF output circuit. Two directional couplers and a photodetector are attached to the output. These components and an RF switching device on the input form a protective network against output transmission line mismatch. The RF switch is activated by the photodetector or the reflected power monitor and must be capable of removing RF drive power from the klystron in less than 10 ms (typically).

In the RF output circuit, the forward power coupler is used to measure the relative power output of the klystron. The reflected power coupler is used to measure the RF power reflected by the output circuit components, or antenna. Damaged components or foreign material in the RF line will increase the RF reflected power. The amount of reflected power should be no more than 5% of the actual forward RF output power of the device in most applications. An interlock monitors the reflected power and removes RF drive to the klystron if the reflected energy reaches an unsafe level. To protect against arcs occurring between the monitor and the output window, a photodetector is placed between the monitor and the window. Light from an arc will trigger the photodetector, which actuates the interlock system and removes RF drive before the window is damaged.

23.5.4 Filament Voltage Control

Extending the life of a microwave tube begins with accurate adjustment of filament voltage. The filament should not be operated at a reduced voltage in an effort to extend tube life. Unlike a thoriated tungsten grid tube, reduced filament voltage may cause uneven emission from the surface of the cathode with little or no improvement in cathode life.

Voltage should be applied to the filament for a specified warm-up period before the application of beam current to minimize thermal stress on the cathode/gun structure. However, voltage normally should not be applied to the filaments for extended periods (2 h or more) if no beam voltage is present. The net rate of evaporation of emissive material from the cathode surface is greater without beam voltage. Subsequent condensation of material on gun components may lead to voltage standoff problems.

23.5.5 Cooling System

The cooling system is vital to any RF generator. In a high-power microwave transmitter, the cooling system may need to dissipate as much as 70% of the input AC power in the form of waste heat. For

TABLE 23.2 Variation of Electrical and Thermal Properties of Common Insulators as a Function of Temperature

	20°C	120°C	260°C	400°C	538°C
Thermal Conductivity[a]					
99.5% BeO	140	120	65	50	40
99.5% Al_2O_3	20	17	12	7.5	6
95.0% Al_2O_3	13.5				
Glass	0.3				
Power Dissipation[b]					
BeO	2.4	2.1	1.1	0.9	0.7
Electrical Resistivity[c]					
BeO	10^{16}	10^{14}	5×10^{12}	10^{12}	10^{11}
Al_2O_3	10^{14}	10^{14}	10^{12}	10^{12}	10^{11}
Glass	10^{12}	10^{10}	10^{8}	10^{6}	
Dielectric Constant[d]					
BeO	6.57	6.64	6.75	6.90	7.05
Al_2O_3	9.4	9.5	9.6	9.7	9.8
Loss Tangent[d]					
BeO	0.00044	0.00040	0.00040	0.00049	0.00080

[a] Heat transfer in $Btu/ft^2/h/°F$.
[b] Dissipation in $W/cm/°C$.
[c] Resistivity in $\Omega \cdot cm$.
[d] At 8.5 GHz.

vapor phase-cooled devices, pure (distilled or demineralized) water must be used. Because the collector is usually only several volts above ground potential, it is generally not necessary to use deionized water.

Excessive dissipation is perhaps the single greatest cause of catastrophic failure in a power tube. The critical points of almost every tube type are the metal-to-ceramic junctions or seals. At temperatures below 250°C these seals remain secure, but above that temperature, the bonding in the seal may begin to disintegrate. Warping of internal structures also may occur at temperatures above the maximum operating level of the device. The result of prolonged overheating is shortened tube life or catastrophic failure. Several precautions are usually taken to prevent damage to tube seals under normal operating conditions. Air directors or sections of tubing may be used to provide spot cooling at critical surface areas of the device. Airflow and waterflow sensors typically prevent operation of the RF system in the event of a cooling system failure.

Temperature control is important for microwave tube operation because the properties of many of the materials used to build a device change with increasing temperature. In some applications, these changes are insignificant. In others, however, such changes can result in detrimental effects, leading to, in the worst case, catastrophic failure. Table 23.2 details the variation of electrical and thermal properties with temperature for various substances.

Water Cooling Systems

A water-cooled tube depends on an adequate flow of fluid to remove heat from the device and transport it to an external heat sink. The recommended flow as specified in the technical data sheet should be maintained at all times when the tube is in operation. Inadequate water flow at high temperature may cause the formation of steam bubbles at the collector surface where the water is in direct contact with the collector. This condition can contribute to premature tube failure.

Circulating water can remove about 1.0 kW/cm^2 of effective internal collector area. In practice, the temperature of water leaving the tube is limited to 70°C to preclude the possibility of spot boiling. The water is then passed through a heat exchanger where it is cooled to 30–40°C before being pumped back to the device.

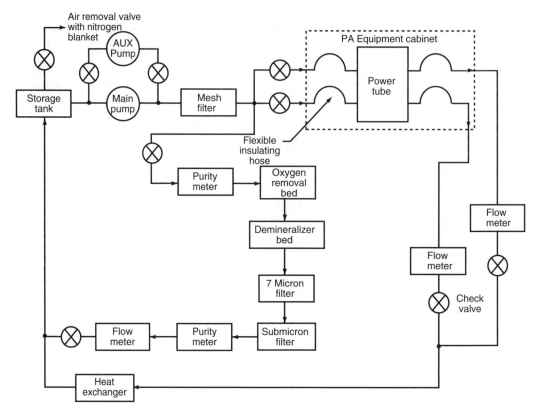

FIGURE 23.33   Functional schematic of a water-cooling system for a high-power amplifier or oscillator.

Cooling System Design

A liquid cooling system consists of the following principal components:

- A source of coolant
- Circulation pump
- Heat exchanger
- Coolant purification loop
- Various connection pipes, valves, and gauges
- Flow interlocking devices (required to ensure coolant flow anytime the equipment is energized)

Such a system is shown schematically in Figure 23.33. In most cases the liquid coolant will be water, however, if there is a danger of freezing, it will be necessary to use an antifreeze solution such as ethylene glycol. In these cases, coolant flow must be increased or plate dissipation reduced to compensate for the poorer heat capacity of the ethylene glycol solution. A mixture of 60% ethylene glycol to 40% water by weight will be about 75% as efficient as pure water as a coolant at 25°C. Regardless of the choice of liquid, the system volume must be maintained above the minimum required to insure proper cooling of the vacuum tube(s).

The main circulation pump must be of sufficient size to ensure necessary flow and pressure as specified on the tube data sheet. A filter screen of at least 60 mesh is usually installed in the pump outlet line to trap any circulating debris that might clog coolant passages within the tube.

The heat exchanger system is sized to maintain the outlet temperature such that the outlet water from the tube at full dissipation does not exceed 70°C. Supplementary coolant courses may be connected in parallel or series with the main supply as long as the maximum outlet temperature is not exceeded.

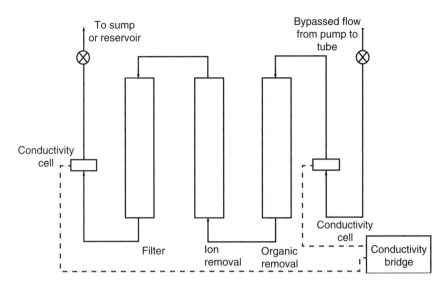

FIGURE 23.34   Typical configuration of a water purification loop.

Valves and pressure meters are installed on the inlet lines to the tube to permit adjustment of flow and measurement of pressure drop, respectively. A pressure meter and check valve are employed in the outlet line. In addition, a flow meter, sized in accordance with the tube data sheet, and a thermometer are included in the outlet line of each coolant course. These flow meters are equipped with automatic interlock switches, wired into the system electrical controls, so that the tube will be completely deenergized in the event of a loss of coolant flow in any one of the coolant passages. In some tubes, filament power alone is sufficient to damage the tube structure in the absence of proper water flow.

The lines connecting the plumbing system to the inlet and outlet ports of tube are made of a flexible insulating material configured so as to avoid excessive strain on the tube flanges. The coolant lines must be of sufficient length to keep electrical leakage below approximately 4 mA at full operating power.

The hoses are coiled or otherwise supported so they do not contact each other or any conducting surface between the high voltage end and ground. Conducting hose *barbs*, connected to ground, are provided at the low-potential end so that the insulating column of water is broken and grounded at the point it exits the equipment cabinet.

Even if the cooling system is constructed using the recommended materials, and is filled with distilled or deionized water, the solubility of the metals, carbon dioxide, and free oxygen in the water make the use of a coolant purification **regeneration loop** essential. The integration of the purification loop into the overall cooling system is shown in Figure 23.34. The regeneration loop typically taps 5–10% of the total cooling system capacity, circulating it through oxygen scavenging and deionization beds and submicron filters before returning it into the main system. The purification loop can theoretically process water to 18-MΩ resistivity at 25°C. In practice, resistivity will be somewhat below this value.

Figure 23.34 shows a typical purification loop configuration. Packaged systems such as the one illustrated are available from a number of manufacturers. In general, such systems consist of replaceable cartridges that perform the filtering, ion exchange, and organic solid removal functions. The system will usually include flow and pressure gauges and valves and conductivity cells for continuous evaluation of the condition of the water and filters.

Water Purity and Resistivity

The purity of the cooling water is an important operating parameter for any water-cooled amplifier or oscillator. The specific **resistivity** typically must be maintained at 1 MΩ · cm minimum at 25°C. Distilled or deionized water should be used and the purity and flow protection periodically checked to ensure against degradation.

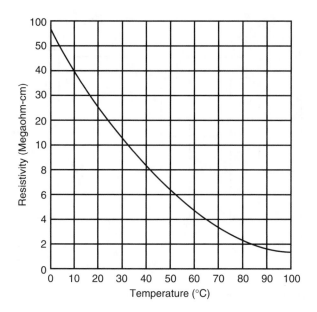

FIGURE 23.35   The effect of temperature on the resistivity of ultrapure water.

Oxygen and carbon dioxide in the coolant will form copper oxide, reducing cooling efficiency, and electrolysis may destroy the coolant passages. In addition, a filter screen should be installed in the tube inlet line to trap any circulating debris that might clog coolant passages within the tube.

After normal operation for an extended period, the cooling system should be capable of holding 3–4 M$\Omega \cdot$ cm until the filter beds become contaminated and must be replaced. The need to replace the filter bed resins is indicated when the purification loop output water falls below 5 M$\Omega \cdot$cm. Although the resistivity measurement is not a test for free oxygen in the coolant, the oxygen filter bed should always be replaced when replacing the deionizing bed.

The resistivity of the coolant is also affected by the temperature of the water. The temperature dependence of the resistivity of pure water is charted in Figure 23.35.

Generally speaking, it is recommended that the coolant water be circulated at all times. This procedure provides the following benefits:

- Maintains high resistivity
- Reduces bacteria growth
- Minimizes oxidation resulting from coolant stagnation

If it is undesirable to circulate the coolant at the regular rate when the tube is deenergized, a secondary circulating pump can be used to move the coolant at a lower rate to purge any air that might enter the system and to prevent stagnation. Typical recommended minimum circulation rates within the coolant lines are as follows:

- 2 m/s during normal operation
- 30 cm/s during standby

The regeneration loop is typically capable of maintaining the cooling system such that the following maximum contaminant levels are not exceeded:

- Copper: 0.05 PPM by weight
- Oxygen: 0.5 PPM by weight
- CO_2: 0.5 PPM by weight
- Total solids: 3 PPM by weight

These parameters represent maximum levels; if the precautions outlined in this section are taken, actual levels will be considerably lower.

If the cooling system water temperature is allowed to reach 50°C, it will be necessary to use cartridges in the coolant regeneration loop that are designed to operate at elevated temperatures. Ordinary cartridges will decompose in high temperature service.

Defining Terms

Arc detector: A device placed within a microwave power tube or within one or more of the external cavities of a microwave power tube whose purpose is to sense the presence of an overvoltage arc.

Beam supply: The high-voltage power supply that delivers operating potential to a microwave power tube.

Body current: A parameter of operation for a microwave power tube that describes the amount of current passing through leakage paths within or around the device, principally through the body of the tube. Some small body current is typical during normal operation. Excessive current, however, can indicate a problem condition within the device or its support systems.

Regeneration loop: A water purification system used to maintain proper conditions of the cooling liquid for a power vacuum tube.

Water resistivity: A measure of the purity of cooling liquid for a power tube, typically measured in $M\Omega \cdot cm$.

References

Crutchfield, E.B. 1992. *NAB Engineering Handbook*, 8th ed. National Association of Broadcasters, Washington DC.

Harper, C.A. 1991. *Electronic Packaging and Interconnection Handbook*. McGraw-Hill, New York.

Kimmel, E. 1983. Temperature sensitive indicators: how they work, when to use them. *Heat Treating* (March).

Spangenberg, K. 1947. *Vacuum Tubes*. McGraw-Hill, New York.

Terman, F.E. 1947. *Radio Engineering*. McGraw-Hill, New York.

Varian. n.d. *Integral Cavity Klystrons for UHF-TV Transmitters*. Varian Associates, Palo Alto, CA.

Varian. 1971. Foaming test for water purity. *Application Engineering Bulletin* No. AEB-26. Varian Associates, Palo Alto, CA, Sept.

Varian. 1977. Water purity requirements in liquid cooling systems. *Application Bulletin* No. 16. Varian Associates, San Carlos, CA, July.

Varian. 1978. Protecting integral-cavity klystrons against water leakage. *Technical Bulletin* No. 3834. Varian Associates, Palo Alto, CA, July.

Varian. 1982. Cleaning and flushing klystron water and vapor cooling systems. *Application Engineering Bulletin* No. AEB-32. Varian Associates, Palo Alto, CA, Feb.

Varian. 1984. *Care and Feeding of Power Grid Tubes*. Varian Associates, San Carlos, CA.

Varian. 1984. Temperature measurements with eimac power tubes. *Application Bulletin* No. 20. Varian Associates, San Carlos, CA, Jan.

Varian. 1991. Water purity requirements in water and vapor cooling systems. *Application Engineering Bulletin* No. AEB-31. Varian Associates, Palo Alto, CA, Sept.

Whitaker, J.C. 1991. *Radio Frequency Transmission Systems: Design and Operation*. McGraw-Hill, New York.

Whitaker, J.C. 1992, *Maintaining Electronic Systems*. CRC Press, Boca Raton, FL.

Whitaker, J.C. *Power Vacuum Tubes Handbook* 2nd ed., CRC Press, Boca Raton, FL.

Further Information

Specific information on the application of microwave tubes can be obtained from the manufacturers of those devices. More general application information can be found in the following publications:

Maintaining Electronic Systems, written by Jerry C. Whitaker, CRC Press, Boca Raton, FL, 1990.

Radio Frequency Transmission Systems: Design and Operation, written by Jerry C. Whitaker, McGraw-Hill, New York, 1991.

Power Vacuum Tubes Handbook 2nd ed., written by Jerry C. Whitaker, CRC Press, Boca Raton, FL.

24

Monolithic Microwave IC Technology

Lawrence P. Dunleavy
Modelithics, Inc.

24.1 Monolithic Microwave Integrated Circuit Technology

24.1.1 MMIC Definition and Concepts

An excellent introduction to monolithic microwave integrated circuit (MMIC) technology is presented by Robert Pucel in 1981 [1]. Pucel went on to assemble a collection of papers on the subject in which he states in the introduction [2]:

> "the monolithic approach is an approach wherein all active and passive circuit elements and interconnections are formed, in situ on or within a semi-insulating semi-conductor substrate by a combination of deposition schemes such as epitaxy, ion implantation, sputtering, and evaporation."

Figure 24.1 [3] is a conceptual MMIC chip illustrating most of the major constituents. These include field effect transistor (FET) active devices, metal–insulator–metal (MIM) capacitors, thin film resistors, spiral strip inductors via hole grounding, and airbridges.

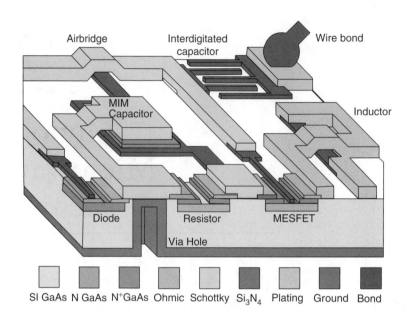

FIGURE 24.1   Three dimensional conceptual illustration of MMIC technology. (From Ladbrooke, P., *MMIC Design GaAs FETs and HEMTs*, Artech House, 1989, p. 46. With permission.)

As implied by the above quote and Figure 24.1, in a MMIC all of the circuit components, including transistors, resistors, capacitors, and interconnecting transmission lines are integrated onto a single semi-insulating/semi-conducting (usually gallium arsenide (GaAs)) substrate. Use of a mask set and a corresponding series of processing steps achieve the integrated circuit fabrication. The mask set can be thought of as a mold. Once the mold has been cast the process can be repeated in a turn-the-crank fashion to batch process tens, hundreds, or thousands of essentially identical circuits on each wafer.

24.1.2 A Brief History of GaAs MMICs

The origin of MMICs may be traced to a 1964 government program at Texas Instruments [4]. A few key milestones are summarized in the following:

1. 1964—U.S. Government funded research program at TI based on Silicon integrated circuit technology [5]:
   - The objective was a transmit/receive module for a phased array radar antenna.
   - The results were disappointing due to the poor semi-insulating properties of silicon.
2. 1968—Mehal and Wacker [6] used semi-insulating GaAs as the substrate with Schottky diodes and Gunn devices as active devices to fabricate an integrated circuit comprising a 94 GHz receiver front-end.
3. 1976—Pengelly and Turner [7] used MESFET devices on GaAs to fabricate an X-band (~10 GHz) amplifier and sparked an intense activity in GaAs MMICs.
4. 1988 (approximately)—U.S. Government's Defense Advanced Research Projects Agency (DARPA, today called ARPA) launched a massive research and development program called the MIMIC program (included Phase I, Phase II, and Phase III efforts) that involved most of the major MMIC manufacturing companies.

In the early 1980s a good deal of excitement was generated and several optimistic projections were made predicting the rapid adoption of GaAs MMIC technology by microwave system designers, with correspondingly large profits for MMIC manufacturers. The reality is that there was a much slower rate of progress to widespread use of MMIC technology, with the majority of the early thrust being provided by the government for defense applications. Still, steady progress was made through the 1980s, and the government's MIMIC program was very successful in allowing companies to develop lower cost design and fabrication techniques to make commercial application of the technology viable. The 1990s have seen good progress towards commercial use with applications ranging from direct broadcast satellite (DBS) TV receivers, to automotive collision avoidance radar, and many wireless communication applications (Cell phones, WLANs, etc.).

In recent years GaAs MMICs are giving ground to SiGe HBT- and Si CMOS-based silicon radio frequency integrated circuits (RFICs), especially at Wireless frequencies below 6 GHz. Since a review of Silicon RFIC technology is outside our scope here, the interested reader is referred to one of the many treatments on these topics [8–10]. As will be discussed later on in this section, GaAs and related III–V compounds like GaN, InP remain strong contenders for switching, power amplifier, and millimeter-wave IC applications.

As a point of terminology clarification, according to Dr. Gailon Brehm of Triquint "Many recent publications have adopted the usage of 'RFIC' to refer to monolithic integrated circuits performing radio frequency functions for high volume applications up to 5 GHz, and have applied this term to such circuits independent of the semiconductor—GaAs or silicon. The term 'MMIC' can be used interchangeably as well [11]."

24.1.3 Hybrid versus MMIC Microwave IC Technologies

The conventional approach to microwave circuit design that MMIC technology competes with, or is used in combination with is called "hybrid microwave integrated circuit," "discrete microwave integrated circuit" technology, or simply MIC technology. In a hybrid MIC the circuit pattern is formed using photolithography. Discrete components are then assembled onto the substrate (e.g., using solder or silver epoxy) then connected using bondwires. In contrast to the batch processing afforded the MMIC approach, MICs has to be assembled with discrete components attached using relatively labor intensive manufacturing methods. Table 24.1 summarizes some of the contrasting features of hybrid and monolithic approaches.

The choice of MMICs versus hybrid approach is mainly a matter of volume requirements. The batch processing of MMICs gives this approach advantages for high volume applications. Significant cost savings can be reaped in reduced assembly labor; however, for MMIC the initial design and mask preparation costs are considerable. The cost of maintaining a MMIC manufacturing facility is also extremely high and this has forced several companies out of the business. A couple of examples are Harris who sold their GaAs operation to Samsung and put their resources on Silicon. Another is AT& T, who also is relying on Silicon for its anticipated microwave IC needs.

The high cost of maintaining a facility can only be offset by high volume production of MMICs. Still this does not prevent companies without MMIC foundries from using the technology, as there are several commercial foundries that offset the costs of maintaining their facilities by manufacturing MMIC chip products for third party companies through a foundry design working relationship.

TABLE 24.1   Features of Hybrid and Monolithic Approaches

Feature	Hybrid	Monolithic
Type of substrate	Insulator	Semiconductor
Passive components	Discrete/deposited	Deposited
Active components	Discrete	Deposited
Interconnects	Deposited and wire-bonded	Deposited
Batch-processed?	No	Yes
Labor intensive/chip	Yes	No

Source: Pucel, R.A., ed., *Monolithic Microwave Integrated Circuits*, IEEE Press, 1985, p. 1. With permission.

A key advantage of MMICs is small size. To give an example, hybrid MIC the size of a business card can easily be reduced to a small chip one or two millimeters on a side. An associated advantage is the ease of integration that allows several functions to be integrated onto a single chip. For example, Anadigics and Raytheon have both manufactured DBS related MMICs, wherein, the functions of amplification (LNA and IF amplifiers), signal generation (VCO), and signal conversion (mixer) and filtering are all accomplished on a 1 × 2 mm, or smaller chip.

In contrast to MMIC, MIC lithography is quite inexpensive and a much smaller scale investment is required to maintain a MIC manufacturing capability. There are also some performance advantages of the hybrid approach. For example, it is much easier to tune or repair a hybrid circuit after fabrication than it is for a MMIC. For this reason, for applications where the lowest noise figure is required, such as in a satellite TV receiver, an individually tuned hybrid LNA may be preferred as the first stage.

Ultimately, there is no such thing as a truly all MMIC system. Monolithic technology can be used to integrate single functions, or several system functions, but cannot sustain a system function in isolation. Usually, a MMIC is packaged along with other circuitry to make a practically useful component, or system. Figure 24.2, a radar module made by Raytheon, exemplifies how the advantages of both MMIC and hybrid approaches are realized in a hybrid connection of MMIC chips. Another example of combined hybrid/MMIC technology is shown in Figure 24.3, in the form of a low noise block downconverter for DBS Applications. An example of combined MMIC and hybrid MIC technologies, this radar module includes several MMIC chips interconnected using microstrip lines. The hybrid substrate (white areas) is an alumina insulating substrate; the traces on the hybrid substrate are microstrip lines (courtesy Raytheon Systems Company).

There are five GaAs MMIC chips that can be recognized as black substrates with gold traces.

FIGURE 24.2 An example of combined MMIC and hybrid MIC technologies, this radar module includes several MMIC chips interconnected using microstrip lines. The hybrid substrate (white areas) is an Alumina insulating substrate; the traces on the hybrid substrate are microstrip lines. (Courtesy Raytheon Systems Company.)

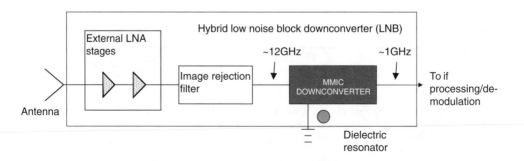

FIGURE 24.3 Application of MMICs in Ku-Band Direct Broadcast Satellite downconverters. The antenna is typically a small parabolic dish.

24.1.4   GaAs MMICs in Comparison to Silicon VLSI Computer Chips

Everyone is familiar with silicon digital IC chips or, at least, the enormous impact silicon-based very large scale integrated (VLSI) circuits have had on the computer industry. Silicon computer chips are digital circuits that contain hundreds to thousands of transistors on each chip. In a digital circuit the transistors are used as switches that are in one of two possible states depending on the "logic" voltage across a pair of terminals. The information processed by a digital circuit consists of a sequence of "1's" and "0's", which translate into logic voltages as the signal passes through the digital IC. Noise distorts the logic waveform in much the same way that it distorts a sinusoidal signal, however, as long as the signal distortion due to noise is not severe, the digital circuitry can assign the correct (discrete) logic levels to the signal as it is processed. Signal interpretation errors that occur due to noise are measured in terms of a bit error rate (BER).

The speed of the digital processing is related to how fast the transistors can switch between one state and another, among other factors. Because of certain material factors, such as electron mobility, digital circuits made on GaAs have been demonstrated to have speed advantages over silicon digital ICs, however, the speed advantages have not been considered by the majority of companies to outweigh the significant economic advantages of well established, lower cost, silicon processing technology.

Because of the large volumes of Silicon chips that have been produced over the last 20 years silicon processing techniques are significantly more established and in many cases standardized as compared to GaAs processing techniques which still vary widely from foundry to foundry. The digital nature of the signals and operation modes of transistors in digital ICs makes uniformity between digital ICs, and even similar ICs made by different manufacturers, much easier to achieve than achieving uniformity with analog GaAs MMICs.

In contrast, GaAs MMICs are analog circuits that usually contain less than 10 transistors on a typical chip. The analog signals (can take on any value between certain limits) processed may, generally, be thought of as combinations of noisy sinusoidal signals. Bias voltages are applied to the transistors in such a way that each transistor will respond in one of several predetermined ways to an applied input signal. One common use for microwave transistors is amplification, whereby the result of a signal passing through the transistor is for it to be boosted by an amount determined by the gain of the transistor.

A complication that arises is that no two transistors are identical in the analog sense. That is, taking gain for example, there will be a statistical distribution of gain for a set of amplifier chips measured on the same GaAs MMIC wafer, a different (wider) set of statistics applies to variations in gain from wafer-to-wafer for the same design. These variations are caused primarily by variations in transistors, but also by variations in other components that make up the MMIC, including MIM capacitors, spiral inductors, film resistors, and transmission line interconnects. Successful foundries are able to control the variations within acceptable limits, in order to achieve a satisfactory yield of chips meeting a customer's requirements; however, translating a MMIC design mask set to a different manufacturing foundry, generally, requires considerable effort on the part of the designer and the foundry.

24.1.5   MMIC Yield and Cost Considerations

Yield is an important concept for MMICs and refers to the percentage of circuits on a given wafer with acceptable performance relative to the total number of circuits fabricated. Since yield may be defined at several points in the MMIC process, it must be interpreted carefully.

- DC yield is the number of circuits whose voltage and currents measured at DC (zero frequency) conditions meets fall within acceptable limits.
- Radio frequency (RF) yield is generally defined as the number of circuits that have acceptable RF/microwave performance when measured "on-wafer," before circuit dicing.
- Packaged RF yield is the final determination of the number of acceptable MMIC products that have been assembled using the fabricated MMIC chips.

If measured in terms of the total number of circuits fabricated each of these yields will be successively lower numbers. For typical foundries DC yields exceed 90% , while packaged yields may be around 50%. Final packaged RF yield depends heavily on the difficulty of the RF specifications, the uniformity of the process (achieved by statistical process control), as well as how sensitive the RF performance of the circuit design is to fabrication variations.

The costs involved with MMICs include

- Material
- Design
- Mask set preparation
- Wafer processing
- Capital equipment
- Testing
- Packaging
- Inspection

A typical wafer run may cost \$20,000–50,000 with \$5,000–10,000 for the mask set alone, and design costs excluded! per-chip MMIC costs are determined by

- Difficulty of design specifications
- Yield
- Material (wafer size and quality)
- Production volume
- Degree of automation

Some 1989–1990 example prices for MMIC chips are as follows [13]:

- 1–5 GHz wideband amplifier—\$30.00
- 2–8 GHz wideband amplifier—\$45.00
- 6–18 GHz wideband amplifier—\$100.00
- DC-12 GHz attenuator—\$60.00
- DBS downconverter chip—\$10.00

In comparison, example prices for 1999 MMIC chips [14]

- DBS downconverter chip < \$2.50
- DC-8 GHz HBT MMIC amp. ~ \$3
- Packaged power FET MMIC \$1 for ~2 GHz
- 4–7 and 8–11 GHz high power/high eff. Power amps. \$85 (2 W)–\$330 (12 W)

As can be seen above, prices vary widely and the key driver that dictates the lower end of the chip prices is the highly competitive price pressure of high volume commercial applications. Prices for low volume specials at high frequency have not moved much since 1989, however, 2006 prices for higher volume chips are about a factor of 2 lower than the above prices [12].

24.1.6 Silicon versus GaAs for Microwave Integrated Circuits

The subject of silicon versus GaAs has been a hotly debated subject since the beginning of the MMIC concept in around 1965 (see Section 24.1.2). Although a comprehensive treatment of this issue is outside of the main scope of interest, for the present treatment, a brief discussion will be given. Two of the main discriminating issues between the technologies are microwave transistor performance and the loss of the semi-conductor when used as a semi-insulating substrate for passive components.

A comparison of relevant physical parameters for silicon and GaAs materials is given in Table 24.2. The dielectric constants of the materials mainly affects the velocity of propagation down transmission line

TABLE 24.2   Properties of GaAs, Silicon, and Common Insulating Substrates

Property	GaAs	Silicon	Semi-Insulating GaAs	Semi-Insulating Silicon	Sapphire	Alumina
Dielectric constant	12.9	11.7	12.9	11.7	11.6	9.7
Thermal conductor (W/cm-K)	0.46 (Fair)	1.45 (Good)	0.46 (Fair)	1.45 (Good)	0.46 (Fair)	0.37 (Fair)
Resistivity (Ω-cm)	—	—	10^7–10^9 (Fair)	10^3–10^5 (Poor)	$>10^{14}$ (Good)	10^{11}–10^{14} (Good)
Elec. mobility	4300* (Best)	700* (Good)	—	—	—	—
Sat. elec. velocity	1.3×10^7 (Best)	9×106 (Good)	—	—	—	—

* At a doping concentration of 10^{17}/cm^3.

Source: Adapted from Pucel, R., ed., *Monolithic Microwave Integrated Circuits*, IEEE Press, 1985, p. 2. With permission.

interconnects, and for the materials compared this parameter is on the same order. For the other factors considered significant differences are observed. The thermal conductivity, a measure of how efficiently the substrate conducts heat (generated by DC currents) away from the transistors, is best for Silicon and one key advantage of the silicon approach. This advantage is offset by silicon's lower mobility and lower resistivity. Mobility in a semiconductor is a measure of how easily electrons can move through the "doped" region of the semiconductor (see discussion of FET operation in the following section). The mobility, as well as the saturated velocity, have a strong influence on the maximum frequency that a microwave transistor can have useful gain.

Turning our attention to passive component operation, GaAs has much better properties for lower loss passive circuit realization. With the exception of a resistor, the ideal passive component is a transmission line, inductor, or capacitor that causes no signal loss. Resistivity is a measure of how "resistant" the substrate is to leakage currents that could flow, for example, from the top conductor of a microstrip line and the ground plane below. Looking first at the properties of the insulators Sapphire and Alumina, the resistivity is seen to be quite high. Semi-insulating GaAs is almost as high, and silicon has the lowest resistivity (highest leakage currents for a given voltage). GaAs has a radiation tolerance advantage as well, although silicon-on-insulator (SOI) has been used with some success in this area at Honeywell and Peregrine [15].

These considerations have led many companies to historically invest heavily in GaAs technology for microwave applications. However, the use of silicon RFICs is now dominant in wireless applications like GPS, Bluetooth, Zig Bee, 802.11a-b-g, WiFi, UWB, and automotive radar, which according to one RFIC design expert are almost 100% Silicon [15].

Also, atleast one company has pushed silicon-based microwave ICs into Ku-band frequencies to compete in the DBS satellite receiver market [16]. Silicon–Germanium Heterojunction Bipolar Transistor Technology is also paving the way for increasing the applicability of silicon technology to even higher frequencies.

To close this section, let us summarize an industry viewpoint on a few areas where GaAs is still better than silicon [17]:

1. RF high power/linearity switches. You can make CMOS switches, but it just is not as good as you can do with GaAs due to the lower breakdown voltage and poor substrate (loss).
2. High power output—once you get above about 1 W, silicon, or SiGe runs out of steam, particularly linear power. Silicon-based PAs have made it into cell phones, but only so far on a limited basis, for example, in Asia where connection over distance is not as much of a concern as in the United States, and hence lower power output requirements.

3. High frequency—GaAs and III–V materials (e.g., AlGas, InP, and related composites) still have an advantage. To be sure, papers reporting high frequency integrated transceiver circuits at 60 GHz and above are starting to appear [18], including an interesting development of a power amplifier at 77 GHz [19].

Generally beyond that, GaAs has higher breakdown, higher substrate isolation (hence better Q inductors), better caps, and higher fTs/fMAXs (for the moment). Silicon has better integration potential, complementary devices (hole mobility in GaAs is very bad) and is available in better mass production capability than GaAs.

It is also worth mentioning that progress continues to be made in both silicon and GaAs (as well as related II–V) technologies and that today's silicon limits tend to be pushed further and further out with each passing year. While, it is likely that GaAs material advantages will always translate into advantages for certain MMIC applications, the limits will continue to be challenged and moved toward higher performance. The other factor that will affect this list in the future is the evolution in radio systems. Today's handset-to-basestation requirements give GaAs an advantage in handset PAs. If system architectures of the future move to shorter distances between basestations, much of the GaAs advantage goes away. In contrast, if direct-to-satellite radios become pervasive, GaAs gains significant advantage. While RF system architectures of the future are not known, they will have an impact on the GaAs–silicon technology use.

24.2 Basic Principles of GaAs MESFETs and HEMTs

24.2.1 Basic MESFET Structure

One of the primary active devices in a GaAs MMIC is the metal-electrode-semiconductor FET, or MESFET. In the past 10 years or so, the heterojunction bipolar transistor (HBT) has emerged as an important alternative, especially for power amplifiers and oscillator applications. The reader is referred to one of the many excellent treatments of HBT device technology for coverage of that topic [21].

The basic construction of a MESFET is shown in Figures 24.4 and 24.5. An "active layer" is first formed on top of a semi-insulating GaAs substrate by intentionally introducing an n-type impurity onto the surface of the GaAs, and isolating specific channel regions. These channel regions are semi-conducting in

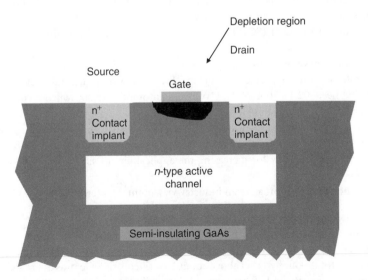

FIGURE 24.4 Cross section of a MESFET transistor. In operation, current I_{ds} flows between the gate and drain terminals through the doped n-type active channel. An AC voltage applied to the gate modulates the size of the depletion region causing the I_{ds} current to be modulated as well. Notice that in a MMIC the doped n-type region is restricted to the transistor region leaving semi-insulating GaAs outside to serve as a passive device substrate.

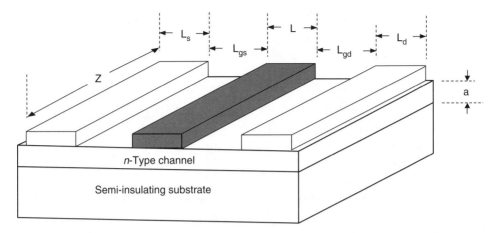

FIGURE 24.5 Aspect view of MESFET. Important dimensions to note are the gate length (L_g) and the gate width (Z). (From J. M. Golio, Ed., Microwave MESFETs and HEMTs, Artech House 1991, p. 28. With permission.)

that they contain free electrons that are available for current flow. When a metal is placed in direct contact with a semi-conductor, as in the case of the gate, a "Schottky diode" is formed. One of the consequences of this is that a natural "depletion region," a region depleted of available electrons, is formed under the gate. A diode allows current flow easily in one direction, while impeding current flow in the other direction. In the case of a MESFET gate, a positive bias voltage between the gate and the source "turns on" the diode and allows current to flow between the gate and the source through the substrate. A negative bias between the gate and the drain "turns off" the diode and blocks current flow, it also increases the depth of the depletion region under the gate.

In contrast to the gate contact, the drain and source contacts are made using what are called "ohmic contacts." In an ohmic contact, current can flow freely in both directions. Whether an ohmic contact or Schottky diode is formed at the metal/semi-conductor interface is determined by the composition of the metal placed on the interface and the doping of the semi-conductor region directly under the metal. The introduction of "pocket $n+$ implants" help form the ohmic contacts in the FET structure illustrated in Figure 24.4. In the absence of the gate, the structure formed by the active channel in combination with the drain and source contacts essentially behave as a resistor obeying ohms law. In fact, this is exactly how one type of GaAs-based resistor commonly used in MMICs is made.

24.2.2 FETs in Microwave Applications

The most common way to operate a MESFET, for example, in an amplifier application, is to ground the source (also called "common source" mode), introduce a positive bias voltage between the drain and source and a negative bias voltage between the gate and source. The positive voltage between the drain and source, V_{ds}, causes current, I_{ds}, to flow in the channel.

As negative bias is applied between the gate and source, V_{gs}, the current, I_{ds}, is reduced as the depletion region extends farther and farther into the channel region. The value of current that flows with zero gate to source voltage is called the saturation current, I_{dss}. Eventually, at a sufficiently large negative voltage the channel is completely depleted of free electrons and the current, I_{ds}, will be reduced to essentially zero. This condition is called "pinch-off." In most amplifier applications the negative gate voltage is set to a "bias condition" between 0 V and the pinch-off voltage, V_{po}.

Figure 24.6 gives a simplified view of a FET configured in an amplifier application. An input sinusoidal signal, $V_{gs}(t)$, is shown offset by a negative DC bias voltage. The sinusoidal variation in, V_{gs}, causes a likewise sinusoidal variation in depth of the depletion region that, in turn, creates a sinusoidal variation (or modulation) in the output current. Amplification occurs because small variations in the, V_{gs}, voltage cause relatively large variations in the output current. By passing the output current through a resistance

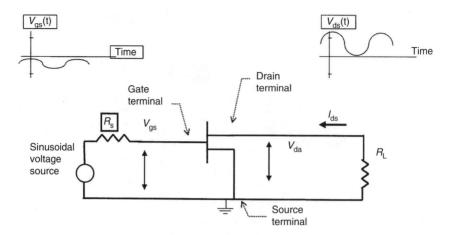

FIGURE 24.6 Simplified schematic of a FET in an amplifier application. Not shown are matching networks needed to match between the source resistance R_s and the FET input impedance, and the load resistance R_L and the FET output impedance. Also omitted are the networks needed to properly apply the DC bias voltages to the device and provide isolation between the RF and DC networks.

RL the voltage waveform, $V_{ds}(t)$, is formed. The, V_{ds}, waveform is shown to have higher amplitude than, V_{gs}, to illustrate the amplification process.

Other common uses, which involve different configurations and biasing arrangements, include use of FETs as the basis for mixers, oscillators (or VCOs).

24.2.3 FET Fabrication Variations and Layout Approaches

Figure 24.7 illustrates a MESFET fabricated with a recessed gate, along with a related type of FET device called a high electron mobility transistor (HEMT) [22]. A recessed gate is used for a number of reasons. First, in processing, it can aid in assuring that the gate stripe is placed in the proper position between the drain and source, and it can also result in better control and uniformity in I_{dss} and V_{po}. A HEMT is a variation of the MESFET structure that, generally, produces a higher performing device, this translates, for example, into a higher gain and a lower noise figure at a given frequency.

In the light of the above cursory understanding of microwave FET structures, some qualitative comments can be made about some of the main factors that cause intended and unintended variations in FET performance. The first factor is the doping profile in the active layer. The doping profile refers to the density of the charge carriers (i.e., electrons) as a function of depth into the substrate. For the simplest type, uniform doping, the density of dopants (intended impurities introduced in the active region) is the same throughout the active region. In practice, there is a natural "tail" of the doping profile that refers to a gradual decrease in doping density as the interface between the active layer and the semi-insulating substrate is approached. One approach to create a more abrupt junction is the so-called "buried p-layer" technique. The buried p-layer influences the distribution of electrons versus depth from the surface of the chip in the "active area" of the chip where the FET devices are made. The idea is to create better definition between where the conducting channel stops and where the non-conducting substrate begins. (More specifically the buried p-layer counteracts the n-type dopants in the "tail" of the doping profile.) The doping profile and density determine the number of charge carriers available for current flow in a given cross section of the active channel. This has a strong influence on the saturation current, I_{dss} and pinch-off voltage, V_{po}. This is also influenced by the depth of the active layer (dimension "a" of Figure 24.5).

Other variables that influence MESFET performance are the gate length and gate width ("L" and "Z" of Figure 24.5). The names for these two parameters are counterintuitive since the gate length refers to

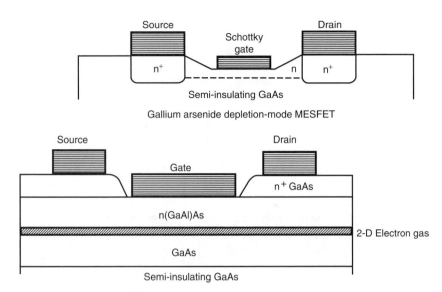

FIGURE 24.7 MESFET (a) and HEMT (b) structures showing "gate recess" structure whose advantages include better control of drain-to-source saturation current I_{dss} and pinch-off voltage V_{po}. (From Goyal, R., Ed., *Monolithic Microwave Integrated Circuits: Technology and Design*, Artech House 1989, p. 113. With permission.)

the shorter of the two dimensions. The gate width and channel depth determine the cross sectional area available for current flow. An increase in gate width increases the value of the saturation current, which translates into the ability to operate the device at higher RF power levels (or AC voltage amplitudes). Typical values for gate widths are in the range of 100 µm for low noise devices to over 2000 µm for high power devices. The gate length is usually the minimum feature size of a device and is the most significant factor in determining the maximum frequency where useful gain can be obtained from a FET; generally, the smaller the gate width, the higher the gain for a given frequency. However, the fabrication difficulty increases, and processing yield decreases, as the gate length is reduced. The difficulty arises from the intricacy in controlling the exact position and length (small dimension) of the gate.

The way a FET is "laid out" can also influence performance. Figure 24.8 shows plan views of common ways by which FETs are actually "laid out" for fabrication [23]. The layout of the FET affects what are called "external parasitics" which are undesired effects that can be modeled as a combination of capacitors, inductors, and resistors added to the basic FET electrical model.

In MMIC fabrication, variations in the most of the above-mentioned parameters are a natural consequence of a real process. These variations cause variations in observed FET performance even for identical microwave FETs made using the same layout geometry on the same wafer. Certainly there are many more subtle factors that influence performance, but the factors considered here should give some intuitive understanding of how unavoidable variations in the physical structure of fabricated FETs cause variations in microwave performance. As previously mentioned, successful GaAs MMIC foundries use statistical process control methods to produce FET devices within acceptable limits of uniformity between devices.

24.3 MMIC Lumped Elements: Resistors, Capacitors, and Inductors

24.3.1 MMIC Resistors

Figure 24.9 shows three common resistor types used in GaAs MMICs. For MMIC resistors the type of resistor material, and the length and width of the resistor determine the value of the resistance [24].

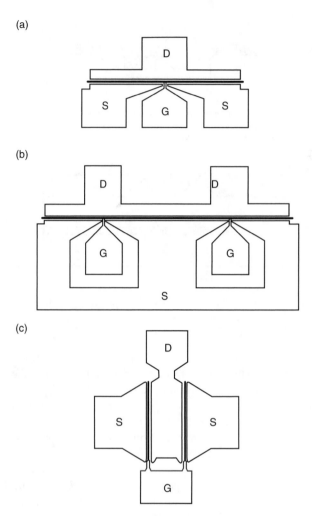

FIGURE 24.8   Three common layout approaches for MESFETs and HEMTs. (From Ladbrooke, P., *MMIC Design: GaAs FETs and HEMTs*, Artech House, 1989, p. 92. With permission.)

In practice, there are also unwanted "parasitic" effects associated with MMIC resistors which can be modeled, generally, as a combination of series inductance, and capacitance to ground, in addition, to the basic resistance of the component.

24.3.2 MMIC Capacitors

The most commonly used type of MMIC capacitor is the metal-insulator-metal capacitor shown in Figure 24.10 [25]. In a MIM capacitor, the value of capacitance is determined from the area of the overlapping metal (smaller dimension of two overlapping plates), the dielectric constant ε of the insulator material, typically silicon nitride, and the thickness of the insulator.

For small value of capacitance, less than 0.5 pF, various arrangements of coupled lines can be used as illustrated in Figure 24.11 [26]. For these capacitors the capacitance is determined from the width and spacing of strips on the surface of the wafer.

At microwave frequencies, "parasitic" effects limit the performance of these capacitors. The two main effects are signal loss due to leakage currents, as measured by the quality factor Q of the capacitor, and a self-resonance frequency, beyond which the component no longer behaves as a capacitor. The final wafer

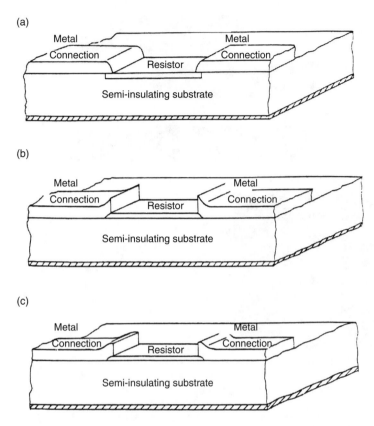

FIGURE 24.9 Common MMIC resistor fabrication approaches: (a) implanted GaAs, (b) mesa-etched/epitaxialy grown GaAs, and (c) thin film (e.g., TaN). (From Goyal, R., Ed., *Monolithic Microwave Integrated Circuits: Technology and Design,* Artech House 1989, p. 342. With permission.)

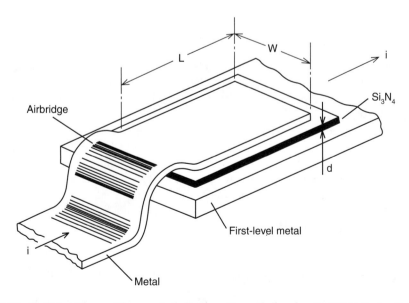

FIGURE 24.10 Metal-insulator-metal conceptual diagram. (From Ladbrooke, P., *MMIC Design GaAs FETs and HEMTs,* Artech House, 1989, p. 46. With permission.)

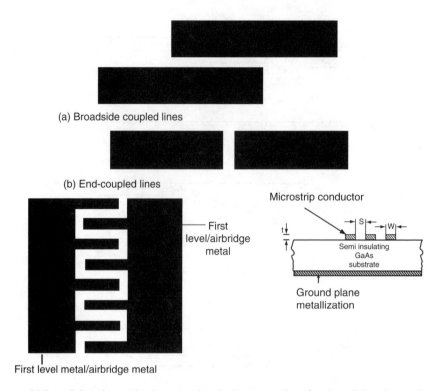

(a) Broadside coupled lines

(b) End-coupled lines

(c) 'Interdigitated capacitor layout and vertical cross section of an interdigitated capacitor.

FIGURE 24.11   MMIC approaches for small valued capacitors. (From Goyal, R., Ed. *Monolithic Microwave Integrated Circuits: Technology and Design*, Artech House, 1989, p. 331. With permission.).

or chip thickness can have a strong influence on these parasitic effects and the associated performance of the capacitors in the circuit. Parasitic effects must be accurately modeled for successful MMIC design usage.

24.3.3 MMIC Inductors

MMIC Inductors are realized with narrow strips of metal on the surface of the chip. Various layout geometries are possible as illustrated in Figure 24.12 [27]. The choice of layout is dictated mainly by the available space and the amount of inductance L that is required in the circuit application, with the spiral inductors providing the highest values.

The nominal value of inductance achievable from strip inductors is determined from the total length, for the simpler, layouts, and by the number of turns, spacing and line width for the spiral inductors.

At microwave frequencies, "parasitic" effects limit the performance of these inductors. The two main effects are signal loss due to leakage currents, as measured by the quality factor Q of the capacitor, and a self-resonance frequency, beyond which the component no longer behaves as an inductor. The final thickness of the substrate influences not only the nominal value of inductance, but also the quality factor and self-resonance frequency. Inductor parasitic effects must be accurately modeled for successful MMIC design usage.

24.3.4 Airbridge Spiral Inductors

Airbridge spiral inductors are distinguished from conventional spiral inductors by the presence of metal traces that make up the inductor suspended from the top of the substrate using MMIC airbridge technology. The MMIC airbridges are, generally, used to allow crossing lines to jump over one another

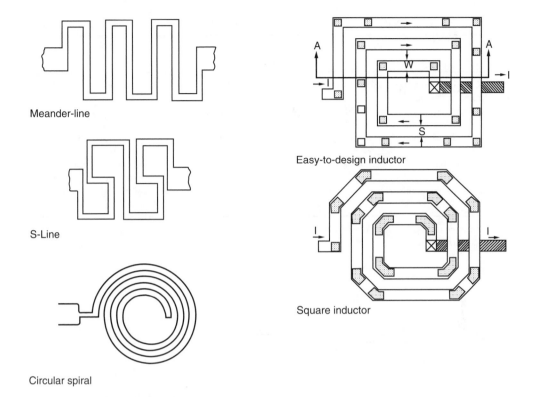

Meander-line

S-Line

Circular spiral

Easy-to-design inductor

Square inductor

FIGURE 24.12 Various MMIC inductor layouts. (From Goyal, R., Ed. *Monolithic Microwave Integrated Circuits: Technology and Design,* Artech House, 1989, pp. 320–325. With permission.).

without touching and are, almost invariably, used in conventional spiral inductors to allow the center of the spiral to be brought through the turns of the spiral inductor for connection to the circuit outside of the spiral. In an airbridge spiral inductor all the turns are suspended off the substrate using a series of airbridges supported by metalized posts. The reason for doing this is to improve inductor performance by reducing loss as well as the effective dielectric constant of the lines that make up the spiral. The latter can have the effect of reducing inter-turn capacitance and increasing the resonant frequency of the inductor. Whether or not airbridge inductors are "worth the effort" is a debatable subject as the airbridge process is an important yield-limiting factor. This means circuit failure due to collapsed airbridges, for example, occur at an increasing rate the more airbridges that are used.

24.3.5 Typical Values for MMIC Lumped Elements

Each MMIC fabrication foundry sets its limits on the geometrical dimensions and range of materials available to the designer in constructing the MMIC lumped elements discussed in the previous section. Accordingly, the range of different resistor, capacitor and inductor values available for design will vary from foundry to foundry. With this understanding a "typical" set of element values associated with MMIC lumped elements are presented in Table 24.3 [28].

24.4 MMIC Processing and Mask Sets

Most common MMIC processes in the industry can be characterized as having 0.25 μm to 0.5 μm gate lengths fabricated on a GaAs wafer whose final thickness is 100 μm, or 4 mils. The backside of the wafer has plated gold; via holes are used to connect from the backside of the wafer to the topside of the wafer.

TABLE 24.3   Ranges of MMIC Lumped Element Values Available to Designer for a "Typical" Foundry Process

Type	Value	Q-Factor (10 GHz)	Dielectric or Metal	Application
INDUCTOR: Single loop, meander line, and so on	0.01–0.5 nH	30–60	Plated gold	Matching
INDUCTOR: Spiral	0.5–10 nH	20–40	Plated gold	Matching, DC power (bias) supply choke
CAPACITOR: Coupled lines	0.001–0.05 pF		Plated or un-plated gold	Matching, RF/DC signal separation
RESISTOR: Thin film	5 Ω to 1 kΩ		NiCr, TaN	DC bias ckts, feedback, matching, stabilization
RESISTOR: GaAs monolithic	10 Ω to 10 kΩ		Implanted or epitaxial GaAs	DC biasing, feedback, matching, stabilization

Source: Adapted from Goyal, R., *Monolithic Microwave Integrated Circuits: Technology and Design,* Artech House, 1989, p. 320. With permission.

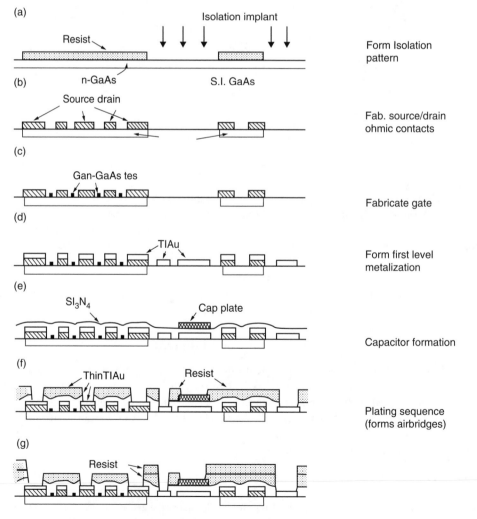

FIGURE 24.13   Conceptual diagrams illustrating flow for typical MMIC process. (From Williams, R., *Modern GaAs Processing Methods,* Artech House, 1990, pp. 11–15. With permission.).

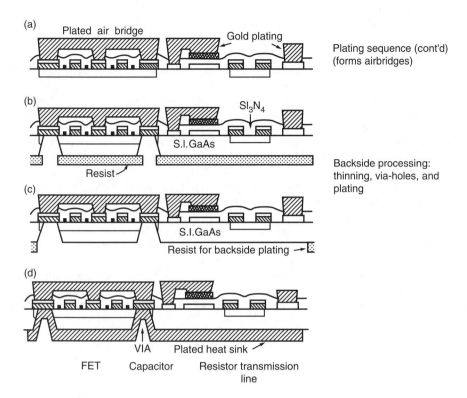

FIGURE 24.13 Continued.

Although specific procedures and steps vary from foundry to foundry, Figure 24.13 [29] illustrates a typical process.

References

1. Pucel, R. A., "Design considerations for monolithic microwave circuits," *IEEE Transactions on Microwave Theory and Techniques*, vol. MTT-29, pp. 513–534, June 1981.
2. Pucel, R.A. ed., *Monolithic Microwave Integrated Circuits*, IEEE Press, 1985, p. 1.
3. Ladbrooke, P., *MMIC Design GaAs FETs and HEMTs*, Artech House, 1989, p. 29.
4. Pucel, R.A. ed., *Monolithic Microwave Integrated Circuits*, IEEE Press, 1985, p. 1–2.
5. Hyltin, T.M. "Microstrip transmission on semiconductor substrates," *IEEE Transactions Microwave Theory and Techniques,* vol. MTT-13, pp. 777–781. Nov. 1965.
6. Mehal, E. and Wacker, R. "GaAs integrated microwave circuits," *IEEE Transactions Microwave Theory and Techniques,* vol. MTT-16, pp. 451–454, July 1968.
7. Pengelly, R. S. and Turner, J.A. "Monolithic broadband GaAs FET amplifiers," *Electronics Letters*, vol. 12, pp. 251–252, May 13, 1976.
8. Cressler, J.D. and Niu, G., *Silicon-Germanium Heterojunction Bipolar Transistors*, Artech House, 2003.
9. T. Lee, *The Design of CMOS Radio-Frequency Integrated Circuits*, Cambridge University Press, 1998.
10. Joseph, A.J., Dunn, J., Freeman, G., Harame, D.L., Coolbaugh, D., Groves, R., Stein, K.J., et al. "Product applications and technology directions with SiGe BiCMOS", *IEEE Journal of Solid-State Circuits*, vol. 38 (9), pp. 1471–1478, Sept. 2003.
11. Brehm, G., Triquint Semiconductor, *personal communication*, Mar., 2006.
12. Pucel, R.A., ed., *Monolithic Microwave Integrated Circuits*, IEEE Press 1985, p. 1.
13. Goyal, R., *Monolithic Microwave Integrated Circuits: Technology and Design*, Artech House, 1989, p. 17.
14. Pengelly R., Cree, and Schmitz, Bert M/A-Com, *personal communication*, Aug., 1999.

15. Wyatt, M., Insyte Corporation, *personal communication*, Feb. 2006.
16. Frank, Mike, Hewlett Packard Company, *personal communication*, Dec. 22, 1995.
17. Kane, B., Insyte Corporation, *personal communication*, Feb., 2006.
18. Floyd, B.A., Reynolds, S.K., Pfeiffer, U.R., Zwick, T., Beukema, T., and Gaucher, B. "SiGe bipolar transceiver circuits operating at 60 GHz,"
19. Komijani, A. and Hajimiri, A., "A wideband 77GHz, 17.5 dBm power amplifier in silicon," *Proceedings of the Custom Integrated Circuits Conference, 2005. IEEE 2005*, 18–21, pp. 566–569, Sept. 2005.
20. Pucel, R., ed., *Monolithic Microwave Integrated Circuits*, IEEE Press, 1985, p. 2.
21. Liu, W. *Fundamentals of III-V Devices: HBTs, MESFETs, and HFETs/HEMTs*, Wiley-Interscience, 1999.
22. Goyal, R., Ed., *Monolithic Microwave Integrated Circuits: Technology and Design*, Artech House 1989, p. 113.
23. Ladbrooke, P., *MMIC Design: GaAs FETs and HEMTs*, Artech House, 1989, p. 92.
24. Goyal, R., Ed., *Monolithic Microwave Integrated Circuits: Technology and Design*, Artech House 1989, p. 342.
25. Ladbrooke, P., *MMIC Design GaAs FETs and HEMTs*, Artech House, 1989, p. 46.
26. Goyal, R., Ed. *Monolithic Microwave Integrated Circuits: Technology and Design*, Artech House, 1989, p. 331.
27. Goyal, R., Ed. *Monolithic Microwave Integrated Circuits: Technology and Design*, Artech House, 1989, pp. 320–325.
28. Goyal, R., *Monolithic Microwave Integrated Circuits: Technology and Design*, Artech House, 1989, p. 320.
29. Williams, R., *Modern GaAs Processing Methods*, Artech House, 1990, pp. 11–15.

25

Bringing RFICs to the Market

John C. Cowles
Analog Devices—Northwest Labs

25.1 Introduction: A Brief History of Radios

The art and science of RFIC design developed over a period of at least 150 years. The fundamental work of the philosopher–scientists of the nineteenth century paved the way with invaluable empirical discoveries related to electricity and magnetism. The identities of these early players are known to us today through the basic units that electrical engineers have been using over the past century: Volta, Ohm, Ampere, Henry, and Faraday. In 1873, Maxwell laid the theoretical foundation for the propagation of electromagnetic energy through space in his famous treatise. Maxwell's equations demonstrated that electric and magnetic fields are inexorably coupled and fueled by charges and currents.

The nineteenth century also brought a flurry of activity as electricity was used in practice to transmit information over a distance, first through wires and then through air. Samuel Morse's invention of the telegraph in 1835, accompanied by his digital, self-named code, and Alexander Graham Bell's invention of the telephone in 1876 demonstrated to the world the potential of electricity as an effective means for communications. In the late 1800s, Hertz proved the existence of electromagnetic waves and was shortly followed by Marconi, who in 1896 demonstrated the first wireless transmission by sending telegraph signals over 2 km from ship to shore. Thus, began the age of wireless transmission.

In the first quarter of the twentieth century, broadcasts began in earnest with Reginald Fesseden getting first credit by broadcasting Christmas carols with a crude form of amplitude modulation in 1906. Shortly thereafter, in the 1920s the first commercial radio and television transmissions began operation. Such broadcasts required radio transmitters and receivers along with the means to carry the information: modulation. The great Edward Armstrong is credited with the invention of frequency modulation, the regenerative receiver and the super-heterodyne receiver, which had been the work-horse until the advent of digital direct-conversion radios. In a matter of 25 years, the world was brought closer together than ever before by radio and television, although only few could afford the equipment for technology since manufacturing had not yet come of age.

The next step involved a confluence of disparate developments that propelled the era of electronics to the next level. Coincident with the advancements in radios and broadcasting was the observation of effects not predicted by the classical sciences, which led to the birth of quantum mechanics. The unraveling of the underlying physical processes in materials and the harnessing of their properties launched the era of vacuum tubes and the seeds for future solid-state electronic devices. As early as 1883, Thomas Edison stumbled on a crude thermionic diode while in 1906 DeForest invented the Audion, the first triode tube amplifier. World War II gave impetus to the development of the RADAR, the magnetron, the klystron and the traveling-wave-tube, further advancing the generation and detection of signals. There was also the appearance of the digital computer in the form of the ENIAC developed by Eckert and Mauchly in 1946, although there was no thought as yet of combining it with a radio. This period of discovery reached a pinnacle with the discovery of the point-contact transistor in 1947 by Shockley, Brattain, and Bardeen. Ironically, their "discovery" of bipolar action was allegedly a fortuitous by-product of a failed experiment in search of a field-effect device. In time, the first true JFETs and MOSFETs followed suit and launched the age of solid-state integrated circuits.

By the mid-1960s there was a growing semiconductor industry led by Jack Kilby of Texas Instruments and Robert Noyce of Fairchild among others. To supplement the technology, Prof. Robert Pederson of Berkeley led the development of the simulation tool SPICE which enabled solid-state circuits to grow in complexity by replacing laborious bread-boarding with computer simulations. As levels of integration increased towards VLSI, the microprocessor and then the personal computer appeared in the market-place, followed by numerous consumer items for entertainment that also relied on digital technology such as video camcorders, DVD/CD/MP3 players, and digital cameras. The regular release of consumer products containing complex digital and analog ICs at affordable prices reflected a mature solid-state electronics industry consisting of robust device technologies, accurate models, sophisticated simulation tools, organized design teams, and a strong sense of entrepreneurship.

Coincident with the growth of consumer-oriented electronics was the development of microwave electronics, primarily driven by defense and satellite industries. These niche applications drove the development of materials, design philosophies, and test techniques required to deliver performance above 1 GHz. A typical early microwave amplifier might consist of a discrete GaAs MESFET or silicon BJT mounted on FR4 board material along with passive matching networks either as printed transmission lines or as mounted components. In the 1980s and 1990s, the fabrication technologies for III–V materials, their modeling and simulation tools had reached the level where multitransistor, multifunction, monolithic ICs could be designed with confidence, the era of microwave (MICs) and millimeter-wave ICs (MMICs). The devices, circuit topologies, spawning testing methods and even the nature of the figures of merit were foreign to traditional analog and digital designers. It would take yet one more link to bring these disparate worlds together to forge the concept of the RFIC.

The confluence of radio architectures, semiconductor technologies, low-cost manufacturing, and a strong consumer electronics market required a key catalyst before launching the age of wireless technology. The spark was simply the latent need for wireless connectivity by the public at large. Prosaic devices as the pager, the garage door opener, and the television remote control were the primitive ancestors of the cellular phone that appeared in the late 1980s along with the Internet. The expectation today is wireless access to digital information anywhere, at any time, with anyone, using whichever access method is available: GSM, UMTS, WiMax, WiFi, and so on. As the public has adapted quickly to the new opportunities afforded by such access to information, the wireless industry has also adapted to create new markets and services. The wireless industry, in cooperation and in competition with wired access methods such as optical-fiber, CATV, and Ethernet, are vying for the delivery of information to the public and enterprise. The distinction between telephony, television, internet, satellite, and computer is quickly vanishing, opening opportunities for service providers to compete for customer access. There is a "productivity cycle" at work, shown in Figure 25.1, in which radio techniques, semiconductor technologies, and bandwidth constraints are in constant flux, one driving the other.

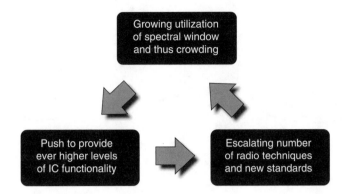

FIGURE 25.1   The productivity cycle drives innovation and progress.

The on-going challenges are perfectly clear:

1. Low cost to reach the widest consumer market
2. Re-configurability and flexibility to adapt to changing channels and standards
3. More efficient use of bandwidth to increase data capacity
4. Higher frequency allocations to increase available bandwidth

Addressing these challenges has required the marriage of classical high-speed analog and microwave designs to deliver timely, robust products at the appropriate integration level, at a competitive price while operating at microwave frequencies. The possibilities are endless as the public demands more access and services which in turn keep the cycle in motion.

25.2   Preliminaries

The RFIC is the building block that links the digital engines that perform high-level mediation of information transfer and processing and the antennas that form the air-interface to the radio. Before the product design is to begin there are many questions, both technical and strategic, to be asked. First of all, one must identify the basic technical function and articulate, as completely as possible at this point, the specifications. In addition to the key metrics, this might include the package with pin descriptions, the proposed technical approach and even a rough die layout, and any applications' details indicating how the part might be used. This information is nothing more than a concept data sheet that describes the product even before committing heavy design and manufacturing resources. It is surprising how many critical decisions can be made in the process of creating such a data sheet.

Underlying the technical requirements captured in the concept data sheet is a set of strategic requirements. As shown in Figure 25.2, there are market forces and corporate realities which must be considered. First of all, is the proposed product of such entrepreneurial vigor that it is likely to create new markets or is it an upgrade of an existing product or an answer to competition? The classic market questions of time-to-market and cost/profit must be clear and understood. If the product has no precedence, one must rely on experience with customers and an anticipation of what customers WILL need. In defining timing and costing, it is essential to understand the target market. The so-called "vertical markets" are often driven by consumer electronics and demand very specific functionality and specifications, low cost, and quick cycle times. The volumes tend to go to a few customers and typically skyrocket for a period of time and then disappear. The "horizontal markets" are driven by a slower-paced infrastructure market that requires high-performance parts but is not under the same time and cost pressure. The customer base tends to be broad with moderate volumes growing systematically over time and holding their value.

The other component required for product development is, of course, corporate resources. This covers the design expertise, the available semiconductor fabrication technologies, the product engineering

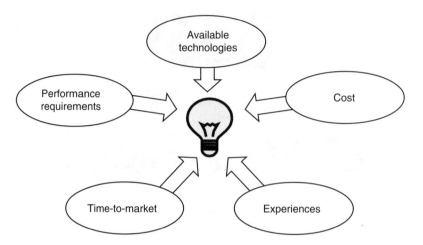

FIGURE 25.2   There are many variables to consider when proposing a product.

required to bring a product to a market release and customer applications' support once the part in released. If any of these components is miscalculated, the product development is likely to fail. It is unlikely that a junior digital designer with access to a low-cost CMOS technology will produce a break-through RF product in GaAs without adequate training in a short TTM scenario. It must be remembered that circuits, in themselves, are not products; products require a team of dedicated specialists to deal with the device modeling, the actual design, the fabrication, the packaging, the evaluation, the reliability, the data sheet, and so on. Any weak link in this chain is likely to be catastrophic.

25.3   The Design Process

The design of a product is built on the foundation established early on during the concept phase. At this point, there is a concept data sheet with preliminary information to guide the design phase and the necessary resources allocated to the task. As noted earlier, this includes the designer's capabilities, the fabrication technology, the development schedule with key milestones, and the greater team responsible for bringing the design into production. The design phase enters its first part in which detailed technical approaches are investigated to validate whether the proposed concept is feasible. Any building block can be implemented in a myriad of ways. Figure 25.3 illustrates a handful of amplifiers whose embodiments are quite distinct. This is where an understanding of the bandwidth, gain, dynamic range, and impedance levels would eliminate many of the candidate approaches. Similar examples for mixers, oscillators, variable gain amplifiers, and power detectors are shown in Figures 25.4–25.7. The designer should be encouraged to consider all options and investigate new topologies as well. The simulation tools allow for "experiments" that can quickly weed out pretenders and, sometimes, even identify an undiscovered star. When topologies and technologies seem to conflict, the option of integration at the package level can also be an option depending on the cost. Figure 25.8 illustrates a common example of a portable handset power amplifier in which a GaAs power stage and a silicon power detector are integrated on a substrate along with passive lumped and distributed components. There might also be extra functionality to improve the efficiency-distortion trade-off and to protect against dangerous loading conditions. This form on integration has become possible due to advances in packaging that have turned expensive, military module technology into low cost consumer-level components.

During this exploratory, feasibility period, the designer would focus on the areas of greatest technical risk and difficulty, leaving the more mundane details for later. Simultaneously, the market feasibility should also be explored further with customer visits, survey of the competition, and preliminary cost analysis. The team then reaches a critical gating point in which the overall feasibility of the product is

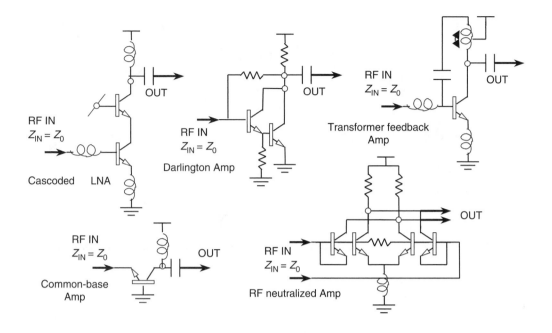

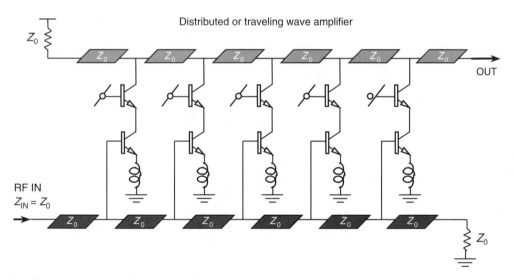

FIGURE 25.3 Which amplifier is appropriate for your application?

reviewed from technical and business points-of-view. From here, there should be an agreement to proceed towards a product, return to feasibility, or stop the work altogether. Note that, as painful as it may be to cancel a development, it is much less painful to do so earlier rather than later. In many cases, the feasibility of a product is not in question, and this phase can be mostly by-passed as long as the goals and requirements are clearly articulated.

A positive feasibility review moves the designer towards a full-blown development. The technical approach is cemented early in this phase with the balance of the time dedicated to fine tuning the design to address robustness and manufacturability. The time spent on these details is paramount to a successful product as it maximizes the production yield, the profits, and customer satisfaction. Simulation sweeps over temperature and over supply and over certain device model parameters quickly reveal weak points in the design. RFIC design has emerged as a symbiosis of high-speed analog with its ties to silicon technology

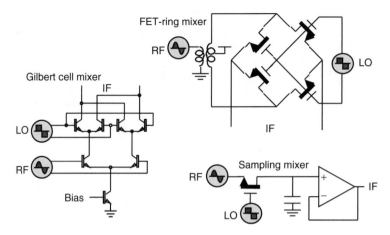

FIGURE 25.4 Mixers have not changed much in 20 years yet few understand the merits of passive and active topologies.

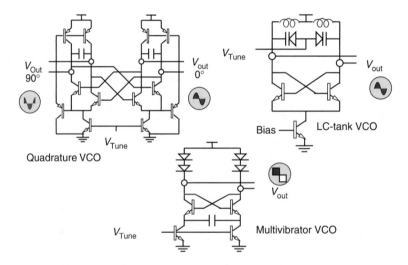

FIGURE 25.5 For a small circuit, oscillators generate more disagreement than any other.

and of more traditional microwave design with its ties to GaAs. Adopting the best aspects of both worlds in design, simulation and layout methodologies is the best toolbox.

Certain practices have long been established in the realm of analog design. Many parameters such as gain are dimensionless and should be defined by ratios of like quantities. The desensitization of such scalar quantities is addressed by use of unit elements of resistors, capacitors, and transistors placed in close proximity and in the same orientation to improve their matching. It is also prudent to use the largest permissible dimensions for these unit elements and surround them with dummies to create a common topography for all. The schematic should give the engineer responsible for the layout the visual clues as to the placement of the unit cells and the dummies. The more clearly the schematic can communicate the intent of the designer, the less is left up to chance. For critical matching, such as in the differential pairs shown in Figure 25.9, components can be connected in a cross-quad arrangement where any temperature or process gradients are spatially averaged. At high frequencies, matching of capacitances in differential circuits may be more important than minimizing them, particularly since minimum sized devices are rarely used due to noise or current handling reasons.

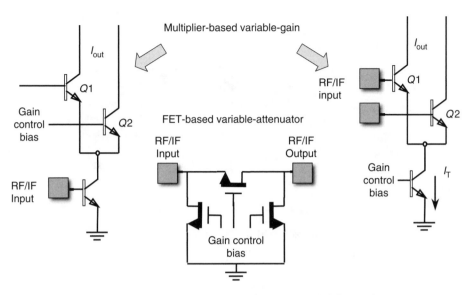

FIGURE 25.6 The classic analog and microwave approaches for gain control differ greatly.

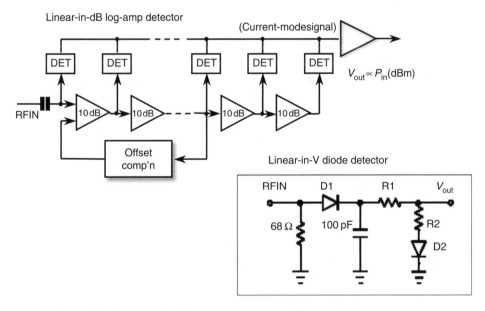

FIGURE 25.7 Power detection has evolved from magic to science in the age of RFICs.

In microwave design, the effect of package and layout parasitics has always had a dramatic impact on the high-frequency performance of circuits. With the increased complexity, integration level and requirements for RFICs, package models and layout parasitic extraction have become critical. Many inexperienced and experienced designers have been caught saying something akin to, "This is a low-frequency circuit so the package model is not important," only to realize later that the 90 GHz transistors in the process were not told about being a low frequency circuit. The accuracy of these parasitic models depends directly on the complexity of the networks that represent them. The iteration between design, package, and layout simulation can be protracted and, at some point, fruitless as many competing effects are present simultaneously. Proper judgment is required to trade-off model accuracy against time to tape-out. It is

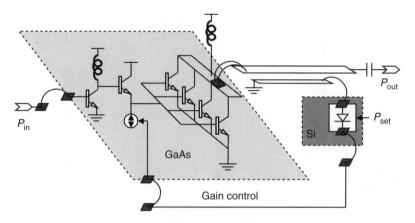

FIGURE 25.8   A cellular PA is often a complex multi-chip system implemented as a low-cost module.

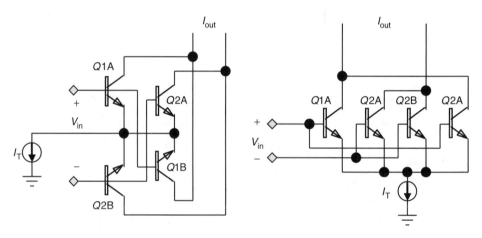

FIGURE 25.9   Cross-quads ubiquitous in analog design have added merit in RFICs. Two forms are shown here.

useful to be able to turn "on" and "off" elements in the package and parasitic network for debugging purposes. This might help identify the need for extra bond-wires or pins for a pesky ground oscillation or the need to balance a pair of differential lines for a lower offset voltage.

The greatest chasm between analog and RF design lies in the design of the bias circuits. While analog IC designers have been using carefully crafted bias currents to achieve a certain precision or stability, microwave designers were just happy to get any current through a device. The era of RFICs has brought the sophistication of biasing to the RF domain. Consider the differential pair in Figure 25.10a. In order to achieve a gain that is independent of temperature, the current must be shaped to be proportional to temperature so that the G_m becomes constant. Now consider a heavily degenerated differential pair in Figure 25.10b. What shape should the current take now that $G_m \sim 1/R$? The answer can be complicated. If constant input capacity or output swing (both related to $I_{Bias}R$) is the important factor, then the current should be made stable with temperature. If the output of a differential pair is resistively loaded to form a CML gate, shown in Figure 25.10c, then the current might be made complementary to temperature (decrease at higher temperatures) since the voltage swing needed to toggle a CML input drops at higher temperatures. There is much to be gained in thinking about the appropriate bias cell. A fair question to pose is whether a formal bias cell is needed or can a simpler bias scheme suffice. The incidental bias for the simple Darlington amplifier in Figure 25.3 might be derived from the supply while the current that defines the reference for a square law detector must be precise to ensure accuracy.

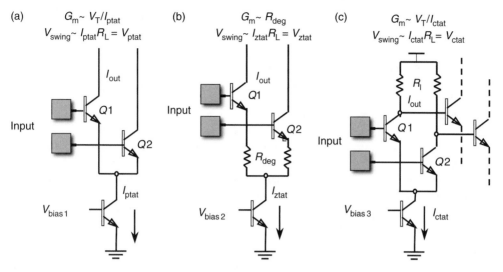

FIGURE 25.10 (a)–(c) Biasing differential pairs requires some forethought on the goals.

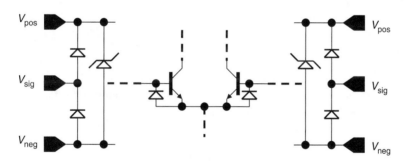

FIGURE 25.11 Typical ESD protection for differential pair signal pins.

An area of robustness that is often neglected is ESD protection. All die and packaged parts are subject to ESD events during handling. It is typical to test every product for a human body model (HBM), machine model (MM), and a charged device model (CDM) up to a certain discharge level. Failing to protect pins can be catastrophic to factory yield and to reliability in the field. Every technology tends to have suggested "good practice," but there is no substitute for thinking ahead. Figure 25.11 shows one example of protecting the input pins of a differential pair. Note that the technology recommendation is to place specially designed ESD diodes from each pin to the supply and ground along with a power ESD diode across the supply. However, it is clear that an ESD event across the input pins would destroy the emitter–base junction that was reverse biased. The solution is to place a diode antiparallel to each emitter–base junction. Under normal operation, these diodes are off; under a pin-to-pin ESD event, the diodes clamp the reverse bias and protect the inputs. It is difficult to anticipate all the possible permutations of ESD events in a complex circuit; here again experience and some forethought are essential.

Once the design is completed, there is usually an event called a design review. The goals of a design review can vary greatly according to the product and the requirements. It is intended to address a broad audience of interested folks ranging from colleagues who share the technical expertise, colleagues from different design disciplines, the marketing team and those who will inherit the silicon and carry it to release and support the customers. The content should address all the attendees' concerns such as achieved technical specification, package and fabrication technologies, circuit techniques of interest and concern, remaining risk factors, robustness, revised cost estimates, schedules, and so on. There will inevitably be a list of actions to investigate and correct if needed. A successful design review hinges on a relevant and participatory audience asking probing questions and giving productive criticism. A passive, uninvolved crowd does the team no favor.

During the design phase, the layout activity should begin. Note that parasitic extraction and validation that the die will fit in the package are necessary to complete the simulation of the RFIC. The greater the realism captured in the schematic, the closer it will be represented in layout. A layout review with a team is just as crucial as the design review. Great patience is required to look for the needle in the haystack. Common topics for discussion are the routing grid of power supplies and ground, critical matching areas, current density along metal lines, ESD protection scheme, unintentional coupling of signals and the bonding diagram associated with the package. As in the design review, there are always issues raised that deserve investigation. The layout and design at this point are inextricably intertwined and tape-out of the mask set only takes place once both are deemed complete by the team. The tape-out authorization is another key gating point where the market and technical merits of the part are once again scrutinized before committing money and resources to mask-making and manufacturing.

25.4 From Silicon to Release

Once the wafers are in process, the team does not go dormant. The time schedules for when material returns must be outlined. The data sheet needs to be updated with a more comprehensive set of specifications, final package, details, key features, and basic connections to the part. In addition to advancing the status of the official data sheet, the update helps define the limits of the test and characterization plans needed to release the part into full production. The evaluation boards can now be designed, go to layout and be reviewed in much the same way the RFIC was scrutinized. The choice of board material, supporting passive components, board traces, proper grounding, supply decoupling and connectors all affect the measurements of the RFIC high-frequency performance.

Once the wafers are finished, the die must be separated, packaged, and mounted on the evaluation board. Characterization and the inevitable debugging phases begin. At this point a full look at the part is critical to help decide whether revisions are necessary. It is common to have parts from wafers that have had certain parameters, such as a resistor value or a doping level, purposefully skewed in order to force certain sensitivities. From nominal and skew material, the robustness of the part to process variations can be inferred. Full temperature and ESD testing complete the characterization. If all seems well at this point, stocking quantities can be generated and the official data sheet can be posted, indicating that the product has been released to the world.

More often, some revisions are necessary. The process of debugging is perhaps the most frustrating and murky of all. In the best of circumstances, the changes may involve only revisions to metallization or to resistors and capacitors. In the worst of cases, a whole new mask set needs to be ordered. It is prudent at this point to re-evaluate the future of this product in terms of time-to-market, consumption of resources, and continued risk. The debug process relies on ingenuity of testing and extreme familiarity with the internal structure of the part. It is absolutely essential for the designer to be present with the test engineer, if they are not one and the same. Quick, resourceful thinking is the only answer here. With some luck and forethought, de-capped ICs, in which the plastic has been etched away, can be probed to gather more information not accessible from the package pins alone.

25.5 Toward the Future

Once the product is released into production, the whole team shifts modes. The applications' support teams go to the field to aid customers and gather feedback to help drive the next development. In the meantime, the design team is already pondering the next concept, tracking the competition, and looking for new product ideas. Just as there is a macro-level productivity engine cycle where markets, technology, and manufacturing chase each other, there is a micro-level cycle in which product definition, design, and release go roundabout. Both cycles yield technological progress, more competitive solutions, and continuous process improvement. This cycle has carried the industry from the first vacuum-tube radios that heated a small room to a video-enabled cellular phone of today which was, coincidentally, inspired by Dick Tracy's watch in the 1930s at the dawn of radios.

Materials Properties

26

Metals

Mike Golio
HVVi Semiconductor

Metals serve many different functions in the realization of RF and microwave products. These functions include

- The wire or guided wave boundary for circuits
- The carrier or structural support for dielectric substrates or semiconductor chips
- The heat sink for devices or circuits, which exhibit high power density
- The reflector element for antennas or screen room applications

Each of these functions results in different requirements that must be met. Thus the optimum metal for each application will vary. Consideration of a wide range of electrical, mechanical, chemical and thermal properties for each metal is needed to choose an appropriate metal for most applications.

26.1 Resistance, Resistivity, and Conductivity

A first-order consideration in the choice of metals for many electrical applications is the electrical resistance of the metal conductor. DC resistance of a metal rod is given by

$$R = \frac{\rho L}{A} \tag{26.1}$$

where R is the resistance of the rod, ρ is the resistivity of the metal, L is the length of the rod and A is the cross-sectional area of the rod. The dc electrical properties of metals are also sometimes discussed in terms of conductivity. Conductivity is the inverse of resistivity given by

$$\sigma = \frac{1}{\rho} \tag{26.2}$$

where σ is the conductivity of the material. For most applications, high conductivity, or conversely low resistivity, is desirable. The resistivity of a number of metals is listed in Table 26.1. The tabulated data

TABLE 26.1 Electrical Resistivity of Several Metals at
Room Temperature in $10^{-8}\,\Omega$m [6–17]

Metal	Electrical Resistivity
Aluminum	2.7
Antimony	38.6
Beryllium	3.6
Brass	5.6
Chromium	12.7
Copper	1.7
Gold	2.2
Kovar	49.0
Lead	21.1
Magnesium	4.4
Manganese	147.2
Molybdenum	5.3
Nickel	7.0
Palladium	10.4
Platinum	10.4
Silver	1.6
Tantalum	13.0
Titanium	44.9
Tungsten	5.4
Vanadium	22.9
Zinc	6.0

TABLE 26.2 Temperature Coefficient of Metal Resistivity
at Room Temperature in (1/K) [6–17]

Metal	Temperature Coefficient of Resistivity
Aluminum	4.38×10^{-3}
Copper	3.92×10^{-3}
Gold	3.61×10^{-3}
Molybdenum	4.60×10^{-3}
Silver	3.71×10^{-3}

represents an average resistivity value from a number of sources. Values for resistivity of a metal can vary significantly because of small impurity differences between samples. Pure Aluminum may exhibit a 5% increase in resistivity with the introduction of only 0.01% impurities, for example. Metals such as Brass and Kovar are alloys and may exhibit even wider variations in properties based on the percentage of elemental metals in the composition. Brass is an alloy of copper and zinc, while Kovar is a nickel–cobalt ferrous alloy.

The resistivity of metal is also a function of temperature. For small perturbations in temperature, this temperature dependence may be characterized by the equation

$$\alpha = \frac{1}{\rho}\frac{\partial \rho}{\partial T} \tag{26.3}$$

where α is the temperature coefficient of resistivity. The measured temperature dependence of resistivity changes as a function of temperature and can vary significantly at temperatures near 0 K, or well above room temperature. For most applications, a low value for the temperature coefficient of resistivity is desirable. Table 26.2 presents the temperature coefficient of resistivity at room temperature for several metals that might be chosen for their low resistivity values [1,3].

26.2 Skin Depth

An electromagnetic field can penetrate into a conductor only a minute distance at microwave frequencies. The field amplitude decays exponentially from its surface value according to

$$A = e^{-x/\delta_s} \tag{26.4}$$

where x is the normal distance into the conductor measured from the surface, and δ_s is the skin depth. The skin depth or depth of penetration into a metal is defined as the distance the wave must travel in order to decay by an amount equal to $e^{-1} = 0.368$ or 8.686 dB. The skin depth δ_s is given by

$$\delta_s = \frac{1}{\sqrt{\pi f \mu \sigma}} \tag{26.5}$$

where f is the frequency, σ is the metal conductivity, and μ is the permeability of the metal given as

$$\mu = \mu_0 \mu_r \tag{26.6}$$

with μ_0 equal to the permeability of free space and μ_r the relative permeability of the metal. For most metals used as conductors for microwave and RF applications, the relative permeability, $\mu_r = 1$. The relative permeability of ferroelectric materials such as iron and steel are typically on the order of several hundred.

Skin depth is closely related to the shielding effectiveness of a metal since the attenuation of electric field strength into a metal can be expressed as Equation 26.4. For static or low frequency fields, the only method of shielding a space is by surrounding it with a high-permeability material. For RF and microwave frequencies, however, a thin sheet or screen of metal serves as an effective shield from electric fields.

Skin depth can be an important consideration in the development of guided wave and reflecting structures for high frequency work. For the best conductors, skin depth is on the order of microns for 1 GHz fields. Since electric fields cannot penetrate very deeply into a conductor, all current is concentrated near the surface. As conductivity or frequency approach infinity, skin depth approaches zero and the current is contained in a narrower and narrower region. For this reason, only the properties of the surface metal affect RF or microwave resistance. A poor conductor with a thin layer of high conductivity metal will exhibit the same RF conduction properties as a solid high conductivity structure [1,3,5].

26.3 Heat Conduction

One of the uses of metal in the development of RF and microwave parts and modules is as a heat spreader. For many applications that involve high power density electronic components, the efficient removal of heat is of great importance in order to preserve component reliability (Chapter 19 of *RF and Microwave Applications and Systems*, a companion to this volume, discusses heat transfer fundamentals and Chapter 20 of the same volume discusses hardware reliability in greater detail).

The one-dimensional heat flow equation that applies to metals (as well as other media) is given as

$$q = kA \frac{\partial T}{\partial x} \tag{26.7}$$

where q is the heat flow, k is the thermal conductivity of the metal, A is the cross-sectional area for heat flow, and $\frac{\partial T}{\partial x}$ the temperature gradient across the metal. For applications where good heat-sinking characteristics are desired, high thermal conductivity, k, is desirable. Table 26.3 lists thermal conductivity values for several important metals.

TABLE 26.3   Thermal Conductivities of Typical Metals (W/m K) at Room Temperature [6–17]

Metal	Thermal Conductivity
Aluminum	216.8
Antimony	24.4
Beryllium	200.5
Brass	105.0
Chromium	93.9
Copper	393.7
Gold	310.0
Kovar	17.1
Lead	34.3
Magnesium	160.3
Manganese	7.8
Molybdenum	138.3
Nickel	91.0
Palladium	71.8
Platinum	70.8
Silver	426.6
Tantalum	58.4
Titanium	21.9
Tungsten	172.3
Vanadium	31.9
Zinc	121.3

26.4   Thermal Expansion

Because RF and microwave components are often exposed to significant temperature excursions, consideration of the coefficient of linear expansion of the metals must be included in making a metal selection for many applications. When temperature is increased, the average distance between atoms is also increased. This leads to an expansion of the whole solid body with increasing temperature. The changes to the linear dimension of the metal can be characterized by

$$\beta = \frac{1}{l}\frac{\Delta l}{\Delta T}$$

where β is called the coefficient of linear expansion, l is the linear dimension of the material, ΔT is the change in temperature and Δl is the change in linear dimension arising from the change in temperature.

Linear expansion properties of metals are important whenever metallic structures are bonded to other materials in an electronic assembly. When two materials with dissimilar thermal expansion characteristics are bonded together, significant stress is experienced during temperature excursions. This stress can result in one or both of the materials breaking or cracking and this can result in degraded electrical performance. The best choice of metals to match thermal linear expansion properties is therefore determined by the thermal coefficient of linear expansion of the material that is used with the metal. Kovar, for example, is often chosen as the metal material of preference for use as a carrier when Alumina dielectric substrates are used to fabricate RF or microwave guided wave elements. Although Kovar is neither a superior electrical conductor nor a superior thermal conductor, its coefficient of linear expansion is a close match to that of the dielectric material, Alumina. Table 26.4 presents the coefficients of linear expansion for several metals as well as other materials that are often used for RF and microwave circuits.

26.5   Chemical Properties

The chemical properties of metals can be especially important in the selection of metals to be used for semiconductor device contacts and integrated circuits. Metal is used extensively in the development of

TABLE 26.4 Thermal Coefficient of Linear Expansion of Some of the Materials Used in Microwave and RF Packaging Applications (at Room Temperature, in $10^{-6}/K$) [6–17]

Material	Thermal Coefficient of Expansion
Dielectrics	
Aluminum nitride	4
Alumina 96%	6
Beryllia	6.5
Diamond	1
Glass-ceramic	4–8
Quartz (fused)	0.54
Metals	
Aluminum	23.1
Antimony	10.9
Beryllium	11.6
Brass	18.9
Chromium	4.9
Copper	16.5
Gold	14.1
Kovar	5.5
Lead	29.1
Magnesium	25.4
Manganese	22.5
Molybdenum	5.0
Nickel	13.0
Palladium	11.5
Platinum	8.9
Silver	18.9
Tantalum	6.4
Titanium	8.7
Tungsten	4.4
Vanadium	8.1
Zinc	30.1
Semiconductors	
GaAs	5.9
Silicon	2.6
Silicon Carbide	2.2

transistors and ICs. Uses include

— The contact material to establish ohmic and rectifying junctions
— The interconnect layers
— The material used to fabricate passive components such as inductors and transmission line segments

For these applications, the chemical properties of the metal when exposed to heat and in contact with the semiconductor material play a significant role in the metal selection criteria. The process of fabricating a semiconductor device often involves hundreds of individual process steps and significant exposure to thermal cycling and temperature ranges that will far exceed the environment extremes experienced by the final device.

For Silicon processes, for example, Aluminum (or its alloys) are often chosen for contacts and interconnects. Aluminum has high conductivity, but it is also chosen because it adheres well to Silicon and Silicon dioxide and because it does not interact significantly with Silicon during the thermal cycling associated with processing. In contrast, Gold also has high conductivity, but its use is typically avoided in Silicon fabrication facilities because Gold forms deep levels (traps) in Silicon that dramatically degrade device performance. Other metals of specific interest in Silicon processing include Gallium and Antiminide, which are often used in the formation of ohmic contacts.

Because interconnects continue to shrink in size as fabricated devices continue to be scaled down, interconnect resistance has begun to pose significant limitations on the levels of integration that can be achieved. One solution to extend these limits is to use Copper rather than Aluminum for interconnect metal. Copper's higher conductivity translates directly into improved interconnect performance. Progress is being made in this area, but the issues that are limiting progress are related to the chemical/thermal budget problems associated with Copper [2,4].

GaAs—the formation of good ohmic contacts and schottky barriers is critical to the fabrication of most GaAs devices. Different metals react chemically in distinct ways when exposed to a GaAs surface at high temperature. Metals such as Gold, Tin and Zinc tend to form ohmic contacts when placed in contact with a GaAs surface. In contrast, Titanium, Platinum, Tungsten, and Nickel normally form Schottky barriers on GaAs. In order to obtain optimum electrical conductivity and still produce good contacts, sandwiched layers of different metals are sometimes used. For example, a thick layer of Gold is often utilized over layers of Titanium and Platinum to produce Schottky barrier contacts. When this technique is employed, the Titanium resting on the surface of the GaAs forms the Schottky barrier, the Platinum serves as a diffusion barrier to keep the Gold and Titanium from diffusing together, and the Gold is used to produce a low resistance connection to the contact pads or remaining IC circuitry.

Certain metals can also react with GaAs to produce undesirable effects. Chromium, for example, produces undesirable deep levels in GaAs that can degrade device performance [2,4].

26.6   Weight

Over the past several decades a dominant trend in the development of electronic circuits has been the continued reduction of size and weight. Although much of this progress has been made possible by the continued scaling of semiconductor devices and ICs, metal portions of many electronic assemblies still dominate the weight of the system. Metal density can be an important factor in choosing metals for certain applications. Table 26.5 presents the density in g/cm^3 for several metals of interest.

TABLE 26.5   Density of Several Metals in g/cm^3 [6–17]

Metal	Density
Aluminum	2.66
Antimony	6.67
Beryllium	1.85
Brass	8.55
Chromium	7.10
Copper	8.93
Gold	19.33
Kovar	8.03
Lead	11.34
Magnesium	1.74
Manganese	7.43
Molybdenum	10.21
Nickel	8.85
Paladium	12.02
Platinum	21.42
Silver	10.49
Tantalum	16.65
Titanium	4.50
Tungsten	19.14
Vanadium	5.70
Zinc	7.12

Bibliography

1. D. Halliday and R. Resnick, *Fundamentals of Physics*, New York: John Wiley & Sons, 1970.
2. D. Schroder, *Semiconductor Material and Device Characterization*, New York: John Wiley & Sons, 1990.
3. M. Plonus, *Applied Electromagnetics*, New York: McGraw-Hill, 1978.
4. D. Elliott, *Integrated Circuit Fabrication Technology*, New York: McGraw-Hill, 1982.
5. R. Collin, *Foundations for Microwave Engineering*, New York: McGraw-Hill, 1966.
6. A. Smith, *Radio Frequency Principles and Applications*, New York: IEEE Press, 1998.
7. J. Cutnell and K. Johnson, *Physics*, 3rd Edn. New York: John Wiley & Sons Inc., 1995.
8. A. James and M. Lord, *Macmillan's Chemical and Physical Data*, London: Macmillan, 1992.
9. G. Kaye and T. Laby, *Tables of Physical and Chemical Constants*, 15th Edn, London: Longman, 1993.
10. G. Samsonov (Ed.), *Handbook of the Physicochemical Properties of the Elements*, New York: IFI-Plenum, 1968.
11. D. Lide (Ed.), *Chemical Rubber Company Handbook of Chemistry and Physics*, 79th Edn, Boca Raton, FL: CRC Press, 1998.
12. H. Ellis (Ed.), *Nuffield Advanced Science Book of Data*, London: Longman, 1972.
13. E. Oberg, F. Jones, H. Horton, and H. Ryffell, *Machinery Handbook*, 24th Edn, New York: Industrial Press Inc, 2000.
14. M. Mitchell, "Electrical resistivity of Beryllium," *Journal of Applied Physics*, Vol. 46, pp. 4742–4746, November 1975.
15. J. Dean (Ed.), *Lange's Handbook of Chemistry*, 14th Edn, New York: McGraw-Hill, 1992.
16. F. Seitz, *The Modern Theory of Solids*, New York: Dover Publications, 1987.
17. W. Jung, F. Schmidt, and G. Danielson, "Thermal conductivity of high-purity vanadium," *Physical Review B (Solid State)*, Vol. 15, Issue 2, pp. 659–665, January 15, 1977.

27

Dielectrics

K. F. Etzold
IBM T. J. Watson Research Center

27.1 Basic Properties

When electrical circuits are built, the physical environment consists of conductors separated from each other by insulating materials. Depending on the application, the properties of the insulating materials are chosen to satisfy requirements of low conductivity, a high or low dielectric permittivity and desirable loss properties. In most cases the high field properties of the materials do not play a role (certainly not in signal processing applications). This is also true with respect to electrical breakdown. However, there are exceptions. For instance, in transmitters the fields can be very large and the choice of insulators has to reflect those conditions. In semiconductor devices, dielectric breakdown of insulating layers can be a problem, but in many cases it is actually tunneling through the dielectric that is responsible for charge transport. In DRAM, the leakage currents in the memory storage capacitor have to be small enough to give a charge half-life of about one second over the full temperature range. In Flash memories, the charge is transferred via tunneling to the controlling site in the FET where it must then remain resident for years. This requires a really good insulator.

In most applications it is desirable that the material properties be independent of the strength of the applied field. This is not always true and there are now materials becoming available that can be used as transmission line phase shifters for steerable antennas, taking advantage of the field dependent dielectric permittivity.

Probably the most important property of dielectrics in circuit applications is the dielectric permittivity, often colloquially called the dielectric constant. This quantity relates the electric field in the material to the free charge on the surfaces. Superficially the connection between this property and the application of dielectrics in a circuit seems rather tenuous. Consider now, however, the capacity C of a device

$$C = \frac{q}{V}, \tag{27.1}$$

where q is the charge on the plates and V the voltage across the device. Usually in a discrete capacitor (understood here as a circuit element) we would like to store as much charge as possible in a given volume. Therefore, if we can somehow double the stored charge, the capacity C will double. What allows us to do this is the choice of the material that fills the device. Thus, the material in this example has twice the charge storage capability or dielectric permittivity. This quantity is typically labeled by the Greek letter epsilon (ε).

Consider, as an example, a parallel plate capacitor. This simple geometry is easily analyzed, and in general we can calculate the capacity of a device exactly if it is possible to calculate the electric field. For the parallel plate device the capacity is

$$C = \varepsilon \frac{A}{d}, \tag{27.2}$$

where A is the area and d is the separation between the plates. If the material between the plates is vacuum, $\varepsilon = 8.85 \times 10^{-12}$ F/m or in units that are a little easier to remember, 8.85 pF/m. The unit of capacity is the Farad. A capacitor with square 1m $\times$ 1m plates and a separation of 1m has a capacity of 8.85 pF, using Equation 27.2. This assumes parallel field lines between the plates that clearly will not be the case here unless a special, guarded geometry is chosen (to guard a circuit, conductors at the same potential are placed nearby. There are no field lines between equipotentials, and thus there are no contributions to the capacity.). We can consider Equation 27.2 for the capacity as a definition of dielectric permittivity. In general, the capacity is proportional to ε and the capacity of a device is given by

$$C = \varepsilon \times \text{Geometry factor}. \tag{27.3}$$

Also for practical applications, the dielectric permittivity is usually specified relative to that of vacuum

$$\varepsilon = \varepsilon_0 \times k. \tag{27.4}$$

Thus, ε_0 is defined as the dielectric permittivity of vacuum and k is the relative dielectric permittivity, most often referred to as simply, but somewhat inaccurately, the dielectric permittivity or also dielectric constant; k of course is dimensionless. As an example, the dielectric permittivity of typical Glass-Epoxy circuit board is 4.5, where this is understood to be the relative dielectric permittivity. On physical grounds all materials have a dielectric permittivity greater than 1. Also, for the same reason the dielectric permittivity of vacuum has no frequency dependence. In Table 27.1, the important properties of selected insulating materials used in engineering application are given.

Note that for crystalline materials the dielectric permittivity is not isotropic, that is, the permittivity depends on the orientation of the crystal axes relative to the electric field direction. The angular variation is directly related to the spatial symmetry of the crystal (see the discussion in Reference 2).

TABLE 27.1 Properties of Some Typical Engineering Insulating Materials

Material	k	Loss	Frequency	Conductivity
Vacuum	1.00	0	All	Zero
Air	1.0006	0		
Glass Vycor 7910	3.8	9.1×10^{-4}		
Glass Corning 0080	6.75	5.8×10^{-2}		
Al_2O_3	8.5	10^{-3}	1 MHz	
Teflon (PTFE)	2.0	2×10^{-4}	1 MHz	10^{-17}
Arlon 25N Circuit board	3.28	2.5×10^{-3}	1 MHz	
Epoxy-Glass circuit board	4.5			
Beryllium oxide	7.35			
Diamond	5.58			10^{-16}
PZT (Lead Zirconium oxide)	~ 1000			
Undoped silicon	11.8			
TaO_5	28			
Quartz (SiO_2)	3.75–4.1	2×10^{-4}		
Mica (Ruby)	6.5–8.7	3.5×10^{-4}		
Water	78.2	0.04	1 MHz	

So far, we have treated dielectrics as essentially perfect and lossless insulators that allow energy storage greater than that of vacuum. However, as can be seen from the fact that the table has a loss factor column, frequency-dependent energy losses will occur in all materials (except again in vacuum) [2]. As could be anticipated, there are a number of loss mechanisms. Perhaps the easiest loss mechanism to deal with conceptually is the low-frequency conductivity of the material. If there is a current flow as a result of an applied field, energy will be dissipated. The dc conductivity can be due to the basic properties of the materials such as the presence of impurities and their associated energy levels. It can also be due to crystalline defects, and conduction can take place along the surfaces of crystallites that constitute many materials (for example ceramics). Most often suitable materials can be chosen selected for the desired properties. In Table 27.1, the conductivity for selected materials is presented. It is seen that some natural and man-made materials have an extremely low conductivity. Indeed, the time constant for loss of charge for a free-floating (disconnected) device can be weeks. The time constant τ (in seconds) of an RC network is defined as

$$\tau = RC, \tag{27.5}$$

where R is the resistance in parallel with the capacitor C (R is in Ohms and C in Farads). Consider Teflon (PTFE). The dc conductivity at room temperature (25°C) is 10^{-17}.

Therefore, for this material the time constant for the charge to decay to $1/e$ of its original value is 23 days using the dielectric permittivity of 2.0 from Table 27.1. This demonstrates that there exist insulating materials for the cabling and supporting structures that are suitable for applications where only the smallest leakage can be tolerated. An example would be systems that measure very small charges such as the quartz-based force transducers and accelerometers.

27.2 Frequency Dependence of the Properties

We need now to consider the behavior at all user frequencies. These extend from dc into the microwave and optical range. In this range a number of different loss mechanisms have to be considered. But, before this can be done, the definition of the dielectric permittivity needs to be extended, so that AC effects can be properly considered. The losses are accounted for if an imaginary component is added to the definition of the dielectric permittivity. Thus,

$$\varepsilon = \varepsilon' - j\varepsilon'', \tag{27.6}$$

where ε' is the real part of the dielectric permittivity and j is the square root of -1. The existence of losses is now explicitly given by the imaginary quantity ε''. It is also obvious that the term "permittivity" as a property of dielectrics is somewhat of a misnomer and is partly historical. In fact both the real and imaginary parts are functions of frequency. As is often the case in describing physical parameters that are complex (in the mathematical sense), the two quantities in Equation 27.6 are actually related. This is a consequence of the fact that the electromagnetic behavior is governed by Maxwell's equations, the absence of discontinuities and also that causality holds, a statement that relates to the time evolution of electric fields. Thus, the so-called Kramers–Kronig equations relate the two quantities [1]

$$\varepsilon'(\omega) = \frac{1}{\pi} \int \frac{\varepsilon''}{x - \omega} dx$$

$$\varepsilon''(\omega) = \frac{1}{\pi} \int \frac{\varepsilon'}{x - \omega} dx \tag{27.7}$$

$$-\infty < \text{limits} < +\infty$$

$$\omega = 2\pi f,$$

where f is the frequency in Hz and where the dependence of the real and imaginary parts of ε on frequency is now explicitly stated and where the integrals are the Cauchy principal values. Thus, we only need to know the real or imaginary parts and the corresponding complimentary part can be calculated. For instance, if we know the real part of the dielectric response over the full frequency range, we can calculate the losses by integration. Even if our knowledge of the real part (capacity) is only over a limited frequency range, the complementary part (loss) can often still be estimated, albeit with reduced accuracy.

Let us now discuss the behavior of dielectrics between the frequency extremes of dc and the optical range. At optical frequencies, there is a relation from Maxwell's equations that relates the optical index of refraction to the dielectric and magnetic properties. Most dielectrics are nonmagnetic, so the magnetic susceptibility is just that of free space, μ_0. The optical index of refraction n is given by

$$n = \sqrt{\frac{\varepsilon}{\mu_0}}. \tag{27.8}$$

This is of interest because the optical index for all materials varies only over a fairly narrow range which also relates to the dielectric properties at high (optical) frequencies. Vacuum has an index of 1. Many insulators have an index near 2, which means that the dielectric permittivity is around 4. High optical indices are rare and one of the highest ones known is that of diamond that has an index of 2.42 (hence the sparkle), and therefore a calculated dielectric permittivity of 5.81. The measured value is 5.58 at low frequencies. Thus, there is slight inconsistency between these values and one would expect a value equal to or slightly higher than 5.58.

Consider now the other end of the frequency range. At low frequencies (Figure 27.1, up to about 1 GHz), the dielectric permittivity for different materials has a much wider range. Here a small diversion is necessary. We need to address the question of what gives rise to dielectric constants greater than 1. The attribute that allows additional charge to be placed on the plates of a capacitor is the (bulk) polarizability of the material or the ability to deposit charge on interfaces associated with the contacts or grain boundaries. Note: Except for amorphous ones, most materials are made up of aggregates of small crystallites. The solid then consists of adjacent crystallites separated by a "grain boundary." The properties of the grain boundary can be vastly different from the properties within a crystallite.

The polarizability can take various forms. It can be in the form of permanent dipoles that are observed, for example, in water or PZT. For many materials there is no permanent polarization, but there is an

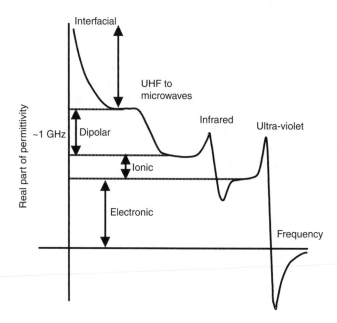

FIGURE 27.1   Dielectric behavior as a function of frequency.

induced polarization that is caused by an external applied field. The latter materials typically have relatively low dielectric constants, up to about 20. Materials that have permanent dipoles can have very high dielectric constants (up to about 20,000 and higher) especially in the presence of cooperative effects. These same cooperative effects imply the presence of a structural phase transition typically in a temperature range of somewhat below room temperature up to several 100°C depending on the material. The inherent (molecular) polarization does not disappear above the phase transition where the materials change from ferroelectric to paraelectric, but the long-range order (domains) that prevails at temperatures below the phase transition temperature (the Curie temperature) disappears, see Chapter 28 on Ferroelectrics. This kind of behavior is typical of PZT and many other perovskites (a perovskite is a crystal with a specific crystal geometry. Many of the materials with this kind of geometry have recently become technologically important; examples are PZT and $BaTiO_3$). Water is an example of a material whose molecules are polarized permanently but that does not have a ferroelectric phase, a phase in which the dipolar vector for a large number of molecules point in the same direction.

Let us now describe the trend of the dielectric permittivity over the entire frequency range. As discussed earlier at very high (optical) frequencies, the dielectric permittivity tends to have its lowest value (up to ~4) and increases as the frequency decreases toward dc. What connects these end points? Where do the changes take place and do they need to be considered in the RF and Microwave regime? The general trend of the dielectric permittivity as a function of frequency is shown in Figure 27.1. As can be seen, there are several regimes in which the dielectric permittivity is changing. At each transition, the interaction mechanism between the field and the material is different.

Consider the low frequency extreme in Figure 27.1. The low frequency rise in the permittivity is attributable to interface effects. These take place at the contacts or in the bulk of the material at the grain boundaries. Charge is able to accumulate at these interfaces and thereby able to contribute to the total capacity or dielectric permittivity. As the frequency is raised, eventually, the system can no longer follow these time-dependent charge fluctuations and the dielectric permittivity settles to a value equal to the one without the interface charges. As a widely discussed phenomenon (papers on ferroelectrics or high permittivity thin films), surface and interface effects can take place in ferroelectric (for example PZT) and nonferroelectric materials such as $BaTiO_3$ or glassy amorphous materials (see Reference 6, Chapter 28). The latter are the materials for chip capacitors.

A manifestation of this low-frequency effect is memory or dielectric relaxation that is attributable to the same interface behavior. This relaxation can readily be observed if a capacitor is charged or discharged over long periods of time. Consider a discharged capacitor that is suddenly connected to a steady state source such as a dc power supply or battery. If we measure the current in this circuit after a long time, we will not find a zero current, as we would expect after the capacitor is fully charged. Rather, a slowly decreasing current is observed in the circuit. This current is not due to dc current leakage or a current variation associated with the RC time permittivity of the Capacitor and the charging circuit. A leakage current would eventually become unvarying and the time constant of the charging RC circuit is much shorter than the observed decay. It is also observed that the capacitor returns some of this low charging current. If the power supply is suddenly removed and the leads of the capacitor are connected to the current meter, we see the initial ordinary discharge from the capacitor. After this discharge is completed a small, slowly decaying current in the opposite direction from the charging current is observed. It is also of course opposite to the residual charging current described above. These relaxation currents are due to charges that slowly accumulate on the various interior interfaces in the device. An example of areas where charge often becomes resident is the interface between the contacts and the dielectric. Of course, the charge is returned when the capacitor is discharged. By way of contrast if there is leakage in a device there is no stored charge.

As a concrete example, consider the variation of the capacity due of a commercial 470 pF chip capacitor as a function of frequency, shown in Figure 27.2. We observe the increase of the capacity due to interface effects as the frequency decreases. However, somewhat unexpectedly, the response does not flatten or drop at high frequencies, as Figure 27.1 implies. Instead, the capacity goes through a minimum and then rises again. This rise is due to a parasitic effect, a resonance in the capacitor and the test jig. The unavoidable

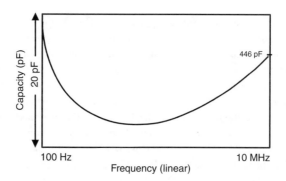

FIGURE 27.2   The "capacity" of a commercial ceramic 470 pF capacitor.

resonant elements are the lead inductance and the inherent capacity of the device. All capacitors exhibit this phenomenon. Indeed, above the resonance the impedance can become positive and the device behaves like an inductor! In bypass applications (i.e., larger value capacitors) this usually requires paralleling capacitors of different values (and therefore different resonance frequencies) to obtain a composite capacitor that has a negative reactance over a wider frequency range. Manufacturers of chip capacitors have also developed devices with a particularly low inductance. In any case, the rise in the response toward high frequencies in Figure 27.2 is not a material property but rather is a consequence of the physical construction of the device. Therefore, the low-frequency behavior of this chip capacitor is comparable to that shown in Figure 27.1, modified for the resonance. As can be seen for the 470 pF capacitor, the interface effect can be neglected at frequencies above 2–3 MHz.

Let us return now to the material-dependent dielectric behavior as a function of frequency. As the frequency is increased, the next polarizability is associated with the material rather than interfaces. Perovskites have Oxygen vacancies that will exhibit a dipolar polarization. This effect can have associated frequencies as low as 100 MHz. It can therefore play a role in chip capacitors, which are typically made from mixed or doped perovskites. Therefore, frequency-dependent effects can appear in these devices at relatively low frequencies. However, it is possible to control the number of vacancies in the manufacturing process, so that in most cases this effect will be small.

As the frequency is increased again, now into the range above 1 GHz, there will be contributions due to the ionic separation of the components of a material. For instance in PZT, the electronic charge cloud associated with a unit cell is slightly distorted. Therefore, the octahedron associated with the unit cell has a dipole moment, which will contribute to the polarization. Again, as the frequency of the applied field is raised, eventually, these charges will no longer be able to follow the external field and the polarization, and therefore the dielectric permittivity will decrease to a new plateau.

The only polarization left at this point is the electronic contribution that is, the distortion induced in the electronic cloud of the material by the external field. Note that this is an induced polarization unlike the permanent polarization in water molecules or PZT. This happens at optical frequencies, hence the example of diamond earlier in the chapter.

The above view of the dielectric behavior of many materials is somewhat idealized. In fact the transition form one regime to another is usually not characterized by a single, well-located (in frequency) transition but is presented to highlight the various mechanisms that enhance the dielectric permittivity. Most often there are multiple mechanisms resulting in a distribution of relaxation times. This in turn causes the transitions (see Figure 27.1) to be smeared out. In fact, the more typical behavior is that of the "Universal Law." Here the dielectric permittivity follows a power law typically with an exponent less than one. In many materials this behavior is observed over many decades of frequency. Thus, the observed response follows

$$\varepsilon(\omega) - \varepsilon_{\infty} = \varepsilon_0 \omega^{(n-1)}, \tag{27.9}$$

with $0 < n < 1$. Interestingly, this relation holds (at least for some perovskites) over as many as 12 orders of magnitude.

27.3 Measurements

The measurement of devices was historically done with a bridge circuit that is an ac modification of the Wheatstone bridge. The method essentially involves manually finding a pair of values representing the capacitance and dissipation factor. This is done by adjusting calibrated resistors to balance a bridge circuit and observing a minimum voltage across the diagonal terminals of the bridge. Because this is a manual method it is slow and tedious, particularly because the two values are interacting, necessitating an adjustment of the first value after the minimum was found for the second value. It is therefore a slowly converging cyclic adjustment. Classical ac bridges are also limited to low frequencies (below approximately 1 MHz).

In recent years, the measurement of capacity and the loss of devices have become significantly easier with the development of fixed-frequency or swept-impedance meters by commercial vendors [4]. At moderate frequencies, up to about 100 MHz, the measurement is typically done as a four terminal measurement. Two terminals deliver a known ac current (I). The complex voltage (V) with its corresponding phase angle relative to the phase of the exciting current across the device is measured with another pair of terminals. The complex impedance (Z) is then given by the AC equivalent of Ohm's law, with complex V and I

$$Z = \frac{V}{I}.$$

(27.10)

The transformations between the measured quantity $|Z|$, associated phase angle Θ, engineering values of the capacity, loss for a series equivalent circuit, and parallel equivalent are given in Table 27.2.

It should be remembered that a lossy two-terminal capacitor necessarily has to be represented either as an ideal capacitor connected in series with its loss equivalent resistor or as the two equivalent elements connected in parallel. In the series column the resistance R and the capacity C are in series. In the parallel column the (equivalent) conductance G and inverse reactance B are connected in parallel. For the parallel connection the measured quantity is typically the admittance Y. The quality factor Q and the dissipation factor D are also included in the table. These transformations allow us to find any of the desired values.

It is possible to force an adjustable dc voltage across the current terminals. This allows a measurement of the capacity of the device as a function of an applied steady-state bias field. This application is particularly important in the analysis of semiconductor devices where the junctions properties have strong field dependencies.

An example is the acquisition of capacity versus voltage or CV curves. Another example would be the characterization of phase shifter materials for phased array antennas, mentioned at the beginning of this article.

At higher frequencies, that is, above about 100 MHz, it not possible to measure accurately the current or voltage, primarily because of the aberrations introduced by the stray capacities and stray inductances in the connections. It is similarly difficult to measure the current and voltages without interactions between each other. Because of this the impedance measurement instead is made indirectly and is based on reflections in a transmission line using directional couplers that are capable of separating traveling waves going in the forward or reverse direction. High frequency network analyzers exist that are designed to measure the

TABLE 27.2 Impedance Conversions for Series and Parallel Connections [4]

Measured Quality	Series Equivalent	Parallel Equivalent
$|Z|$	$\sqrt{R^2 + X^2}$	
$|Y|$		$\sqrt{G^2 + B^2}$
Ω	$\mathrm{Tan}^{-1}(X/R)$	$\mathrm{Tan}^{-1}(B/G)$
C	$-1/(\omega X)$	B/ω
Q	$|X|/R$	$|B|/G$
D	$R/|X|$	$G/|B|$

transmission and reflection on a 50 Ω transmission line over a very wide range of frequencies. Thus, if a complex impedance terminates such a line, the impedance of the terminating device can be determined from the reflected signals. In fact the complex reflection coefficient Gamma (Γ) is given by

$$\Gamma = \frac{Z - Z_0}{Z + Z_0}, \tag{27.11}$$

where Z is the reflecting impedance and Z_0 is the impedance of the instrument and transmission line (usually 50 Ω). This can be inverted to solve for the unknown impedance

$$Z = Z_0 \frac{\Gamma + 1}{\Gamma - 1}. \tag{27.12}$$

Both the four terminal and the network analyzer impedance measurement methods place the sample at the terminals of the respective analyzer. In fact, it is often desirable to make an *in situ* measurement requiring cables. An example would be the characterization of devices that are being probed on a Silicon wafer [3]. To accommodate this requirement, many instruments have the capability to move the reference plane from the terminals to the measurement site. An accurate, purely resistive 50 Ω termination is placed at the actual test site. The resulting data are used to calibrate the analyzer. An internal built-in algorithm does a little more than to simply remeasure the calibration resistor. Two other measurements are also necessary to compensate for imperfections in the network analyzer and the cabling. The compensation is made for an imperfect directional coupler and direct reflection or imperfect impedance match at the sample site. This is accomplished with additional measurements with an open circuit and a short circuit in addition to the 50 Ω standard at the sample site. Using this data one can recalculate the value of the unknown at the end of the cable connections to remove the stray impedances. Similar compensations can be made for the four terminal measurements at lower frequencies.

References

1. A.K. Jonscher, *Dielectric Relaxation In Solids*, Chelsea Dielectric Press, London, 1983.
2. C. Kittel, *Introduction to Solid State Physics*, 3rd edn. John Wiley and Sons, New York, 1967.
3. J.D. Baniecki, *Dielectric Relaxation of Barium Strontium Titanate and Application to Thin Films for DRAM Capacitors*. PhD Dissertation, Columbia University, 1999.
4. Hewlett Packard, Instruction Manuals for the HP4194A Impedance and Gain/Phase Analyzer and HP8753C Network Analyzer with the 85047A S-Parameter Test Set.

28

Ferroelectrics and Piezoelectrics

K. F. Etzold
IBM T. J. Watson Research Center

28.1 Introduction

Piezoelectric materials have been used extensively in actuator and ultrasonic receiver applications, while ferroelectric materials (in thin film form) have recently received much attention for their potential use in nonvolatile (NV) memory applications. We will discuss the basic concepts in the use of these materials, highlight their applications, and describe the constraints limiting their uses. This chapter emphasizes properties that need to be understood for the effective use of these materials but are often very difficult to research. Among the properties that are discussed are hysteresis and domains.

Ferroelectric and piezoelectric materials derive their properties from a combination of structural and electrical properties. As the name implies, both types of materials have electric attributes. A large number of materials which are ferroelectric are also piezoelectric. However, the converse is not true. Pyroelectricity (heat to electric field conversion) is closely related to ferroelectric and piezoelectric properties via the symmetry properties of the crystals.

Examples of the classes of materials that are technologically important are given in Table 28.1. It is apparent that many materials exhibit electric phenomena which can be attributed to ferroelectric, piezoelectric, and electret behavior. It is also clear that vastly different materials (organic and inorganic) can exhibit ferroelectricity or piezoelectricity, and many have actually been commercially exploited for these properties.

As shown in Table 28.1, there are two dominant classes of ferroelectric materials, ceramics and organics. Both classes have important applications of their piezoelectric properties. To exploit the ferroelectric

TABLE 28.1   Ferroelectric, Piezoelectric, and Electrostrictive Materials

Type	Material Class	Example	Applications
Electret	Organic	Waxes	No recent
Electret	Organic	Fluorine based	Microphones
Ferroelectric	Organic	PVF2	No known
Ferroelectric	Organic	Liquid crystals	Displays
Ferroelectric	Ceramic	PZT thin film	NV-memory
Piezoelectric	Organic	PVF2	Transducer
Piezoelectric	Ceramic	PZT	Transducer
Piezoelectric	Ceramic	PLZT	Optical
Piezoelectric	Single crystal	Quartz	Freq. control
Piezoelectric	Single crystal	LiNbO3	SAW devices
Electrostrictive	Ceramic	PMN	Actuators

property, recently a large effort has been devoted to producing thin films of PZT (Lead [**Pb**] **Z**irconate **T**itanate) on various substrates for silicon-based memory chips for nonvolatile storage. In these devices, data is retained in the absence of external power as positive and negative polarization. Organic materials have not been used for their ferroelectric properties but have seen extensive application as electrets (the ability to retain a permanent charge). Liquid crystals in display applications are used for their ability to rotate the plane of polarization of light but not for their ferroelectric attribute.

It should be noted that the prefix ferro refers to the permanent nature of the electric polarization in analogy with the magnetization in the magnetic case. It does not imply the presence of iron, even though the root of the word suggests it. The root of the word piezo means pressure; hence the original meaning of the word piezoelectric implied "pressure electricity" the generation of an electric field from applied pressure. This early definition ignores the fact that these materials are reversible, allowing the generation of mechanical motion by applying a field.

28.2   Mechanical Characteristics

Materials are acted on by forces (stresses) and the resulting deformations are called strains. An example of a strain due to a force on a material is the change of dimension parallel and perpendicular to the applied force. It is useful to introduce the coordinate system and the numbering conventions which are used when discussing these materials. Subscripts 1, 2, and 3 refer to the x, y, and z directions respectively. Displacements are vectors and have single indices associated with their direction. If the material has a preferred axis, such as the poling direction in PZT, that axis is designated the z or 3 axis. Stresses and strains are tensors and require double indices such as xx or xy. To make the notation less cluttered and confusing, contracted notation has been defined. The following mnemonic rule is used to reduce the double index to a single one:

$$
\begin{array}{ccc}
1 & 6 & 5 \\
xx & xy & xz \\
 & 2 & 4 \\
 & yy & yz \\
 & & 3 \\
 & & zz
\end{array}
$$

This rule can be thought of as a matrix with the diagonal elements having repeated indices in the expected order, then continuing the count in a counterclockwise direction. Note that xy = yx, and so forth so that subscript 6 applies equally to xy and yx.

TABLE 28.2 Properties of Well-Known PZT Formulations (Based on the Original Navy Designations)

	Units	PZT4	PZT5A	PZT5H	PZT8
ε_{33}	—	1300	1700	3400	1000
d_{33}	10^{-2}A/V	289	374	593	225
d_{13}	10^{-2}A/V	−123	−171	−274	−297
d_{15}	10^{-2}A/V	496	584	741	330
g_{33}	10^{-3}Vm/N	26.1	24.8	19.7	25.4
k_{33}	—	0.70	0.705	0.752	0.64
Tc	°C	328	365	193	300
Q	—	500	75	65	1000
Rho	g/cm^3	7.5	7.75	7.5	7.6
Application	—	High signal	Medium signal	Receiver	Highest signal

Any mechanical object is governed by the well-known relationship between stress and strain,

$$S = sT \tag{28.1}$$

where S is the strain (relative elongation), T is the stress (force per unit area), and s contains the coefficients connecting the two. All quantities are tensors; S and T are second rank, and s is fourth rank. Note, however, that usually contracted notation is used so that the full complement of subscripts is not visible.

PZT converts electrical fields into mechanical displacements and vice versa. The connection between the two is via the d and g coefficients. The d coefficients give the displacement when a field is applied (motion transmitter), while the g coefficients give the field across the device when a stress is applied (motion receiver). The electrical effects are added to the basic Equation 28.1 such that

$$S = sT + dE \tag{28.2}$$

where E is the electric field and d is the tensor which contains the electromechanical coupling coefficients. The latter parameters are reported in Table 28.2 for representative materials. Expanded, the matrix Equation 28.1, becomes Equation 28.3. The exact form of Equation 28.3 depends on the crystalline symmetry; here the equation is given for a polarized ceramic material like PZT.

$$
\begin{bmatrix} S_1 \\ S_2 \\ S_3 \\ S_4 \\ S_5 \\ S_6 \end{bmatrix} =
\begin{bmatrix} s_{11} & s_{12} & s_{13} & & & \\ s_{12} & s_{11} & s_{13} & & 0 & \\ s_{13} & s_{13} & s_{33} & & & \\ & & & s_{44} & & \\ & 0 & & & s_{44} & \\ & & & & & 2(s_{11}-s_{12}) \end{bmatrix}
\begin{bmatrix} T_1 \\ T_2 \\ T_3 \\ T_4 \\ T_5 \\ T_6 \end{bmatrix} +
\begin{bmatrix} 0 & 0 & d_{13} \\ 0 & 0 & d_{13} \\ 0 & 0 & d_{33} \\ 0 & d_{15} & 0 \\ d_{13} & 0 & 0 \\ 0 & 0 & 0 \end{bmatrix}
\begin{bmatrix} E_1 \\ E_2 \\ E_3 \end{bmatrix}
\tag{28.3}
$$

Note that T and E are shown as column vectors for typographical reasons; they are in fact row vectors. This equation shows explicitly the stress–strain relation and the effect of the electromechanical conversion.

A similar equation applies when the material is used as a receiver:

$$\mathbf{E} = -\mathbf{g}T + (\boldsymbol{\varepsilon}^{T})^{-1}\mathbf{D} \tag{28.4}$$

where T is the transpose and D the electric displacement. The matrices are not fully populated for any material. Whether a coefficient is nonzero depends on the crystalline symmetry. For PZT, a ceramic which is given a preferred direction by the poling operation (the z-axis), only d_{33}, d_{13}, and d_{15} are nonzero. Also, again by symmetry, $d_{13} = d_{23}$ and $d_{15} = d_{25}$.

28.3 Applications

Historically the material which was used earliest for its piezoelectric properties was single-crystal quartz. Crude sonar devices were built by Langevin using quartz transducers, but the most important application for Quartz was, and still is, frequency control. Crystal oscillators are today at the heart of every clock that does not derive its frequency reference from the ac power line. They are also used in every color television set and personal computer. In these applications at least one (or more) "quartz crystal" controls frequency or time. This explains the label "quartz" which appears on many clocks and watches. The use of quartz resonators for frequency control relies on another unique property. Not only is the material piezoelectric (which allows the excitation of mechanical vibrations), but the material has also a very high mechanical "Q" or quality factor ($Q > 100,000$). The actual value depends on the mounting details, whether the crystal is in a vacuum, and other details. Compare this value to a Q for PZT between 75 and 1000. The Q factor is a measure of the rate of decay of the vibration and thus the mechanical losses of an excitation with no external drive. A high Q leads to a very sharp resonance and thus tight frequency control. For frequency control it has been possible to find orientations of cuts of quartz that reduce the influence of temperature on the vibration frequency.

Ceramic materials of the PZT family have also found increasingly important applications. The piezoelectric but not the ferroelectric property of these materials is made use of in transducer applications. PZT has a very high efficiency (electric energy to mechanical energy coupling factor k) and can generate high-amplitude ultrasonic waves in water or solids. The coupling factor is defined by

$$k^2 = \frac{\text{Energy stored mechanically}}{\text{Total energy stored electrically}} \tag{28.5}$$

Typical values of k_{33} are 0.7 for PZT 4 and 0.09 for quartz, showing that PZT is a much more efficient transducer material than quartz. Note that the energy is a scalar; the subscripts are assigned by finding the energy conversion coefficient for a specific vibrational mode and field direction and selecting the subscripts accordingly. Thus k_{33} refers to the coupling factor for a longitudinal (thickness) mode driven by a longitudinal field.

Probably the most important applications of PZT today are based on ultrasonic echo ranging. Sonar uses the conversion of electrical signals to mechanical displacement as well as the reverse transducer property, which is to generate electrical signals in response to a stress wave. Medical diagnostic ultrasound and nondestructive testing systems devices rely on these same properties. Actuators have also been built but a major obstacle is the small displacement that can conveniently be generated. Even then, the required drive voltages are typically hundreds of volts and the displacements are only a few hundred angstroms. For PZT the strain in the z-direction due to an applied field in the z-direction is (no stress, $T = 0$)

$$s_3 = d_{33} E_3 \tag{28.6}$$

or

$$s_3 = \frac{\Delta d}{d} = d_{33} \frac{V}{d} \tag{28.7}$$

where s is the strain, E the electric field, and V the potential; d_{33} is the coupling coefficient that connects the two. Thus

$$\Delta d = d_{33} V \tag{28.8}$$

Note that this expression is independent of the thickness d of the material but this is true only when the applied field is parallel to the displacement. Let the applied voltage be 100 V and let us use PZT8 for which d_{33} is 225 (from Table 28.2). Hence $\Delta d = 225$ Å or 2.25 Å/V, a small displacement indeed. We also note that Equation 28.6 is a special case of Equation 28.2 with the stress equal to zero. This is the situation when an actuator is used in a force-free environment, for example, as a mirror driver. This arrangement results in the maximum displacement. Any forces that tend to oppose the free motion of the PZT will subtract from the available displacement with the reduction given by the normal stress–strain relation, Equation 28.1.

It is possible to obtain larger displacements with mechanisms that exhibit mechanical gain, such as laminated strips (similar to bimetallic strips). The motion can then be up to about 1 mm but at a cost of a reduced available force. An example of such an application is the video head translating device to provide tracking in VCRs.

There is another class of ceramic materials that recently has become important. PMN (Lead [**Pb**] **M**agnesium **N**iobate), typically doped with approximately 10% Lead Titanate) is an electrostrictive material which has seen applications where the absence of hysteresis is important. For example, deformable mirrors require repositioning of the reflecting surface to a defined location regardless of whether the final position is above or below the initial position.

Electrostrictive materials exhibit a strain that is quadratic as a function of the applied field. Producing a displacement requires an internal polarization. Because the latter polarization is induced by the applied field and is not permanent, as it is in the ferroelectric materials, electrostrictive materials have essentially no hysteresis. Unlike PZT, electrostrictive materials are not reversible; PZT will change shape on application of a field and generate a field when a strain is induced. Electrostrictive materials only change shape on application of a field and, therefore, cannot be used as receivers. PZT has inherently large hysteresis because of the domain nature of the polarization.

Organic electrets (materials that carry a permanent net charge) have important applications in self-polarized condenser (or capacitor) microphones where the required electric bias field in the gap is generated by the diaphragm material rather than by an external power supply.

28.4 Structure of Ferroelectric and Piezoelectric Materials

Ferroelectric materials have, as their basic building block, atomic groups which have an associated permanent electric field, either as a result of their structure or as result of distortion of the charge clouds that make up the groups. In the first case, the field arises from an asymmetric placement of the individual ions in the group (these groupings are called unit cells). In the second case, the electronic cloud is moved with respect to the ionic core. If the group is distorted permanently, a permanent electric field can be associated with each group. We can think of these distorted groups as represented by electric dipoles, defined as two equal but opposite charges which are separated by a small distance. Electric dipoles are similar to magnetic dipoles, which have the familiar north and south poles. The external manifestation of a magnetic dipole is a magnetic field and that of an electric dipole is an electric field.

Figure 28.1a represents a hypothetical slab of material in which the dipoles are perfectly arranged. In actual materials the atoms are not as uniformly arranged, but, nevertheless, from this model there would

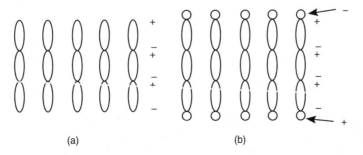

(a)   (b)

FIGURE 28.1 Charge configurations in ferroelectric model materials: (a) uncompensated and (b) compensated dipole arrays.

be a very strong field emanating from the surface of the crystal. The common observation, however, is that the fields are either absent or very weak. This effective charge neutrality arises from the fact that there are free, mobile charges available in the material, which can be attracted to the surfaces. The polarity of the mobile charges is opposite to the charge of the free dipole end. The added charges on the two surfaces generate their own field, equal and opposite to the field due to the internal dipoles. Thus the effect of the internal field is canceled and the external field is zero, as if no charges were present at all (Figure 28.1b).

In ferroelectric materials a crystalline asymmetry exists, which allows electric dipoles to form. If the dipoles are absent the internal field disappears. Consider an imaginary horizontal line drawn through the middle of a dipole. We can see readily that the dipole is not symmetric about that line. The asymmetry thus requires that there be no center of inversion when a material is in the ferroelectric state.

All ferroelectric and piezoelectric materials have phase transitions at which the material changes crystalline symmetry. For example, in PZT there is a change from tetragonal or rhombohedral symmetry to cubic as the temperature is increased. The temperature at which the material changes crystalline phases is called the Curie temperature, T_C. For typical PZT compositions the Curie temperature is between 250 and 450°C (see Table 28.2).

A consequence of a phase transition is that a rearrangement of the lattice takes place when the material is cooled through the transition. Intuitively we would expect that the entire crystal assumes the same orientation throughout as the temperature passes through the transition. By orientation we mean the direction of the preferred crystalline axis (say the tetragonal axis). Experimentally it is found, however, that the material breaks up into smaller regions in which the preferred direction and thus the polarization is uniform. Note that cubic materials have no preferred direction. In tetragonal crystals the polarization points along the c-axis (the longer axis) whereas in rhombohedral lattices the polarization is along the body diagonal. The volume in which the preferred axis is pointing in the same direction is called a domain and the border between the regions is called a domain wall. The energy of the multidomain state is slightly lower than the single-domain state and is thus the preferred configuration. The direction of the polarization changes by either 90° or 180° as we pass from one uniform region to another. Thus the domains are called 90° and 180° domains. Whether an individual crystallite or grain consists of a single domain depends on the size of the crystallite and external parameters such as strain gradients, impurities, and so forth. It is also possible that a domain extends beyond the grain boundary and encompasses two or more grains of the crystal.

28.5 Ferroelectric Materials

PZT ($PbZr_xTi_{(1-x)}O_3$) is an example of a ceramic material that is ferroelectric. We will use PZT as a prototype system for many of the ferroelectric attributes to be discussed. The concepts, of course, have general validity. The structure of this material is ABO_3 where A is Lead and B is one or the other atoms, Ti or Zr. This material consists of many randomly oriented crystallites that vary in size between approximately 10 nm to several microns depending on the preparation method. The crystalline symmetry of the material is determined by the magnitude of the parameter x. The material changes from rhombohedral to tetragonal symmetry when $x > 0.48$. This transition is almost independent of temperature. The line which divides the two phases is called a morphotropic phase boundary (MPB) (change of symmetry as a function of composition only). Commercial materials are made with $x \sim 0.48$, where the d and g sensitivity of the material is maximum. It is clear from Table 28.2 that there are other parameters which can be influenced as well. Doping the material with donors or acceptors often changes the properties dramatically. Thus niobium is important to obtain higher sensitivity and resistivity and to lower the Curie temperature. PZT typically is a p-type conductor and niobium will significantly decrease the conductivity because of the electron which Nb^{5+} contributes to the lattice. The Nb ion substitutes for the B-site ion Ti^{4+} or Zr^{4+}. The resistance to depolarization (the so-called hardness of the material) is affected by iron doping. Many other dopants and admixtures have been used, often in very exotic combinations to affect aging, sensitivity, and

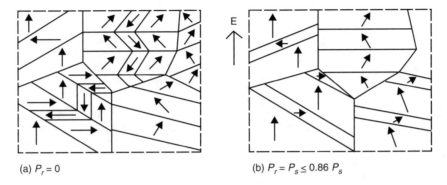

FIGURE 28.2 Domains in PZT, as prepared (a) and poled (b).

so forth. Hardness is a definition giving the relative resistance to depolarization. It should not be confused with mechanical hardness.

The designations used in Table 28.2 reflect very few of the many combinations that have been developed. The PZT designation types were originated by the U.S. Navy to reflect certain property combinations. These can also be obtained with different combinations of compositions and dopants. The examples given in the table are representative of typical PZT materials, but today essentially all applications have their own custom formulation. The name PZT has become generic for the Lead Zirconate Titanates and does not reflect Navy or proprietary designations.

Real materials consist of large numbers of unit cells, and the manifestation of the individual charged groups is an internal and an external electric field when the material is stressed. Internal and external refer to inside and outside of the material. The interaction of an external electric field with a charged group causes a displacement of certain atoms in the group. The macroscopic manifestation of this is a displacement of the surfaces of the material if the groups are similarly oriented, that is, if the material is poled. This motion is called the piezoelectric effect, the conversion of an applied field into a corresponding displacement.

When PZT ceramic material is prepared, the crystallites and domains are randomly oriented, and therefore the material does not exhibit any piezoelectric behavior (Figure 28.2a). The random nature of the displacements for the individual crystallites causes the net displacement to average to zero when an external field is applied. The tetragonal axis has three equivalent directions 90° apart and the material can be poled by reorienting the polarization of the domains into a direction nearest the applied field. When a sufficiently high field is applied, some but not all of the domains will be rotated toward the electric field through the allowed angle 90° or 180°. If the field is raised further, eventually all domains will be oriented as close as possible to the direction of the field. Note however, that the polarization of each domain will not point exactly in the direction of the field (Figure 28.2b). At this point, no further domain motion is possible and the material is saturated. As the field is reduced, the majority of domains retain the orientation they had with the field on leaving the material in an oriented state, which now has a net polarization. This operation is called poling and is accomplished by raising the temperature to about 150°C, usually in oil to avoid breakdown. Raising the temperature lowers the coercive field, E_c and the poling field of about 30–60 kV/cm is applied for several minutes. The temperature is then lowered but it is not necessary to keep the applied field turned on during cooling because the domains will not spontaneously re-randomize.

28.6 Electrical Characteristics

A piezoelectric component often has a very simple geometric shape, especially when it is prepared for measurement purposes. There will be mechanical resonances associated with the major dimensions of a sample piece. The resonance spectrum will be more or less complicated, depending on the shape of

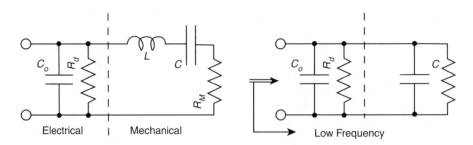

FIGURE 28.3 Equivalent circuit for a piezoelectric resonator. The reduction of the equivalent circuit at low frequencies is shown on the right.

a sample piece. If the object has a simple shape (say a circular disc) then some of the resonances will be well separated from each other and can be associated with specific vibrations and dimensions (modes). Each of these resonances has an electrical equivalent, and inspection of the equivalent circuit shows that there will be a resonance (minimum impedance) and an antiresonance (maximum impedance). Thus an impedance plot can be used to determine the frequencies and also the coupling constants and mechanical parameters for the various modes.

Consider the equivalent circuit for a slab of ferroelectric material. In Figure 28.3, the circuit shows a mechanical (acoustic) component and the static or clamped capacity C_0 (and the dielectric loss R_d) which are connected in parallel [3,4]. The acoustic components are due to their motional or mechanical equivalents, the compliance (capacity, C) and the mass (inductance, L). There will be mechanical losses, which are indicated in the mechanical branch by R_m. The electrical branch has the clamped capacity C_0 and a dielectric loss (R_d), distinct from the mechanical losses. This configuration will have a resonance that is usually assumed to correspond to the mechanical thickness mode but can represent other modes as well. This simple model does not show the many other modes a slab (or rod) of material will have. Thus transverse, plate, and flexural modes are present. Each can be represented by its own combination of L, C, and R_m. The presence of a large number of modes often causes difficulties in characterizing the material since some parameters must be measured either away from the resonances or from clean, nonoverlapping resonances. For instance, the clamped capacity (or clamped dielectric constant) of a material is measured at high frequencies where there are usually a large number of modes present. For an accurate measurement these must be avoided and often a low-frequency measurement is made in which the material is physically clamped to prevent motion. This yields the static, nonmechanical capacity, C_0. The circuit can be approximated at low frequencies by ignoring the inductor and redefining R and C. Thus, the coupling constant can be extracted from the value of C and C_0. From the previous definition of k we find

$$k^2 = \frac{\text{Energy stored mechanically}}{\text{Total energy stored electrically}} = \frac{CV^2/2}{(C + C_0)V^2/2} = \frac{1}{\dfrac{C_0}{C} + 1} \tag{28.9}$$

It requires charge to rotate or flip a domain. Thus, there is charge flow associated with the rearrangement of the polarization in the ferroelectric material. If a bipolar, repetitive signal is applied to a ferroelectric material, its hysteresis loop is traced out and the charge in the circuit can be measured using the Sawyer Tower circuit (Figure 28.4) [7]. In some cases the drive signal to the material is not repetitive and only a single cycle is used. In that case the starting point and the end point do not have the same polarization value and the hysteresis curve will not close on itself due to other dielectric charging phenomena.

The charge flow through the sample is due to the rearrangement of the polarization vectors in the domains (the polarization) and contributions from the static capacity and losses (C_0 and R_d in Figure 28.3). The charge into the test device is integrated by the measuring capacitor, which is in series with the sample. The measuring capacitor is sufficiently large to avoid a significant voltage loss. The polarization is plotted on a X-Y oscilloscope or plotter against the applied voltage and therefore the applied field.

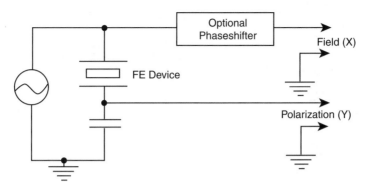

FIGURE 28.4   Sawyer Tower circuit.

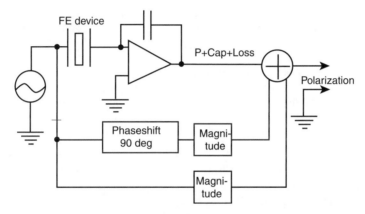

FIGURE 28.5   Modern hysteresis circuit. An op amp is used to integrate the charge into the device; loss and static capacitance compensation are included.

Ferroelectric and piezoelectric materials are lossy. This will distort the shape of the hysteresis loop and can even lead to incorrect identification of materials as ferroelectric when they merely have nonlinear conduction characteristics. A resistive component (from R_d in Figure 28.3) will introduce a phase shift in the polarization signal. Thus the display has an elliptical component, which looks like the beginnings of the opening of a hysteresis loop. However, if the horizontal signal has the same phase shift, the influence of this lossy component is eliminated, because it is in effect subtracted. Obtaining the exact match is the function of the optional phase shifter, and in the original Sawyer Tower circuits a bridge was constructed, which had a second measuring capacitor in the comparison arm (identical to the one in series with the sample). The phase was then matched with adjustable high-voltage components that match C_0 and R_d [2,3,8].

This design is inconvenient to implement and modern Sawyer Tower circuits have the capability to shift the reference phase either electronically or digitally to compensate for the loss and static components. A contemporary version, which has compensation and no voltage loss across the integrating capacitor, is shown in Figure 28.5. The op-amp integrator provides a virtual ground at the input, reducing the voltage loss to negligible values. The output from this circuit is the sum of the polarization and the capacitive and loss components. These later contributions can be canceled using a purely real (resistive) and a purely imaginary (capacitive, 90° phaseshift) compensation component proportional to the drive across the sample. Both need to be scaled (magnitude adjustments) to match them to the device being measured and then have to be subtracted (adding negatively) from the output of the op amp. The remainder signal is the polarization. The hysteresis response for typical ferroelectrics is frequency dependent, and traditionally the reported values of the polarization are measured at 50 or 60 Hz.

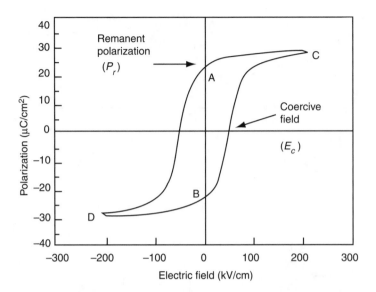

FIGURE 28.6 Idealized hysteresis curve for typical PZT materials. Many PZT materials display offsets from the origin and have asymmetries with respect to the origin. The curve shows how the remanent polarization (P_r) and the coercive field (E_c) are defined. While the loop is idealized, the values given for the polarization and field are realistic for typical PZT materials.

The improved version of the Sawyer Tower (Figure 28.5) circuit allows the cancellation of C_0 and R_d and the losses, thus determining the mechanically active components. This is important in the development of materials for ferroelectric memory applications. It is far easier to judge the squareness of the loop when the inactive components are canceled. Also, by calibrating the "magnitude controls" the values of the inactive components can be read off directly. In typical measurements the sample resonance is far above the test frequencies used (50 or 60 Hz), so ignoring the inductance in the equivalent circuit is justified. The circuit shown in Figure 28.5 is an analog configuration. In commercial testers, available now, the compensation is done digitally, but the measurement principle is the same. The charge is still acquired with an op amp but the compensating values are obtained digitally and are used to extract the equivalent mechanical parameters. Representative, idealized data from hysteresis loop is shown in Figure 28.6 for PZT.

The measurement of the dielectric constant and the losses is usually very straightforward. A slab with a circular or other well-defined cross section is prepared, electrodes are applied, and the capacity and loss are measured (usually as a function of frequency). The relative dielectric permittivity k is found by rearranging the capacitor equation Equation 28.10,

$$C = \varepsilon_0 k \frac{A}{t} \qquad (28.10)$$

where A is the area of the device and t the thickness. In this definition (also used in Table 28.2) k is the relative dielectric permittivity and ε_0 is the permittivity of vacuum. The subscripts in the table indicate to measuring field direction and the poling direction. Until recently, the relative dielectric permittivity, like the polarization, was measured at 50 or 60 Hz because the test voltage could easily be obtained with a transformer from the power line. Today the dielectric parameters are typically specified at 1 kHz or higher, which is possible because impedance analyzers with high-frequency capability are readily available and for much of the device work the tests are done on thin films. To avoid low-frequency anomalies, even higher frequencies such as 1 MHz are often selected. This is especially required when evaluating PZT or nonferroelectric high k thin films. Low-frequency anomalies can be present in the actual devices but are not modeled in the equivalent circuit (Figure 28.3). These are typically due to interface layers between the dielectric or ferroelectric material and the contact material, often but not always a metal. These layers will cause both the resistive and reactive components to rise at low frequencies producing readings that are not representative of the dielectric material properties (see Chapter 27 of this book).

28.7 Ferroelectric and High Epsilon Thin Films

While PZT and other ferroelectric (FE) bulk materials have had major commercial importance, thin films prepared from these materials have only recently been the focus of significant research efforts. In this section the material properties and process issues will be discussed. Because of the potentially large payoff, major efforts have been directed at developing the technologies for depositing thin films of ferroelectric and nonferroelectric but high k (high dielectric permittivity) thin films [6].

A recent trend has been the sharply increasing density of dynamic random access memory (DRAM). The data storage capacitor in these devices is becoming a major limiting factor because its dielectric has to be very thin in order to achieve the desired capacitance values to yield, in turn, a sufficient signal for the storage cell. It is often also desirable to have nonvolatile operation (no data loss on power loss). These two desires have, probably more than anything else, driven the development of high k and FE thin films. Of course, these are not the only applications of FE films. Table 28.3 lists the applications of FE (nonvolatile, NV) and high k films (volatile) and highlights which of the properties are important for their use. It is seen that the NV memory application is very demanding. Satisfying all these requirements simultaneously has produced significant challenges in the manufacture of these films.

Perhaps the least understood and to some extent still unsolved problem is that of fatigue and aging. In nonvolatile memory applications the polarization represents the memory state of the material (up = bit 1; down = bit 0). In use the device can be switched at the clock rate, say 100 MHz. Thus for a lifetime of 5 years the material must withstand $\sim 10^{16}$ polarization reversals or large field reversals. Typical candidate materials for ferroelectric memory applications are PZTs with the ratio of zirconium to titanium adjusted to yield the maximum dielectric constant and polarization. This maximum will be near the MPB for PZT. Small quantities of other materials can be added, such as Lanthanum or Niobium to modify optical or switching characteristics. The Sol-Gel method discussed below is particularly suitable for incorporating these additives. Devices made from materials at the current state of the art lose a significant portion of their polarization after 10^{10}–10^{12} cycles, rendering them useless for their intended memory use because of the associated signal loss. This is a topic of intensive investigation and only one proprietary material has emerged which might be more suitable for memory use than PZT. Strontium Bismuth Titanate (SBT) has better fatigue characteristics than PZT but large scale memories, comparable to volatile DRAM have not been developed yet, and so far NV applications have been small scale like, for instance, for smart cards.

High k nonferroelectric materials are of great interest for DRAM applications [9,10]. As an example, major efforts are extant to produce thin films of mixtures of barium and strontium titanate (BST). Dielectric constants of 600 and above have been achieved (compared to 4–6 for silicon oxides and nitrides).

TABLE 28.3 Material Properties and Applications Areas

Type	Ferro-electric	Epsilon	Polarization	Coercive Field	Leakage	Aging	Electro-Optical	Electromechanical
NV RAM	X		X	X	X	X		
DRAM		X			X	X		
Actuator				X				X
Display	X				X	X	X	
Optical modulator	X				X	X	X	

In applications for FE films, significant opportunities also exist for electro-optical modulators for fiber-optic devices and light valves for displays. Another large scale application is actuators and sensors. For the latter the electromechanical conversion property is used and large values of d_{33} (the electromechanical conversion coefficient) are desirable. However, economically the importance of all other applications are, and probably will be in the foreseeable future, less significant than that of memory devices if successful large scale devices can be developed.

Integration of ferroelectric or nonferroelectric materials with silicon devices and substrates has proved to be very challenging. Contacts and control of the crystallinity and crystal size and the stack structure of the capacitor device are the principal issues. In both volatile and nonvolatile memory cells the dielectric material interacts with the silicon substrate. Thus an appropriate barrier layer must be incorporated, which at the same time is a suitable substrate on which to grow the dielectric or ferroelectric films. A typical device structure starts with an oxide layer (SiO_x) on the silicon substrate followed by a thin titanium layer that prevents diffusion of the final substrate layer, platinum (the actual growth substrate and contact material). Ruthenium based barrier layer materials have also been used as a dual purpose contact and seed layer. These are also perovskites, structurally similar to the FE or high epsilon films.

While the major effort in developing thin films for data storage has been for the films described above, a significant effort has recently been made to produce low k films. In contemporary logic chips the device speed is not only determined by the speed of the active devices but also by the propagation delay in the interconnect wiring. To reduce the delay the conductivity of the metal must be increased (hence the use copper as conductors) and the dielectric separating the conductors must have as low as possible dielectric permittivity. Materials for this purpose are now being developed in thin film form.

Significant differences have been observed in the quality of the films depending on the nature of the substrate. The quality can be described by intrinsic parameters such as the crystallinity (i.e., the degree to which noncrystalline phases are present). The uniformity of the orientation of the crystallites also seems to play a role in determining the electrical properties of the films. In the extreme case of perfect alignment of the crystallites of the film with the single-crystal substrate an epitaxial film is obtained. These films tend to have the best electrical properties.

In addition to amorphous material, other crystalline but nonferroelectric phases can be present. An example is the pyrochlore phase in PZT. These phases often form incidentally to the growth process of the desired film and usually degrade one or more of the desired properties of the film (for instance the dielectric constant). The pyrochlore and other oxide materials can accumulate between the metal electrode and the desired PZT or BST layer. The interface layer is then electrically in series with the desired dielectric layer and degrades its properties. The apparent reduction of the dielectric constant that is often observed in these films as the thickness is reduced can be attributed to the presence of these low dielectric constant layers.

There are many growth methods for these films. Table 28.4 lists the most important techniques along with some of the critical parameters. Wet methods use metal organic compounds in liquid form. In the Sol-Gel process the liquid is spun onto the substrate. The wafer is then heated, typically to a lower,

TABLE 28.4 Deposition Methods for PZT and Perovskites

Type	Process	Rate nm/min	Substrate Temperature	Anneal Temperature	Target/Source
Wet	Sol-Gel	100 nm/coat	RT	450–750	Metal organic
Wet	MOD	300 nm/coat	RT	500–750	Metal organic
Dry	RF sputter	.5–5	RT-700	500–700	Metals and oxides
Dry	Magnetron sputter	5–30	RT-700	500–700	Metals and oxides
Dry	Ion beam sputter	2–10	RT-700	500–700	Metals and oxides
Dry	Laser sputter	5–100	RT-700	500–700	Oxide
Dry	MOCVD	5–100	400–800	500–700	MO vapor and carrier gas

intermediate temperature (around 300°C) to remove the organic components. This spin-on and heat process is repeated until the desired thickness is reached. At this temperature only an amorphous film forms consisting of the desired film oxides composition, but is not yet reacted (calcined). The wafer is then heated to between 500 and 700°C usually in oxygen and the actual chemical reaction and crystal growth takes place. Instead of simple long term heating (order of hours) rapid thermal annealing (RTA) is often used. In this process the sample is only briefly exposed to the elevated temperature, usually by a scanning infrared beam. It is in the transition between the low decomposition temperature and the firing temperature that the pyrochlore tends to form. At the higher temperatures the more volatile components have a tendency to evaporate, thus producing a chemically unbalanced compound which also has a great propensity to form one or more of the pyrochlore phases. In the case of PZT, 5–10% excess lead is usually incorporated which helps to form the desired perovskite material and compensates for the loss. In preparing Sol-Gel films it is generally easy to prepare the compatible liquid compounds of the major constituents and the dopants. The composition is then readily adjusted by appropriately changing the ratio of the constituents. Very fine quality films have been prepared by this method, including epitaxial films.

Current semiconductor technology is tending toward dry processing. Thus, in spite of the advantages of the Sol-Gel method, other methods using physical vapor deposition (PVD) are being investigated. These methods use energetic beams or a plasma to move the constituent materials from the target to the heated substrate. The compound then forms in situ on the heated wafer (500°C). Even then, however, a subsequent anneal is often required. With PVD methods it is much more difficult to change the composition since now the oxide or metal ratios of the target have to be changed or dopants have to be added. This can involve the fabrication of a new target for each composition ratio. MOCVD is an exception here; the ratio is adjusted by regulating the carrier gas flow. However, the equipment is very expensive and the substrate temperatures tend to be high (up to 800°C, uncomfortably high for semiconductor device processing, where active circuitry is usually located below the memory film). Laser sputtering is very attractive and it has produced very fine films. The disadvantage is that the films are defined by the plume which forms when the laser beam is directed at the source. This produces only small areas of good films and scanning methods need to be developed to cover full-size silicon wafers. Debris is also a significant issue in laser deposition. However, it is a convenient method to produce films quickly and with a small investment. In the long run MOCVD or Sol-Gel will probably evolve as the method of choice for realistic DRAM devices with state of the art densities.

Defining Terms

A-site: Many ferroelectric materials are oxides with a chemical formula ABO_3. The A-site is the crystalline location of the A atom.

B-site: Analogous to the definition of the A-site.

Coercive field: When a ferroelectric material is cycled through the hysteresis loop the coercive field is the electric field value at which the polarization is zero. A material has a negative and a positive coercive field and these are usually, but not always, equal in magnitude to each other.

Crystalline phase: In crystalline materials the constituent atoms are arranged in regular geometric ways; for instance in the cubic phase the atoms occupy the corners of a cube (edge dimensions ~2–5 A for typical oxides).

Curie temperature: The temperature at which a material spontaneously changes its crystalline phase or symmetry. Ferroelectric materials are often cubic above the Curie temperature, and tetragonal or rhombohedral below.

Domain: Domains are portions of a material in which the polarization is uniform in magnitude and direction. A domain can be smaller, larger, or equal in size to a crystalline grain.

Electret: A material which is similar to ferroelectrics but charges are macroscopically separated and thus are not structural. In some cases the net charge in the electrets is not zero, for instance when an implantation process was used to embed the charge.

Electrostriction: The change in size of a nonpolarized, dielectric material when it is placed in an electric field.

Ferroelectric: A material with permanent charge dipoles which arise from asymmetries in the crystal structure.

Hysteresis: When the electric field is raised in a ferroelectric material the polarization lags behind. When the field across the material is cycled the hysteresis loop is traced out by the polarization.

Morphotropic phase boundary (MPB): Materials which have a MPB assume a different crystalline phase depending on the composition of the material. The MPB is sharp (a few percent in composition) and separates the phases of a material. It is approximately independent of temperature in PZT.

Piezoelectric: A material which exhibits an external electric field when a stress is applied to the material, and a charge flow proportional to the strain is observed when a closed circuit is attached to electrodes on the surface of the material.

PLZT: A PZT material with a lanthanum doping or admixture (up to approximately 15% concentration). The lanthanum occupies the A-site. It is transparent and the index can be controlled electrically. Its principal application is for optical devices.

PMN: Generic name for electrostrictive materials of the Lead (Pb) Magnesium Niobate family.

Polarization: The polarization is the amount of charge associated with the dipolar or free charge in a ferroelectric or an electret, respectively. For dipoles the direction of the polarization is the direction of the dipole. The polarization is equal to the external charge which must be supplied to the material to produce a polarized state from a random state (twice that amount is necessary to reverse the polarization). The statement is rigorously true if all movable charges in the material are reoriented (i.e., saturation can be achieved).

Poling: After manufacture the domains are randomly oriented and the material exhibits no piezoelectricity. The domains are reoriented as much as possible in the direction of the applied field during the poling operation. The direction of the poling field is defined as the 3 or z-direction.

PVF2: An organic polymer which can be ferroelectric. The name is an abbreviation for polyvinyledene difluoride.

PZT: Generic name for piezoelectric materials of the Lead (Pb) Zirconate Titanate family.

Remanent polarization: The residual or remanent polarization of a material after an applied field is reduced to zero. If the material was saturated, the remanent value is usually referred to as the polarization, although even at smaller fields a (smaller) polarization remains.

Related Topics

SAW Material Properties, Piezoelectric Excitation, Material Properties Conducive for Smart Material Applications.

References

1. J. C. Burfoot and G. W. *Taylor, Polar Dielectrics and Their Applications*, Berkeley: University of California Press, 1979.
2. H. Diamant, K. Drenck, and R. Pepinsky, Bridge for accurate measurement of ferroelectric hysteresis, *Rev. Sci. Instrum.*, Vol. 28, p. 30, 1957.
3. T. Hueter and R. Bolt, *Sonics*, New York: John Wiley and Sons, 1954.
4. B. Jaffe, W. Cook, and H. Jaffe, *Piezoelectric Ceramics*, London: Academic Press, 1971.
5. M. E. Lines and A. M. Glass, *Principles and Applications of Ferroelectric Materials*, Oxford: Clarendon Press, 1977.
6. R. A. Roy and K. F. Etzold, Ferroelectric film synthesis, past and present: a select review, *Mater. Res. Soc. Symp. Proc.*, Vol. 200, p. 141, 1990.
7. C. B. Sawyer and C. H. Tower, Rochelle salt as a dielectric, *Phys. Rev.*, Vol. 35, p. 269, 1930.

8. Z. Surowiak, J. Brodacki, and H. Zajosz, Electronic system for investigation of the electrical hysteresis of ferroelectric thin films with high dielectric losses, *Rev. Sci. Instrum.*, Vol. 49, p. 1351, 1978.

9. D. E. Kotecki *et al*, "(BaSr)TiO$_3$ dielectrics for future stacked-capacitor DRAM," *IBM J. of Res. Develop.*, Vol. 43, p. 367, 1999.

10. A. Grill and V. Patel, "Characteristics of low-k and ultralow-k PECVD deposited SiCOH films," *Appl. Phys. Lett.*, Vol. 79, No. 6, 2001.

Further Information

1. IEEE Transactions on Ultrasonics, Ferroelectrics, and Frequency Control (UFFC).
2. IEEE Proceedings of International Symposium on the Application of Ferroelectrics (ISAF) (these symposia are held at irregular intervals).
3. Materials Research Society, Symposium Proceedings, Vols. 191, 200, and 243 (this society holds symposia on ferroelectric materials at irregular intervals).
4. K.-H. Hellwege, Ed., Landolt-Bornstein: Numerical Data and Functional Relationships in Science and Technology, New Series, Gruppe III, Vols. 11 and 18, Berlin: Springer-Verlag, 1979 and 1984 (these volumes have elastic and other data on piezoelectric materials).
5. *American Institute of Physics Handbook*, 3rd ed., New York: McGraw-Hill, 1972.

29

Material Properties of Semiconductors

Mike Harris
Georgia Tech Research Institute

29.1 Semiconductors

Semiconductor is a class of materials that can generally be defined as having an electrical resistivity in the range of 10^{-2}–10^9 Ω-cm.[1] Addition of a very small amount of impure atoms can make a large change in the conductivity of the semiconductor material. This unique materials property makes all semiconductor devices and circuits possible. The amount of free charge in the semiconductor and the transport characteristics of the charge within the crystalline lattice determine the conductivity. Device operation is governed by the ability to generate, move, and remove free charge in a controlled manner. Material characteristics vary widely in various semiconductors and only certain materials are suitable for use in the fabrication of microwave and RF devices.

Bulk semiconductors, for microwave and RF applications, include germanium (Ge), silicon (Si), silicon carbide (SiC), gallium arsenide (GaAs), and indium phosphide (InP). Electronic properties of these materials determine the appropriate frequency range for a particular material. Epitaxial layers of other important materials are grown on these host substrates to produce higher performance devices that overcome basic materials limitations of homogeneous semiconductors. These specialized semiconductors include silicon germanium, gallium nitride, aluminum gallium arsenide, and indium gallium arsenide, among others. Many of the advanced devices described in Chapter 6 are made possible by the optimized properties of these materials. Through the use of "bandgap engineering," many of the material compromises that limit electronic performance can be overcome with these heterostructures.

Electron transport properties determine, to a large extent, the frequency at which various semiconductors are used. On the basis of maturity and cost, silicon will dominate all applications in which it can satisfy the performance requirements. Figure 29.1 is a plot showing the general range of frequencies over which semiconductor materials are being used for integrated circuit applications. It should be noted that the boundary tends to shift to the right with time as new device and fabrication technologies emerge. Discrete devices may be used outside these ranges for specific applications.

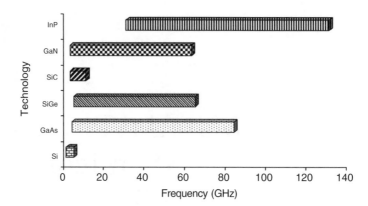

FIGURE 29.1   Frequency range for semiconductor materials.

This section provides information on important materials properties of semiconductors used for microwave and RF applications. Basic information about each semiconductor is presented followed by tables of electronic properties, thermal properties, and mechanical properties. In order to use these materials in microwave and RF applications, devices and circuits must be fabricated. Device fabrication requires etching and deposition of metal contacts and this section provides a partial list of etchants for these semiconductors and provides a list of metallization systems that have been used to successfully produce components. Detailed information on electronic properties of materials can be found in the *Handbook of Electronic and Photonic Materials*, Springer (2006) edited by Kasap and Capper.[2]

29.1.1   Silicon

Silicon is an elemental semiconductor that is by far the best known and most mature. A measure of the maturity of silicon is the fact that it is available in 300-mm (12") diameter wafers. Single crystal silicon wafers are made from electronic grade silicon, which is one of the most refined materials in the world, having an impurity level of no more than one part per billion. Silicon and germanium have the same crystal structure as diamond. In this structure, each atom is surrounded by four nearest neighbor atoms forming a tetrahedron, as shown in Figure 29.2. All of the atoms in the diamond lattice are silicon. Silicon is a "workhorse" material at lower frequencies; however, its electron transport properties and low bulk resistivity limit its application in integrated circuit form to frequencies typically below 4 GHz. Silicon-based discrete devices such as PIN diodes find application at higher frequencies.

Si wafers are available with dopant atoms that make them conductive. Wafers having phosphorus impurities are n-type containing excess electrons. Boron-doped wafers are p-type and have an excess of holes. Flats are cut on the wafers to distinguish between conductivity types and crystal orientation, as shown in Figure 29.3.[3] Silicon is an indirect bandgap material meaning that when an electron and hole recombine, the energy produced is dissipated as a lattice vibration. This should be compared to material like GaAs that is a direct gap semiconductor. When an electron and hole recombine in GaAs, energy is released in the form of light.

29.1.2   Silicon–Germanium

Carrier mobility (the velocity of the carriers to the applied electric field) for both electrons and holes in Si is rather small, and the maximum velocity that these carriers can attain is limited to about 1×10^7 cm/s under normal conditions. Since the speed of a device ultimately depends on how fast the carriers can be forced through the material under practical operating voltages, Si can be regarded as a somewhat "slow" semiconductor and not well-suited for high-frequency devices.[4]

Cost, maturity, and volume production attributes of silicon made it compelling to explore ways to improve silicon's high-frequency performance. The idea of using silicon–germanium (SiGe) alloys to

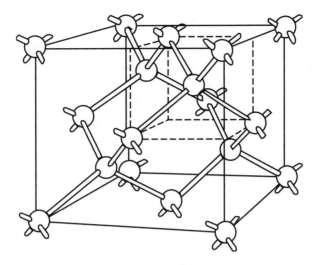

FIGURE 29.2 Cyrstalline structure of Si.

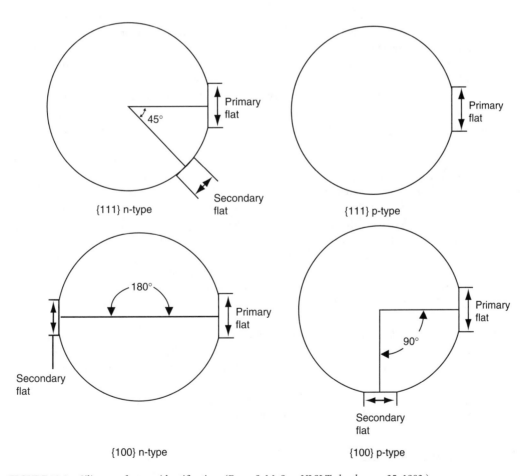

FIGURE 29.3 Silicon wafer type identification. (From S. M. Sze, *VLSI Technology*, p. 35, 1983.)

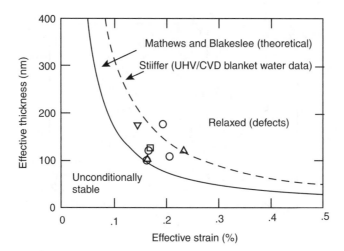

FIGURE 29.4   Strain of SiGe films grown on Si as a function of thickness.

bandgap engineer Si devices dates back to the 1960s; however, synthesis of defect-free SiGe films proved difficult. Device-quality SiGe films were not successfully produced until the 1980s.

While Si and Ge can be combined to produce a chemically stable alloy, their lattice constants differ by roughly 4%. As a result, SiGe alloys grown on Si substrates are compressively strained with the SiGe film adopting the underlying Si lattice constant. These SiGe-strained layers are subject to a fundamental stability criterion limiting their thickness for a given Ge concentration.[5] Figure 29.4 shows this stability diagram, which plots effective film thickness as a function of effective film strain (i.e., Ge content). Deposited SiGe films that lie below the stability curve are thermodynamically stable, and can be processed using conventional furnace, rapid-thermal annealing, or ion implantation without generating defects. Deposited SiGe films that lie above the stability curve are "metastable" and will relax to their natural lattice.

The compressive strain associated with SiGe alloys produces additional bandgap shrinkage, and the net result is a bandgap reduction of approximately 7.5 meV for each 1% of Ge added by mole percent. This Ge-induced "band offset" occurs predominantly in the valence band, making it conducive for use in n-p-n bipolar transistors. In addition, the compressive strain lifts the conduction and valence band degeneracies at the band extremes, effectively reducing the density of states and improving the carrier mobilities through a reduction in carrier scattering. Because a practical SiGe film must be very thin to remain stable, it is a natural candidate for use in the base region of a bipolar transistor which, by definition, must be thin to achieve high-frequency performance. A SiGe heterojunction bipolar transistor (SiGe HBT) contains an n-Si/p-SiGe emitter–base (EB) heterojunction and a p-SiGe/n-Si base–collector heterojunction and is the first practical bandgap-engineered Si-based transistor. Figure 29.5 is a cross-sectional diagram of the material stack used to fabricate an SiGe transistor.[6]

SiGe material and device technology is clearly suitable for microwave and millimeter wave applications. Scaling is used to achieve higher frequency performance, and as a result, the voltage levels are reduced to below 2 V. This limits the output power that SiGe can provide at high frequency. Fourth generation SiGe technology has demonstrated room temperature F_t of over 200 GHz and up to 510 GHz at 4.5K.[4] Since Si–Ge alloy is used in localized regions for this technology and in very thin layers governed by the stability curves described above, the bulk properties of this material system are the same as for Si.

29.1.3 Gallium Arsenide

Silicon is the most widely used semiconductor to make electronic devices; however, there are compounds that perform functions beyond the physical limits of the electronic properties of silicon. There are many

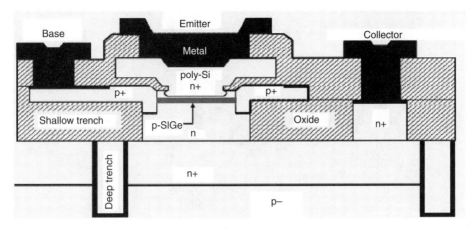

FIGURE 29.5 Cross-sectional diagram of a SiGe HBT showing the thin SiGe base region. (From R. Krithivasan, et al., *IEEE Electron Device Lett.*, Vol. 27, No. 7, pp. 567–569, July 2006.)

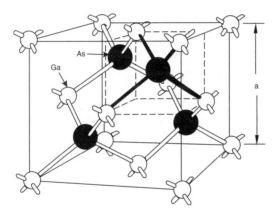

FIGURE 29.6 Zincblende crystalline structure of gallium arsenide (GaAs).

different kinds of compound semiconductor materials, but the most common material combinations used in microwave and RF applications come from the group III and group V elements.

GaAs has a zincblende crystalline structure and is one of the most important compound semiconductors. Zincblende consists of two interpenetrating face-centered cubic (fcc) sublattices as seen in Figure 29.6. One sublattice is displaced by one-fourth of a lattice parameter in each direction from the other sublattice, so that each site of one sublattice is tetrahedrally coordinated with sites from the other sublattice. That is, each atom is at the center of a regular tetrahedron formed by four atoms of the opposite type. When the two sublattices have the same type of atom, the zincblende lattice becomes the diamond lattice as shown above for silicon. Other examples of compound semiconductors with the zincblende lattice include InP and SiC.

GaAs wafers, for microwave and RF applications, are available in 4 inch and 6 inch diameters. Six-inch wafers are currently used for high-speed digital and wireless applications and will be used for higher frequency applications as demand increases. Figure 29.7 shows the standard wafer orientation for semi-insulating GaAs. The front face is the (100) direction or 2 degrees off (100) toward the [110] direction. Figure 29.8 shows the different edge profiles that occur when etching GaAs. These profiles are a function of the crystal orientation as seen in the diagram.[7]

Figure 29.9 is a plot of the bandgap energy of various semiconductors as a function of temperature.[8] GaAs has a bandgap energy at room temperature of 1.43 eV compared to 1.12 eV for silicon. This means that the intrinsic carrier concentration of GaAs can be very low compared to silicon. Since the intrinsic

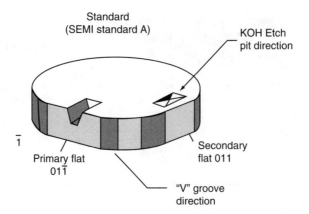

FIGURE 29.7   Standard wafer orientation for semi-insulating GaAs.

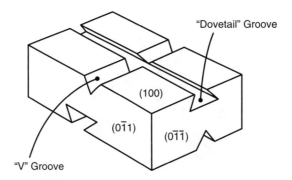

FIGURE 29.8   Orientation-dependent etching profiles of GaAs. (From S. M. Sze, *VLSI Technology*, p. 35, 1983.)

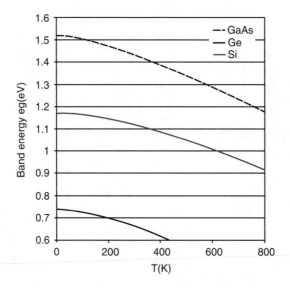

FIGURE 29.9   Energy bandgaps of GaAs, Si, and Ge as a function of temperature. (From S. M. Sze, *Physics of Semiconductors*, p. 15, 1981.)

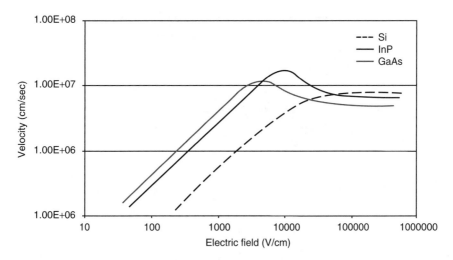

FIGURE 29.10 Electron velocity as a function of electric field for common semiconductors. (From I. Bahl and P. Bhartia, *Microwave Solid State Circuit Design*, p. 307, 1988.)

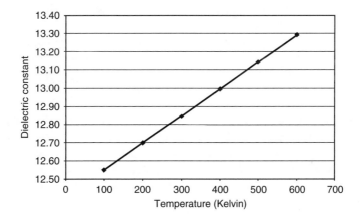

FIGURE 29.11 Relative dielectric constant of GaAs as a function of temperature.

carrier concentration in GaAs is less than that in silicon, higher resistivity substrates are available in GaAs. High resistivity substrates are desirable since they allow the active region of a device to be isolated using a simple ion implantation or mesa etching. Availability of high resistivity GaAs substrates is one reason that this material has found such widespread use for microwave and wireless applications.

The ability to move charge is determined by the transport characteristics of the material. This information is most often presented in the charge carrier velocity electric field characteristic as shown in Figure 29.10.[5] For low values of electric field, the carrier velocity is linearly related to the electric field strength. The proportionality constant is the mobility and this parameter is important in determining the low field operation of a device. Generally, a high value of mobility is desired for optimum device performance. Since the mobility of electrons in GaAs is about six times that of silicon, GaAs is a more attractive material for high-frequency RF and high-speed digital applications. This mobility advantage along with the availability of high resistivity substrates makes GaAs the preferred and most widely used semiconductor material for these applications.[9]

The relative dielectric constant of GaAs is of critical importance in the design of monolithic microwave integrated circuits (MMICs). This parameter is used to determine the width of transmission lines used to interconnect transistors. The characteristic impedance of the transmission line is a function of the relative dielectric constant, the substrate thickness, and the width of the line. Figure 29.11 shows how

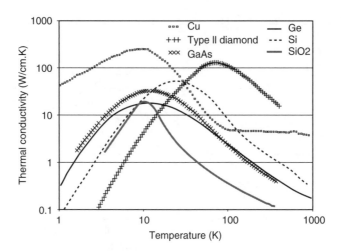

FIGURE 29.12   Thermal conductivity of various materials as a function of temperature. (From S. M. Sze, *Physics of Semiconductors*, p. 15, 1981.)

the relative dielectric constant behaves with temperature. Microwave and RF designs must accommodate this variation over the expected temperature range of operation.

Reliable design and use of GaAs-based MMICs depends on keeping the device channel temperature below a certain absolute maximum level. Channel temperature is defined by the equation,

$$T_{ch} = T_{sink} + \theta \times P \tag{29.1}$$

where T_{ch} is the channel temperature in (K), and T_{sink} is the heat sink temperature in (K). θ is the thermal resistance defined as L/kA, where L is the thickness of the GaAs substrate, k is the thermal conductivity of GaAs and A is the area of the channel. P is the power dissipated (W). Thermal conductivity is a bulk material property that varies with temperature. Figure 29.12 shows the variation in thermal conductivity as a function of temperature for various semiconductors. At temperatures on the order of 100 K, the thermal conductivity of GaAs approaches that of copper.

29.1.4   II-V Heterostructures

A new class of high performance materials, based on GaAs and InP, has been developed using advanced epitaxial processes such as molecular beam epitaxy (MBE). These materials have heterojunctions that are formed between semiconductors having different compositions and bandgaps. The bandgap discontinuity can provide significant improvement in the transport properties of the material and allows optimization of device characteristics not possible with homojunctions. In FET devices, the current density can be high while retaining high electron mobility. This is possible because the free electrons are separated from their donor atoms and are confined to a region where the lattice scattering is minimal. Table 29.1 lists the most common heterostructures for microwave and RF applications. Material compositions vary and are designated by mole fraction using subscripts. In order to match the lattice spacing of GaAs, $Al_{0.25}Ga_{0.75}As$

TABLE 29.1   Common Heterostructures Used for Microwave and RF Applications

$Al_xGa_{1-x}As/GaAs$
$Al_xIn_{x-1}As/InGaAs$
$Al_xGa_{1-x}As/In_yGa_{1-y}As/GaAs$
InGaAs/InP
$Al_xIn_{x-1}As/InP$

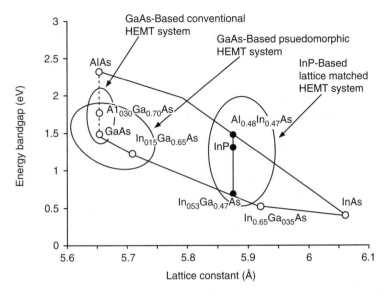

FIGURE 29.13 Energy bandgap and associated lattice constants for II-V heterostructures.

	Thickness	Dopant	Doping
N⁺ GaAs cap	500 Å	Si	$5.0 \times 10^{18} \text{cm}^{-3}$
i AlGaAs donor	300Å	None	·········
Si Planar doping	········	Si	$5.0 \times 10^{12} \text{cm}^{-2}$
i AlGaAs spacer	20 Å	None	··········
i In GaAs channel	120 Å	None	··········
i GaAs spacer			
Si Planar doping	········	Si	$1.2 \times 10^{12} \text{cm}^{-2}$
S/L Buffer	1000Å	None	··········
i GaAs Buffer	5000Å	None	··········
GaAs substrate			

FIGURE 29.14 Double pulsed doped pseudomorphic HEMT layer structure. (Courtesy of Quantum Epitaxial Designs. With permission.)

is routinely used as shown in the diagram of Figure 29.13. This diagram also shows the compounds that are lattice-matched to InP. When a compound is formed with materials that have different lattice spacing, there is strain in the material and under certain conditions, improved performance is possible. This materials system is called pseudomorphic. Figure 29.14 is a cross-sectional diagram of a double pulsed doped pseudomorphic layer structure used for microwave power transistors.

29.1.5 Indium Phosphide

Indium phosphide is an important compound semiconductor for microwave and RF devices due to its physical and electronic properties. Some of the most important properties of InP are high peak electron velocity, high electric field breakdown, and relatively high thermal conductivity. Three-inch diameter bulk, semi-insulating InP wafers are available and 4-inch diameter wafers are being validated. InP has a zincblende crystalline structure like GaAs and its lattice constant is 5.8687 Å compared to 5.6532 Å for GaAs. This materials property is important in the growth of heterostructures discussed below. Bulk InP is used in the fabrication of optoelectronic devices but is not used directly for microwave and RF applications. In microwave and RF devices, InP is used as a substrate for epitaxial growth to support

GaInAs/AlInAs pseudomorphic high electron mobility transistors (HEMTs). This material system has proved to be a more desirable choice for microwave power amplifiers and millimeter wave low noise amplifiers.

A fundamental design goal, for microwave and RF device engineers, is to achieve the best electron transport properties possible within the reliability, breakdown and leakage requirements of the application. Transition from GaAs/AlGaAs HEMTs to pseudomorphic GaAs/InGaAs/AlGaAs HEMTs resulted in significant improvements in device capability for both low noise and power applications.[10] This was due primarily to the increased electron velocity associated with the smaller electron effective mass in InGaAs compared to GaAs. However, the InAs lattice parameter of about 0.606 nm is considerably larger than the GaAs lattice constant of 0. 565 nm, and due to strain effects the compositional limit of pseudomorphic InGaAs on GaAs substrates is limited to about $x = 0.30$. One method to achieve higher InAs content in the InGaAs channel is to use an InP substrate with a lattice constant of 0.587 nm. This newer generation of HEMT devices uses $In_{0.53}Ga_{0.47}As$ channels lattice matched to InP substrates. InP-based pseudomorphic HEMTs with $In_xGa_{1-x}As$ channel compositions of up to about $x = 0.80$ have achieved improvements in performance capability for both low noise and power amplifier applications compared to the best GaAs based devices.

InP-based MMICs require semi-insulating substrates with resistivities from 10^{-6} to $10^{-8}\Omega$ cm. Achieving such resistivities, in nominally undoped crystals, requires that the residual donor concentration to be reduced by a factor of at least 10^6 from current values. This is not practical and an alternate method is required that employs acceptor doping to compensate the residual donors. In principle, any acceptor can compensate the donors. However, because bulk InP crystals commonly have a short range variation of at least 5% in donor and acceptor concentration, the maximum resistivity that can be obtained by compensation with shallow acceptors in the absence of p-n junction formation, is only about 15 Ω cm. Resistivities in the semi-insulating range are usually obtained by doping with a deep acceptor such as iron (Fe). Fe is, by far, the most widely used deep acceptor to produce semi-insulating substrates. As a substitutional impurity, Fe is generally stable under normal device operating temperatures. At high temperatures there is concern of possible diffusion of Fe into the epitaxial layer leading to adverse effects on the devices. Diffusion studies of Fe in InP at a temperature of 650°C for 5 h indicated virtually no change from the control sample.[11]

Use of InP materials is sometimes restricted by its cost. InP substrates are significantly more costly than those made from GaAs, and even more so when compared to silicon. In addition, the technology of InP substrate manufacturing is much more difficult than that of GaAs or silicon. This situation is not simply the result of lower market demand but is linked fundamentally to the high vapor pressure of phosphorus that creates an obstacle to the synthesis of single crystal boules of larger diameters. While 8 inch silicon substrates and 6 inch GaAs substrates are the rule in commercial fabrication, InP substrates are still primarily 2 inch. Three-inch diameter wafers are becoming available and 4 inch wafers are being validated. New concepts of single crystal growth may provide larger diameter wafers at low cost leading to wider acceptance of this compound semiconductor for microwave and RF applications.

29.1.6 Silicon Carbide

Silicon carbide possesses many intrinsic material properties that make it ideal for a wide variety of high power, high temperature, and high-frequency electronic device applications. In comparison with Si and GaAs, SiC has greater than 2.5x larger bandgap, 3x greater thermal conductivity, and a factor of 10 larger breakdown electric field.[12] These characteristics enable SiC devices to operate at higher temperatures and power levels with lower on-resistances and in harsh environments inaccessible to other semiconductors. However, electron transport properties may limit SiC applications to frequencies less than 10 GHz. While excellent prototype SiC devices have been demonstrated, SiC devices will not be widely available until material quality improves and material cost drops.[13]

SiC belongs to a class of semiconductors commonly known as "wide bandgap," which makes it, among other things, less sensitive to increased temperatures. Properly designed and fabricated SiC devices should operate at 500°C or higher, a realm that Si does not even approach. Furthermore, the thermal conductivity

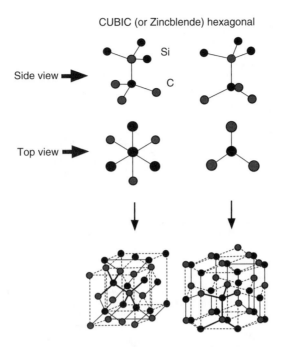

FIGURE 29.15 Difference between the cubic and hexagonal polytypes of SiC. (From Virgil B. Shields, 1994.)

of SiC exceeds even that of copper; and heat produced by a device is therefore quickly dissipated. The inertness of SiC to chemical reaction implies that devices have the potential to operate even in the most caustic of environments. SiC is extremely hard and is best known as the grit coating on sandpaper. This hardness again implies that SiC devices can operate under conditions of extreme pressure. SiC is extremely radiation-hard and can be used close to reactors or for space electronic hardware. Properties of particular importance to the microwave and RF device design engineer are high electric field strength and relatively high saturation drift velocity.

SiC exists not as a single crystal type, but as a whole family of crystals known as polytypes. Each crystal structure has its own unique electrical and optical properties.[14,15] Polytypes differ not in the relative numbers of Si and C atoms, but in the arrangement of these atoms in layers, as illustrated in Figure 29.15. The polytypes are named according to the periodicity of these layers; for example, one of the most common polytypes is called 6H. This means a hexagonal type lattice with an arrangement of six different Si + C layers before the pattern repeats itself. In total, more than 200 different polytypes of SiC have been shown to exist, some with patterns that do not repeat for hundreds of layers. The exact physical properties of SiC depend on the crystal structure adopted, some of the most common structures used are 6H, 4H, and 3C, the final of these structures is the one cubic form of SiC.

SiC is also emerging as a substrate material that may meet the challenging requirements for GaN growth.[16,17] Type 6H-SiC is lattice matched to within 3.5% of GaN, compared to 16% for sapphire.[18] SiC has a thermal conductivity that is over ten times higher than sapphire.

Figure 29.16 compares the energy band and lattice constants of the conventional III-V semiconductor material systems shown in the circle and the wide gap semiconductors. As shown in Figure 29.16, SiC is reasonably lattice matched to GaN and makes a good candidate for a substrate material.

29.1.7 Gallium Nitride

Gallium nitride in the wurtzite form (2H polytype) has a bandgap of 3.45 eV (near UV region) at room temperature. It also forms a continuous range of solid solutions with AlN (6.28 eV) and a discontinuous range of solid solutions with InN (1.95 eV). The wide bandgap, the heterojunction capability, and the strong atomic bonding of these materials make them good candidates for RF and microwave devices.

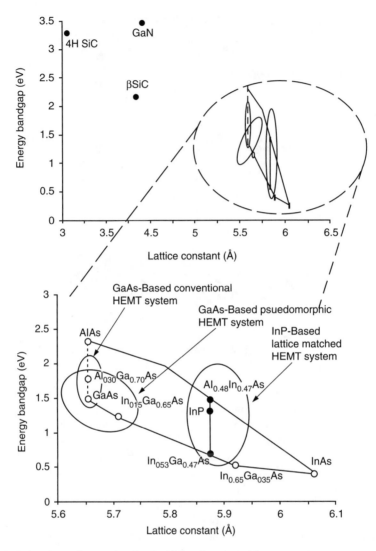

FIGURE 29.16   Comparison of conventional and wide bandgap materials.

Other pertinent device-related parameters include a good thermal conductivity of 1.5 W/cm-K, a type I heterojunction with AlN and AlGaN alloys, large band discontinuities with resultant large interface carrier concentrations, and a large breakdown voltage.

Intrinsic material properties of GaN combined with heterostructure designs are producing revolutionary improvement in output power for microwave devices. GaN-based FET devices have demonstrated power densities of greater than 4 W/mm of gate width compared to 1 W/mm for the best GaAs-based devices.[19] GaN has a bandgap of 3.45 eV compared to 1.43 eV for GaAs. This property leads to orders of magnitude lower than thermal leakage. GaN has a thermal conductivity of almost three times higher than GaAs, a parameter that is critically important for power amplifier applications. The dielectric breakdown strength of GaN is a factor of two greater than GaAs further supporting the consideration of this material for microwave power amplifier applications. A key limitation in the development of GaN devices is the fact that there is no native GaN substrate. Currently, GaN must be deposited on host substrates such as sapphire or SiC.

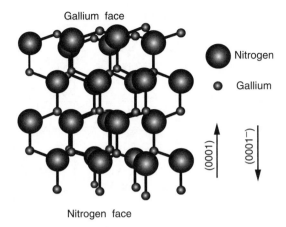

FIGURE 29.17   Wurtzite structure of GaN (Hellman 20).

GaN, AlN, and InGaN have a polar wurtzite structure, as shown in Figure 29.17, and epitaxial films of these materials typically grow along the polar axis. Although the polarity of these nitrides has been studied by a number of techniques, many results in the literature are in conflict.[20] The wurtzite lattice can be considered as two interpenetrating hexagonal close-packed lattices. The wurtzite structure has a tetrahedral arrangement of four equidistant nearest neighbors, similar to the GaAs zincblende structure.

Electron transport in GaN is governed by the electron mobility.[21] Low field electron mobility, in bulk wurtzite GaN, is limited by charged states that are dislocation related, as well as by isolated donor ions and phonons. With no dislocations or donor ions, the $300°K$ phonon-limited mobility is near 2000 cm^2/V-s. Combined scattering effects reduce the mobility for both highly doped and lightly doped material for a given dislocation density. Dislocation density depends on the nucleation methods used at growth initiation, usually ranging from $3 \times 10^9/cm^2$ to $3 \times 10^{10}/cm^2$.[22]

The average drift velocity at high fields in bulk GaN, or in GaN MODFETs, is expected to be above 2×10^7 cm/s. These high fields extend increasingly toward the drain with higher applied drain-source voltage. The average transit velocity is $>1.25 \times 10^7$ cm/s when only the effective gate length is considered—without the extended high-field region. This result is consistent with an average drift velocity of $\sim 2 \times 10^7$ cm/s over an extended region.

Growth of GaN and its related compounds is dominated by metalorganic vapor phase epitaxy (MOVPE) or metalorganic chemical vapor deposition (MOCVD), techniques with obvious advantages in fields where high throughput is required. MBE is also used to grow GaN films. The III-nitride community grows films of GaN and related nitride materials using heteroepitaxial growth routes because of the lack of bulk substrates of these materials. This results in films containing dislocations because of the mismatches in the lattice parameters and the coefficients of thermal expansion between the buffer layer and the film and/or the buffer layer and the substrate. These high concentrations of dislocations may also limit the performance of devices.

Advanced growth procedures, including selective area growth (SAG) and lateral epitaxial overgrowth (LEO) techniques for GaN deposition, are being used specifically to significantly reduce the dislocation density. The latter technique involves the initial vertical growth of a configuration and material composition through windows etched in an SiO_2 or Si_3N_4 mask previously deposited on an underlying GaN seed layer and the subsequent lateral and vertical growth of a film over and adjacent to the mask. Reduction in the number of defects in the material and growth of GaN on larger diameter wafers will lead to the production of GaN-based devices and circuits for microwave and higher frequency applications.

TABLE 29.2   Selected Material Properties of Semiconductors for Microwave and RF Applications

Property	Si	β-SiC	InP	GaAs	GaN
Atoms/cm^3	5.0×10^{22}	6.53×10^{22}	3.95×10^{22}	4.43×10^{22}	4.96×10^{22}
Atomic weight	28.09	40.1	72.90	72.32	41.87
Breakdown field (V/cm)	3×10^5 [37]	4×10^5 [29]	5×10^5 [30]	6×10^5	$> 10 \times 10^5$
Crystal structure	Diamond	Zincblende	Zincblende	Zincblende	Zincite
Density (g/cm^3)	2.3283	3.21	4.81	5.3176	6.1
Dielectric constant	11.8	9.75 [23] 9.66 [24]	12.4	12.5	9 [40]
Effective mass m*/m$_0$ Electron	1.1	.37 3C-SiC [25] .45 6H-SiC [26]	.067	.067	.22 [40, 41]
Electron affinity, eV	4.05 [37]	---	4.38 [30]	4.07 [30]	3.4 [39]
Energy gap (eV) at 300K	1.107	2.403 3C-SiC 3.101 6H-SiC	1.29 [37]	1.35 [37]	3.34 [42]
Intrinsic carrier concentration (cm^{-3})	1.45×10^{10}	3×10^6 3C-SiC [27] 10^{15}-10^{16} 6H-SiC [27]	1.6×10^7	1.8×10^6	3-6×10^9
Lattice constant (Å)	5.431 [37]	4.348 [37]	5.860[37]	5.651[38]	3.190 [43]
Linear coeff. of thermal expansion (10^{-6} K^{-1})	2.49	5.48 [28]	4.6	5.4	5.6[33]
Melting point (K)	1685	3070 [29]	1335[37]	1511 [37]	1500 [37]
Electron mobility μ$_n$ (cm^2/V-S)	1900	1000 3C-SiC [30] 600 6H-SiC [30]	4600	8800	1000[44]
Holes mobility μ$_p$ (cm^2/V-S)	500	40 3C-SiC [30] 40 6H-SiC [30]	150	400	30 [43]
Optical phonon energy (eV)	.063 eV [37]	---	.043 [37]	.035 [37]	.912 [45]
Refractive index	3.42	2.65 3C-SiC [9] 2.72 6H-SiC [10]	3.1 [37]	3.66 [37]	2.7 [46] (at bandedge)
Resistivity, intrinsic (Ω-cm)	1000 [37]	150 [33]	8.2×10^7 [37]	3.8×10^8 [37]	$>10^{13}$ [33]
Specific heat (J/kg-K)	702	640 [34]	310[37]	325[38]	847.39 [47]
Thermal conductivity at 300K (Watt/cm-K)	1.24	3.2 3C-SiC [35] 4.9 6H-SiC [36]	.77 [37]	.56 [37]	1.3 [48]

TABLE 29.3   World Wide Web Sites for
Semiconductor Material Properties

http://www.ioffe.rssi.ru/SVA/NSM/Nano/index.html
http://mems.isi.edu/mems/materials/
http://www.webelements.com/webelements/elements/
http://www.sensors-research.com/senres/materials.htm
http://nsr.mij.mrs.org/
http://nina.ecse.rpi.edu/shur/nitride.htm

29.2   Selected Material Properties

Table 29.2 contains the most common material properties on semiconductors used for microwave and RF applications. Sources of information in this table include textbooks, articles, and World Wide Web sites. Web sites are a convenient way to find materials data and several of the sites used to prepare Table 29.2 are listed in Table 29.3. Most of these sites include detailed references that will provide the user with more extensive information. When using these material properties in the design of components, it is

TABLE 29.4 Etching Processes for Various Semiconductors

Etchant	Substrate and Conditions	Applications	Etch Rate	Reference
15 HNO3, 5 CH3COOH, 2 HF (planar)	Si	For general etching	Variant	49
110 ml CH3COOH, 100 ml HNO3, 50 ml HF, 3 g I2 (iodine etch)	Si	For general etching	Variant	50
KOH solutions; hydrazine (a) KOH (3–50%) (b) 60 hydrazine, 40 H2O	(100) Si 70–90°C, SiO2 unmasked 110°C, 10 min, unmasked	For texturing and V grooving solar etch (no texturizing) for diffuse-reflectivity texturizing	Variant	51
28% by weight KOH at 90–95°C	[100] Si, with oxide mask	V-shaped channel etching for optical waveguides	~2 microns/min	52
3CH3OH, 1H3PO4, 1H2O2	[110], [100], Ga [111]GaAs	Preferential structural etching	~2 microns/min, except Ga[111] reduced twofold	53
3 ml HPO4, 1 ml H2O2, 50 ml H2O	GaAs, 300 K with Shipley 1813 mask	Wet etch	~.1 microns/min	
BCL3 Gas, reactive ion etch	GaN, BCL3, 150 W, 10 mTorr	Slow etching, used to obtain a vertical profile for mesa isolation	~.02 microns/min	

TABLE 29.5 Metallization Systems for Ohmic and Schottky Barrier Contacts

Contact	SiC	GaAs	GaN	InP
Ohmic contact mettalization	Ta, Al/Ti, Ni, Ti, Mo, Ta[55]	Au/Ge/Ni, In, Sn, Ag and Ag alloys, Au, Ni[57]	Ti/Al/Ni/Aun-GaN Ni/Aup-GaN[58]	Au and Au alloys, near noble transition metals (Co, Ni, Ti, Pd, Pt)[54]
Schottky gate mettalization	Au and Au alloys[55]	Ti/Pt/Au multilayer structures, Au alloys[56]	Ni/Au/Ti n-GaN Aup-GaN[58]	For AlInP/InP 15% Al superlattice, 20% Al quantum well[57]

recommended that the user examines the temperature and frequency behavior of the specific property before finalizing the design.

29.2.1 Etching Processes for Semiconductors

Table 29.4 lists typical etching processes for the various semiconductors described in this section.

29.2.1.1 Ohmic and Schottky Contacts

Table 29.5 is a list of metallizations that have been used successfully to fabricate low resistance ohmic contacts for drain and source contacts in FET devices and low leakage Schottky barrier contacts for metal semiconductor diode gates.

References

1. P. Y. Yu and M. Cardona, *Fundamentals of Semiconductors*, New York, Springer-Verlag, p. 1, 1999.
2. S. Kasap and P. Cappers, Eds., *Handbook of Electronic and Photonic Materials*, New York, Springer, 2006.
3. S. M. Sze, *VLSI Technology*, New York, McGraw-Hill, p. 35, 1983.
4. J. D. Cressler, "SiGe HBT technology: A new contender for Si-based RF and microwave circuit applications," *IEEE Trans. Microw. Theory Tech.*, Vol. 46, No. 5, pp. 572–589, May 1998.

5. R. Krithivasan, Y. Lu, J. D. Cressler, J.-S. Rieh, M. H. Khater, D. Ahlgren, and G. Freeman, "Half-Terahertz operation of SiGe HBTs," *IEEE Electron Device Lett.*, Vol. 27, No.7, pp. 567–569, July 2006.

6. J. D. Cressler, "Overview slides of Cressler's SiGe research program," http://users.ece.gatech.edu/~cressler/research/research.html.

7. Sumitomo Electric III-V Semiconductors Specifications, p. 32.

8. S. M. Sze, *Physics of Semiconductors*, New York, John-Wiley & Sons, p. 15, 1981.

9. I. Bahl and P. Bhartia, *Microwave Solid State Circuit Design*, New York, John Wiley & Sons, p. 307, 1988.

10. T. P. Pearsall, *Properties, Processing and Applications of Indium Phosphide*, H. Eisele and G. I., Haddad, Eds. INSPEC, London, p. 40, 2000.

11. A. Katz, *Indium Phosphide and Related Materials: Processing, Technology and Devices*, Byrne, E. K. and Katz, A., Eds. Boston, Artech House, p. 169, 1992.

12. M. S. Shur, S. L. Rumyantsev, and M. E. Levinshtein, Eds., *SiC Materials and Devices*, Vol. 1, World Sci. 2006.

13. G. R. Brandes, "Growth of SiC Boules and Epitaxial Films for Next Generation Electronic Devices," *American Physical Society Centennial Meeting*, March 20–26, Atlanta, 1999.

14. W. Von-munch, "Silicon Carbide," in *Physik der Elemete der IV. Gruppe und der III-V* Verbindungen, K.-H. Hellwege, Ed., Berlin, Heidelberg: Springer-Verlag, pp. 132–142, 1982.

15. J. A. Powell, P. Pirouz, and W. J. Choyke, "Growth and characterization of silicon carbide polytypes for electronic applications," in *Semiconductor Interfaces, Microstructures, and Devices: Properties and Applications*, Z.C. Feng, Ed., Bristol, United Kingdom: Institute of Physics Publishing, pp. 257–293, 1993.

16. S. N. Mohammad, A. A. Salvador, and H. Morkoc, "Emerging gallium nitride based devices," *Proceedings of the IEEE*, Vol. 83, No. 10, pp. 1306–1355, October 1995.

17. M. E. Lin, B. Sverdlov, G. L. Zhou, and H. Morkoc, "A comparative study of GaN epilayers grown on sapphire and SiC substrates by plasma-assisted molecular-beam epitaxy," *Appl. Phys. Lett.* Vol. 62, No. 26, pp. 3479–3481, June 28, 1993.

18. M. E. Lin, Z. Ma, F. Y. Huang, Z. F. Fan, L. H. Allen, and H. Morkov, "Low resistance ohmic contacts on wide band-gap GaN," *Appl. Phys. Lett.* Vol. 64, pp. 1003–1005, 1994.

19. S. Sheppard, K. Doverspike, M. Leonard, W. Pribble, S. Allen, and J. Palmour, "Improved operation of GaN/AlGaN HEMTS on silicon carbide," *International Conference on Silicon Carbide and Related Materials* 1999, paper 349.

20. E. S. Hellman, "The polarity of GaN: a critical review," *Mat. Res. Soc. Internet J.*, Vol. 3.

21. S. K. O'Leary, B. E. Foutz, M. S. Shur, and L. F. Eastman, "Electron transport within the III-V, nitride semiconductors, GaN, AlN, and InN: A Monte Carlo analysis," in *Handbook of Electronic and Optoelectronic Materials*, S. Kasap and P. Capper Eds., Springer, 2006.

22. L. Eastman, K. Chu, W. Schaff, M. Murphy, M. Weimann, and T. Eustis, "High frequency AlGaN/GaN MODFET's," *Mat. Res. Soc. Internet J.*, Vol. 2.

23. L. Patrick and W. J. Choyke, *Phys. Rev. J.*, Vol. 2, p. 2255, 1977.

24. W. J. Choyke, *NATO ASI Ser. E, Appl. Sci.*, Vol. 185, and references therein, 1990.

25. A. Suzuki, A. Ogura, K. Furukawa, Y. Fujii, M. Shigeta, and S. Nakajima, *J. Appl. Phys.*, Vol. 64, p. 2818, 1988.

26. B. W. Wessels and J. C. Gatos, *J. Phys. Chem. Solids*, Vol. 15, p. 83, 1977.

27. M. M. Anikin, A. S. Zubrilov, A. A. Lebedev, A. M. Sterlchuck, and A. E. Cherenkov, *Fitz. Tekh. Polurpovdn.*, Vol. 24, p. 467, 1992.

28. *CRC Materials Science and Engineering Handbook*, p. 304.

29. *CRC Handbook of Chemistry and Physics*, pp. 12–98, 1999–2000 edition.

30. http://vshields.jpl.nasa.gov/windchime.html.

31. P. T. B. Schaffer and R. G. Naum, *Journal of Optical Society of America*, Vol. 59, p. 1498, 1969.

32. P. T. B. Schaffer, *Appl. Opt.*, Vol. 10, p. 1034, 1977.

33. M. N. Yoder, *Wide Bandgap Semiconductor Materials and Devices*, Vol. 43, No. 10, p. 1634, 1966.

34. E. L. Kern, H. W. Hamill, H. W. Deem, and H. D. Sheets, *Mater Res. Bull.*, Vol. 4, p. 107, 1969.

35. D. Morelli, J. Hermans, C. Bettz, W. S. Woo, G. L. Harris, and C. Taylor, *Inst. Phys. Conf. Ser.*, Vol. 137, pp. 313–316, 1990.

36. I. I. Parafenova, Y. M. Tairov, and V. F. Tsvetkov, *Sov. Physics-Semiconductors*, Vol. 24 No. 2, pp. 158–161, 1990.
37. http://www.ioffe.rssi.ru/SVA/NSM/Nano/index.html.
38. C. W. Garland and K. C. Parks, *J. Appl. Phys.*, Vol. 33, p. 759, 1962.
39. M. C. Benjamin, C. Wang, R. F. Davis, and R. J. Nemanich, *Appl. Phys. Lett.*, Vol. 64, p. 3288, 1994.
40. S. N. Mohammad and H. Morkoc, *Prog. Quant. Electron.*, Vol. 20, p. 361, 1996.
41. A. S. Barker and M. Ilegems, *Phys. Rev. B*, Vol 7, p. 743, 1973.
42. H. P. Maruska and J. J. Tietjen, *Appl. Phys. Lett.*, Vol. 15, p. 327, 1969.
43. M. S. Shur and A. M. Khan, *Mat. Res. Bull.*, Vol. 22 No. 2, p. 44, 1977.
44. U. V. Bhapkar and M. S. Shur, *J. Appl. Phys.*, Vol. 82 No. 4, p. 1649, 1997.
45. V. W. L. Chin, T. L. Tansley, and T. Osotchan, *J. Appl. Phys.*, Vol. 75, p. 7365, 1994.
46. A. Billeb, W. Grieshhaber, D. Stocker, E. F. Schubert, and R. F. Karlicek, *Appl. Phys. Lett.*, Vol. 69, p. 2953, 1996.
47. V. I. Koshchenko, Ya. Kh. Grinberg, and A. F. Demidienko, *Inorg. Mat.*, Vol. 11, pp. 1550-1553, 1984.
48. E. K. Sichel and J. I. Pankove, *J. Phys. Chem. Solids*, Vol. 38, p. 330, 1977.
49. W. R. Ruynun, *Semiconductor Measurements and Instrumentation*, Chaps. 1, 2, 7 and 9. McGraw-Hill, New York, p. 129, Table 7.3, 1975.
50. "Integrated Circuit Silicon Device Technology; X-Chemical Metallurgical Properties of Silicon," ASD-TDR-63-316, Vol. X, AD 625, 985. Research Triangle Inst., Research Triangle Park, North Carolina.
51. C. R. Baroana and W. Brandhorst, *IEEE Photovoltaic Spec. Conf. Proc.*, Scottsdale, AZ., pp. 44–48 (1975).
52. http://peta.ee.cornell.edu/jay/res/vgroove/
53. J. L. Merz and R. A. Logan, *J. Appl. Phys.*, Vol. 47, p. 3503, 1976.
54. T. P. Pearshall, "Processing Technologies" in *Properties, Processing and Applications of Indium Phosphide*, A. Katz and T. P. Pearshall, Eds. Inspec, London, p. 246, 2000.
55. G. L. Harris, SiC Devices and Ohmic Contacts, *Properties of SiC*, G. L. Harris, G. Kelner, and M. S. Shur, Eds. Inspec, London, pp. 233–243, 1995.
56. M. Misssous, "Interfaces and Contacts," *Properties of Gallium Arsenide*, D. V. Morgan and J. Wood, Inspec, London, p. 386–387, 1990.
57. J. Lammasniemi, K. Tappura, and K. Smekalin, *Appl. Phys. Lett.*, Vol. 65, No. 20 pp. 2574–2575, 1998.
58. J. H. Edgar, S. Strite, I. Akasaki, H. Amano, and C. Wetzel, "Specifications, Characterisation and Applications of GaN Based Devices," *Properties, Processing and Applications of Gallium Nitride and Related Semiconductors*, S. E. Mohney, Inspec, London, pp. 491–496, 1999.

APPENDIX A
Mathematics, Symbols, and Physical Constants

Greek Alphabet

Greek Letter		Greek Name	English Equivalent	Greek Letter		Greek Name	English Equivalent
A	α	Alpha	a	N	ν	Nu	n
B	β	Beta	b	Ξ	ξ	Xi	x
Γ	γ	Gamma	g	O	o	Omicron	ŏ
Δ	δ	Delta	d	Π	π	Pi	p
E	ε	Epsilon	ĕ	P	ρ	Rho	r
Z	ζ	Zeta	z	Σ	σ	Sigma	s
H	η	Eta	ē	T	τ	Tau	t
Θ	θ ϑ	Theta	th	Y	υ	Upsilon	u
I	ι	Iota	i	Φ	ϕ φ	Phi	ph
K	κ	Kappa	k	X	χ	Chi	ch
Λ	λ	Lambda	l	Ψ	ψ	Psi	ps
M	μ	Mu	m	Ω	ω	Omega	ō

International System of Units (SI)

The International System of units (SI) was adopted by the 11th General Conference on Weights and Measures (CGPM) in 1960. It is a coherent system of units built form seven *SI base units,* one for each of the seven dimensionally independent base quantities: they are the meter, kilogram, second, ampere, kelvin, mole, and candela, for the dimensions length, mass, time, electric current, thermodynamic temperature, amount of substance, and luminous intensity, respectively. The definitions of the SI base units are given below. The *SI derived units* are expressed as products of powers of the base units, analogous to the corresponding relations between physical quantities but with numerical factors equal to unity.

In the International System there is only one SI unit for each physical quantity. This is either the appropriate SI base unit itself or the appropriate SI derived unit. However, any of the approved decimal prefixes, called *SI prefixes,* may be used to construct decimal multiples or submultiples of SI units.

It is recommended that only SI units be used in science and technology (with SI prefixes where appropriate). Where there are special reasons for making an exception to this rule, it is recommended always to define the units used in terms of SI units. This section is based on information supplied by IUPAC.

Definitions of SI Base Units

Meter—The meter is the length of path traveled by light in vacuum during a time interval of 1/299 792 458 of a second (17th CGPM, 1983).

Kilogram—The kilogram is the unit of mass; it is equal to the mass of the international prototype of the kilogram (3rd CGPM, 1901).

Second—The second is the duration of 9 192 631 770 periods of the radiation corresponding to the transition between the two hyperfine levels of the ground state of the cesium-133 atom (13th CGPM, 1967).

Ampere—The ampere is that constant current which, if maintained in two straight parallel conductors of infinite length, of negligible circular cross-section, and placed 1 meter apart in vacuum, would produce between these conductors a force equal to 2×10^{-7} newton per meter of length (9th CGPM, 1948).

Kelvin—The kelvin, unit of thermodynamic temperature, is the fraction 1/273.16 of the thermodynamic temperature of the triple point of water (13th CGPM, 1967).

Mole—The mole is the amount of substance of a system which contains as many elementary entities as there are atoms in 0.012 kilogram of carbon-12. When the mole is used, the elementary entities must be specified and may be atoms, molecules, ions, electrons, or other particles, or specified groups of such particles (14th CGPM, 1971).

Examples of the use of the mole:

1 mol of H_2 contains about 6.022×10^{23} H_2 molecules, or 12.044×10^{23} H atoms

1 mol of HgCl has a mass of 236.04 g

1 mol of Hg_2Cl_2 has a mass of 472.08 g

1 mol of Hg_2^{2+} has a mass of 401.18 g and a charge of 192.97 kC

1 mol of $Fe_{0.91}S$ has a mass of 82.88 g

1 mol of e^- has a mass of 548.60 µg and a charge of -96.49 kC

1 mol of photons whose frequency is 10^{14} Hz has energy of about 39.90 kJ

Candela—The candela is the luminous intensity, in a given direction, of a source that emits monochromatic radiation of frequency 540×10^{12} hertz and that has a radiant intensity in that direction of (1/683) watt per steradian (16th CGPM, 1979).

Names and Symbols for the SI Base Units

Physical Quantity	Name of SI Unit	Symbol for SI Unit
Length	Meter	m
Mass	Kilogram	kg
Time	Second	s
Electric Current	Ampere	A
Thermodynamic temperature	Kelvin	K
Amount of substance	Mole	mol
Luminous intensity	Candela	cd

SI Derived Units with Special Names and Symbols

Physical Quantity	Name of SI Unit	Symbol for SI Unit	Expression in Terms of SI Base Units	
Frequency[1]	Hertz	Hz	s^{-1}	
Force	Newton	N	$m\ kg\ s^{-2}$	
Pressure, stress	Pascal	Pa	$N\ m^{-2}$	$= m^{-1}\ kg\ s^{-2}$
Energy, work, heat	Joule	J	$N\ m$	$= m^2\ kg\ s^{-2}$
Power, radiant flux	Watt	W	$J\ s^{-1}$	$= m^2\ kg\ s^{-3}$
Electric charge	Coulomb	C	$A\ s$	
Electric potential, electromotive force	Volt	V	$J\ C^{-1}$	$= m^2\ kg\ s^{-3}\ A^{-1}$

Physical Quantity	Name of SI Unit	Symbol for SI Unit	Expression in Terms of SI Base Units	
Electric resistance	Ohm	Ω	$V\ A^{-1}$	$= m^2\ kg\ s^{-3}\ A^{-2}$
Electric conductance	Siemens	S	Ω^{-1}	$= m^{-2}\ kg^{-1}\ s^3\ A^2$
Electric capacitance	Farad	F	$C\ V^{-1}$	$= m^{-2}\ kg^{-1}\ s^4\ A^2$
Magnetic flux density	Tesla	T	$V\ s\ m^{-2}$	$= kg\ s^{-2}\ A^{-1}$
Magnetic flux	Weber	Wb	$V\ s$	$= m^2\ kg\ s^{-2}\ A^{-1}$
Inductance	Henry	H	$V\ A^{-1}\ s$	$= m^2\ kg\ s^{-2}\ A^{-2}$
Celsius temperature[2]	Degree Celsius	°C	K	
Luminous flux	Lumen	lm	cd sr	
Illuminance	Lux	lx	$cd\ sr\ m^{-2}$	
Activity (Radioactive)	Becquerel	Bq	s^{-1}	
Absorbed dose (of radiation)	Gray	Gy	$J\ kg^{-1}$	$= m^2\ s^{-2}$
Dose equivalent (dose equivalent index)	Sievert	Sv	$J\ kg^{-1}$	$= m^2\ s^{-2}$
Plane angle	Radian	rad	1	$= m\ m^{-1}$
Solid angle	Steradian	sr	1	$= m^2\ m^{-2}$

[1]For radial (circular) frequency and for angular velocity the unit rad s^{-1}, or simply s^{-1}, should be used, and this may not be simplified to Hz. The unit Hz should be used only for frequency in the sense of cycles per second.

[2]The Celsius temperature θ is defined by the equation:

$$\theta/°C = T/K - 273.15$$

The SI unit of Celsius temperature interval is the degree Celsius, °C, which is equal to the kelvin, K. °C should be treated as a single symbol, with no space between the ° sign and the letter C. (The symbol °K, and the symbol °, should no longer be used.)

Units in Use Together with the SI

These units are not part of the SI, but it is recognized that they will continue to be used in appropriate contexts. SI prefixes may be attached to some of these units, such as milliliter, ml; millibar, mbar; megaelectronvolt, MeV; kilotonne, ktonne.

Physical Quantity	Name of Unit	Symbol for Unit	Value in SI Units
Time	Minute	min	60 s
Time	Hour	h	3600 s
Time	Day	d	86 400 s
Plane angle	Degree	°	$(\pi/180)$ rad
Plane angle	Minute	′	$(\pi/10\ 800)$ rad
Plane angle	Second	″	$(\pi/648\ 000)$ rad
Length	Ångstrom[1]	Å	10^{-10} m
Area	Barn	b	$10^{-28}\ m^2$
Volume	Litre	l, L	$dm^3 = 10^{-3}\ m^3$
Mass	Tonne	t	$Mg = 10^3$ kg
Pressure	Bar[1]	bar	$10^5\ Pa = 10^5\ N\ m^{-2}$
Energy	Electronvolt[2]	$eV\ (= e \times V)$	$\approx 1.60218 \times 10^{-19}$ J
Mass	Unified atomic mass unit[2,3]	$u\ (= m_a(^{12}C)/12)$	$\approx 1.66054 \times 10^{-27}$ kg

[1]The ångstrom and the bar are approved by CIPM for "temporary use with SI units," until CIPM makes a further recommendation. However, they should not be introduced where they are not used at present.

[2]The values of these units in terms of the corresponding SI units are not exact, since they depend on the values of the physical constants e (for the electronvolt) and N_a (for the unified atomic mass unit), which are determined by experiment.

[3]The unified atomic mass unit is also sometimes called the dalton, with symbol Da, although the name and symbol have not been approved by CGPM.

Physical Constants

General

Equatorial radius of the earth = 6378.388 km = 3963.34 miles (statute).

Polar radius of the earth, 6356.912 km = 3949.99 miles (statute).

1 degree of latitude at 40° = 69 miles.

1 international nautical mile = 1.15078 miles (statute) = 1852 m = 6076.115 ft.

Mean density of the earth = 5.522 g/cm^3 = 344.7 lb/ft^3.

Constant of gravitation (6.673 ± 0.003) × 10^{-8} cm^3 gm^{-1} s^{-2}.

Acceleration due to gravity at sea level, latitude 45° = 980.6194 cm/s^2 = 32.1726 ft/s^2.

Length of seconds pendulum at sea level, latitude 45° = 99.3575 cm = 39.1171 in.

1 knot (international) = 101.269 ft/min = 1.6878 ft/s = 1.1508 miles (statute)/h.

1 micron = 10^{-4} cm.

1 ångstrom = 10^{-8} cm.

Mass of hydrogen atom = (1.67339 ± 0.0031) × 10^{-24} g.

Density of mercury at 0°C = 13.5955 g/ml.

Density of water at 3.98°C = 1.000000 g/ml.

Density, maximum, of water, at 3.98°C = 0.999973 g/cm^3.

Density of dry air at 0°C, 760 mm = 1.2929 g/l.

Velocity of sound in dry air at 0°C = 331.36 m/s – 1087.1 ft/s.

Velocity of light in vacuum = (2.997925 ± 0.000002) × 10^{10} cm/s.

Heat of fusion of water 0°C = 79.71 cal/g.

Heat of vaporization of water 100°C = 539.55 cal/g.

Electrochemical equivalent of silver 0.001118 g/s international amp.

Absolute wavelength of red cadmium light in air at 15°C, 760 mm pressure = 6438.4696 Å.

Wavelength of orange-red line of krypton 86 = 6057.802 Å.

π Constants

$$\pi = 3.14159\ 26535\ 89793\ 23846\ 26433\ 83279\ 50288\ 41971\ 69399\ 37511$$
$$1/\pi = 0.31830\ 98861\ 83790\ 67153\ 77675\ 26745\ 02872\ 40689\ 19291\ 48091$$
$$\pi^2 = 9.8690\ \ 44010\ 89358\ 61883\ 44909\ 99876\ 15113\ 53136\ 99407\ 24079$$
$$\log_e\pi = 1.14472\ 98858\ 49400\ 17414\ 34273\ 51353\ 05871\ 16472\ 94812\ 91531$$
$$\log_{10}\pi = 0.49714\ 98726\ 94133\ 85435\ 12682\ 88290\ 89887\ 36516\ 78324\ 38044$$
$$\log_{10}\sqrt{2}\,\pi = 0.39908\ 99341\ 79057\ 52478\ 25035\ 91507\ 69595\ 02099\ 34102\ 92128$$

Constants Involving *e*

$$e = 2.71828\ 18284\ 59045\ 23536\ 02874\ 71352\ 66249\ 77572\ 47093\ 69996$$
$$1/e = 0.36787\ 94411\ 71442\ 32159\ 55237\ 70161\ 46086\ 74458\ 11131\ 03177$$
$$e^2 = 7.38905\ 60989\ 30650\ 22723\ 04274\ 60575\ 00781\ 31803\ 15570\ 55185$$
$$M = \log_{10}e = 0.43429\ 44819\ 03251\ 82765\ 11289\ 18916\ 60508\ 22943\ 97005\ 80367$$
$$1/M = \log_e10 = 2.30258\ 50929\ 94045\ 68401\ 79914\ 54684\ 36420\ 67011\ 01488\ 62877$$
$$\log_{10}M = 9.63778\ 43113\ 00536\ 78912\ 29674\ 98645\ -10$$

Numerical Constants

$$\sqrt{2} = 1.41421\ 35623\ 73095\ 04880\ 16887\ 24209\ 69807\ 85696\ 71875\ 37695$$
$$3\sqrt{2} = 1.25992\ 10498\ 94873\ 16476\ 72106\ 07278\ 22835\ 05702\ 51464\ 70151$$
$$\log_e2 = 0.69314\ 71805\ 59945\ 30941\ 72321\ 21458\ 17656\ 80755\ 00134\ 36026$$
$$\log_{10}2 = 0.30102\ 99956\ 63981\ 19521\ 37388\ 94724\ 49302\ 67881\ 89881\ 46211$$
$$\sqrt{3} = 1.73205\ 08075\ 68877\ 29352\ 74463\ 41505\ 87236\ 69428\ 05253\ 81039$$
$$\sqrt[3]{3} = 1.44224\ 95703\ 07408\ 38232\ 16383\ 10780\ 10958\ 83918\ 69253\ 49935$$
$$\log_e3 = 1.09861\ 22886\ 68109\ 69139\ 52452\ 36922\ 52570\ 46474\ 90557\ 82275$$
$$\log_{10}3 = 0.47712\ 12547\ 19662\ 43729\ 50279\ 03255\ 11530\ 92001\ 28864\ 19070$$

Symbols and Terminology for Physical and Chemical Quantities

Name	Symbol	Definition	SI Unit
		Classical Mechanics	
Mass	m		kg
Reduced mass	μ	$\mu = m_1 m_2/(m_1 + m_2)$	kg
Density, mass density	ρ	$\rho = M/V$	kg m^{-3}
Relative density	d	$d = \rho/\rho^\theta$	1
Surface density	ρ_A, ρ_S	$\rho_A = m/A$	kg m^{-2}
Specific volume	v	$v = V/m = 1/\rho$	m^3 kg^{-1}
Momentum	p	$p = mv$	kg m s^{-1}
Angular momentum, action	L	$l = r \times p$	J s
Moment of inertia	I, J	$I = \Sigma m_i r_i^2$	kg m^2
Force	F	$F = dp/dt = ma$	N
Torque, moment of a force	$T, (M)$	$T = r \times F$	N m
Energy	E		J
Potential energy	E_p, V, Φ	$E_p = -\int F \cdot ds$	J
Kinetic energy	E_k, T, K	$e_k = (1/2)mv^2$	J
Work	W, w	$w = \int F \cdot ds$	J
Hamilton function	H	$H(q, p)$ $= T(q, p) + V(q)$	J
Lagrange function	L	$L(q, \dot{q})$ $T(q, \dot{q}) - V(q)$	J
Pressure	p, P	$p = F/A$	Pa, N m^{-2}
Surface tension	γ, σ	$\gamma = dW/dA$	N m^{-1}, J m^{-2}
Weight	$G, (W, P)$	$G = mg$	N
Gravitational constant	G	$F = Gm_1 m_2/r^2$	N m^2 kg^{-2}
Normal stress	σ	$\sigma = F/A$	Pa
Shear stress	τ	$\tau = F/A$	Pa
Linear strain, relative elongation	ε, e	$\varepsilon = \Delta l/l$	1
Modulus of elasticity, Young's modulus	E	$E = \sigma/\varepsilon$	Pa
Shear strain	γ	$\gamma = \Delta x/d$	1
Shear modulus	G	$G = \tau/\gamma$	Pa
Volume strain, bulk strain	θ	$\theta = \Delta V/V_0$	1
Bulk modulus, compression modulus	K	$K = -V_0(dp/dV)$	Pa
Viscosity, dynamic viscosity	η, μ	$\tau_{xz} = \eta(dv_x/dz)$	Pa s
Fluidity	ϕ	$\phi = 1/\eta$	m kg^{-1} s
Kinematic viscosity	v	$v = \eta/\rho$	m^2 s^{-1}
Friction coefficient	$\mu, (f)$	$F_{frict} = \mu F_{norm}$	1
Power	P	$P = dW/dt$	W
Sound energy flux	P, P_a	$P = dE/dt$	W
Acoustic factors			
Reflection factor	ρ	$\rho = P_r/P_0$	1
Acoustic absorption factor	$\alpha_a, (\alpha)$	$\alpha_a = 1 - \rho$	1
Transmission factor	τ	$\tau = P_{tr}/P_0$	1
Dissipation factor	δ	$\delta = \alpha_a - \tau$	1

Fundamental Physical Constants

Summary of the 1986 Recommended Values
of the Fundamental Physical Constants

Quantity	Symbol	Value	Units	Relative Uncertainty (ppm)
Speed of light in vacuum	c	299 792 458	ms^{-1}	(exact)
Permeability of vacuum	μ_o	$4\pi \times 10^{-7}$	N A^{-2}	
		$= 12.566\ 370614\ ...$	10^{-7} N A^{-2}	(exact)
Permittivity of vacuum	ε_o	$1/\mu_o c^2$		
		$= 8.854\ 187\ 817\ ...$	10^{-12} F m^{-1}	(exact)
Newtonian constant of gravitation	G	6.672 59(85)	10^{-11} m^3 kg^{-1} s^{-2}	128
Planck constant	h	6.626 0755(40)	10^{-34} J s	0.60
$h/2\pi$	$\hbar$	1.054 572 66(63)	10^{-34} J s	0.60
Elementary charge	e	1.602 177 33(49)	10^{-19} C	0.30
Magnetic flux quantum, $h/2e$	Φ_o	2.067 834 61(61)	10^{-15} Wb	0.30
Electron mass	m_e	9.109 3897(54)	10^{-31} kg	0.59
Proton mass	m_p	1.672 6231(10)	10^{-27} kg	0.59
Proton–electron mass ratio	m_p/m_e	1836.152701(37)		0.020
Fine-structure constant, $\mu_o ce^2/2h$	α	7.297 353 08(33)	10^{-3}	0.045
Inverse fine-structure constant	α^{-1}	137 035 9895(61)		0.045
Rydberg constant, $m_e c\alpha^2/2h$	R_∞	10 973 731.534(13)	m^{-1}	0.0012
Avogadro constant	N_A, L	6.022 1367(36)	10^{23} mol^{-1}	0.59
Faraday constant, $N_A e$	F	96 485.309(29)	C mol^{-1}	0.30
Molar gas constant	R	8.314 510(70)	J mol^{-1} K^{-1}	8.4
Boltzmann constant, R/N_A	k	1.380 658(12)	10^{-23} J K^{-1}	8.5
Stafan–Boltzmann constant, $(\pi^2/60)k^4/\hbar^3 c^2$	σ	5.670 51(19)	10^{-8} W m^{-2} K^{-4}	34
Non-SI units used with SI				
Electronvolt, $(e/C)J = \{e\}J$	eV	1.602 17733(40)	10^{-19} J	0.30
(Unified) atomic mass unit, $1\ u = m_u = 1/12m(^{12}C)$	u	1.660 5402(10)	10^{-27} kg	0.59

Note: An abbreviated list of the fundamental constants of physics and chemistry based on a least-squares adjustment with 17 degrees of freedom. The digits in parentheses are the one-standard-deviation uncertainty in the last digits of the given value. Since the uncertainties of many entries are correlated, the full covariance matrix must be used in evaluating the uncertainties of quantities computed from them.

PERIODIC TABLE OF THE ELEMENTS

New Notation
Previous IUPAC Form
CAS Version

Key to Chart

Atomic Number →	**50**	← Oxidation States (+2 +4)
Symbol →	**Sn**	
1995 Atomic Weight →	118.710	
	-18-18-4 ←	Electron Configuration

Group / Element	Atomic No.	Atomic Weight	Oxidation States	Electron Config.	Shell
H (IA, 1)	1	1.00794	+1, -1	1	K
He (VIIIA, 18)	2	4.002602	0	2	K
Li (IA, 1)	3	6.941	+1	2-1	K-L
Be (IIA, 2)	4	9.012182	+2	2-2	
B (IIIA, 13)	5	10.811	+3	2-3	
C (IVA, 14)	6	12.0107	+2 +4 -4	2-4	
N (VA, 15)	7	14.00674	+1 +2 +3 +4 +5 -1 -2 -3	2-5	
O (VIA, 16)	8	15.9994	-2	2-6	
F (VIIA, 17)	9	18.9984032	-1	2-7	
Ne (VIIIA, 18)	10	20.1797	0	2-8	
Na (IA, 1)	11	22.989770	+1	2-8-1	K-L-M
Mg (IIA, 2)	12	24.3050	+2	2-8-2	
Al (IIIA, 13)	13	26.981538	+3	2-8-3	
Si (IVA, 14)	14	28.0855	+2 +4 -4	2-8-4	
P (VA, 15)	15	30.973761	+3 +5 -3	2-8-5	
S (VIA, 16)	16	32.066	+4 +6 -2	2-8-6	
Cl (VIIA, 17)	17	35.4527	+1 +5 +7 -1	2-8-7	
Ar (VIIIA, 18)	18	39.948	0	2-8-8	
K (IA, 1)	19	39.0983	+1	2-8-8-1	-L-M-N
Ca (IIA, 2)	20	40.078	+2	2-8-8-2	
Sc (IIIB, 3)	21	44.955910	+3	2-8-9-2	
Ti (IVB, 4)	22	47.867	+2 +3 +4	2-8-10-2	
V (VB, 5)	23	50.9415	+2 +3 +4 +5	2-8-11-2	
Cr (VIB, 6)	24	51.9961	+2 +3 +6	2-8-13-1	
Mn (VIIB, 7)	25	54.938049	+2 +3 +4 +7	2-8-13-2	
Fe (VIII, 8)	26	55.845	+2 +3	2-8-14-2	
Co (VIII, 9)	27	58.933200	+2 +3	2-8-15-2	
Ni (VIII, 10)	28	58.6934	+2 +3	2-8-16-2	
Cu (IB, 11)	29	63.546	+1 +2	2-8-18-1	
Zn (IIB, 12)	30	65.39	+2	2-8-18-2	
Ga (IIIA, 13)	31	69.723	+3	2-8-18-3	
Ge (IVA, 14)	32	72.61	+2 +4	2-8-18-4	
As (VA, 15)	33	74.92160	+3 +5 -3	2-8-18-5	
Se (VIA, 16)	34	78.96	+4 +6 -2	2-8-18-6	
Br (VIIA, 17)	35	79.904	+1 +5 -1	2-8-18-7	
Kr (VIIIA, 18)	36	83.80	0	2-8-18-8	
Rb (IA, 1)	37	85.4678	+1	2-8-18-8-1	-M-N-O
Sr (IIA, 2)	38	87.62	+2	2-8-18-8-2	
Y (IIIB, 3)	39	88.90585	+3	2-8-18-9-2	
Zr (IVB, 4)	40	91.224	+4	2-8-18-10-2	
Nb (VB, 5)	41	92.90638	+3 +5	2-8-18-12-1	
Mo (VIB, 6)	42	95.94	+6	2-8-18-13-1	
Tc (VIIB, 7)	43	(98)	+4 +6 +7	2-8-18-13-2	
Ru (VIII, 8)	44	101.07	+3	2-8-18-15-1	
Rh (VIII, 9)	45	102.90550	+3	2-8-18-16-1	
Pd (VIII, 10)	46	106.42	+2 +4	2-8-18-18-0	
Ag (IB, 11)	47	107.8682	+1	2-8-18-18-1	
Cd (IIB, 12)	48	112.411	+2	2-8-18-18-2	
In (IIIA, 13)	49	114.818	+3	2-8-18-18-3	
Sn (IVA, 14)	50	118.710	+2 +4	2-8-18-18-4	
Sb (VA, 15)	51	121.760	+3 +5 -3	2-8-18-18-5	
Te (VIA, 16)	52	127.60	+4 +6 -2	2-8-18-18-6	
I (VIIA, 17)	53	126.90447	+1 +5 +7 -1	2-8-18-18-7	
Xe (VIIIA, 18)	54	131.29	0	2-8-18-18-8	
Cs (IA, 1)	55	132.90545	+1	2-8-18-18-8-1	-N-O-P
Ba (IIA, 2)	56	137.327	+2	2-8-18-18-8-2	
La* (IIIB, 3)	57	138.9055	+3	2-8-18-18-9-2	
Hf (IVB, 4)	72	178.49	+4	2-8-18-32-10-2	
Ta (VB, 5)	73	180.9479	+5	2-8-18-32-11-2	
W (VIB, 6)	74	183.84	+6	2-8-18-32-12-2	
Re (VIIB, 7)	75	186.207	+4 +6 +7	2-8-18-32-13-2	
Os (VIII, 8)	76	190.23	+3 +4	2-8-18-32-14-2	
Ir (VIII, 9)	77	192.217	+3 +4	2-8-18-32-15-2	
Pt (VIII, 10)	78	195.078	+2 +4	2-8-18-32-17-1	
Au (IB, 11)	79	196.96655	+1 +3	2-8-18-32-18-1	
Hg (IIB, 12)	80	200.59	+1 +2	2-8-18-32-18-2	
Tl (IIIA, 13)	81	204.3833	+1 +3	2-8-18-32-18-3	
Pb (IVA, 14)	82	207.2	+2 +4	2-8-18-32-18-4	
Bi (VA, 15)	83	208.98038	+3 +5	2-8-18-32-18-5	
Po (VIA, 16)	84	(209)	+2 +4	2-8-18-32-18-6	
At (VIIA, 17)	85	(210)	+3 +5 -1	2-8-18-32-18-7	
Rn (VIIIA, 18)	86	(222)	0	2-8-18-32-18-8	-O-P-Q
Fr (IA, 1)	87	(223)	+1	2-8-18-32-18-8-1	-N-O-P
Ra (IIA, 2)	88	(226)	+2	2-8-18-32-18-8-2	
Ac** (IIIB, 3)	89	(227)	+3	2-8-18-32-18-9-2	
Rf (IVB, 4)	104	(261)	+3	2-8-18-32-32-10-2	
Db (VB, 5)	105	(262)		2-8-18-32-32-11-2	
Sg (VIB, 6)	106	(266)		2-8-18-32-32-12-2	
Bh (VIIB, 7)	107	(264)		2-8-18-32-32-13-2	
Hs (VIII, 8)	108	(269)		2-8-18-32-32-14-2	
Mt (VIII, 9)	109	(268)		2-8-18-32-32-15-2	
Uun (VIII, 10)	110	(271)		2-8-18-32-32-16-2	-O-P-Q
Uuu (IB, 11)	111	(272)			
Uub (IIB, 12)	112				

* Lanthanides

Element	Atomic No.	Atomic Weight	Oxidation States	Electron Config.
Ce	58	140.116	+3 +4	-19-9-2
Pr	59	140.90765	+3 +4	-21-8-2
Nd	60	144.24	+3	-22-8-2
Pm	61	(145)	+3	-23-8-2
Sm	62	150.36	+2 +3	-24-8-2
Eu	63	151.964	+2 +3	-25-8-2
Gd	64	157.25	+3	-25-9-2
Tb	65	158.92534	+3	-27-8-2
Dy	66	162.50	+3	-28-8-2
Ho	67	164.93032	+3	-29-8-2
Er	68	167.26	+3	-30-8-2
Tm	69	168.93421	+3	-31-8-2
Yb	70	173.04	+2 +3	-32-8-2
Lu	71	174.967	+3	-32-9-2

** Actinides

Element	Atomic No.	Atomic Weight	Oxidation States	Electron Config.
Th	90	232.0381	+4	-18-10-2
Pa	91	231.03588	+4 +5	-20-9-2
U	92	238.0289	+3 +4 +5 +6	-21-9-2
Np	93	(237)	+3 +4 +5 +6	-22-9-2
Pu	94	(244)	+3 +4 +5 +6	-24-8-2
Am	95	(243)	+3 +4 +5 +6	-25-8-2
Cm	96	(247)	+3	-25-9-2
Bk	97	(247)	+3 +4	-27-8-2
Cf	98	(251)	+3 +4	-28-8-2
Es	99	(252)	+3	-29-8-2
Fm	100	(257)	+3	-30-8-2
Md	101	(258)	+2 +3	-31-8-2
No	102	(259)	+2 +3	-32-8-2
Lr	103	(262)	+3	-32-9-2

The new IUPAC format numbers the groups from 1 to 18. The previous IUPAC numbering system and the system used by Chemical Abstracts Service (CAS) are also shown. For radioactive elements that do not occur in nature, the mass number of the most stable isotope is given in parentheses.

References
1. G. J. Leigh, Editor, *Nomenclature of Inorganic Chemistry*, Blackwell Scientific Publications, Oxford, 1990.
2. *Chemical and Engineering News*, 63(5), 27, 1985.
3. Atomic Weights of the Elements, 1995, *Pure & Appl. Chem.*, 68, 2339, 1996.

Electrical Resistivity

Electrical Resistivity of Pure Metals

The first part of this table gives the electrical resistivity, in units of 10^{-8} Ω m, for 28 common metallic elements as a function of temperature. The data refer to polycrystalline samples. The number of significant figures indicates the accuracy of the values. However, at low temperatures (especially below 50 K) the electrical resistivity is extremely sensitive to sample purity. Thus the low-temperature values refer to samples of specified purity and treatment.

The second part of the table gives resistivity values in the neighborhood of room temperature for other metallic elements that have not been studied over an extended temperature range.

Electrical Resistivity in 10^{-8} Ω m

T/K	Aluminum	Barium	Beryllium	Calcium	Cesium	Chromium	Copper
1	0.000100	0.081	0.0332	0.045	0.0026		0.00200
10	0.000193	0.189	0.0332	0.047	0.243		0.00202
20	0.000755	0.94	0.0336	0.060	0.86	0.00280	
40	0.0181	2.91	0.0367	0.175	1.99		0.0239
60	0.0959	4.86	0.067	0.40	3.07		0.0971
80	0.245	6.83	0.075	0.65	4.16		0.215
100	0.442	8.85	0.133	0.91	5.28	1.6	0.348
150	1.006	14.3	0.510	1.56	8.43	4.5	0.699
200	1.587	20.2	1.29	2.19	12.2	7.7	1.046
273	2.417	30.2	3.02	3.11	18.7	11.8	1.543
293	2.650	33.2	3.56	3.36	20.5	12.5	1.678
298	2.709	34.0	3.70	3.42	20.8	12.6	1.712
300	2.733	34.3	3.76	3.45	21.0	12.7	1.725
400	3.87	51.4	6.76	4.7		15.8	2.402
500	4.99	72.4	9.9	6.0		20.1	3.090
600	6.13	98.2	13.2	7.3		24.7	3.792
700	7.35	130	16.5	8.7		29.5	4.514
800	8.70	168	20.0	10.0	34.6	5.262	
900	10.18	216	23.7	11.4		39.9	6.041

T/K	Gold	Hafnium	Iron	Lead	Lithium	Magnesium	Manganese
1	0.0220	1.00	0.0225		0.007	0.0062	7.02
10	0.0226	1.00	0.0238		0.008	0.0069	18.9
20	0.035	1.11	0.0287		0.012	0.0123	54
40	0.141	2.52	0.0758		0.074	0.074	116
60	0.308	4.53	0.271		0.345	0.261	131
80	0.481	6.75	0.693	4.9	1.00	0.557	132
100	0.650	9.12	1.28	6.4	1.73	0.91	132
150	1.061	15.0	3.15	9.9	3.72	1.84	136
200	1.462	21.0	5.20	13.6	5.71	2.75	139
273	2.051	30.4	8.57	19.2	8.53	4.05	143
293	2.214	33.1	9.61	20.8	9.28	4.39	144
298	2.255	33.7	9.87	21.1	9.47	4.48	144
300	2.271	34.0	9.98	21.3	9.55	4.51	144
400	3.107	48.1	16.1	29.6	13.4	6.19	147
500	3.97	63.1	23.7	38.3		7.86	149
600	4.87	78.5	32.9			9.52	151
700	5.82		44.0			11.2	152
800	6.81		57.1			12.8	
900	7.86					14.4	

(continued)

T/K	Molybdenum	Nickel	Palladium	Platinum	Potassium	Rubidium	Silver
1	0.00070	0.0032	0.0200	0.002	0.0008	0.0131	0.00100
10	0.00089	0.0057	0.0242	0.0154	0.0160	0.109	0.00115
20	0.00261	0.0140	0.0563	0.0484	0.117	0.444	0.0042
40	0.0457	0.068	0.334	0.409	0.480	1.21	0.0539
60	0.206	0.242	0.938	1.107	0.90	1.94	0.162
80	0.482	0.545	1.75	1.922	1.34	2.65	0.289
100	0.858	0.96	2.62	2.755	1.79	3.36	0.418
150	1.99	2.21	4.80	4.76	2.99	5.27	0.726
200	3.13	3.67	6.88	6.77	4.26	7.49	1.029
273	4.85	6.16	9.78	9.6	6.49	11.5	1.467
293	5.34	6.93	10.54	10.5	7.20	12.8	1.587
298	5.47	7.12	10.73	10.7	7.39	13.1	1.617
300	5.52	7.20	10.80	10.8	7.47	13.3	1.629
400	8.02	11.8	14.48	14.6			2.241
500	10.6	17.7	17.94	18.3			2.87
600	13.1	25.5	21.2	21.9			3.53
700	15.8	32.1	24.2	25.4			4.21
800	18.4	35.5	27.1	28.7			4.91
900	21.2	38.6	29.4	32.0			5.64

T/K	Sodium	Strontium	Tantalum	Tungsten	Vanadium	Zinc	Zirconium
1	0.0009	0.80	0.10	0.00016		0.0100	0.250
10	0.0015	0.80	0.102	0.000137	0.0145	0.0112	0.253
20	0.016	0.92	0.146	0.00196	0.039	0.0387	0.357
40	0.172	1.70	0.751	0.0544	0.304	0.306	1.44
60	0.447	2.68	1.65	0.266	1.11	0.715	3.75
80	0.80	3.64	2.62	0.606	2.41	1.15	6.64
100	1.16	4.58	3.64	1.02	4.01	1.60	9.79
150	2.03	6.84	6.19	2.09	8.2	2.71	17.8
200	2.89	9.04	8.66	3.18	12.4	3.83	26.3
273	4.33	12.3	12.2	4.82	18.1	5.46	38.8
293	4.77	13.2	13.1	5.28	19.7	5.90	42.1
298	4.88	13.4	13.4	5.39	20.1	6.01	42.9
300	4.93	13.5	13.5	5.44	20.2	6.06	43.3
400		17.8	18.2	7.83	28.0	8.37	60.3
500		22.2	22.9	10.3	34.8	10.82	76.5
600		26.7	27.4	13.0	41.1	13.49	91.5
700		31.2	31.8	15.7	47.2		104.2
800		35.6	35.9	18.6	53.1		114.9
900			40.1	21.5	58.7		123.1

Electrical Resistivity of Pure Metals (continued)

Element	*T*/K	Electrical Resistivity $10^{-8}\ \Omega$ m
Antimony	273	39
Bismuth	273	107
Cadmium	273	6.8
Cerium	290–300	82.8
Cobalt	273	5.6
Dysprosium	290–300	92.6
Erbium	290–300	86.0
Europium	290–300	90.0
Gadolinium	290–300	131
Gallium	273	13.6
Holmium	290–300	81.4
Indium	273	8.0
Iridium	273	4.7
Lanthanum	290–300	61.5
Lutetium	290–300	58.2
Mercury	273	94.1
Neodymium	290–300	64.3
Niobium	273	15.2
Osmium	273	8.1
Polonium	273	40
Praseodymium	290–300	70.0
Promethium	290–300	75
Protactinium	273	17.7
Rhenium	273	17.2
Rhodium	273	4.3
Ruthenium	273	7.1
Samrium	290–300	94.0
Scandium	290–300	56.2
Terbium	290–300	115
Thallium	273	15
Thorium	273	14.7
Thulium	290–300	67.6
Tin	273	11.5
Titanium	273	39
Uranium	273	28
Ytterbium	290–300	25.0
Yttrium	290–300	59.6

Electrical Resistivity of Selected Alloys

Values of the resistivity are given in units of 10^{-8} Ω m. General comments in the preceding table for pure metals also apply here.

	273 K	293 K	300 K	350 K	400 K		273 K	293 K	300 K	350 K	400 K
		Alloy—Aluminum-Copper						Alloy—Copper-Nickel			
Wt % Al						**Wt % Cu**					
99[a]	2.51	2.74	2.82	3.38	3.95	99[c]	2.71	2.85	2.91	3.27	3.62
95[a]	2.88	3.10	3.18	3.75	4.33	95[c]	7.60	7.71	7.82	8.22	8.62
90[b]	3.36	3.59	3.67	4.25	4.86	90[c]	13.69	13.89	13.96	14.40	14.81
85[b]	3.87	4.10	4.19	4.79	5.42	85[c]	19.63	19.83	19.90	2032	20.70
80[b]	4.33	4.58	4.67	5.31	5.99	80[c]	25.46	25.66	25.72	26.12[aa]	26.44[aa]
70[b]	5.03	5.31	5.41	6.16	6.94	70[i]	36.67	36.72	36.76	36.85	36.89
60[b]	5.56	5.88	5.99	6.77	7.63	60[i]	45.43	45.38	45.35	45.20	45.01
50[b]	6.22	6.55	6.67	7.55	8.52	50[i]	50.19	50.05	50.01	49.73	49.50
40[c]	7.57	7.96	8.10	9.12	10.2	40[c]	47.42	47.73	47.82	48.28	48.49
30[c]	11.2	11.8	12.0	13.5	15.2	30[i]	40.19	41.79	42.34	44.51	45.40
25[f]	16.3[aa]	17.2	17.6	19.8	22.2	25[c]	33.46	35.11	35.69	39.67[aa]	42.81[aa]
15[h]	—	12.3	—	—	—	15[c]	22.00	23.35	23.85	27.60	31.38
19[g]	10.8[aa]	11.0	11.1	11.7	12.3	10[c]	16.65	17.82	18.26	21.51	25.19
5[e]	9.43	9.61	9.68	10.2	10.7	5[c]	11.49	12.50	12.90	15.69	18.78
1[b]	4.46	4.60	4.65	5.00	5.37	1[c]	7.23	8.08	8.37	10.63[aa]	13.18[aa]
		Alloy—Aluminum-Magnesium						Alloy—Copper-Palladium			
Wt % Al						**Wt % Cu**					
99[c]	2.96	3.18	3.26	3.82	4.39	99[c]	2.10	2.23	2.27	2.59	2.92
95[c]	5.05	5.28	5.36	5.93	6.51	95[c]	4.21	4.35	4.40	4.74	5.08
90[c]	7.52	7.76	7.85	8.43	9.02	90[c]	6.89	7.03	7.08	7.41	7.74
85	—	—	—	—	—	85[c]	9.48	9.61	9.66	10.01	10.36
80	—	—	—	—	—	80[c]	11.99	12.12	12.16	12.51[aa]	12.87
70	—	—	—	—	—	70[c]	16.87	17.01	17.06	17.41	17.78
60	—	—	—	—	—	60[c]	21.73	21.87	21.92	22.30	22.69
50	—	—	—	—	—	50[c]	27.62	27.79	27.86	28.25	28.64
40	—	—	—	—	—	40[c]	35.31	35.51	35.57	36.03	36.47
30	—	—	—	—	—	30[c]	46.50	46.66	46.71	47.11	47.47
25	—	—	—	—	—	25[c]	46.25	46.45	46.52	46.99[aa]	47.43[aa]
15	—	—	—	—	—	15[c]	36.52	36.99	37.16	38.28	39.35
10[b]	17.1	17.4	17.6	18.4	19.2	10[c]	28.90	29.51	29.73	31.19[aa]	32.56[aa]
5[b]	13.1	13.4	13.5	14.3	15.2	5[c]	20.00	20.75	21.02	22.84[aa]	24.54[aa]
1[a]	5.92	6.25	6.37	7.20	8.03	1[c]	11.90	12.67	12.93[aa]	14.82[aa]	16.68[aa]
		Alloy—Copper-Gold						Alloy—Copper-Zinc			
Wt % Cu						**Wt % Cu**					
99[c]	1.73	1.86[aa]	1.91[aa]	2.24[aa]	2.58[aa]	99[b]	1.84	1.97	2.02	2.36	2.71
95[c]	2.41	2.54[aa]	2.59[aa]	2.92[aa]	3.26[aa]	95[b]	2.78	2.92	2.97	3.33	3.69
90[c]	3.29	4.42[aa]	3.46[aa]	3.79[aa]	4.12[aa]	90[b]	3.66	3.81	3.86	4.25	4.63
85[c]	4.20	4.33	4.38[aa]	4.71[aa]	5.05[aa]	85[b]	4.37	4.54	4.60	5.02	5.44
80[c]	5.15	5.28	5.32	5.65	5.99	80[b]	5.01	5.19	5.26	5.71	6.17
70[c]	7.12	7.25	7.30	7.64	7.99	70[b]	5.87	6.08	6.15	6.67	7.19
60[c]	9.18	9.13	9.36	9.70	10.05	60	—	—	—	—	—
50[c]	11.07	11.20	11.25	11.60	11.94	50	—	—	—	—	—
40[c]	12.70	12.85	12.90[aa]	13.27[aa]	13.65[aa]	40	—	—	—	—	—
30[c]	13.77	13.93	13.99[aa]	14.38[aa]	14.78[aa]	30	—	—	—	—	—
25[c]	13.93	14.09	14.14	14.54	14.94	25	—	—	—	—	—
15[c]	12.75	12.91	12.96[aa]	13.36[aa]	13.77	15	—	—	—	—	—
10[c]	10.70	10.86	10.91	11.31	11.72	10	—	—	—	—	—
5[c]	7.25	7.41[aa]	7.46	7.87	8.28	5	—	—	—	—	—
1[c]	3.40	3.57	3.62	4.03	4.45	1	—	—	—	—	—

Alloy—Gold-Palladium

Wt % Au	273 K	293 K	300 K	350 K	400 K
99[c]	2.69	2.86	2.91	3.32	3.73
95[c]	5.21	5.35	5.41	5.79	6.17
90[i]	8.01	8.17	8.22	8.56	8.93
85[b]	10.50[aa]	10.66	10.72[aa]	11.100[aa]	11.48[aa]
80[b]	12.75	12.93	12.99	13.45	13.93
70[c]	18.23	18.46	18.54	19.10	19.67
60[b]	26.70	26.94	27.01	27.63[aa]	28.23[aa]
50[a]	27.23	27.63	27.76	28.64[aa]	29.42[aa]
40[a]	24.65	25.23	25.42	26.74	27.95
30[b]	20.82	21.49	21.72	23.35	24.92
25[b]	18.86	19.53	19.77	21.51	23.19
15[a]	15.08	15.77	16.01	17.80	19.61
10[a]	13.25	13.95	14.20[aa]	16.00[aa]	17.81[aa]
5[a]	11.49[aa]	12.21	12.46[aa]	14.26[aa]	16.07[aa]
1[a]	10.07	10.85[aa]	11.12[aa]	12.99[aa]	14.80[aa]

Alloy—Iron-Nickel

Wt % Fe	273 K	293 K	300 K	350 K	400 K
99[a]	10.9	12.0	12.4	—	18.7
95[c]	18.7	19.9	20.2	—	26.8
90[c]	24.2	25.5	25.9	—	33.2
85[c]	27.8	29.2	29.7	—	37.3
80[c]	30.1	31.6	32.2	—	40.0
70[b]	32.3	33.9	34.4	—	42.4
60[c]	53.8	57.1	58.2	—	73.9
50[d]	28.4	30.6	31.4	—	43.7
40[d]	19.6	21.6	22.5	—	34.0
30[c]	15.3	17.1	17.7	—	27.4
25[b]	14.3	15.9	16.4	—	25.1
15[c]	12.6	13.8	14.2	—	21.1
10[c]	11.4	12.5	12.9	—	18.9
5[c]	9.66	10.6	10.9	—	16.1[aa]
1[b]	7.17	7.94	8.12	—	12.8

Alloy—Gold-Silver

Wt % Au	273 K	293 K	300 K	350 K	400 K
99[b]	2.58	2.75	2.80[aa]	3.22[aa]	3.63[aa]
95[a]	4.58	4.74	4.79	5.19	5.59
90[j]	6.57	6.73	6.78	7.19	7.58
85[j]	8.14	8.30	8.36[aa]	8.75	9.15
80[j]	9.34	9.50	9.55	9.94	10.33
70[j]	10.70	10.86	10.91	11.29	11.68[aa]
60[j]	10.92	11.07	11.12	11.50	11.87
50[j]	10.23	10.37	10.42	10.78	11.14
40[j]	8.92	9.06	9.11	9.46[aa]	9.81
30[a]	7.34	7.47	7.52	7.85	8.19
25[a]	6.46	6.59	6.63	6.96	7.30[aa]
15[a]	4.55	4.67	4.72	5.03	5.34
10[a]	3.54	3.66	3.71	4.00	4.31
5[i]	2.52	2.64[aa]	2.68[aa]	2.96[aa]	3.25[aa]
1[b]	1.69	1.80	1.84[aa]	2.12[aa]	2.42[aa]

Alloy—Silver-Palladium

Wt % Ag	273 K	293 K	300 K	350 K	400 K
99[b]	1.891	2.007	2.049	2.35	2.66
95[b]	3.58	3.70	3.74	4.04	4.34
90[b]	5.82	5.94	5.98	6.28	6.59
85[k]	7.92[aa]	8.04[aa]	8.08	8.38[aa]	8.68[aa]
80[k]	10.01	10.13	10.17	10.47	10.78
70[k]	14.53	14.65	14.69	14.99	15.30
60[i]	20.9	21.1	21.2	21.6	22.0
50[k]	31.2	31.4	31.5	32.0	32.4
40[m]	42.2	42.2	42.2	42.3	42.3
30[b]	40.4	40.6	40.7	41.3	41.7
25[k]	36.67[aa]	37.06	37.19	38.1[aa]	38.8[aa]
15[i]	27.08[aa]	26.68[aa]	27.89[aa]	29.3[aa]	30.6[aa]
10[i]	21.69	22.39	22.63	24.3	25.9
5[b]	15.98	16.72	16.98	18.8[aa]	20.5[aa]
1[a]	11.06	11.82	12.08[aa]	13.92[aa]	15.70[aa]

[a] Uncertainty in resistivity is ± 2%.
[b] Uncertainty in resistivity is ± 3%.
[c] Uncertainty in resistivity is ± 5%.
[d] Uncertainty in resistivity is ± 7% below 300 K and ± 5% at 300 and 400 K.
[e] Uncertainty in resistivity is ± 7%.
[f] Uncertainty in resistivity is ± 8%.
[g] Uncertainty in resistivity is ± 10%.
[h] Uncertainty in resistivity is ± 12%.
[i] Uncertainty in resistivity is ± 4%.
[j] Uncertainty in resistivity is ± 1%.
[k] Uncertainty in resistivity is ± 3% up to 300 K and ± 4% above 300 K.
[m] Uncertainty in resistivity is ± 2% up to 300 K and ± 4% above 300 K.
[a] Crystal usually a mixture of α-hep and fcc lattice.
[aa] In temperature range where no experimental data are available.

Resistivity of Selected Ceramics (Listed by Ceramic)

Ceramic	Resistivity (Ω-cm)
Borides	
Chromium diboride (CrB_2)	21×10^{-6}
Hafnium diboride (HfB_2)	$10–12 \times 10^{-6}$ at room temp.
Tantalum diboride (TaB_2)	68×10^{-6}
Titanium diboride (TiB_2) (polycrystalline)	
85% dense	$26.5–28.4 \times 10^{-6}$ at room temp.
85% dense	9.0×10^{-6} at room temp.
100% dense, extrapolated values	$8.7–14.1 \times 10^{-6}$ at room temp.
	3.7×10^{-6} at liquid air temp.
Titanium diboride (TiB_2) (monocrystalline)	
Crystal length 5 cm, 39 deg. and 59 deg. orientation with respect to growth axis	$6.6 \pm 0.2 \times 10^{-6}$ at room temp.
Crystal length 1.5 cm, 16.5 deg. and 90 deg. orientation with respect to growth axis	$6.7 \pm 0.2 \times 10^{-6}$ at room temp.
Zirconium diboride (ZrB_2)	9.2×10^{-6} at 20°C
	1.8×10^{-6} at liquid air temp.
Carbides: boron carbide (B_4C)	0.3–0.8

Dielectric Constants

Dielectric Constants of Solids

These data refer to temperatures in the range 17–22°C.

Material	Freq. (Hz)	Dielectric Constant	Material	Freq. (Hz)	Dielectric Constant
Acetamide	4×10^8	4.0	Diphenylmethane	4×10^8	2.7
Acetanilide	—	2.9	Dolomite $\perp$ optic axis	10^8	8.0
Acetic acid (2°C)	4×10^8	4.1	Dolomite ‖	10^8	6.8
Aluminum oleate	4×10^8	2.40	Ferrous oxide (15°C)	10^8	14.2
Ammonium bromide	10^8	7.1	Iodine	10^8	4
Ammonium chloride	10^8	7.0	Lead acetate	10^8	2.6
Antimony trichloride	10^8	5.34	Lead carbonate (15°C)	10^8	18.6
Apatite $\perp$ optic axis	3×10^8	9.50	Lead chloride	10^8	4.2
Apatite ‖ optic axis	3×10^8	7.41	Lead monoxide (15°C)	10^8	25.9
Asphalt	$<3 \times 10^4$	2.68	Lead nitrate	6×10^7	37.7
Barium chloride (anhyd.)	6×10^7	11.4	Lead oleate	4×10^8	3.27
Barium chloride ($2H_3O$)	6×10^7	9.4	Lead sulfate	10^4	14.3
Barium nitrate	6×10^7	5.9	Lead sulfide (15°C)	16^8	17.9
Barium sulfate (15°C)	10^8	11.40	Malachite (mean)	10^{12}	7.2
Beryl $\perp$ optic axis	10^4	7.02	Mercuric chloride	10^8	3.2
Beryl ‖ optic axis	10^4	6.08	Mercurous chloride	10^8	9.4
Calcite $\perp$ optic axis	10^4	8.5	Naphthalene	4×10^8	2.52
Calcite ‖ optic axis	10^4	8.0	Phenanthrene	4×10^8	2.80
Calcium carbonate	10^4	6.14	Phenol (10°C)	4×10^8	4.3
Calcium fluoride	10^4	7.36	Phosphorus, red	10^8	4.1
Calcium sulfate ($2H_2O$)	10^4	5.66	Phosphorus, yellow	10^8	3.6
Cassiterite $\perp$ optic axis	10^{12}	23.4	Potassium aluminum sulfate	10^8	3.8
Cassiterite ‖ optic axis	10^{12}	24	Potassium carbonate (15°C)	10^8	5.6
d-Cocaine	5×10^8	3.10	Potassium chlorate	6×10^7	5.1
Cupric oleate	4×10^8	2.80	Potassium chloride	10^4	5.03
Cupric oxide (15°C)	10^8	18.1	Potassium chromate	6×10^7	7.3
Cupric sulfate (anhyd.)	6×10^7	10.3	Potassium iodide	6×10^7	5.6
Cupric sulfate ($5H_2O$)	6×10^7	7.8	Potassium nitrate	6×10^7	5.0
Diamond	10^8	5.5	Potassium sulfate	6×10^7	5.9

Material	Freq. (Hz)	Dielectric Constant	Material	Freq. (Hz)	Dielectric Constant
Quartz ⊥ optic axis	3×10^7	4.34	Sodium carbonate ($10H_2O$)	6×10^7	5.3
Quartz ‖ optic axis	3×10^7	4.27	Sodium chloride	10^4	6.12
Resorcinol	4×10^8	3.2	Sodium nitrate	—	5.2
Ruby ⊥ optic axis	10^4	13.27	Sodium oleate	4×10^8	2.75
Ruby ‖ optic axis	10^4	11.28	Sodium perchlorate	6×10^7	5.4
Rutile ⊥ optic axis	10^8	86	Sucrose (mean)	3×10^8	3.32
Rutile ‖ optic axis	10^8	170	Sulfur (mean)	—	4.0
Selenium	10^8	6.6	Thallium chloride	10^4	46.9
Silver bromide	10^4	12.2	*p*-Toluidine	4×10^8	3.0
Silver chloride	10^4	11.2	Tourmaline ⊥ optic axis	10^4	7.10
Silver cyanide	10^4	5.6	Tourmaline ‖ optic axis	10^4	6.3
Smithsonite ⊥ optic axis	10^{12}	9.3	Urea	4×10^8	3.5
Smithsonite ‖ optic axis	10^{10}	9.4	Zircon ⊥, ‖	10^8	12
Sodium carbonate (anhyd.)	6×10^7	8.4			

Dielectric Constants of Ceramics

Material	Dielectric Constant 10^4 Hz	Dielectric strength Volts/mil	Volume Resistivity Ohm-cm (23°C)	Loss Factor[a]
Alumina	4.5–8.4	40–160	10^{11}–10^{14}	0.0002–0.01
Corderite	4.5–5.4	40–250	10^{12}–10^{14}	0.004–0.012
Forsterite	6.2	240	10^{14}	0.0004
Porcelain (dry process)	6.0–8.0	40–240	10^{12}–10^{14}	0.0003–0.02
Porcelain (wet process)	6.0–7.0	90–400	10^{12}–10^{14}	0.006–0.01
Porcelain, zircon	7.1–10.5	250–400	10^{13}–10^{15}	0.0002–0.008
Steatite	5.5–7.5	200–400	10^{13}–10^{15}	0.0002–0.004
Titanates (Ba, Sr, Ca, Mg, and Pb)	15–12.000	50–300	10^8–10^{13}	0.0001–0.02
Titanium dioxide	14–110	100–210	10^{13}–10^{18}	0.0002–0.005

Dielectric Constants of Glasses

Type	Dielectric Constant At 100 MHz (20°C)	Volume Resistivity (350°C megohm-cm)	Loss Factor[a]
Corning 0010	6.32	10	0.015
Corning 0080	6.75	0.13	0.058
Corning 0120	6.65	100	0.012
Pyrex 1710	6.00	2,500	0.025
Pyrex 3320	4.71	—	0.019
Pyrex 7040	4.65	80	0.013
Pyrex 7050	4.77	16	0.017
Pyrex 7052	5.07	25	0.019
Pyrex 7060	4.70	13	0.018
Pyrex 7070	4.00	1,300	0.0048
Vycor 7230	3.83	—	0.0061
Pyrex 7720	4.50	16	0.014
Pyrex 7740	5.00	4	0.040
Pyrex 7750	4.28	50	0.011
Pyrex 7760	4.50	50	0.0081
Vycor 7900	3.9	130	0.0023
Vycor 7910	3.8	1,600	0.00091
Vycor 7911	3.8	4,000	0.00072
Corning 8870	9.5	5,000	0.0085
G. E. Clear (silica glass)	3.81	4,000–30,000	0.00038
Quartz (fused)	3.75 4.1 (1 MHz)	—	0.0002 (1 MHz)

[a] Power factor × dielectric constant equals loss factor.

Properties of Semiconductors
H. Mike Harris *Georgia Tech Research Institute*
Semiconducting Properties of Selected Materials

Substance	Minimum Energy Gap (eV) R.T.	Minimum Energy Gap (eV) 0 K	dE_g / dT × 10^4 eV/°C	dE_g / dP × 10^6 eV·cm²/kg	Density of States eleCtron Effective Mass m_{d_n} (m_o)	Electron Mobility and Temperature Dependence μ_n (cm²/V·s)	$-x$	Density of States Hole Effec- Tive Mass m_{d_p} (m_o)	Hole Mobility and Temperature Dependence μ_p (cm²/V·s)	$-x$
Si	1.107	1.153	−2.3	−2.0	1.1	1,900	2.6	0.56	500	2.3
Ge	0.67	0.744	−3.7	±7.3	0.55	3,80	1.66	0.3	1,820	2.33
αSn	0.08	0.094	−0.5		0.02	2,500	1.65	0.3	2,400	2.0
Te	0.33				0.68	1,100		0.19	560	
III–V Compounds										
AlAs	2.2	2.3				1,200			420	
AlSb	1.6	1.7	−3.5	−1.6	0.09	2..	1.5	0.4	500	1.8
GaP	2.24	2.40	−5.4	−1.7	0.35	300	1.5	0.5	150	1.5
GaAs	1.35	1.53	−5.0	+9.4	0.068	9,000	1.0	0.5	500	2.1
GaSb	0.67	0.78	−3.5	+12	0.050	5,000	2.0	0.23	1,400	0.9
InP	1.27	1.41	−4.6	+4.6	0.067	5,000	2.0		200	2.4
InAs	0.36	0.43	−2.8	+8	0.022	33,000	1.2	0.41	460	2.3
InSb	0.165	0.23	−2.8	+15	0.014	78,000	1.6	0.4	750	2.1
II–VI Compounds										
ZnO	3.2		−9.5	+0.6	0.38	180	1.5			
ZnS	3.54		−5.3	+5.7		180			5 (400°C)	
ZnSe	2.58	2.80	−7.2	+6		540			28	
ZnTe	2.26			+6		340			100	
CdO	2.5 ± 0.1		−6		0.1	120				
CdS	2.42		−5	+3.3	0.165	400		0.8		
CdSe	1.74	1.85	−4.6		0.13	650	1.0	0.6		
CdTe	1.44	1.56	−4.1	+8	0.14	1,200		0.35	50	
HgSe	0.30				0.030	20,000	2.0			
HgTe	0.15		−1		0.017	25,000		0.5	350	
Halite Structure Compounds										
PbS	0.37	0.28	+4		0.16	800		0.1	1,000	2.2
PbSe	0.26	0.16	+4		0.3	1,500		0.34	1,500	2.2
PbTe	0.25	0.19	+4	−7	0.21	1,600		0.14	750	2.2
Others										
ZnSb	0.50	0.56			0.15	10				1.5
CdSb	0.45	0.57	−5.4		0.15	300			2,000	1.5
Bi_2S_3	1.3					200			1,100	
Bi_2Se_3	0.27					600			675	
Bi_2Te_3	0.13		−0.95		0.58	1,200	1.68	1.07	510	1.95
Mg_2Si		0.77	−6.4		0.46	400	2.5		70	
Mg_2Ge		0.74	−9			280	2		110	
Mg_2Sn	0.21	0.33	−3.5		0.37	320			260	
Mg_3Sb_2		0.32				20			82	
Zn_3As_2	0.93					10	1.1		10	
Cd_3As_2	0.55				0.046	100,000	0.88			
GaSe	2.05		3.8						20	
GaTe	1.66	1.80	−3.6			14	−5			
InSe	1.8					9000				
TlSe	0.57		−3.9		0.3	30		0.6	20	1.5
$CdSnAs_2$	0.23				0.05	25,000	1.7			
Ga_2Te_3	1.1	1.55	−4.8							
α-In_2Te_3	1.1	1.2			0.7				50	1.1
β-In_2Te_3	1.0								5	
$Hg_5In_2Te_8$	0.5								11,000	
SnO_2									78	

Band Properties of Semiconductors

Part A. Data on Valence Bands of Semiconductors (Room Temperature)

Substance	Band curvature effective mass (expressed as fraction of free electron mass)				Measured Light Hole Mobility (cm²/V·s)
	Heavy Holes	Light Holes	"Split-off" Band Holes	Energy Separation of "Split-off" Band (eV)	
Semiconductors with Valence Bands Maximum at the Center of the Brillouin Zone ("F")					
Si	0.52	0.16	0.25	0.044	500
Ge	0.34	0.043	0.08	0.3	1,820
Sn	0.3				2,400
AlAs					
AlSb	0.4			0.7	550
GaP				0.13	100
GaAs	0.8	0.12	0.20	0.34	400
GaSb	0.23	0.06		0.7	1,400
InP				0.21	150
InAs	0.41	0.025	0.083	0.43	460
InSb	0.4	0.015		0.85	750
CdTe	0.35				50
HgTe	0.5				350

Semiconductors with Multiple Valence Band Maxima

Substance	Number of Equivalent Valleys and Directions	Band Curvature Effective Masses		Anisotropy $K = m_L/m_T$	Measured (light) Hole Mobility cm²/V·s
		Longitudinal m_L	Transverse m_T		
PbSe	4 "L" [111]	0.095	0.047	2.0	1,500
PbTe	4 "L" [111]	0.27	0.02	10	750
Bi_2Te_3	6	0.207	~0.045	4.5	515

Part B. Data on Conduction Bands of Semiconductors (Room Temperature Data)

Single Valley Semiconductors

Substance	Energy Gap (eV)	Effective Mass (m_o)	Mobility (cm²/V·s)
GaAs	1.35	0.067	8,500
InP	1.27	0.067	5,000
InAs	0.36	0.022	33,000
InSb	0.165	0.014	78,000
CdTe	1.44	0.11	1,000

Multivalley Semiconductors

Substance	Energy Gap	Number of equivalent valleys and direction	Band curvature effective mass		Anisotropy $K = m_L/m_T$
			Longitudinal m_L	Transverse m_T	
Si	1.107	6 in [100] "Δ"	0.90	0.192	4.7
Ge	0.67	4 in [111] at "L"	1.588	~0.0815	19.5
GaSb	0.67	as Ge	~1.0	~0.2	~5
PbSe	0.26	4 in [111] at "L"	0.085	0.05	1.7
PbTe	0.25	4 in [111] at "L"	0.21	0.029	5.5
Bi_2Te_3	0.13	6			~0.05

References

1. L. Patrick and W. J. Choyke, *Physics Review Journal*, vol. 2, p. 2255, 1977.
2. W. J. Choyke, NATO ASI Ser. E, Applied Sciences, vol. 185, and references therein, 1990.
3. A. Suzuki, A. Ogura K. Furukawa, Y. Fuji, M. Shigeta and S. Nakajima, *Journal of Applied Physics*, vol. 64, p. 2818, 1988.
4. B. W. Wessels and J. C. Gatos, *Journal of Physics Chemical Solids*, vol. 15, p. 83, 1977.
5. M. M. Anikin, A. S. Zubrilov, A. A. Lebedev, A. M. Sterlchuck, and A. E. Cherenkov, *Fitz. Tekh. Polurpovdn.*, vol. 24, p. 467, 1992.
6. *CRC Materials Science and Engineering Handbook*, p. 304.
7. D. R. Lide, Properties of Semiconductors, *CRC Handbook of Chemistry and Physics*, pp. 12–98, 1999–2000 ed. CRC Press 2000.
8. http://vshields.jpl.nasa.gov/windchime.html
9. P. T. B. Schaffer and R. G. Naum, *Journal of Optical Society of America*, vol. 59, p. 1498, 1969.
10. P. T. B. Schaffer, *Applied Optics*, vol. 10, p. 1034, 1977.
11. M. N. Yoder, *Wide Bandgap Semiconductor Materials and Devices*, vol. 43, no.10, p.1634, 1966.
12. E. L. Kern, H. W. Hamill, H. W. Deem and H. D. Sheets, *Mater. Res. Bull.*, vol. 4, p. 107, 1969.
13. D. Morelli, J. Hermans, C. Bettz, W. S. Woo, G. L. Harris and C. Taylor, *Inst. Physics Conf. Ser.*, vol. 137, p. 313–6, 1990.
14. I. I. Parafenova, Y. M. Tairov and V. F. Tsvetkov, Sov. *Physics-Semiconductors*, vol. 24, no. 2, pp. 158–61, 1990.
15. http://www.ioffe.rssi.ru/SVA/NSM/Nano/index.html
16. C.W. Garland and K. C. Parks, *Journal Applied Physics*, vol. 33, p. 759, 1962.
17. http://mems.isi.edu/mems/materials/measurements.cgi?MATTAG=galliumarsenidegaas-bulk&PAGE_SIZE=20
18. M. C. Benjamin, C. Wang, R. F. Davis and R. J. Nemanich, *Applied Phys. Letter*, vol. 64, p. 3288, 1994.
19. S. N. Mohammad and H. Morkoc, *Prog. Quant. Electron.* vol. 20, p. 361, 1996.
20. A. S. Barker and M. Ilegems *Phys. Rev. B*, vol. 7, p. 743, 1973.
21. H. P. Maruska and J. J. Tietjen, A*ppl. Phys. Lett.* vol. 15, p. 327, 1969.

Resistance of Wires

The following table gives the approximate resistance of various metallic conductors. The values have been computed from the resistivities at 20°C, except as otherwise stated, and for the dimensions of wire indicated. Owing to differences in purity in the case of elements and of composition in alloys, the values can be considered only as approximations.

B. & S. Gauge	Diameter		B. & S. gauge	Diameter	
	mm	mills 1 mil = .001 in		mm	mills 1 mil = .001 in
10	2.588	101.9	26	0.4049	15.94
12	2.053	80.81	27	0.3606	14.20
14	1.628	64.08	28	0.3211	12.64
16	1.291	50.82	30	0.2546	10.03
18	1.024	40.30	32	0.2019	7.950
20	0.8118	31.96	34	0.1601	6.305
22	0.6438	25.35	36	0.1270	5.000
24	0.5106	20.10	40	0.07987	3.145

B. & S. No.	Ohms per cm	Ohms per ft	B. & S. No.	Ohms per cm	Ohms per ft
Advance (0°C) $Q = 48. \times 10^{-6}$ ohm cm			Brass $Q = 7.00 \times 10^{-6}$ ohm cm		
10	.000912	.0278	10	.000133	.00406
12	.00145	.0442	12	.000212	.00645
14	.00231	.0703	14	.000336	.0103
16	.00367	.112	16	.000535	.0163
18	.00583	.178	18	.000850	.0259
20	.00927	.283	20	.00135	.0412
22	.0147	.449	22	.00215	.0655
24	.0234	.715	24	.00342	.104
26	.0373	1.14	26	.00543	.166
27	.0470	1.43	27	.00686	.209
28	.0593	1.81	28	.00864	.263
30	.0942	2.87	30	.0137	.419
32	.150	4.57	32	.0219	.666
34	.238	7.26	34	.0348	1.06
36	.379	11.5	36	.0552	1.68
40	.958	29.2	40	.140	4.26
Aluminum $Q = 2.828 \times 10^{-6}$ ohm cm			Climax $Q = 87. \times 10^{-6}$ ohm cm		
10	.0000538	.00164	10	.00165	.0504
12	.0000855	.00260	12	.00263	.0801
14	.000136	.00414	14	.00418	.127
16	.000216	.00658	16	.00665	.203
18	.000344	.0105	18	.0106	.322
20	.000546	.0167	20	.0168	.512
22	.000869	.0265	22	.0267	.815
24	.00138	.0421	24	.0425	1.30
26	.00220	.0669	26	.0675	2.06
27	.00277	.0844	27	.0852	2.60
28	.00349	.106	28	.107	3.27
30	.00555	.169	30	.171	5.21
32	.00883	.269	32	.272	8.28
34	.0140	.428	34	.432	13.2
36	.0223	.680	36	.687	20.9
40	.0564	1.72	40	1.74	52.9
Constantan (0°C) $Q = 44.1 \times 10^{-6}$ ohm cm			Excello $Q = 92. \times 10^{-6}$ ohm cm		
10	.000838	.0255	10	.00175	.0533
12	.00133	.0406	12	.00278	.0847
14	.00212	.0646	14	.00442	.135
16	.00337	.103	16	.00703	.214
18	.00536	.163	18	.0112	.341
20	.00852	.260	20	.0178	.542
22	.0135	.413	22	.0283	.861
24	.0215	.657	24	.0449	1.37
26	.0342	1.04	26	.0714	2.18
27	.0432	1.32	27	.0901	2.75
28	.0545	1.66	28	.114	3.46
30	.0866	2.64	30	.181	5.51
32	.138	4.20	32	.287	8.75
34	.219	6.67	34	.457	13.9
36	.348	10.6	36	.726	22.1
40	.880	26.8	40	1.84	56.0

B. & S. No.	Ohms per cm	Ohms per ft	B. & S. No.	Ohms per cm	Ohms per ft
Copper, annealed $Q = 1.724 \times 10^{-6}$ ohm cm			German silver $Q = 33. \times 10^{-6}$ ohm cm		
10	.0000328	.000999	10	.000627	.0191
12	.0000521	.00159	12	.000997	.0304
14	.0000828	.00253	14	.00159	.0483
16	.000132	.00401	16	.00252	.0768
18	.000209	.00638	18	.00401	.122
20	.000333	.0102	20	.00638	.194
22	.000530	.0161	22	.0101	.309
24	.000842	.0257	24	.0161	.491
26	.00134	.0408	26	.0256	.781
27	.00169	.0515	27	.0323	.985
28	.00213	.0649	28	.0408	1.24
30	.00339	.103	30	.0648	1.97
32	.00538	.164	32	.103	3.14
34	.00856	.261	34	.164	4.99
36	.0136	.415	36	.260	.794
40	.0344	1.05	40	.659	20.1
Eureka (0°C) $Q = 47. \times 10^{-6}$ ohm cm			Gold $Q = 2.44 \times 10^{-6}$ ohm cm		
10	.000893	.0272	10	.0000464	.00141
12	.00142	.0433	12	.0000737	.00225
14	.00226	0.688	14	.000117	.00357
16	.00359	.109	16	.000186	.00568
18	.00571	.174	18	.000296	.00904
20	.00908	.277	20	.000471	.0144
22	.0144	.440	22	.000750	.0228
24	.0230	.700	24	.00119	.0363
26	.0365	1.11	26	.00189	.0577
27	.0460	1.40	27	.00239	.0728
28	.0580	1.77	28	.00301	.0918
30	.0923	2.81	30	.00479	.146
32	.147	4.47	32	.00762	.232
34	.233	7.11	34	.0121	.369
36	.371	11.3	36	.0193	.587
40	.938	28.6	40	.0487	1.48
Iron $Q = 10. \times 10^{-6}$ ohm cm			Manganin $Q = 44. \times 10^{-6}$ ohm cm		
10	.000190	.00579	10	.000836	.0255
12	.000302	.00921	12	.00133	.0405
14	.000481	.0146	14	.00211	.0644
16	.000764	.0233	16	.00336	.102
18	.00121	.0370	18	.00535	.163
20	.00193	.0589	20	.00850	.259
22	.00307	.0936	22	.0135	.412
24	.00489	.149	24	.0215	.655
26	.00776	.237	26	.0342	1.04
27	.00979	.299	27	.0431	1.31
28	.0123	.376	28	.0543	1.66
30	.0196	.598	30	.0864	2.63
32	.0312	.952	32	.137	4.19
34	.0497	1.51	34	.218	6.66
36	0.789	2.41	36	.347	10.6
40	.200	6.08	40	.878	26.8

B. & S. No.	Ohms per cm	Ohms per ft	B. & S. No.	Ohms per cm	Ohms per ft
Lead $Q = 22. \times 10^{-6}$ ohm cm			Molybdenum $Q = 5.7 \times 10^{-6}$ ohm cm		
10	.000418	.0127	10	.000108	.00330
12	.000665	.0203	12	.000172	.00525
14	.00106	.0322	14	.000274	.00835
16	.00168	.0512	16	.000435	.0133
18	.00267	.0815	18	.000693	.0211
20	.00425	.130	20	.00110	.0336
22	.00676	.206	22	.00175	.0534
24	.0107	.328	24	.00278	.0849
26	.0171	.521	26	.00443	.135
27	.0215	.657	27	.00558	.170
28	.0272	.828	28	.00704	.215
30	.0432	1.32	30	.0112	.341
32	.0687	2.09	32	.0178	.542
34	.109	3.33	34	.0283	.863
36	.174	5.29	36	.0450	1.37
40	.439	13.4	40	.114	3.47
Magnesium $Q = 4.6 \times 10^{-6}$ ohm cm			Monel Metal $Q = 42. \times 10^{-6}$ ohm cm		
10	.0000874	.00267	10	.000798	.0243
12	.000139	.00424	12	.00127	.0387
14	.000221	.00674	14	.00202	.0615
16	.000351	.0107	16	.00321	.0978
18	.000559	.0170	18	.00510	.156
20	.000889	.0271	20	.00811	.247
22	.00141	.0431	22	.0129	.393
24	.00225	.0685	24	.0205	.625
26	.00357	.109	26	.0326	.994
27	.00451	.137	27	.0411	1.25
28	.00568	.173	28	.0519	1.58
30	.00903	.275	30	.0825	2.51
32	.0144	.438	32	.131	4.00
34	.0228	.696	34	.209	6.36
36	.0363	1.11	36	.331	10.1
40	.0918	2.80	40	.838	25.6
*Nichrome $Q = 150. \times 10^{-6}$ ohm cm			Silver (18°C) $Q = 1.629 \times 10^{-6}$ ohm cm		
10	.0021281	.06488	10	.0000310	.000944
12	.0033751	.1029	12	.0000492	.00150
14	.0054054	.1648	14	.0000783	.00239
16	.0085116	.2595	16	.000124	.00379
18	.0138383	.4219	18	.000198	.00603
20	.0216218	.6592	20	.000315	.00959
22	.0346040	1.055	22	.000500	.0153
24	.0548088	1.671	24	.000796	.0243
26	.0875760	2.670	26	.00126	.0386
28	.1394328	4.251	27	.00160	.0486
30	.2214000	6.750	28	.00201	.0613
32	.346040	10.55	30	.00320	.0975
34	.557600	17.00	32	.00509	.155
36	.885600	27.00	34	.00809	.247
38	1.383832	42.19	36	.0129	.392
40	2.303872	70.24	40	.0325	.991

B. & S. No.	Ohms per cm	Ohms per ft	B. & S. No.	Ohms per cm	Ohms per ft
Nickel $Q = 7.8 \times 10^{-6}$ ohm cm			Steel, piano wire (0°C) $Q = 11.8 \times 10^{-6}$ ohm cm		
10	.000148	.00452	10	.000224	.00684
12	.000236	.00718	12	.000357	.0109
14	.000375	.0114	14	.000567	.0173
16	.000596	.0182	16	.000901	.0275
18	.000948	.0289	18	.00143	.0437
20	.00151	.0459	20	.00228	.0695
22	.00240	.0730	22	.00363	.110
24	.00381	.116	24	.00576	.176
26	.00606	.185	26	.00916	.279
27	.00764	.233	27	.0116	.352
28	.00963	.294	28	.0146	.444
30	.0153	.467	30	.0232	.706
32	.0244	.742	32	.0368	1.12
34	.0387	1.18	34	.0586	1.79
36	.0616	1.88	36	.0931	2.84
40	.156	4.75	40	.236	7.18
Platinum $Q = 10. \times 10^{-6}$ ohm cm			Steel, invar (35% Ni) $Q = 81. \times 10^{-6}$ ohm cm		
10	.000190	.00579	10	.00154	.0469
12	.000302	.00921	12	.00245	.0746
14	.000481	.0146	14	.00389	.119
16	.000764	.0233	16	.00619	.189
18	.00121	.0370	18	.00984	.300
20	.00193	.0589	20	.0156	.477
22	.00307	.0936	22	.0249	.758
24	.00489	.149	24	.0396	1.21
26	.00776	.237	26	.0629	1.92
27	.00979	.299	27	.0793	2.42
28	.0123	.376	28	.100	3.05
30	.0196	.598	30	.159	4.85
32	.0312	.952	32	.253	7.71
34	.0497	1.51	34	.402	12.3
36	.0789	2.41	36	.639	19.5
40	.200	6.08	40	1.62	49.3
Tantalum $Q = 15.5 \times 10^{-6}$ ohm cm			Tungsten $Q = 5.51 \times 10^{-6}$ ohm cm		
10	.000295	.00898	10	.000105	.00319
12	.000468	.0143	12	.000167	.00508
14	.000745	.0227	14	.000265	.00807
16	.00118	.0361	16	.000421	.0128
18	.00188	.0574	18	.000669	.0204
20	.00299	.0913	20	.00106	.0324
22	.00476	.145	22	.00169	.0516
24	.00757	.231	24	.00269	.0820
26	.0120	.367	26	.00428	.130
27	.0152	.463	27	.00540	.164
28	.0191	.583	28	.00680	.207
30	.0304	.928	30	.0108	.330
32	.0484	1.47	32	.0172	.524
34	.0770	2.35	34	.0274	.834
36	.122	3.73	36	.0435	1.33
40	.309	9.43	40	.110	3.35

B. & S. No.	Ohms per cm	Ohms per ft	B. & S. No.	Ohms per cm	Ohms per ft
Tin $Q = 11.5 \times 10^{-6}$ ohm cm			Zinc (0°C) $Q = 5.75 \times 10^{-6}$ ohm cm		
10	.000219	.00666	10	.000109	.00333
12	.000348	.0106	12	.000174	.00530
14	.000553	.0168	14	.000276	.00842
16	.000879	.0268	16	.000439	.0134
18	.00140	.0426	18	.000699	.0213
20	.00222	.0677	20	.00111	.0339
22	.00353	.108	22	.00177	.0538
24	.00562	.171	24	.00281	.0856
26	.00893	.272	26	.00446	.136
27	.0113	.343	27	.00563	.172
28	.0142	.433	28	.00710	.216
30	.0226	.688	30	.0113	.344
32	.0359	1.09	32	.0180	.547
34	.0571	1.74	34	.0286	.870
36	.0908	2.77	36	.0454	1.38
40	.230	7.00	40	.115	3.50

Credits

Except for the Properties of Semiconductors section, material in Appendix A was reprinted from the following sources:

D. R. Lide, Ed., *CRC Handbook of Chemistry and Physics,* 76th ed., Boca Raton, Fla.: CRC Press, 1992: International System of Units (SI), conversion constants and multipliers (conversion of temperatures), symbols and terminology for physical and chemical quantities, fundamental physical constants, classification of electromagnetic radiation.

W. H. Beyer, Ed., *CRC Standard Mathematical Tables and Formulae,* 29th ed., Boca Raton, Fla.: CRC Press, 1991: Greek alphabet, conversion constants and multipliers (recommended decimal multiples and submultiples, metric to English, English to metric, general, temperature factors), physical constants, series expansion, integrals, the Fourier transforms, numerical methods, probability, positional notation.

R. J. Tallarida, *Pocket Book of Integrals and Mathematical Formulas,* 2nd ed., Boca Raton, Fla.: CRC Press, 1991: Elementary algebra and geometry; determinants, matrices, and linear systems of equations; trigonometry; analytic geometry; series; differential calculus; integral calculus; vector analysis; special functions; statistics; tables of probability and statistics; table of derivatives.

J. F. Pankow, *Aquatic Chemistry Concepts,* Chelsea, Mich.: Lewis Publishers, 1991: Periodic table of the elements.

J. Shackelford and W. Alexander, Eds., *CRC Materials Science and Engineering Handbook,* Boca Raton, Fla.: CRC Press, 1992: Electrical resistivity of selected alloy cast irons, resistivity of selected ceramics.

APPENDIX B
Microwave Engineering Appendix

John P. Wendler
Tyco Electronics
Wireless Network Solutions

Attenuator Design Values

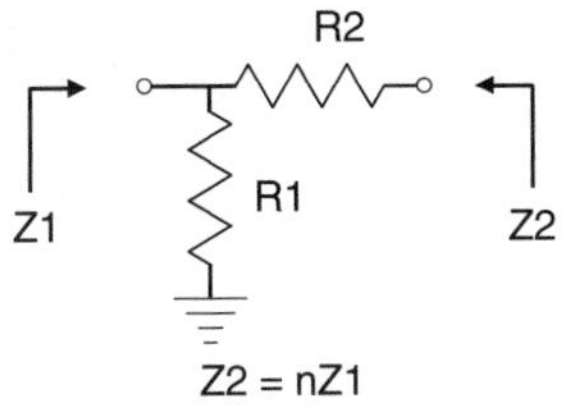

FIGURE B.1 Equivalent circuit for a minimum loss pad.

TABLE 1 Minimum Loss Matching Pad Resistance Values as a Function of Transformation Ratio For Z1 = 1 Ohm, 50 Ohms, and 75 Ohms

n Z2/Z1	r1 Z1 = 1	r2 Z1 = 1	R1 Z1 = 50	R2 Z1 = 50	R1 Z1 = 75	R2 Z1 = 75	Loss [dB]
1.1	3.3166	0.3317	165.8	16.6	248.7	24.9	2.7
1.2	2.4495	0.4899	122.5	24.5	183.7	36.7	3.8
1.3	2.0817	0.6245	104.1	31.2	156.1	46.8	4.5
1.4	1.8708	0.7483	93.5	37.4	140.3	56.1	5.2
1.5	1.7321	0.8660	86.6	43.3	129.9	65.0	5.7
1.6	1.6330	0.9798	81.6	49.0	122.5	73.5	6.2
1.7	1.5584	1.0909	77.9	54.5	116.9	81.8	6.6
1.8	1.5000	1.2000	75.0	60.0	112.5	90.0	7.0
1.9	1.4530	1.3077	72.6	65.4	109.0	98.1	7.3
2.0	1.4142	1.4142	70.7	70.7	106.1	106.1	7.7
2.1	1.3817	1.5199	69.1	76.0	103.6	114.0	8.0
2.2	1.3540	1.6248	67.7	81.2	101.6	121.9	8.2
2.3	1.3301	1.7292	66.5	86.5	99.8	129.7	8.5
2.4	1.3093	1.8330	65.5	91.7	98.2	137.5	8.7
2.5	1.2910	1.9365	64.5	96.8	96.8	145.2	9.0
2.6	1.2748	2.0396	63.7	102.0	95.6	153.0	9.2
2.7	1.2603	2.1424	63.0	107.1	94.5	160.7	9.4
2.8	1.2472	2.2450	62.4	112.2	93.5	168.4	9.6
2.9	1.2354	2.3473	61.8	117.4	92.7	176.1	9.8
3.0	1.2247	2.4495	61.2	122.5	91.9	183.7	10.0
3.1	1.2150	2.5515	60.7	127.6	91.1	191.4	10.1

TABLE 1 (continued)

n Z2/Z1	r1 Z1 = 1	r2 Z1 = 1	R1 Z1 = 50	R2 Z1 = 50	R1 Z1 = 75	R2 Z1 = 75	Loss [dB]
3.2	1.2060	2.6533	60.3	132.7	90.5	199.0	10.3
3.3	1.1978	2.7550	59.9	137.7	89.8	206.6	10.5
3.4	1.1902	2.8566	59.5	142.8	89.3	214.2	10.6
3.5	1.1832	2.9580	59.2	147.9	88.7	221.9	10.8
3.6	1.1767	3.0594	58.8	153.0	88.3	229.5	10.9
3.7	1.1706	3.1607	58.5	158.0	87.8	237.1	11.0
3.8	1.1650	3.2619	58.2	163.1	87.4	244.6	11.2
3.9	1.1597	3.3630	58.0	168.2	87.0	252.2	11.3
4.0	1.1547	3.4641	57.7	173.2	86.6	259.8	11.4
4.1	1.1500	3.5651	57.5	178.3	86.3	267.4	11.6
4.2	1.1456	3.6661	57.3	183.3	85.9	275.0	11.7
4.3	1.1415	3.7670	57.1	188.3	85.6	282.5	11.8
4.4	1.1376	3.8678	56.9	193.4	85.3	290.1	11.9
4.5	1.1339	3.9686	56.7	198.4	85.0	297.6	12.0
4.6	1.1304	4.0694	56.5	203.5	84.8	305.2	12.1
4.7	1.1271	4.1701	56.4	208.5	84.5	312.8	12.2
4.8	1.1239	4.2708	56.2	213.5	84.3	320.3	12.3
4.9	1.1209	4.3715	56.0	218.6	84.1	327.9	12.4
5.0	1.1180	4.4721	55.9	223.6	83.9	335.4	12.5
5.1	1.1153	4.5727	55.8	228.6	83.6	343.0	12.6
5.2	1.1127	4.6733	55.6	233.7	83.5	350.5	12.7
5.3	1.1102	4.7739	55.5	238.7	83.3	358.0	12.8
5.4	1.1078	4.8744	55.4	243.7	83.1	365.6	12.9
5.5	1.1055	4.9749	55.3	248.7	82.9	373.1	13.0
5.6	1.1034	5.0754	55.2	253.8	82.8	380.7	13.1
5.7	1.1013	5.1759	55.1	258.8	82.6	388.2	13.2
5.8	1.0992	5.2764	55.0	263.8	82.4	395.7	13.3
5.9	1.0973	5.3768	54.9	268.8	82.3	403.3	13.3
6.0	1.0954	5.4772	54.8	273.9	82.2	410.8	13.4
6.1	1.0937	5.5776	54.7	278.9	82.0	418.3	13.5
6.2	1.0919	5.6780	54.6	283.9	81.9	425.9	13.6
6.3	1.0903	5.7784	54.5	288.9	81.8	433.4	13.6
6.4	1.0887	5.8788	54.4	293.9	81.6	440.9	13.7
6.5	1.0871	5.9791	54.4	299.0	81.5	448.4	13.8
6.6	1.0856	6.0795	54.3	304.0	81.4	456.0	13.9
6.7	1.0842	6.1798	54.2	309.0	81.3	463.5	13.9
6.8	1.0828	6.2801	54.1	314.0	81.2	471.0	14.0
6.9	1.0814	6.3804	54.1	319.0	81.1	478.5	14.1
7.0	1.0801	6.4807	54.0	324.0	81.0	486.1	14.1
7.1	1.0789	6.5810	53.9	329.1	80.9	493.6	14.2
7.2	1.0776	6.6813	53.9	334.1	80.8	501.1	14.3
7.3	1.0764	6.7816	53.8	339.1	80.7	508.6	14.3
7.4	1.0753	6.8819	53.8	344.1	80.6	516.1	14.4
7.5	1.0742	6.9821	53.7	349.1	80.6	523.7	14.5
7.6	1.0731	7.0824	53.7	354.1	80.5	531.2	14.5
7.7	1.0720	7.1826	53.6	359.1	80.4	538.7	14.6
7.8	1.0710	7.2829	53.6	364.1	80.3	546.2	14.6
7.9	1.0700	7.3831	53.5	369.2	80.3	553.7	14.7
8.0	1.0690	7.4833	53.5	374.2	80.2	561.2	14.8
8.1	1.0681	7.5835	53.4	379.2	80.1	568.8	14.8
8.2	1.0672	7.6837	53.4	384.2	80.0	576.3	14.9
8.3	1.0663	7.7840	53.3	389.2	80.0	583.8	14.9
8.4	1.0654	7.8842	53.3	394.2	79.9	591.3	15.0
8.5	1.0646	7.9844	53.2	399.2	79.8	598.8	15.0
8.6	1.0638	8.0846	53.2	404.2	79.8	606.3	15.1

TABLE 1 (continued)

n Z2/Z1	r1 Z1 = 1	r2 Z1 = 1	R1 Z1 = 50	R2 Z1 = 50	R1 Z1 = 75	R2 Z1 = 75	Loss [dB]
8.7	1.0630	8.1847	53.1	409.2	79.7	613.9	15.2
8.8	1.0622	8.2849	53.1	414.2	79.7	621.4	15.2
8.9	1.0614	8.3851	53.1	419.3	79.6	628.9	15.3
9.0	1.0607	8.4853	53.0	424.3	79.5	636.4	15.3
9.1	1.0599	8.5855	53.0	429.3	79.5	643.9	15.4
9.2	1.0592	8.6856	53.0	434.3	79.4	651.4	15.4
9.3	1.0585	8.7858	52.9	439.3	79.4	658.9	15.5
9.4	1.0579	8.8859	52.9	444.3	79.3	666.4	15.5
9.5	1.0572	8.9861	52.9	449.3	79.3	674.0	15.6
9.6	1.0565	9.0863	52.8	454.3	79.2	681.5	15.6
9.7	1.0559	9.1864	52.8	459.3	79.2	689.0	15.7
9.8	1.0553	9.2865	52.8	464.3	79.1	696.5	15.7
9.9	1.0547	9.3867	52.7	469.3	79.1	704.0	15.7
10.0	1.0541	9.4868	52.7	474.3	79.1	711.5	15.8

$$R_1 = \frac{Z_1(n+\sqrt{n^2-n})}{n-1+\sqrt{n^2-n}} \qquad R_2 = Z_1\sqrt{n^2-n} \qquad \frac{P_O}{P_A} = \frac{1}{n}\left(\frac{1}{1+\sqrt{\frac{1}{n}}}\right)^2$$

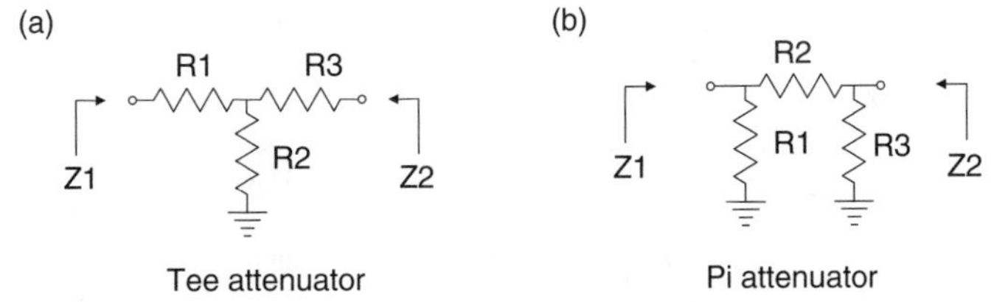

FIGURE B.2 (a) Equivalent circuit for a Tee attenuator; (b) Equivalent circuit for a Pi attenuator.

TABLE 2 Tee- and Pi-Pad Resistor Values for Zo = 1 Ohm and Zo = 50 Ohms

Loss [dB]	Voltage Atten	r1, r3, g1, g3 Z1 = Z2 = 1	r2, g2 Z1 = Z2 = 1	Tee R1, R3 Z1 = Z2 = 50	Tee R2 Z1 = Z2 = 50	Pi R1, R3 Z1 = Z2 = 50	Pi R2 Z1 = Z2 = 50
0.1	0.98855	0.0058	86.8570	0.3	4342.8	8686.0	0.6
0.2	0.97724	0.0115	43.4256	0.6	2171.3	4343.1	1.2
0.3	0.96605	0.0173	28.9472	0.9	1447.4	2895.6	1.7
0.4	0.95499	0.0230	21.7071	1.2	1085.4	2171.9	2.3
0.5	0.94406	0.0288	17.3622	1.4	868.1	1737.7	2.9
0.6	0.93325	0.0345	14.4650	1.7	723.2	1448.2	3.5
0.7	0.92257	0.0403	12.3950	2.0	619.7	1241.5	4.0
0.8	0.91201	0.0460	10.8420	2.3	542.1	1086.5	4.6
0.9	0.90157	0.0518	9.6337	2.6	481.7	966.0	5.2
1.0	0.89125	0.0575	8.6667	2.9	433.3	869.5	5.8
1.2	0.87096	0.0690	7.2153	3.4	360.8	725.0	6.9
1.4	0.85114	0.0804	6.1774	4.0	308.9	621.8	8.1
1.6	0.83176	0.0918	5.3981	4.6	269.9	544.4	9.3
1.8	0.81283	0.1032	4.7911	5.2	239.6	484.3	10.4
2.0	0.79433	0.1146	4.3048	5.7	215.2	436.2	11.6

TABLE 2 (continued)

Loss [dB]	Voltage Atten	r1, r3, g1, g3 Z1 = Z2 = 1	r2, g2 Z1 = Z2 = 1	Tee R1, R3 Z1 = Z2 = 50	Tee R2 Z1 = Z2 = 50	Pi R1, R3 Z1 = Z2 = 50	Pi R2 Z1 = Z2 = 50
2.2	0.77625	0.1260	3.9062	6.3	195.3	396.9	12.8
2.4	0.75858	0.1373	3.5735	6.9	178.7	364.2	14.0
2.6	0.74131	0.1486	3.2914	7.4	164.6	336.6	15.2
2.8	0.72444	0.1598	3.0490	8.0	152.5	312.9	16.4
3.0	0.70795	0.1710	2.8385	8.5	141.9	292.4	17.6
3.2	0.69183	0.1822	2.6539	9.1	132.7	274.5	18.8
3.4	0.67608	0.1933	2.4906	9.7	124.5	258.7	20.1
3.6	0.66069	0.2043	2.3450	10.2	117.3	244.7	21.3
3.8	0.64565	0.2153	2.2144	10.8	110.7	232.2	22.6
4.0	0.63096	0.2263	2.0966	11.3	104.8	221.0	23.8
4.2	0.61660	0.2372	1.9896	11.9	99.5	210.8	25.1
4.4	0.60256	0.2480	1.8921	12.4	94.6	201.6	26.4
4.6	0.58884	0.2588	1.8028	12.9	90.1	193.2	27.7
4.8	0.57544	0.2695	1.7206	13.5	86.0	185.5	29.1
5.0	0.56234	0.2801	1.6448	14.0	82.2	178.5	30.4
5.5	0.53088	0.3064	1.4785	15.3	73.9	163.2	33.8
6.0	0.50119	0.3323	1.3386	16.6	66.9	150.5	37.4
6.5	0.47315	0.3576	1.2193	17.9	61.0	139.8	41.0
7.0	0.44668	0.3825	1.1160	19.1	55.8	130.7	44.8
7.5	0.42170	0.4068	1.0258	20.3	51.3	122.9	48.7
8.0	0.39811	0.4305	0.9462	21.5	47.3	116.1	52.8
8.5	0.37584	0.4537	0.8753	22.7	43.8	110.2	57.1
9.0	0.35481	0.4762	0.8118	23.8	40.6	105.0	61.6
9.5	0.33497	0.4982	0.7546	24.9	37.7	100.4	66.3
10.0	0.31623	0.5195	0.7027	26.0	35.1	96.2	71.2
10.5	0.29854	0.5402	0.6555	27.0	32.8	92.6	76.3
11.0	0.28184	0.5603	0.6123	28.0	30.6	89.2	81.7
11.5	0.26607	0.5797	0.5727	29.0	28.6	86.3	87.3
12.0	0.25119	0.5985	0.5362	29.9	26.8	83.5	93.2
12.5	0.23714	0.6166	0.5025	30.8	25.1	81.1	99.5
13.0	0.22387	0.6342	0.4714	31.7	23.6	78.8	106.1
13.5	0.21135	0.6511	0.4425	32.6	22.1	76.8	113.0
14.0	0.19953	0.6673	0.4156	33.4	20.8	74.9	120.3
14.5	0.18836	0.6830	0.3906	34.1	19.5	73.2	128.0
15.0	0.17783	0.6980	0.3673	34.9	18.4	71.6	136.1
15.5	0.16788	0.7125	0.3455	35.6	17.3	70.2	144.7
16.0	0.15849	0.7264	0.3251	36.3	16.3	68.8	153.8
16.5	0.14962	0.7397	0.3061	37.0	15.3	67.6	163.3
17.0	0.14125	0.7525	0.2883	37.6	14.4	66.4	173.5
17.5	0.13335	0.7647	0.2715	38.2	13.6	65.4	184.1
18.0	0.12589	0.7764	0.2558	38.8	12.8	64.4	195.4
18.5	0.11885	0.7875	0.2411	39.4	12.1	63.5	207.4
19.0	0.11220	0.7982	0.2273	39.9	11.4	62.6	220.0
19.5	0.10593	0.8084	0.2143	40.4	10.7	61.8	233.4
20.0	0.10000	0.8182	0.2020	40.9	10.1	61.1	247.5
20.5	0.09441	0.8275	0.1905	41.4	9.5	60.4	262.5
21.0	0.08913	0.8363	0.1797	41.8	9.0	59.8	278.3
21.5	0.08414	0.8448	0.1695	42.2	8.5	59.2	295.0
22.0	0.07943	0.8528	0.1599	42.6	8.0	58.6	312.7
22.5	0.07499	0.8605	0.1508	43.0	7.5	58.1	331.5
23.0	0.07079	0.8678	0.1423	43.4	7.1	57.6	351.4
23.5	0.06683	0.8747	0.1343	43.7	6.7	57.2	372.4
24.0	0.06310	0.8813	0.1267	44.1	6.3	56.7	394.6
24.5	0.05957	0.8876	0.1196	44.4	6.0	56.3	418.2
25.0	0.05623	0.8935	0.1128	44.7	5.6	56.0	443.2

TABLE 2 (continued)

Loss [dB]	Voltage Atten	r1, r3, g1, g3 Z1 = Z2 = 1	r2, g2 Z1 = Z2 = 1	Tee R1, R3 Z1 = Z2 = 50	Tee R2 Z1 = Z2 = 50	Pi R1, R3 Z1 = Z2 = 50	Pi R2 Z1 = Z2 = 50
26.0	0.05012	0.9045	0.1005	45.2	5.0	55.3	497.6
27.0	0.04467	0.9145	0.0895	45.7	4.5	54.7	558.6
28.0	0.03981	0.9234	0.0797	46.2	4.0	54.1	627.0
29.0	0.03548	0.9315	0.0711	46.6	3.6	53.7	703.7
30.0	0.03162	0.9387	0.0633	46.9	3.2	53.3	789.8
31.0	0.02818	0.9452	0.0564	47.3	2.8	52.9	886.3
32.0	0.02512	0.9510	0.0503	47.5	2.5	52.6	994.6
33.0	0.02239	0.9562	0.0448	47.8	2.2	52.3	1116.1
34.0	0.01995	0.9609	0.0399	48.0	2.0	52.0	1252.5
35.0	0.01778	0.9651	0.0356	48.3	1.8	51.8	1405.4
36.0	0.01585	0.9688	0.0317	48.4	1.6	51.6	1577.0
37.0	0.01413	0.9721	0.0283	48.6	1.4	51.4	1769.5
38.0	0.01259	0.9751	0.0252	48.8	1.3	51.3	1985.5
39.0	0.01122	0.9778	0.0224	48.9	1.1	51.1	2227.8
40.0	0.01000	0.9802	0.0200	49.0	1.0	51.0	2499.8
41.0	0.00891	0.9823	0.0178	49.1	0.9	50.9	2804.8
42.0	0.00794	0.9842	0.0159	49.2	0.8	50.8	3147.1
43.0	0.00708	0.9859	0.0142	49.3	0.7	50.7	3531.2
44.0	0.00631	0.9875	0.0126	49.4	0.6	50.6	3962.1
45.0	0.00562	0.9888	0.0112	49.4	0.6	50.6	4445.6

Note: P_i values are duals of Tee values.

$$a = \sqrt{\frac{P_{z_2}}{P_{z_1}}} \qquad R_{1T} = \left(\frac{2}{(1-a^2)} - 1\right) Z_1 - \frac{2a}{(1-a^2)} \sqrt{Z_1 Z_2}$$

$$R_{2T} = 2\sqrt{Z_1 Z_2} \frac{a}{(1-a^2)}$$

$$R_{3T} = \left(\frac{2}{(1-a^2)} - 1\right) Z_1 - \frac{2a}{(1-a^2)} \sqrt{Z_1 Z_2}$$

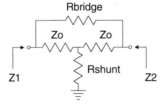

Bridged-tee attenuator
Z1=Z2=Zo

FIGURE B.3 Equivalent circuit for a Bridged-T attenuator.

TABLE 3 Bridged-T Attenuator Resistance Values for Zo = 1 Ohm, 50 Ohms, 75 Ohms

Loss [dB]	Voltage Atten	Bridge Arm Z1 = Z2 = 1	Shunt Arm Z1 = Z2 = 1	Bridge Arm Z1 = Z2 = 50	Shunt Arm Z1 = Z2 = 50	Bridge Arm Z1 = Z2 = 75	Shunt Arm Z1 = Z2 = 75
0.1	0.98855	0.0116	86.3599	0.6	4318.0	6477.0	0.9
0.2	0.97724	0.0233	42.9314	1.2	2146.6	3219.9	1.7
0.3	0.96605	0.0351	28.4558	1.8	1422.8	2134.2	2.6
0.4	0.95499	0.0471	21.2186	2.4	1060.9	1591.4	3.5
0.5	0.94406	0.0593	16.8766	3.0	843.8	1265.7	4.4
0.6	0.93325	0.0715	13.9822	3.6	699.1	1048.7	5.4
0.7	0.92257	0.0839	11.9151	4.2	595.8	893.6	6.3

TABLE 3 (continued)

Loss [dB]	Voltage Atten	Bridge Arm Z1 = Z2 = 1	Shunt Arm Z1 = Z2 = 1	Bridge Arm Z1 = Z2 = 50	Shunt Arm Z1 = Z2 = 50	Bridge Arm Z1 = Z2 = 75	Shunt Arm Z1 = Z2 = 75
0.8	0.91201	0.0965	10.3650	4.8	518.3	777.4	7.2
0.9	0.90157	0.1092	9.1596	5.5	458.0	687.0	8.2
1.0	0.89125	0.1220	8.1955	6.1	409.8	614.7	9.2
1.2	0.87096	0.1482	6.7498	7.4	337.5	506.2	11.1
1.4	0.85114	0.1749	5.7176	8.7	285.9	428.8	13.1
1.6	0.83176	0.2023	4.9440	10.1	247.2	370.8	15.2
1.8	0.81283	0.2303	4.3428	11.5	217.1	325.7	17.3
2.0	0.79433	0.2589	3.8621	12.9	193.1	289.7	19.4
2.2	0.77625	0.2882	3.4692	14.4	173.5	260.2	21.6
2.4	0.75858	0.3183	3.1421	15.9	157.1	235.7	23.9
2.6	0.74131	0.3490	2.8656	17.4	143.3	214.9	26.2
2.8	0.72444	0.3804	2.6289	19.0	131.4	197.2	28.5
3.0	0.70795	0.4125	2.4240	20.6	121.2	181.8	30.9
3.2	0.69183	0.4454	2.2450	22.3	112.2	168.4	33.4
3.4	0.67608	0.4791	2.0872	24.0	104.4	156.5	35.9
3.6	0.66069	0.5136	1.9472	25.7	97.4	146.0	38.5
3.8	0.64565	0.5488	1.8221	27.4	91.1	136.7	41.2
4.0	0.63096	0.5849	1.7097	29.2	85.5	128.2	43.9
4.2	0.61660	0.6218	1.6082	31.1	80.4	120.6	46.6
4.4	0.60256	0.6596	1.5161	33.0	75.8	113.7	49.5
4.6	0.58884	0.6982	1.4322	34.9	71.6	107.4	52.4
4.8	0.57544	0.7378	1.3554	36.9	67.8	101.7	55.3
5.0	0.56234	0.7783	1.2849	38.9	64.2	96.4	58.4
5.5	0.53088	0.8836	1.1317	44.2	56.6	84.9	66.3
6.0	0.50119	0.9953	1.0048	49.8	50.2	75.4	74.6
6.5	0.47315	1.1135	0.8981	55.7	44.9	67.4	83.5
7.0	0.44668	1.2387	0.8073	61.9	40.4	60.5	92.9
7.5	0.42170	1.3714	0.7292	68.6	36.5	54.7	102.9
8.0	0.39811	1.5119	0.6614	75.6	33.1	49.6	113.4
8.5	0.37584	1.6607	0.6021	83.0	30.1	45.2	124.6
9.0	0.35481	1.8184	0.5499	90.9	27.5	41.2	136.4
9.5	0.33497	1.9854	0.5037	99.3	25.2	37.8	148.9
10.0	0.31623	2.1623	0.4625	108.1	23.1	34.7	162.2
10.5	0.29854	2.3497	0.4256	117.5	21.3	31.9	176.2
11.0	0.28184	2.5481	0.3924	127.4	19.6	29.4	191.1
11.5	0.26607	2.7584	0.3625	137.9	18.1	27.2	206.9
12.0	0.25119	2.9811	0.3354	149.1	16.8	25.2	223.6
12.5	0.23714	3.2170	0.3109	160.8	15.5	23.3	241.3
13.0	0.22387	3.4668	0.2884	173.3	14.4	21.6	260.0
13.5	0.21135	3.7315	0.2680	186.6	13.4	20.1	279.9
14.0	0.19953	4.0119	0.2493	200.6	12.5	18.7	300.9
14.5	0.18836	4.3088	0.2321	215.4	11.6	17.4	323.2
15.0	0.17783	4.6234	0.2163	231.2	10.8	16.2	346.8
15.5	0.16788	4.9566	0.2018	247.8	10.1	15.1	371.7
16.0	0.15849	5.3096	0.1883	265.5	9.4	14.1	398.2
16.5	0.14962	5.6834	0.1759	284.2	8.8	13.2	426.3
17.0	0.14125	6.0795	0.1645	304.0	8.2	12.3	456.0
17.5	0.13335	6.4989	0.1539	324.9	7.7	11.5	487.4
18.0	0.12589	6.9433	0.1440	347.2	7.2	10.8	520.7
18.5	0.11885	7.4140	0.1349	370.7	6.7	10.1	556.0
19.0	0.11220	7.9125	0.1264	395.6	6.3	9.5	593.4
19.5	0.10593	8.4406	0.1185	422.0	5.9	8.9	633.0
20.0	0.10000	9.0000	0.1111	450.0	5.6	8.3	675.0
20.5	0.09441	9.5925	0.1042	479.6	5.2	7.8	719.4
21.0	0.08913	10.2202	0.0978	511.0	4.9	7.3	766.5
21.5	0.08414	10.8850	0.0919	544.3	4.6	6.9	816.4

TABLE 3 (continued)

Loss [dB]	Voltage Atten	Bridge Arm Z1 = Z2 = 1	Shunt Arm Z1 = Z2 = 1	Bridge Arm Z1 = Z2 = 50	Shunt Arm Z1 = Z2 = 50	Bridge Arm Z1 = Z2 = 75	Shunt Arm Z1 = Z2 = 75
22.0	0.07943	11.5893	0.0863	579.5	4.3	6.5	869.2
22.5	0.07499	12.3352	0.0811	616.8	4.1	6.1	925.1
23.0	0.07079	13.1254	0.0762	656.3	3.8	5.7	984.4
23.5	0.06683	13.9624	0.0716	698.1	3.6	5.4	1047.2
24.0	0.06310	14.8489	0.0673	742.4	3.4	5.1	1113.7
24.5	0.05957	15.7880	0.0633	789.4	3.2	4.8	1184.1
25.0	0.05623	16.7828	0.0596	839.1	3.0	4.5	1258.7
26.0	0.05012	18.9526	0.0528	947.6	2.6	4.0	1421.4
27.0	0.04467	21.3872	0.0468	1069.4	2.3	3.5	1604.0
28.0	0.03981	24.1189	0.0415	1205.9	2.1	3.1	1808.9
29.0	0.03548	27.1838	0.0368	1359.2	1.8	2.8	2038.8
30.0	0.03162	30.6228	0.0327	1531.1	1.6	2.4	2296.7
31.0	0.02818	34.4813	0.0290	1724.1	1.5	2.2	2586.1
32.0	0.02512	38.8107	0.0258	1940.5	1.3	1.9	2910.8
33.0	0.02239	43.6684	0.0229	2183.4	1.1	1.7	3275.1
34.0	0.01995	49.1187	0.0204	2455.9	1.0	1.5	3683.9
35.0	0.01778	55.2341	0.0181	2761.7	0.9	1.4	4142.6
36.0	0.01585	62.0957	0.0161	3104.8	0.8	1.2	4657.2
37.0	0.01413	69.7946	0.0143	3489.7	0.7	1.1	5234.6
38.0	0.01259	78.4328	0.0127	3921.6	0.6	1.0	5882.5
39.0	0.01122	88.1251	0.0113	4406.3	0.6	0.9	6609.4
40.0	0.01000	99.0000	0.0101	4950.0	0.5	0.8	7425.0
41.0	0.00891	111.2018	0.0090	5560.1	0.4	0.7	8340.1
42.0	0.00794	124.8925	0.0080	6244.6	0.4	0.6	9366.9
43.0	0.00708	140.2538	0.0071	7012.7	0.4	0.5	10519.0
44.0	0.00631	157.4893	0.0063	7874.5	0.3	0.5	11811.7
45.0	0.00562	176.8279	0.0057	8841.4	0.3	0.4	13262.1

Return Loss, Reflection Coefficient, VSWR, and Mismatch Loss

TABLE 4   Conversion Between Return Loss, Reflection Coefficient, VSWR, and Mismatch Loss

Return Loss [dB]	Reflection Coefficient (Rho)	VSWR ( ):1	Mismatch Loss [dB]	Return Loss [dB]	Reflection Coefficient (Rho)	VSWR ( ):1	Mismatch Loss [dB]
Infinite	0.0000	1.00	0.00	33.00	0.0224	1.05	0.00
50.00	0.0032	1.01	0.00	32.00	0.0251	1.05	0.00
49.00	0.0035	1.01	0.00	31.00	0.0282	1.06	0.00
48.00	0.0040	1.01	0.00	30.00	0.0316	1.07	0.00
47.00	0.0045	1.01	0.00	29.00	0.0355	1.07	0.01
46.00	0.0050	1.01	0.00	28.00	0.0398	1.08	0.01
45.00	0.0056	1.01	0.00	27.00	0.0447	1.09	0.01
44.00	0.0063	1.01	0.00	26.00	0.0501	1.11	0.01
43.00	0.0071	1.01	0.00	25.00	0.0562	1.12	0.01
42.00	0.0079	1.02	0.00	24.00	0.0631	1.13	0.02
41.00	0.0089	1.02	0.00	23.00	0.0708	1.15	0.02
40.00	0.0100	1.02	0.00	22.00	0.0794	1.17	0.03
39.00	0.0112	1.02	0.00	21.00	0.0891	1.20	0.03
38.00	0.0126	1.03	0.00	20.00	0.1000	1.22	0.04
37.00	0.0141	1.03	0.00	19.50	0.1059	1.24	0.05
36.00	0.0158	1.03	0.00	19.00	0.1122	1.25	0.06
35.00	0.0178	1.04	0.00	18.50	0.1189	1.27	0.06
34.00	0.0200	1.04	0.00	18.00	0.1259	1.29	0.07
17.50	0.1334	1.31	0.08	6.02	0.5000	3.00	1.25
17.00	0.1413	1.33	0.09	6.00	0.5012	3.01	1.26
16.50	0.1496	1.35	0.10	5.80	0.5129	3.11	1.33
16.00	0.1585	1.38	0.11	5.60	0.5248	3.21	1.40
15.50	0.1679	1.40	0.12	5.40	0.5370	3.32	1.48
15.00	0.1778	1.43	0.14	5.20	0.5495	3.44	1.56
14.50	0.1884	1.46	0.16	5.11	0.5556	3.50	1.60
14.00	0.1995	1.50	0.18	5.00	0.5623	3.57	1.65
13.50	0.2113	1.54	0.20	4.80	0.5754	3.71	1.75
13.00	0.2239	1.58	0.22	4.60	0.5888	3.86	1.85
12.50	0.2371	1.62	0.25	4.44	0.6000	4.00	1.94
12.00	0.2512	1.67	0.28	4.40	0.6026	4.03	1.96
11.50	0.2661	1.73	0.32	4.20	0.6166	4.22	2.08
11.00	0.2818	1.78	0.36	4.00	0.6310	4.42	2.20
10.50	0.2985	1.85	0.41	3.93	0.6364	4.50	2.25
10.00	0.3162	1.92	0.46	3.80	0.6457	4.64	2.34
9.80	0.3236	1.96	0.48	3.60	0.6607	4.89	2.49
9.60	0.3311	1.99	0.50	3.52	0.6667	5.00	2.55
9.54	0.3333	2.00	0.51	3.40	0.6761	5.17	2.65
9.40	0.3388	2.03	0.53	3.20	0.6918	5.49	2.83
9.20	0.3467	2.06	0.56	3.00	0.7079	5.85	3.02
9.00	0.3548	2.10	0.58	2.80	0.7244	6.26	3.23
8.80	0.3631	2.14	0.61	2.60	0.7413	6.73	3.46
8.60	0.3715	2.18	0.65	2.40	0.7586	7.28	3.72
8.40	0.3802	2.23	0.68	2.20	0.7762	7.94	4.01
8.20	0.3890	2.27	0.71	2.00	0.7943	8.72	4.33
8.00	0.3981	2.32	0.75	1.80	0.8128	9.69	4.69
7.80	0.4074	2.37	0.79	1.74	0.8182	10.00	4.81
7.60	0.4169	2.43	0.83	1.60	0.8318	10.89	5.11
7.40	0.4266	2.49	0.87	1.40	0.8511	12.44	5.60
7.36	0.4286	2.50	0.88	1.20	0.8710	14.50	6.17
7.20	0.4365	2.55	0.92	1.00	0.8913	17.39	6.87
7.00	0.4467	2.61	0.97	0.80	0.9120	21.73	7.74
6.80	0.4571	2.68	1.02	0.60	0.9333	28.96	8.89
6.60	0.4677	2.76	1.07	0.40	0.9550	43.44	10.56
6.40	0.4786	2.84	1.13	0.20	0.9772	86.86	13.47
6.20	0.4898	2.92	1.19	0.00	1.0000	Infinite	Infinite

Notes:

1. Return Loss = -20*log(|Rho|)
2. Mismatch Loss = -10*log(1-|Rho|^2)
3. VSWR = (1+|Rho|)/(1-|Rho|)

Waveguide Components

TABLE 5 Waveguide Performance and Dimensions

EIA WR-	Mil-W-85E RG( )/U	TE10 Cutoff Frequency [GHz]	Recommended Frequency Range Min [GHz]	Max [GHz]	Theoretical Attenuation Fmin [dB/100']	Fmax [dB/100']	Inside Dimensions a [inches]	b [inches]	Tolerance ± [inches]	Outside Dimensions [inches]	[inches]	Tolerance ± [inches]	Wall Thickness [Inches]	Material	Contact Flange	Choke Flange	Cover Flange	Hole Pattern Figure
3	139	173.5726	220.00	325.00	503.90	352.59	0.0340	0.0170	0.00020	4.156	(Diameter)	0.001	0.040	Silver				
4	137	137.2434	170.00	260.00	371.25	246.94	0.0430	0.0215	0.00020	3.156	(Diameter)	0.001	0.040	Silver				
5	135	115.7151	140.00	220.00	303.47	190.96	0.0510	0.0255	0.00025	2.156	(Diameter)	0.001	0.040	Silver				
7	136	90.7918	110.00	170.00	210.19	133.39	0.0650	0.0325	0.00025	1.156	(Diameter)	0.001	0.040	Silver				
8	138	73.7683	90.00	140.00	151.38	97.26	0.080	0.040	0.0003	0.156	(Diameter)	0.001	0.040	Silver				
10		59.0147	75.00	110.00			0.100	0.050	0.0005	0.180	0.130	0.002	0.040					
12		48.3727	60.00	90.00	122.71	82.37	0.122	0.061	0.0005	0.202	0.141	0.002	0.040	Brass				
12	99	48.3727	60.00	90.00	77.46	51.99	0.122	0.061	0.0005	0.202	0.141	0.002	0.040	Silver			387	1
15		39.8748	50.00	75.00	89.78	61.41	0.148	0.074	0.0010	0.228	0.154	0.002	0.040	Brass				
15	98	39.8748	50.00	75.00	56.67	38.76	0.148	0.074	0.0010	0.228	0.154	0.002	0.040	Silver			385	1
19		31.3908	40.00	60.00			0.188	0.094	0.0010	0.268	0.174	0.002	0.040					
22		26.3458	33.00	50.00	48.33	32.89	0.224	0.112	0.0010	0.304	0.192	0.002	0.040	Brass				
22	97	26.3458	33.00	50.00	30.50	20.76	0.224	0.112	0.0010	0.304	0.192	0.002	0.040	Silver			383	1
28		21.0767	26.50	40.00	28.00	19.20	0.280	0.140	0.0015	0.360	0.220	0.002	0.040	Aluminum				1,2
28		21.0767	26.50	40.00	34.32	23.53	0.280	0.140	0.0015	0.360	0.220	0.002	0.040	Brass				1,2
28	96	21.0767	26.50	40.00	21.66	14.85	0.280	0.140	0.0015	0.360	0.220	0.002	0.040	Silver		600	599	1,2
34		17.3573	22.00	33.00			0.340	0.170	0.0020	0.420	0.250	0.003	0.040					
42	121	14.0511	18.00	26.50	16.86	12.40	0.420	0.170	0.0020	0.500	0.250	0.003	0.040	Aluminum	425	598	597	1,2
42	53	14.0511	18.00	26.50	20.66	15.19	0.420	0.170	0.0020	0.500	0.250	0.003	0.040	Brass	425	596	595	1,2
42	66	14.0511	18.00	26.50	13.04	9.59	0.420	0.170	0.0020	0.500	0.250	0.003	0.040	Silver	425	596A	595	1,2
51		11.5715	15.00	22.00			0.510	0.255	0.0025	0.590	0.335	0.003	0.040					
62		9.4879	12.40	18.00	7.88	5.80	0.622	0.311	0.0025	0.702	0.391	0.003	0.040	Aluminum				2
62	91	9.4879	12.40	18.00	9.66	7.11	0.622	0.311	0.0025	0.702	0.391	0.003	0.040	Brass		541	419	2
62	107	9.4879	12.40	18.00	6.10	4.49	0.622	0.311	0.0025	0.702	0.391	0.003	0.040	Silver				2
75		7.8686	10.00	15.00			0.750	0.375	0.003	0.850	0.475	0.003	0.050					
90	67	6.5572	8.20	12.40	5.29	3.66	0.900	0.400	0.003	1.000	0.500	0.003	0.050	Aluminum		136A	135	2
90	52	6.5572	8.20	12.40	6.48	4.49	0.900	0.400	0.003	1.000	0.500	0.003	0.050	Brass		40A	39	2
112	68	5.2598	7.05	10.00	3.39	2.63	1.122	0.497	0.004	1.250	0.625	0.004	0.064	Aluminum		137A	138	2
112	51	5.2598	7.05	10.00	4.15	3.23	1.122	0.497	0.004	1.250	0.625	0.004	0.064	Brass		52A	51	2
137	106	4.3014	5.85	8.20	2.42	1.91	1.372	0.622	0.004	1.500	0.750	0.004	0.064	Aluminum		440A	441	2
137	50	4.3014	5.85	8.20	2.96	2.34	1.372	0.622	0.004	1.500	0.750	0.004	0.064	Brass		343A	344	2
159		3.7116	4.90	7.05			1.590	0.795	0.004	1.718	0.923	0.004	0.064	Brass				2

TABLE 5 (continued) Waveguide Performance and Dimensions

EIA WR-	Mil-W-85E RG( )/U	TE10 Cutoff Frequency [GHz]	Recommended Frequency Range Min [GHz]	Max [GHz]	Theoretical Attenuation Fmin [dB/100']	Fmax [dB/100']	Inside Dimensions a [inches]	b [Inches]	Tolerance ± [Inches]	Outside Dimensions [Inches]	[Inches]	Tolerance ± [Inches]	Wall Thickness [Inches]	Material	Contact Flange	Choke Flange	Cover Flange	Hole Pattern Figure
187	95	3.1525	3.95	5.85	1.70	1.18	1.872	0.872	0.005	2.000	1.000	0.005	0.064	Aluminum		406A	407	
187	49	3.1525	3.95	5.85	2.09	1.45	1.872	0.872	0.005	2.000	1.000	0.005	0.064	Brass		148B	149A	
229		2.5771	3.30	4.90			2.290	1.145	0.005	2.418	1.273	0.005	0.064					
284	75	2.0780	2.60	3.95	0.91	0.62	2.840	1.340	0.005	3.000	1.500	0.005	0.080	Aluminum		585	584	
284	48	2.0780	2.60	3.95	1.11	0.76	2.840	1.340	0.005	3.000	1.500	0.005	0.080	Brass		54B	53	
340	113	1.7357	2.20	3.30	0.65	0.45	3.400	1.700	0.005	3.560	1.860	0.005	0.080	Aluminum			554	
340	112	1.7357	2.20	3.30	0.80	0.56	3.400	1.700	0.005	3.560	1.860	0.005	0.080	Brass			553	
430	105	1.3724	1.70	2.60	0.48	0.32	4.300	2.150	0.005	4.460	2.310	0.005	0.080	Aluminum			437A	
430	104	1.3724	1.70	2.60	0.59	0.39	4.300	2.150	0.005	4.460	2.310	0.005	0.080	Brass			435A	
510		1.1572	1.45	2.20			5.100	2.550	0.005	5.260	2.710	0.005	0.080					
650	103	0.9079	1.12	1.70	0.26	0.17	6.500	3.250	0.005	6.660	3.410	0.005	0.080	Aluminum			418A	
650	69	0.9079	1.12	1.70	0.32	0.21	6.500	3.250	0.005	6.660	3.410	0.005	0.080	Brass			417A	
770	205	0.7664	0.96	1.45	0.20	0.13	7.700	3.850	0.005	7.950	4.100	0.005	0.125	Aluminum				
975	204	0.6053	0.75	1.12	0.14	0.09	9.750	4.875	0.010	10.000	5.125	0.010	0.125	Aluminum				
1150	203	0.5132	0.64	0.96	0.11	0.07	11.500	5.750	0.015	11.750	6.000	0.015	0.125	Aluminum				
1500	202	0.3934	0.49	0.75	0.07	0.05	15.000	7.500	0.015	15.250	7.750	0.015	0.125	Aluminum				
1800	201	0.3279	0.41	0.63	0.05	0.04	18.000	9.000	0.020	18.250	9.250	0.020	0.125	Aluminum				
2100		0.2810	0.35	0.53	0.04	0.03	21.000	10.500	0.020	21.250	10.750	0.020	0.125	Aluminum				
2300		0.2566	0.32	0.49	0.04	0.03	23.000	11.500	0.020	23.250	11.750	0.020	0.125	Aluminum				

Notes:

1. Conductivity of 63.0e6 Mhos/Meter used for Silver.
2. Conductivity of 37.7e6 Mhos/Meter used for Aluminum.
3. Conductivity of 25.1e6 Mhos/Meter used for Brass.
4. Loss is inversely proportional to the square root of conductivity.

Source:

1. Balanis, C.A., *Advanced Engineering Electromagnetics*, John Wiley & Sons, New York, 1990.
2. Catalog, Microwave Development Labs, Natick, MA.
3. Catalog, Aerowave Inc, Medford, MA.
4. Catalog, Formcraft Tool Co, Chicago, IL.
5. Catalog, Penn Engineering Components, No. Hollywood, CA.

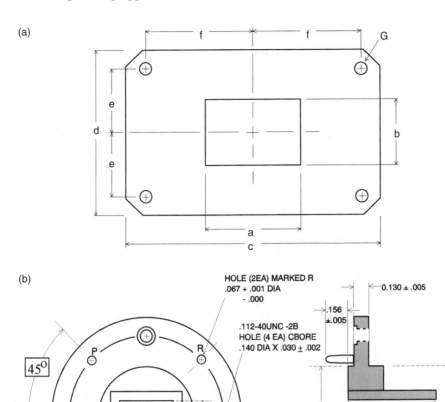

FIGURE B.4 (a) Standard waveguide flange dimensions (rectangular flange); (b) Standard waveguide flange dimensions (circular flange).

TABLE 6   Waveguide Flange Dimensions

			Rectangular Flange Dimensions				
WR-()	a [inches] (Ref.)	b [inches] (Ref.)	c [inches] ±0.015	d [inches] ±0.015	e [inches] ±0.005	f [inches] ±0.005	G [inches-dia] ±0.0015
28	0.280	0.140	0.750	1.750	0.265	0.250	0.1175
42	0.420	0.170	0.875	0.875	0.335	0.320	0.1175
51	0.510	0.255	1.313	1.313	0.497	0.478	0.1445
62	0.620	0.311	1.313	1.313	0.478	0.497	0.1445
75	0.750	0.375	1.500	2.500	0.560	0.520	0.1445
90	0.900	0.400	1.625	1.625	0.610	0.640	0.1705
112	1.122	0.497	1.875	1.875	0.676	0.737	0.1705

			Circular Flange Dimensions			
WR-()	a [inches] (Ref.)	b [inches] (Ref.)	c [inches] +0.000 −0.002	d [inches] BSC.	e [inches] ±0.005	f [inches] ±0.005
10	0.1	0.05	0.75	0.5625	0.375	0.312
12	0.122	0.061	0.75	0.5625	0.375	0.312
15	0.148	0.074	0.75	0.5625	0.375	0.312
19	0.188	0.094	1.125	0.9375	0.5	0.468
22	0.224	0.112	1.125	0.9375	0.5	0.468
28	0.28	0.14	1.125	0.9375	0.5	0.468
42	0.42	0.17	1.125	0.9375	0.625	0.625

Coaxial Cables

TABLE 7   Flexible Coax Specifications

RG-()/U	Mil-C-17/()	Zo	Loss dB/100' @1 MHz	Loss dB/100' @10 MHz	Loss dB/100' @100 MHz	Loss dB/100' @1000 MHz	Center Conductor	Outer Conductor	Jacket	Outside Diameter [Inches]	Dielectric	Velocity Factor [%]	Capacitance pF/foot	Dielectric Core Diameter [inches]	Maximum Voltage [RMS]
8		52	0.16	0.56	1.9	7.4	#13 Stranded Bare Copper .058 Dia	Bare Copper Braid, 97%	Black PVC	0.405	Polyethylene	66	29.5	0.285	3700
8A	[3] 163	52	0.16	0.56	1.9	7.4	#13 Stranded Bare Copper .058 Dia	Bare Copper Braid, 97%	Black PVC, Noncontaminating	0.405	Polyethylene	66	29.5	0.285	3700
9		51	0.18	0.62	2.1	8.2	#13 Stranded Silver Coated Copper .086 Dia	Double Braid, Silver Coated Inner, Bare Copper Outer, 97%	Gray PVC, Noncontaminating	0.42	Polyethylene	66	30	0.28	3700
9B	[4] 164	50					#13 Stranded Silver Coated Copper .089 Dia	Double Braid, Silver Coated Inner, Bare Copper Outer, 97%	PVC	0.36	Polyethylene	66		0.285	3701
11	[6]	75	0.19	0.66	2	7.1	#18 Stranded Tinned Copper .048	Bare Copper Braid, 97%	Black PVC	0.405	Flame Retardant Semi-Foam Polyethylene	66	20.5	0.285	300
11A		75	0.19	0.66	2	8.5	#18 Stranded Tinned Copper .048	Bare Copper Braid, 97%	Black PVC, Noncontaminating	0.405	Polyethylene	66	20.5	0.285	3700
58	[28]	50	0.42	1.5	5.4	22.8	#20 Tinned Copper .035 Dia	Tinned Copper Braid, 95%	Black PVC, Noncontaminating	0.193	Polyethylene	66	30.8	0.116	1400
58A		50	0.42	1.5	5.4	22.8	#20 Solid Bare Copper .035 Dia	Tinned Copper Braid, 95%	Black PVC	0.193	Polyethylene	66	30.8	0.116	1400
58C	155	50	0.42	1.5	5.4	22.8	#20 Tinned Copper .035 Dia	Tinned Copper Braid, 95%	Black PVC, Noncontaminating	0.193	Polyethylene	66	30.8	0.116	1400
59	[29]	75	0.6	1.1	3.4	12	#23 Solid Bare Copper Covered Steel .023 Dia	Bare Copper Braid, 95%	Black PVC, Noncontaminating	0.241	Polyethylene	66	20.5	0.146	1700
59B		75	0.6	1.1	3.4	12	#23 Solid Bare Copper Covered Steel .023 Dia	Bare Copper Braid, 95%	Black PVC, Noncontaminating	0.241	Polyethylene	66	20.5	0.146	1700
62A	[30]	93	0.25	0.85	2.7	8.7	#22 Solid Bare Copper Covered Steel .023 Dia	Bare Copper Braid, 95%	Black PVC, Noncontaminating	0.242	Semi-solid Polyethylene	84	13.5	0.146	750

TABLE 7 (continued)

RG-()/U	Mil-C-17/()	Zo	Loss dB/100' @1 MHz	Loss dB/100' @10 MHz	Loss dB/100' @100 MHz	Loss dB/100' @1000 MHz	Center Conductor	Outer Conductor	Jacket	Outside Diameter [Inches]	Dielectric	Velocity Factor [%]	Capacitance pF/foot	Dielectric Core Diameter [inches]	Maximum Voltage [RMS]
62B	[91] 97	93	0.31	0.9	2.9	11	#24 Solid Bare Copper Covered Steel .025 Dia	Bare Copper Braid, 95%	Black PVC, Noncontaminating	0.242	Semi-solid Polyethylene	84	13.5	0.146	750
63	[31]	125	0.19	0.52	1.5	5.8	#22 Solid Bare Copper Covered Steel .025 Dia	Bare Copper Braid, 97%	Black PVC, Noncontaminating	0.405	Semi-solid Polyethylene	84	9.7	0.285	750
71	[90]	93	0.25	0.85	2.7	8.7	#22 Solid Bare Copper Covered Steel .025 Dia	Double Braid, Tinned Copper Outer, Bare Copper Inner, 98%	Black Polyethylene	0.245	Semi-solid Polyethylene	84	13.5	0.146	750
122	[54] 157	50	0.4	1.7	7	29	#22 Stranded Tinned Copper .030 Dia	Tinned Copper Braid, 95%	Black PVC, Noncontaminating	0.16	Polyethylene	66	30.8	0.096	1400
141	[59] 170	50					#18 Solid Silver Coated Copper Covered Steel .037 Dia	Silver Coated Copper Braid, 94%	Fluorinated Ethylene-Propylene	0.17	TFE Teflon	69.5		0.116	1400
141A		50	0.34	1.1	3.9	13.5	#18 Solid Silver Coated Copper Covered Steel .037 Dia	Silver Coated Copper Braid, 94%	Tinted Brown Fiberglass	0.187	TFE Teflon	69.5	29.2	0.116	1400
142	[60] 158	50	0.34	1.1	3.9	13.5	#18 Solid Silver Coated Copper Covered Steel .037 Dia	Double Silver Coated Copper Braid, 94%	Tinted Brown Fluorinated Ethylene-Propylene	0.187	TFE Teflon	69.5	29.2	0.116	1400
174	[119] 173	50	1.9	3.3	8.4	34	#26 Stranded Bare Copper Covered Steel .019 Dia	Tinned Copper Braid, 90%	Black PVC Jacket	0.11	Polyethylene	66	30.8	0.059	1100
178	[93] 169	50					#30 Stranded Silver Coated Copper Covered Steel .012 Dia	Silver Coated Copper Braid, 96%	Fluorinated Ethylene-Propylene	0.071	TFE Teflon	69.5		0.033	750
178B		50	2.6	5.6	14	46	#30 Solid Silver Coated Copper Covered Steel .012 Dia	Silver Coated Copper Braid, 96%	White Fluorinated Ethylene-Propylene	0.071	TFE Teflon	69.5	29.2	0.033	750

RG	Spec						Conductor	Braid	Jacket		Dielectric				
179	[94]	75	3	5.3	10	24	#30 Solid Silver Coated Copper Covered Steel .012 Dia	Silver Coated Copper Braid, 95%	Tinted Brown Fluorinated Ethylene-Propylene	0.1	TFE Teflon	69.5	19.5	0.062	900
180	[95]	95	2.4	3.3	5.7	17	#30 Solid Silver Coated Copper Covered Steel .012 Dia	Silver Coated Copper Braid, 95%	Tinted Brown Fluorinated Ethylene-Propylene	0.141	TFE Teflon	69.5	15.4	0.102	1100
187	[68] 94	75					#30 Solid Silver Coated Copper Covered Steel .012 Dia	Silver Coated Copper Braid, 92.3%	Fluorinated Ethylene-Propylene	0.1	TFE Teflon	69.5		0.063	900
212	[73] 162	50	0.26	0.83	2.7	9.8	#15.5 Solid Silver Coated Copper .0556 Dia	Double Silver Coated Copper Braid, 95%	Black PVC Noncontaminating	0.332	Polyethylene	66	30.8	0.185	2200
213	[74] 163	50	0.18	0.62	2.1	8.2	#13 Stranded Bare Copper .089 Dia	Bare Copper Braid 97%	Black PVC Noncontaminating	0.405	Polyethylene	66	30.8	0.285	3700
214	[75] 164	50	0.17	0.55	1.9	8	#13 Stranded Silver Coated Copper .089 Dia	Double Silver Coated Copper Braid, 97%	Black PVC Noncontaminating	0.425	Polyethylene	66	30.8	0.285	3700
216	[77]	75	0.19	0.66	2	7.1	#18 Stranded Tinned Copper .048	Double Bare Copper Braid 95%	Black PVC Noncontaminating	0.425	Polyethylene	66	20.5	0.285	3700
223	[84]	50	0.35	1.2	4.1	14.5	#19 Solid Silver Coated Copper .034 Dia	Double Silver Coated Copper Braid, 95%	Black PVC Noncontaminating	0.212	Polyethylene	66	30.8	0.117	1700
303	[111] 170	50	0.34	1.1	3.9	13.5	#18 Solid Silver Coated Copper Covered Steel .037 Dia	Silver Coated Copper Braid, 95%	Tinted Brown Fluorinated Ethylene-Propylene	0.17	TFE Teflon	69.5	29.2	0.116	1400
316	[113] 172	50	1.2	2.7	8.3	29	#26 Stranded Silver Coated Copper Covered Steel .020 Dia	Silver Coated Copper Braid, 95%	White Fluorinated Ethylene-Propylene	0.098	TFE Teflon	69.5	29.2	0.06	900

Note: Mil-C-17/() part numbers were revised. Initial specification numbers are shown in brackets, current specification numbers are unbracketed.
Source:
Mil-C-17G
Mil-C-17G Supplement 1
Belden Master Catalog, Belden Wire & Cable Co, Richmond, IN.

TABLE 8 Semi-Rigid Coax Specifications

RG-()/U	Mil-C-17 Part Number M17/	Zo	Loss [dB/100 ft] Power [W] @500 MHz	Loss [dB/100 ft] Power [W] @1 GHz	Loss [dB/100 ft] Power [W] @3 GHz	Loss [dB/100 ft] Power [W] @5 GHz	Loss [dB/100 ft] Power [W] @10 GHz	Loss [dB/100 ft] Power [W] @18 GHz	Loss [dB/100 ft] Power [W] @20 GHz	Center Conductor Material	Center Conductor Diameter	Outer Conductor	Outside Diameter [Inches]	Dielectric	Dielectric Constant (1 GHz)	Max Capacitance pF/foot	Dielectric Diameter [inches]	Maximum Voltage (60Hz) [RMS]
	154-00001	50 ± 3.0	42 / 14	60 / 10	100 / 6	140 / 4.5	190 / 3.1		280 / 2	Silver Plated Copper Coated Steel	0.008 ±0.0005	Copper	0.034 ±0.001	Solid PTFE	2.03	29.9	0.026 ±0.001	750
	154-00002	50 ± 3.0	42 / 14	60 / 10	100 / 6	140 / 4.5	190 / 3.1		280 / 2	Silver Plated Copper Coated Steel	0.008 ±0.0005	Tin-Plated Copper	0.034 ±0.002	Solid PTFE	2.03	29.9	0.026 ±0.001	750
	151-00001	50 ± 2.5	28 / 45	40 / 32	70 / 18	90 / 13	130 / 9		190 / 65	Silver Plated Copper Coated Steel	0.0113 ±0.0005	Copper	0.047 ±0.001	Solid PTFE	2.03	29.9	0.037 ±0.001	1000
	151-00002	50 ± 2.5	28 / 45	40 / 32	70 / 18	90 / 13	130 / 9		190 / 6.5	Silver Plated Copper Coated Steel	0.0113 ±0.0005	Tin-Plated Copper	0.047 +0.002 -0.001	Solid PTFE	2.03	29.9	0.037 ±0.001	1000
405	133-RG-405	50 ± 1.5	15 / 180	22 / 130		50 / 54	80 / 35		130 / 20	Silver Plated Copper Coated Steel	0.0201 ±0.0005	Copper	0.0865 ±0.001	Solid PTFE	2.03	29.9	0.066 ±0.002	5000
	133-00001	50 ± 1.5	15 / 180	22 / 130		50 / 54	80 / 35		130 / 20	Silver Plated Copper Coated Steel	0.0201 ±0.0005	Tin-Plated Copper	0.0865 +0.002 -0.001	Solid PTFE	2.03	29.9	0.066 ±0.002	5000
	133-00002	50 ± 1.5	15 / 180	22 / 130		50 / 54	80 / 35		130 / 20	Silver Plated Copper Coated Steel	0.0201 ±0.0005	Copper	0.0865 ±0.001	Solid PTFE	2.03	29.9	0.066 ±0.002	5000
	133-00003	50 ± 1.5	15 / 180	22 / 130		50 / 54	80 / 35		130 / 20	Silver Plated Copper	0.0201 ±0.0005	Tin-Plated Copper	0.0865 +0.002 -0.001	Solid PTFE	2.03	29.9	0.066 ±0.002	5000
	133-00004	50 ± 1.5	15 / 180	22 / 130		50 / 54	80 / 35		130 / 20	Silver Plated Nickel Copper Coated Steel	0.0201 ±0.0005	Copper	0.0865 ±0.001	Solid PTFE	2.03	29.9	0.066 ±0.002	5000
	133-00005	50 ± 1.5	15 / 180	22 / 130		50 / 54	80 / 35		130 / 20	Silver Plated Nickel Copper Coated Steel	0.0201 ±0.0005	Tin-Plated Copper	0.0865 +0.002 -0.001	Solid PTFE	2.03	29.9	0.066 ±0.002	5000
	133-0006	50 ± 1.5	15 / 180	22 / 130		50 / 54	80 / 35		130 / 20	Silver Plated Copper Coated Steel	0.0201 ±0.0005	Copper	0.0865 ±0.001	Solid PTFE	2.03	29.9	0.066 ±0.002	5000

Part No.	Impedance							Shield		Conductor		Dielectric				
133-00007	50 ± 1.5	15 / 180	22 / 130			50 / 54	80 / 35	130 / 20	Silver Plated Copper Coated Steel	0.0201 ±0.0005	Tin-Plated Copper	0.086 +0.0021 -0.001	Solid PTFE	2.03	29.9	0.066 ±0.002	5000
133-00008	50 ± 1.5	15 / 180	22 / 130			50 / 54	80 / 35	130 / 20	Silver Plated Copper	0.0201 ±0.0005	Copper	0.0865 ±0.001	Solid PTFE	2.03	29.9	0.066 ±0.002	5000
133-00009	50 ± 1.5	15 / 180	22 / 130			50 / 54	80 / 35	130 / 20	Silver Plated Copper	0.0201 ±0.0005	Tin-Plated Copper	0.0865 +0.002 -0.001	Solid PTFE	2.03	29.9	0.066 ±0.002	5000
133-00010	50 ± 1.5	15 / 180	22 / 130			50 / 54	80 / 35	130 / 20	Silver Plated Nickel Copper Coated Steel	0.0201 ±0.0005	Copper	0.086 ±0.001	Solid PTFE	2.03	29.9	0.066 ±0.002	5000
133-00011	50 ± 1.5	15 / 180	22 / 130			50 / 54	80 / 35	130 / 20	Silver Plated Nickel Copper Coated Steel	0.0201 ±0.0005	Tin-Plated Copper	0.086 +0.002 -0.001	Solid PTFE	2.03	29.9	0.066 ±0.002	5000
130-RG-402 (402)	50 ± 1.0	8 / 600	12 / 450	21 / 250	29 / 180	45 / 120	70 / 70	Silver Plated Copper Coated Steel	0.0362 ±0.0007	Copper	0.141 ±0.001	Solid PTFE	2.03	29.9	0.1175 ±0.001	5000
130-00001	50 ± 1.0	8 / 600	12 / 450	21 / 250	29 / 180	45 / 120	70 / 70	Silver Plated Copper Coated Steel	0.0362 ±0.0007	Tin-Plated Copper	0.141 +0.002 -0.001	Solid PTFE	2.03	29.9	0.1175 ±0.001	5000
130-00002	50 ± 1.0	8 / 601	12 / 451	21 / 251	29 / 181	45 / 121	70 / 71	Silver Plated Nickel Copper Coated Steel	0.0362 ±0.0007	Copper	0.141 ±0.001	Solid PTFE	2.03	29.9	0.1175 ±0.001	5000
130-00003	50 ± 1.0	8 / 602	12 / 452	21 / 252	29 / 182	45 / 122	70 / 72	Silver Plated Nickel Copper Coated Steel	0.0362 ±0.0007	Tin-Plated Copper	0.141 +0.002 -0.001	Solid PTFE	2.03	29.9	0.1175 ±0.001	5000
130-00004	50 ± 1.0	8 / 603	12 / 453	21 / 253	29 / 183	45 / 123	70 / 73	Silver Plated Copper Coated Steel	0.0362 ±0.0007	Copper	0.141 ±0.001	Solid PTFE	2.03	29.9	0.1175 ±0.001	5000
130-00005	50 ± 1.0	8 / 604	12 / 454	21 / 254	29 / 184	45 / 124	70 / 74	Silver Plated Copper Coated Steel	0.0362 ±0.0007	Tin-Plated Copper	0.141 +0.002 -0.001	Solid PTFE	2.03	29.9	0.1175 ±0.001	5000
130-00006	50 ± 1.0	8 / 605	12 / 455	21 / 255	29 / 185	45 / 125	70 / 75	Silver Plated Nickel Copper Coated Steel	0.0362 ±0.0007	Copper	0.141 ±0.001	Solid PTFE	2.03	29.9	0.1175 ±0.001	5000

TABLE 8 (continued)

RG-(/)U	Mil-C-17 Part Number M17/	Zo	Loss [dB/100 ft] Power [W] @500 MHz	Loss [dB/100 ft] Power [W] @1 GHz	Loss [dB/100 ft] Power [W] @3 GHz	Loss [dB/100 ft] Power [W] @5 GHz	Loss [dB/100 ft] Power [W] @10 GHz	Loss [dB/100 ft] Power [W] @18 GHz	Loss [dB/100 ft] Power [W] @20 GHz	Center Conductor Material	Center Conductor Diameter	Outer Conductor	Outside Diameter [Inches]	Dielectric	Dielectric Constant (1 GHz)	Max Capacitance pF/foot	Dielectric Diameter [inches]	Maximum Voltage (60Hz) [RMS]
	130-00007	50 ± 1.0	8 606	12 456	21 256	29 186	45 126		70 76	Silver Plated Nickel Copper Coated Steel	0.0362 ±0.0007	Tin-Plated Copper	0.141 +0.002 −0.001	Solid PTFE	2.03	29.9	0.1175 ±0.001	5000
	130-00008	50 ± 1.0	8 607	12 457	21 257	29 187	45 127		70 77	Silver Plated Copper Coated Steel	0.0362 ±0.0007	Aluminum	0.141 ±0.001	Solid PTFE	2.03	29.9	0.1175 ±0.001	5000
	130-00009	50 ± 1.0	8 608	12 458	21 258	29 188	45 128		70 78	Silver Plated Copper Coated Steel	0.0362 ±0.0007	Aluminum	0.141 ±0.001	Solid PTFE	2.03	29.9	0.1175 ±0.001	5000
	130-00010	50 ± 1.0	8 609	12 459	21 259	29 189	45 129		70 79	Silver Plated Nickel Copper Coated Steel	0.0362 ±0.0007	Aluminum	0.141 ±0.001	Solid PTFE	2.03	29.9	0.1175 ±0.001	5000
	130-00011	50 ± 1.0	8 610	12 460	21 260	29 190	45 130		70 80	Silver Plated Copper Coated Steel	0.0362 ±0.0007	Aluminum	0.141 ±0.001	Solid PTFE	2.03	29.9	0.1175 ±0.001	5000
	130-00012	50 ± 1.0	8 611	12 461	21 261	29 191	45 131		70 81	Silver Plated Copper Coated Steel	0.0362 ±0.0007	Silver Plated Copper	0.141 +0.002 −0.001	Solid PTFE	2.03	29.9	0.1175 ±0.001	5000
	130-00013	50 ± 1.0	8 612	12 462	21 262	29 192	45 132		70 82	Silver Plated Nickel Copper Coated Steel	0.0362 ±0.0007	Silver Plated Copper	0.141 +0.002 −0.001	Solid PTFE	2.03	29.9	0.1175 ±0.001	5000
401	129-RG-401	50 ± 0.5	4.5 1900	7.5 1400	11 750		33 350	48 200	— —	Silver Plated Copper	0.0641 ±0.001	Copper	0.250 ±0.001	Solid PTFE	2.03	29.9	0.209 ±0.002	7500
	129-0001	50 ± 0.5	4.5 1900	7.5 1400	11 750		33 350	48 200	— —	Silver Plated Copper	0.0641 ±0.001	Tin-Plated Copper	0.250 +0.002 −0.001	Solid PTFE	2.03	29.9	0.209 ±0.002	7500

Notes: Attenuation/Power Ratings are maximum values for families.
Sources: Mil-C-17/130E
Semi-Rigid Coaxial Cable Catalog, Micro-Coax Components, Inc, Collegeville, PA.

Guide To Use of Dissimilar Metals In Sea Water, Marine Atmosphere, and Industrial Atmosphere

Code	Metals	A	B	C	D	E	F	G	H	I	J	K	L	M	N	O	P	Q	R	S	T
A	Magnesium	G																			
B	Zinc Zinc Coating		G																		
C	Cadmium Beryllium			G																	
D	Aluminum, Aluminum-Mg, Aluminum-Zn				G																
E	Aluminum-Copper					G															
F	Steels - Carbon, Low Alloy						G														
G	Lead							G													
H	Tin, Tin–Lead Indium								G												
I	Stainless Steels - Martensitic, Ferritic									G											
J	Chromium, Molybdenum, Tungsten										G										
K	Stainless Steels - Austenitic, Type PH Super Strength Heat Resistant											G									
L	Brass-Lead, Bronze												G								
M	Brass - Low Copper, Bronze - Low Copper													G							
N	Brass - High Copper, Bronze - High Copper														G						
O	Copper - High Nickel, Monel															G					
P	Nickel, Cobalt																G				
Q	Titanium																	G			
R	Silver																		G		
S	Palladium, Rhodium, Gold, Platinum																			G	
T	Graphite																				G

Sea Water

| Marine
Atmosphere | Industrial
Atmosphere |

Example: Can High-Copper Brass be put safely in direct contact with Gold?
High-Copper Brass has code letter M, and Gold has code letter T.
Find the intersection of Row M and Column T, to find that these metals are incompatible in salt water and marine atmosphere, but compatible in an industrial atmosphere.

Notes:
1. Metals are ranked from most anodic (A) to most cathodic (T).
2. "G" indicates bare identical metal couple compatible in sea water, marine atmosphere, and industrial atmosphere.
3. I indicates joined metals incompatible without appropriate protective coatings in the indicated environment.
4. C indicates joined metals are compatible in the indicated environment.
5. Reference: MIL-STD-889B.

Single Sideband and Image Reject Mixers

TABLE 10 Maximum Tolerable Phase Error (Degrees) for Single Sideband and Image Reject Mixers as a Function of Suppression and Amplitude Imbalance

Amplitude Imbalance [dB]	Suppression												
	−10 dBc	−13 dBc	−15 dBc	−17 dBc	−20 dBc	−23 dBc	−25 dBc	−27 dBc	−30 dBc	−33 dBc	−35 dBc	−37 dBc	−40 dBc
0.00	35.10	25.24	20.17	16.08	11.42	8.10	6.44	5.12	3.62	2.56	2.04	1.62	1.15
0.05	35.10	25.24	20.16	16.08	11.42	8.09	6.43	5.10	3.61	2.54	2.01	1.58	1.10
0.10	35.09	25.23	20.16	16.07	11.40	8.07	6.40	5.07	3.56	2.48	1.93	1.48	0.94
0.15	35.08	25.22	20.14	16.05	11.38	8.04	6.36	5.02	3.48	2.37	1.78	1.28	0.58
0.20	35.08	25.21	20.13	16.03	11.35	7.99	6.30	4.94	3.37	2.20	1.55	0.94	
0.25	35.06	25.19	20.10	16.00	11.30	7.93	6.22	4.84	3.23	1.97	1.20		
0.30	35.05	25.17	20.07	15.96	11.25	7.86	6.13	4.72	3.04	1.63	0.49		
0.35	35.03	25.14	20.04	15.92	11.19	7.77	6.01	4.57	2.79	1.12			
0.40	35.01	25.11	20.00	15.87	11.12	7.66	5.87	4.38	2.48				
0.45	34.99	25.07	19.96	15.81	11.03	7.54	5.72	4.17	2.08				
0.50	34.96	25.04	19.91	15.75	10.94	7.40	5.53	3.91	1.50				
0.55	34.93	24.99	19.85	15.68	10.84	7.25	5.32	3.61					
0.60	34.90	24.95	19.79	15.60	10.72	7.07	5.08	3.25					
0.65	34.87	24.90	19.73	15.51	10.60	6.88	4.81	2.80					
0.70	34.83	24.84	19.65	15.42	10.46	6.66	4.50	2.21					
0.75	34.79	24.78	19.58	15.32	10.31	6.42	4.13	1.32					
0.80	34.75	24.72	19.49	15.21	10.15	6.16	3.70						
0.85	34.70	24.65	19.41	15.10	9.97	5.86	3.18						
0.90	34.66	24.58	19.31	14.98	9.78	5.53	2.52						
0.95	34.61	24.50	19.21	14.84	9.58	5.16	1.53						
1.00	34.55	24.42	19.10	14.70	9.35	4.73							
1.10	34.44	24.24	18.87	14.40	8.86	3.65							
1.20	34.31	24.05	18.62	14.06	8.28	1.83							
1.30	34.17	23.84	18.34	13.67	7.61								
1.40	34.02	23.61	18.03	13.25	6.80								
1.50	33.86	23.35	17.69	12.78	5.82								
1.60	33.68	23.08	17.32	12.25	4.53								
1.70	33.50	22.79	16.92	11.67	2.52								
1.80	33.30	22.48	16.48	11.01									
1.90	33.09	22.14	16.00	10.28									
2.00	32.86	21.78	15.49	9.44									
2.10	32.63	21.39	14.93	8.47									
2.20	32.37	20.98	14.31	7.31									
2.30	32.11	20.54	13.64	5.87									
2.40	31.83	20.07	12.90	3.81									
2.50	31.54	19.56	12.08										
2.60	31.23	19.03	11.17										
2.70	30.90	18.45	10.13										
2.80	30.56	17.83	8.93										
2.90	30.21	17.17	7.48										
3.00	29.83	16.46	5.59										
3.10	29.44	15.68	2.41										
3.20	29.03	14.84											
3.30	28.60	13.92											
3.40	28.16	12.91											
3.50	27.69	11.77											
3.60	27.19	10.47											
3.70	26.68	8.94											
3.80	26.14	7.02											
3.90	25.57	4.24											
4.00	24.98												
4.10	24.35												
4.20	23.69												
4.30	23.00												
4.40	22.27												
4.50	21.49												
4.60	20.67												
4.70	19.79												
4.80	18.86												
4.90	17.85												
5.00	16.76												

Notes: Example: An image reject mixer requires 25 dB of unwanted image suppression, and a maximum amplitude imbalance of 1 dB is expected between the channels. Looking at the intersection of the 1 dB row and −25 dBc column shows that a maximum interchannel phase imbalance of 9.35 degrees is tolerable.
Suppression [dBc] = $10 \log ((1 - 2 a \cos(p) + a^2) / (1 + 2 a \cos(p) + a^2))$ where a is the voltage ratio amplitude imbalance and p is the phase imbalance.

Power, Voltage and Decibels

TABLE 11   Power, Voltage, dB Conversion Table

dB	Power Ratio	Voltage Ratio	dB	Power Ratio	Voltage Ratio	dB	Power Ratio	Voltage Ratio	dB	Power Ratio	Voltage Ratio
0	1	1	9	7.943282347	2.818382931	34	2511.886432	50.11872336	60	1000000	1000
0.1	1.023292992	1.011579454	9.5	8.912509381	2.985382619	35	3162.27766	56.23413252	61	1258925.412	1122.018454
0.2	1.047128548	1.023292992	10	10	3.16227766	36	3981.071706	63.09573445	62	1584893.192	1258.925412
0.3	1.071519305	1.035142167	11	12.58925412	3.548133892	37	5011.872336	70.79457844	63	1995262.315	1412.537545
0.4	1.096478196	1.047128548	12	15.84893192	3.981071706	38	6309.573445	79.43282347	64	2511886.432	1584.893192
0.5	1.122018454	1.059253725	13	19.95262315	4.466835922	39	7943.282347	89.12509381	65	3162277.66	1778.27941
0.6	1.148153621	1.071519305	14	25.11886432	5.011872336	40	10000	100	66	3981071.706	1995.262315
0.7	1.174897555	1.083926914	15	31.6227766	5.623413252	41	12589.25412	112.2018454	67	5011872.336	2238.721139
0.8	1.202264435	1.096478196	16	39.81071706	6.309573445	42	15848.93192	125.8925412	68	6309573.445	2511.886432
0.9	1.230268771	1.109174815	17	50.11872336	7.079457844	43	19952.62315	141.2537545	69	7943282.347	2818.382931
1	1.258925412	1.122018454	18	63.09573445	7.943282347	44	25118.86432	158.4893192	70	10000000	3162.27766
1.5	1.412537545	1.188502227	19	79.43282347	8.912509381	45	31622.7766	177.827941	71	12589254.12	3548.133892
2	1.584893192	1.258925412	20	100	10	46	39810.71706	199.5262315	72	15848931.92	3981.071706
2.5	1.77827941	1.333521432	21	125.8925412	11.22018454	47	50118.72336	223.8721139	73	19952623.15	4466.835922
3	1.995262315	1.412537545	22	158.4893192	12.58925412	48	63095.73445	251.1886432	74	25118864.32	5011.872336
3.5	2.238721139	1.496235656	23	199.5262315	14.12537545	49	79432.82347	281.8382931	75	31622776.6	5623.413252
4	2.511886432	1.584893192	24	251.1886432	15.84893192	50	100000	316.227766	76	39810717.06	6309.573445
4.5	2.818382931	1.678804018	25	316.227766	17.7827941	51	125892.5412	354.8133892	77	50118723.36	7079.457844
5	3.16227766	1.77827941	26	398.1071706	19.95262315	52	158489.3192	398.1071706	78	63095734.45	7943.282347
5.5	3.548133892	1.883649089	27	501.1872336	22.38721139	53	199526.2315	446.6835922	79	79432823.47	8912.509381
6	3.981071706	1.995262315	28	630.9573445	25.11886432	54	251188.6432	501.1872336	80	100000000	10000
6.5	4.466835922	2.11348904	29	794.3282347	28.18382931	55	316227.766	562.3413252	81	125892541.2	11220.18454
7	5.011872336	2.238721139	30	1000	31.6227766	56	398107.1706	630.9573445	82	158489319.2	12589.25412
7.5	5.623413252	2.371373706	31	1258.925412	35.48133892	57	501187.2336	707.9457844	83	199526231.5	14125.37545
8	6.309573445	2.511886432	32	1584.893192	39.81071706	58	630957.3445	794.3282347	84	251188643.2	15848.93192
8.5	7.079457844	2.66072506	33	1995.262315	44.66835922	59	794328.2347	891.2509381	85	316227766	17782.7941

dB	Power Ratio	Voltage Ratio	dB	Power Ratio	Voltage Ratio	dB	Power Ratio	Voltage Ratio	dB	Power Ratio	Voltage Ratio
86	398107170.6	19952.62315	−5.5	0.281838293	0.530884444	−35	0.000316228	0.017782794	−69	1.25893E-07	0.000354813
87	501187233.6	22387.21139	−6	0.251188643	0.501187234	−36	0.000251189	0.015848932	−70	0.0000000	0.000316228
88	630957344.5	25118.86432	−6.5	0.223872114	0.473151259	−37	0.000199526	0.014125375	−71	7.94328E-08	0.000281838
89	794328234.7	28183.82931	−7	0.199526231	0.446683592	−38	0.000158489	0.012589254	−72	6.30957E-08	0.000251189
90	1000000000	31622.7766	−7.5	0.177827941	0.421696503	−39	0.000125893	0.011220185	−73	5.01187E-08	0.000223872
91	1258925412	35481.33892	−8	0.158489319	0.398107171	−40	0.0001	0.01	−74	3.98107E-08	0.000199526
92	1584893192	39810.71706	−8.5	0.141253754	0.375837404	−41	7.94328E-05	0.008912509	−75	3.16228E-08	0.000177828
93	1995262315	44668.35922	−9	0.125892541	0.354813389	−42	6.30957E-05	0.007943282	−76	2.51189E-08	0.000158489
94	2511886432	50118.72336	−9.5	0.112201845	0.334965439	−43	5.01187E-05	0.007079458	−77	1.99526E-08	0.000141254
95	3162277660	56234.13252	−10	0.1	0.316227766	−44	3.98107E-05	0.006309573	−78	1.58489E-08	0.000125893
96	3981071706	63095.73445	−11	0.079432823	0.281838293	−45	3.16228E-05	0.005623413	−79	1.25893E-08	0.000112202
97	5011872336	70794.57844	−12	0.063095734	0.251188643	−46	2.51189E-05	0.005011872	−80	0.00000001	0.0001
98	6309573445	79432.82347	−13	0.050118723	0.223872114	−47	1.99526E-05	0.004466836	−81	7.94328E-09	8.91251E-05
99	7943282347	89125.09381	−14	0.039810717	0.199526231	−48	1.58489E-05	0.003981072	−82	6.30957E-09	7.94328E-05
100	10000000000	100000	−15	0.031622777	0.177827941	−49	1.25893E-05	0.003548481	−83	5.01187E-09	7.07946E-05
0	1	1	−16	0.025118864	0.158489319	−50	0.00001	0.003162278	−84	3.98107E-09	6.30957E-05
−0.1	0.977237221	0.988553095	−17	0.019952623	0.141253754	−51	7.94328E-06	0.002818383	−85	3.16228E-09	5.62341E-05
−0.2	0.954992586	0.977237221	−18	0.015848932	0.125892541	−52	6.30957E-06	0.002511886	−86	2.51189E-09	5.01187E-05
−0.3	0.933254301	0.966050879	−19	0.012589254	0.112201845	−53	5.01187E-06	0.002238721	−87	1.99526E-09	4.46684E-05
−0.4	0.912010839	0.954992586	−20	0.01	0.1	−54	3.98107E-06	0.001995262	−88	1.58489E-09	3.98107E-05
−0.5	0.891250938	0.944060876	−21	0.007943282	0.089125094	−55	3.16228E-06	0.001778279	−89	1.25893E-09	3.54813E-05
−0.6	0.87096359	0.933254301	−22	0.006309573	0.079432823	−56	2.51189E-06	0.001584893	−90	0.000000001	3.16228E-05
−0.7	0.851138038	0.922571427	−23	0.005011872	0.070794578	−57	1.99526E-06	0.001412538	−91	7.94328E-10	2.81838E-05
−0.8	0.831763771	0.912010839	−24	0.003981072	0.063095734	−58	1.58489E-06	0.001258925	−92	6.30957E-10	2.51189E-05
−0.9	0.812830516	0.901571138	−25	0.003162278	0.056234133	−59	1.25893E-06	0.001122018	−93	5.01187E-10	2.23872E-05
−1	0.794328235	0.891250938	−26	0.002511886	0.050118723	−60	0.000001	0.001	−94	3.98107E-10	1.99526E-05
−1.5	0.707945784	0.841395142	−27	0.001995262	0.044668359	−61	7.94328E-07	0.000891251	−95	3.16228E-10	1.77828E-05
−2	0.630957344	0.794328235	−28	0.001584893	0.039810717	−62	6.30957E-07	0.000794328	−96	2.51189E-10	1.58489E-05
−2.5	0.562341325	0.749894209	−29	0.001258925	0.035481339	−63	5.01187E-07	0.000707946	−97	1.99526E-10	1.41254E-05
−3	0.501187234	0.707945784	−30	0.001	0.031622777	−64	3.98107E-07	0.000630957	−98	1.58489E-10	1.25893E-05
−3.5	0.446683592	0.668343918	−31	0.000794328	0.028183829	−65	3.16228E-07	0.000562341	−99	1.25893E-10	1.12202E-05
−4	0.398107171	0.630957344	−32	0.000630957	0.025118864	−66	2.51189E-07	0.000501187	−100	1E-10	0.00001
−4.5	0.354813389	0.595662144	−33	0.000501187	0.022387211	−67	1.99526E-07	0.000446684			
−5	0.316227766	0.562341325	−34	0.000398107	0.019952623	−68	1.58489E-07	0.000398107			

Notes:
1. Multiply by appropriate factor.
2. Use voltage factor to convert S-parameters.

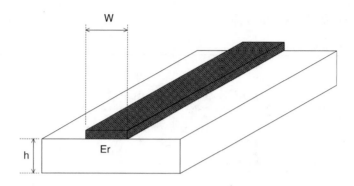

FIGURE B.5

TABLE 12a   Zero Thickness Microstrip Dimensions, Effective Dielectric Constant, and PUL Capacitance and Inductance for Er = 2.2

Zo [Ohms]	W/h	Keff	pF/cm	nH/mm	Zo [Ohms]	W/h	Keff	pF/cm	nH/mm
1	250.3363	2.1861	49.3192	0.0493	57	2.5199	1.8500	0.7959	2.5860
2	123.5739	2.1728	24.5846	0.0983	58	2.4514	1.8471	0.7816	2.6294
3	81.4007	2.1601	16.3417	0.1471	59	2.3854	1.8443	0.7678	2.6727
4	60.3537	2.1480	12.2218	0.1955	60	2.3218	1.8416	0.7544	2.7160
5	47.7490	2.1364	9.7510	0.2438	61	2.2603	1.8389	0.7415	2.7592
6	39.3615	2.1253	8.1046	0.2918	62	2.2010	1.8362	0.7290	2.8024
7	33.3816	2.1146	6.9294	0.3395	63	2.1437	1.8336	0.7170	2.8456
8	28.9051	2.1044	6.0485	0.3871	64	2.0883	1.8310	0.7053	2.8887
9	25.4299	2.0946	5.3639	0.4345	65	2.0347	1.8285	0.6939	2.9318
10	22.6550	2.0851	4.8166	0.4817	66	1.9829	1.8259	0.6829	2.9749
11	20.3889	2.0760	4.3692	0.5287	67	1.9327	1.8235	0.6723	3.0179
12	18.5040	2.0673	3.9967	0.5755	68	1.8919	1.8214	0.6620	3.0612
13	16.9121	2.0589	3.6817	0.6222	69	1.8447	1.8190	0.6520	3.1042
14	15.5503	2.0508	3.4120	0.6688	70	1.7990	1.8166	0.6423	3.1471
15	14.3723	2.0429	3.1784	0.7152	71	1.7548	1.8143	0.6328	3.1900
16	13.3435	2.0354	2.9743	0.7614	72	1.7119	1.8120	0.6236	3.2329
17	12.4375	2.0280	2.7943	0.8075	73	1.6704	1.8097	0.6147	3.2757
18	11.6337	2.0210	2.6344	0.8536	74	1.6301	1.8075	0.6060	3.3186
19	10.9159	2.0141	2.4915	0.8994	75	1.5910	1.8053	0.5976	3.3614
20	10.2711	2.0075	2.3631	0.9452	76	1.5531	1.8031	0.5894	3.4041
21	9.6889	2.0010	2.2469	0.9909	77	1.5163	1.8010	0.5814	3.4469
22	9.1607	1.9948	2.1414	1.0365	78	1.4806	1.7988	0.5736	3.4896
23	8.6793	1.9887	2.0452	1.0819	79	1.4458	1.7968	0.5660	3.5322
24	8.2389	1.9828	1.9571	1.1273	80	1.4121	1.7947	0.5586	3.5749
25	7.8346	1.9771	1.8761	1.1726	81	1.3793	1.7926	0.5514	3.6175
26	7.4621	1.9715	1.8014	1.2177	82	1.3474	1.7906	0.5443	3.6601
27	7.1178	1.9661	1.7323	1.2628	83	1.3164	1.7886	0.5375	3.7027
28	6.7988	1.9608	1.6682	1.3078	84	1.2862	1.7867	0.5308	3.7453
29	6.5024	1.9557	1.6085	1.3528	85	1.2568	1.7847	0.5243	3.7878

TABLE 12a (continued)

Zo [Ohms]	W/h	Keff	pF/cm	nH/mm	Zo [Ohms]	W/h	Keff	pF/cm	nH/mm
30	6.2263	1.9507	1.5529	1.3976	86	1.2282	1.7828	0.5179	3.8303
31	5.9686	1.9458	1.5010	1.4424	87	1.2003	1.7809	0.5117	3.8728
32	5.7274	1.9410	1.4523	1.4871	88	1.1732	1.7791	0.5056	3.9152
33	5.5013	1.9364	1.4066	1.5318	89	1.1467	1.7772	0.4996	3.9577
34	5.2890	1.9319	1.3636	1.5763	90	1.1210	1.7754	0.4938	4.0001
35	5.0892	1.9274	1.3231	1.6208	91	1.0959	1.7736	0.4882	4.0424
36	4.9009	1.9231	1.2849	1.6653	92	1.0714	1.7718	0.4826	4.0848
37	4.7231	1.9189	1.2488	1.7096	93	1.0476	1.7700	0.4772	4.1272
38	4.5550	1.9147	1.2146	1.7539	94	1.0243	1.7683	0.4719	4.1695
39	4.3959	1.9107	1.1822	1.7982	95	1.0016	1.7665	0.4667	4.2118
40	4.2450	1.9067	1.1515	1.8424	96	0.9795	1.7658	0.4617	4.2552
41	4.1018	1.9028	1.1223	1.8865	97	0.9579	1.7651	0.4569	4.2987
42	3.9658	1.8990	1.0945	1.9306	98	0.9369	1.7644	0.4521	4.3421
43	3.8363	1.8953	1.0680	1.9746	99	0.9163	1.7637	0.4475	4.3855
44	3.7129	1.8917	1.0427	2.0186	100	0.8962	1.7629	0.4429	4.4289
45	3.5953	1.8881	1.0185	2.0625	101	0.8767	1.7621	0.4384	4.4722
46	3.4831	1.8846	0.9955	2.1064	102	0.8575	1.7613	0.4340	4.5154
47	3.3758	1.8811	0.9734	2.1502	103	0.8389	1.7605	0.4297	4.5586
48	3.2733	1.8778	0.9523	2.1940	104	0.8206	1.7596	0.4255	4.6018
49	3.1751	1.8745	0.9320	2.2378	105	0.8028	1.7588	0.4213	4.6449
50	3.0811	1.8712	0.9126	2.2814	106	0.7855	1.7579	0.4172	4.6880
51	2.9909	1.8680	0.8939	2.3251	107	0.7685	1.7570	0.4132	4.7310
52	2.9044	1.8649	0.8760	2.3687	108	0.7519	1.7561	0.4093	4.7740
53	2.8214	1.8618	0.8588	2.4122	109	0.7357	1.7552	0.4054	4.8169
54	2.7416	1.8587	0.8422	2.4557	110	0.7198	1.7543	0.4016	4.8599
55	2.6648	1.8558	0.8262	2.4992	111	0.7043	1.7534	0.3979	4.9027
56	2.5910	1.8528	0.8108	2.5426	112	0.6892	1.7524	0.3943	4.9456
113	0.6744	1.7515	0.3907	4.9884	132	0.4484	1.7330	0.3327	5.7964
114	0.6600	1.7505	0.3871	5.0312	133	0.4389	1.7321	0.3301	5.8387
115	0.6459	1.7496	0.3837	5.0739	134	0.4296	1.7311	0.3275	5.8810
116	0.6320	1.7486	0.3803	5.1166	135	0.4206	1.7302	0.3250	5.9232
117	0.6185	1.7477	0.3769	5.1593	136	0.4117	1.7292	0.3225	5.9654
118	0.6053	1.7467	0.3736	5.2020	137	0.4030	1.7283	0.3201	6.0076
119	0.5924	1.7457	0.3704	5.2446	138	0.3945	1.7273	0.3177	6.0498
120	0.5798	1.7447	0.3672	5.2872	139	0.3862	1.7264	0.3153	6.0920
121	0.5675	1.7438	0.3640	5.3298	140	0.3781	1.7254	0.3130	6.1342
122	0.5554	1.7428	0.3609	5.3723	141	0.3702	1.7245	0.3107	6.1763
123	0.5436	1.7418	0.3579	5.4148	142	0.3624	1.7236	0.3084	6.2185
124	0.5321	1.7408	0.3549	5.4573	143	0.3547	1.7226	0.3062	6.2606
125	0.5208	1.7399	0.3520	5.4998	144	0.3473	1.7217	0.3039	6.3027
126	0.5097	1.7389	0.3491	5.5422	145	0.3400	1.7208	0.3018	6.3448
127	0.4989	1.7379	0.3462	5.5846	146	0.3328	1.7199	0.2996	6.3868
128	0.4884	1.7369	0.3434	5.6270	147	0.3259	1.7190	0.2975	6.4289
129	0.4780	1.7360	0.3407	5.6694	148	0.3190	1.7181	0.2954	6.4709
130	0.4679	1.7350	0.3380	5.7118	149	0.3123	1.7172	0.2934	6.5130
131	0.4580	1.7340	0.3353	5.7541	150	0.3058	1.7163	0.2913	6.5550

Notes: Calculation of W/H has an error of less than 1%.

Source: Gupta, K.C., Garg, R., Bahl, I., Bhartia, P., *Microstrip Lines and Slotlines,* 2nd Ed., Artech House, Norwood, MA, 1996, 103.

TABLE 12b   Zero Thickness Microstrip Dimensions, Effective Dielectric Constant, and PUL Capacitance and Inductance for Er =3.78

Zo [Ohms]	W/h	Keff	pF/cm	nH/mm	Zo [Ohms]	W/h	Keff	pF/cm	nH/mm
1	190.5772	3.7382	64.4927	0.0645	26	5.4498	3.1668	2.2831	1.5433
2	93.9049	3.6989	32.0763	0.1283	27	5.1890	3.1537	2.1940	1.5994
3	61.7511	3.6619	21.2770	0.1915	28	4.9474	3.1410	2.1113	1.6553
4	45.7086	3.6271	15.8817	0.2541	29	4.7230	3.1287	2.0345	1.7110
5	36.1035	3.5942	12.6477	0.3162	30	4.5140	3.1167	1.9629	1.7666
6	29.7137	3.5632	10.4941	0.3778	31	4.3190	3.1051	1.8961	1.8221
7	25.1593	3.5337	8.9578	0.4389	32	4.1365	3.0938	1.8335	1.8775
8	21.7508	3.5059	7.8070	0.4997	33	3.9655	3.0827	1.7747	1.9327
9	19.1053	3.4794	6.9133	0.5600	34	3.8050	3.0720	1.7195	1.9878
10	16.9935	3.4542	6.1994	0.6199	35	3.6539	3.0616	1.6676	2.0428
11	15.2694	3.4301	5.6162	0.6796	36	3.5116	3.0514	1.6185	2.0976
12	13.8357	3.4072	5.1309	0.7389	37	3.3773	3.0414	1.5722	2.1524
13	12.6252	3.3853	4.7210	0.7978	38	3.2504	3.0317	1.5284	2.2070
14	11.5899	3.3643	4.3702	0.8566	39	3.1302	3.0222	1.4869	2.2616
15	10.6946	3.3442	4.0666	0.9150	40	3.0164	3.0130	1.4475	2.3160
16	9.9129	3.3249	3.8014	0.9732	41	2.9083	3.0039	1.4101	2.3703
17	9.2247	3.3064	3.5678	1.0311	42	2.8056	2.9951	1.3745	2.4246
18	8.6143	3.2885	3.3605	1.0888	43	2.7080	2.9864	1.3406	2.4787
19	8.0694	3.2714	3.1754	1.1463	44	2.6150	2.9780	1.3082	2.5327
20	7.5800	3.2549	3.0090	1.2036	45	2.5263	2.9697	1.2774	2.5867
21	7.1382	3.2389	2.8586	1.2607	46	2.4417	2.9615	1.2479	2.6406
22	6.7375	3.2235	2.7222	1.3175	47	2.3609	2.9536	1.2197	2.6943
23	6.3724	3.2086	2.5978	1.3743	48	2.2836	2.9458	1.1927	2.7480
24	6.0386	3.1942	2.4840	1.4308	49	2.2097	2.9381	1.1669	2.8016
25	5.7321	3.1803	2.3794	1.4871	50	2.1389	2.9306	1.1421	2.8552
51	2.0711	2.9233	1.1183	2.9086	101	0.5140	2.7126	0.5439	5.5488
52	2.0060	2.9160	1.0954	2.9620	102	0.5008	2.7099	0.5383	5.6008
53	1.9435	2.9089	1.0734	3.0152	103	0.4878	2.7071	0.5328	5.6529
54	1.8840	2.9020	1.0523	3.0685	104	0.4752	2.7043	0.5274	5.7048
55	1.8268	2.8952	1.0320	3.1217	105	0.4630	2.7016	0.5222	5.7568
56	1.7719	2.8886	1.0124	3.1748	106	0.4511	2.6989	0.5170	5.8086
57	1.7190	2.8820	0.9935	3.2278	107	0.4394	2.6961	0.5119	5.8605
58	1.6682	2.8756	0.9752	3.2807	108	0.4281	2.6934	0.5069	5.9123
59	1.6192	2.8693	0.9577	3.3336	109	0.4171	2.6907	0.5020	5.9640
60	1.5720	2.8631	0.9407	3.3865	110	0.4064	2.6880	0.4972	6.0157
61	1.5265	2.8569	0.9243	3.4392	111	0.3960	2.6853	0.4924	6.0674
62	1.4826	2.8509	0.9084	3.4919	112	0.3858	2.6826	0.4878	6.1190
63	1.4402	2.8450	0.8931	3.5446	113	0.3759	2.6800	0.4832	6.1706
64	1.3993	2.8392	0.8782	3.5971	114	0.3663	2.6774	0.4788	6.2221
65	1.3597	2.8334	0.8638	3.6496	115	0.3569	2.6747	0.4744	6.2736
66	1.3215	2.8278	0.8499	3.7021	116	0.3477	2.6721	0.4701	6.3251
67	1.2845	2.8222	0.8364	3.7545	117	0.3388	2.6696	0.4658	6.3765
68	1.2488	2.8167	0.8233	3.8068	118	0.3301	2.6670	0.4616	6.4279
69	1.2142	2.8113	0.8106	3.8591	119	0.3217	2.6644	0.4575	6.4793
70	1.1807	2.8060	0.7982	3.9113	120	0.3134	2.6619	0.4535	6.5307
71	1.1482	2.8008	0.7862	3.9635	121	0.3054	2.6594	0.4496	6.5820
72	1.1168	2.7956	0.7746	4.0156	122	0.2976	2.6569	0.4457	6.6333
73	1.0863	2.7905	0.7633	4.0676	123	0.2900	2.6544	0.4418	6.6845
74	1.0568	2.7855	0.7523	4.1196	124	0.2826	2.6520	0.4381	6.7358
75	1.0282	2.7805	0.7416	4.1716	125	0.2754	2.6496	0.4344	6.7870
76	1.0004	2.7756	0.7312	4.2235	126	0.2683	2.6472	0.4307	6.8382
77	0.9735	2.7737	0.7215	4.2776	127	0.2615	2.6448	0.4271	6.8893
78	0.9474	2.7717	0.7120	4.3316	128	0.2548	2.6424	0.4236	6.9405
79	0.9220	2.7696	0.7027	4.3855	129	0.2483	2.6401	0.4201	6.9916

TABLE 12b (continued)

Zo [Ohms]	W/h	Keff	pF/cm	nH/mm	Zo [Ohms]	W/h	Keff	pF/cm	nH/mm
80	0.8974	2.7675	0.6936	4.4393	130	0.2420	2.6378	0.4167	7.0427
81	0.8735	2.7653	0.6848	4.4929	131	0.2358	2.6355	0.4134	7.0938
82	0.8503	2.7630	0.6762	4.5465	132	0.2298	2.6332	0.4101	7.1449
83	0.8277	2.7606	0.6677	4.6000	133	0.2239	2.6309	0.4068	7.1959
84	0.8058	2.7582	0.6595	4.6534	134	0.2182	2.6287	0.4036	7.2469
85	0.7845	2.7557	0.6514	4.7067	135	0.2126	2.6265	0.4004	7.2980
86	0.7639	2.7532	0.6436	4.7599	136	0.2072	2.6243	0.3973	7.3490
87	0.7438	2.7506	0.6359	4.8130	137	0.2019	2.6221	0.3943	7.3999
88	0.7242	2.7481	0.6284	4.8660	138	0.1968	2.6200	0.3912	7.4509
89	0.7052	2.7454	0.6210	4.9190	139	0.1917	2.6179	0.3883	7.5019
90	0.6868	2.7428	0.6138	4.9718	140	0.1869	2.6158	0.3853	7.5528
91	0.6688	2.7401	0.6068	5.0246	141	0.1821	2.6137	0.3825	7.6037
92	0.6513	2.7374	0.5999	5.0773	142	0.1775	2.6116	0.3796	7.6546
93	0.6344	2.7347	0.5931	5.1300	143	0.1729	2.6096	0.3768	7.7055
94	0.6178	2.7320	0.5865	5.1826	144	0.1685	2.6076	0.3741	7.7564
95	0.6018	2.7292	0.5801	5.2351	145	0.1642	2.6056	0.3713	7.8073
96	0.5861	2.7265	0.5737	5.2875	146	0.1600	2.6036	0.3687	7.8582
97	0.5709	2.7237	0.5675	5.3399	147	0.1560	2.6017	0.3660	7.9091
98	0.5561	2.7209	0.5615	5.3922	148	0.1520	2.5998	0.3634	7.9599
99	0.5417	2.7182	0.5555	5.4444	149	0.1481	2.5979	0.3608	8.0108
100	0.5277	2.7154	0.5497	5.4966	150	0.1444	2.5960	0.3583	8.0616

Notes: Calculation of W/H has an error of less than 1%.

Source: Gupta, K.C., Garg, R., Bahl, I., Bhartia, P., *Microstrip Lines and Slotlines,* 2nd Ed., Artech House, Norwood, MA, 1996, 103.

TABLE 12c　Zero Thickness Microstrip Dimensions, Effective Dielectric Constant, and PUL Capacitance and Inductance for Er =5.75

Zo [Ohms]	W/h	Keff	pF/cm	nH/mm	Zo [Ohms]	W/h	Keff	pF/cm	nH/mm
1	154.1545	5.6626	79.3758	0.0794	55	1.3098	4.1200	1.2310	3.7239
2	75.8057	5.5818	39.4035	0.1576	56	1.2664	4.1088	1.2074	3.7864
3	49.7547	5.5068	26.0920	0.2348	57	1.2248	4.0978	1.1846	3.8488
4	36.7611	5.4372	19.4449	0.3111	58	1.1847	4.0869	1.1627	3.9112
5	28.9839	5.3723	15.4628	0.3866	59	1.1461	4.0763	1.1415	3.9734
6	23.8117	5.3116	12.8127	0.4613	60	1.1089	4.0658	1.1210	4.0355
7	20.1263	5.2548	10.9235	0.5352	61	1.0731	4.0555	1.1012	4.0976
8	17.3689	5.2014	9.5094	0.6086	62	1.0386	4.0453	1.0821	4.1596
9	15.2296	5.1512	8.4118	0.6814	63	1.0054	4.0353	1.0636	4.2214
10	13.5223	5.1037	7.5357	0.7536	64	0.9733	4.0305	1.0464	4.2859
11	12.1289	5.0589	6.8204	0.8253	65	0.9423	4.0265	1.0297	4.3507
12	10.9706	5.0163	6.2257	0.8965	66	0.9124	4.0222	1.0136	4.4153
13	9.9929	4.9759	5.7236	0.9673	67	0.8836	4.0178	0.9979	4.4797
14	9.1570	4.9375	5.2942	1.0377	68	0.8557	4.0132	0.9827	4.5439
15	8.4343	4.9008	4.9229	1.1077	69	0.8288	4.0084	0.9679	4.6080
16	7.8035	4.8659	4.5987	1.1773	70	0.8028	4.0035	0.9535	4.6719
17	7.2484	4.8324	4.3133	1.2466	71	0.7777	3.9985	0.9394	4.7357
18	6.7562	4.8004	4.0602	1.3155	72	0.7534	3.9933	0.9258	4.7993
19	6.3169	4.7697	3.8342	1.3841	73	0.7299	3.9881	0.9125	4.8628
20	5.9225	4.7403	3.6312	1.4525	74	0.7072	3.9828	0.8996	4.9261
21	5.5666	4.7119	3.4479	1.5205	75	0.6853	3.9774	0.8870	4.9893
22	5.2438	4.6847	3.2817	1.5883	76	0.6641	3.9720	0.8747	5.0524
23	4.9499	4.6584	3.1302	1.6559	77	0.6435	3.9665	0.8628	5.1153
24	4.6812	4.6331	2.9916	1.7232	78	0.6236	3.9609	0.8511	5.1781
25	4.4346	4.6087	2.8644	1.7902	79	0.6044	3.9554	0.8397	5.2408

TABLE 12c (continued)

Zo [Ohms]	W/h	Keff	pF/cm	nH/mm	Zo [Ohms]	W/h	Keff	pF/cm	nH/mm
26	4.2075	4.5851	2.7471	1.8571	80	0.5858	3.9498	0.8287	5.3034
27	3.9979	4.5623	2.6388	1.9237	81	0.5678	3.9442	0.8178	5.3659
28	3.8037	4.5402	2.5384	1.9901	82	0.5503	3.9386	0.8073	5.4283
29	3.6233	4.5187	2.4451	2.0563	83	0.5334	3.9329	0.7970	5.4905
30	3.4554	4.4980	2.3581	2.1223	84	0.5171	3.9273	0.7870	5.5527
31	3.2988	4.4778	2.2769	2.1881	85	0.5012	3.9217	0.7771	5.6148
32	3.1523	4.4583	2.2010	2.2538	86	0.4859	3.9161	0.7676	5.6768
33	3.0151	4.4393	2.1297	2.3193	87	0.4710	3.9105	0.7582	5.7387
34	2.8863	4.4208	2.0628	2.3846	88	0.4566	3.9049	0.7490	5.8005
35	2.7652	4.4028	1.9997	2.4497	89	0.4427	3.8994	0.7401	5.8623
36	2.6511	4.3853	1.9403	2.5147	90	0.4292	3.8938	0.7314	5.9239
37	2.5434	4.3682	1.8842	2.5795	91	0.4161	3.8883	0.7228	5.9855
38	2.4417	4.3516	1.8311	2.6441	92	0.4034	3.8829	0.7144	6.0470
39	2.3455	4.3353	1.7808	2.7087	93	0.3912	3.8774	0.7063	6.1085
40	2.2543	4.3195	1.7331	2.7730	94	0.3793	3.8720	0.6983	6.1699
41	2.1678	4.3040	1.6878	2.8373	95	0.3677	3.8667	0.6904	6.2312
42	2.0856	4.2889	1.6448	2.9014	96	0.3565	3.8614	0.6828	6.2925
43	2.0075	4.2741	1.6037	2.9653	97	0.3457	3.8561	0.6753	6.3537
44	1.9289	4.2588	1.5645	3.0288	98	0.3352	3.8508	0.6679	6.4148
45	1.8591	4.2449	1.5272	3.0926	99	0.3250	3.8456	0.6607	6.4759
46	1.7926	4.2312	1.4916	3.1562	100	0.3152	3.8405	0.6537	6.5369
47	1.7291	4.2178	1.4576	3.2198	101	0.3056	3.8354	0.6468	6.5979
48	1.6684	4.2048	1.4250	3.2832	102	0.2963	3.8304	0.6400	6.6588
49	1.6104	4.1919	1.3938	3.3464	103	0.2874	3.8254	0.6334	6.7197
50	1.5549	4.1794	1.3638	3.4096	104	0.2787	3.8204	0.6269	6.7806
51	1.5017	4.1671	1.3351	3.4727	105	0.2702	3.8155	0.6205	6.8414
52	1.4507	4.1550	1.3076	3.5356	106	0.2620	3.8107	0.6143	6.9022
53	1.4018	4.1431	1.2811	3.5985	107	0.2541	3.8059	0.6082	6.9629
54	1.3549	4.1315	1.2556	3.6612	108	0.2464	3.8011	0.6022	7.0236
109	0.2389	3.7964	0.5963	7.0842	130	0.1255	3.7101	0.4942	8.3525
110	0.2317	3.7918	0.5905	7.1449	131	0.1217	3.7065	0.4902	8.4127
111	0.2247	3.7872	0.5848	7.2055	132	0.1180	3.7030	0.4863	8.4729
112	0.2179	3.7827	0.5792	7.2660	133	0.1144	3.6996	0.4824	8.5331
113	0.2113	3.7782	0.5738	7.3265	134	0.1110	3.6962	0.4786	8.5933
114	0.2049	3.7738	0.5684	7.3871	135	0.1076	3.6928	0.4748	8.6535
115	0.1987	3.7694	0.5631	7.4475	136	0.1044	3.6895	0.4711	8.7137
116	0.1927	3.7651	0.5580	7.5080	137	0.1012	3.6863	0.4675	8.7739
117	0.1869	3.7608	0.5529	7.5684	138	0.0982	3.6830	0.4639	8.8341
118	0.1813	3.7566	0.5479	7.6288	139	0.0952	3.6799	0.4603	8.8943
119	0.1758	3.7524	0.5430	7.6892	140	0.0923	3.6767	0.4569	8.9544
120	0.1705	3.7483	0.5382	7.7496	141	0.0896	3.6737	0.4534	9.0146
121	0.1653	3.7443	0.5334	7.8099	142	0.0869	3.6706	0.4500	9.0748
122	0.1603	3.7403	0.5288	7.8703	143	0.0842	3.6676	0.4467	9.1350
123	0.1555	3.7363	0.5242	7.9306	144	0.0817	3.6647	0.4434	9.1952
124	0.1508	3.7324	0.5197	7.9909	145	0.0792	3.6618	0.4402	9.2553
125	0.1462	3.7286	0.5153	8.0512	146	0.0768	3.6589	0.4370	9.3155
126	0.1418	3.7248	0.5109	8.1115	147	0.0745	3.6561	0.4339	9.3757
127	0.1375	3.7210	0.5066	8.1717	148	0.0723	3.6533	0.4308	9.4359
128	0.1334	3.7173	0.5024	8.2320	149	0.0701	3.6505	0.4277	9.4961
129	0.1294	3.7137	0.4983	8.2922	150	0.0680	3.6478	0.4247	9.5562

Notes: Calculation of W/H has an error of less than 1%.

Source: Gupta, K.C., Garg, R., Bahl, I., Bhartia, P., *Microstrip Lines and Slotlines,* 2nd Ed., Artech House, Norwood, MA, 1996, 103.

TABLE 12d Zero Thickness Microstrip Dimensions, Effective Dielectric Constant, and PUL Capacitance and Inductance for Er = 9.4

Zo [Ohms]	W/h	Keff	pF/cm	nH/mm	Zo [Ohms]	W/h	Keff	pF/cm	nH/mm
1	120.1221	9.2047	101.2010	0.1012	24	3.4066	7.1750	3.7229	2.1444
2	58.8859	9.0280	50.1125	0.2004	25	3.2161	7.1309	3.5630	2.2269
3	38.5353	8.8676	33.1101	0.2980	26	3.0409	7.0885	3.4157	2.3090
4	28.3901	8.7212	24.6268	0.3940	27	2.8791	7.0475	3.2797	2.3909
5	22.3208	8.5871	19.5493	0.4887	28	2.7293	7.0079	3.1537	2.4725
6	18.2864	8.4635	16.1735	0.5822	29	2.5902	6.9697	3.0366	2.5538
7	15.4132	8.3493	13.7691	0.6747	30	2.4609	6.9326	2.9276	2.6348
8	13.2647	8.2433	11.9712	0.7662	31	2.3403	6.8967	2.8258	2.7156
9	11.5985	8.1445	10.5771	0.8567	32	2.2275	6.8619	2.7306	2.7961
10	10.2695	8.0521	9.4653	0.9465	33	2.1220	6.8281	2.6413	2.8764
11	9.1854	7.9655	8.5584	1.0356	34	2.0229	6.7952	2.5574	2.9564
12	8.2846	7.8841	7.8050	1.1239	35	1.9221	6.7606	2.4780	3.0356
13	7.5248	7.8074	7.1695	1.2116	36	1.8364	6.7301	2.4037	3.1152
14	6.8754	7.7348	6.6264	1.2988	37	1.7556	6.7004	2.3336	3.1947
15	6.3143	7.6661	6.1571	1.3853	38	1.6792	6.6715	2.2673	3.2740
16	5.8248	7.6009	5.7477	1.4714	39	1.6070	6.6434	2.2045	3.3530
17	5.3942	7.5389	5.3875	1.5570	40	1.5386	6.6159	2.1449	3.4319
18	5.0126	7.4798	5.0682	1.6421	41	1.4738	6.5891	2.0884	3.5105
19	4.6722	7.4234	4.7833	1.7268	42	1.4122	6.5629	2.0346	3.5890
20	4.3668	7.3694	4.5276	1.8110	43	1.3537	6.5373	1.9834	3.6673
21	4.0913	7.3178	4.2968	1.8949	44	1.2981	6.5122	1.9346	3.7454
22	3.8416	7.2683	4.0876	1.9784	45	1.2451	6.4877	1.8880	3.8233
23	3.6143	7.2207	3.8971	2.0616	46	1.1946	6.4638	1.8436	3.9010
47	1.1465	6.4403	1.8011	3.9786	99	0.1530	5.8352	0.8139	7.9771
48	1.1005	6.4173	1.7604	4.0560	100	0.1472	5.8268	0.8052	8.0518
49	1.0567	6.3948	1.7215	4.1332	101	0.1417	5.8184	0.7966	8.1265
50	1.0148	6.3728	1.6841	4.2103	102	0.1365	5.8102	0.7883	8.2012
51	0.9747	6.3596	1.6494	4.2901	103	0.1314	5.8021	0.7801	8.2758
52	0.9364	6.3507	1.6165	4.3711	104	0.1265	5.7942	0.7720	8.3504
53	0.8997	6.3412	1.5849	4.4519	105	0.1217	5.7864	0.7642	8.4251
54	0.8646	6.3312	1.5543	4.5323	106	0.1172	5.7787	0.7565	8.4997
55	0.8309	6.3208	1.5248	4.6124	107	0.1128	5.7712	0.7489	8.5742
56	0.7987	6.3100	1.4963	4.6923	108	0.1086	5.7638	0.7415	8.6488
57	0.7678	6.2989	1.4687	4.7719	109	0.1046	5.7565	0.7342	8.7234
58	0.7382	6.2875	1.4421	4.8512	110	0.1006	5.7493	0.7271	8.7979
59	0.7098	6.2759	1.4163	4.9303	111	0.0969	5.7423	0.7201	8.8725
60	0.6826	6.2641	1.3914	5.0091	112	0.0933	5.7354	0.7133	8.9470
61	0.6564	6.2521	1.3673	5.0877	113	0.0898	5.7286	0.7065	9.0216
62	0.6313	6.2400	1.3439	5.1661	114	0.0864	5.7219	0.6999	9.0961
63	0.6072	6.2278	1.3213	5.2443	115	0.0832	5.7154	0.6934	9.1707
64	0.5841	6.2156	1.2994	5.3223	116	0.0801	5.7090	0.6871	9.2452
65	0.5619	6.2032	1.2781	5.4001	117	0.0771	5.7026	0.6808	9.3197
66	0.5406	6.1909	1.2575	5.4777	118	0.0743	5.6964	0.6747	9.3943
67	0.5201	6.1786	1.2375	5.5552	119	0.0715	5.6903	0.6687	9.4688
68	0.5004	6.1662	1.2181	5.6325	120	0.0688	5.6843	0.6627	9.5434
69	0.4814	6.1539	1.1992	5.7096	121	0.0662	5.6785	0.6569	9.6179
70	0.4632	6.1417	1.1809	5.7866	122	0.0638	5.6727	0.6512	9.6925
71	0.4457	6.1295	1.1631	5.8634	123	0.0614	5.6670	0.6456	9.7670
72	0.4289	6.1173	1.1458	5.9401	124	0.0591	5.6615	0.6401	9.8416
73	0.4127	6.1053	1.1290	6.0166	125	0.0569	5.6560	0.6346	9.9162
74	0.3972	6.0933	1.1127	6.0931	126	0.0548	5.6506	0.6293	9.9907
75	0.3822	6.0814	1.0968	6.1694	127	0.0527	5.6454	0.6241	10.0653
76	0.3679	6.0696	1.0813	6.2456	128	0.0508	5.6402	0.6189	10.1399
77	0.3540	6.0579	1.0662	6.3217	129	0.0489	5.6351	0.6138	10.2146
78	0.3407	6.0464	1.0516	6.3977	130	0.0471	5.6301	0.6088	10.2892
79	0.3279	6.0350	1.0373	6.4736	131	0.0453	5.6252	0.6039	10.3638

TABLE 12d (continued)

Zo [Ohms]	W/h	Keff	pF/cm	nH/mm	Zo [Ohms]	W/h	Keff	pF/cm	nH/mm
80	0.3156	6.0236	1.0233	6.5494	132	0.0436	5.6204	0.5991	10.4385
81	0.3038	6.0125	1.0098	6.6251	133	0.0420	5.6157	0.5943	10.5131
82	0.2924	6.0014	0.9965	6.7007	134	0.0404	5.6111	0.5897	10.5878
83	0.2815	5.9905	0.9836	6.7762	135	0.0389	5.6065	0.5850	10.6625
84	0.2709	5.9797	0.9710	6.8517	136	0.0375	5.6021	0.5805	10.7372
85	0.2608	5.9691	0.9588	6.9271	137	0.0361	5.5977	0.5761	10.8119
86	0.2510	5.9586	0.9468	7.0024	138	0.0347	5.5934	0.5717	10.8867
87	0.2416	5.9482	0.9351	7.0777	139	0.0334	5.5892	0.5673	10.9614
88	0.2326	5.9380	0.9237	7.1529	140	0.0322	5.5850	0.5631	11.0362
89	0.2239	5.9280	0.9125	7.2281	141	0.0310	5.5809	0.5589	11.1110
90	0.2155	5.9180	0.9016	7.3032	142	0.0298	5.5770	0.5547	11.1858
91	0.2074	5.9083	0.8910	7.3782	143	0.0287	5.5730	0.5507	11.2606
92	0.1997	5.8986	0.8806	7.4532	144	0.0276	5.5692	0.5467	11.3354
93	0.1922	5.8891	0.8704	7.5281	145	0.0266	5.5654	0.5427	11.4103
94	0.1850	5.8798	0.8605	7.6031	146	0.0256	5.5617	0.5388	11.4852
95	0.1781	5.8706	0.8507	7.6779	147	0.0247	5.5581	0.5350	11.5600
96	0.1715	5.8616	0.8412	7.7528	148	0.0237	5.5545	0.5312	11.6350
97	0.1651	5.8526	0.8319	7.8276	149	0.0229	5.5510	0.5274	11.7099
98	0.1589	5.8439	0.8228	7.9023	150	0.0220	5.5476	0.5238	11.7848

Notes: Calculation of W/H has an error of less than 1%.

TABLE 12e   Zero Thickness Microstrip Dimensions, Effective Dielectric Constant, and PUL Capacitance and Inductance for Er = 9.8

Zo [Ohms]	W/h	Keff	pF/cm	nH/mm	Zo [Ohms]	W/h	Keff	pF/cm	nH/mm
1	117.6023	9.5914	103.3045	0.1033	57	0.7347	6.5379	1.4963	4.8615
2	57.6329	9.4030	51.1424	0.2046	58	0.7059	6.5254	1.4691	4.9421
3	37.7044	9.2322	33.7840	0.3041	59	0.6783	6.5128	1.4428	5.0224
4	27.7700	9.0767	25.1237	0.4020	60	0.6519	6.5000	1.4174	5.1025
5	21.8272	8.9344	19.9408	0.4985	61	0.6265	6.4870	1.3927	5.1824
6	17.8771	8.8035	16.4952	0.5938	62	0.6022	6.4740	1.3689	5.2621
7	15.0641	8.6827	14.0413	0.6880	63	0.5788	6.4609	1.3458	5.3415
8	12.9606	8.5706	12.2066	0.7812	64	0.5564	6.4477	1.3234	5.4208
9	11.3295	8.4662	10.7841	0.8735	65	0.5349	6.4346	1.3017	5.4999
10	10.0285	8.3688	9.6496	0.9650	66	0.5142	6.4214	1.2807	5.5788
11	8.9673	8.2775	8.7244	1.0557	67	0.4944	6.4082	1.2603	5.6575
12	8.0857	8.1917	7.9558	1.1456	68	0.4753	6.3951	1.2405	5.7360
13	7.3419	8.1109	7.3075	1.2350	69	0.4570	6.3820	1.2213	5.8144
14	6.7063	8.0345	6.7535	1.3237	70	0.4395	6.3690	1.2026	5.8927
15	6.1572	7.9622	6.2749	1.4118	71	0.4226	6.3561	1.1844	5.9708
16	5.6782	7.8937	5.8573	1.4995	72	0.4064	6.3432	1.1668	6.0488
17	5.2568	7.8285	5.4900	1.5866	73	0.3908	6.3305	1.1497	6.1266
18	4.8834	7.7664	5.1644	1.6733	74	0.3758	6.3179	1.1330	6.2043
19	4.5503	7.7071	4.8738	1.7595	75	0.3614	6.3053	1.1168	6.2820
20	4.2515	7.6505	4.6131	1.8452	76	0.3476	6.2929	1.1010	6.3595
21	3.9820	7.5963	4.3778	1.9306	77	0.3343	6.2807	1.0857	6.4368
22	3.7377	7.5443	4.1645	2.0156	78	0.3215	6.2685	1.0707	6.5141
23	3.5154	7.4944	3.9703	2.1003	79	0.3092	6.2565	1.0561	6.5913
24	3.3122	7.4464	3.7926	2.1846	80	0.2974	6.2447	1.0419	6.6684
25	3.1259	7.4002	3.6296	2.2685	81	0.2860	6.2329	1.0281	6.7455

TABLE 12e (continued)

Zo [Ohms]	W/h	Keff	pF/cm	nH/mm	Zo [Ohms]	W/h	Keff	pF/cm	nH/mm
26	2.9545	7.3557	3.4795	2.3522	82	0.2751	6.2214	1.0146	6.8224
27	2.7962	7.3128	3.3408	2.4355	83	0.2646	6.2100	1.0015	6.8993
28	2.6497	7.2713	3.2124	2.5185	84	0.2545	6.1987	0.9887	6.9760
29	2.5138	7.2312	3.0930	2.6012	85	0.2448	6.1876	0.9762	7.0528
30	2.3873	7.1923	2.9819	2.6837	86	0.2355	6.1766	0.9640	7.1294
31	2.2693	7.1547	2.8781	2.7659	87	0.2265	6.1658	0.9520	7.2060
32	2.1591	7.1182	2.7811	2.8478	88	0.2179	6.1552	0.9404	7.2825
33	2.0558	7.0827	2.6901	2.9295	89	0.2096	6.1447	0.9291	7.3590
34	1.9590	7.0483	2.6046	3.0109	90	0.2016	6.1344	0.9180	7.4355
35	1.8614	7.0124	2.5237	3.0916	91	0.1939	6.1242	0.9071	7.5118
36	1.7777	6.9805	2.4480	3.1727	92	0.1865	6.1142	0.8965	7.5882
37	1.6987	6.9494	2.3766	3.2535	93	0.1794	6.1044	0.8862	7.6645
38	1.6241	6.9191	2.3090	3.3342	94	0.1726	6.0947	0.8760	7.7408
39	1.5535	6.8896	2.2450	3.4146	95	0.1660	6.0852	0.8661	7.8170
40	1.4867	6.8609	2.1843	3.4948	96	0.1597	6.0758	0.8565	7.8932
41	1.4233	6.8328	2.1266	3.5749	97	0.1536	6.0666	0.8470	7.9693
42	1.3632	6.8053	2.0718	3.6547	98	0.1478	6.0575	0.8377	8.0455
43	1.3060	6.7785	2.0197	3.7343	99	0.1422	6.0486	0.8286	8.1216
44	1.2517	6.7523	1.9699	3.8138	100	0.1368	6.0398	0.8198	8.1977
45	1.1999	6.7266	1.9225	3.8931	101	0.1316	6.0312	0.8111	8.2737
46	1.1507	6.7015	1.8772	3.9721	102	0.1266	6.0227	0.8026	8.3498
47	1.1037	6.6770	1.8339	4.0510	103	0.1217	6.0143	0.7942	8.4258
48	1.0588	6.6529	1.7924	4.1298	104	0.1171	6.0062	0.7860	8.5018
49	1.0161	6.6293	1.7527	4.2083	105	0.1127	5.9981	0.7780	8.5778
50	0.9752	6.6149	1.7158	4.2895	106	0.1084	5.9902	0.7702	8.6538
51	0.9361	6.6054	1.6810	4.3722	107	0.1043	5.9824	0.7625	8.7298
52	0.8988	6.5953	1.6474	4.4545	108	0.1003	5.9748	0.7549	8.8057
53	0.8630	6.5846	1.6150	4.5365	109	0.0965	5.9673	0.7476	8.8817
54	0.8289	6.5735	1.5837	4.6182	110	0.0928	5.9600	0.7403	8.9576
55	0.7961	6.5619	1.5536	4.6996	111	0.0893	5.9527	0.7332	9.0336
56	0.7648	6.5501	1.5245	4.7807	112	0.0859	5.9456	0.7262	9.1095
113	0.0826	5.9387	0.7194	9.1855	132	0.0396	5.8280	0.6101	10.6295
114	0.0795	5.9318	0.7126	9.2614	133	0.0381	5.8232	0.6052	10.7056
115	0.0765	5.9251	0.7060	9.3374	134	0.0366	5.8185	0.6005	10.7818
116	0.0736	5.9185	0.6996	9.4133	135	0.0352	5.8139	0.5958	10.8579
117	0.0708	5.9120	0.6932	9.4893	136	0.0339	5.8093	0.5912	10.9341
118	0.0681	5.9057	0.6870	9.5652	137	0.0326	5.8049	0.5866	11.0102
119	0.0655	5.8994	0.6808	9.6412	138	0.0314	5.8005	0.5821	11.0864
120	0.0630	5.8933	0.6748	9.7172	139	0.0302	5.7962	0.5777	11.1626
121	0.0606	5.8873	0.6689	9.7931	140	0.0290	5.7920	0.5734	11.2389
122	0.0583	5.8814	0.6631	9.8691	141	0.0279	5.7879	0.5691	11.3151
123	0.0561	5.8756	0.6574	9.9451	142	0.0269	5.7838	0.5649	11.3914
124	0.0540	5.8699	0.6517	10.0211	143	0.0258	5.7799	0.5608	11.4677
125	0.0519	5.8643	0.6462	10.0971	144	0.0249	5.7760	0.5567	11.5440
126	0.0499	5.8588	0.6408	10.1731	145	0.0239	5.7722	0.5527	11.6203
127	0.0480	5.8534	0.6354	10.2492	146	0.0230	5.7684	0.5487	11.6966
128	0.0462	5.8482	0.6302	10.3252	147	0.0221	5.7647	0.5448	11.7730
129	0.0445	5.8430	0.6250	10.4013	148	0.0213	5.7611	0.5410	11.8494
130	0.0428	5.8379	0.6200	10.4773	149	0.0205	5.7576	0.5372	11.9258
131	0.0411	5.8329	0.6150	10.5534	150	0.0197	5.7541	0.5334	12.0022

Notes: Calculation of W/H has an error of less than 1%.

Source: Gupta, K.C., Garg, R., Bahl, I., Bhartia, P., *Microstrip Lines and Slotlines,* 2nd Ed., Artech House, Norwood, MA, 1996, 103.

TABLE 12f Zero Thickness Microstrip Dimensions, Effective Dielectric Constant, and PUL Capacitance and Inductance for Er = 11.6

Zo [Ohms]	W/h	Keff	pF/cm	nH/mm	Zo [Ohms]	W/h	Keff	pF/cm	nH/mm
1	107.9253	11.3278	112.2672	0.1123	26	2.6229	8.5446	3.7502	2.5351
2	52.8209	11.0843	55.5270	0.2221	27	2.4782	8.4928	3.6003	2.6246
3	34.5132	10.8654	36.6506	0.3299	28	2.3444	8.4426	3.4615	2.7138
4	25.3888	10.6674	27.2364	0.4358	29	2.2202	8.3942	3.3325	2.8026
5	19.9316	10.4873	21.6044	0.5401	30	2.1047	8.3473	3.2124	2.8912
6	16.3052	10.3226	17.8617	0.6430	31	1.9970	8.3019	3.1003	2.9794
7	13.7232	10.1712	15.1973	0.7447	32	1.8884	8.2543	2.9948	3.0667
8	11.7929	10.0313	13.2059	0.8452	33	1.7964	8.2125	2.8967	3.1545
9	10.2964	9.9016	11.6625	0.9447	34	1.7101	8.1718	2.8045	3.2420
10	9.1031	9.7809	10.4321	1.0432	35	1.6291	8.1324	2.7178	3.3293
11	8.1299	9.6682	9.4289	1.1409	36	1.5527	8.0939	2.6361	3.4163
12	7.3215	9.5625	8.5958	1.2378	37	1.4807	8.0565	2.5589	3.5031
13	6.6397	9.4632	7.8932	1.3340	38	1.4128	8.0201	2.4859	3.5897
14	6.0573	9.3696	7.2931	1.4294	39	1.3485	7.9846	2.4168	3.6759
15	5.5541	9.2812	6.7747	1.5243	40	1.2877	7.9499	2.3513	3.7620
16	5.1153	9.1975	6.3226	1.6186	41	1.2301	7.9161	2.2890	3.8478
17	4.7294	9.1180	5.9249	1.7123	42	1.1754	7.8830	2.2299	3.9335
18	4.3875	9.0424	5.5725	1.8055	43	1.1235	7.8507	2.1735	4.0189
19	4.0826	8.9703	5.2581	1.8982	44	1.0742	7.8192	2.1199	4.1041
20	3.8091	8.9016	4.9760	1.9904	45	1.0273	7.7883	2.0687	4.1890
21	3.5625	8.8358	4.7215	2.0822	46	0.9827	7.7654	2.0207	4.2758
22	3.3390	8.7728	4.4908	2.1736	47	0.9402	7.7532	1.9762	4.3653
23	3.1357	8.7124	4.2807	2.2645	48	0.8998	7.7401	1.9334	4.4545
24	2.9499	8.6543	4.0887	2.3551	49	0.8612	7.7262	1.8922	4.5432
25	2.7796	8.5984	3.9125	2.4453	50	0.8244	7.7117	1.8526	4.6315
51	0.7893	7.6966	1.8145	4.7195	101	0.0957	6.9814	0.8726	8.9017
52	0.7558	7.6810	1.7778	4.8072	102	0.0917	6.9719	0.8635	8.9837
53	0.7238	7.6650	1.7425	4.8945	103	0.0880	6.9625	0.8545	9.0657
54	0.6933	7.6487	1.7084	4.9816	104	0.0844	6.9534	0.8458	9.1477
55	0.6641	7.6322	1.6755	5.0683	105	0.0809	6.9444	0.8372	9.2297
56	0.6362	7.6154	1.6438	5.1548	106	0.0776	6.9356	0.8287	9.3116
57	0.6095	7.5985	1.6131	5.2410	107	0.0744	6.9269	0.8205	9.3936
58	0.5840	7.5815	1.5835	5.3270	108	0.0714	6.9185	0.8124	9.4756
59	0.5596	7.5643	1.5549	5.4127	109	0.0684	6.9102	0.8044	9.5576
60	0.5363	7.5472	1.5273	5.4982	110	0.0656	6.9020	0.7967	9.6396
61	0.5139	7.5301	1.5005	5.5835	111	0.0630	6.8940	0.7890	9.7216
62	0.4925	7.5129	1.4747	5.6686	112	0.0604	6.8862	0.7815	9.8036
63	0.4721	7.4959	1.4496	5.7535	113	0.0579	6.8786	0.7742	9.8857
64	0.4525	7.4789	1.4253	5.8382	114	0.0555	6.8710	0.7670	9.9677
65	0.4337	7.4620	1.4018	5.9227	115	0.0533	6.8637	0.7599	10.0498
66	0.4158	7.4452	1.3790	6.0071	116	0.0511	6.8565	0.7530	10.1318
67	0.3986	7.4286	1.3569	6.0913	117	0.0490	6.8494	0.7461	10.2139
68	0.3821	7.4121	1.3355	6.1753	118	0.0470	6.8425	0.7394	10.2960
69	0.3663	7.3957	1.3147	6.2592	119	0.0450	6.8357	0.7329	10.3781
70	0.3512	7.3795	1.2945	6.3430	120	0.0432	6.8290	0.7264	10.4602
71	0.3367	7.3635	1.2749	6.4266	121	0.0414	6.8225	0.7201	10.5423
72	0.3228	7.3477	1.2558	6.5101	122	0.0397	6.8161	0.7138	10.6245
73	0.3095	7.3321	1.2373	6.5935	123	0.0381	6.8099	0.7077	10.7067
74	0.2967	7.3166	1.2193	6.6768	124	0.0365	6.8037	0.7017	10.7888
75	0.2845	7.3014	1.2018	6.7600	125	0.0350	6.7977	0.6957	10.8711
76	0.2728	7.2864	1.1847	6.8430	126	0.0336	6.7919	0.6899	10.9533
77	0.2616	7.2716	1.1682	6.9260	127	0.0322	6.7861	0.6842	11.0355
78	0.2508	7.2570	1.1520	7.0089	128	0.0309	6.7804	0.6786	11.1178
79	0.2405	7.2426	1.1363	7.0918	129	0.0296	6.7749	0.6730	11.2001
80	0.2306	7.2285	1.1210	7.1745	130	0.0284	6.7695	0.6676	11.2824

TABLE 12f (continued)

Zo [Ohms]	W/h	Keff	pF/cm	nH/mm	Zo [Ohms]	W/h	Keff	pF/cm	nH/mm
81	0.2211	7.2146	1.1061	7.2572	131	0.0273	6.7642	0.6622	11.3647
82	0.2121	7.2009	1.0916	7.3398	132	0.0261	6.7590	0.6570	11.4471
83	0.2033	7.1874	1.0774	7.4224	133	0.0251	6.7539	0.6518	11.5294
84	0.1950	7.1741	1.0636	7.5049	134	0.0240	6.7489	0.6467	11.6118
85	0.1870	7.1611	1.0501	7.5873	135	0.0231	6.7440	0.6417	11.6943
86	0.1793	7.1483	1.0370	7.6697	136	0.0221	6.7392	0.6367	11.7767
87	0.1719	7.1357	1.0242	7.7520	137	0.0212	6.7345	0.6318	11.8592
88	0.1649	7.1233	1.0117	7.8343	138	0.0203	6.7300	0.6271	11.9417
89	0.1581	7.1111	0.9994	7.9166	139	0.0195	6.7255	0.6223	12.0242
90	0.1516	7.0992	0.9875	7.9988	140	0.0187	6.7210	0.6177	12.1067
91	0.1454	7.0874	0.9758	8.0810	141	0.0179	6.7167	0.6131	12.1893
92	0.1394	7.0759	0.9645	8.1632	142	0.0172	6.7125	0.6086	12.2718
93	0.1337	7.0646	0.9533	8.2453	143	0.0165	6.7084	0.6042	12.3545
94	0.1282	7.0535	0.9424	8.3274	144	0.0158	6.7043	0.5998	12.4371
95	0.1230	7.0426	0.9318	8.4095	145	0.0152	6.7003	0.5955	12.5197
96	0.1179	7.0319	0.9214	8.4916	146	0.0146	6.6964	0.5912	12.6024
97	0.1131	7.0214	0.9112	8.5736	147	0.0140	6.6926	0.5870	12.6851
98	0.1085	7.0111	0.9013	8.6556	148	0.0134	6.6889	0.5829	12.7679
99	0.1040	7.0010	0.8915	8.7377	149	0.0128	6.6852	0.5788	12.8506
100	0.0998	6.9911	0.8820	8.8197	150	0.0123	6.6817	0.5748	12.9334

Notes: Calculation of W/H has an error of less than 1%.

Source: Gupta, K.C., Garg, R., Bahl, I., Bhartia, P., *Microstrip Lines and Slotlines,* 2nd Ed., Artech House, Norwood, MA, 1996, 103.

TABLE 12g   Zero Thickness Microstrip Dimensions, Effective Dielectric Constant, and PUL Capacitance and Inductance for Er =11.9

Zo [Ohms]	W/h	Keff	pF/cm	nH/mm	Zo [Ohms]	W/h	Keff	pF/cm	nH/mm
1	106.5298	11.6168	113.6899	0.1137	55	0.6453	7.8084	1.6947	5.1265
2	52.1270	11.3637	56.2223	0.2249	56	0.6179	7.7908	1.6626	5.2139
3	34.0530	11.1365	37.1049	0.3339	57	0.5918	7.7731	1.6316	5.3009
4	25.0454	10.9312	27.5710	0.4411	58	0.5667	7.7554	1.6016	5.3878
5	19.6583	10.7446	21.8678	0.5467	59	0.5428	7.7375	1.5726	5.4744
6	16.0785	10.5741	18.0780	0.6508	60	0.5199	7.7197	1.5446	5.5607
7	13.5298	10.4175	15.3802	0.7536	61	0.4980	7.7019	1.5176	5.6469
8	11.6245	10.2730	13.3640	0.8553	62	0.4771	7.6841	1.4914	5.7328
9	10.1474	10.1390	11.8014	0.9559	63	0.4571	7.6664	1.4660	5.8186
10	8.9696	10.0144	10.5558	1.0556	64	0.4379	7.6488	1.4414	5.9041
11	8.0091	9.8981	9.5403	1.1544	65	0.4195	7.6313	1.4176	5.9895
12	7.2113	9.7891	8.6970	1.2524	66	0.4020	7.6140	1.3946	6.0748
13	6.5385	9.6867	7.9859	1.3496	67	0.3852	7.5968	1.3722	6.1598
14	5.9637	9.5902	7.3784	1.4462	68	0.3691	7.5797	1.3505	6.2447
15	5.4672	9.4991	6.8538	1.5421	69	0.3536	7.5628	1.3295	6.3295
16	5.0342	9.4128	6.3961	1.6374	70	0.3389	7.5462	1.3090	6.4142
17	4.6534	9.3309	5.9937	1.7322	71	0.3247	7.5297	1.2892	6.4987
18	4.3160	9.2531	5.6370	1.8264	72	0.3112	7.5134	1.2699	6.5831
19	4.0152	9.1789	5.3189	1.9201	73	0.2982	7.4973	1.2511	6.6674
20	3.7454	9.1081	5.0334	2.0134	74	0.2858	7.4814	1.2329	6.7516
21	3.5020	9.0404	4.7759	2.1062	75	0.2739	7.4658	1.2152	6.8356
22	3.2816	8.9755	4.5424	2.1985	76	0.2625	7.4504	1.1980	6.9196
23	3.0810	8.9134	4.3298	2.2905	77	0.2516	7.4352	1.1812	7.0035
24	2.8977	8.8536	4.1355	2.3820	78	0.2411	7.4202	1.1649	7.0873
25	2.7297	8.7961	3.9572	2.4732	79	0.2311	7.4055	1.1490	7.1711

TABLE 12g (continued)

Zo [Ohms]	W/h	Keff	pF/cm	nH/mm	Zo [Ohms]	W/h	Keff	pF/cm	nH/mm
26	2.5751	8.7408	3.7930	2.5641	80	0.2215	7.3910	1.1335	7.2547
27	2.4324	8.6874	3.6413	2.6545	81	0.2123	7.3767	1.1185	7.3383
28	2.3004	8.6359	3.5009	2.7447	82	0.2035	7.3627	1.1038	7.4218
29	2.1779	8.5861	3.3704	2.8345	83	0.1950	7.3489	1.0895	7.5053
30	2.0640	8.5378	3.2489	2.9240	84	0.1869	7.3353	1.0755	7.5887
31	1.9578	8.4911	3.1355	3.0132	85	0.1791	7.3220	1.0619	7.6721
32	1.8513	8.4425	3.0288	3.1014	86	0.1717	7.3089	1.0486	7.7554
33	1.7607	8.3995	2.9295	3.1902	87	0.1646	7.2960	1.0356	7.8386
34	1.6756	8.3577	2.8362	3.2787	88	0.1577	7.2834	1.0230	7.9219
35	1.5956	8.3171	2.7485	3.3669	89	0.1512	7.2709	1.0106	8.0051
36	1.5204	8.2776	2.6658	3.4549	90	0.1449	7.2588	0.9985	8.0882
37	1.4494	8.2391	2.5877	3.5426	91	0.1389	7.2468	0.9868	8.1713
38	1.3824	8.2016	2.5139	3.6301	92	0.1331	7.2350	0.9752	8.2544
39	1.3190	8.1651	2.4440	3.7173	93	0.1276	7.2235	0.9640	8.3375
40	1.2591	8.1295	2.3777	3.8043	94	0.1223	7.2122	0.9530	8.4206
41	1.2023	8.0947	2.3147	3.8910	95	0.1172	7.2011	0.9422	8.5036
42	1.1484	8.0607	2.2548	3.9775	96	0.1124	7.1902	0.9317	8.5866
43	1.0973	8.0275	2.1979	4.0639	97	0.1077	7.1795	0.9214	8.6696
44	1.0487	7.9951	2.1436	4.1499	98	0.1033	7.1691	0.9113	8.7526
45	1.0026	7.9633	2.0918	4.2358	99	0.0990	7.1588	0.9015	8.8356
46	0.9586	7.9500	2.0446	4.3263	100	0.0949	7.1487	0.8919	8.9185
47	0.9168	7.9367	1.9994	4.4167	101	0.0909	7.1388	0.8824	9.0015
48	0.8770	7.9226	1.9560	4.5067	102	0.0872	7.1292	0.8732	9.0845
49	0.8391	7.9078	1.9143	4.5962	103	0.0836	7.1197	0.8641	9.1674
50	0.8029	7.8923	1.8742	4.6854	104	0.0801	7.1104	0.8552	9.2504
51	0.7684	7.8762	1.8356	4.7743	105	0.0768	7.1012	0.8466	9.3333
52	0.7354	7.8598	1.7984	4.8628	106	0.0736	7.0923	0.8380	9.4163
53	0.7040	7.8429	1.7626	4.9510	107	0.0705	7.0835	0.8297	9.4992
54	0.6740	7.8258	1.7280	5.0389	108	0.0676	7.0749	0.8215	9.5822
109	0.0648	7.0665	0.8135	9.6652	130	0.0266	6.9243	0.6752	11.4107
110	0.0621	7.0583	0.8056	9.7481	131	0.0255	6.9190	0.6698	11.4940
111	0.0595	7.0502	0.7979	9.8311	132	0.0245	6.9138	0.6645	11.5774
112	0.0571	7.0423	0.7903	9.9141	133	0.0235	6.9086	0.6592	11.6608
113	0.0547	7.0345	0.7829	9.9971	134	0.0225	6.9036	0.6541	11.7442
114	0.0524	7.0269	0.7756	10.0801	135	0.0216	6.8987	0.6490	11.8276
115	0.0503	7.0195	0.7685	10.1632	136	0.0207	6.8939	0.6440	11.9110
116	0.0482	7.0122	0.7615	10.2462	137	0.0198	6.8892	0.6391	11.9945
117	0.0462	7.0050	0.7546	10.3293	138	0.0190	6.8845	0.6342	12.0780
118	0.0443	6.9980	0.7478	10.4123	139	0.0182	6.8800	0.6294	12.1615
119	0.0424	6.9911	0.7412	10.4954	140	0.0174	6.8756	0.6247	12.2451
120	0.0407	6.9844	0.7346	10.5785	141	0.0167	6.8713	0.6201	12.3287
121	0.0390	6.9778	0.7282	10.6617	142	0.0160	6.8670	0.6156	12.4123
122	0.0374	6.9714	0.7219	10.7448	143	0.0154	6.8628	0.6111	12.4959
123	0.0358	6.9651	0.7157	10.8280	144	0.0147	6.8588	0.6067	12.5795
124	0.0343	6.9589	0.7096	10.9112	145	0.0141	6.8548	0.6023	12.6632
125	0.0329	6.9528	0.7036	10.9944	146	0.0135	6.8509	0.5980	12.7469
126	0.0316	6.9469	0.6978	11.0776	147	0.0130	6.8471	0.5938	12.8306
127	0.0302	6.9411	0.6920	11.1608	148	0.0124	6.8433	0.5896	12.9144
128	0.0290	6.9354	0.6863	11.2441	149	0.0119	6.8396	0.5855	12.9982
129	0.0278	6.9298	0.6807	11.3274	150	0.0114	6.8361	0.5814	13.0820

Notes: Calculation of W/H has an error of less than 1%.

Source: Gupta, K.C., Garg, R., Bahl, I., Bhartia, P., *Microstrip Lines and Slotlines,* 2nd Ed., Artech House, Norwood, MA, 1996, 103.

TABLE 12h　Zero Thickness Microstrip Dimensions, Effective Dielectric Constant, and PUL Capacitance and Inductance for Er =12.88

Zo [Ohms]	W/h	Keff	pF/cm	nH/mm	Zo [Ohms]	W/h	Keff	pF/cm	nH/mm
1	102.3159	12.5596	118.2135	0.1182	24	2.7402	9.5011	4.2841	2.4676
2	50.0315	12.2746	58.4323	0.2337	25	2.5791	9.4384	4.0991	2.5619
3	32.6633	12.0197	38.5483	0.3469	26	2.4309	9.3779	3.9288	2.6559
4	24.0084	11.7903	28.6339	0.4581	27	2.2942	9.3197	3.7715	2.7494
5	18.8329	11.5823	22.7043	0.5676	28	2.1677	9.2635	3.6258	2.8427
6	15.3941	11.3928	18.7648	0.6755	29	2.0503	9.2091	3.4905	2.9355
7	12.9460	11.2191	15.9610	0.7821	30	1.9316	9.1518	3.3637	3.0273
8	11.1161	11.0591	13.8660	0.8874	31	1.8324	9.1020	3.2463	3.1197
9	9.6977	10.9111	12.2425	0.9916	32	1.7397	9.0536	3.1365	3.2117
10	8.5668	10.7736	10.9487	1.0949	33	1.6528	9.0068	3.0335	3.3035
11	7.6446	10.6455	9.8939	1.1972	34	1.5714	8.9612	2.9369	3.3950
12	6.8787	10.5255	9.0182	1.2986	35	1.4948	8.9170	2.8459	3.4862
13	6.2329	10.4130	8.2799	1.3993	36	1.4228	8.8739	2.7602	3.5772
14	5.6812	10.3071	7.6492	1.4993	37	1.3549	8.8320	2.6792	3.6678
15	5.2048	10.2071	7.1046	1.5985	38	1.2908	8.7911	2.6027	3.7582
16	4.7893	10.1125	6.6296	1.6972	39	1.2302	8.7513	2.5302	3.8484
17	4.4239	10.0228	6.2119	1.7952	40	1.1729	8.7125	2.4614	3.9383
18	4.1003	9.9376	5.8418	1.8927	41	1.1186	8.6745	2.3962	4.0280
19	3.8117	9.8565	5.5117	1.9897	42	1.0672	8.6375	2.3341	4.1174
20	3.5529	9.7791	5.2155	2.0862	43	1.0184	8.6014	2.2751	4.2066
21	3.3196	9.7051	4.9483	2.1822	44	0.9721	8.5791	2.2205	4.2989
22	3.1082	9.6342	4.7061	2.2778	45	0.9281	8.5645	2.1693	4.3928
23	2.9159	9.5663	4.4856	2.3729	46	0.8863	8.5487	2.1202	4.4863
47	0.8465	8.5321	2.0731	4.5794	99	0.0845	7.6726	0.9333	9.1471
48	0.8086	8.5147	2.0278	4.6720	100	0.0808	7.6620	0.9233	9.2331
49	0.7726	8.4967	1.9843	4.7643	101	0.0774	7.6516	0.9136	9.3192
50	0.7383	8.4781	1.9425	4.8562	102	0.0740	7.6415	0.9040	9.4052
51	0.7056	8.4591	1.9023	4.9478	103	0.0709	7.6315	0.8946	9.4912
52	0.6744	8.4397	1.8635	5.0390	104	0.0678	7.6218	0.8855	9.5773
53	0.6447	8.4201	1.8263	5.1299	105	0.0649	7.6123	0.8765	9.6633
54	0.6163	8.4002	1.7903	5.2206	106	0.0621	7.6030	0.8677	9.7494
55	0.5893	8.3802	1.7557	5.3109	107	0.0595	7.5938	0.8591	9.8354
56	0.5634	8.3601	1.7223	5.4010	108	0.0569	7.5849	0.8506	9.9215
57	0.5388	8.3399	1.6900	5.4908	109	0.0545	7.5761	0.8423	10.0076
58	0.5152	8.3198	1.6588	5.5804	110	0.0521	7.5675	0.8342	10.0937
59	0.4928	8.2996	1.6288	5.6697	111	0.0499	7.5592	0.8262	10.1798
60	0.4713	8.2796	1.5997	5.7588	112	0.0477	7.5509	0.8184	10.2659
61	0.4508	8.2596	1.5716	5.8477	113	0.0457	7.5429	0.8107	10.3521
62	0.4312	8.2397	1.5443	5.9365	114	0.0437	7.5350	0.8032	10.4382
63	0.4125	8.2200	1.5180	6.0250	115	0.0418	7.5273	0.7958	10.5244
64	0.3946	8.2005	1.4925	6.1133	116	0.0400	7.5198	0.7885	10.6106
65	0.3775	8.1811	1.4678	6.2015	117	0.0383	7.5124	0.7814	10.6968
66	0.3611	8.1619	1.4439	6.2895	118	0.0367	7.5052	0.7744	10.7830
67	0.3455	8.1429	1.4207	6.3774	119	0.0351	7.4981	0.7676	10.8693
68	0.3306	8.1242	1.3982	6.4652	120	0.0336	7.4912	0.7608	10.9556
69	0.3163	8.1057	1.3763	6.5528	121	0.0321	7.4844	0.7542	11.0419
70	0.3026	8.0874	1.3551	6.6402	122	0.0308	7.4778	0.7477	11.1282
71	0.2895	8.0694	1.3346	6.7276	123	0.0294	7.4713	0.7413	11.2146
72	0.2770	8.0517	1.3146	6.8148	124	0.0282	7.4649	0.7350	11.3009
73	0.2651	8.0342	1.2952	6.9020	125	0.0270	7.4587	0.7288	11.3873
74	0.2536	8.0169	1.2763	6.9890	126	0.0258	7.4526	0.7227	11.4737
75	0.2427	8.0000	1.2579	7.0759	127	0.0247	7.4467	0.7167	11.5602
76	0.2323	7.9833	1.2401	7.1628	128	0.0236	7.4409	0.7109	11.6466
77	0.2222	7.9669	1.2227	7.2496	129	0.0226	7.4352	0.7051	11.7331
78	0.2127	7.9507	1.2058	7.3363	130	0.0217	7.4296	0.6994	11.8197
79	0.2035	7.9348	1.1894	7.4229	131	0.0207	7.4241	0.6938	11.9062

TABLE 12h (continued)

Zo [Ohms]	W/h	Keff	pF/cm	nH/mm	Zo [Ohms]	W/h	Keff	pF/cm	nH/mm
80	0.1947	7.9192	1.1734	7.5095	132	0.0198	7.4188	0.6883	11.9928
81	0.1864	7.9039	1.1578	7.5960	133	0.0190	7.4136	0.6829	12.0794
82	0.1783	7.8889	1.1425	7.6825	134	0.0182	7.4085	0.6775	12.1660
83	0.1707	7.8741	1.1277	7.7689	135	0.0174	7.4035	0.6723	12.2527
84	0.1633	7.8596	1.1133	7.8552	136	0.0166	7.3986	0.6671	12.3393
85	0.1563	7.8453	1.0992	7.9415	137	0.0159	7.3938	0.6621	12.4261
86	0.1496	7.8314	1.0854	8.0278	138	0.0152	7.3891	0.6570	12.5128
87	0.1431	7.8176	1.0720	8.1140	139	0.0146	7.3845	0.6521	12.5996
88	0.1370	7.8042	1.0589	8.2002	140	0.0140	7.3800	0.6473	12.6863
89	0.1311	7.7910	1.0461	8.2864	141	0.0134	7.3756	0.6425	12.7732
90	0.1255	7.7781	1.0336	8.3725	142	0.0128	7.3713	0.6378	12.8600
91	0.1201	7.7654	1.0215	8.4587	143	0.0122	7.3671	0.6331	12.9469
92	0.1149	7.7529	1.0095	8.5448	144	0.0117	7.3630	0.6286	13.0338
93	0.1100	7.7407	0.9979	8.6309	145	0.0112	7.3590	0.6241	13.1207
94	0.1052	7.7288	0.9865	8.7169	146	0.0107	7.3551	0.6196	13.2077
95	0.1007	7.7171	0.9754	8.8030	147	0.0103	7.3512	0.6152	13.2946
96	0.0964	7.7056	0.9645	8.8890	148	0.0098	7.3475	0.6109	13.3817
97	0.0922	7.6944	0.9539	8.9751	149	0.0094	7.3438	0.6067	13.4687
98	0.0883	7.6833	0.9435	9.0611	150	0.0090	7.3402	0.6025	13.5558

Notes: Calculation of W/H has an error of less than 1%.

Source: Gupta, K.C., Garg, R., Bahl, I., Bhartia, P., *Microstrip Lines and Slotlines,* 2nd Ed., Artech House, Norwood, MA, 1996, 103.

TABLE 12i Zero Thickness Microstrip Dimensions, Effective Dielectric Constant, and PUL Capacitance and Inductance for Er =35

Zo [Ohms]	W/h	Keff	pF/cm	nH/mm	Zo [Ohms]	W/h	Keff	pF/cm	nH/mm
1	61.2525	33.5453	193.1947	0.1932	57	0.1141	20.3209	2.6380	8.5709
2	29.6179	32.3412	94.8478	0.3794	58	0.1063	20.2652	2.5890	8.7093
3	19.1302	31.3265	62.2321	0.5601	59	0.0990	20.2113	2.5417	8.8477
4	13.9141	30.4569	46.0216	0.7363	60	0.0923	20.1591	2.4961	8.9860
5	10.8011	29.7005	36.3573	0.9089	61	0.0860	20.1087	2.4521	9.1243
6	8.7367	29.0345	29.9561	1.0784	62	0.0801	20.0598	2.4096	9.2626
7	7.2699	28.4417	25.4132	1.2452	63	0.0746	20.0126	2.3686	9.4010
8	6.1757	27.9094	22.0275	1.4098	64	0.0695	19.9670	2.3289	9.5393
9	5.3291	27.4273	19.4101	1.5722	65	0.0648	19.9228	2.2906	9.6776
10	4.6555	26.9878	17.3286	1.7329	66	0.0604	19.8801	2.2534	9.8160
11	4.1073	26.5845	15.6351	1.8918	67	0.0562	19.8389	2.2175	9.9543
12	3.6529	26.2124	14.2315	2.0493	68	0.0524	19.7990	2.1827	10.0928
13	3.2705	25.8674	13.0501	2.2055	69	0.0488	19.7604	2.1490	10.2312
14	2.9445	25.5460	12.0424	2.3603	70	0.0455	19.7232	2.1163	10.3697
15	2.6635	25.2453	11.1732	2.5140	71	0.0424	19.6872	2.0846	10.5082
16	2.4189	24.9629	10.4161	2.6665	72	0.0395	19.6525	2.0538	10.6468
17	2.2042	24.6968	9.7510	2.8180	73	0.0368	19.6189	2.0239	10.7855
18	2.0144	24.4452	9.1623	2.9686	74	0.0343	19.5865	1.9949	10.9242
19	1.8358	24.1924	8.6351	3.1173	75	0.0319	19.5551	1.9667	11.0630
20	1.6891	23.9716	8.1658	3.2663	76	0.0298	19.5249	1.9394	11.2018
21	1.5569	23.7610	7.7427	3.4145	77	0.0277	19.4957	1.9127	11.3407
22	1.4372	23.5597	7.3594	3.5619	78	0.0258	19.4674	1.8869	11.4796
23	1.3284	23.3669	7.0106	3.7086	79	0.0241	19.4402	1.8617	11.6187
24	1.2292	23.1820	6.6918	3.8545	80	0.0224	19.4139	1.8372	11.7578
25	1.1385	23.0043	6.3995	3.9997	81	0.0209	19.3884	1.8133	11.8970

TABLE 12i (continued)

Zo [Ohms]	W/h	Keff	pF/cm	nH/mm	Zo [Ohms]	W/h	Keff	pF/cm	nH/mm
26	1.0554	22.8334	6.1304	4.1442	82	0.0195	19.3639	1.7900	12.0362
27	0.9790	22.6972	5.8857	4.2907	83	0.0181	19.3402	1.7674	12.1755
28	0.9087	22.6289	5.6670	4.4429	84	0.0169	19.3173	1.7453	12.3149
29	0.8438	22.5532	5.4624	4.5939	85	0.0157	19.2952	1.7238	12.4544
30	0.7840	22.4719	5.2708	4.7437	86	0.0147	19.2739	1.7028	12.5940
31	0.7286	22.3863	5.0911	4.8925	87	0.0137	19.2533	1.6823	12.7336
32	0.6774	22.2977	4.9222	5.0403	88	0.0127	19.2334	1.6624	12.8733
33	0.6300	22.2069	4.7633	5.1873	89	0.0119	19.2142	1.6429	13.0131
34	0.5861	22.1148	4.6136	5.3333	90	0.0111	19.1957	1.6238	13.1530
35	0.5453	22.0220	4.4724	5.4787	91	0.0103	19.1778	1.6052	13.2929
36	0.5074	21.9291	4.3390	5.6233	92	0.0096	19.1605	1.5871	13.4329
37	0.4723	21.8365	4.2128	5.7673	93	0.0089	19.1439	1.5693	13.5730
38	0.4397	21.7445	4.0933	5.9107	94	0.0083	19.1278	1.5520	13.7132
39	0.4093	21.6536	3.9800	6.0535	95	0.0078	19.1122	1.5350	13.8535
40	0.3811	21.5638	3.8724	6.1959	96	0.0072	19.0972	1.5184	13.9938
41	0.3549	21.4755	3.7702	6.3378	97	0.0067	19.0828	1.5022	14.1342
42	0.3305	21.3889	3.6730	6.4792	98	0.0063	19.0688	1.4863	14.2747
43	0.3078	21.3039	3.5805	6.6203	99	0.0059	19.0553	1.4708	14.4152
44	0.2867	21.2208	3.4923	6.7610	100	0.0055	19.0422	1.4556	14.5559
45	0.2670	21.1396	3.4081	6.9014	101	0.0051	19.0297	1.4407	14.6966
46	0.2487	21.0603	3.3278	7.0416	102	0.0047	19.0175	1.4261	14.8374
47	0.2317	20.9831	3.2510	7.1814	103	0.0044	19.0058	1.4118	14.9782
48	0.2158	20.9078	3.1776	7.3211	104	0.0041	18.9945	1.3978	15.1191
49	0.2010	20.8346	3.1073	7.4605	105	0.0038	18.9836	1.3841	15.2601
50	0.1873	20.7635	3.0399	7.5998	106	0.0036	18.9730	1.3707	15.4012
51	0.1745	20.6944	2.9753	7.7388	107	0.0033	18.9629	1.3575	15.5423
52	0.1625	20.6272	2.9134	7.8778	108	0.0031	18.9530	1.3446	15.6835
53	0.1514	20.5621	2.8539	8.0166	109	0.0029	18.9436	1.3319	15.8247
54	0.1411	20.4989	2.7967	8.1553	110	0.0027	18.9344	1.3195	15.9661
55	0.1314	20.4377	2.7418	8.2939	111	0.0025	18.9256	1.3073	16.1074
56	0.1224	20.3784	2.6889	8.4324	112	0.0023	18.9170	1.2954	16.2489
113	0.0022	18.9088	1.2836	16.3904	132	0.0006	18.7969	1.0956	19.0896
114	0.0020	18.9009	1.2721	16.5320	133	0.0005	18.7928	1.0872	19.2321
115	0.0019	18.8932	1.2608	16.6736	134	0.0005	18.7889	1.0790	19.3747
116	0.0018	18.8858	1.2497	16.8153	135	0.0005	18.7851	1.0709	19.5173
117	0.0016	18.8786	1.2387	16.9571	136	0.0004	18.7815	1.0629	19.6600
118	0.0015	18.8717	1.2280	17.0989	137	0.0004	18.7779	1.0551	19.8027
119	0.0014	18.8651	1.2175	17.2407	138	0.0004	18.7745	1.0473	19.9454
120	0.0013	18.8587	1.2071	17.3826	139	0.0003	18.7713	1.0397	20.0882
121	0.0012	18.8524	1.1970	17.5246	140	0.0003	18.7681	1.0322	20.2310
122	0.0012	18.8465	1.1870	17.6666	141	0.0003	18.7650	1.0248	20.3738
123	0.0011	18.8407	1.1771	17.8087	142	0.0003	18.7621	1.0175	20.5167
124	0.0010	18.8351	1.1675	17.9508	143	0.0003	18.7592	1.0103	20.6596
125	0.0009	18.8297	1.1580	18.0930	144	0.0002	18.7565	1.0032	20.8026
126	0.0009	18.8245	1.1486	18.2352	145	0.0002	18.7538	0.9962	20.9456
127	0.0008	18.8195	1.1394	18.3775	146	0.0002	18.7513	0.9893	21.0886
128	0.0008	18.8146	1.1304	18.5198	147	0.0002	18.7488	0.9825	21.2316
129	0.0007	18.8100	1.1215	18.6622	148	0.0002	18.7464	0.9758	21.3747
130	0.0007	18.8055	1.1127	18.8046	149	0.0002	18.7441	0.9692	21.5178
131	0.0006	18.8011	1.1041	18.9471	150	0.0002	18.7419	0.9627	21.6609

Notes: Calculation of W/H has an error of less than 1%.

Source: Gupta, K.C., Garg, R., Bahl, I., Bhartia, P., *Microstrip Lines and Slotlines,* 2nd Ed., Artech House, Norwood, MA, 1996, 103.

TABLE 12j   Zero Thickness Microstrip Dimensions, Effective Dielectric Constant, and PUL Capacitance and Inductance for Er =85

Zo [Ohms]	W/h	Keff	pF/cm	nH/mm	Zo [Ohms]	W/h	Keff	pF/cm	nH/mm
1	38.5925	79.6824	297.7560	0.2978	26	0.3739	51.7462	9.2288	6.2387
2	18.3705	75.6651	145.0765	0.5803	27	0.3349	51.4124	8.8583	6.4577
3	11.6859	72.5009	94.6738	0.8521	28	0.3000	51.0886	8.5150	6.6757
4	8.3710	69.9235	69.7318	1.1157	29	0.2688	50.7757	8.1961	6.8930
5	6.3982	67.7680	54.9189	1.3730	30	0.2409	50.4745	7.8994	7.1095
6	5.0937	65.9270	45.1398	1.6250	31	0.2159	50.1850	7.6226	7.3253
7	4.1695	64.3277	38.2191	1.8727	32	0.1935	49.9076	7.3640	7.5407
8	3.4821	62.9184	33.0733	2.1167	33	0.1734	49.6422	7.1218	7.7557
9	2.9517	61.6613	29.1034	2.3574	34	0.1554	49.3886	6.8947	7.9702
10	2.5309	60.5285	25.9513	2.5951	35	0.1393	49.1467	6.6813	8.1845
11	2.1895	59.4981	23.3904	2.8302	36	0.1249	48.9161	6.4804	8.3986
12	1.8946	58.5089	21.2622	3.0618	37	0.1119	48.6966	6.2911	8.6125
13	1.6651	57.6611	19.4840	3.2928	38	0.1003	48.4877	6.1124	8.8263
14	1.4696	56.8730	17.9682	3.5218	39	0.0899	48.2891	5.9435	9.0400
15	1.3012	56.1365	16.6614	3.7488	40	0.0806	48.1003	5.7835	9.2537
16	1.1551	55.4455	15.5236	3.9740	41	0.0723	47.9210	5.6319	9.4673
17	1.0274	54.7949	14.5245	4.1976	42	0.0648	47.7509	5.4881	9.6810
18	0.9153	54.4535	13.6748	4.4306	43	0.0581	47.5893	5.3514	9.8947
19	0.8165	54.1607	12.9202	4.6642	44	0.0521	47.4361	5.2213	10.1085
20	0.7290	53.8383	12.2376	4.8950	45	0.0467	47.2907	5.0975	10.3224
21	0.6514	53.4976	11.6179	5.1235	46	0.0418	47.1529	4.9794	10.5364
22	0.5825	53.1467	11.0534	5.3498	47	0.0375	47.0222	4.8667	10.7505
23	0.5211	52.7921	10.5375	5.5743	48	0.0336	46.8984	4.7590	10.9648
24	0.4664	52.4383	10.0645	5.7972	49	0.0301	46.7810	4.6561	11.1792
25	0.4176	52.0888	9.6297	6.0185	50	0.0270	46.6697	4.5575	11.3938
51	0.0242	46.5643	4.4631	11.6085	101	0.0001	44.8028	2.2106	22.5503
52	0.0217	46.4644	4.3726	11.8234	102	0.0001	44.7963	2.1888	22.7720
53	0.0195	46.3698	4.2857	12.0385	103	0.0001	44.7901	2.1674	22.9936
54	0.0175	46.2801	4.2023	12.2538	104	0.0001	44.7842	2.1464	23.2153
55	0.0156	46.1952	4.1221	12.4693	105	0.0001	44.7787	2.1258	23.4371
56	0.0140	46.1148	4.0449	12.6849	106	0.0001	44.7734	2.1056	23.6589
57	0.0126	46.0386	3.9707	12.9008	107	0.0001	44.7685	2.0858	23.8808
58	0.0113	45.9665	3.8992	13.1168	108	0.0000	44.7638	2.0664	24.1027
59	0.0101	45.8982	3.8302	13.3330	109	0.0000	44.7593	2.0474	24.3247
60	0.0091	45.8335	3.7637	13.5495	110	0.0000	44.7551	2.0287	24.5467
61	0.0081	45.7722	3.6996	13.7661	111	0.0000	44.7511	2.0103	24.7688
62	0.0073	45.7142	3.6376	13.9829	112	0.0000	44.7473	1.9923	24.9908
63	0.0065	45.6592	3.5777	14.1999	113	0.0000	44.7437	1.9745	25.2130
64	0.0059	45.6072	3.5198	14.4170	114	0.0000	44.7404	1.9572	25.4351
65	0.0052	45.5579	3.4638	14.6344	115	0.0000	44.7371	1.9401	25.6573
66	0.0047	45.5112	3.4095	14.8519	116	0.0000	44.7341	1.9233	25.8795
67	0.0042	45.4670	3.3570	15.0696	117	0.0000	44.7312	1.9068	26.1018
68	0.0038	45.4252	3.3061	15.2875	118	0.0000	44.7285	1.8906	26.3241
69	0.0034	45.3856	3.2568	15.5056	119	0.0000	44.7259	1.8746	26.5464
70	0.0030	45.3481	3.2089	15.7238	120	0.0000	44.7235	1.8589	26.7688
71	0.0027	45.3126	3.1625	15.9422	121	0.0000	44.7212	1.8435	26.9911
72	0.0024	45.2789	3.1174	16.1607	122	0.0000	44.7190	1.8284	27.2135
73	0.0022	45.2471	3.0736	16.3794	123	0.0000	44.7169	1.8135	27.4360
74	0.0020	45.2169	3.0311	16.5982	124	0.0000	44.7149	1.7988	27.6584
75	0.0018	45.1884	2.9897	16.8172	125	0.0000	44.7131	1.7844	27.8809
76	0.0016	45.1614	2.9495	17.0363	126	0.0000	44.7113	1.7702	28.1034
77	0.0014	45.1358	2.9104	17.2556	127	0.0000	44.7097	1.7562	28.3259
78	0.0013	45.1115	2.8723	17.4750	128	0.0000	44.7081	1.7425	28.5484
79	0.0011	45.0886	2.8352	17.6946	129	0.0000	44.7066	1.7289	28.7710
80	0.0010	45.0669	2.7991	17.9142	130	0.0000	44.7052	1.7156	28.9936

TABLE 12j (continued)

Zo [Ohms]	W/h	Keff	pF/cm	nH/mm	Zo [Ohms]	W/h	Keff	pF/cm	nH/mm
81	0.0009	45.0463	2.7639	18.1340	131	0.0000	44.7038	1.7025	29.2162
82	0.0008	45.0268	2.7296	18.3539	132	0.0000	44.7026	1.6896	29.4388
83	0.0007	45.0084	2.6962	18.5739	133	0.0000	44.7014	1.6768	29.6614
84	0.0007	44.9909	2.6636	18.7941	134	0.0000	44.7002	1.6643	29.8840
85	0.0006	44.9744	2.6317	19.0143	135	0.0000	44.6992	1.6519	30.1067
86	0.0005	44.9587	2.6007	19.2347	136	0.0000	44.6981	1.6398	30.3293
87	0.0005	44.9439	2.5704	19.4551	137	0.0000	44.6972	1.6278	30.5520
88	0.0004	44.9299	2.5408	19.6757	138	0.0000	44.6963	1.6160	30.7747
89	0.0004	44.9166	2.5118	19.8963	139	0.0000	44.6954	1.6043	30.9974
90	0.0003	44.9040	2.4836	20.1171	140	0.0000	44.6946	1.5929	31.2201
91	0.0003	44.8921	2.4560	20.3379	141	0.0000	44.6938	1.5816	31.4429
92	0.0003	44.8808	2.4290	20.5588	142	0.0000	44.6931	1.5704	31.6656
93	0.0002	44.8701	2.4026	20.7798	143	0.0000	44.6924	1.5594	31.8883
94	0.0002	44.8600	2.3767	21.0009	144	0.0000	44.6917	1.5486	32.1111
95	0.0002	44.8504	2.3515	21.2220	145	0.0000	44.6911	1.5379	32.3339
96	0.0002	44.8414	2.3267	21.4432	146	0.0000	44.6905	1.5273	32.5567
97	0.0002	44.8328	2.3025	21.6645	147	0.0000	44.6899	1.5169	32.7794
98	0.0001	44.8247	2.2788	21.8859	148	0.0000	44.6894	1.5067	33.0022
99	0.0001	44.8170	2.2556	22.1073	149	0.0000	44.6889	1.4966	33.2250
100	0.0001	44.8097	2.2329	22.3288	150	0.0000	44.6884	1.4866	33.4478

Notes: Calculation of W/H has an error of less than 1%.

Source: Gupta, K.C., Garg, R., Bahl, I., Bhartia, P., *Microstrip Lines and Slotlines*, 2nd Ed., Artech House, Norwood, MA, 1996, 103.

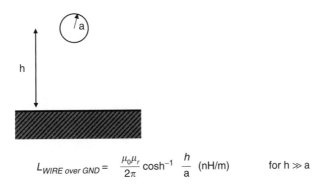

$$L_{WIRE\ over\ GND} = \frac{\mu_0 \mu_r}{2\pi} \cosh^{-1} \frac{h}{a} \quad (nH/m) \qquad \text{for } h \gg a$$

FIGURE B.6 Inductance of a wire over a ground plane. When consistent units are used for the wire radius, a, and the height over the ground plane, h, the formula provides an estimate of the inductance per unit length in (nH/m).

Bondwires, Ribbons, Mesh

The assembly of semiconductor and hybrid integrated circuits for microwave and millimeter-wave frequencies generally requires the use of gold bondwires, ribbons, or mesh. The impedance of the interconnection must be accounted for in a good design. Unfortunately, there is no single accepted electrical model. The complexity required of the model will depend on the frequency of operation and the general impedance levels of the circuits being connected. At low frequencies and moderate to high impedances, the connection is frequently modeled as an inductor (sometimes in series with a resistor); at high frequencies, a full 3-D electromagnetic simulation may be required for accurate results. At intermediate points it may be modeled as a high impedance transmission line or as a lumped LC circuit. Note that the resistances of the rf interconnects should be included in the design of extremely low-noise circuits as they will affect the noise figure. In connecting a semiconductor die to package leads, it may also be necessary to model the mutual inductances and interlead capacitances in addition to the usual self inductances and shunt capacitance. Figure 13 illustrates one method of modeling bond wire inductance that has been shown adequate for many microwave applications. More sophisticated methods of modeling bond wires, ribbon or mesh are described in the references.

References

1. Grover, F.W., *Inductance Calculations*, Dover Publications, New York, NY Available through Instrument Society of America, Research Triangle Park, NC.
2. Caulton, M., Lumped Elements in Microwave Circuits, from Advances in Microwaves 1974, Academic Press, NY 1974, pp. 143–202.
3. Wadell, B. C., Transmission Line Design Handbook, Artech House, Boston, MA 1991, pp. 151–155.
4. Terman, F.E., *Radio Engineer's Handbook*, McGraw-Hill, 1943 pp. 48–51.
5. Caverly, R. H., "Characteristic Impedance of Integrated Circuit Bond Wires," IEEE Transactions on Microwave Theory and Techniques, Vol MTT-34, No. 9, September, 1986, pp. 982–984.
6. Kuester, E. F., Chang, D. C., "Propagating Modes Along a Thin Wire Located Above a Grounded Dielectric Slab," IEEE Transactions on Microwave Theory and Techniques, Vol MTT-25, No. 12, December, 1977, pp. 1065–1069.
7. Mondal, J. P. "Octagonal spiral inductor measurement and modelling for MMIC applications," Int. J. Electronics, 1990, Vol 68, No. 1, pp. 113–125.
8. MIL-STD-883E, "Test Method Standard, Microcircuits" US Department of Defense.
9. Gupta, K. C., Garg, E., Bahl, I., Bhartia, P., *Microstrip Lines and Slothlines*, 2nd Ed., Artech House, Norwood, MA, 1996, 103.

Index